6th International Conference on Mechanical Engineering, Materials Science and Civil Engineering (ICMEMSCE 2018)

IOP Conference Series: Materials Science and Engineering
Volume 542

Xiamen, China
21 – 22 December 2018

ISBN: 978-1-5108-9087-9
ISSN: 1757-8981

Printed from e-media with permission by:

Curran Associates, Inc.
57 Morehouse Lane
Red Hook, NY 12571

Some format issues inherent in the e-media version may also appear in this print version.

No changes have been made to the content of these proceedings. There may be changes to pagination and minor adjustments for aesthetics.

Printed by Curran Associates, Inc. (2019)

For permission requests, please contact the Institute of Physics at the address below.

Institute of Physics
Dirac House, Temple Back
Bristol BS1 6BE UK

Phone: 44 1 17 929 7481
Fax: 44 1 17 920 0979

techtracking@iop.org

Additional copies of this publication are available from:

Curran Associates, Inc.
57 Morehouse Lane
Red Hook, NY 12571 USA
Phone: 845-758-0400
Fax: 845-758-2633
Email: curran@proceedings.com
Web: www.proceedings.com

Table of contents

Volume 542

2018 the 6th International Conference on Mechanical Engineering, Materials Science and Civil Engineering

21–22 December 2018, Xiamen, China

Accepted papers received: 25 April 2019

Preface

Papers

IOP Conf. Series: Materials Science and Engineering **542** (2019) 011001 doi:10.1088/1757-899X/542/1/011001

Preface

Dear Distinguished Authors and Guests,

The Organizing Committee warmly welcomes you to 2018 International Conference on Mechanical Engineering, Materials Science and Civil Engineering (ICMEMSCE2018), held on 21th-22th, December, 2018, Xiamen, China.

The aim of ICMEMSCE2018 is to present the latest research and results of scientists (professors, students, PhD Students, engineers, and post-doc scientist) related to Mechanical Engineering and Materials topics. This conference provides opportunities for the different areas delegates to exchange new ideas and application experiences face to face, to establish business or research relations and to find global partners for future collaboration.

After the peer-review process, the submitted papers were selected on the basis of originality, significance, and clarity for the purpose of the conference. The selected papers and additional late-breaking contributions to be presented as lectures will make an exciting technical program. The conference program is extremely rich, featuring high-impact presentation. We hope that the conference results constituted significant contribution to the knowledge in these up to date scientific field.

The proceeding records the fully refereed papers presented at the conference. The main conference themes and tracks are Mechanical Engineering and Composite materials, nano-materials, bio-materials, Material processing, molding technology, Advanced Materials Science, Materials Processing Technology, Electrical Engineering, Vehicle and Motor, Metal Alloys and Simulation. Hopefully, all participants and other interested readers benefit scientifically from the proceedings and also find it stimulating in the process.

On behalf of the organizing committee, I would like to especially thank all the editors from IOP Conference Series for their great support to ICMEMSCE2018. Without their excellent editorial work, ICMEMSCE2018 will not be published so timely and successfully.

Finally we wish all the authors and attendees of ICMEMSCE2018 a unique, rewarding

IOP Conf. Series: Materials Science and Engineering **542** (2019) 011001 doi:10.1088/1757-899X/542/1/011001

and enjoyable memory at ICMEMSCE2018 in Xiamen, China. We look forward to your participation in the 7th ICMEMSCE in 2019.

With our warmest regards,

Prof. JingWei Zhao

Conference Organizing Chair

December 25th 2018

IOP Conf. Series: Materials Science and Engineering **542** (2019) 011002 doi:10.1088/1757-899X/542/1/011002

ICMEMSCE2018 Committee

Conference Chairs

Prof. JingWei Zhao, University of wollongong, Australia

Technical Program Committee Chairs

Hyun-Do YUN, Chungnam National University, Korea

THANH-LAM NGUYEN, Lac Hong University, Vietnam

Mou Chen, Nanjing University of Aeronautics and Astronautics, China

Keishi Matsuda, Matsuyama University, Japan

International Program Committee Chairs（America）

Jeremy (Zheng) Li, School of Engineering, University of Bridgeport, America (Guest editor for 3rd ICMEMSCE2014)

International Program Committee Chairs（Japan）

Ikuo IHARA, Nagaoka University of Technology, Japan (Guest editor for 2nd ICMEMSCE2013)

International Program Committee Chairs（Malaysia）

Norhaliza Abdul Wahab, University Teknologi Malaysia, Malaysia

Abdul L Mohd Tobi, University Tun Hussein Onn Malaysia, Malaysia

Aishah Najiah Dahnel, International Islamic University Malaysia, Malaysia

Ahmad Baharuddin Abdullah, University Sains Malaysia, Malaysia

Toh Tien Choon, University Tunku Abdul Rahman, Malaysia

Goh KaiChen, University Tun Hussein Onn Malaysia, Malaysia

Yusri Yusof, University Tun Hussein Onn Malaysia, Malaysia

Mohammad Jawaid, University Putra Malaysia, Malaysia

International Program Committee Chairs（Srilanka）

K.M. Muditha Dassanayake, University of Peradeniya, Srilanka

International Program Committee Chairs（Slovak Republic）

Michal Kelemen, Technical University of Kosice, Slovak Republic

International Program Committee Chairs （South Korea）

Sangyoon Lee, Konkuk University, South Korea

International Program Committee Chairs（Hungarian）

László Horváth, óbuda University, Hungarian

International Program Committee Chairs（Thailand）

Achanai Buasri, Silpakorn University, Thailand

International Program Committee Chairs（Taiwan）

Ming-Hung Shu, National Kaosiung University of Applied Sciences, Taiwan

Chih-Ta Tsai, National Cheng Kung Universit, Taiwan

International Program Committee Chairs（Iran ）

Seyed Hashem Eslami, University of Tehran, Iran

International Program Committee Chairs（Brazil）

Walter Jesus Paucar Casas, Universidade Federal do Rio Grande do Sul, Brazil

International Program Committee Chairs（India）

Sandip Patil, Bharat Forge Ltd, Pune, India

Prof.Vijayan Gurumurthy Iyer, Department of Civil Engineering, Koneru Lakshmaiah Education Foundation (KLEF) Deemed To Be University

Dr. Subhasis Roy,Department of Chemical Engineering University of Calcutta

International Program Committee Chairs (Turkey)

Osman Adiguzel, Firat University, Turkey

PINAR TERZIOGLU, Mugla Sitki Kocman University University,Turkey

International Program Committee Chairs (Poland)

Magorzata asiulewicz – Kaczmarek, Poznan University of Technology

Miroslaw Kwiatkowski, AGH University of Science and Technology

Krzysztof Kuś, University of Warmia and Mazury in Olsztyn

Grzegorz Woroniak, Bialystok University of Technology

International Program Committee Chair (Romanian)

Stefan Talu, University of Cluj-Napoca, Romanian

IOP Conf. Series: Materials Science and Engineering **542** (2019) 011002 doi:10.1088/1757-899X/542/1/011002

International Program Committee Chair (UK)

Yingang Du, Anglia Ruskin University, Chelmsford

International Program Committee Chair (Portugal)

F.Pacheco Torgal, University of Minho

International Program Committee Chair (Czech Republic)

Fatma Yalcinkaya, Technical University of Liberec

International Program Committee Chair (Paris)

GHORBEL Elhem, University of Cergy-Pontoise

Peer review statement

All papers published in this volume of *IOP Conference Series: Materials Science and Engineering* have been peer reviewed through processes administered by the proceedings Editors. Reviews were conducted by expert referees to the professional and scientific standards expected of a proceedings journal published by IOP Publishing.

IOP Conf. Series: Materials Science and Engineering **542** (2019) 012001 doi:10.1088/1757-899X/542/1/012001

Experiment Study on Sound Properties of Carbon Fiber Composite Material

Zhong Yan, Zhou Pu, Feng Haijun ,and Zhang Yi

Shanghai Marine Equipment Research Institute, Shanghai, China
E-mail: zhongyan704@163.com

Abstract. In this paper, the sound absorption coefficient and transmission loss of carbon fiber composite material are tested by stand wave tube equipment, and the factors affecting the performance of the carbon fiber composite material are analyzed. The results show that the sandwich structure can effectively improve the sound absorption coefficient and transmission loss; increasing material thickness appropriately can also improve acoustic performance of the carbon fiber composite material. The carbon fiber composite material with reasonable structure design can be used as acoustic material in the field of vibration and noise reduction.

1. Introduction

Carbon fiber composite material is a kind of excellent mechanical material, with high strength, high modulus, low density, low temperature resistance, small linear expansion coefficient and a series of advantages, which is get increasingly attention and application in aerospace, building reinforcement, upgrade industries such as new energy vehicles[1]. As the application of carbon fiber composites expands to the field of vibration and noise reduction, new requirements are proposed for the acoustic performance of carbon fiber composites[2].

In this paper, acoustic properties of carbon fiber composite materials are tested by standing wave tube method and the affect factors are analyzed. The conclusion of the experiment can provide a scientific basis for the application of vibration damping noise reduction.

2. Experimental Study on Sound Absorption Performance

2.1 Test Method

The sound absorption coefficient is a physical parameter reflecting the sound absorption performance of sound absorbing materials. There are two methods for measuring the sound absorption performance of materials, reverberation chamber method and standing wave tube method[3]. Standing wave tube method can measure absorption coefficient and specific acoustic impedance of the normal incident sound wave .The method has the following characteristics: small specimen area, easy installation and measurement, high measurement accuracy and low cost. The sound absorption coefficient measured by this method can also be used to compare the sound absorption performance of different materials and the same materials under different conditions[4-6]. Therefore, the standing wave tube method is adopted to test the sound absorption performance of the material in this paper.

2.2 Test Results and Analysis

The measurement of sound absorption coefficient is carried out on B&K4206 standard impedance tube. The internal diameter of impedance tube is 100mm and test frequency range is 64~1600Hz.

The materials used in the test are shown in table 1, and the test results are shown.

Table 1 List of Acoustic Performance Test Materials

Material	Thickness /mm	Remark
Steel Plate	11	contrast- material
Carbon Fiber with E51 Resin	11	
Carbon fiber with E51 Resin	22	
Carbon Fiber with 3D Fabric Sandwich	11	
Carbon Fiber with Honeycomb Sandwich	11	

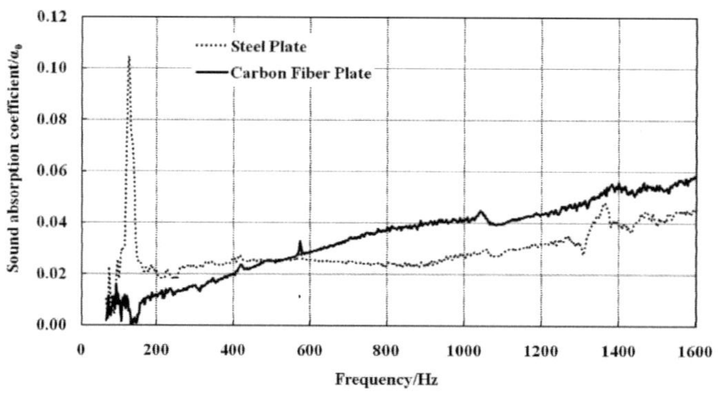

Figure1. Comparison of Carbon fiber and Steel On Sound Absorption Coefficient

As shown as Figure 1, the absorption coefficient of carbon fiber and steel of same thickness is in the same order of magnitude, the absorption coefficient of carbon fiber is slightly lower than steel in low frequency range, slightly higher than the steel in high frequency. The overall sound absorption coefficients is below 0.1 .That is to say simple structure of carbon fiber composites has no obvious advantage in sound absorption performance. To improve the sound absorption performance of carbon fiber composites, the structural design and internal quality distribution should be optimized.

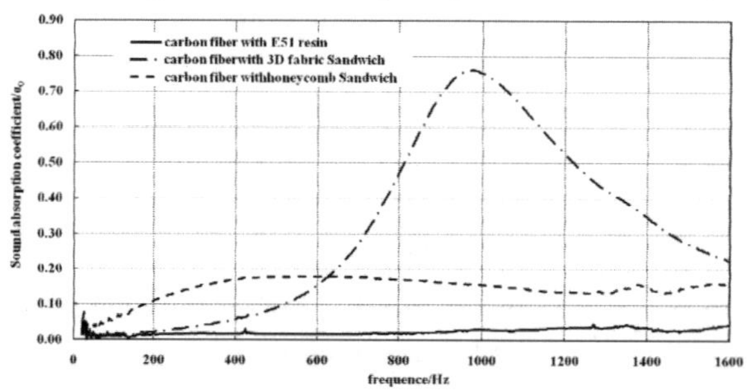

Figure2. Influence of Different Structures on Sound Absorption Coefficient

IOP Conf. Series: Materials Science and Engineering **542** (2019) 012001 doi:10.1088/1757-899X/542/1/012001

As shown as Figure 2, the sound absorption coefficient can be significantly increased by use the 3D fabric sandwich and honeycomb sandwich structure. The sound absorption coefficient of carbon fiber with 3D fabric sandwich above 600Hz is above 0.2, and the peak value reaches about 0.75, appearing around 1000Hz. This is because the 3D fabric sandwich changes the mass distribution characteristics, increases the flow resistance of the material, strengthens the reflection effect in the material internal, and improves the sound absorption performance of carbon fiber composite material.

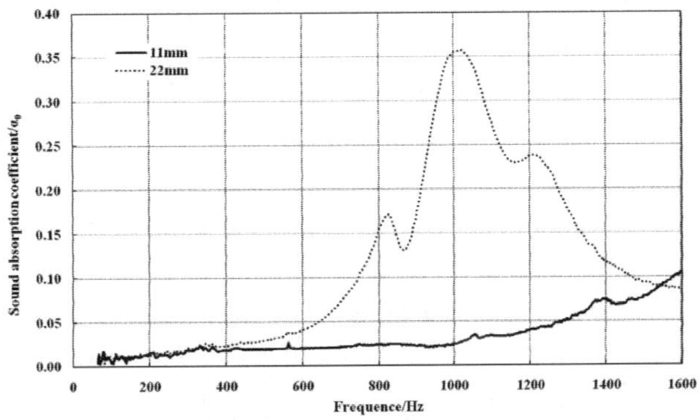

Figure3. Comparison of Different Thickness On Sound Absorption Coefficient

As shown as Figure 3, with the increase of thickness, the sound absorption coefficient in medium and high frequencies is greatly increased, and the sound absorption coefficient of low frequencies is also increased slightly. Because with the increase of the thickness, the propagation distance of the sound waves increases, more energy is consumed. So in the whole frequency range, especially in high frequencies, the sound absorption coefficient is increased significantly, the absorption property is improved. But when the material thickness increases to a certain extent, the change of sound absorption coefficient caused by thickness will be small.

3. Experimental Study on Acoustic Insulation Performance

3.1 Test Method

Sound insulation is the most commonly used physical quantity in the evaluation index of sound insulation performance. The sound insulation measurement methods have reverberation chamber method and four- microphone method[7-8]. The reverberation chamber method need special test environment and instruments, the test cycle is too long to meet the laboratory research. The four-microphone method is high-precision and convenient to test and study the performance of laboratory acoustic materials. Therefore, the four- microphone method is adopted in this paper.

Figure 4 shows the measurement principle of four-microphone method. The signals produced by signal generator, are converted into sound waves through the speaker and enter the sound source tube. In the tube plane sound wave is generated. The plane waves are divided into two parts when meet test samples, one part is launched, forming plane sound waves; the other part pass through the samples, enter into the receiving tube and form reflected plane sound wave. Plane sound waves are also divided into two parts when meeting the end of sound absorption, one part is absorbed, the other part is reflected, and forming reflected plane sound waves. Two microphones are placed before and after the test sample to measure the sound pressure at the position.

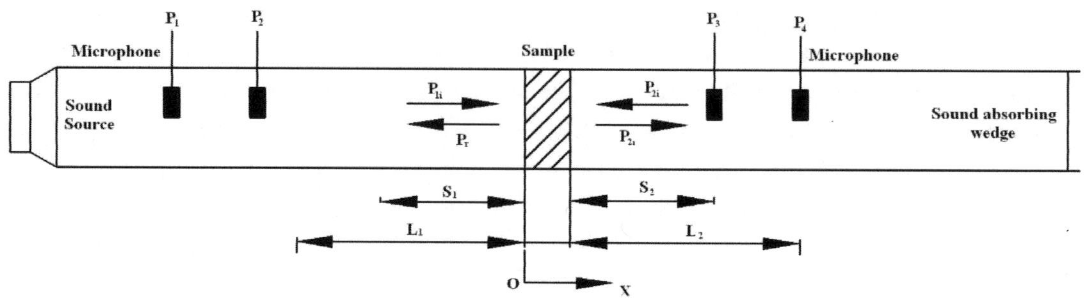

Figure4. Schematic Diagram of Measurement System

3.2 Test Results and Analysis

Sound insulation measurement system is based on the absorption coefficient test device and increases transmission tube and sound absorption terminals. This make two microphones in the incident tube two microphone and two microphones in transmission tube measurements at the same time, it can separate two columns of wave in the incident tube and transmission tube, which have the opposite direction in the propagation. So the transmission loss is calculated conveniently. The test results are shown in Figure 5 and Figure 6.

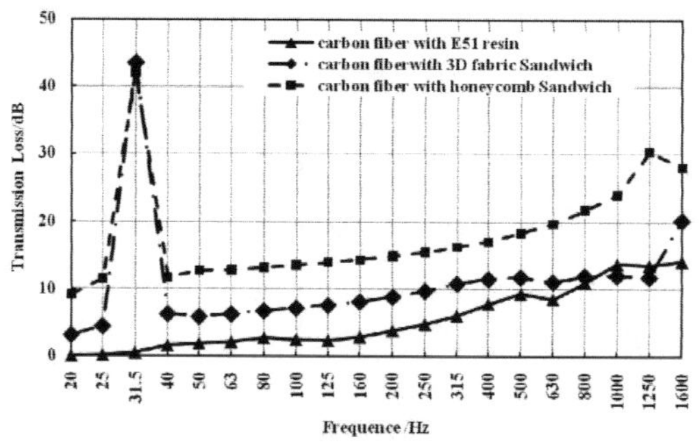

Figure5. Influence of Different Structures on Transmission Loss

As shown as Figure 5, 3D fabric sandwich and honeycomb sandwich structure can greatly improve the sound insulation performance of carbon fiber composite materials. The peak value of transmission loss appears at 31.5Hz, reaching over 40dB. This is because the sandwich structure divides the material into several closed cavities, which hinder sound waves propagation and heat exchange greatly. These cavity walls can also reflect sound waves effectively. Therefore the sandwich structure has good isolation effect, especially for lower frequencies. The sandwich structure can be adopted in the sound insulation engineering.

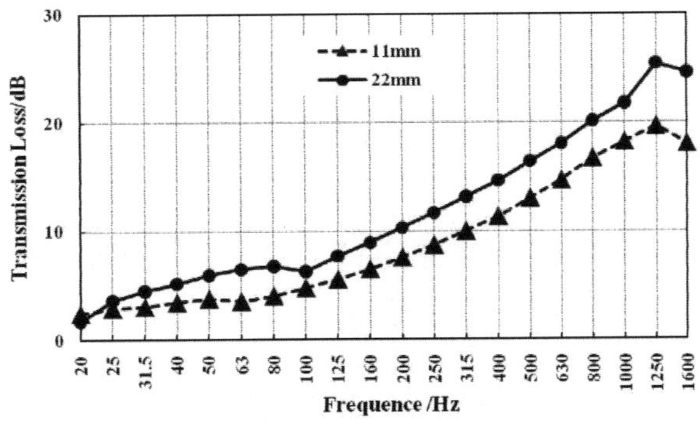

Figure6. Comparison of Different Thickness on Transmission Loss

As shown as Figure6., increasing the thickness can improve the sound insulation within a certain frequency range. This is because with the increase of thickness, the resistance to sound wave strengthens gradually. Sound waves lose more energy as they travel, the transmission energy decreases gradually, the transmission loss of the plate increases gradually. So in order to get better acoustic noise reduction, the thickness of carbon fiber composite materials can be appropriately increased.

4. Conclusion

In this paper, the sound absorption coefficient and insulation of carbon fiber composites are tested by standing wave tube method, and the influencing factors of material acoustics performance are analyzed. The following conclusions can be drawn:

(1) The carbon fiber sandwich structure can change the acoustics and dynamic characteristics of composite materials, increase the flow resistance of the material, strengthen the internal reflection effect of the sound waves in the material, and improve the sound absorption performance of materials.

(2) The sandwich structure has good sound insulation effect, especially for low frequency noise.

(3) Increasing of material thickness can improve the sound absorption coefficient in medium and high frequencies; it also can increase sound transmission loss on all frequencies.

(4) The carbon fiber composite materials designed with reasonable structure can be used as acoustic materials to reduce vibration and noise.

References

[1] Zhang Dingjin and Chen Hong, 2015 *Advanced Materials Industry*, **5**, 31-35.

[2] Wang Guan, Ruan Zhuqing, 2015 *Technical Acoustics*, **34**, 453-454.

[3] Maa Dahyou, 2002 *Noise and Vibration Control Engineering*, 427-431.

[4] Qu Bo and Zhu Beili, 2012 *Noise and Vibration Control*, **12**, 44-46.

[5] MinQi, Peng Feng, YinYa, 2010 *Acta Acoustic*, **29**, 321-330.

[6] MaYanhong, 2009 *Journal of Beijing University of Aeronautics and Astronautics*, **35**, 653-656

[7] Yuelu Ren, 2013 *Materials Letters*, **91**, 242-244.

[8] Thilagavathi T, 2010 *Journal of Industrial Textiles*, **39**, 267-278.

IOP Conf. Series: Materials Science and Engineering **542** (2019) 012002 doi:10.1088/1757-899X/542/1/012002

Research on influence of admixtures to the plastic shrinkage cracking of mortar

Yiping Ma[1], Han Zhang[1], Shaotong Yu[1], Chang Tan[1,2] and Xiaojie Yang[1]

1 Key Laboratory of Advanced Civil Engineering Materials of Education Ministry, Tongji University, Shanghai, China
2 Wuhan Chengxing United Technology Development Co., Ltd., Wuhan, China
E-mail: ypma1239@sina.com

Abstract. In this paper, the influence of admixture on plastic shrinkage cracking was studied. The results show that the plastic tensile strength increases while plastic shrinkage stress decreases as the content of air-entraining agent grows to 150×10^{-6} wt.%, and the anti-cracking index increases as a whole. Water reducing agent, water retaining agent, and shrinkage reducing agent all have the effect of improving plastic shrinkage cracking of fresh cement mortar. The constitutive relations between the admixtures content and anti-cracking index have been built.

1. Introduction

Plastic shrinkage cracking of cement-based materials is a common engineering construction problem which has a great impact on its performance. As an important influencing factor for plastic shrinkage and cracking of cement-based materials, admixture has also been studied [1-2]. Andreas Leemann [1] had studied the effect of shrinkage reducing admixture and paraffin-based curing compound to fresh concrete in plastic state, the results showed that the effectiveness in reducing plastic shrinkage cracking. On the other hand, it has been noted that the addition of SRA delays the time of cracking. Many researchers have proposed models to simulate plastic shrinkage and try to predict the plastic shrinkage cracking. A model called PShC Severity Model was proposed by Boshoff et al [3]. The model was based on the volume of water that evaporates from the concrete between the placing and the initial setting time of the concrete. The author [4] has studied the plastic shrinkage and cracking mechanism of cement-based materials and established the correlation between the cracking index and the plastic shrinkage cracking of cement-based materials.

Based on the existing research methods of the plastic shrinkage constitutive relationship of cement-based materials, this study explores the influence of the types and contents of various commonly used cement admixtures on the plastic shrinkage constitutive relationship of cement-based materials.

2. Experiment

2.1. Raw materials

Ordinary Portland Cement (P·O 42.5) was used in this experiment. The fineness modulus of sand with a 5 mm maximum particle size was 2.6, and the water used in the experiment was tap water. Air-entraining agent and water retaining agent were sodium dodecyl benzene sulfonate and methyl cellulose, respectively. These reagents were manufactured by Sinopharm Chemical Reagent Co., Ltd and all of their purity levels were analytical. The polycarboxylate water reducing agent (powder agent) was used. The composition of the shrinkage reducing agent was fatty alcohol-polyoxyethylene ether.

2.2. Research methods of constitutive relations in plastic shrinkage cracking

The flat method was used to test the cement mortar plastic capillary shrinkage stress and the plastic tensile strength of mortar was tested in this experiment using briguette mould designed by the authors. The schematic diagram of the test device is shown in figure 1, figure 2 and figure 3. The specific operation method of this device can be found in the author's previous article [5].

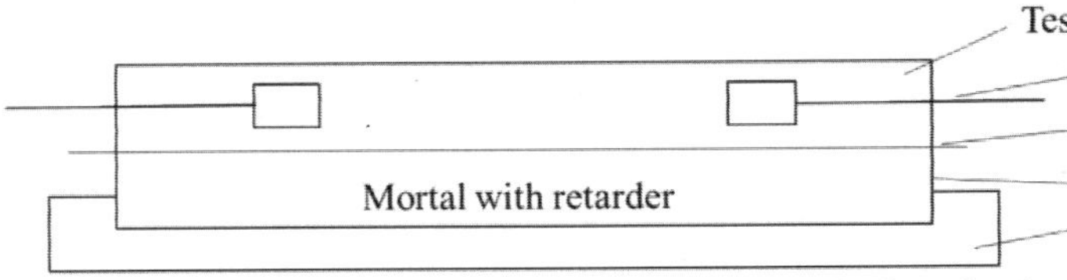

Figure 1. Sketch of measuring principle for plastic shrinkage stress

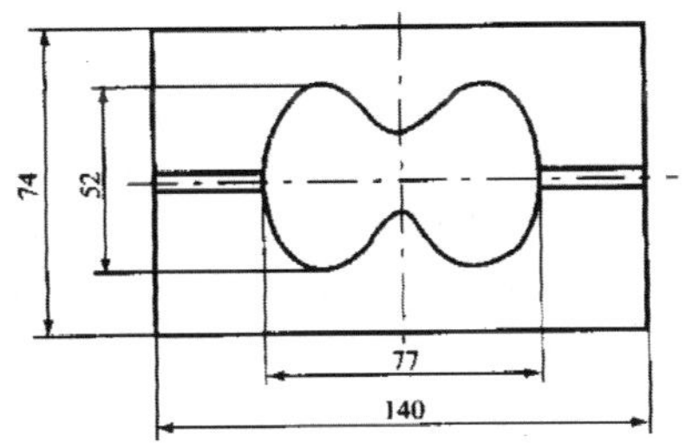

Figure 2. Schema of the briguette mould

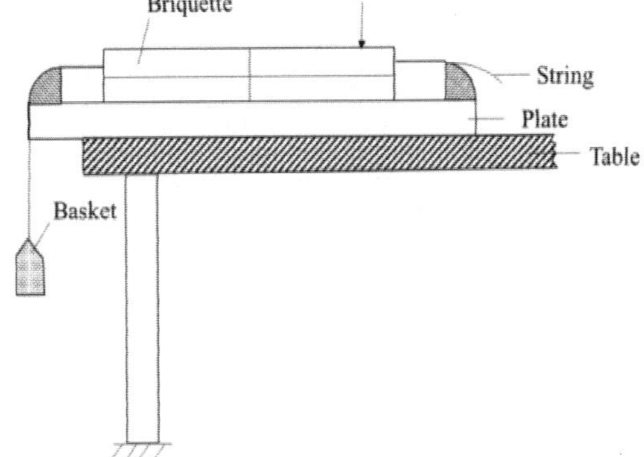

Figure 3. Schema of the test set-up of plastic tensile strength of mortal

3. Results and discussion

3.1. Effect of air-entraining agent on plastic shrinkage cracking of mortar[6]

The mortar consists of cement, sand and water in a weight ratio of 1: 1: 0.5. Adding air-entraining agent in a ratio of 0.005 wt.%~0.025 wt.% to the mortar according to GB50119-2013.

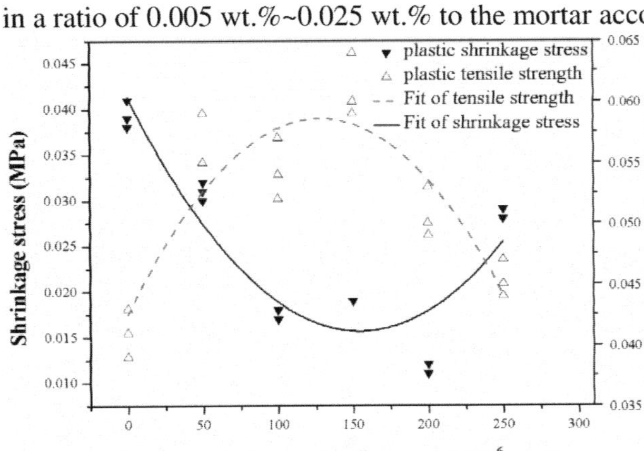

Figure 4. Relationship between plastic shrinkage index and air-entraining agent

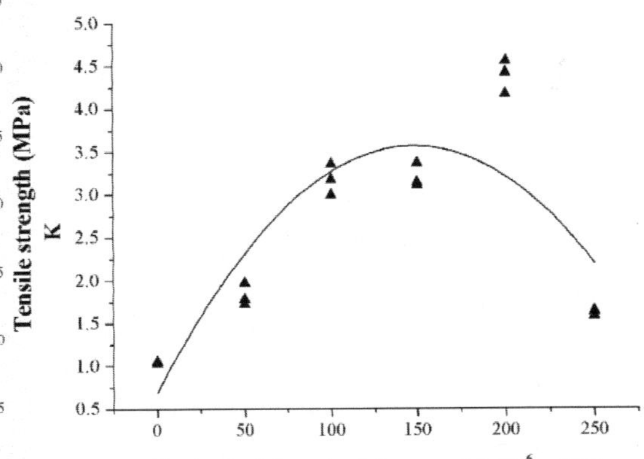

Figure 5. Relationship between anti-cracking index K and content of air-entraining agent

The effect of air-entraining agent content on plastic shrinkage stress and plastic tensile strength is presented in figure 4. The correlation coefficients (R^2) of plastic shrinkage stress and plastic tensile strength fitting are respectively 0.83 and 0.74. It is shown that the shrinkage stress will decrease with the increase of air-entraining agent content in the initial stage. The minimum of shrinkage stress is 0.0108MPa when the dosage is 200×10^{-6} wt.%. After that, the shrinkage stress will increase with the increase of air-entraining agent content. As can be seen from figure 4, the tensile strength of mortar will increase first and then reduces with the air-entraining agent content increased.

The relationship between the anti-cracking index K and the dosage of the air-entraining agent is shown in figure 5. The correlation coefficient (R^2) is 0.68. It is seen that with the increase of the amount of air-entraining agent, the anti-cracking index K of the mortar increases first and then decreases, which is similar to the trend of the tensile strength. The maximum value of K will achieve when the air-entraining agent dosage is 200×10^{-6} wt.%.

We believe that the phenomenon mentioned above after introduction of air-entraining agent can be explained in the following three aspects. (1) As a surfactant, the air-entraining agent reduces the surface tension of water which leads to the reduction of the shrinkage stress of capillary concave surface. At the same time, the bubbles introduced by air-entraining agent impede the rate of moisture migration, resulting in a decrease in the rate of capillary concave surfaces formation. (2) The increase in tensile strength is due to the increase of viscosity in the internal porosity of the liquid caused by the air-entraining agent. In addition, the micro-bubble introduced by air-entraining agent has a certain mechanical strength. When the dosage of air-entraining agent exceeds a certain amount, the original micro-bubbles connect with each other to form a large air bubble, which reduces the overall mechanical strength, thereby decreasing the tensile strength. (3) The moisture migration is impeded by the bubbles and the formation rate of capillary concave surface reduced owing to the introduction of air-entraining agent, which resulting in the extension of cracking time.

3.2. Effect of water reducing agent on plastic shrinkage cracking of mortar

The mortar consists of cement, sand and water in a weight ratio of 1: 1: 0.45. The temperature, humidity and wind speed are set to 23°C, 45% and 3.5m/s, respectively. Adding water reducing agent in a ratio of 0.01wt.%~0.08 wt.% to the mortar.

The influence of the water reducing agent content on plastic shrinkage stress and tensile strength of mortar is presented in figure 6. The correlation coefficient (R^2) of tensile strength fitting is 0.75. With the increase of the amount of water reducing agent, the plastic tensile strength tends to decrease gradually, while the plastic shrinkage stress is basically unchanged. Figure 7 is the fitting curve of water reducing agent content and anti-cracking index. It is known that the anti-cracking index tends to decrease as the amount of water reducing agent increases. The fitting equation of water reducing agent content and the mortar anti-cracking index is shown in equation (1). J indicates the amount of water reducing agent, and the correlation coefficient (R^2) is 0.58.

$$K=-2.28J+1.05 \tag{1}$$

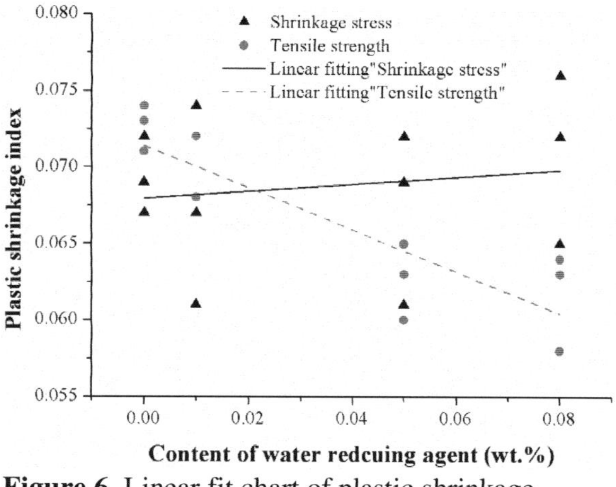

Figure 6. Linear fit chart of plastic shrinkage index and content of water reducing agent

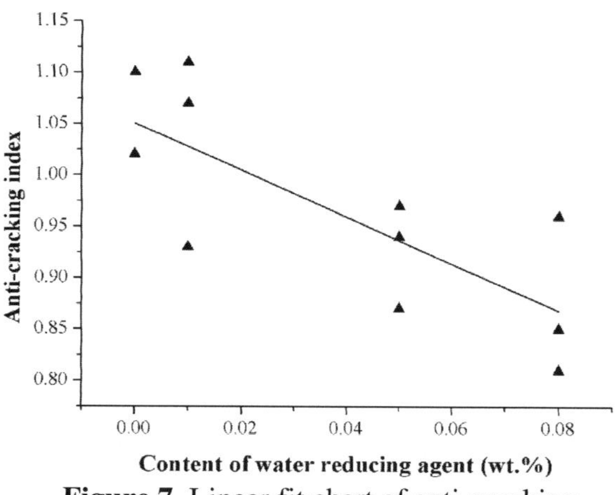

Figure 7. Linear fit chart of anti-cracking index and content of water reducing agent

In order to verify the reliability of the equation, we still choose a new content of water reducing agent for verification. The temperature, humidity and wind speed are set to 23°C, 45% and 3.5m/s, respectively. The water reducing agent dosage is 0.06 wt.%. The value of the anti-cracking index calculated by equation (1) is 0.91, so we predicted that the mortar will crack. In the verification experiment, the mortar cracked at 218min and the actual anti-cracking index was 0.85. The relative error between the actual measured value and the theoretical calculated is 7.1%.

3.3. Effect of water retaining agent on plastic shrinkage cracking of cement mortar

Adding water retaining agent in a ratio of 0.02 wt.%、0.05 wt.% and 0.1 wt.% to the mortar (Water-cement ratio is 0.45, Cement sand ratio is 1:1). The temperature, humidity and wind speed are set to 23°C, 60% and 3.5m/s, respectively.

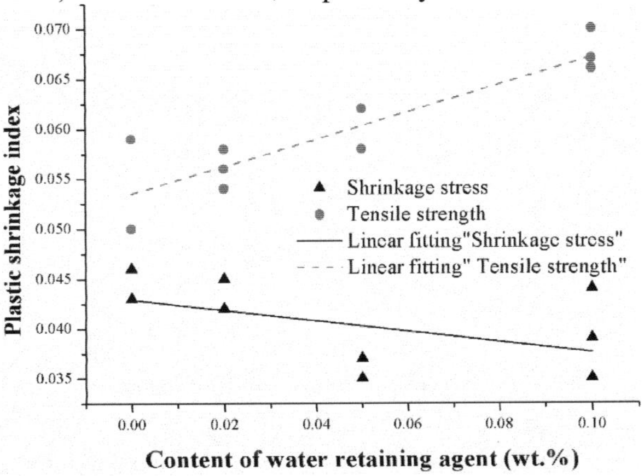

Figure 8. Linear fit chart of plastic shrinkage index and content of water retaining agent

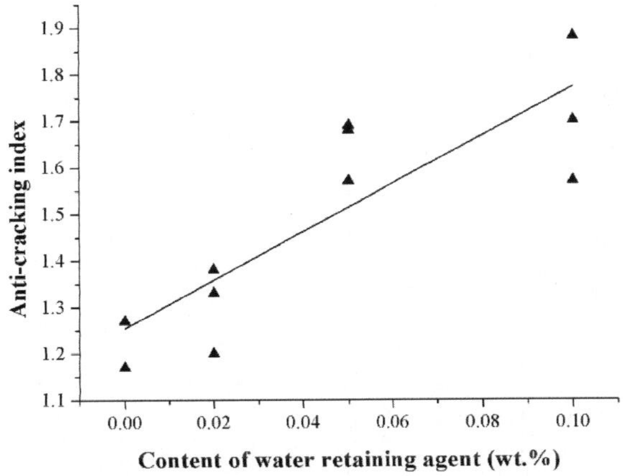

Figure 9. Linear fit chart of anti-cracking index and content of water retaining agent

The effect of water retaining agent on the plastic shrinkage stress and plastic tensile strength is presented in figure 8. The correlation coefficient (R^2) of tensile strength fitting is 0.77. Figure 9 is the relationship between the water retaining agent dosage and anti-cracking index K. As shown in figure 8, the plastic shrinkage stress of the mortar decreases with the addition of water retaining agent while the plastic tensile strength increases gradually. From figure 9, the anti-cracking index increases with the addition of water retaining agent. The plastic shrinkage cracking of mortar specimens occurs if the content of water retaining agent is less than 0.05 wt.%. The fitting equation of water retaining agent content and mortar anti-cracking index is shown in equation (2). *B* indicates the water retaining agent content, and the correlation coefficient (R^2) is 0.69.

$$K=5.18B+1.25 \tag{2}$$

A new experiment was carried out to verify the equation (2). The temperature, humidity and wind speed are set to 23°C, 60% and 3.5m/s, respectively. The dosage of water retaining agent is 0.07 wt.%. The value of the anti-cracking index calculated by the equation (2) is 1.61. In the verification experiment, the mortar did not crack and the actual anti-cracking index is 1.55. The relative error between the actual measured value and the theoretical calculated value is 3.9%.

3.4. Effect of shrinkage reducing agent on plastic shrinkage cracking of cement mortar

Adding shrinkage reducing agent in a ratio of 0.01 wt.%~0.1 wt.% to the mortar (Water-cement ratio is 0.45, Cement sand ratio is 1:1). The plastic shrinkage stress and tensile strength of the test mortar are presented in figure 10. The correlation coefficients (R^2) of plastic shrinkage stress and plastic tensile strength fitting are respectively 0.92 and 0.91. Figure 11 is the relationship between the anti-cracking index and the amount of shrinkage reducing agent. The correlation coefficient (R^2) is 0.86. From figure 10, we can see that the plastic shrinkage stress and the plastic tensile strength decrease with the addition of the shrinkage reducing agent. As is shown in figure 11, the anti-cracking index increases linearly with the increase of the shrinkage reducing agent content.

On one hand, the incorporation of shrinkage reducing agent reduces the surface tension of the water on the mortar surface, thereby reducing the capillary shrinkage stress caused by water evaporation. In addition, due to the incorporation of shrinkage reducing agent, the mortar mixture has a certain expansion, which compensates for the plastic shrinkage of mortal. So, the plastic shrinkage stress would reduce with the addition of shrinkage reducing agent. On the other hand, the solidification and hardening process of the mortar is delayed owing to the water retention and retarding effect of shrinkage reducing agent which leads to the reduction of plastic tensile strength. A certain amount of bleed air was observed in the experiment that resulted in a certain expansion of the specimen, increased porosity in the slurry and decreased the cohesiveness of the mortar, which can also be seen as a cause of a significant reduction in plastic tensile strength.

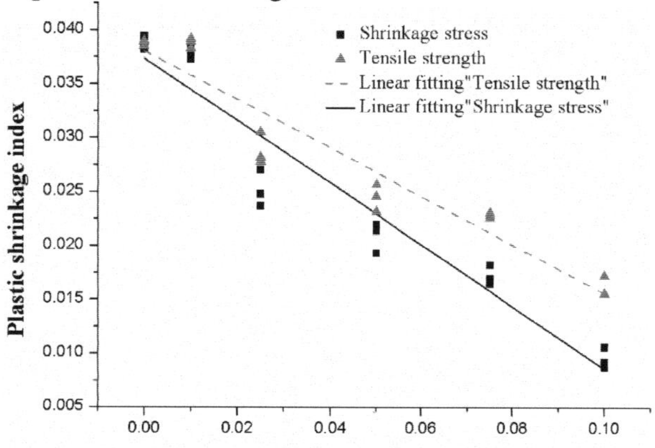

Figure 10. Linear fit chart of plastic shrinkage index and content of shrinkage reducing agent

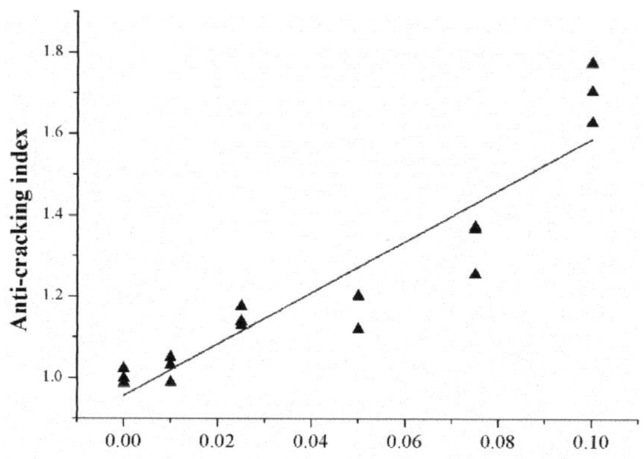

Figure 11. Linear fit chart of anti-cracking index and content of shrinkage reducing agent

4. Conclusions

(1) With the increase of air-entraining agent dosage, the anti-cracking index of mortar increases first and then decreases. In the range of experimental parameters, the best anti-cracking effect can be achieved with about the dosage of 200×10^{-6} wt.%, while the more accurate value should be further studied.
(2) The anti-cracking index of mortar decreases with the addition of water reducing agent.
(3) With the increase of the amount of water retaining agent, the anti-cracking index shows an increasing trend, and shrinkage cracks can be significantly reduced.
(4) After the addition of shrinkage reducing agent, the anti-cracking index increases significantly, which indicates the improvement of anti-cracking performance about the mortal.

Acknowledgments

The authors were grateful for the financial supports form the project of Thirteen Five National Key Research and Development Programs (2016YFB0303200).

References

[1] Leemann A, Nygaard P and Lura P 2014 *Cement and Concrete Composites.* (**46**)**:**1-7.
[2] Wyrzykowski M, Trtik P, Münch B, *et al.* 2015 *Cement and Concrete Research.* (**73**):238-245.
[3] Boshoff W P and Combrinck R 2013 *Cement and Concrete Research.* (**48**):34-39.
[4] Ma Y P, Zhu B R and Tan M H 2003 *Journal of Building Materials.* **6** (1):20-24. (in Chinese)
[5] Ma Y P and Wang P M 2006 *Journal of Building Materials.* **9** (5):533-537. (in Chinese)
[6] Ma Y P, Tan C, Yang X J, *et al.* 2014 *Materials review.* (**28**):342-348. (in Chinese)

Effect of three-step mixing technology based on vibratory mixing on properties of high-strength concrete

Kaiyin Zhao[1], Lijun Zhao[1*], Shanshan Liu[1], Jinru Hou[1]

[1] Key Laboratory of Highway Construction Technology and Equipment of Ministry of Education, Chang'an University, Xi'an 710064, China
E-mail: zhaolj@chd.edu.cn

Abstract: A large number of experimental studies have shown that both vibratory mixing and multi-step mixing processes can improve the workability of concrete and improve the strength and durability of high-strength concrete(HSC). In this paper, for C60 HSC, two kinds of mixing methods: conventional compulsory mixing and vibratory mixing, two kinds of mixing technology: one-step mixing technology and three-step mixing technology are used to carry out contrast test. The purpose is to study the influence of mixing method and mixing technology on the performance of HSC. The research results show that compared with the one-step mixing technology, the three-step mixing technology can significantly reduce the plastic viscosity of the concrete, improve its fluidity, and significantly improve the compressive strength at the age of and after 28 days and 56 days under the condition of the conventional compulsory mixing and the vibratory mixing. And the study found that the most suitable mixing method and mixing technology for C60 HSC are vibratory mixing method and three-step mixing technology.

***Key words*:** vibratory mixing; HSC; three-step mixing; mixing technology

1. Introduction

Because of its high strength, small deformation and good durability, HSC has been widely used in high-rise building structure, large-span bridge structure and some special structures, which is more suitable for the development of modernization than ordinary concrete. A large number of experiments

*Corresponding author. E-mail: zhaolj@chd.edu.cn
The project supported by Natural Science Basic Research Plan in Shaanxi Province of China (Program No. 2017JM5118).

[1-5] show that the vibratory mixing and three-step mixing technology can improve the performance of HSC, but how to combine the two methods has not been studied yet. Therefore, in order to study the effects of different mixing methods and technologies on the performance of HSC, two kinds of mixing methods (conventional compulsory mixing and vibratory mixing) and two kinds of mixing technologies (one-step mixing technology and three-step mixing technology), were combined and compared various scheme tests. The results show that both under the condition of the conventional compulsory mixing and the vibratory mixing, compared with the one-step mixing technology, three-step mixing technology can significantly reduce the plastic viscosity of concrete, improve its fluidity, and obviously enhance the compressive strength of HSC at 28 days and 56 days, and vibratory mixing method and three-step mixing technology are the optimal combination of mixing method and mixing technology.

2. Experiment details

2.1. Equipment and Materials

The test mixer was a 60L twin-shaft vibrating mixer (No vibratory mixing and Vibratory mixing).

(1)P.O52.5R ordinary Portland cements;

(2)The poly carboxylate superplasticizer with high water reduction rate is selected in this experiment. The dosage is 0.7～1.8% , and 1.3% is selected;

(3)The coarse aggregate includes two grades in 5～10mm and 10～20mm. In this experiment, the amount of each of the two materials is calculated according to 10~20mm : 5~10mm=6 : 4.

2.2. Mix proportion design

The compressive strength grade of HSC is designed as C60, the sand ratio is 33%, and the water-cement ratio is 0.33. According to the raw material test results, the design ratio of HSC is shown in Table 1.

Table 1. The concrete mix proportion.

Materials	Water /(kg/m^3)	Cement /(kg/m^3)	Fly ash /(kg/m^3)	Water reducer /(kg/m^3)	Coarse aggregate /(kg/m^3)	Fine aggregate /(kg/m^3)
Amount /(kg/m^3)	165	500	0	5	1229	606
Mix proportion	0.33	1	0	0.01	2.458	1.212

2.3. Mixing methods

As a kind of effective method to strengthen the concrete mixing process, vibratory mixing is the most representative of the research done by Professor Feng Zhongxu of Chang'an University in China. The research group has done a lot of experimental research on the influence of vibratory mixing on concrete performance [3][8-10]. The research found that: (1) vibratory mixing can significantly increase the air content of concrete; (2) vibratory mixing can improve the interface between cement stone and coarse aggregate and the morphology of cement hydration slurry in concrete, so that its strength and durability can be greatly improved.

In this paper, HSC is mixed by vibratory mixing and conventional compulsory mixing. The effects of different mixing methods on the performance of HSC are compared and analyzed. The vibration parameters are selected as shown in Table 2.

Table2. The vibration parameters

Number	Amplitude/mm	Vibration acceleration/g
1	0	0
2	0.82	4
3	0.82	8
4	1.36	4
5	1.36	8

2.4. Mixing technology

The one-step mixing technology and the three-step mixing technology are adopted in this experiment. Under the same raw material and mixing ratio, the performance of fresh concrete and the compressive strength of concrete after 28 days and 56days were compared and analyzed between the two mixing technologies.

The one-step mixing technology is that cement, water, water reducer, coarse aggregate and fine aggregate are added at the same time. The mixing time is 150 seconds.

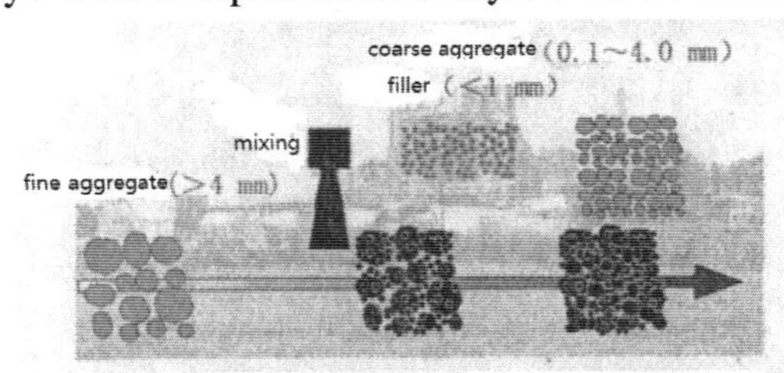

Figure 1. Asphalt mixture mixing method

Three-step mixing technology used in this paper is an improved multi-step mixing technology. The improvement idea is derived from a new method used by Hu Guiling [6] for mixing asphalt mixture, as shown in Figure 1.Three-step mixing technology is that coarse aggregate, 30%water and 30% water reducer are added leading to the formation of a thin water film on the surface of coarse aggregate during the first step of mixing. At the second step of mixing, cement is added slowly (Feeding duration is 10 seconds) leading to the surface of the wet stone wrapped evenly by the cement. At the third step of mixing, the remaining water, water reducer and fine aggregate are added to complete the concrete mixing process. The specific process is shown in Figure 2.

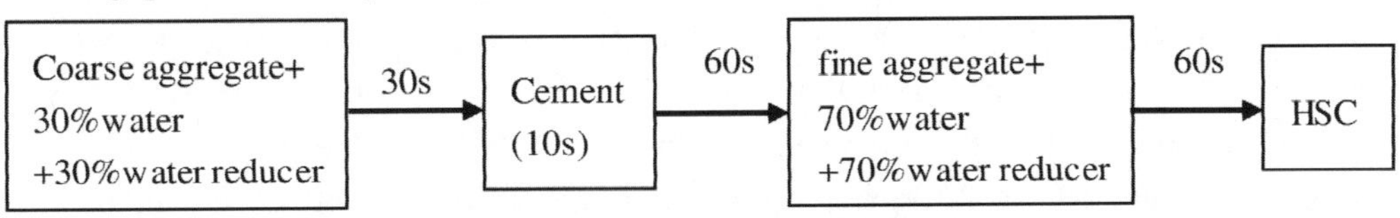

Figure 2. Three-step mixing

(a) Stones after mixing for 30 seconds **(b)** After adding cement for 60s

Figure 3 Three-step mixing process

3. Test results and discussions

3.1. The rheology of fresh concrete

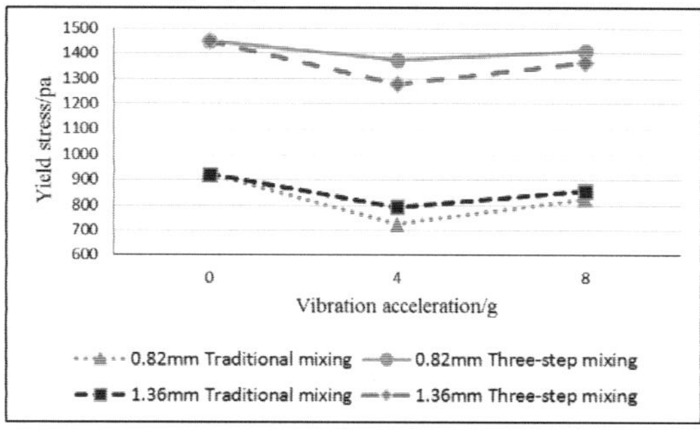

Figure4 Relationship between Plastic viscosity and Vibration acceleration under different amplitude

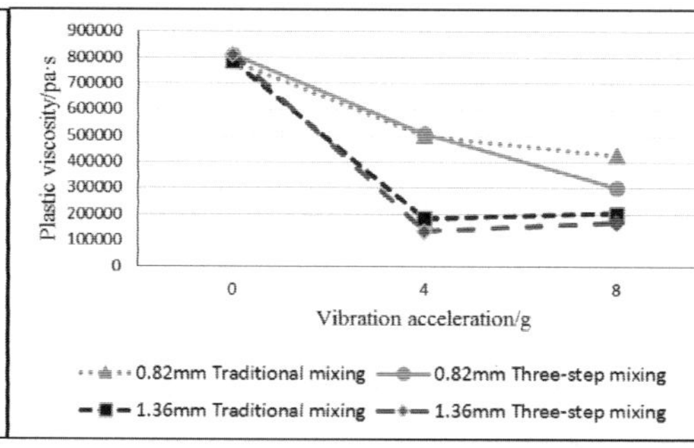

Figure5 Relationship between Yield stress and Vibration acceleration under different amplitude

Figure 4 show the relationship between yield stress of HSC and vibration acceleration at different amplitude. Analysis of the test results found that vibration mixing can significantly reduce the yield stress and plastic viscosity of concrete compared with ordinary mixing. Compared with the one-step mixing technology, the three-step mixing technology does not reduce the yield stress of the HSC, which is related to the amount of water used in the third mixing process. The reason for this result is that the water addition amount in the third step is 70%, the 70% water added time is shorter, and another reason is that a large amount of water is easily leaked in the batch mixer, resulting in the lower water distribution uniformity. Figure 5 show that compared with the one-step mixing technology, three-step mixing technology reduces the plastic viscosity of the HSC, which is reduced by 25.9%.

Studies [12] have shown that the yield stress and plastic viscosity of HSC are related to unit water consumption. In this experiment, the slump of the three-step mixing technology is reduced because of the reduction of the unit water consumption, which leads to the increase in the yield stress. However, the "shell-forming" process of the cement-coated stone method increase the amount of slurry between the aggregates, and the internal movement of the concrete is more easy, which greatly reduces the plastic viscosity of the HSC and improves the fluidity

IOP Conf. Series: Materials Science and Engineering **542** (2019) 012003 doi:10.1088/1757-899X/542/1/012003

of the HSC.

3.2. The air content of fresh concrete

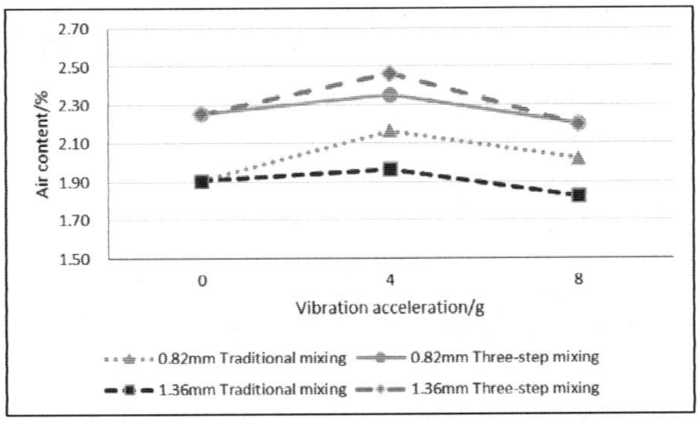

Figure6 Relationship between Air content and Vibration acceleration under different amplitudes.

As shown in Figure 6, the vibratory mixing can increase the air content of the HSC, and as the vibration acceleration increases, the concrete gas content quickly increases to a maximum value (up to 2.5%), and as the vibration acceleration continues to increase, the air content of the HSC decreases. Studies have shown that [13-14] gas content plays an important role in the durability of concrete, especially for freezing. Figure6 show that both vibratory mixing method and three-step mixing technology can increase the air content of HSC. And the air content is the highest and durability is the best when the amplitude is 1.36 mm, the vibration acceleration is 4 g and the mixing technology is three-step mixing technology.

3.3. Compressive strength of HSC

The standard test block of 100 mm × 100 mm × 100 mm was prepared by the one-step mixing technology and the three-step mixing technology, then the standard curing was carried out, and the compressive strength test was carried out separately. The test results are shown in Tables 3.

Table3. Compressive strength test results

Mixing technology	Mixing method			Compressive strength of hardened concrete at 28 days		Compressive strength of hardened concrete at 56 days	
		Amplitude /mm	Vibration acceleration/g	$\bar{f}$/Mpa	Cv	$\bar{f}$/Mpa	Cv
One-step mixing technology	Conventional compulsory mixing	0	0	60.76	0.067	63.84	0.066
	Vibratory mixing	0.82	4	64.57	0.111	72.27	0.061
			8	66.38	0.064	77.34	0.096
		1.36	4	61.15	0.106	75.45	0.125
			8	63.46	0.076	77.16	0.123
Three-step mixing technology	Conventional compulsory mixing	0	0	63.52	0.081	68.14	0.088
	Vibratory mixing	0.82	4	65.07	0.079	74.73	0.055
			8	66.55	0.088	75.45	0.097
			4	67.86	0.036	77.94	0.041

IOP Conf. Series: Materials Science and Engineering **542** (2019) 012003 doi:10.1088/1757-899X/542/1/012003

1.36	8	65.18	0.065	75.16	0.052

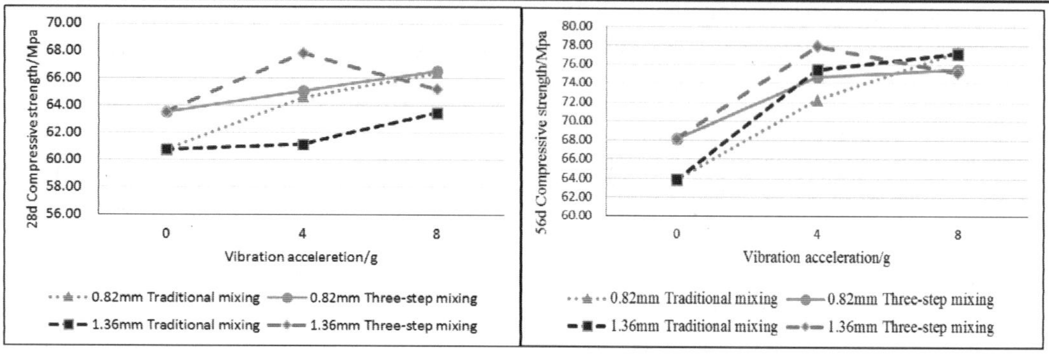

(a) 28days and 56days compressive strength – vibration acceleration

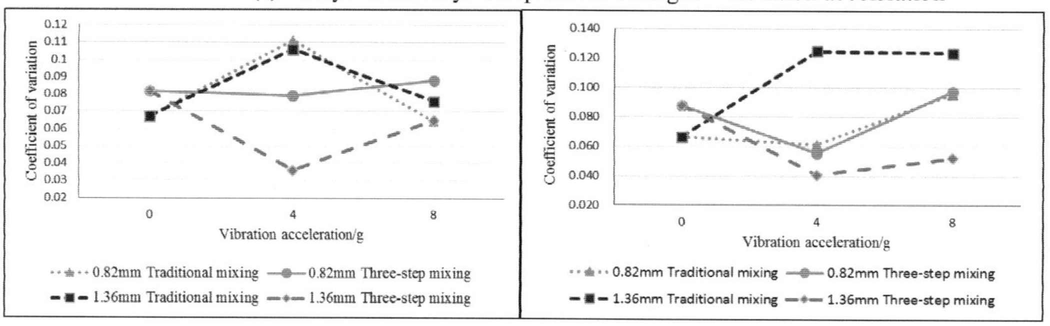

(b) Coefficient of variation – vibration acceleration

Figure 7 Relationship between Compressive strength (coefficient of variation) and Vibration acceleration under different amplitudes

Figure 7 shows the relationship between the compressive strength and the coefficient of variation and the vibration acceleration of HSC at 28 and 56 days.

(1)Figure7 (a) shows that compared with the one-step mixing technology, three-step mixing technology both increased the compressive strength of concrete at 28days and 56days by 2.9% ~ 10.9% the under the condition of the conventional compulsory mixing method and the vibratory mixing method.;

(2) Figure7 (b) shows that under conventional compulsory mixing method, the coefficient of variation of compressive strength of three-step mixing technology at 28 days and 56 days are greater than those of one-step mixing technology; under vibration mixing method, the coefficient of variation of compressive strength of three-step mixing technology can be reduced at 28 days and 56 days. And when the amplitude is 1.36mm, the vibration acceleration is 4g, the coefficient of variation is the smallest.

4. Conclusion

(1) Under the conventional compulsory mixing method, compared with the one-step mixing technology, the three-step mixing technology can obviously reduce the plastic viscosity of HSC, improve its fluidity, workability, and significantly improve the compressive strength of HSC at 28 days and 56 days.

(2) Under the vibratory mixing method, compared with the one-step mixing technology, the three-step mixing technology obviously reduces the plastic viscosity and improves the

fluidity of concrete, and the compressive strength of HSC at 28 days is obviously improved. The optimum vibration parameter is that the vibration acceleration. is 4g, the amplitude is 1.36mm.

(3) The yield stress of HSC (HSC) is not reduced by the three-step mixing technology either in the conventional compulsory mixing method and in the vibratory mixing method, which is related to the leakage of water in the mixer and the unit water consumption in every step, which needs to be further verified by a large number of subsequent tests.

Acknowledgements
The project was supported by Natural Science Basic Research Plan in Shaanxi Province of China (Program No. 2017JM5118). The project was also supported by China Postdoctoral Science Foundation (No.: 2018M633443) and the Special Fund for Basic Scientific Research of Central Colleges, Chang'an University (No.: 310826171002).

Reference

[1] TATTERSALL G. H, BAKER P. H. The effect of vibration on the rheological properties of fresh concrete[J] *Magazine of Concrete Research*, 1988,40(143):79-88

[2] Peng Y. Study on Mixing Technology of Mixing HSC with Double Screw Shaft Mixer [D].*Chang'an University*, 2014.

[3] Weiming Q, Shaomin Song. Effect of Vibratory Mixing on Performance of High Strength Fluid Concrete [J]. *Magazine of Concrete World*, 2017(12):90-95.

[4] Huaying Z, Zhongjian Sun. Application Research of Pure Pulp Wrap Process in Formulating HSC [J]. *Journal of Foshan University of Science and Technology (Natural Science Edition)*, 2006(02):14-16.

[5] Xianghong F.. Influence by Multi-step Mixing Technology to the Strength and Frost-resistance of High Performance Concrete [J]. *Magazine of Concrete*,2014(06):157-160.

[6] Guiling H. The new Mixing Method of Asphalt Mixture[J]. *Magazine of Asphalt*, 2011, 25（06）,46-48.

[7] Chengcheng S, Dong Y. Study on the Influence of Secondary Mixing Process on the Strength of Recycled Concrete [J]. *Magazine of Concrete*,2008(06):125-128.

[8] Zhongxu F. Research Progress on Mixing Theory and Equipment [J]. *Magazine of Construction Machinery*, 2014（5）: 1-7.

[9] Liangqi Z. Research on Vibratory Mixing Mechanism and Industrial Application of Concrete [D].Xian: *Chang'an University*, 2013

[10] Shaojie Y, Shaomin . Effect of Vibratory Mixing on Performance of Large Fluid Concrete [J]. *Magazine of Concrete*,2017(08):152-155.

[11] Lin L. Experimental Method Based on Rheological Properties of Large Fluid Concrete [D].*Chongqing University*,2015

[12] FEYS D, VERHOEVEN R, DE SCHUTTER G. Why is fresh self-compacting concrete shear Thickening ? [J]. *Cement and Concrete Research*, 2009, 39(6): 510-23.

[13] Bassuoni M T, Nehdi M L. The Case for Air-entrainment in High-performance Concrete[J]. *Proceedings of the Institution of Civil Engineers, Structures and Buildings*, 2005,158(SB5):311-319

[14] Ksaibati, K; Zeng, M; Sellers, F R; Dolan, C W. Air Change in Hydraulic Concrete due to Pumping [J]. *Transportation Research Record*, 2003,18(34): 85-92

IOP Conf. Series: Materials Science and Engineering **542** (2019) 012004 doi:10.1088/1757-899X/542/1/012004

Corrosion of aluminium in water-in-diesel-emulsion fuel

D Chandran, M Khalid, R Raviadaran and PR Jagadish

Graphene & Advanced 2D Materials Research Group (GAMRG), School of Science and Technology, Sunway University, Malaysia
Email: dchandran@sunway.edu.my

Abstract. This study aimed to determine the corrosion characteristics of water-in-diesel emulsion fuel towards aluminium. Investigation was performed by immersing aluminium coupons in emulsified diesel for 240 hours, 480 hours, 720 hours, 960 hours and 1200 hours at 25 °C. For comparison, additional aluminium coupons were immersed in diesel for 1200 hours at 25 °C. Span 80 and Tween 80 surfactants were used to prepare the emulsified diesel which consisted of 5 vol% of water and 95 vol% of diesel. Mass loss of the coupons were measured to calculate the corrosion rate. Water content, total acid number, density and viscosity of the fuel samples were measured before and after immersion. The corrosion rate of aluminium coupons exposed to emulsified diesel reduced by 97.6% between 240 hours and 1200 hours of immersion duration. The aluminium coupons immersed in emulsified diesel had 2.2 times higher corrosion rate than in diesel. The higher total acid number of emulsified diesel by 4.7 times than diesel is suggested to have influenced this. To benefit the potential reduction in harmful exhaust emission through the adoption of emulsified diesel, studies towards producing less acidic stable emulsified diesel are essential.

1. Introduction

Reduction in the harmful emission produced by vehicles are absolutely necessary to address the declining air quality. Declined air quality are among the major reason leading to respiratory diseases and may also cause neurological disorders. With regard to this, research works have demonstrated the possibilities of reducing harmful exhaust emission from diesel engine by adopting water-in-diesel emulsion fuel (emulsified diesel). Ithnin et al. [1] reported reduced nitrogen oxide and particulate matter by 41% and 35%, respectively, for using emulsified diesel with 20 vol% of water as compared to diesel. In another study, the authors reported reduction in nitrogen oxide and particulate matter by 32% and 16%, respectively, with the use of emulsified diesel consisting of 5 vol% of water as compared to neat diesel [2]. Several other studies have also exhibited similar outcome [3-5]. However, limited information is available till present concerning the compatibility of emulsified diesel with fuel delivery materials. Use of incompatible fuel could significantly reduce the components lifespan culminating in stalling the engine due to fuel leakage [6, 7]. This study therefore investigated the corrosive effect of emulsified diesel on aluminium (common material for high pressure fuel pump) under immersion investigation.

2. Materials and methods

Aluminium coupons with a diameter of 0.02 m, thickness of 0.002 m and an elemental composition of 99.59 wt% of aluminium, 0.28 wt% of iron, 0.10 wt% of silicon and 0.03 wt% of zinc were used in this study. Emulsified diesel was prepared using not additized diesel, distilled water prepared from TKA Smart2Pure and surfactants (Span 80 and Tween 80). The emulsion consisted of 5 vol% of water, 1.5 vol% of Span 80, 1.5 vol% of Tween 80 and 92 vol% of diesel. The emulsion was prepared using Silverstone LM-5 high shear mixer at a rotational speed of 3500 revolutions/min. The immersion investigations were carried out by employing ASTM G1 and ASTM G31 standard methods. Each coupon was immersed in 38 ml of emulsified diesel according to SAE J1747 standard method, stored in a closed and dark environment at 25 °C for 240 hours, 480 hours, 720 hours, 960 hours and 1200 hours. Additionally, coupons were also immersed in diesel for 1200 hours at 25 °C for comparisons. For each condition, an additional coupon was analysed as duplicate. The mass of the coupons was measured before and after the immersion. Details on coupons preparation/cleaning have been reported earlier [8]. The corrosion rate was calculated using the average mass loss. The water content, total acid number (TAN), viscosity and density of the fuels were measured before and after the immersion. The specific details concerning the analysis procedure and utilized equipment have been reported elsewhere [9, 10].

3. Results and discussion

As shown in Figure 1, the corrosion rate of aluminium exposed to emulsified diesel reduced by up to 97.6% with increasing immersion duration from 240 hours to 1200 hours. For the comparative corrosion rate of aluminium in emulsified diesel and diesel after 1200 hours of immersion at 25 °C, 2.2 times higher corrosion rate occurred to the coupon exposed to emulsified diesel at 0.000249 mm/yr than to diesel at 0.000116 mm/yr. Comparisons to the existing studies were not made possible due to the limited information available in this area. The trend of reducing corrosion rate with increasing immersion duration nevertheless corroborated with the study reported by Chandran et al. [9] in which the authors reported reduced corrosion rate of copper due to biodiesel exposure by 2.8% after 540 hours as compared to 108 hours of immersion duration. Fazal et al. [11] also reported likewise in which the corrosion rate of copper reduced by 19.5% due to biodiesel exposure between 1200 hours and 2880 hours of immersion duration.

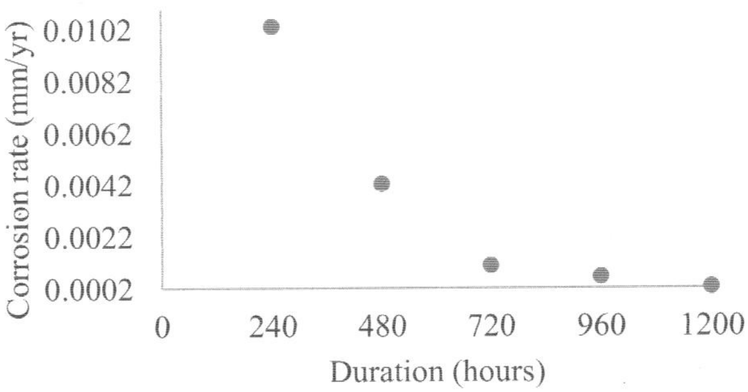

Figure 1. Corrosion rate of aluminium in emulsified diesel with respect to immersion duration at 25 °C.

Figure 2 shows the TAN of emulsified diesel due to aluminium coupon exposure for 240 hours, 480 hours, 720 hours, 960 hours and 1200 hours. The TAN of the emulsified diesel is observed to reduce with increasing immersion duration. 12% reduction in TAN was measured between 240 hour and 1200 hours of immersion duration. As shown in Figure 3, the water content of the emulsified diesel remained similar (< 0.03% change) after exposed to aluminium coupon for 240 hours, 480 hours, 720 hours, 960 hours and 1200 hours. From Figure 4, it is noticed that the viscosity of the emulsified diesel increased with increasing immersion duration of aluminium coupons. The viscosity value increased by 20.3% between 240 hour and 1200 hours of immersion duration. As shown in Figure 5, the density of the emulsified diesel is observed to increase with increasing immersion duration. The density of the emulsified diesel increased by 0.7% between 240 hours and 1200 hours of aluminium coupon exposure.

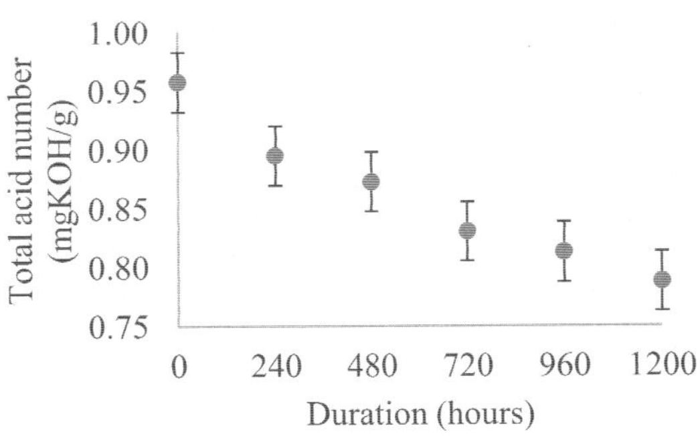

IOP Conf. Series: Materials Science and Engineering **542** (2019) 012004 doi:10.1088/1757-899X/542/1/012004

Figure 2. Changes in total acid number of emulsified diesel exposed to aluminium with respect to immersion duration.

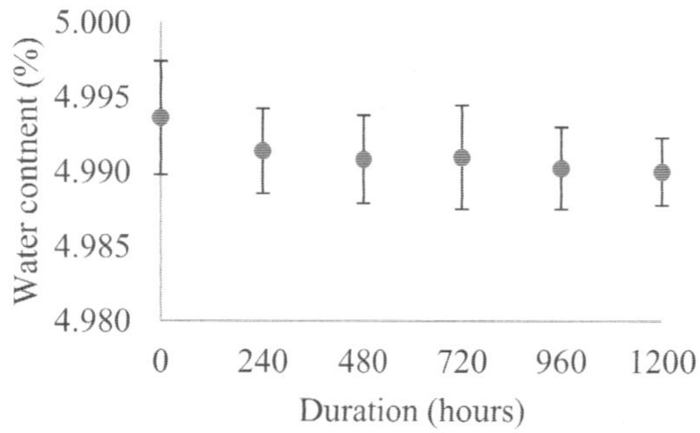

Figure 3. Changes in water content of emulsified diesel exposed to aluminium with respect to immersion duration.

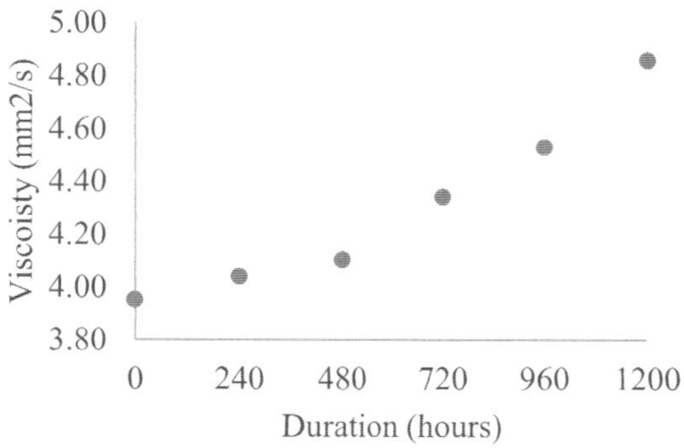

Figure 4. Changes in viscosity of emulsified diesel exposed to aluminium with respect to immersion duration.

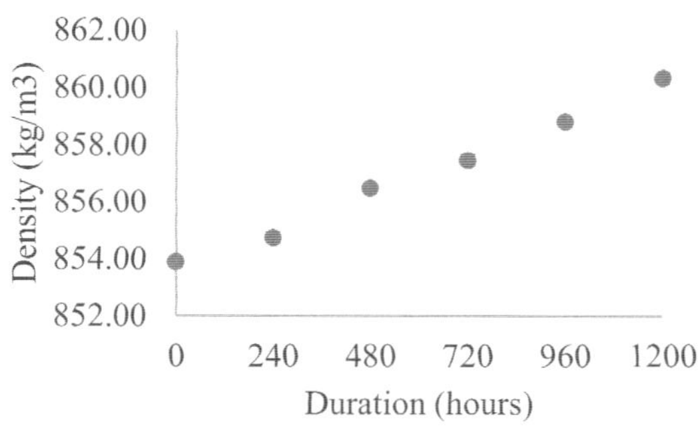

Figure 5. Changes in density of emulsified diesel exposed to aluminium with respect to immersion duration.

As shown in Table 1, the prepared emulsified diesel has 4.7 times higher TAN than diesel. The surfactants Span 80 and Tween 80 which has a TAN of 13.90 mgKOH/g and 2.65 mgKOH/g, respectively, used in the preparation of emulsified diesel is suggested to have influenced the rise. After exposed to aluminium, the TAN reduced by 18% for emulsified diesel, while it increased by 39% for the diesel. The reduction in TAN for the emulsified diesel suggest the involvement of the acidic substances in the corrosion process. The prepared emulsified diesel has 768 times higher water content than diesel. The ~5% water content in the emulsified diesel reflects the 5 vol% of water used to produce the emulsified diesel. The water content for both diesel and emulsified diesel did not change after exposed to aluminium coupon. The prepared emulsified diesel has 23% higher viscosity value than diesel. The presence of water in the form of emulsion in

IOP Conf. Series: Materials Science and Engineering **542** (2019) 012004 doi:10.1088/1757-899X/542/1/012004

diesel could have caused this rise. Existing studies have reported up to 50% rise in viscosity value to formation of emulsified diesel with 5 vol% of water [12, 13]. 18% increase in the viscosity value occurred for the emulsified diesel after exposed to aluminium while it did not change for diesel. The increase in the water droplets size could have influenced this rise for the emulsified diesel. The emulsified diesel has 2% higher density than diesel. The presence of 5 vol% of water in the emulsified diesel could have caused this rise since water has a higher density value by 19% than diesel. Existing studies have reported increased density value by up to 3.5% for emulsified diesel with 5 vol% of water than diesel [12, 13]. After exposed to aluminium,
< 1% increase in density occurred for emulsified diesel while it remained close to as-received for diesel.

Table 1. Changes in the properties of diesel and emulsified diesel due to exposure to aluminium.

	Diesel				Emulsified diesel			
	Before		After		Before		After	
	Mean	SD	Mean	SD	Mean	SD	Mean	SD
TAN (mgKOH/g)	0.21	1.29E-02	0.29	5.77E-03	0.96	5.00E-03	0.79	9.57E-03
Water content (%)	0.007	5.77E-04	0.008	5.77E-04	4.994	7.63E-03	4.990	4.54E-03
Viscosity (mm2/s)	3.358	7.41E-04	3.361	7.39E-04	3.952	7.41E-04	4.859	1.56E-03
Density (kg/m3)	838.1	1.26E-01	838.7	3.51E-01	853.9	3.30E-01	860.3	3.50E-01

By relating the properties of the fuels to the corrosion rate, the increased acidity of the emulsified diesel as compared to diesel is suggested to have influenced the higher corrosion rate. This suggestion corroborates to the trend of reduced corrosion rate with reduced TAN observed in this study, which also agrees to an existing study [14]. Emulsified diesel with lower TAN should therefore be formulated in order to prevent accelerated metal corrosion. The effect of water towards corrosion rate can be ruled out in this study since free water did not form during the investigation. Nevertheless, taking into account the de-mulsification of emulsified diesel which could culminate in the formation of free water [12], stable emulsified diesel are equivalently necessary to prevent accelerated metal corrosion.

4. Conclusion

The corrosive characteristics of emulsified diesel was investigated in this work. The corrosion rate of aluminium coupons exposed to emulsified diesel reduced by 97.6% with increasing immersion duration from 240 hours to 1200 hours. 2.2 times higher corrosion rate occurred for the aluminium coupons exposed to emulsified diesel than to diesel. Considering the influence of water could be overruled since free water did not form during the investigations, the higher acidic value of emulsified diesel by 4.7 times than diesel is suggested to have influenced the higher corrosion rate for the aluminium exposed to emulsified diesel than diesel. The surfactants used to formulate stable water-in-diesel emulsion to prevent free water formation are suggested to have increased the acidic value of the emulsified diesel thus adversely affected the metal corrosion. Studies towards producing stable emulsified diesel with lower acidic value are required to prevent metals failure due to accelerated corrosion. More studies in the area are required to establish the compatibility of emulsified diesel with fuel delivery materials.

5. References

[1] Ithnin A M, Ahmad M A, Bakar M A A, Rajoo S and Yahya W J 2015 Combustion performance and emission analysis of diesel engine fuelled with water-in-diesel emulsion fuel made from low-grade diesel fuel *Energy Convs. Manage.* **90** 375-82

[2] Ithnin A M, Yahya W J, Ahmad M A, Ramlan N A, Kadir H A, Sidik N A C and Koga T 2018 Emulsifier-free Water-in-Diesel emulsion fuel: Its stability behaviour, engine performance and exhaust emission *Fuel* **215** 454-62

[3] Hasannuddin A, Yahya W, Sarah S, Ithnin A, Syahrullail S, Sugeng D, Razak I, Fatah A A, Aqma W and Rahman A 2018 Performance, emissions and carbon deposit characteristics of diesel engine operating on emulsion fuel *Energy* **142** 496-506

[4] Ogunkoya D, Li S, Rojas O J and Fang T 2015 Performance, combustion, and emissions in a diesel engine operated with fuel-in-water emulsions based on lignin *Appl. Energy* **154** 851-61

[5] Perumal V and Ilangkumaran M 2018 Water emulsified hybrid pongamia biodiesel as a modified fuel for the experimental analysis of performance, combustion and emission characteristics of a direct injection diesel engine *Renew. Energy* **121** 623-31

[6] Chandran D, Ng H, Harrison L, Gan S, Choo Y and Jahis S 2016 Compatibility of biodiesel fuel with metals and elastomers in fuel delivery system of a diesel engine *J. Oil Palm Res.* **28** 64-73

[7] Chandran D 2017 Experimental investigation into the physico-chemical properties changes of palm biodiesel under common rail diesel engine operation for the elucidation of metal corrosion and elastomer degradation in fuel delivery system (Doctoral dissertation). University of Nottingham

[8] Chandran D, Gan S, Lau H L N, Raviadaran R, Salim M and Khalid M 2018 Critical relationship between biodiesel fuel properties and degradation of fuel delivery materials of a diesel engine *Therm. Sci. Eng. Prog.* **7** 20-6

[9] Chandran D, Ng H K, Lau H L N, Gan S and Choo Y M 2016 Investigation of the effects of palm biodiesel dissolved oxygen and conductivity on metal corrosion and elastomer degradation under novel immersion method *Appl. Therm. Eng.* **104** 294-308

[10] Chandran D, Ng H K, Lau H L N, Gan S and Choo Y M 2017 Deterioration of palm biodiesel fuel under common rail diesel engine operation *Energy* **120** 854-63

[11] Fazal M A, Haseeb A S M A and Masjuki H H 2013 Corrosion mechanism of copper in palm biodiesel *Corrosion Sci.* **67** 50-9

[12] Ismael M A, Heikal M R, Aziz A R A, Syah F and Crua C 2018 The effect of fuel injection equipment on the dispersed phase of water-in-diesel emulsions *Appl. Energy* **222** 762-71

[13] Elsanusi O A, Roy M M and Sidhu M S 2017 Experimental investigation on a diesel engine fueled by diesel-biodiesel blends and their emulsions at various engine operating conditions *Appl. Energy* **203** 582-93

[14] Tsuchiya T, Shiotani H, Goto S, Sugiyama G and Maeda A 2006 Japanese standards for diesel fuel containing 5% FAME: Investigation of acid generation in FAME blended diesel fuels and its impact on corrosion *SAE Technical Paper, 2006-01-3303*

Acknowledgements
The authors acknowledge the financial support provided by Sunway University through the grant INT-2018-SST-RCNMET-05 to facilitate this study.

Measurement and analysis of backlash on harmonic drive

Bin Wang[1], Jikui Liu[2] and Chunjie Wang[1]

1 School of Mechanical Engineering and Automation, Beihang University, Beijing, China
2 Beijing Institute of Control Engineering, Beijing, China
E-mail: binw502@126.com

Abstract. Harmonic drive transmissions are widely used in robotics, space manipulation systems because of their excellent mechanical properties including low backlash, high speed reduction ratio and compact size. Analytical investigation and modeling of backlash and hysteresis for harmonic drive are reported, but no thorough understanding of backlash as well as its attributes is available in literature. In this research, the experimental and analytical studies of backlash are conducted. First, the definition of backlash according to China National Standard (CNS) is given, which also matches the requirements in application. A dedicated apparatus was designed to measure the backlash. The results of this paper offer an insight into backlash of harmonic drive and the results are valuable in the mechanical design of harmonic drive gears as well as the dynamic modeling and precision control of harmonic drive systems.

1. Introduction

Harmonic gear drive transmissions are compact, low-backlash, high-ratio, high-resolution rotary motion transmissions. Military and civil mechanism benefit from these attributes[1]. In space applications, high-stability SADA (Solar Array Drive Assembly) and high-precision antenna GDA (Gimbal Drive Assembly) take advantage of the low-backlash characteristic of harmonic drive to reduce the disturbance caused by vibration of large flexible components. The value of backlash in harmonic drive transmission required is less than 3 arc minutes or even 1 arc minute.

The positional errors of harmonic drive transmission can be simply divided into two categories, the kinematic error, which refers to pointing error when the input rotation is unidirectional, and backlash error when reversing rotation is included. Emelyanov [2] carried out a mathematical analysis on the kinematic error and concluded that the error was due to physical imperfections and assembly tolerance of the three principle components in harmonic gear. The results in the report revealed that the kinematic error periodically occurred twice per wave generator rotation. Besides, dynamic models of harmonic drive transmission for various applications were built [3] [4][5], which all reduced the maximum pointing error through implementing compensation.

Though the kinematic error in harmonic drive has been adequately studied, the literature known on backlash error is relatively few, possibly due to the difficult access to sufficient experimental data which will be explored later. Because of the complex source of backlash, we adopt experiment measurement in this paper to investigate the characterization of backlash. When appropriate, the seemingly random test data is explained analytically.

2. Measurement of backlash

Even though few experimental researches were obtained from known literature, the definition of backlash is generally agreed According to CNS GB/T 12601-90, the backlash of harmonic drive which is also called lost motion is the angle lag of the output shaft when the input shaft change the rotational direction in working condition. Hence the harmonic gear should be driven at input end with backlash measured at the output end, as shown in Figure 1. Furthermore, the term "in working condition" in the definition indicated that appropriate inertia or load of torque should be included at the output end. In this section we present a precise characterization on the contribution of backlash and emphasize our research on one particular aspect.

2.1. Mechanism of backlash

Through analysis of transmission principle, the backlash of harmonic drive transmission could be decomposed into two dominant components including lost motion caused by clearance and lost motion caused by elastic deformation. The source of clearance mainly includes tooth clearance between circular spline and flexspline, the clearance from Oldham coupling in the wave generator, while the elastic deformation is mainly due to flexspline under load of torque. The above process could be shown in Figure 2 by analysing the transmission chain. The notation $\Delta\Phi$ is the summation of clearance distributed among the transmission chain. Kt is the torsional stiffness, a variable which increases with increasing load. The research in our experiment in the following sections focus on lost motion caused by clearance.

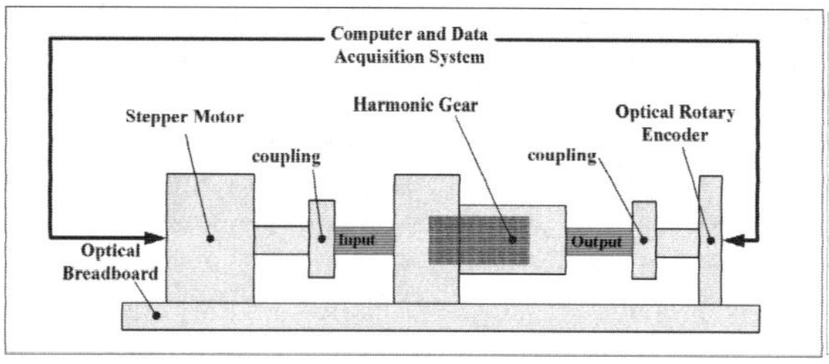

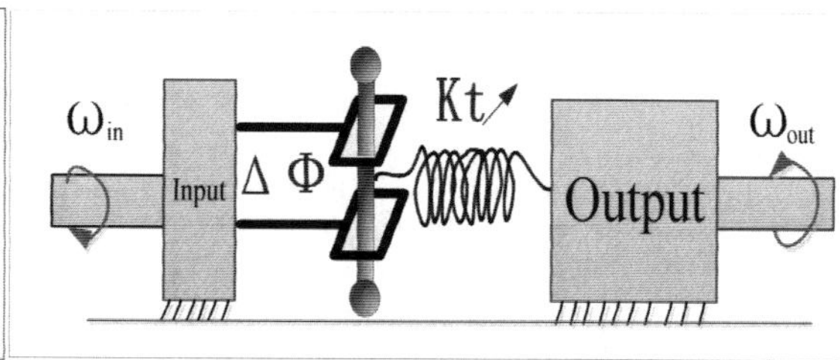

Figure 1. Backlash measurement apparatus **Figure 2.** Harmonic drive transmission model

A method of calculation of tooth clearance for involute profile tooth in design state[6] is shown in Figure 3. A coordinate system {0xyz} is fixed to the center of wave generator with y coaxial with major axis, and {0x1y1z1} is fixed to a meshing flexspline tooth with y1 coaxial with its symmetric line. Since the minimal clearance occurs at tooth top of the flexspline (point K1), it is chosen to calculate the clearance. The coordinates of point K1 and K2 could be defined by the following expression:

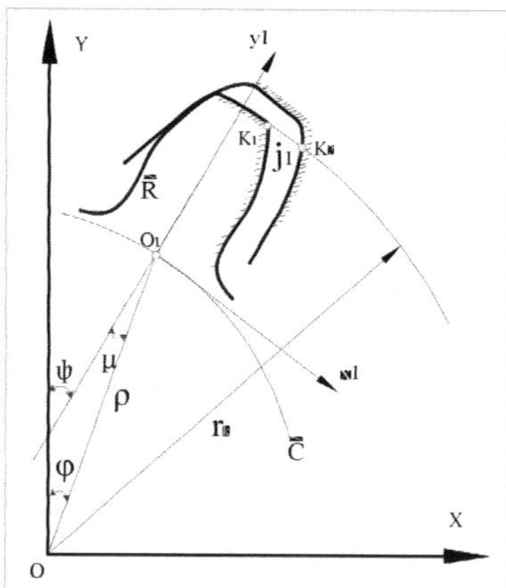

Figure 3. Theoretic clearance calculation diagram

The theoretic clearance in design state can thus be given by:

$$\mathrm{j} \approx \sqrt{(X_{k2} - X_{k1})^2 + (Y_{k2} - Y_{k1})^2} \tag{1}$$

IOP Conf. Series: Materials Science and Engineering **542** (2019) 012005 doi:10.1088/1757-899X/542/1/012005

The center angel or backlash can be expressed as:

$$\Delta = \frac{j}{\rho} \tag{2}$$

2.2. Measurement apparatus and procedure

The harmonic drive studied in our research is a cup type of 80 size, as shown in Figure 4, and its parameters are listed in table 1. Figure 5 and 6 are photos of the backlash measuring apparatus as well as its profile. In order to minimize the effect of uncertain factors, all testing samples were carefully prepared in assembling except for intended purpose.

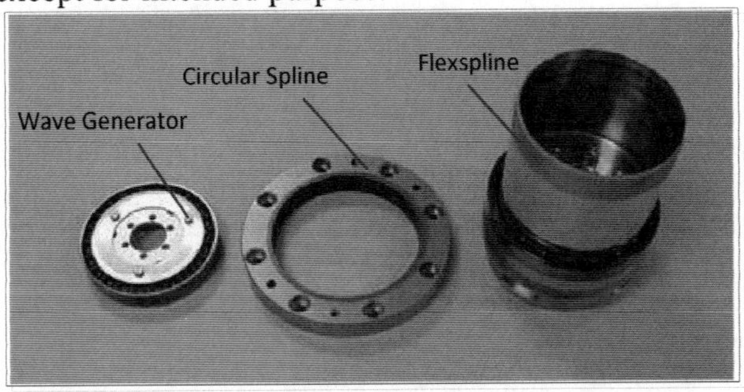

Figure 4. Three main components of harmonic gear

Table 1. Parameter of harmonic drive

Item	Value	Unit
Ratio	160:1	
Design backlash	1	arc minute
Flexspline dimension	80	mm
Modulus	0.25	mm

Figure 5. View of measurement apparatus

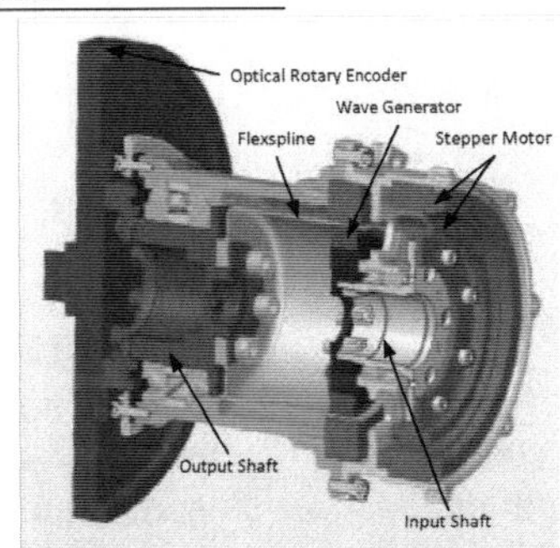

Figure 6. Profile of backlash measurement apparatus

3. Characteristic of backlash

3.1. General characteristics

The test and analysis of harmonic drive backlash is a hard task due to the phenomenon that units from the same batches may show completely different error signatures and even the same unit shows different backlash in different assembly. What makes it worse is that the test method restricts the number of test data, which makes it difficult to analyze its discipline. However, it is generally agreed that, the backlash is involved with the following sources:

- machining error of gear tooth of flexspline and circular spline;
- machining error of cam profile;

- assemble error of the three components;
- clearance of coupling;
- and flexibility of flexspline and other parts.

Because the harmonic drive is rigidly connected to the input and output shaft and the test is carried out in quasi-static without load of torque, the effects of the last two terms can be removed. In fact, their influences are either relatively clear or sufficiently studied. In the subsequent sections, the machining error and assembly tolerance which have important influences on backlash are experimentally tested using the apparatus described above, and the results are analysed to give a physical explanation.

3.2. Assembly dependence

Another important reason for low backlash characteristic of harmonic drive besides multi-teeth engagement is the existing angle between flexspline and circular spline along rotation axis, as shown in Figure 7. The angle causes higher depth of meshing at outer area of the spline, which results in smaller backlash of harmonic gear with no or little load. When the load of torque increases, the cup flexes to allow more width of flexspline to engage with the circular spline, and the stiffness increases, which explained the variable torsional stiffness discussed in section mechanism of backlash.

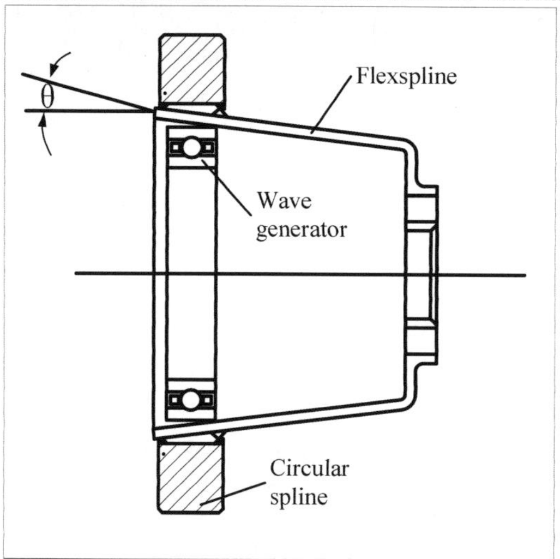

Figure 7. Harmonic drive cross section

Increasing the dimension of the major axis of wave generator could increase the meshing depth of tooth and consequently decrease the backlash or increase the torsional stiffness of harmonic drive. Johnson and Head [7] increased the torsional stiffness of harmonic drive by using wave generator of different major diameter.

Increase in insert depth of wave generator into flexspline could similarly achieve the same objective. Harmonic drive with wave generator inserted at different axial position were tested, one of the results were shown in Figure 8. The values on the two curves were tested at the same 26 points uniformly distributed on an output revolution. To enable a visual comparison and minimize the disturbance of irregularity, the results were sorted in ascending order. As shown in figure, with 3 more millimeters into the flexspline along axis, the backlash is obviously reduced, especially at positions of large backlash.

3.3. Life dependence

The backlash requirement is to be satisfied throughout the life, and in systems with backlash compensation, the value should remain stable during the whole life.

The wear contains two opposite aspects on backlash: 1) wear of tooth crest and tooth surface could lead to increases of clearance; 2) accumulations of debris in wave generator and tooth slot decrease the backlash. Both the phenomenon has been found in the life test of five units. The accumulations of debris especially metal debris caused zero backlash or even minus value backlash, and abnormal wear were found in subsequent test after disassembly. For the units which passed the life test of about 4 million revolutions at the input end, variation of backlash is shown in Figure 9.

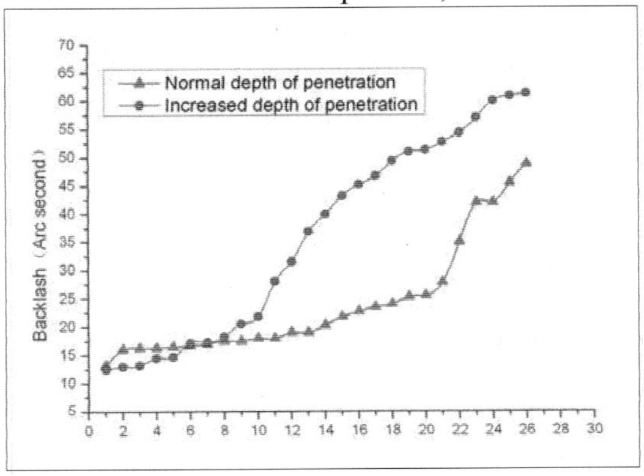

Figure 8. Dependence of backlash on penetration depth of wave generator

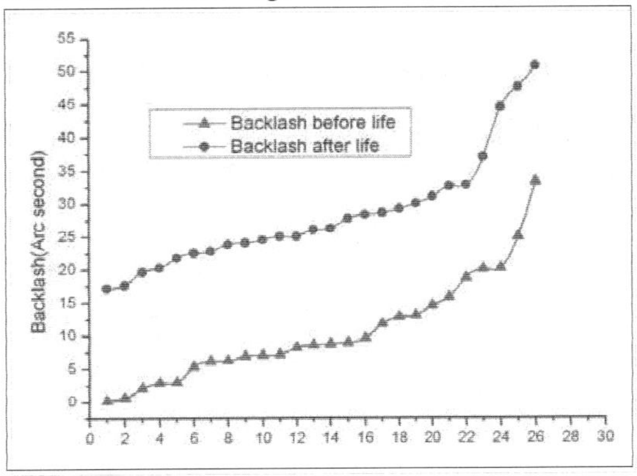

Figure 9. Dependence of backlash on life

4. Conclusions

In this paper new experimental measurement results and analysis were presented which give new insight into the behavior of backlash in harmonic drives. A dedicated backlash measurement apparatus as well as a test procedure were proposed which conformed to application requirements. Extensive measurements were carried out using the apparatus. Besides, we characterized the dependence of the backlash on assembly and life, revealing the mechanism of backlash under different conditions. The measurement results obtained in this paper are very useful in mechanical design of harmonic drive gear and precision dynamical control systems involving backlash compensation.

5. References

[1]. Ivanov M 1987 *The Harmonic Drive*. Beijing: Defense Industry Press.
[2]. Emelyanov A 1983 Calculation of the kinematic error of a harmonic gear transmission taking into account the compliance of elements. *Soviet Engineering Research* 3(7): 7-10.
[3]. Preissner C, Royston TJ and Shu D 2012 A high-fidelity harmonic drive model. *ASME Journal of Dynamic Systems, Measurement, and Control* 134(1): 011002 1-3.
[4]. Preissner C, Shu D and Royston TJ 2007 Experimental investigation and model development for a harmonic drive transmission. *Proc. of SPIE* (California, USA, 28 August 2007). Vol. 6665 (6665-0P).
[5]. Tuttle TD 1992 Understanding and modeling the behavior of a harmonic drive gear transmission. Report,
[6]. Shen YW and Ye QT 1985 Theory and d Schempf H 1990 Comparative Design, Modeling, and Control Analysis of Robotic Transmissions. *Woods Hole Oceanographic Institution MIT Report,* USA, August.
[7]. Johnson MR, Gehling R and Head R 2006 Life test failure of harmonic gears in a Two-axis Gimbal for the Mars Reconnaissance Orbiter Spacecraft. *Jet Propulsion Laboratory Report*, USA, May.

IOP Conf. Series: Materials Science and Engineering **542** (2019) 012006 doi:10.1088/1757-899X/542/1/012006

Seismic response analysis and damping method of spherical liquid storage tank

Junqi Chen[1], Qingjun Xian[1] and Peng Zhang[1]

1 School of Civil Engineering and Architecture, Henan University of Technology, Zhengzhou, China
E-mail: junqihit@163.com

Abstract. Earthquake is a natural disaster with outburst and ruinous in the world. The damage of spherical liquid storage tank due to earthquake can result in serious deaths and economic loss. In this thesis, the analysis of seismic response of spherical tank, which considering interior pressure, gravity, liquid sloshing and earthquake, is carried out. Seismic response analysis shows that the spherical tank rotates horizontally while rotating a small amount around the horizontal axis, mainly around the vertical axis. Time-history analysis of the tank under passive control is performed. The spherical tank model with dissipative bracing system installed has a very good damping effect by comparing the seismic responses of controlled and uncontrolled structures.

1. Introduction

Earthquake is a natural disaster with outburst and ruinous in the world. The damage of spherical liquid storage tank due to earthquake can result in serious deaths and economic loss [1-2]. Especially leak of flammability and exploded-prone medium deposited in tanks can cause secondly disaster, which led to the serious destroy of the survival environment. The spherical liquid-storage tank is developing with high parameter and huge, analyze the seismic response of it and reduce seismic damage to the least has always become one of the most important researches, which becomes more outstanding in currently study of seismic protection of petrochemical facilities.

Structural vibration control technology is a burgeoning technology to take precautions against and to reduce natural calamities, apply which to seismic protection of petrochemical facilities is an innovative attempt [3-4]. In this task, the analysis of seismic response of spherical tank considering that it effect on interior pressure, gravity, liquid sloshing and earthquake. Then, applying passive control technologies of fluid damper to the support system of the tank, which is base on passive energy dissipated control technology, and compare the seismic response in order to investigate the effectiveness of the seismic protection system [5-7].

2. Description of structure

The structure analyzed in the present study, shown in Fig. 1, is a typical spherical liquid storage tank with a volume of 1000m^3. The contained liquid is assumed to be refrigerated polypropylene with a density of 480kg/m3. The sphere has an inner diameter of 12.3m, an average shell thickness of 34mm and is constructed from a steel plate with Elastic Modulus of 2.1GPa, Poisson's ratio of 0.3 and density of 7850kg/m^3. The equator of the sphere is 8.0m above ground. The sphere is supported by then circular columns with an outer diameter of 486mm and a wall thickness of 10mm. The spherical tank is laterally strengthened by pairs of diagonal braces between columns, i.e. 11 pairs in total, as shown in Fig. 1. The braces are made of steel pole with diameter of 56mm, the mechanical characteristics of which are the same as those of the spherical container. Each brace is pin-connected to the adjacent columns, at 0.5m and 6m above the bottom level of the columns. All braces are considered to be tension-only elements since they buckle at very low compression levels.

Figure 1. A typical spherical liquid storage tank.

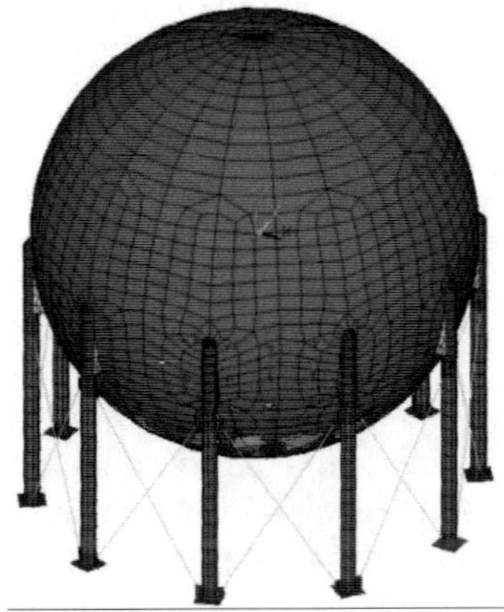

Figure 2. Finite element model of the tank.

3. Seismic response analysis of spherical liquid storage tanks

3.1. Finite element model

The numerical analysis of the spherical storage tank structure is performed on the basis of detailed finite element model developed with the help of the routines available in the ANSYS Finite Element program, as shown in Fig. 2. The spherical shell and the supporting columns are modelled by 3310 four-node shell elements (SHELL181) with six DOFs per node. The eight node solid fluid element (FLUID80) with three DOFs per node has been chosen to model the incompressible fluid content. A total of 1023 or 1793 FLUID80 elements are used, respectively, for 89% tank fullness considered in this work.

In order to satisfy the continuity conditions between the fluid and solid media at the spherical boundary, the coincident nodes of the fluid and shell elements are constrained to be coupled in the direction normal to the interface, while relative movements are allowed to occur in the tangential directions. The uniaxial "tension only" behaviour of the braces is simulated by means of the 3-D spar elements LINK10, which feature a bilinear stiffness matrix, i.e. the stiffness is removed if the element goes into compression.

The above finite element model of spherical tank is numerically analyzed by means of a full transient non-linear analysis. The governing equations of motion can be expressed in matrix form as

$$[M]\ddot{\vec{u}} + [C]\dot{\vec{u}} + [K]\vec{u} = \vec{F}(t) \tag{1}$$

with $[M]$, $[C]$ and $[K]$ being the mass, damping and stiffness matrices of the structure, $\vec{u}$, $\dot{\vec{u}}$ and $\ddot{\vec{u}}$ being the displacement, velocity and acceleration of the structure, and $\vec{F}(t)$ of the load applied to the structure. Eq. (1) is integrated directly in time using the Newmark-β method.

3.2. Seismic wave selection

The selected Tianjin wave is suitable for the third and fourth types of sites. The duration is taken from the first 10 seconds of the Tianjin wave record. The reading interval is 0.02 seconds. The time history is shown in Fig. 3.

In this paper, the seismic fortification intensity to be 8 degrees, the peak value of X to Tianjin wave acceleration is adjusted to 0.20g, the corresponding Y-direction peak is 0.29g, and the Z-direction peak is 0.20g.

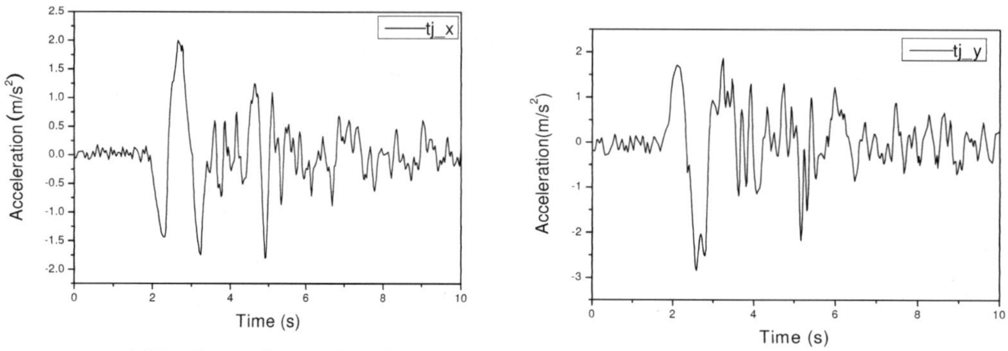

a) North-south acceleration b) East-west acceleration

Figure 3. Acceleration time history curve of Tianjin wave.

3.3. Time history analysis of 89% full spherical tank

Enter the three-direction seismic wave. Fig. 4 and Fig. 5 show the x-direction and y-direction displacement time history of the apex and bottom points of the spherical shell. The maximum x-direction displacement of the apex (up) and bottom (low) of the spherical tank is 53mm and 50mm, respectively, and the maximum difference in the time course is 3mm. The maximum y-direction displacement of the apex (up) and bottom (low) of the spherical tank is 18 mm and 17 mm, respectively, and the maximum difference in the time course is 1 mm. It shows that the spherical tank rotates horizontally while rotating a small amount around the horizontal axis, mainly around the y axis.

Fig. 6 shows the horizontal displacement cloud of the spherical tank when the seismic wave is input for 5.3s while the maximum displacement occurs. The equivalent stress of the spherical shell, the tensile stress of the rod and the vertical reaction force at the bottom of the column also appear at this time. Fig. 7 shows the stress distribution of the spherical shell at this moment. In the position where the pillar and the spherical shell are connected, the stress level is high because the displacement of the casing is resisted to satisfy deformation coordination.

4. Damping method study

4.1. Finite element model

A viscous damper is a device that uses viscous fluid to dissipate energy through an orifice or a gap between a cylinder and a piston to generate viscous damping [8-9]. Replace all tie rods in the original structure with a viscous damper. In this paper, only the simplest model is used for analysis. Therefore, the linear viscous damper is selected, and its viscous damping coefficient is 600kN/(m/s).

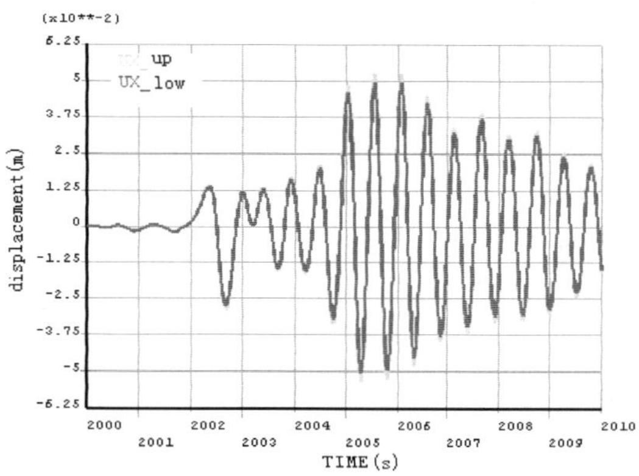

Figure 4. X-direction displacement time history of the apex and bottom points of the spherical shell.

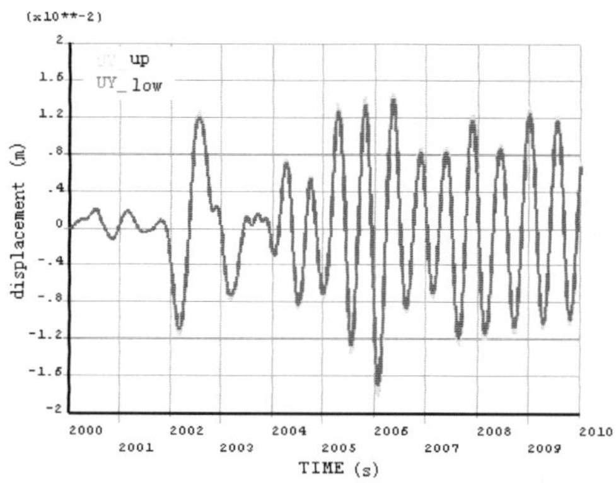

Figure 5. Y-direction displacement time history of the apex and bottom points of the spherical shell.

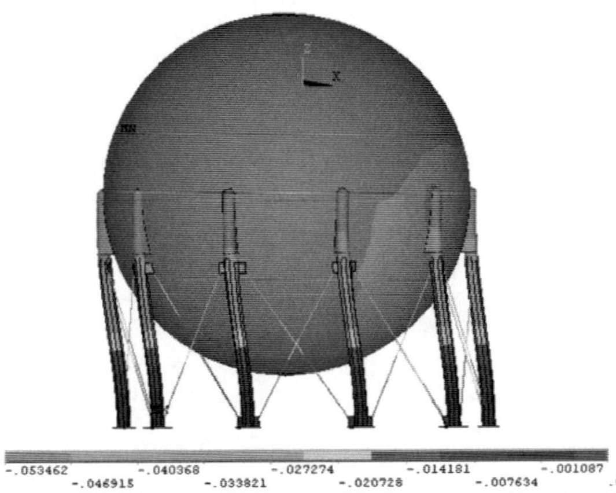

Figure 6. The horizontal displacement cloud at 5.3s.

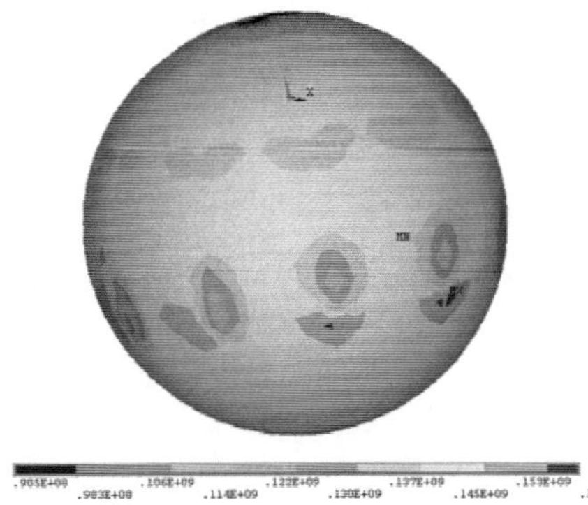

Figure 7. The stress distribution cloud at 5.3s.

The viscous fluid damper devices are modelled using the 1-D non-linear damper elements COMBIN37. The above finite element model of spherical tank is numerically analyzed by means of a full transient non-linear analysis. The governing equations of motion can be expressed in matrix form as

$$[M]\ddot{\vec{u}} + [C]\dot{\vec{u}} + [K]\vec{u} = [M][\ell]\{\ddot{u}_g(t)\} \tag{2}$$

with $[\ell]$ being an influence coefficient matrix, and $\{\ddot{u}_g(t)\}$ the ground acceleration. Eq. (2) is integrated directly in time using the Newmark-β method. The nonlinear, introduced by the "tension-only" behaviour of the conventional braces and by the viscous fluid damper devices, require an iterative solution with incremental load steps. In the present study the iterative Newton-Raphson approach is employed.

The seismic response of the spherical liquid storage tank with and without the dissipative bracing system is investigated by performing two types of analyses: time domain analysis. The problem is solved for 89% fullness of tank.

The seismic response of the spherical tank structure with a dissipative bracing system is compared to that of the tank with conventional bracing in order to investigate the effectiveness of the seismic protection system. The time history analysis corresponds to the TJ earthquake ground motion, the ground acceleration of which has a duration of 10sec and a peak ground acceleration (PGA) of 0.4g, applied in horizontal x-direction of the tank.

4.2. Result analysis

The time variation of the displacement in the x-direction at the top of the sphere, the total base shear and the vertical fluid displacement at the intersection of the fluid free surface are shown in Figs 8, 9 and 10. Figs 8 and 9 reveal a significant reduction in horizontal movement and base shear of the tank, implying the effectiveness of the seismic protection system.

More specifically in Fig. 8, the maximum horizontal displacements at the top of the sphere are reduced from 10.6cm when conventional braces are used to 3cm when dissipative ones are employed. Similarly in Fig 9, the total base shear peak responses are reduced from 401kN to 196kN.

On the other hand, the sloshing vertical displacements of the fluid content, shown in Fig. 10, remain almost the same, as expected, for both bracing systems. This is due to the fact that the natural frequency of sloshing liquid mass (2.3sec) is well apart from the dominant natural frequencies of the structure and corresponds to negligible portion of the seismic excitation in Fig. 11 the hysteretic loop of the viscous damper had effective absorbed the energy of earthquakes.

IOP Conf. Series: Materials Science and Engineering **542** (2019) 012006 doi:10.1088/1757-899X/542/1/012006

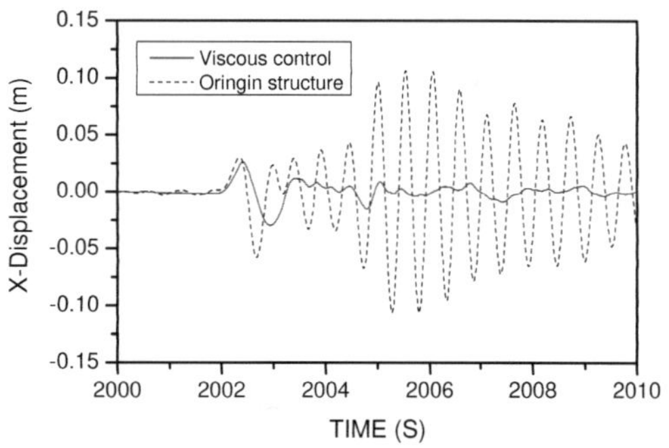

Figure 8. Horizontal displacement at the top.

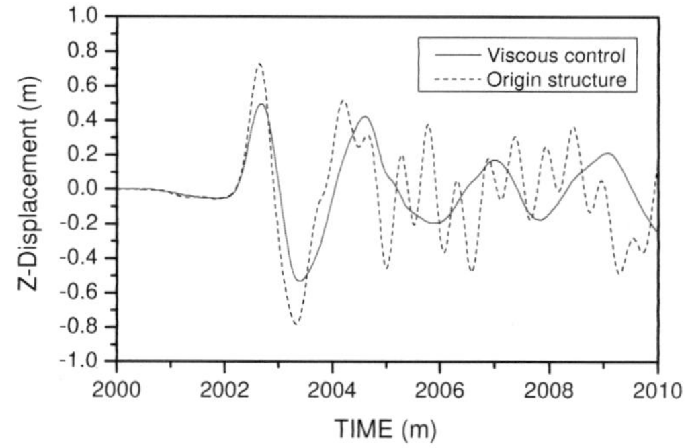

Figure 9. Total base shear force of a 89% full spherical.

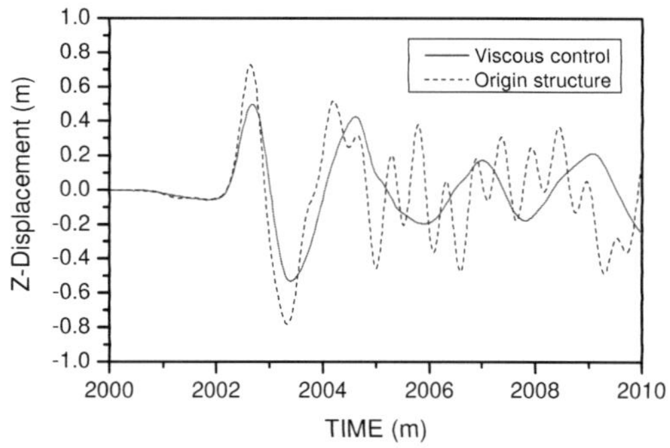

Figure 10. Vertical movement of the fluid surface.

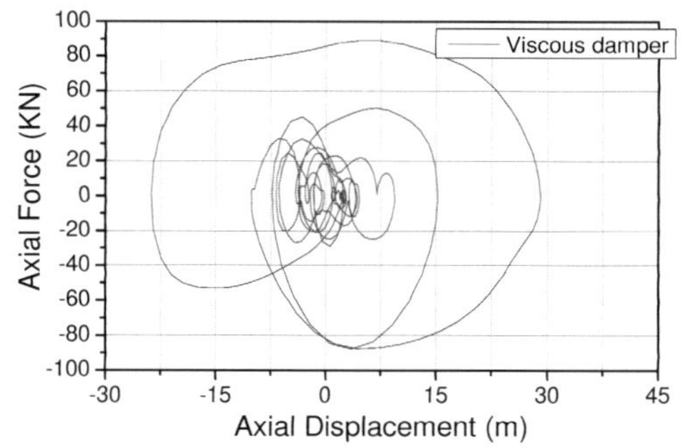

Figure 11. Hysteretic loop of the viscous damper.

5. Conclusions

Seismic response analysis shows that the spherical tank rotates horizontally while rotating a small amount around the horizontal axis, mainly around the y axis. Time-history analysis of 89% of the tanks under passive control is performed. The spherical tank model with two dampers installed has a very good damping effect by comparing the seismic responses of controlled and uncontrolled structures.

References

[1] Leon G S and Kausel E A 1986 *J. Struct. Eng.* **112** 1-18.
[2] Shibata H 1999 *J. Press. Vess-T* **121** 30-36.
[3] Pasman H J, Rogers W J, and Mannan M S 2017 *Safety Sci.* **99** 140-55.
[4] Curadelli O 2011 *Eng. Struct.* **33** 2662-67.
[5] Paolacci F, Giannini R, and De Angelis M 2013 *J. loss Prevent. Proc.* **26** 924-35.
[6] Curadelli O, Ambrosini D, Mirasso A, and Amani M 2010 *J. Fluid Struct.* **26** 148-59.
[7] Soni D P, Mistry B B, and Panchal V R 2011 *Nucl. Eng. and des.* **241** 700-13.
[8] Lay K S 1993 *J. Eng. Mech.* **119** 1747-61.
[9] Tsyss V G, and Sergaeva M Y 2015 *Procedia Eng.* **113** 395-401.

Acknowledgments

This work was financially supported by Key Research Project of Colleges and Universities in Henan Province(No.15A560019), Zhengzhou Scientific and Technology Project Plan(No.20140719) and Henan University of Technology Fund Project of High-level Talent(No.2014BS002).

Determination of optimum gear ratios of a three stage bevel helical gearbox

Vu Ngoc Pi[1], Tran Thi Phuong Thao[1], Tran Thi Hong[2], Nguyen Khac Tuan[1], Le Xuan Hung[1], Luu Anh Tung[1]

1 Thai Nguyen University of Technology, Thai Nguyen city, Vietnam
2 Nguyen Tat Thanh University, Ho Chi Minh city, Vietnam.
E-mail: vungocpi@tnut.edu.vn.

Abstract. This paper presents a study on the determination of optimum gear ratios of a three stage bevel helical gearbox. In the study, for finding the optimum gear ratios, an optimization problem was conducted. In the optimization problem, the cross sectional area of the gearbox was chosen as the objective function. Also, the influence of input factors including the total gearbox ratio, the coefficient of the face width of the bevel and the helical gear sets, the allowable contact stress and the output torque were considered. evaluating the effect of these parameters on the optimum gear ratios, an "experiment" was designed and a computer program was constructed to perform the "experiment". From the results of the study, the effect of the input factors on the optimum gear ratios were learned and several models for determination of the optimum gear ratios were proposed.
Keywords: Gear ratio, optimum gearbox design, bevel helical gearbox.

1. Introduction

In optimum gearbox design, the determination of the optimum gear ratios is significant. The reason is that the dimension, the mass, and the cost of a gearbox depend powerfully on the gear ratios.
For bevel helical gearboxes, so far, there have been several studies on the determination of the gear ratios. For two-stage bevel helical gearboxes, V.N. Kudreavtev et al [1] introduced a graph method (figure 1) for determining the gear ratio for a two-stage bevel helical gearbox. According to this method, we must select the coefficient large in value ($\lambda c^3 = 0.6$ to 4). Therefore, it is very complex and almost impossible to get the optimum value. Later, G. Milou et al. [2] reported "a practical method" in which the gear ratios were found in the tabulated form from practical data.
A more useful method for determining optimum gear ratios is the model method. In this method, models for finding the optimum gear were suggested for different objectives. For two state bevel helical gearboxes V.N. Pi [3] proposed a model for the minimum height of the gearbox. For three stage bevel helical gearboxes, the minimum length of the gearbox was chosen as the objective of the optimization problem [4]. Recently, V.N. Pi et al. [5] conducted a study on the determination of the optimum gear ratios of a mechanical system using a V-belt drive and a two-stage bevel helical gearbox in order to get the minimum cross-sectional dimension of the system.

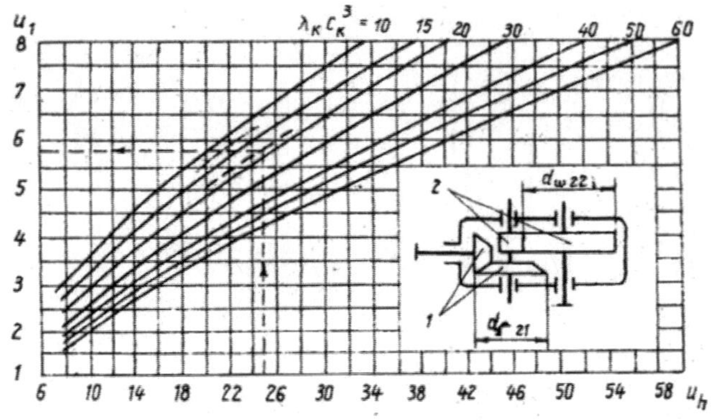

Figure 1. The gear ratio of the bevel gear set vs the total gearbox ratio [1].

This paper introduces a study for determination of the optimum gear ratios of a three stage bevel helical gearbox with the objective is the minimum surface of the cross section of the gearbox. Also, the effect of the input parameters on the optimum gear ratios was evaluated.

2. Optimization problem

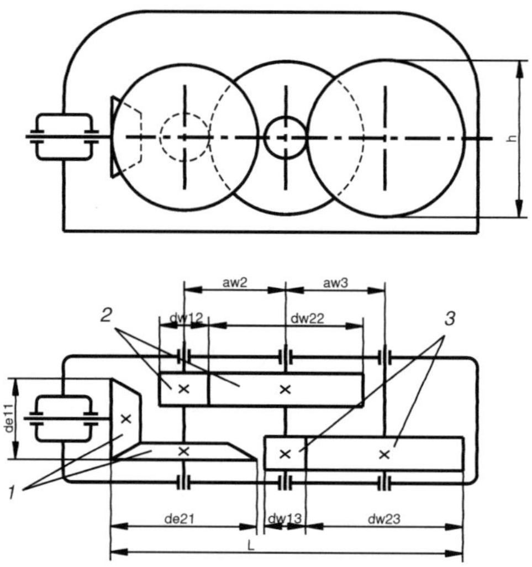

Figure 2. Calculation schema.

From figure 2, the surface of cross section of the gearbox is determined by:

$$A = L \cdot h \tag{1}$$

Where, L and h are determined as (see Figure 2):

$$L = d_{e21}/2 + a_{w2} + a_{w3} + d_{w23}/2 \tag{2}$$

$$h = \max\left(d_{e21}, d_{w22}, d_{w23}\right) \tag{3}$$

In the above equations, a_{w2}, a_{w3}, d_{w22} and d_{w23} are the center distance and the pitch diameters of the second and the third stage, respectively.

For helical gear sets, d_{w22} and d_{w23} can be determined as [6]:

$$d_{w22} = 2 \cdot a_{w2} \cdot u_2 / \left(u_2 + 1\right) \tag{4}$$

$$d_{w23} = 2 \cdot a_{w3} \cdot u_3 / \left(u_3 + 1\right) \tag{5}$$

Where, u_2 and u_3 are the gear ratios of the second and the third stage.

Thus, the optimization problem is defined as:

$$\text{minimize } A = L \cdot h \tag{6}$$

With the following constraints:

$$1 \le u_2 \le 9 \tag{7}$$
$$1 \le u_3 \le 9$$

From (2), (3), (4), (5) and (6), it is clear that for solving the optimization problem it is necessary to determine d_{e21}, a_{w2} and a_{w3}.

2.1. Determining the driven diameter of the bevel gear set

Based on the pitting resistance of the straight bevel gear set, the external cone distance R_e is determined by [6]:

$$R_e = k_R \cdot \left(u_1^2+1\right)^{1/2} \cdot \left(T_{11} \cdot K_{H\beta 1} / \left[\left(1-k_{be}\right) \cdot k_{be} \cdot u_1 \cdot \left[\sigma_{H1}\right]^2\right]\right)^{1/3} \tag{8}$$

In which, k_R is the coefficient which depends on the gear material and the gear type; For steel straight bevel gearings $k_R = 50$ ($MP_a^{1/3}$); $k_{be} = 0.25...0.3$ is the coefficient of the face width; $[\sigma_{H1}]$ is the allowable contact stress (MPa); u_1 is the gear ratio of the bevel gear set; $K_{H\beta 1}$ is the contact load ratio for pitting resistance of the bevel gear set. From the tabulated data in [6], the following regression equations were found for determination of $K_{H\beta 1}$ (with the coefficient of determination $R^2 = 1$):

$$K_{H\beta 1} = 0.25 \cdot k^2 + 0.2 \cdot k + 1.02 \tag{9}$$

Wherein, $k = k_{be} \cdot u_1 / \left(2 - k_{be}\right)$.

From the condition of the moment equilibrium of the mechanic system with three gear sets we have:

$$T_{11} = T_{out} / \left(u_g \cdot \eta_{bg} \cdot \eta_{hg} \cdot \eta_b^3\right) \tag{10}$$

Where, T_{out} is the output torque (N.mm); u_g is the total ratio of the gearbox; η_{bg} is the bevel gear transmission efficiency ($\eta_{bg} = 0.95...0.97$ [6]); η_{hg} is the helical gear transmission efficiency ($\eta_{hg} = 0.96...0.98$ [6]); η_b is the transmission efficiency of a pair of rolling bearings ($\eta_b = 0.99...0.995$ [6]). Choosing $\eta_{bg} = 0.96$, $\eta_{hg} = 0.97$, $\eta_b = 0.992$ and substituting them into (10) give:

$$T_{11} = 1.101 \cdot T_{out} / u_g \tag{11}$$

Substituting $k_R = 50$ and (11) into (8) gets

$$R_e = 51.6296 \cdot \left(u_1^2+1\right)^{1/2} \cdot \left(T_{out} \cdot k_{H\beta 1} / \left[\left(1-k_{be}\right) \cdot k_{be} \cdot u_1 \cdot u_g \cdot \left[\sigma_H\right]^2\right]\right)^{1/3} \tag{12}$$

The outer pitch diameter of the pinion then can be calculated by [6]:

$$d_{e11} = 2 \cdot R_e . \left(1+u_1^2\right)^{-1/2} \tag{13}$$

2.2. Determining the center distance of the second stage

The center distance of the second stage a_{w2} is calculated by [6]:

$$a_{w2} = k_m \cdot \left(u_2+1\right) \cdot \left(\frac{T_{12} \cdot k_{H\beta}}{\left[\sigma_H\right]^2 \cdot u_2 \cdot \psi_{ba2}}\right)^{1/3} \tag{14}$$

In which, $K_{H\beta}$ is the contact load ratio for pitting resistance for the second stage of the gearbox $k_{H\beta} = 1.02 \div 1.28$ [6]. Therefore, we can chose $k_{H\beta} = 1.1$; $[\sigma_H]$ - allowable contact stress (MPa); In practice, $[\sigma_H] = 350...410$ (MPa); k_m is the material coefficient; As the gear material is steel, k_m =43 [6]; ψ_{ba2} is the coefficient of the wheel face width of the second stage; $\psi_{ba2} = 0.3...0.35$ [6];

From the moment equilibrium condition of the mechanic system with two helical gear sets the following equation is given:

$$T_{out} = T_{12} \cdot \eta_{hg}^2 \cdot \eta_b^3 \cdot u_2 \cdot u_3 \quad (15)$$

Wherein, η_{hg} is the helical gear transmission efficiency (η_{hg} is from 0.96 to 0.98 [6]); η_b is the transmission efficiency of a pair of rolling bearings (η_{be} is from 0.99 to 0.995 [6]). Choosing $\eta_{hg} = 0.97$ and $\eta_b = 0.992$ then substituting them into (15) gives

$$T_{12} = 1.0887 \cdot T_{out} / (u_2 \cdot u_3) \quad (16)$$

Substituting (16) and $k_{H\beta} = 1.1$ into (14) with the note that $u_2 \cdot u_3 = u_g / u_1$ gets:

$$a_{w2} = 45.6635 \cdot (u_2 + 1) \cdot \left(\frac{T_{out} \cdot u_1}{[\sigma_H]^2 \cdot u_g \cdot u_2 \cdot \psi_{ba2}} \right)^{1/3} \quad (17)$$

The pitch diameter of the second stage then is calculated by [6]:

$$d_{w22} = 2 \cdot a_{w2} \cdot u_2 / (u_2 + 1) \quad (18)$$

2.3. Determining the center distance of the third stage

The center distance of the third stage a_{w3} can be determined by the following equation [6]:

$$a_{w3} = K_m \cdot (u_3 + 1) \cdot \left(\frac{T_{13} \cdot k_{H\beta}}{[\sigma_H]^2 \cdot u_3 \cdot \psi_{ba3}} \right)^{1/3} \quad (19)$$

Also, for the third stage we have

$$T_{out} = T_{13} \cdot \eta_{hg} \cdot \eta_b^2 \cdot u_3 \quad (20)$$

Choosing $\eta_{hg} = 0.97$ and $\eta_b = 0.992$ as in section 2.2 gives

$$T_{12} = 1.0476 \cdot T_{out} / u_3 \quad (21)$$

Substituting (21), k_m=43 and $k_{H\beta} = 1.1$ (as in section 2.2) into (19) gets:

$$a_{w3} = 45.0814 \cdot (u_3 + 1) \cdot \left(\frac{T_{out}}{[\sigma_H]^2 \cdot u_3^2 \cdot \psi_{ba3}} \right)^{1/3} \quad (22)$$

The pitch diameter of the third stage then is determined by [6]:

$$d_{w23} = 2 \cdot a_{w3} \cdot u_3 / (u_3 + 1) \quad (23)$$

2.4. Experimental work

To investigate the influence of the input parameters on the optimum gear ratios, an "experiment" was designed and performed. For this experiment, a 2-level full factorial design was selected. Also, 6 input parameters were chosen for the exploring (table 1). Therefore, the design was arranged with $2^6 = 64$ number of experiments. To perform the experiment, a computer program was constructed based on equations (6) and (7). Table 2 shows the various levels of input parameters and the output responses (the optimum gear ratios of the first and the second stage u_1, u_2).

Table 1. Input parameters.

Factor	Code	Unit	Low	High
Total gearbox ratio	u_g	-	40	100
Coefficient of the face width of bevel gear set	Kbc	-	0.25	0.3
Coefficient of wheel face width of stage 2	x_{ba2}	-	0.3	0.35
Coefficient of wheel face width of stage 3	x_{ba3}	-	0.35	0.4
Allowable contact stress	AS	MPa	350	420
Output torque	T_{out}	Nmm	10^5	10^7

Table 2. Experimental plans and output response.

StdOrder	RunOrder	CenterPt	Blocks	u_g	K_{be}	Xba2	Xba3	AS (MPa)	Tout (Nm)	u_1	u_2
33	1	1	1	40	0.25	0.3	0.35	350	10000	4.26	4.17
15	2	1	1	40	0.3	0.35	0.4	350	100	4.06	4.32
58	3	1	1	100	0.25	0.3	0.4	420	10000	5.19	6.42
10	4	1	1	100	0.25	0.3	0.4	350	100	5.19	6.42
62	5	1	1	100	0.25	0.35	0.4	420	10000	5.67	6.39
61	6	1	1	40	0.25	0.35	0.4	420	10000	4.06	4.32
20	7	1	1	100	0.3	0.3	0.35	420	100	5.75	6.3
…											
24	63	1	1	100	0.3	0.35	0.35	420	100	6.25	6.27
64	64	1	1	100	0.3	0.35	0.4	420	10000	5.51	6.51

3. Optimization results and discussions

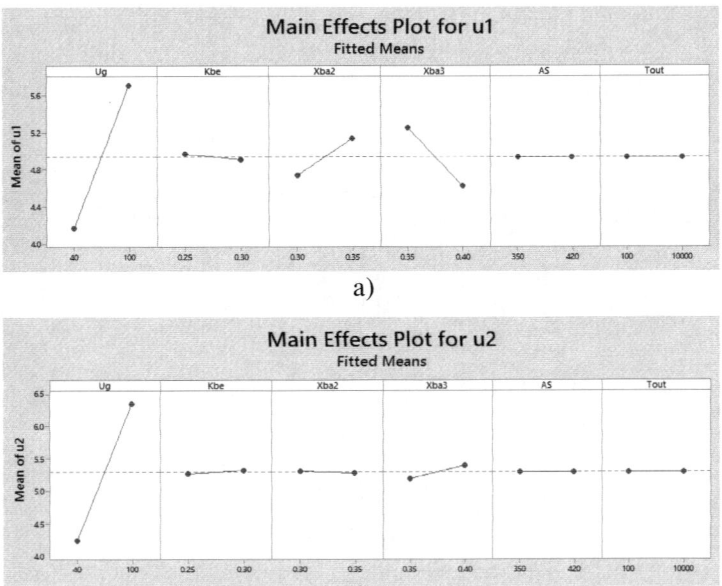

a)

b)

Figure 3. Main effects plot for u_1 and u_2 .

IOP Conf. Series: Materials Science and Engineering **542** (2019) 012007 doi:10.1088/1757-899X/542/1/012007

To evaluate the effect of input factors on the response and the relative strength of the effect, figure 3 shows the main effect of each parameter on the optimum gear ratios of the bevel gear set u_1 (figure 3.a) and the second stage u_2 (figure 3 b). From the figure, both u_1 and u_2 increase significantly with the increase of the total gearbox ratio u_g. Besides, they also were affected by the coefficient of wheel face width of the bevel gear and the helical gear sets (k_{ba}, ψ_{ba1} and ψ_{ba2}). Furthermore, it was found that the allowable contact stress and the output torque do not affect the optimum gear ratios.

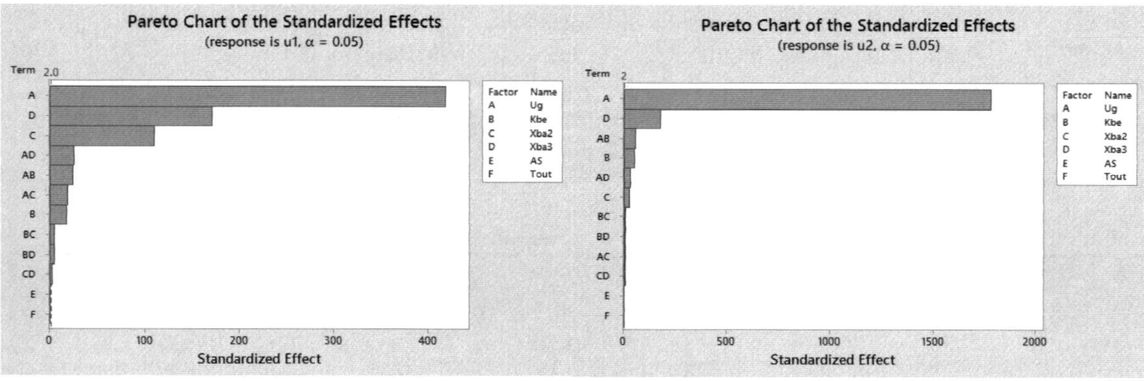

a) b)

Figure 4. Pareto Chart of the Standardized Effects for u_1 and u_2.

Figure 4 shows the Pareto chart of the standardized effects from the largest to the smallest one. From this figure, the bars that represent parameters including the total gearbox ratio (factor A), the coefficient of wheel face width of the bevel gear (factor B) and the helical gear sets (factors C and D) and the interactions between them cross the reference line. Consequently, these factors are statistically significant at the 0.05 level with the response model.

Figure 5 shows the Normal Plot of the standardized effects for u_1 (figure 5.a) and u_2 (figure 5.b). From the figure, the total gearbox ratio (factor A) is the most significant factor for both u_1 and u_2. Also, it has a positive standardized effect for both u_1 and u_2. Besides, k_{be} and ψ_{ba3} (factor B and D) have a negative standardized effect on u_1 but they have a positive standardized effect on u_2. In addition, ψ_{ba2} (factor C) has a positive standardized effect on u_1 while it has a negative standardized effect for u_2.

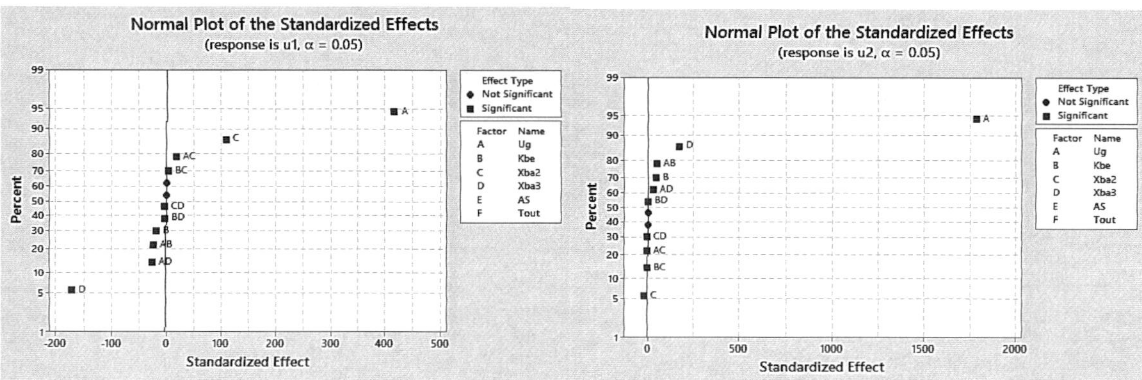

a) b)

Figure 5. Normal Plot for u_1 and u_2.

IOP Conf. Series: Materials Science and Engineering **542** (2019) 012007 doi:10.1088/1757-899X/542/1/012007

Coded Coefficients

Term	Effect	Coef	SE Coef	T-Value	P-Value	VIF
Constant		4.94272	0.00185	2675.50	0.000	
Ug	1.54546	0.77273	0.00185	418.28	0.000	1.00
Kbe	-0.06355	-0.03178	0.00185	-17.20	0.000	1.00
Xba2	0.40564	0.20282	0.00185	109.79	0.000	1.00
Xba3	-0.63234	-0.31617	0.00185	-171.14	0.000	1.00
AS	0.00000	0.00000	0.00185	0.00	1.000	1.00
Tout	-0.00000	-0.00000	0.00185	-0.00	1.000	1.00
Ug*Kbe	-0.08850	-0.04425	0.00185	-23.95	0.000	1.00
Ug*Xba2	0.06688	0.03344	0.00185	18.10	0.000	1.00
Ug*Xba3	-0.09141	-0.04570	0.00185	-24.74	0.000	1.00
Kbe*Xba2	0.01690	0.00845	0.00185	4.57	0.000	1.00
Kbe*Xba3	-0.01397	-0.00698	0.00185	-3.78	0.000	1.00
Xba2*Xba3	-0.00975	-0.00488	0.00185	-2.64	0.011	1.00

Model Summary

S	R-sq	R-sq(adj)	R-sq(pred)
0.0147792	99.98%	99.97%	99.96%

a)

Coded Coefficients

Term	Effect	Coef	SE Coef	T-Value	P-Value	VIF
Constant		5.29687	0.00059	9022.33	0.000	
Ug	2.09625	1.04812	0.00059	1785.30	0.000	1.00
Kbe	0.056250	0.028125	0.000587	47.91	0.000	1.00
Xba2	-0.026250	-0.013125	0.000587	-22.36	0.000	1.00
Xba3	0.206250	0.103125	0.000587	175.66	0.000	1.00
AS	0.000000	0.000000	0.000587	0.00	1.000	1.00
Tout	-0.000000	-0.000000	0.000587	-0.00	1.000	1.00
Ug*Kbe	0.063750	0.031875	0.000587	54.29	0.000	1.00
Ug*Xba2	-0.003750	-0.001875	0.000587	-3.19	0.002	1.00
Ug*Xba3	0.033750	0.016875	0.000587	28.74	0.000	1.00
Kbe*Xba2	-0.003750	-0.001875	0.000587	-3.19	0.002	1.00
Kbe*Xba3	0.003750	0.001875	0.000587	3.19	0.002	1.00
Xba2*Xba3	-0.003750	-0.001875	0.000587	-3.19	0.002	1.00

Model Summary

S	R-sq	R-sq(adj)	R-sq(pred)
0.0046967	100.00%	100.00%	100.00%

b)

Figure 6. Estimated Effects and Coefficients for u_1 and u_2.

Figure 6 presents the estimated effects and coefficients for u_1 (figure 6 a) and u_2 (figure 6 b). It can be seen from the figure 6 that factors which significantly affect a response with P-values lower than 0.05 include the total gearbox ratio u_g, the coefficient of wheel face width of the bevel gear set and the helical gear sets (k_{be}, ψ_{ba2} and ψ_{ba3}) and their interactions. Consequently, the following equations can be found for the relation between the optimum gear ratios and the significant effect factors:

$$u_1 = 2.98 + 0.05034 \cdot u_g + 2.65 \cdot k_{be} + 4.2 \cdot \psi_{ba2} - 2.77 \cdot \psi_{ba3} - 0.059 \cdot u_g \cdot k_{be} + 0.04459 \cdot u_g \cdot \psi_{ba2} - \\ -0.06094 \cdot u_g \cdot \psi_{ba3} + 13.52 \cdot k_{be} \cdot \psi_{ba2} - 11.7 \cdot k_{be} \cdot \psi_{ba3} - 7.8 \cdot \psi_{ba2} \cdot \psi_{ba3} \tag{24}$$

$$u_2 = 2.193 + 0.015625 \cdot u_g - 2 \cdot k_{be} + 1.6 \cdot \psi_{ba2} + 2.7 \cdot \psi_{ba3} + 0.0425 \cdot u_g \cdot k_{be} - 0.0025 \cdot u_g \cdot \psi_{ba2} + \\ +0.0225 \cdot u_g \cdot \psi_{ba3} - 3 \cdot k_{be} \cdot \psi_{ba2} + 3 \cdot k_{be} \cdot \psi_{ba3} - 3 \cdot \psi_{ba2} \cdot \psi_{ba3} \tag{25}$$

The above estimated models fit the data very well because the adj-R_2 and pred-R_2 are in the high values (figure 6).

Equations (24) and (25) are used to calculate the optimum gear ratios of the first and the second stages u_1 and u_2. After that, the optimum gear ratio of the third stage is determined by $u_3 = u_g / (u_1 \cdot u_2)$.

4. Conclusions

A study on calculation of the optimum gear ratios of a three stage bevel helical gearbox for getting the minimum cross sectional surface of the gearbox was conducted. In the study, the influence of the input parameters including the total gearbox ratio, the wheel face width coefficients of the bevel gear set and the helical gear sets, the allowable contact stress and the output torque was investigated. In addition, some models for determination of the optimum partial gear ratios for getting the minimum surface cross section of the gearbox were proposed. The optimum gear ratios can be determined simply by using these explicit models.

Acknowledgements

The work described in this paper was supported by Thai Nguyen University of Technology for a scientific project.

References

[1] Kudreavtev V N, Gierzaves I A and Glukharev E G 1971 *Design and calculus of gearboxes* (Sankt Petersburg: Mashinostroenie Publishing)

[2] Milou G, Dobre G, Visa S and Vitila H 1996 *Optimal Design of Two Step Gear Units, regarding the Main Parameters* (VDI Berichte No 1230) p 227

[3] Pi V N 2000, A new and effective method for optimal calculation of total transmission ratio of two step bevel - helical gearboxes, *International colloquium on Mechanics of Solids, Fluids, Structures & Interaction* (Nha Trang: Vietnam) pp 716- 719

[4] Pi Vu N 2008 A study on optimal calculation of partial transmission ratios of three-step bevel helical gearboxes, *International Workshop on Advanced Computing and Applications* (Ho Chi Minh city: Vietnam) pp 277-286

[5] Pi V N, Cam N T H and Tuan N K 2016 Optimum calculation of partial transmission ratios of mechanical driven systems using a V-belt and two-step bevel helical gearbox, *Journal of Environmental Science and Engineering A 5* pp 566-570

[6] Chat T and Uyen L V 1998 *Design and calculus of Mechanical Transmissions* (Ha Noi: Educational Republishing House)

Influence of dressing parameters on surface roughness of workpiece for grinding hardened 9XC tool steel

Hoang Xuan Tu[1, 2], Le Phuong Thao[1], Tran Thi Hong[3], Nguyen Thi Thanh Nga[1], Do Duc Trung[4], Jun Gong [2, 5] and Vu Ngoc Pi[1, *]

[1]Thai Nguyen University of Technology, Thai Nguyen city, Vietnam
[2]School of Mechanical and Electrical Engineering, Lanzhou University of Technology, Lanzhou, 730050, China
[3]Nguyen Tat Thanh University, Ho Chi Minh city, Vietnam
[4]Hanoi University of Industry, Vietnam
[5]Key Laboratory of Digital Manufacturing Technology and Application, Ministry of Education of China, Lanzhou, 730050, China

Abstract. The quality of final machining surfaces in a grinding process, such as the surface roughness and the accuracy depends on the grinding and dressing parameters. In order to improve the surface smoothness and precise tolerances, many grinding and dressing parameters need to be optimized. However, several dressing parameters, i.e., the rough dressing, the final dressing, and the dressing without depth of cut affected on the surface roughness have not yet been mentioned in the grinding process. This paper presents an optimization of these parameters on the surface roughness for grinding the hardened 9XC steel using grinding wheel Cn80MV1 400x40x203. An experimental system was set up to evaluate those parameters. The optimal values of these parameters were determined by using Minitab® 16 for analyzing data from the experiment.

1. Introduction

Grinding is one of the most important methods used in a machining process. The grinding process is widely used for the final shaping of elements that require very smooth surfaces and high precise tolerances. Due to this reason, the number of grinding machines can be accounted up to 30% in manufacturing industry and up to 60% in various special industries, i.e., a ball bearing manufacturing process [1].

In order to improve the smoothness of surfaces and precise tolerances, there are numerous researches focused on optimization of grinding and dressing parameters. The Taguchi method has been used to optimize the surface roughness of AISI 1040 steel plates using EKR46K grinding wheels [2]. In this research, the optimal parameters, such as the cutting speed, the rate of feed, and the depth of cut were determined by minimizing the surface roughness. To continue the minimization of the surface roughness, Ganesan et al. [3] also used the Taguchi method, ANOVA, and the regression analysis to predict the optimal value of the cylindrical grinding parameters, i.e., the cutting speed, the feed rate, and the depth of cut. In order to improve the surface finish and the surface characteristics, a research proposed the effect of grinding parameters on surface finish of EN8 steel [4]. The optimization of the ceramic grinding process using the genetic algorithm has been developed by Agarwal [5]. This work also suggested the effect of several parameters including the depth of cut, the

* Corresponding author: vungocpi@tnut.edu.vn

table feed, the size, and the density of grit on the metal removal rate, the surface roughness, and subsurface damages. The aim of this work is to obtain the best metal removal rate and surface finish and to decrease subsurface damages. Rabiey et al. [6] presented the influence of dressing parameters to grinding forces, the workpiece roughness, and the wheel wear of hybrid bond CBN grinding wheels. The multi-objective function, which depends on dressing parameters, such as the dressing speed, the dressing depth, and the dressing cross-feed, was exhibited to minimize the tangential grinding force and the surface roughness [7]. The result of this research has been given for optimizing the grinding process. Another optimization of dressing parameters in the cylindrical grinding is presented in [8]. Based on the results of this research, a model was established for simulating the wheel lifetime, the cutting ability, the production rate, and cutting forces during the grinding process. Dressing processes were displayed for cylindrical grinding using CBN and Seeded Gel hybrid wheels to compute the optimal dressing parameters [9]. In addition, Mohite et al. [10] proposed a method to determine the optimal dressing parameters, i.e., the dressing depth of cut, the dressing of cross feed rate, the drag angle of dresser, and the number of passes of dresser in order to minimize the surface roughness of EN19 steel bar by using the CNC cylindrical grinding machine.

From the brief literature review, it can be noticed that the quality of surfaces and the accuracy of elements in the grinding process have been affected by grinding and dressing parameters. However, little attention has been paid to the effect of the rough dressing, the final dressing and the dressing without depth of cut on the roughness of surfaces. This paper, thus, presents an optimization of these parameters applied for grinding hardened 9XC steel to improve the surface roughness of the ground parts.

2. Experimental design

This section presents an experiment to determine the influence of dressing parameters, such as the rough dressing, the final dressing and the dressing without depth of cut on the surface roughness and cutting forces of the hardened 9XC steel. Figure 1 shows the experiment system. The details of the experimental system and cutting parameters are shown as follows:

Experimental system:
Cylindrical grinder: CONDO - Hi-450 HTS (Japan);
Grinding wheel: Cn80MV1 400x40x203, 35 m/s (Vietnam);
Dresser: Multi-point - 3908-0088C; Type 2; diamond group: APC4; diamond mass: 1 carat (Russia);
Surface roughness tester: Mitutoyo 178-923-2A, SJ-201 (Japan);
Workpiece: hardened 9XC steel with 22 mm in diameter and 170 mm in length, 58-62 HRC, and surface roughness is not larger than 4 μm.

Cutting parameters:
- Depth of cut: t = 0.01 mm;
- Longitudinal feed rate: S_d = 1 (m/min);
- Cross-feed rate: S_n=0.01 (mm/stroke);
- Velocity of wheel: V_d = 30 (m/s).

In this study, Box-Behnken design was selected for the experiment with three types of dressing modes including coarse dressing, fine dressing, and non-feeding dressing (or dressing without depth of cut). The experimental matrix was designed by Minitab® 16 and was shown in Table 1. The surface roughness of the ground parts was measured. Figure 2 presents the relationship between the number of the ground parts and the surface roughness with five experimental runs. From this results, with the requirement of the surface roughness (in this case $R_a \leq 0.4$ μm), we can define the maximum number of ground parts with each experimental run. As a results, the number of ground parts is shown in Figure.3.

Figure 1. Experimental system

3. Experimental results and analysis

Table 1. Experimental matrix with dressing modes

Runs	Coarse dressing depth (mm)	Fine dressing depth (mm)	Non-feeding dressing (times)
1	0.07	0.03	4
2	0.06	0.02	4
3	0.06	0.02	2
4	0.07	0.02	3
5	0.06	0.01	3
6	0.06	0.03	3
7	0.07	0.02	3
8	0.08	0.01	3
9	0.07	0.02	3
10	0.07	0.01	2
11	0.07	0.03	2
12	0.07	0.01	4
13	0.08	0.03	3
14	0.08	0.02	4
15	0.08	0.02	2

IOP Conf. Series: Materials Science and Engineering **542** (2019) 012008 doi:10.1088/1757-899X/542/1/012008

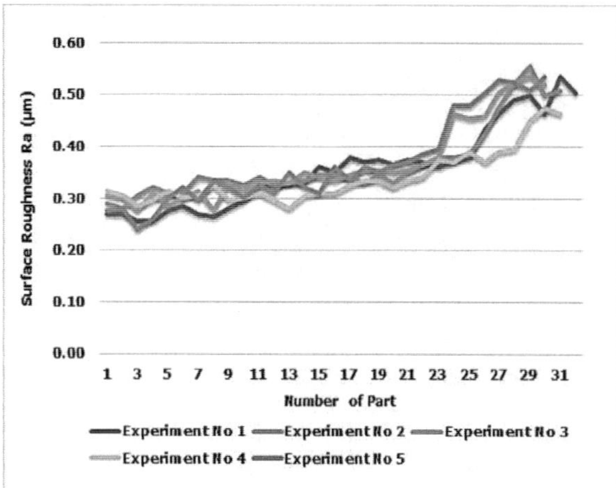

Figure 2a. Surface roughness after grinding

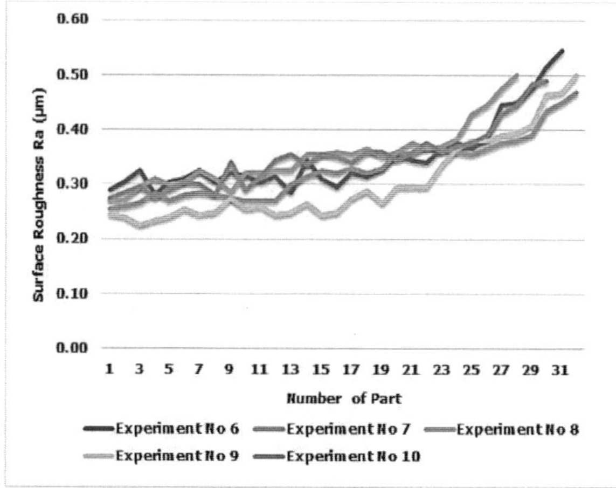

Figure 2b. Surface roughness after grinding

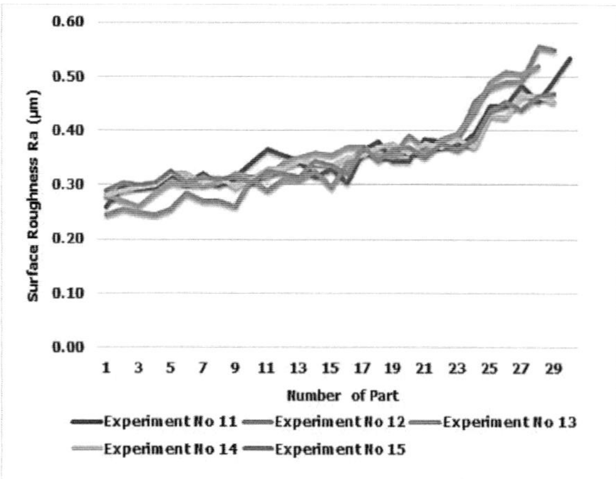

Figure 2c. Surface roughness after grinding

The wheel life T of each experiment can be calculated by the following equation [1]:

$$T = \frac{95 * n}{60} \ (min) \quad (1)$$

where, n is the number of ground parts. The values of the wheel life of each experimental run are presented in Table 2.

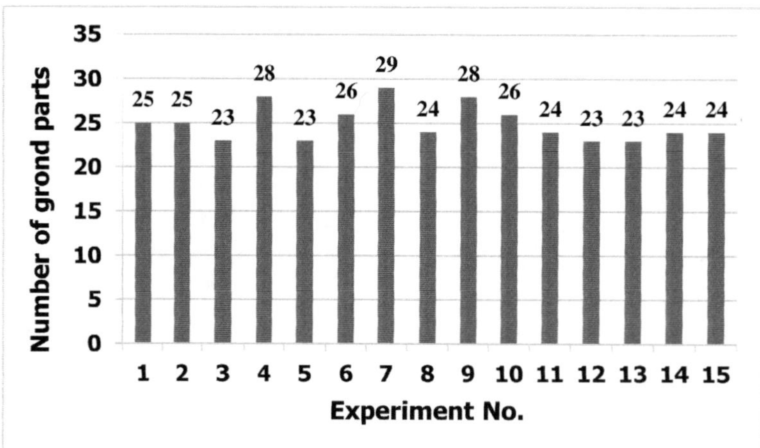

Figure 3. Number of ground parts

IOP Conf. Series: Materials Science and Engineering **542** (2019) 012008 doi:10.1088/1757-899X/542/1/012008

Table 2. The wheel life of each experiment

Runs (P_i)	Number of ground parts (-)	Wheel life (min)
1	25	39.58
2	25	39.58
3	23	36.42
4	28	44.33
5	23	36.42
6	26	41.17
7	29	45.92
8	24	38.00
9	28	44.33
10	26	41.17
11	24	38.00
12	23	36.42
13	23	36.42
14	24	38.00
15	24	38.00

```
Estimated Regression Coefficients for Y(T)

Term                                Coef  SE Coef       T      P
Constant                         42.2222   0.8531  49.495  0.000
Rough dressing                   -1.9792   0.8303  -2.384  0.063
Finish dressing                  -5.6736   1.4904  -3.807  0.013
Dressing without depth of cut     1.5833   0.8303   1.907  0.115
Rough dressing*Rough dressing    -3.8264   0.7056  -5.423  0.003
Finish dressing*Finish dressing  -3.0347   0.7056  -4.301  0.008
Dressing without depth of cut*   -3.0347   0.7056  -4.301  0.008
  Dressing without depth of cut
Rough dressing*Finish dressing   -1.5833   0.6779  -2.335  0.067
Rough dressing*                  -0.7917   0.6779  -1.168  0.296
  Dressing without depth of cut
Finish dressing*                  1.5833   0.6779   2.335  0.067
  Dressing without depth of cut


S = 1.35589     PRESS = 124.094
R-Sq = 93.45%  R-Sq(pred) = 11.61%  R-Sq(adj) = 81.67%
```

Figure 4: Analysis results by Minitab® 16

The influence of the dressing parameters on the wheel life can be described by the following equation:

$$Y(T_{Ra}) = \beta_0 + \beta_1 x_1 + \beta_2 x_2 + \beta_3 x_3 + x_1^2 + x_2^2 + x_3^2 + \beta_{12} x_1 x_2 + \beta_{13} x_1 x_3 + \beta_{23} x_2 x_3, \quad (2)$$

In which, x_1, x_2, and x_3 are the fine dressing, the coarse dressing and non-feeding dressing, respectively. $Y(T_{Ra})$ is the objective function of the regression equation; β_i are impact factors of variables x_i.

The coefficients in Equation 2 are determined by using Minitab® software 16 to analyze the data in Table 2. Thus, Equation 2 is rewritten as:

$$Y(T_{Ra}) = 42.2222\text{-}1.9792.x_1\text{-}5.6736x_2 + 0.5833x_3\text{-}3.8264x_1^2\text{-}3.0347x_2^2\text{-}3.0347x_3^2\text{-}1.5833x_1x_2\text{-} \\ \text{-}0.7917x_1x_3 + 1.5833x_2x_3 \quad (3)$$

IOP Conf. Series: Materials Science and Engineering **542** (2019) 012008 doi:10.1088/1757-899X/542/1/012008

Response Optimization

```
Parameters

      Goal     Lower  Target  Upper  Weight  Import
Y(T)  Maximum     36      46     46       1       1

Starting Point
Rough dressing                     =  0.06
Finish dressing                    =  0.01
Dressing without depth of cut      =  2

Global Solution
Rough dressing                     =  0.0692929
Finish dressing                    =  0.0209091
Dressing without depth of cut      =  3.03030

Predicted Responses

Y(T)  =  44.8943 ,  desirability =  0.889432

Composite Desirability = 0.889432
```

Optimization Plot

Figure 5. Optimum values of parameters.

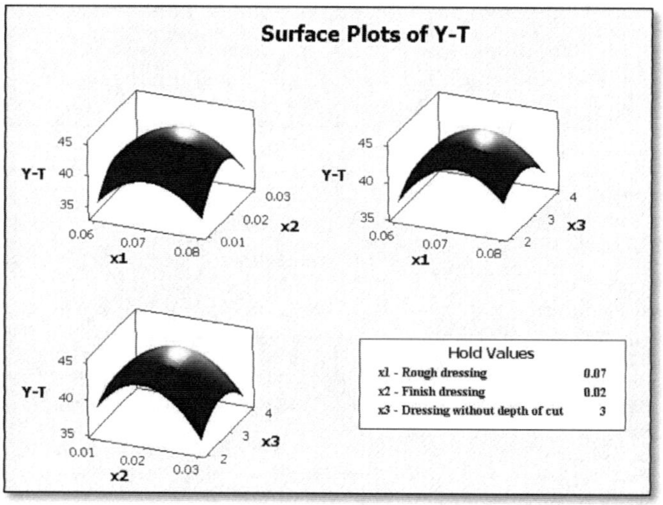

Figure 6. Interaction effect of dressing factors

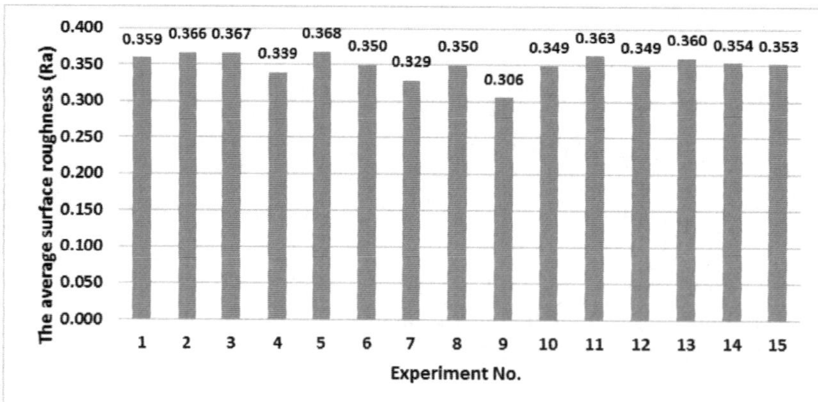

Figure 7. The average surface roughness

The optimal values of the dressing parameters can also be obtained by using Minitab® 16 as seen in Figure 5. Figure 6 shows the effect of the dressing parameters on the wheel life. From these figures, the optimal values of the dressing parameters are achieved as follows: the coarse dressing depth of 0.07 (mm); the fine dressing depth of 0.02 (mm); the number of non-feeding dressing times is 3.

4. Conclusions

This paper addressed an experimental study on the optimum dressing parameters in grinding 9CrSi tool steel using grinding wheel Cn80MV1 400x40x203 for getting the minimum surface roughness.
- Three types of dressing modes including the coarse dressing, the fine dressing and the non-feeding dressing were investigated.
- The influence of the dressing parameters on the wheel life was described.
- The optimum dressing parameters for getting the minimum surface roughness as follows: The coarse dressing depth of 0.07 (mm); the fine dressing depth of 0.02 (mm); the number of non-feeding dressing times is 3.

Acknowledgements
The work was supported by Thai Nguyen University of Technology for a scientific project.

References

[1] Bành Tiến Long, Trần Thế Lục and Trần Sỹ Túy 2001 *Principle of Material Machining*, Science and Technology HaNoi (in Vietname language).
[2] Mustafa Kemal Külekcý 2013 *Analysis of Process Parameters for a Surface-Grinding Process Based on the Taguchi Method* Materials and technology 47 pp 105–109.
[3] M Ganesan, S Karthikeyan and N Karthikeyan 2014 *Prediction and Optimization of Cylindrical Grinding Parameters for Surface Roughness Using Taguchi Method* Journal of Mechanical and Civil Engineering pp 39–46.
[4] D K Patel, D Goyal and Bs Pabla 2018 *Optimization of parameters in cylindrical and surface grinding for improved surface finish* Royal Society open science 5 pp 1–10.
[5] Agarwal S 2016 *Optimizing machining parameters to combine high productivity with high surface integrity in grinding silicon carbide ceramics* Ceramics International 42 pp 6244–6262.
[6] Rabiey M, Walter C, Kuster F, Stirnimann, J, Pude, F and Wegener K 2012 *Dressing of Hybrid Bond CBN Wheels Using Short-Pulse Fiber Laser* Journal of Mechanical Engineering, 58 pp 462–469.
[7] Baseri H 2012 *Simulated annealing based optimization of dressing conditions for increasing the grinding performance* The International Journal of Advanced Manufacturing Technology 59 pp 531–538.
[8] Aleksandrova I 2016 *Optimization of the dressing parameters in cylindrical grinding based on a generalized utility function* Chinese Journal of Mechanical Engineering 29 pp 63–73.
[9] Z Prusak, J A Webster and I D Marinescu 2010 *Influence of dressing parameters on grinding performance of CBN/ Seeded Gel hybrid wheels in cylindrical grinding* International Journal of Production Research 35 pp 2899–2915.
[10] Dadaso D M, Neeraj T, Sandeep S and Udayshankar M 2017 *Modelling and Optimization of Dressing Parameters of CNC Cylindrical Grinding Wheel for Minimum Surface Roughness* International Journal of Engineering Research and General Science 5 pp102–111.

IOP Conf. Series: Materials Science and Engineering **542** (2019) 012009 doi:10.1088/1757-899X/542/1/012009

Methods for measuring residual contact pressure of hydraulic expansion joints in heat exchangers

Jingyu Jiang[1, a], Lianfa Yang[2, b*], Jun Wei[3, c], Junjie Han[4, d] and Jinjie Huang[5, e]

[1,2,3,4,5]School of Mechanical and Electrical Engineering, Guilin University of Electronic Technology, Guilin, Guangxi, 541004
E-mail:[a]j_jingyu@163.com,[b]y-lianfa@163.com, [c]weijun6268@163.com, [d]1029314600@qq.com, [e]JinjieHuang231@126.com

Abstract. The hydraulic expansion of the heat exchanger tube and tubesheet is an important method in the uniform expansion. Compared with other methods, the advantages of hydraulic expansion are high productivity, low labor intensity and reliable expansion performance. This paper describes several common methods for the determination of the residual contact pressure between the tube and tubesheet, and describes the principle, characteristics and the current status of the measurement at home and abroad. This paper also summarizes the development of future methods for measuring the residual contact pressure in the hydraulic expansion joint.

1. Introduction

Hydraulic expansion [1] is a method for the expansion of heat exchanger tubes using direct hydraulic pressure to expand the tube forming the tube-to-tubesheet joint. The heat exchanger tube expands outward under the pressure of liquid media, after removing the liquid media in the tube, the springback quantity of the tube is much smaller than that of the tubesheet hole due to plastic deformation. Meanwhile, the tubesheet hole is in the elastic range and will immediately springback. The contact stress in the tube-to-tubesheet joint is sufficient for the heat exchanger device to obtain enough tightness and pull-out strength, which means that the tube can fit with the tubesheet hole closely. However, the main indicators for measuring the quality of hydraulic expansion are sealing performance and pull-out performance, which are associated with the residual contact pressure. Hence, residual contact pressure is one of the important mechanical parameters to indicate the quality of the hydraulic expansion joint.

This paper discusses a variety of methods for measuring the residual contact pressure in the hydraulic expansion joint that can be divided into the following three types: theoretical and computational methods, finite element analysis methods and experimental methods.

2. Theoretical and computational methods

The residual contact pressure of hydraulic expansions can be determined by theoretical calculations.

Simplified models are usually established for analyses and computation.

In 1943, Goodier et al. [2] applied the elastic-plastic theory to simplify the tubesheet into an infinite plane with holes. The residual stress and residual deformation were preliminarily studied, and then the residual contact pressure was obtained by establishing the curve.

In 1976, Krips et al. [1] used finite circular rings for structural representation of the tubesheet, and proposed a concept of an equivalent sleeve. The ring structure is denoted as an equivalent sleeve, and a formula for the residual contact pressure was proposed:

$$\frac{p_c^R}{p_i - p_c} = \left(1 + \left(\frac{k_s^2-1}{k_t^2-1}\right)\left[\frac{k_t^2(1-\mu_s)+(1+\mu_s)}{k_s^2(1+\mu_s)+(1-\mu_s)}\right]\right)^{-1} \quad (1)$$

Where p_c^R and p_i respectively indicate the residual contact pressure and the expansion pressure, μ_s is the Poisson's ratio of the tubesheet material, K_t and K_s respectively indicate the ratios of outer and inner radii of the tube and tubesheet (the following parameters are the same). Krips et al. assumed that the tube and tubesheet are ideal elastic-plastic with the same elastic modulus. When the elastic moduli of the tube and tubesheet differ greatly, the formula (1) is no longer applicable.

In 1991, Yokell [3] assumed that the tube and tubesheet is ideal elastic-plastic and the thickness of the outer tube wall is infinite. Yokell established a formula for the residual contact pressure:

$$p_c^R = p_i[1 - \left(\frac{r_i}{r_o}\right)^2]\frac{2}{\sqrt{3}}\delta_{st}\ln(\frac{r_o}{r_i}) \quad (2)$$

Where r_i and r_o respectively indicate the inner and outer radii of the heat exchanger tube, δ_{st} is the yield strength of the tube (the following parameters are the same). The formula (2) is simple and it indicates that the residual contact pressure is associated with the yield limit, inner and outer diameters of the tube. However, the shortcomings of the formula (2) are the same as those of the formula (1), and the result of formula (2) is always smaller than that of the formula (1).

In 1992, Chaaban et al. [4] utilized statistical methods and established an empirical equation for the outer diameter of the equivalent sleeve. A more practical empirical formula was proposed, and the stress in the transition zone was analyzed by theoretical and computational methods.

In 1995, Kohlpaintner [5] assumed the heat exchanger tube to be a 19-hole tubesheet model with triangular hole patterns, and established a more perfect equation for the equivalent sleeve outer diameter. The empirical formula of the equivalent sleeve diameter was obtained to calculate the residual stress by analyzing the mechanical properties of the perforated tubesheet.

In 1996, Huigeng Y et al. [6] assumed that there is no strain enhancement in the tube during the expansion. Elastic-plastic analysis method was utilized to establish a two-dimensional single-tube model. The residual contact pressure p_c^R was given by:

$$p_c^R = (1 - 2c)p_i - \delta_{st}\ln K_t \quad (3)$$

Where c is associated with parameters such as E_t, E_s, K_t, K_s, μ_s, μ_t, etc. E_t and E_s respectively indicate the elastic modulus of the tube and tubesheet. μ_t indicates the Poisson's ratio of the tube.

In 2001, Huigeng Y et al. [7] realized that the elastic-plastic theory utilized for hydraulic expansion analysis would generate large errors. The calculation of the deformation curve and the slope after the drawing was performed and the value of residual contact pressure was calculated. The accuracy of this method is related to the material and the degree of deformation of the tube.

In 2007, Laghzale N E et al. [8] proposed a new elastic-plastic model which can accurately predict the residual contact pressure. With the single-tube sleeve model, they assumed that the model of the tube and tubesheet is ideal. Meanwhile, the strain hardening was not neglected.

IOP Conf. Series: Materials Science and Engineering **542** (2019) 012009 doi:10.1088/1757-899X/542/1/012009

In 2010, Chen G et al. [9] focused on the analysis of two theoretical and computational models for the residual contact pressure: an ideal elastic-plastic model and a power hardening material model. The two models of the tube were numerically simulated, and the results of the numerical analysis were compared with those of the theoretical model in this analysis.

In summary, the formulas proposed by Krips, Yokell, and Huigeng Y et al. are derived on the basis of the ideal elastic-plastic model. Hence, the calculated value of the residual contact pressure is smaller than the actual one. Although the formula established by Huigeng Y et al. is simple and requires fewer known parameters for materials, however, the effect of gaps is neglected. It is more accurate that Laghzale N E et al. took the strain hardening of the material into account.

3. Finite Element Methods

The finite element method (FEM) is a numerical method for solving problems of engineering and mathematical physics. Within the linear and nonlinear finite element analysis, the residual contact pressure can be obtained.

In 1984, Scott et al. [10] used stress corrosion, X-ray diffraction and other methods to study the residual stress and residual deformation in the expansion joint. Scott studied the stress state at the contact surface of the expansion by the finite element analysis and studied the change in strength and length of this joint during hydraulic expansion.

In 1992, Chaaban et al. [4] utilized finite element methods to perform finite element analysis. Referring to Figure 1, the 7-hole plane stress model and the axisymmetric model were used to determine the residual contact pressure, and the residual stress in the transition zone was analyzed.

In 2003, Feng W H et al. [11] simulated the hydraulic expansion process by the nonlinear finite element analysis. Referring to Figure 2, the 1/12 of the finite circular disk with 19 holes was taken as the computation model for the tube and the tubesheet. ANSYS was used for further processing, it was found that the residual contact pressure was unevenly distributed along the axis of the tube. The quantitative relationship between the residual contact pressure and the material, geometry and expansion pressure was determined. However, the quantitative relationship between pull-out stress, joint sealing pressure and residual contact pressure remains to be studied.

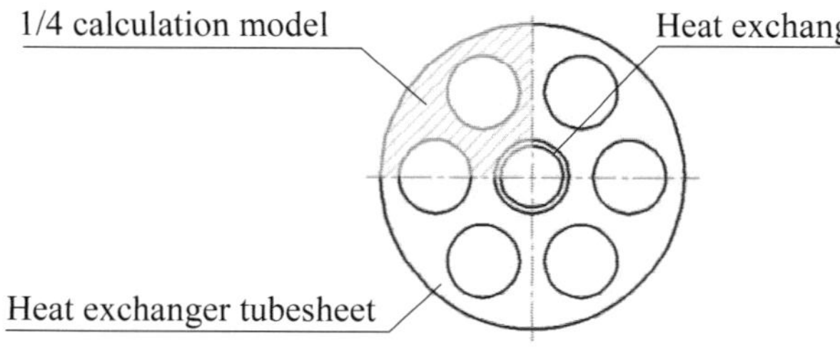

Figure 1. Finite circular disk with 7 holes

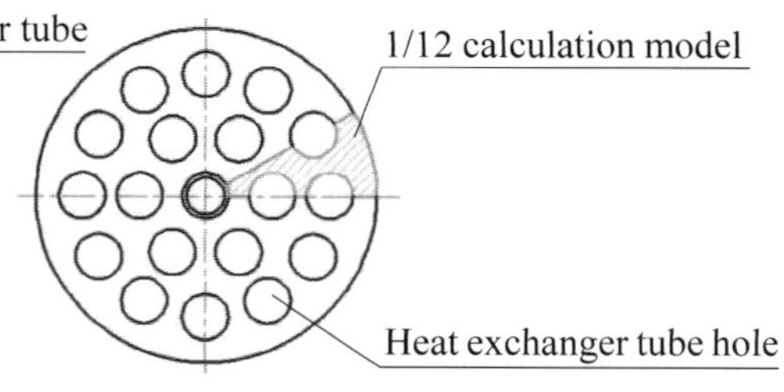

Figure 2. Finite circular disk with 19 holes

In 2012, Zheng W et al. [12] simulated the expansion process of AP1000 heat exchanger tube by the finite element analysis. Since adopting the finite circular disk with 19 holes needs a large amount of calculations, the 1/4 of a finite circular disk with 7 holes was adopted for analysis shown in Figure 1. ANSYS was used to solve geometric nonlinear problems such as large deformation, large strain and stress rigidification. And the residual contact pressure increases with expansion pressure in the range of 250-270 MPa and reaches the maximum when the expansion pressure reaches 270 MPa was summarized.

In summary, the model adopted by Scott and Chaaban et al.is a single-tube model, which is different from the models adopted by Feng W H and Zheng W et al. ANSYS was found a utilization in processing the distribution of residual contact pressure during the expansion process.

4. Experimental methods

During the hydraulic expansion, the pressure is evenly distributed on the inner wall of the tube. Hence, not only the value of the residual contact pressure is calculated, but also the circumferential strain on the outer surface of the single-tube model is measured.

In 1996, Huigeng Y, L Ge et al. [13, 14] conducted further research and got a theoretical solution of the residual contact pressure on the basis of the principle of unloading and related parameters. Huigeng Y, L Ge et al. adopted Elastic unloading methods to verify the correctness of the theoretical solution of the residual contact pressure.

In 2003, Yi Lu et al. [15] used electrical resistance strain gauges to measure the circumferential strain on the outer surface of a single-tube model, as shown in Figure 3. Yi Lu proposed a new method for measuring the residual contact pressure, by which the relevant parameters of the entire measurement system can be calibrated. Hence, the influence of various factors was eliminated. In addition, the accuracy of the algorithm proposed from Yi Lu was verified in the literature [7].

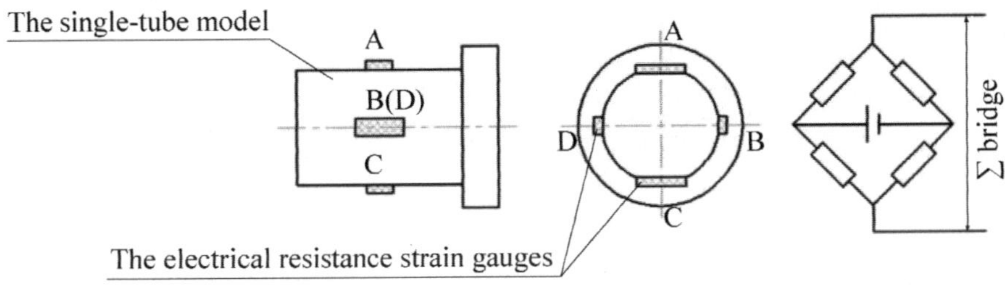

Figure 3. The strain gauge distribution and full-bridge connection

In summary, the errors between the experimental results from Huigeng Y, L Ge et al. and the theoretical results are within a reasonable range, hence, the methods with resistance strain gauges are feasible. According to Yi Lu, the factors such as the structural size of the model, the accuracy of the attachment position of the strain gauge and the sensitivity coefficient of the strain gauge are eliminated.

5. Conclusions

Theoretical analyses and calculations for different types of tubes are needed for the methods for measuring the residual contact pressure in the hydraulic expansion joint, and the finite element simulation software should be used for analyses to improve efficiency and accuracy. It is also necessary to conduct experiments to further verify whether the experimental and theoretical values are within reasonable errors. Some of the hydraulic expansions mentioned in this article are liquid bag expansions. Hence, further researches on pure hydraulic expansions are needed. Most of the objects simulated in the existing literatures are single-tube models, however, multi-tube expansions are required in actual productions. In addition, the residual contact pressure after multi-tube expansion is different from that after single-tube expansion, and it is worth further study.

Acknowledgements

This work was financially supported by the National Natural Science Foundation of China (51564007), Natural Science Foundation of Guangxi Province (2017GXNSFAA198133), and Guangxi Key Laboratory of Manufacturing System & Advanced Manufacturing Technology (14-045-15-005Z).

References

[1] Krips H and Podhorsky M. Hydraulic Expansion—A New Procedure for Fastening Tubes[J]. *VGB Kraftswerkstech,* 1976, 56(7): 456-64.

[2] Goodier J N and Schoessow G J. The holding power and hydraulic tightness of expanded tube joints: analysis of the stress and deformation. Trans [J]. *ASME J. Press. Ves. Tech*, 1943, 65(7):489-96

[3] Yokell S. Heat-exchanger tube-to-tubesheet expanded joints[J]. *Trans. of ASME, J. of Pressure Vessel and Technology*, 1991, 113: 341-4.

[4] Chaaban A, Ma H and Bazergui A. Tube-Tubesheet Joint: A Proposed Equation for the Equivalent Sleeve Diameter Used in the Single-Tube Model[J]. *Journal of Pressure Vessel Technology*, 1992, 114(1):19-22.

[5] Kohlpaintner W R. Calculation of Hydraulically Expanded Tube-to-Tubesheet Joints[J]. *Journal of Pressure Vessel Technology*, 1995, 117:1(1):24-30.

[6] Huigeng Y, Bingsheng Z and Letong G, et al. Research on Hydraulically Expanding of Heat Exchangers (1) Determination of the Expanding Pressure[J]. *China Pressure Vessel Technology*,1996, 13(2): 126-30.

[7] Huigeng Y and Cuiju Y. Engineering Graphic Estimation of the Residual Contact Pressure of the Hydraulic Expanding Tubes of Heat Exchangers[J]. *Chemical Engineering & Machinery*, 2001, 28(4): 211-4.

[8] Laghzale N E and Bouzid A H. Analytical Modeling of Hydraulically Expanded Tube-To-Tubesheet Joints[J]. *Journal of Pressure Vessel Technology*, 2007, 131(1):233-42.

[9] Chen G and Zhang J. Theoretical Calculation Method and Analysis on Residual Kiss Pressure during Hydraulic Tube Expansion [J]. *Power Station Auxiliary Equipment*, 2010, 3: 002.

[10] Scott D A and Wolgemuth G A, Aikin J A. Hydraulically expended tube-to-tubesheet joints [J]. *Trans ASME J Press Ves Tech*, 1984,106(2): 104-9

[11] Feng W H and Fu S Z. Residual contact pressure in hydraulically expanded tube-tubesheet joints of heat exchangers[J]. *Journal of Nanjing University*, 2003.

[12] Zheng W. Analysis of residual contact pressure during hydraulic expanded tube-to-tubesheet joint[J]. *Modern Manufacturing Engineering*, 2012.

[13] Huigeng Y, Bingsheng Z and Letong G, et al. Research on Hydraulic Expanding of Heat Exchangers (2) The Residual Contact Pressure and Friction Coefficient[J]. *Pressure Vessel Technology*, 1996.

[14] The Residual Contact Pressure Measuring of Hydraulically Expanded Tube-to-tubesheet Joints of Heat Exchangers[J]. *Journal of Jiangsu Institute of Petrochemical Technology*, 1996.

[15] Yi Lu and Yan H. A New Method for Measuring the Residual Contact Pressure of Hydraulic Expansion[J]. *Chemical Engineering & Machinery*, 2003

IOP Conf. Series: Materials Science and Engineering **542** (2019) 012010 doi:10.1088/1757-899X/542/1/012010

Detection methods for residual stress of assembled camshaft

Jinjie Huang[1,a], Lianfa Yang[2,b*], Jun Wei[3,c], Jianping Ma[4,d] and Jingyu jiang[5,e]

[1,2,3,4,5]School of Mechanical & Electrical Engineering, Guilin University of Electronic Technology, Guilin, Guangxi, 541004, China

E-mail: [a]JinjieHuang231@126.com, [b]y-lianfa@163.com, [c]weijun6268@163.com, [d]majianping1229@163.com, [e]j_jingyu@163.com

Abstract. Compared with the traditional integrated camshaft, the assembled camshaft has the advantages of lowering production cost, reducing engine weight and improving engine performance, with a broad development prospect. The assembled camshaft is connected by different processes will cause residual stress which will seriously affect the mechanical properties of camshaft. Therefore, the measurement of residual stress is of great importance to the safety and reliability of assembled camshaft. This paper mainly introduces three detection methods currently applied to the residual stress of the assembled camshaft, namely X-ray diffraction method, Hole-Drilling method and Finite element simulation.

1. Introduction

Including a shaft, several cams and other components, a assembled camshaft is connected by various processes, which can be manufactured according to different performance requirements to better exert the performance of the material. After this camshaft is assembled, the residual stress is generated on its surface and inside. The working cam can be easily induced cracks by this stress, which is a great threat to the safety and reliability of camshaft operation. Therefore, studying on the detection of residual stress of assembled camshaft is necessary. Due to the geometries of the camshafts of different manufacturing processes of the same type are relatively close, the residual stress measurement methods of the conventional camshaft are also applicable to the assembled camshaft.

The paper mainly describes three methods currently applied to the detection of residual stress in camshafts, namely X-ray diffraction method, Hole-Drilling method and Finite element simulation. The first two detection methods mainly use specific experimental equipments to measure residual stress, while the last method mainly operates ANSYS, ABAQUS and other software to achieve detection.

2. X-ray diffraction method

In 1929, Russian scholars proposed the basic principle of X-ray stress measurement. Due to the effects of residual stress, the lattice spacing inside the camshaft is changed. Meanwhile, the macroscopic strain is generated. X-ray diffraction technique can be used to calculate the lattice strain by measuring the diffraction on angle of its internal lattice (Figure. 1). The macroscopic strain of the camshaft can be obtained by using the elastic mechanics. According to the consistency of the lattice strain and the macroscopic strain, the macroscopic stress can be calculated. With this technique, since ordinary

X-rays have limited penetration into the metal layer, it is necessary to perform surface depth removal using electrolytic polishing to perform deeper residual stress measurement, but the deviation caused by the removal layer must be corrected.

In 1956, Moore M G et al. [1] took the elastic theory formula to derive modified formulas for the residual stress of the removed layer based on some simple geometric assumptions. However, these formulas did not apply to complex geometries, such as cam lobes and gears, etc. In 2007, Ricardo CLA et al. [2] presented a new stress correction algorithm based on research of literature [1], which considered the effects of X-ray absorption and mechanical equilibrium conditions. This algorithm utilized the stress profile model of the fifth order polynomial to solve the problem of oscillation of the polynomial at a greater depth, and improved the reliability of residual stress measurement.

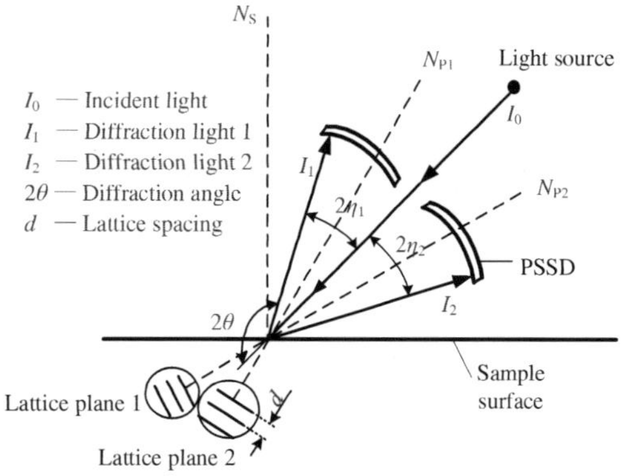

Figure. 1 X-Ray Diffraction Stress Measurement Geometry

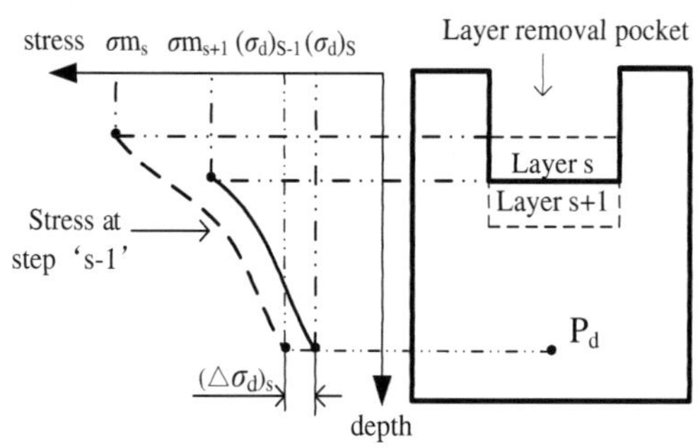

Figure. 2 Schematic diagram of residual stress layer removal process

In 1996, Lambda R. [3] proposed a FEA matrix relaxation correction technique by some ideal assumptions and the idea of layer removal, which was used to correct layer removal in X-ray residual stress measurement. This technique can be taken to create 2-D or 3-D models based on different samples and polished pockets to make stress calculations simple, and is applicable to relatively complex geometries. However, this technique is ineffective for the assumption of simple geometric and symmetric stress fields. In 2012, Savaria V et al. [4] improved Lambda R's technique by using the average of the residual stress values measured at the top of the upper and lower removal layers (Figure. 2) as the calculated value of the correction formula. This improved technique can be used to keep accuracy in high stress gradient zones and it is suitable for more geometric models.

In 2018, LEVIEIL B. et al. [5] utilized X-ray diffraction to measure the residual stress of bars, compared and analyzed the correction methods in the literatures [1,3,4]. The study found that the correction results of the above correction methods proved that the measurement accuracy of the X-ray method is improved. The authors showed the importance of the material retention elasticity after layer removal on the measurement accuracy of the three methods, and found that the elasticity had no effect on the above two finite element correction methods [3,4] below 4 mm on the surface.

3. Hole-Drilling method

The Hole-Drilling method (Blind hole method) was proposed by German scholars in 1934. The specific principle of the method is as follows. Three strain gauges are attached to the surface of the workpiece where residual stress is to be tested and a hole is drilled at the intersection of their neutral lines. Since the stress is redistributed in the vicinity of the hole after drilling, the strain gauge is strained by the stress release, finally according to the stress-strain relationship, residual stress of this point can be calculated, including the assumed two principal stresses σ_1, σ_2 and a main direction angle θ, as shown in Figure. 3.

In 1978, Sandifer J P [6] performed force analysis around the blind hole and calculated the final residual stress by loop iteration. The author through the results of the measurement found that the eccentricity of the borehole will affect the measurement accuracy of the blind hole method, as shown in Figure. 4. But the force analysis process of this method is more complicated, the loop iteration is not easy to converge. In 1994, Liu B et al. [7] through calibration experiments found that plastic deformation occurs at the edge of the hole when using the drilling method to measure residual stress, and proposed a correction formula for the amount of plastic deformation.

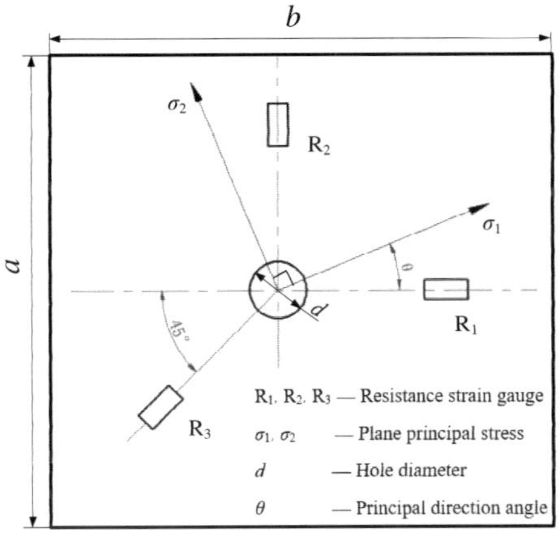

Figure. 3 Drilling method application schematic

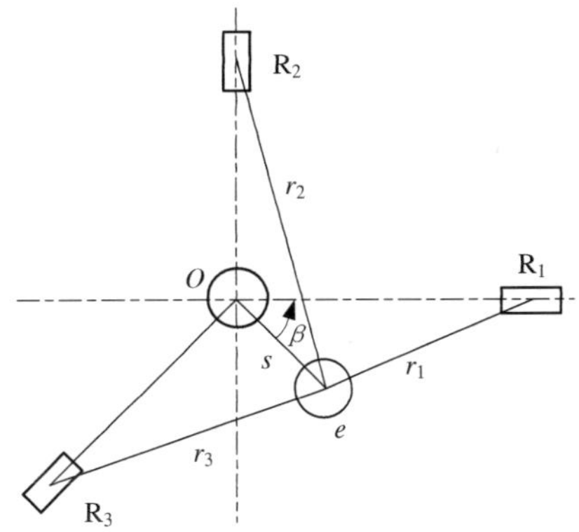

Figure. 4 Schematic diagram of the positional relationship between borehole and strain gauge

In 2012, Liu L L et al. [8] put forward a method of directly correcting the release coefficient by using the finite element method, ignored the influence of plastic deformation on the edge of the hole, which can correct the error caused by the eccentricity of the borehole. Through analysis, it is found that when $D/d \geq 16$ (D — outer diameter of material), the Hole-Drilling method can be applied to determine the residual stress of the tube parts.

In 2014, Von Mirbach D [9] recognized that the plastic strain at the hole edge, which has a great influence on the measurement accuracy of the residual stress of the borehole method. The author used the loop iterative algorithm to derive the calculation and presented the method of adaptive calibration functions. The load type of the material and the performance of the elastoplastic material were taken into account in the method which reduced the calibration procedure and shortened the correction time of the residual stress to some extent.

4. Finite element simulation

Many scholars have employed finite element numerical simulation to observe and study the residual stress distribution of the entire assembled camshaft. The Finite element method is a discretized numerical solution commonly used to analyze and solve complex practical engineering problems.

In 2006, Bayrakceken H et al. [10] utilized the finite element method to determine the residual stress concentration of the camshaft fracture region, including the dangerous zone with high stress concentration, maximum stress and minimum stress. Firstly, the camshaft model is established, then according to the actual force condition, the boundary condition of the camshaft position and the load such as torsion and pressurization are set, and the angular velocity of the camshaft rotation is considered. Finally, the entire working process of the camshaft is simulated.

In 2011, Qiao J et al. [11] employed finite element software to simulate the expansion process of the steel ball of the assembled camshaft, and obtained the residual contact stress distribution of the camshaft. Firstly, the author simplified the cam in the model into a thick-walled cylinder, treated the expansion process as an axisymmetric problem, and then fine-divided the mesh of the shaft and cams joints, and finally the assembly process was simulated. In 2015, Yang S et al. [12] also utilized ANSYS software to analyze the stress deformation of this connection process, and got the distribution of residual stress on the inner wall of the shaft tube.

In 2015, Liu G et al. [13] simulated the hydraulic connection process of the assembled camshaft by finite element method, and achieved the residual stress distribution of the axial and circumferential directions of the camshaft. The author designed the inner hole of the cam to be elliptical, and selected a 1/4 cam section shape to establish a model base on the symmetrical shape. The simulation process assumesd that the shaft is in a thin-walled mode, hydraulic pressure conditions were set.

In 2017, Scherzer R et al. [14] performed a finite element simulation on the Presta joining process of the assembled camshaft, which gained the residual stress distribution after joining. The simulation process was divided into two steps. Firstly, the rolling manufacturing process of the shaft was to simulate, secondly, the interference assembly of shaft and cam was to carry out. The scholar set the mesh of the assembly part to be fine, and regarded the cam as a cylinder treatment. The smallest model section was selected for simulation based on the cyclic symmetry of the cylinder contour.

5. Conclusion

The following are the summaries of three detection methods for residual stress of camshaft.

(1) Through the continuous improvement of the correction method, the accuracy of measuring the residual stress of the camshaft by X-ray diffraction method has been greatly improved. M. Moore's correction method is only suitable for some simple symmetrical geometry, and Ricardo CLA improved the reliability of this method. Although FEA matrix relaxation method has some limitations, it is more versatile and accurate than M. Moore's method.

(2) When measuring the residual stress of the camshaft by the blind hole method, it is necessary to satisfy the axial diameter ratio (D/d). The measurement accuracy of this method is mainly affected by two factors, namely the plastic deformation of the hole edge and the eccentricity of the hole. Many scholars have done a lot of research on these two factors, and the measurement accuracy is getting higher and higher. However, the accuracy of the method required for drilling is relatively high, and it will cause certain damage to the sample.

(3) Finite element simulation has a distinct advantage of a comprehensive observation and analysis of the distribution of residual stress in assembled camshaft. Properly setting the grid size of the camshaft model and selecting the model section can greatly improve the efficiency of calculation.

Acknowledgements

This work was financially supported by the National Natural Science Foundation of China (51564007), Natural Science Foundation of Guangxi Province (2017GXNSFAA198133), and Guangxi Key Laboratory of Manufacturing System & Advanced Manufacturing Technology (14-045-15-005Z).

IOP Conf. Series: Materials Science and Engineering **542** (2019) 012010 doi:10.1088/1757-899X/542/1/012010

References

[1] Moore M G, Evans W P. Mathematical correction for stress in removed layers in X-ray diffraction residual stress analysis[C], *a meeting of the SAE Residual Stress Committee, Pittsburgh*, 1956.

[2] Ricardo C L A, D'Incau M, Scardi P. Revision and extension of the standard laboratory technique for X-ray diffraction measurement of residual stress gradients[J]. *Journal of Applied Crystallography*, 2007, 40(4):675–683.

[3] Lambda R. Diffraction notes no.17. Finite element correction for stress relaxation in complex geometries. *Cincinnati, OH, USA*; 1996. (513) 561-0883.

[4] Savaria V, Bridier F, Bocher P. Computational quantification and correction of the errors induced by layer removal for subsurface residual stress measurements[J]. *International Journal of Mechanical Sciences*, 2012, 64(1):184-195.

[5] Levieil B, Bridier F, Doudard C, et al. User Influence on Two Complementary Residual Stress Determination Methods: Contour Method and Incremental X-Ray Diffraction[J]. *Experimental Mechanics*, 2018, 56(9):1-12.

[6] Sandifer J P, Bowie G E. Residual stress by blind-hole method with off-center hole[J]. *Experimental Mechanics*, 1978, 18(5):173-179.

[7] Liu B, Jiao J, Li G. Plastic Released Strain in Determining Residual Welding Stresses by Hole-Drilling Method[J]. *Journal of Dalian Railway Institute*, 1994.

[8] Liu L L, Fan L X, Dong X H. Study on Measurement of Residual Stresses in Forged Barrel Surface[J]. *Acta Armamentarii*, 2012, 33(6):712-717.

[9] Von Mirbach D. Hole-Drilling Method for Residual Stress Measurement - Consideration of Elastic-Plastic Material Properties[J]. *Materials Science Forum*, 2013, 768-769:174-181.

[10] Bayrakceken H, Ucun I, Tasgetiren S. Fracture analysis of a camshaft made from nodular cast iron[J]. *Engineering Failure Analysis*, 2006, 13(8):1240-1245.

[11] Qiao J, Kou S Q, Kong Q L, et al. Analysis of hollow camshaft assembled process on expanding tube connection by pushing through a ball[J]. *Journal of Jilin University*, 2011, 41:115-120.

[12] Yang S, Zhao D, Zeng G, et al. Stress and Deformation Analysis of the Ball through Assembled Camshaft based on ANSYS[J]. *Journal of Mechanical Transmission*, 2015.

[13] Liu G, Lin J, Yuan S. Mechanism of Hydrojoining and Approach to Increase Torsion Strength of Assembled Camshafts[J]. *Journal of Manufacturing Science & Engineering*, 2015,137(5).

[14] Scherzer R, Silbermann C B, Landgraf R, et al. FE-simulation of the Presta joining process forassembled camshafts - modelling of the joining process[J] *Materials Science and Engineering* , 2017, 181(1):012030.

IOP Conf. Series: Materials Science and Engineering **542** (2019) 012011 doi:10.1088/1757-899X/542/1/012011

Effect of cohesion on internal stability of bilinear geosynthetic-reinforced slopes

Gongquan Zhang[1], Xiaobo Ruan[1], Jun Yan[1] and Dayong Zhu[2]

1 School of Automotive and Transportation Engineering, Hefei University of Technology, Hefei, China
2 School of Civil Engineering, Hefei University of Technology, Hefei, China
E-mail: xiaoboruan@163.com

Abstract. This paper presents a limit equilibrium (LE) approach to analyze the internal stability of the bilinear geosynthetic-reinforced slopes comprised of cohesive reinforced fill. This LE analysis uses a top-down log spiral mechanism to formulate the resultant reinforcement force in the lower tier required for internal design of such reinorced slopes. The presented formulation considers both cohesion and depth of crack, and then the required unfactored reinforcement strength is defined and used for subsequent analysis. Results are presented in the form of stability charts, enabling quick assessment of reinforcement strength required for internal stability. In addition, the effect of cohesion and depth of crack on the critical slip surface is discussed, respectively.

1. Introduction

In this paper, a bilinear geosynthetic-reinforced slop can be defined as a geosynthetic-reinforced slope with zero batter (i.e., vertical) in the upper while another geosynthetic-reinforced slope has a batter greater than zero (non-vertical) in the lower. This reinforced slope can yield the same right-of-way as a single equivalent, yet less steep slope. For analysis of bilinear reinforced slopes, Ruan et al. (2015)[1] conducted a rigorous limit equilibrium (LE) analysis exploring the impact of such reinforced slopes on the maximum reinforcement force required for global stability. Subsequently, considering the effect seismic load, Ruan et al. (2017) [2] extended the formulation in the paper of Ruan et al. (2015) [1].

China has a vast territory, and the soil quality varies greatly. In the absence of high quality filling areas, the one-sided pursuit of filler quality may lead to excessive investment. In engineering practice, Yang et al. (2012) [3] has used the cohesive material as the reinforced fill, but the related engineering experience is still immature. Therefore, the study of the relevant characteristics of reinforced earth structures with cohesive materials is helpful to broaden its scope of application, and achieve the purpose of saving project cost and improving economic benefits. In particular, it is necessary to study the effect of cohesion of internal stability of bilinear geosynthetic-reinforced slopes.

The objective of this study is to formulate the resultant reinforcement force required for internal stability of bilinear geosynthetic-reinforced slopes in the lower tier through extending the methods of Ruan et al. (2015) [1] combined with Vahedifard, et al. (2014) [4]. Presented formulation considers the effect of cohesion and depth of crack. In addition, it is assumed that the foundation soil is competent.

2. Formulation

The outlined analysis assumes log spiral slip surfaces as part of the LE formulation - refer to Figure 1 for notation and convention. In Figure 1, the resisting forces, T_1 and T_2, are the resultant force of all reinforcement layers for the upper tier and the lower tier, respectively. The driving force, W, is the weight of the entire failure mass. The line of action of T_1, D_1, is measured from the bottom of the upper tier while the lines of action of T_2, D_2, is measured from the bottom of the lower tier.

The location of the resultant reinforcement force is not known and must be assumed. For the current design guidelines of Mechanically Stabilized Earth walls [5], for a horizontal and surcharge-free crest subjected to static conditions, the height of the resultant is one third of the height of the wall. With assumed surcharge or seismicity, the elevation of resultant goes up. Therefore, it is reasonable to assume D_1 and D_2, to act at $H_1/3$ and $H_2/2$, respectively.

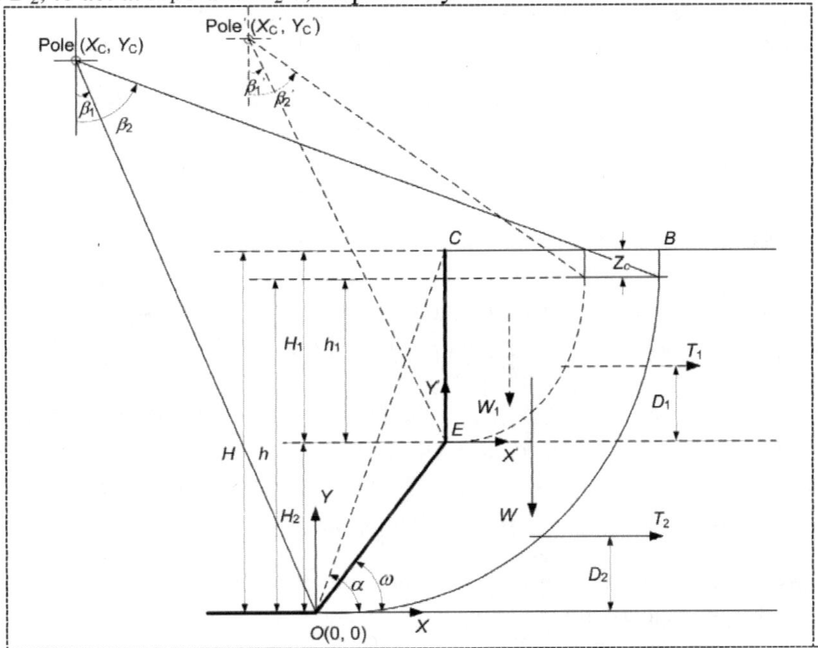

Figure 1. Notation and convention for the presented LE approach..

For completeness, the expression for the resultant resisting force in the upper tier, T_1, is reproduced here from Vahedifard, et al. (2014) [4]

$$
\begin{aligned}
T_1 = \Bigg\{ & \gamma_d \int_{\beta_1'}^{\beta_2'} \left(A' e^{-\psi\beta'} \cos\beta' - A' e^{-\psi\beta_2'} \cos\beta_2' \right) \left(A' e^{-\psi\beta'} \sin\beta' \right) \left(A' e^{-\psi\beta'} \right) \left(\cos\beta' - \psi \sin\beta' \right) d\beta' \\
& - c \int_{\beta_1'}^{\beta_2'} \left(A' e^{-\psi\beta'} \cos\beta' \right) \left(A' e^{-\psi\beta'} \right) \left(\cos\beta' - \psi \sin\beta' \right) d\beta' - c \int_{\beta_1'}^{\beta_2'} \left(A' e^{-\psi\beta'} \sin\beta' \right) \left(A' e^{-\psi\beta'} \right) \\
& \times \left(\sin\beta' + \psi \cos\beta' \right) d\beta' + \frac{1}{2} \gamma_d Z_c \left[\left(A' e^{-\psi\beta_2'} \sin\beta_2' \right)^2 - \left(A' e^{-\psi\beta_1'} \sin\beta_1' \right)^2 \right] \Bigg\} \\
& / \left(A' e^{-\psi\beta_1'} \cos\beta_1' - D_1 \right)
\end{aligned} \tag{1}
$$

where, γ_d is the unit weight of the reinforced soil; c is the cohesion of reinforced fill; β_1' and β_2' are angles at points where the log spiral slip surface enters and exits the upper tier; β' is the angle in polar coordinates defined relative to Cartesian coordinate system translated to Pole' (X_C', Y_C') from the origin E – Fig. 1; A' is log spiral constant, i.e., $h_1 / \left[\exp(-\psi\beta_1') \cos\beta_1' - \exp(-\psi\beta_2') \cos\beta_2' \right]$. where, $h_1 = H_1 - Z_c$, H_1 is height of the upper tier and Z_c is the depth of crack, $\psi = \tan\phi_d$, and ϕ_d is the design internal angle of friction.

The resistive moment component due to normal and shear stress distributions along the log spiral at a LE state vanishes since its elemental resultant force goes through the pole. Consequently, at a LE state the resisting and driving moments are equal as shown:

$$M_W = M_{T1} + M_{T2} \tag{2}$$

where, M_W is the driving moment about the pole and can be calculated using W multiplied by its corresponding leverage arm; M_{T1} and M_{T2} are the resisting moments and can be determined using T_1 and T_2 multiplied by their respective leverage arms, respectively.

Using Eqns. 1 and 2 one can solve T_2 which is as follows

$$\begin{aligned} T_2 = \Big\{ & \gamma_d \int_{\beta_1}^{\beta_2} \left(Ae^{-\psi\beta}\cos\beta - Ae^{-\psi\beta_2}\cos\beta_2\right)\left(Ae^{-\psi\beta}\sin\beta\right)\left(Ae^{-\psi\beta}\right)(\cos\beta - \psi\sin\beta)d\beta \\ & - C\int_{\beta_1}^{\beta_2}\left(Ae^{-\psi\beta}\cos\beta\right)\left(Ae^{-\psi\beta}\right)(\cos\beta - \psi\sin\beta)d\beta - C\int_{\beta_1}^{\beta_2}\left(Ae^{-\psi\beta}\sin\beta\right)\left(Ae^{-\psi\beta}\right) \\ & \times(\sin\beta + \psi\cos\beta)d\beta + \frac{1}{2}\gamma_d Z_c\left[\left(Ae^{-\psi\beta_2}\sin\beta_2\right)^2 - \left(Ae^{-\psi\beta_1}\sin\beta_1 + \cot\alpha\right)^2\right] \\ & - (\gamma_d h_1 H\cot\alpha)\left(Ae^{-\psi\beta_1}\sin\beta_1 + H\cot\alpha/2\right) - (\gamma_d H_2 H\cot\alpha/2)\left(Ae^{-\psi\beta_1}\sin\beta_1\right. \\ & \left. + H\cot\alpha/3\right) - T_1\left(Ae^{-\psi\beta_1}\cos\beta_1 - H_2 - D_1\right)\Big\} / \left(Ae^{-\psi\beta_1}\cos\beta_1 - D_2\right) \end{aligned} \tag{3}$$

where, H_2 and H are heights of the lower tier and the bilinear slope, respectively; β_1 and β_2 are angles of points where the log spiral enters and exits the bilinear slope – Fig. 1; α is the angle of the bilinear slope; β is the angle in polar coordinates defined relative to Cartesian coordinate system translated to Pole (X_C, Y_C) from the origin O (0, 0); A is log spiral constant, i.e., $h/[\exp(-\psi\beta_1)\cos\beta_1 - \exp(-\psi\beta_2)\cos\beta_2]$, where $h=H-Z_c$.

For a dimensionless analysis using T_1 and T_2, one can respectively define K_{T1} and K_{T2} as $2T_1/(\gamma_d H^2)$ and $2T_2/(\gamma_d H^2)$, respectively

3. Results

3.1. Stability charts

When α=70°, Z_c=0 and ϕ_d=34°, Figure 2 shows that K_{T1} increases as λ increases for different values of c while K_{T2} decreases with an increase in λ. In particular, K_{T2} is equal to zero when the value of $c/\gamma_d H$ arrives at 0.075 regardless of the value of λ.

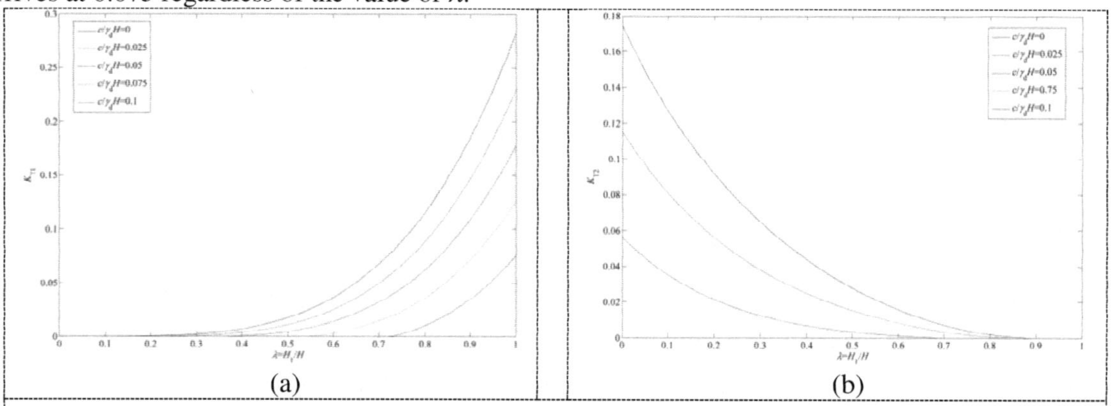

(a) (b)

Figure 2. Stability charts for different c and λ when α=70°: (*a*) K_{T1}, and (*b*) K_{T2}.

When α=80°, Z_c=0 and ϕ_d=34°, Figure 3 shows that K_{T1} increases as λ increases for different values of c while K_{T2} decreases with an increase in λ. In particular, K_{T2} is equal to zero when the value of $c/\gamma_d H$ arrives at 0.1 and λ is more than 0.06.

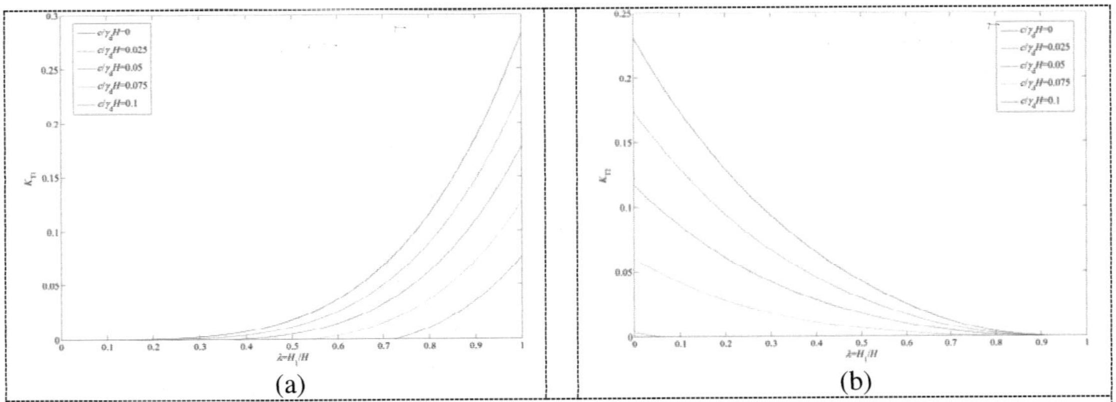

Figure 3. Stability charts for different c and λ when α=80°: (a) K_{T1}, and (b) K_{T2}.

3.2. Critical slip surfaces

It is noted that 'critical slip surfaces' means that these surfaces produce maximum T_1 and T_2 forces; hence, they govern design. In Figure 4, traces of critical slip surfaces are presented for different values of c while λ is equal to 0.6, α is equal to 70° and 80°, $c/\gamma_d H$ is 0, 0.025, and 0.05, and ϕ_d is equal to 34°. With the increase of the value of $c/\gamma_d H$, the critical slip surface moves towards the facing in the upper while moves away from the facing in the lower. In Figure 4(a), when $c/\gamma_d H$=0, K_{T1}=0.0366, and K_{T2}=0.0158, when $c/\gamma_d H$=0.025, K_{T1}=0.0253, and K_{T2}=0.0248, and when $c/\gamma_d H$=0.05, K_{T1}=0.0142, and K_{T2}=0.0013. In Figure 4(b), when $c/\gamma_d H$=0, K_{T1}=0.0366, and K_{T2}=0.0158, when $c/\gamma_d H$=0.025, K_{T1}=0.0253, and K_{T2}=0.0168, and when $c/\gamma_d H$=0.05, K_{T1}=0.0142, and K_{T2}=0.0099.

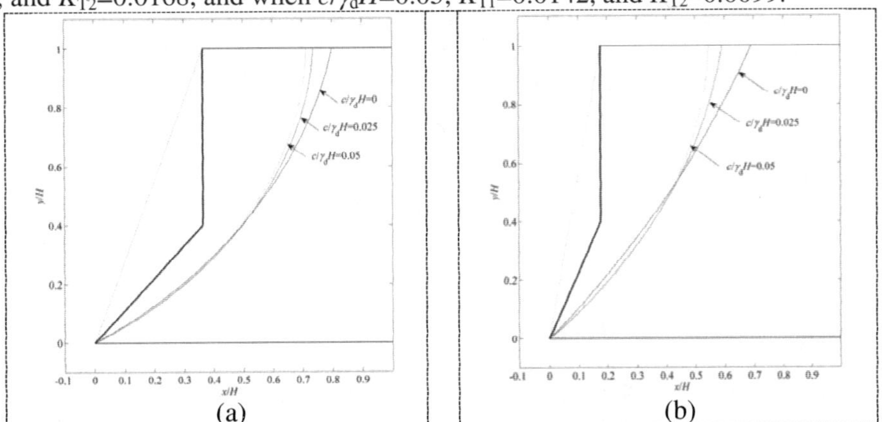

Figure 4. Critical slip surface under different c for λ=0.6: (a) α=70°, and (b) α=80°.

In Figure 5, traces of critical slip surfaces are presented for different values of Z_c while λ is equal to 0.6, α is equal to 70°, $c/\gamma_d H$ is 0.025, and ϕ_d is equal to 34°. With the increase of the value of Z_c/H, the critical slip surface moves towards the facing in the upper while moves away from the facing in the lower. When Z_c/H=0, K_{T1}=0.0253, and K_{T2}=0.0084. When Z_c/H=0.025, K_{T1}=0.0236, and K_{T2}=0.0083. When Z_c/H=0.05, K_{T1}=0.0218, and K_{T2}=0.0082.

IOP Conf. Series: Materials Science and Engineering **542** (2019) 012011 doi:10.1088/1757-899X/542/1/012011

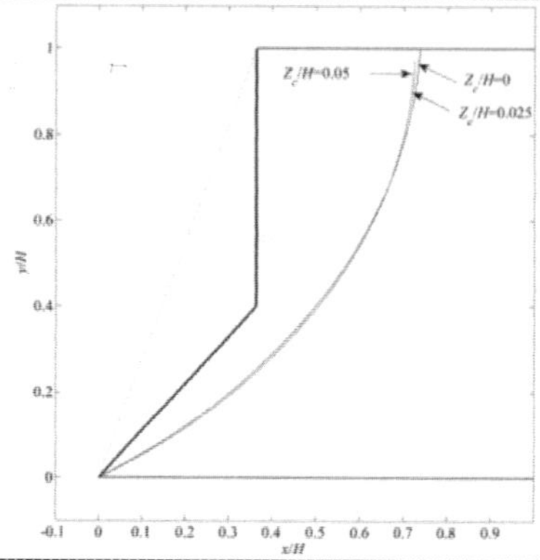

Figure 5. Critical slip surface under different Z_c for α=70° and λ=0.6.

4. Conclusion

In the view of the fact that a cohesive material may be used as a reinforced fill, the LE approach using a top-down log spiral mechanism is presented to determine the internal stability analysis of the bilinear geosynthetic-reinforced slope in this paper. The formulation is then described, which it will result in the resultant reinforcement force required for the internal design of the bilinear geosynthetic-reinforced slope in the upper and lower tiers, respectively.

From stability charts, it is seen that the resultant reinforcement force in the upper tier increases as the ratio of the height of the upper tier to the height of the slope increases for different values of the cohesion while the resultant reinforcement force in the lower tier decreases in the same case. In addition, from critical slip surfaces, with the increase of the value of cohesion or depth of crack, the critical slip surface moves towards the facing in the upper while moves away from the facing in the lower, and the resultant reinforcement force in whether the upper or lower tier decreases. In a word, the presence of cohesion can produce design redundancy.

Acknowledgments

This study was sponsored by the Natural Science Foundation of Anhui Province (Grant No. 1608085QE121), the Fundamental Research Funds for the Central Universities (Grant No. JZ2015HGBZ0458), and the China Postdoctoral Science Foundation (Grant No. 2016M592048).

References

[1] Ruan, X., Leshchinsky, D., and Leshchinsky, B.A. 2015. Global stability of bilinear reinforced slopes. Transportation Infrastructure Geotechnology, 2(1), 34-46.

[2] Ruan, X., Guo, X., Luo, Y.-S., and Sun, S. 2017. Seismic design of bilinear geosynthetic-reinforced slopes. Soil Dynamics and Earthquake Engineering, 100, 454-457.

[3] Yang, G., Liu, H., Lv, P., and Zhang, B. 2012. Geogrid-reinforced lime-treated cohesive soil retaining wall: Case study and implications. Geotextiles and Geomembranes, 35, 112-118.

[4] Vahedifard, F., Leshchinsky, B.A., Sehat, S., Leshchinsky, D. 2014. Impact of cohesion on seismic design of geosynthetic-reinforced earth structures. Journal of Geotechnical and Geoenvironmental Engineering, 140(6), 04014016.

[5] Federal Highway Association (FHWA). 2009. Design and construction of mechanically stabilized earth walls and reinforced soil slopes, FHWA-NHI-10-024, Berg, R.R., Christopher, B.R., Samtani, N.C., eds. Washington, DC.

Numerical simulation and microgravity experiment of fluid flow in the vane type tank

Lei Chen[1, 2, 3], Jintao Liu[1, 2], Wen Li[1, 2], Can Yao[1, 2] and Honglai Zhu[1, 2]

[1] Beijing Institute of Control Engineering, Beijing, 100190, China
[2] Beijing Engineering Research Center of Efficient and Green Aerospace Propulsion Technology, Beijing, 100190, China

[3] E-mail: chenlei1340@126.com

Abstract. Propellant tank is an important component of spacecraft, which determines the life of spacecraft. As a key unit for storage and management fluid on the spacecraft, the vane type tank can provide liquid without gas for thrusters. And the propellant management device(PMD) is the core component in the tank. In the paper, numerical simulation and microgravity experiment have been carried out to study the performance of propellant management. By using a VOF two-phase flow model, the fluid behaviour in tank is numerically simulated. By microgravity drop tower tests, the fluid distribution is obtained actually. The results from numerical simulation are in good agreement with ones from the microgravity drop tower tests. It shows that the PMD of the vane type tank can availably achieve the separation between liquid and gas interface and providing liquid without gas in microgravity environment.

1. Introduction

Propellant tank is an important component of spacecraft, which determines the life of spacecraft. At present, the main propellant tank of spacecraft is surface tension tank. As a key unit for storage and management fluid on the spacecraft, the vane type tank can provide liquid without gas for thrusters. And the core part of the vane type tank is the propellant management device(PMD). The principal advantages of the vanes type tank are light, reliable, repeatable, slosh suppression. The vane type tank with big vanes is one of the most advanced new type propellant tanks, which can guide and sponge fluid[1~5]. The working environment of the vane type tank is the microgravity environment, in which the gravity is not the dominant factor. Therefore the study of the fluid behavior in microgravity environment is crucial. The microgravity drop tower test can be used to study the fluid behavior of the vane type tank, but the experiment verification of the vane type in microgravity environment couldn't be given for a long time. Therefore numerical simulation is necessary for the study of the vane type tank. In this paper, by using a VOF two-phase flow model, the fluid behavior in microgravity environment in tank is numerically simulated to research the performance of the van type tank. Microgravity drop tower tests are carried out. Through comparative the results of numerical simulation and experiment, the fluid distributing rule is obtained[4~6].

2. Numerical simulation

2.1. Volume of fluid method

In this paper, by using a VOF two-phase flow model, the fluid flow characteristics in the tank are numerically simulated[7~10]. The volume of fluid (VOF) method is a free-surface modeling technique for tracking and locating the free surface. It belongs to the class of Eulerian methods which are characterized by a mesh that is either stationary or is moving in a certain prescribed manner to accommodate the evolving shape of the interface. VOF is an advection scheme—a numerical recipe that allows the programmer to track the shape and position of the interface. It is a scalar function, defined as the integral of a fluid's characteristic function in the control volume, namely the volume of a computational grid cell. Due to laminar flow generally in vane type tank, the basic equation of VOF model comprise of physical equation, continuity equation and momentum equation.

2.1.1. Physical equation. The physical property of fluid is determined by volume fraction of different phases in mixed fluid, and the physical equation express physical property of different volume fraction. There is only two-phase mixed flow in the tank, so density properties equation of mixed fluid is given below.

$$\rho = \alpha_1 \rho_1 + \alpha_2 \rho_2 , \quad \alpha_1 + \alpha_2 = 1 \tag{1}$$

$\rho$is the density of mixed fluid, $\alpha_1$and α_2 are the volume fractions of the two phase, ρ_1 and ρ_2 are the densities of the two phase, which are given values.

2.1.2. Continuity equation. The continuity equation for the mixture is

$$\frac{\partial}{\partial t}(\rho) + \frac{\partial}{\partial x_i}(\rho u_i) = R \tag{2}$$

$u_i$is velocity of mixed fluid, R is the source term.

2.1.3. Momentum equation. The momentum equation of mixed fluid is

$$\frac{\partial}{\partial t}\rho u_j + \frac{\partial}{\partial x_j}\rho u_i u_j = -\frac{\partial p}{\partial x_j} + \mu \frac{\partial}{\partial x_j}\left(\frac{\partial u_i}{\partial x_j} + \frac{\partial u_j}{\partial x_i}\right) + \rho g_j \tag{3}$$

p is the pressure of the tank, u_j and $u_i$are velocity of liquid phase and gas phase respectively, x_j and x_i are liquid phase position and gas phase position respectively, t is the time, g_j is microgravity acceleration, and μ is coefficient of viscosity. As the effect of surface tension,

$$\frac{\partial p}{\partial x_j} = -\sigma \frac{\partial}{\partial x_j}(\frac{1}{r}) \tag{4}$$

$\sigma$is the coefficient of surface tension, r is radius, According to equation (3) and (4), the momentum equation of mixed fluid is

$$\frac{\partial}{\partial t}\rho u_j + \frac{\partial}{\partial x_j}\rho u_i u_j = \sigma \frac{\partial}{\partial x_j}(\frac{1}{R}) + \mu \frac{\partial}{\partial x_j}\left(\frac{\partial u_i}{\partial x_j} + \frac{\partial u_j}{\partial x_i}\right) + \rho g_j \tag{5}$$

2.2. Numerical model of the tank

In this paper, the vane type tank shown in figure 1 mainly comprises of inside and outside blade which are both eight. The volume of tank is 4L, and the inside diameter is 170mm. These blades are used for transferring and storing propellant.

According to the model of the tank, the mesh of the tank is divided using block hexahedron gird method. In order to mesh the tank simply, the 1/8 tank is divided into three parts. The grid numbers of the model is 2.3 million. The numerical model of the tank is shown in figure 2.

IOP Conf. Series: Materials Science and Engineering **542** (2019) 012012 doi:10.1088/1757-899X/542/1/012012

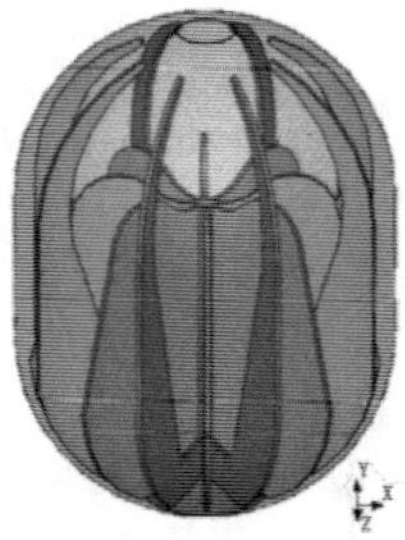

Figure 1. The model of the tank.

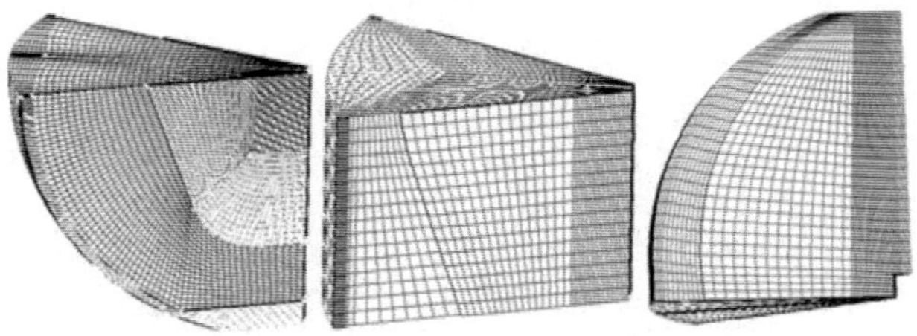

Figure 2. The mesh of the tank.

The commercial CFD code FLUENT is used to perform the simulations. The boundary conditions of all the walls are set to solid wall. The outlet is set to pressure-outlet and the inlet is set to velocity-inlet. The computational domain includes liquid and gas. The first phase is set to liquid which is MMH, and the second phase is set to gas which is air. In this paper, the reorientation process with filling ratio 5%, 40%, 60% are simulated, while the microgravity acceleration is 1×10^{-5} g and the direction is for -Y. The reorientation process with filling ratio 5% and 60% are simulated, while the microgravity acceleration is 1×10^{-3} g and the direction is for +X.

2.3. Result of numerical simulation

The reorientation process with filling ratio 5%, 40% and 60% are simulated to adequately research the characteristic of the tank PMD, while the microgravity acceleration is 1×10^{-5} g and the direction is for -Y. The fluid distributions are shown in figure 3, 4, and 5. The blue color is gas, the red color is liquid.

Figure 3 shows the fluid reorientation with filling ratio 5% in the tank.

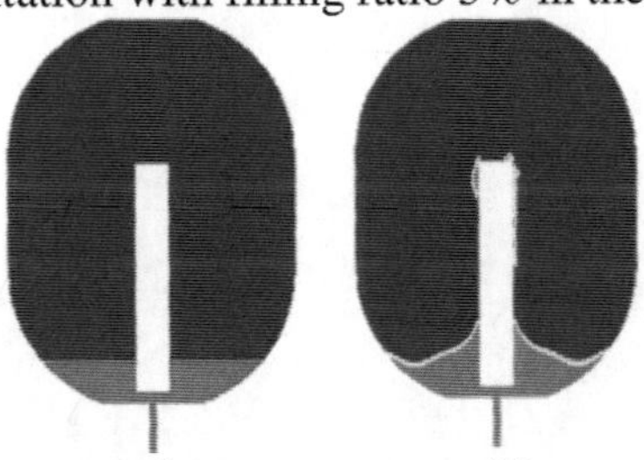

t=0s t=10s

Fig.3 Fluid reorientation in the tank with filling ratio 5%

Figure 4 shows the fluid reorientation with filling ratio 40% in the tank.

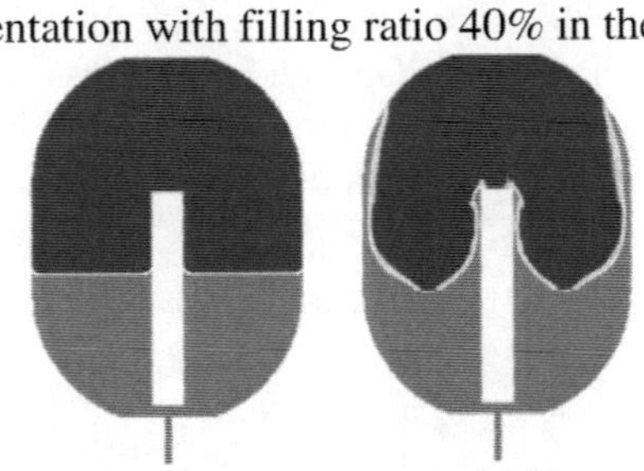

t=0s t=10s

Fig.4 Fluid reorientation in the tank with fill ratio 40%

Figure 5 shows the fluid reorientation with filling ratio 60% in the tank.

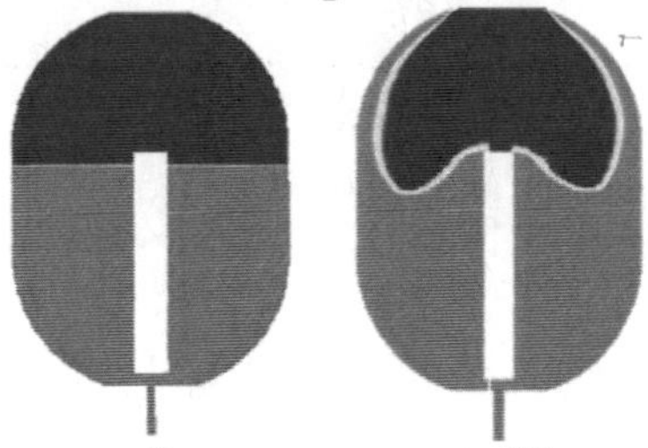

t=0s t=10s

Fig.5 Fluid reorientation in the tank with fill ratio 60%

Seeing from the figure 3, 4 and 5, in microgravity environment, the liquid flow along the inside and outside vanes due to the force of surface tension, and the height of the liquid could reach the top of the blade column. Gas and liquid is without mixing in the process of reorientation, and concave surface is formed between the blades. In the end, the liquid position around the PMD in the tank to cover the liquid outlet, and the gas position in the top of the tank. The result indicates that the vane type PMD can availably achieve the separation between liquid and gas interface and providing liquid without gas to thrusters, while the microgravity acceleration is 1×10^{-5} g and the direction is for -Y.

The reorientation process with filling ratio 5% and 60% are simulated to adequately research the characteristic of the tank PMD, while the microgravity acceleration is 1×10^{-3} g and the direction is for +X. The fluid distributions are shown in figure 6 and 7.

Figure 6 shows the fluid reorientation with filling ratio 5% under lateral microgravity acceleration.

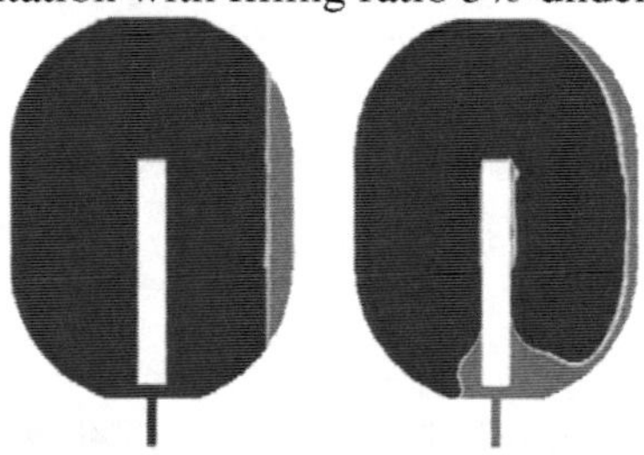

t=0s t=18s

Fig.6 Fluid reorientation in the tank with fill ratio 5% under lateral microgravity acceleration

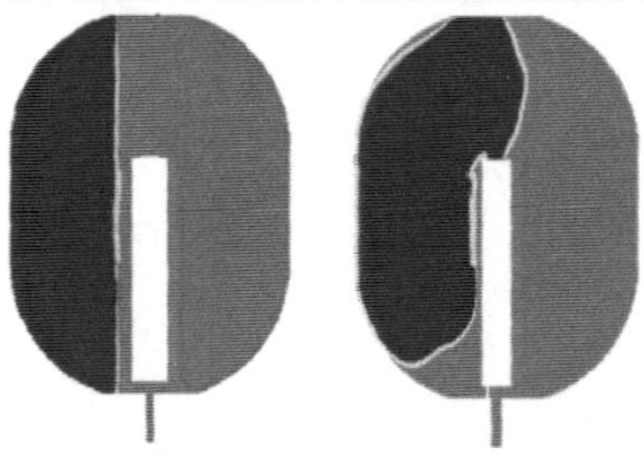

t=0s t=18s

Fig.7 Fluid reorientation in the tank with fill ratio 60% under lateral microgravity acceleration

Seeing from the figure 6 and 7, in the microgravity acceleration which is 1×10^{-3} g and the direction is for +X, the liquid flow along the vanes. Gas and liquid is without mixing in the process of reorientation. In the end, the liquid position around the PMD in the tank to cover the liquid outlet. The result indicates that the vane type PMD can availably achieve the separation between liquid and gas interface and providing liquid without gas to thruster, while the microgravity acceleration is 1×10^{-5} g and the direction is for -Y.

3. Microgravity test

3.1. Test system

Microgravity drop tower tests which are most usual in the means of microgravity tests can provide short microgravity time, but can supply relatively low microgravity level. In the paper, the microgravity tests have been carried out in the hundred meters drop tower of the National Microgravity Laboratory. The free fall experiment facility of the drop tower provides microgravity time which is about 3.5s, and the facility comprises of the falling module, the deceleration and recovery system, the release system, the control system, the measurement system, and auxiliary equipment and so on. According to the requirement of microgravity drop tower experiment research, the test system is built, which consists of abbreviate models, the test bracket, the lighting device and the picture acquisition device and so on. In the system, the high resolution camera with CCD is using for recording fluid climbing process, distribution and reorientation process in abbreviate models.

3.2. Result of the test

In the test, the drop tower experiments with filling ratio 5%, 40% and 60% in the tank which is in forward direction have been carried out. The fluid distributions are shown in figure 8,9 and 10.

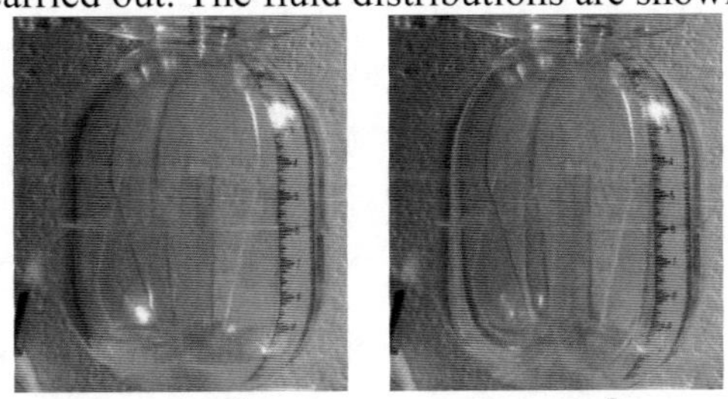

t=0s t=3s

Figure 8. Fluid reorientation with filling ratio 5% in forward direction.

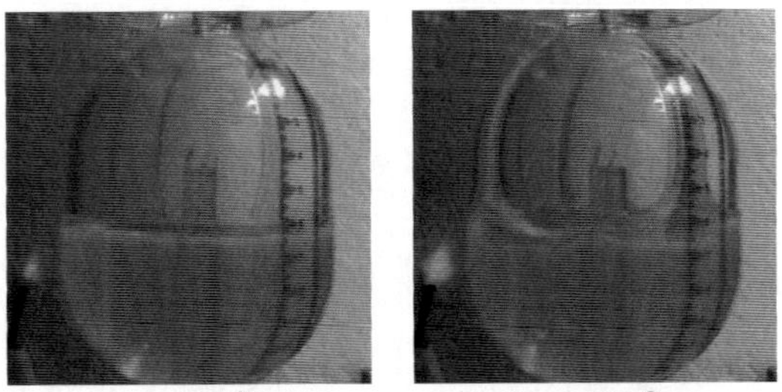

t=0s t=3s

Figure 9. Fluid reorientation with filling ratio 40% in forward direction.

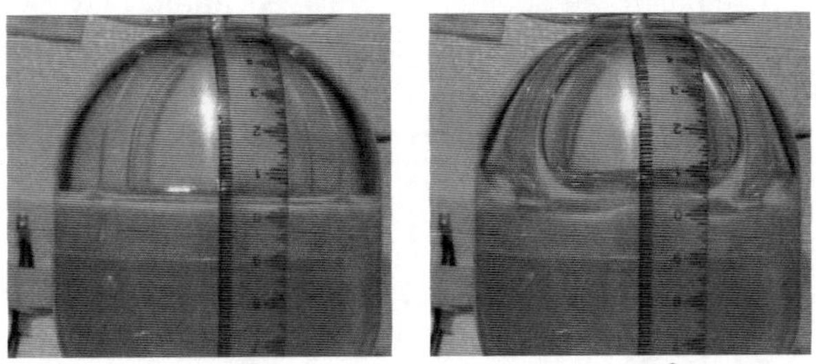

t=0s t=3s

Figure 10. Fluid reorientation with filling ratio 60% in forward direction.

The drop tower experiments with filling ratio 5% and 60% in the tank which is in lateral direction have been carried out. The fluid distributions are shown in figure 11 and 12.

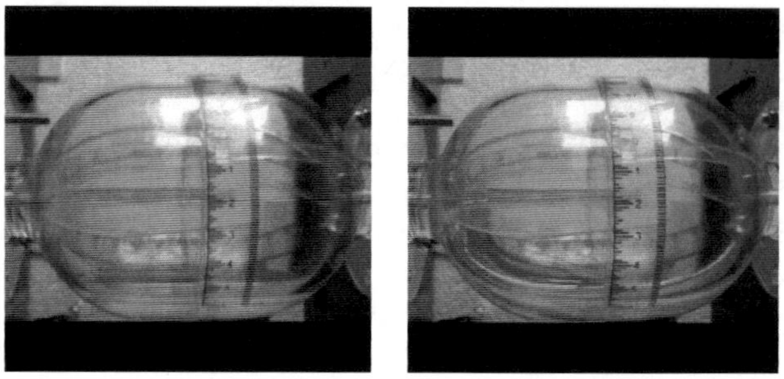

Figure 11. Fluid reorientation with filling ratio 5% in lateral direction.

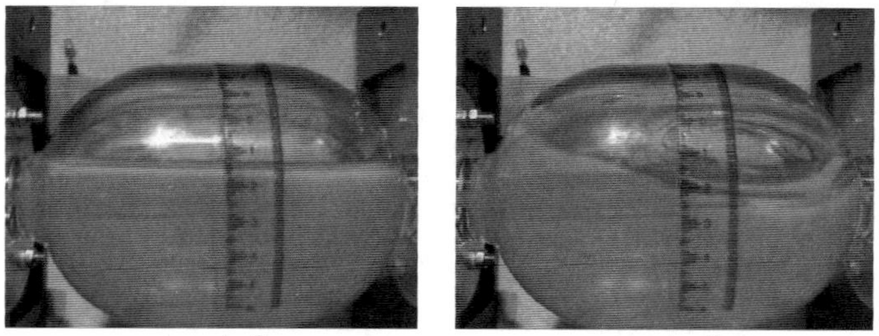

Figure 12. Fluid reorientation with filling ratio 60% in lateral direction.

Seeing from these figures, in the drop tower experiments, by using of surface tension, the fluid rapid climb along the tap region between vanes and inner walls of the tank. And a concave liquid surface between vanes and inner walls is shaped under microgravity environment. The liquid mainly reserves in the bottom of tank and the region between vanes and inner walls and covers the liquid outlet of the tank. The experiment results indicate that the vane type tank has good and initiative fluid orbital management ability and can availably separate between fluid and gas, and provide propellant without gas for thrusters.

4. Conclusion

Numerical simulation and experiment of the flow in vane type with microgravity have been carried out. The fluid distributing rule has been obtained. The results from numerical simulation are in good agreement with ones from the microgravity drop tower tests. The results indicate that the PMD of the vane type tank can availably achieve the separation between liquid and gas interface and providing liquid without gas. Gas and liquid is without mixing in the process of reorientation. The liquid positioned around the PMD in the tank to cover the liquid outlet, and the gas positioned in the top of the tank. The vane type tank has good performance of propellant management in microgravity environment.

References

[1] Dipprey N F and Rotenberger S J 2003 Orbital express propellant resupply servicing *The 39th Joint Propulsion Conference and Exhibit* (2003)20-23
[2] JaekleJr D E 1991 Propellant management device conceptual design and analysis: vanes *The 27th Joint Propulsion Conf.* (1991)24-26
[3] Hu Q, Li Y, et al.2013 *Aero Control and Application.* **39**(3) 58-62
[4] S. Dominick and J. Tegart 1994 Orbital Test Results of a vaned Liquid Acquisition Device *The 30th AIAA Joint Propulsion Conference and Exhibit* (1994) 27-29
[5] David J C and Timothy A M 1997 Vented Tank Resupply Experiment - Flight Test Results *The 33rd Joint Propulsion Conference and Exhibit* (1997) 6-9
[6] Zhuang B T, Li Y, et al. 2014 *Aero Control and Application* **40**(1) 27-30
[7] Sharipov F and Kalempa D 2002 Gaseous Mixture Flow through a Long Tube at Arbitrary Knudsen Numbers *Journal of Vacuum Science and Technology* **20**(3)814-822.

[8] Hirt C W Nichols B D 1981 Volume of fluid (VOF) method for the dynamics of free boundaries *Journal of Computational Physics* **39** 201-225
[9] Tsai W, Yue D K 1996 Computation of Nonlinear Free-Surface Flows *Annual Review of Fluid Mechanics* **28** 249-278
[10] Shin S and Lee W I 2000 Finite element analysis of incompressible viscous flow with moving free surface by selective volume of fluid method *International Journal of Heat and Fluid Flow* **21** 197-206

IOP Conf. Series: Materials Science and Engineering **542** (2019) 012013 doi:10.1088/1757-899X/542/1/012013

The state-of-art and prospect of contactless torque measurement methods

Jun Liu[1, a], LianFa Yang[2, b] and JianPing Ma[3, c]

[1, 2, 3]Faculty of Mechanical & Electrical Engineering ,Guilin University of Electronic Technology, Guilin, Guangxi, 541004,China
Email:[a]LiuJun2333@icloud.com,[b]y-lianfa@163.com, [c]majianping1229@163.com

Abstract. Torsional vibration is an important factor in the transmission process of mechanical shaft. The torque is also a part of the important parameters for working state of the mechanical shaft. In recent years, research on the measurement method of contactless torque of mechanical transmission shaft has been studied by many scholars. The contactless torque measurement methods have been continuously improved to achieve high precision and high reliability.The methods of contactless torque measurement is presented. The advantages of contactless torque measurement methods are introduced. The development of contactless torque measurement methods in the industry is discussed. Finally, the prospect of contactless torque measurement methods is prospected.

1.Introduction

Torque is one of the important parameters of a mechanical drive shaft. The value of torque is closely related to the performance of mechanical power, energy consumption, life and safety performance. Therefore, the torque can be measured accurately, timely and reliably, and the working state of the rotating shaft can be analyzed. This is a key function for automatic detection and automatic control of mechanical equipment[1].

Torque measurements can be divided into contact and non-contact. Contact torque measurement technology was used at an early stage. However, with the development of the industry, in some large-scale work, the rotating shaft is particularly large, and the torque received is also large. At this time, the normal contact sensor cannot be used to detect the torque of the shaft. Because of the torque sensor can be damage by the large tension. If a large contact sensor is designed, which will increase the cost. [2]RG Karpov（1966）[3] used existing contact torque measurement technology to achieve the goal of contactless torque measurement technology.A new measurement field for torque is being developed by RG Karpov.

The main aim of this paper starts classifies the contactless torque measurement technology, and then

IOP Conf. Series: Materials Science and Engineering **542** (2019) 012013 doi:10.1088/1757-899X/542/1/012013

briefly describes the state-of-art of contactless torque measurement in recent years. Finally, this paper summarizes the future development trends and provides a reference for further optimization of design and practical use of contactless torque sensors.

2.Contactless torque measurement method

Contactless torque measurement techniques can be divided into two categories: tension measurement torque and twist angle measurement torque.

2.1. Tension measurement torque

tension measurement is a conventional measure of torque .When the shaft is subjected to torque, the stress and strain values on the surface of the shaft will changes. At this time, the stress and strain values on the surface of the shaft are measured.Then the corresponding measured values are substituted into the corresponding mechanical formulas to calculate.Finally, the value of the torque on the rotating shaft is obtained.

New torque sensor can detect rotational speed and torque at the same time is designed based on amorphous alloy by Shi Yan-ping in Huaihai institute of technology (2011)[4], as shown in figure 1. Excitation current of a certain frequency flows through the sensor excitation winding. When there is no twist, the reluctance of the surface DE1 is equal to the reluctance of DE2, and the magnetic flux passing through the winding N2 is equal to them. Since the two coils are connected in reverse, the value of the sensor output is zero. When there is load torque, tensile stress and compressive stress of the alloy surface are generated. According to the piezomagnetic effect, the change in the magnetic resistance of the magnetic circuit causes a change in the magnetic flux, which Causes the induced voltage of the sense winding to change. The induced voltage of the E1 pole coil N2 increases, and the induced voltage of the E2 pole coil N2 decreases. Because the two coils are differentially connected, the output voltages are added to each other. After calibration, there is a corresponding formula that can be used to calculate the torque.

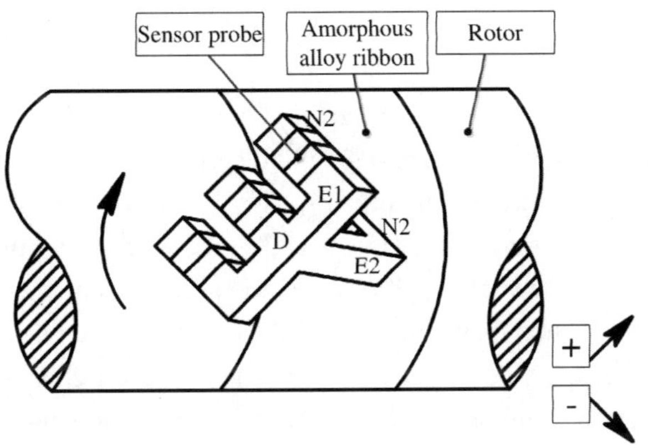

Figure 1. Structure schematic diagram of amorphous alloy torque sensor

An improved dynamic torque measurement method based on strain gauge is designed by Hilal Muftah(2013) [5]. The working principle of the metallic strain gauge torque transducer is to measure the strain in a flexible section of the shaft. In this improved method, a solid square element is used instead of a solid circular element as the sensing element. In this case, four strain gauges are attached into the rotating shaft as a Wheatstone bridge circuit. The first two strain gauges are fixed at ±45° on one side of the shaft axis and the other two strain gauges are also prepared at the same way on the other side of the shaft. In this case, shear strain occurs when torque is applied to the shaft. Shear strain occurs when the torque is applied to the shaft. Therefore tensile strain, which is measured by one pair of gauges, will increase the circuit resistance which, on the other hand, will be decreased by the other pair due to compression strain. Formerly the measured torque can be read from a measuring device as output voltage.

Shengnan Fu(2014)present a torque measurement method based on fiber Bragg grating (FBG) sensor [6]. The two gratings FBG1 and FBG2 are pasted in parallel on the surface of the shaft, as shown in figure 2. When the shaft is subjected to torque, the same strain occurs in both gratings. But the magnitude of their strain is the opposite. The wavelength will also change correspondingly when the grating is strained. Torque can be measured based on the wavelength changes of the two gratings FB1 and FBG2.

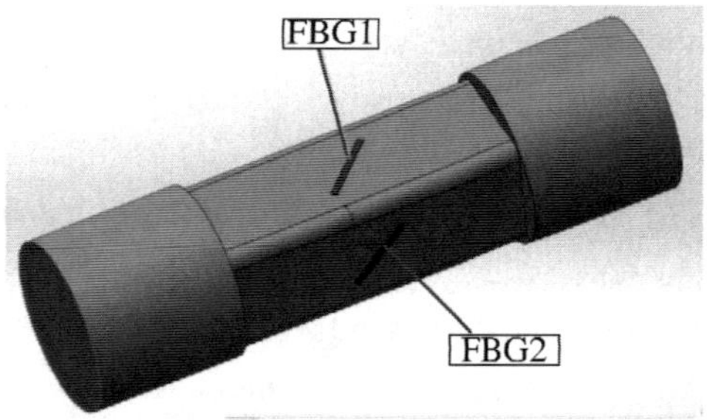

Figure 2. The design of the shaft and situation of FBGs bonded on the planes

Muhammad Nazar Ul Islam(2016)present a new method based on strain measurement torque [7]. There are two identical sealed chambers as chambers A and B, as shown in figure 3. Each chamber is connected to the port of the differential pressure sensor. When the shaft is subjected to torque, the torsional deformation of the shaft causes a difference in volume between the two internal sealed chambers. The value of the differential pressure measured by the sensor, The pressure difference is calculated by a mathematical formula to obtain the value of the torque.

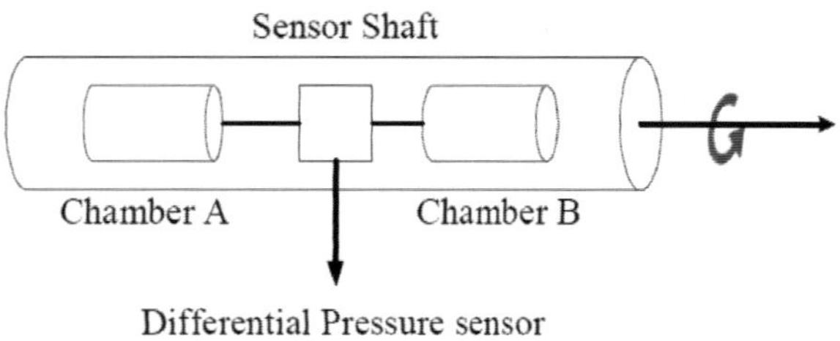

Figure 3. Abstract representation of the proposedmethod for torque measurement[7].

Jacques Cousteau S. Borges(2017) present a torque based on the Hall effect is measured by two self-measured magnets with two flux densities attached to the shaft[8]. The magnet is placed at an angle of 45° to the busbar on the surface of the rotating shaft, which is the maximum shear stress at the position where the shaft surface is subjected to torque. The shaft is deformed by the torque to cause a change in the distance between the magnets, and the magnetic field generated between the two magnets also changes accordingly. The equivalent moment can be calculated using the magnetic dipole model.

2.2. Twist angle measurement torque

Twist angle measurement torque is also a conventional torque measurement method. When the shaft is subjected to the torque, the shaft will produce a torsion angle. At this time, two points on one of the busbars on the surface of the rotating shaft are measured. Then the relative angular difference between the two points is measured. Finally, after the data processing, the corresponding torque value can be

IOP Conf. Series: Materials Science and Engineering **542** (2019) 012013 doi:10.1088/1757-899X/542/1/012013

obtained.

A contactless torque sensor is designed by RG Kärpov(1966)[3], as shown in figure 4. Two soft magnetic material sprocket wheels are mounted on both ends of the rotating shaft, and the detector is also made of soft magnetic material. When the rotating shaft is rotated by the torque, the distance between the probe and the tooth tip or the root is detected by two magnetic pole detectors, and the detector generates different pulses to measure the torque value.

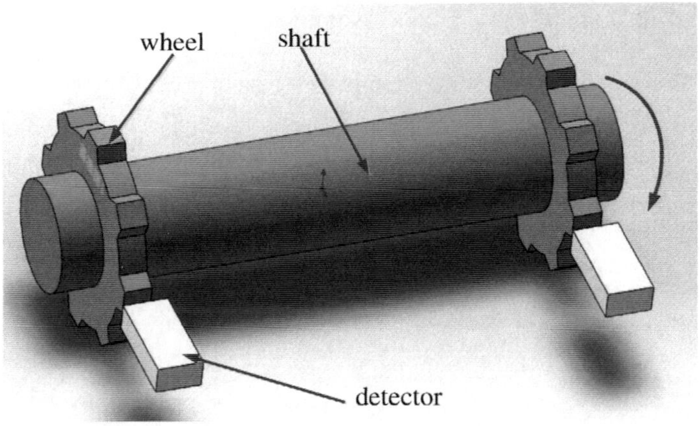

Figure 4. Magnetic pole detection torque sensor structure diagram

A new type of angle measurement torque sensor is designed by Yu Hong Lin(2009)[9]. A new type of non-contact torque sensor is formed by a special annular space array and a dedicated magnetoelectric detector. When the rotating shaft is subjected to the torque to generate a torsion angle, the magnetoelectric detector will move relative to the annular space array. The change in the magnetic medium between the coils causes a change in the mutual inductance, and the amplitude of the output voltage also changes. The measured electrical signal is the conversion of the torque signal.

Alberto Garinei(2017) present a non-contact torque measurement method based on laser speckle contrast method [10]. The measuring principle is that a laser beam passes through a beam splitter to obtain two beams, and the two beams respectively illuminate two different points on the same busbar on the same axis surface. The beam incident on the surface of the shaft is diffusely reflected, and the beam reflected back and diffuses forms a speckle. Speckle distribution varies with angular phase shift. The diffusely reflected beam is directed into the fiber. Electrical signal obtained by coupling an optical fiber to a photodiode. The two electrical signals are processed separately to obtain the measured torque value.

D Zappalá(2018) present a non-intrusive torque measurement of a rotating shaft using optical sensing of a zebra strip [11]. A pair of zebra strips with a relative angular phase of zero are attached to the ends of the shaft. The distance between the two measured zebra strips is as far as possible from each other. The optical sensors for detecting the two zebra strips are mounted on a fixed support. The principle of the proposed method is to quantify the shaft relative twist angle by measuring the phase difference between the two pulse signals and thence deriving the applied torque, from a known torque twist relationship.

3.Conclusion

The non-contact measurement of torque is moving towards digital, versatile and intelligent development. With the increasing degree of automation of various systems and the complexity of information, more and more information needs to be accessed, which not only puts higher requirements on the accuracy, reliability and response speed of the sensor. A sensor and system with standard output form needs to be developed. The torque measuring device is the following development trend:

- Torque measuring device develops to contactless measurement.
- The standardized package provides a standardized solution for the use of torque measurement.
- The production process continues to improve, and the measurement accuracy and response speed are gradually improved.
- The product volume continues to shrink, and the price ratio is significantly improved.

Acknowledgements

This work was financially supported by the National Natural Science Foundation of China (51564007), Natural Science Foundation of Guangxi Province (2017GXNSFAA198133), and Guangxi Key Laboratory of Manufacturing System & Advanced Manufacturing Technology (14-045-15-005Z).

References

[1] Hao Zhao. Present Situation and Development Review of Torque Measurement[J].*Applied Mechanics and Materials*, 2013, **Vol.422**:141-145.

[2] DengQuan Wang, Ming Yang and Lin Ye. Study of non-contact torque measurement of rotating shaft[J]. *Electronic Measurement Technology | Electr Measur Technol,* 2010, **(6)**:8-12.

[3] R G Karpov. Contactless transducer for measuring torque[J]. *Measurement Techniques*, 1966, **Vol.9**:491-495.

[4] YanPing Shi, QingGui Zhou and JiPing Zhou. Research on New Speed and Torque Sensor Based on Amorphous Alloy[J]. *Chinese Internal Combustion Engine Engineering,* 2011, **32(3)**:68-73.

[5] M Hilal Muftah, S Mohamed Haris, K Petroczki and E Awad Khidir. An Improved Strain Gauge-Based Dynamic Torque Measurement Method[J].*International Journal of Circuits,Systems and Signal Processing,*2013

[6] Shengnan Fu, Yinguo Huang and Xinghua Li. Experimental Study of Torque Measurement Based On FBG[J]. *The Proceedings of the Second International Conference on Communications, Signal Processing, and Systems,* 2014, **Vol.246**:255-261.

[7] Islam M N U, Cheng P and Oelmann B. Method of torque measurement based on volumetric strain[C]// *Society of Instrument and Control Engineers of Japan. IEEE,* 2016:116-123.

[8] Borges J C S, Deus D B B D and Filho A C L, et al. New Contactless Torque Sensor Based on the Hall Effect[J]. *IEEE Sensors Journal,* 2017, **PP(99)**:1-1.

[9] HongLin Yu, Jun Yu and AnGuo He. Design and electromagnetic analysis of ring-space array torque sensor[J]. *Chinese Journal of Scientific Instrument,* 2009, **(8)**:1687.

[10] Garinei A and Marsili R. Development of a non-contact torque transducer based on the laser speckle contrast method[J]. 2017, **6(2)**:253-258.

[11] Zappalá D,Bezziccheri M and Crabtree. Non-intrusive torque measurement for rotating shafts using optical sensing of zebra-tapes.[J]. *Measurement science and technology,* 2018.

Simulation Analysis of Ground Infrared Background Based on Numerical Fitting Model

Wang Chengcheng[1], Sun tao[1], Wang yan[2]

1 Naval Aeronautical University, Shandong, Yantai, China;
2 National University of Defense Technology, Hunan, Changsha, China
E-mail: 497990518@qq.com

Abstract. In view of the poor real-time and authenticity of the ground scene generation in the infrared scene simulation, a method of ground infrared background generation based on numerical fitting model is proposed. Combining with the main influencing factors of ground infrared scene, a numerical fitting model of surface temperature field of typical ground material is established. On the basis of the texture of the ground material area, the infrared texture of the ground is dynamically generated in the shader by combining the numerical fitting model of the temperature field and the texture mask generated by the multi-spectral image method. The simulation results show that the numerical fitting model of ground temperature field is in good agreement with the measured data, and the generated infrared texture is more realistic, which meets the real-time and authenticity requirements of infrared scene simulation system.

1. Introduction

Infrared modeling and Simulation of ground background is an important part of infrared battlefield environment simulation. The ground includes natural ground and artificial ground. There are many factors affecting the ground temperature, and the composition of the ground material is more complex, which brings a certain degree of difficulty to the ground infrared scene simulation. The generation of ground infrared texture based on visible remote sensing image or aerial image is a research direction in recent years [1-3]. According to the meteorological conditions and the measured data sets of several typical ground material temperatures, this paper establishes the engineering calculation model of ground temperature field, and then uses multi-spectral image fusion method to generate realistic ground infrared texture.

2. Analysis and modeling of ground temperature field

The theoretical modeling of ground temperature field is a complex problem. In addition to the thermal properties of the ground material, the ground temperature is also affected by many environmental

IOP Conf. Series: Materials Science and Engineering **542** (2019) 012014 doi:10.1088/1757-899X/542/1/012014

factors [4-7], such as solar radiation, weather, temperature, wind speed and air relative humidity (RH). In theoretical research, the ground temperature is often calculated by the theoretical model shown in Formula (1).

$$\begin{cases} \eta c \dfrac{\partial T}{\partial t} = \dfrac{\partial}{\partial x}(k \dfrac{\partial T}{\partial z}) \\ E_s + E_c + E_g + q + EL + G = 0 \\ T \mid_{z=z_0} = T_0 \end{cases} \quad (1)$$

In the upper formula, t is time, C is specific heat, is the thermal conductivity of the material, is solar irradiance, is temperature, is sky irradiance, is the surface radiation, is sensible heat transfer, EL is latent heat transfer, G is downward heat transfer to the surface.

Based on the measured results of temperature, wind speed, solar hourly radiation, atmospheric humidity and ground temperature, a data sample set is established. The mathematical expression of the final fitting model is

$$T = a + bT_a + cH^f + dR_{sun}^g + he^{-RH} \quad (2)$$

Among them, the atmospheric temperature, H convective heat transfer coefficient, solar hourly radiation, RH for atmospheric humidity, a, b, c, d, f, g, h for the fitting coefficient, the value is determined according to season and material. The ground is used as a diffuser, and the infrared radiance of the ground to the detector can be calculated according to formula (3).

$$I_{ground,\lambda_1-\lambda_2} = \frac{\tau}{\pi}(E_{ground,\lambda_1-\lambda_2} + \rho_g(E_{sun,\lambda_1-\lambda_2} + E_{sky,\lambda_1-\lambda_2})) \quad (3)$$

Among them, the upper and lower limits of the working band and the band of the detector are in the middle band. Limited in length, table 1 gives the fitting coefficients of several typical material floors in spring.

Table 1. Fitting coefficient of model for spring

Numerical	asphalt	concrete	pasture	soil (dry)	marble
a	-182.56	-258.379	-63.48	-169.628	-171.91
b	1.565	1.791	1.152	1.4943	1.627
c	-0.822	-1.10	-0.432	-0.455	-9.04
d	8.4×10^{-9}	6.6×10^{-9}	0.84×10^{-8}	1.02×10^{-8}	9.2×10^{-9}
f	0.5	0.5	0.5	0.5	0.5
g	3	3	3	3	3
h	-0.052	-0.047	-0.345	-0.195	-0.045

3. Research on realistic generation technology of infrared texture on the ground

In this paper, the multi spectral image fusion method is used to generate the reality of the infrared texture of the ground. The method uses the texture mask generated from the visible spectrum image to fuse the pixel gray level to realize the reality of the infrared image. The method is as follows:

- Transforming visible light into gray image and finding the gray mean of the same material area.

- Subtract the gray mean of different regions of the visible light gray image to get the texture mask of the region.
- According to formula (2) and formula (3), the temperature and infrared radiance of each region are calculated, and then the infrared texture image of the ground is obtained.
- Texture mask is multiplied by a texture depth adjustment factor, which is used to adjust the depth of mask.
- The sum of infrared gray and mask gray at the same texture coordinate point is taken as the gray value of the final texture pixel. Figure 1 shows the comparison between the actual ground infrared image and the simulated ground infrared image.

（a） infrared real photo (b) visible light real image (c) to simulate ground image

Figure 1. Realistic IR texture of ground

In Figure 1, the ground is cement road and Dali slate. Because the temperature of cement pavement is lower than that of marble surface, the gray level of cement pavement in the image is lower than that of marble surface. In addition, the detail of the simulated ground image is clearer than that of the real image. This is mainly due to the fact that the texture mask is based on the visible image, and the resolution of the visible image is higher than that of the infrared detector.

4. Simulation results evaluation

The accuracy of ground temperature field modeling is evaluated by comparing the predicted results with the measured ones. Figure2 shows the comparison of the model temperature prediction results with the measured data of several typical materials. Table 2 gives the corresponding mean square error evaluation. In Figure 2, the ground is horizontal, and the numerical label indicates the simulation date. 1 is July 2, the weather is cloudy and clear, 2 is May 8, the weather is clear and cloudless.

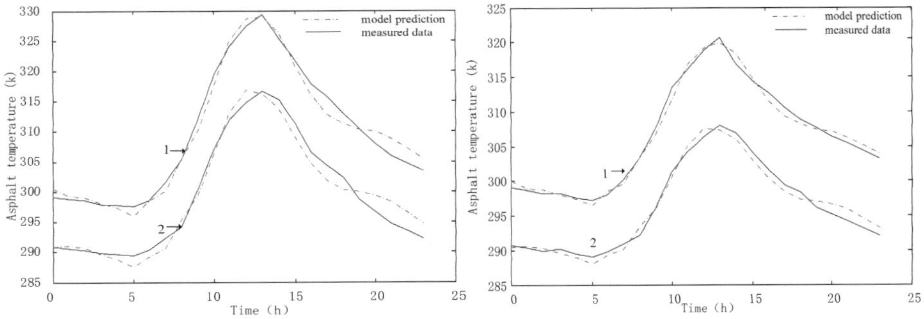

（a） asphalt （b） concrete

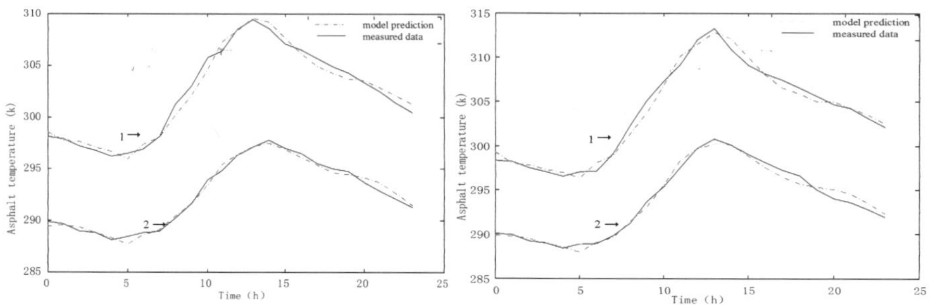

(c) pasture (d) soil

Figure 2. Forecast result of ground temperature by model and real measurement

Table 2. MSE of values between value of model and real measurement

Date	Asphalt	Soncrete	Pasture	Soil
1	1.52K	0.80K	0.70K	0.62K
2	1.79K	0.96K	0.53K	0.37K

According to Figure 2 and Table 2, the predicted results of the ground temperature field model are in good agreement with the measured data, and the accuracy can meet the requirements of Engineering calculation. Obviously, the calculation speed of the ground temperature field model is obviously faster than that of the theoretical model based on the finite difference method.

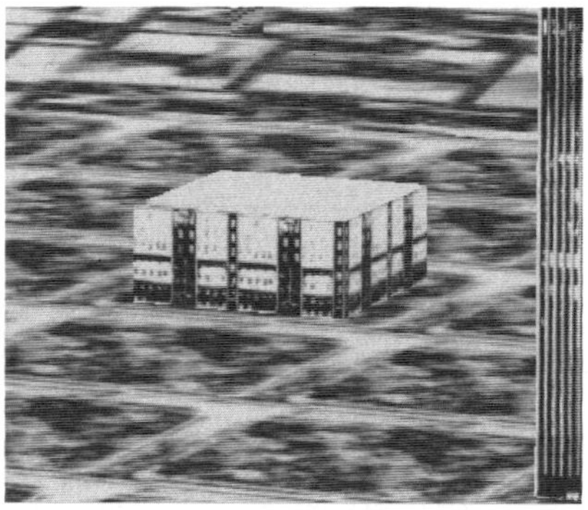

Figure 3. Simulation of IR ground

Figure 3 shows the final 3D scene simulation of the ground infrared scene. The ground texture details in the map are generated based on the texture mask generated from the 15m visible light remote sensing image. The ground material is mainly grassland and asphalt pavement. Ground buildings as part of the ground are rendered together with the ground in the simulation. Because the ground temperature is higher at noon, the infrared gray level of the ground and buildings is obviously higher than that of the ground and buildings in the morning. The frame rate of the scene is kept at around 60Frame/s because of the use of shaders to render the scene grayscale.

5. Concluding remarks

Aiming at the problem that the real-time and authenticity of ground scene dynamic generation in infrared scene simulation are poor, a method of ground infrared background dynamic generation based on numerical fitting model is proposed. Firstly, the engineering calculation model of surface temperature field of typical materials is established based on meteorological conditions and ground temperature data sets. Then, the infrared gray level of ground texture is calculated in real time by Colorer according to the texture and texture mask of the material area on the ground. The simulation results show that the method based on the numerical fitting model is feasible and the computational speed meets the requirements of dynamic real-time simulation. The generated infrared texture has a strong sense of reality. The method has a strong practical significance in engineering.

6. References

[1] Cheng Qiong. Visual image conversion to infrared image simulation . Wuhan: Huazhong University of Science and Technology, 2008.

[2] Ge Xiaoqing. Infrared and visible image fusion research. Chongqing: Chongqing University, 2010.

[3] M P Lvesque. Generation of synthetic IR sky background images. background 209.

[4] Ruan Kun, Fan Yin, Li Huijun, et al. Cloud model design based on midpoint migration algorithm [J].Journal of PLA University of Science and Technology: Natural Science Edition, 2003,4(1): 99-102.

[5] D A Vaitekunas, K Alexan, O E Lawrence, et al.SHIPIR/NTSC: A naval ship IR signature countermeasure and threat engagement simulator.Proceeding of SPIE.1996:41-424.

[6] Ren Haigang, Chen Hanping, Xu Zhu. Sea surface emissivity model in infrared simulation. Infrared and laser engineering, 2007, 36 (7): 36-38.

[7] Ren Haigang, Xingqiang, Chen Hanping. Sea surface radiation model in infrared simulation. Infrared and laser engineering, 2006, 35 (4): 19-22.

Acknowledgments

First and foremost, I would like to show my deepest gratitude to mysupervisor, professorXie Xiaofang, a respectable, responsible and resourceful scholar,who has provided me with valuable guidance in the writing of the study. My sincere appreciation also goes to the co-workers from Naval Aeronautical University, who participated this study with great cooperation. Last but not least, I' d like to thank all my friends, for their encouragement and support.

Reaction Torque of Gas Foil Journal Bearing with Misalignment

Liu Zhenxing[1,2] , Wei Yanming[1,2], Wang Ge[1,2], Cong Yuntian[1,2] and Chen Lei[1,2]

1 Beijing Institute of Control Engineering, Beijing, PRC
2 Beijing Engineering Research Centre of Efficient and Green Aerospace Propulsion Technology, Beijing, PRC
E-mail: liuzhenxing_888@126.com

Abstract. In a rotor gas foil bearing system, zero misalignment can usually not be ensured for assembly errors or creep deformation of the base. The tilting attitude of the journal makes lubricant performance worse. To study the performance of a gas foil journal bearing with misalignment, effect of misalignment on gas film thickness is introduced into the lubrication model. The two typical misalignment attitudes of the journal were defined as radial tilting and tangential tilting. The results show that reaction torques increase with the tilting angle of the journal. A larger eccentricity leads to a larger reaction torque. This work provides a reference for prediction about the self-action of gas foil journal bearing for misalignment.

1. Introduction

Gas foil bearing have a broad application prospect in high speed rotating machinery for its compactness, endurance of foreign matter, improvement of damping. It is difficult to ensure zero misalignment in a rotor foil bearing system for assembly errors or creep deformation of the base. Despite of the ability to accommodate misalignments, the tilting attitudes of the journal change the gas film thickness distribution, making lubricant performance worse.

Many studies were about the performance of gas foil bearings with perfect alignment. Heshmat et al.[1, 2] made early applications of numerical method on gas foil bearing analysis. Based on different numerical methods, many factors were added into model to predict the performance of gas bearings. On structure modeling, uniform stiffness model was used in Heshmat's early work[1, 2]. Further, top foil can be treated as beam element[3-6], plate element[7, 8] or shell element[9, 10]. Friction effect and bearing sleeve was also taken into consideration [11-14]. Besides, considering real gas property, variations of gas physical parameters were introduced in. Peng[15] accounted for viscosity-temperature characteristic, and compared numerical results with experimental study. Park et al.[16] involved rarefaction gas coefficients into the lubrication model for foil thrust bearings. Conboy and Wright[21] and Kim[25] presented a foil thrust bearing model with S-CO2 gas lubrication.

Performances of gas foil journal bearings under misalignment operation conditions were studies experimentally. Howard[17] measured the temperature raise and vibration on a rotor foil bearing system test rig with misalignment. In his another work [18], vibration of this rotor system was analyzed by plotting the waterfall diagram under different angular misalignments, wearness of the tested bearing was observed. Feng [19] conducted a series of static and dynamic load tests to measure the structural stiffness and equivalent viscous damping of the prototype multi-cantilever foil bearing with misalignment, find a high level of misalignment can lead to larger static and dynamic bearing stiffness as well as to larger equivalent viscous damping. Gad[20] evaluated the effects of static and dynamic angular misalignments on the performance of a gas foil thrust bearing.

Most of studies about gas foil journal bearings with misalignment were based on experiments. In this paper, effect of misalignment on gas film thickness is introduced into the lubrication model. The two typical misalignment attitudes of the journal were defined as radial tilting and tangential tilting. The reaction torques caused by misalignment are calculated by solving the pressure distribution of gas film numerically. This work provides a reference for the prediction about the self-action of gas foil journal bearings for misalignment.

2. Theoretical model

As shown in Figure 1a), coordinate frame *o-xyz* is set in the centre of the bearing. θ is the circumferential coordinate. R is the radius of the rotor journal, and h is the film local thickness, c is the bearing clearance, L is the bearing length. A general attitude of the journal respect with the bearing is shown in Figure 1b), the translational motion of the journal is described according to the radial motion of the middle section (z=0) centre shown in Figure 1a).

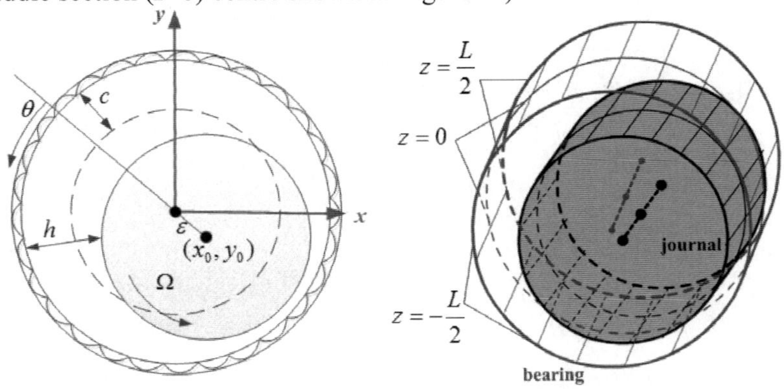

a) Geometry in the middle section (z=0) b) Misalignment in a foil journal bearing

Figure 1 Configuration of gas foil journal bearing

2.1. Hydrodynamic Lubrication Equation

Under laminar flow assumption, Reynolds equation in cylindrical frame is given below.

$$12\frac{\partial(\rho h)}{\partial t}+6\Omega\frac{\partial(\rho h)}{\partial\theta}=\frac{1}{R^2}\frac{\partial}{\partial\theta}\left(\frac{\rho h^3}{\mu}\frac{\partial p}{\partial\theta}\right)+\frac{\partial}{\partial z}\left(\frac{\rho h^3}{\mu}\frac{\partial p}{\partial z}\right) \tag{1}$$

Where, h is the gas film thickness, c is bearing clearance, $u(\theta,z)$ is the deflection of the foil. Gas film thickness $h=h_0+u$ consists of the geometry thickness h_0 and foil deflection u.

Neglecting the transient item, the non-dimensional form of the Eqn.(1) is

$$\frac{\partial}{\partial\theta}\left(\frac{\bar{\rho}\bar{h}^3}{\bar{\mu}}\frac{\partial\bar{p}}{\partial\theta}\right)+\left(\frac{2R}{L}\right)^2\frac{\partial}{\partial\bar{z}}\left(\frac{\bar{\rho}\bar{h}^3}{\bar{\mu}}\frac{\partial\bar{p}}{\partial\bar{z}}\right)=\Lambda\frac{\partial(\bar{\rho}\bar{h})}{\partial\theta} \tag{2}$$

The non-dimensional variables are listed below.

$$\bar{\rho}=\frac{\rho}{\rho_0},\quad \bar{h}=\frac{h}{c}=\frac{h_0}{c}+\bar{u}(\theta,\bar{z}),\quad \bar{u}(\theta,\bar{z})=\frac{u(\theta,z)}{c}$$
$$\varepsilon=\frac{e}{c},\quad \bar{\mu}=\frac{\mu}{\mu_0},\quad \bar{p}=\frac{p}{p_a},\quad \bar{z}=\frac{2z}{L},\quad \bar{t}=t\cdot\Omega,\quad \Lambda=\frac{6\mu_0\Omega}{p_a}\left(\frac{R}{c}\right)^2 \tag{3}$$

Where ρ_0,μ_0,p_0 are the density, viscosity, pressure of the ambient gas, respectively. L is the axial length of the bearing. ε is the eccentricity ratio. Λ is the non-dimensional bearing number.

2.2. Structural model

A uniform stiffness model of the foil structure based on Heshmat's[1] is adopted to acquire the deflection of the foil structure, as shown in Figure 2. Assume that the stiffness of the foil is taken to be uniformly distributed and constant throughout the bearing surface. The overall foil structure stiffness is mainly contributed by bump foil stiffness.

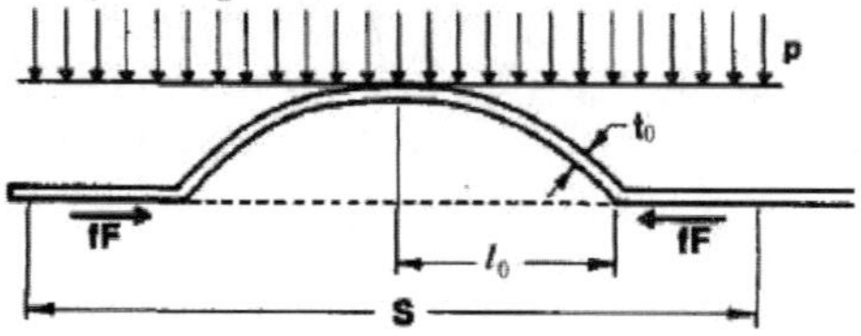

Figure 2 Configuration of bump foil[2]

Under the assumptions above, the relationship between the point pressure and deflection is given by

$$\bar{p}-1=\frac{1}{\bar{\alpha}}\bar{u}+\bar{C}_f\frac{\partial\bar{u}}{\partial\bar{t}} \tag{4}$$

Where, $\bar{\alpha},\bar{C}_f$ are the non-dimensional compliance and dumping coefficients of structure, written as[1]

$$\bar{\alpha}=\frac{1}{k_b}\frac{p_a s}{c},\quad \bar{C}_f=\frac{C_f}{p_a/(c\Omega)} \tag{5}$$

Where s denotes spacing between adjacent bumps for a given radial section, k_b is the stiffness of the bump foil, $k_b=\left[0.5E_b/(1-v_b^2)\right]\cdot(t_0/l_0)^3$ with E_b,v_b are the modulus, posisson ratio of the bump foil material, t_b,l_0 are the thickness and length of half bump sector.

2.3. *Gas film thickness*

Gas film thickness consists of the geometry clearance and foil deflection, the film thickness in the middle section (z=0) is given by

$$h=c+e\cos\theta+u(\theta,z=0) \tag{6}$$

For the journal is tilting, the attitude of the axial line can by define by the tilting angle θ_x and θ_y as shown in Figure 3. The positions of the journal center in different axial sections can be described by the axial distance and tilting angles.

$$(x,y,z)=(x_0,y_0,0)+(-z\tan\theta_y,z\tan\theta_x,z) \tag{7}$$

Where (x_0,y_0) is the position of the journal center in the middle section (z=0) described in Figure 1a). Thus, the film thickness is written as

$$h=c-y\cos\theta+x\cos\theta-z\tan\theta_y,z\tan\theta_x+u(\theta,z) \tag{8}$$

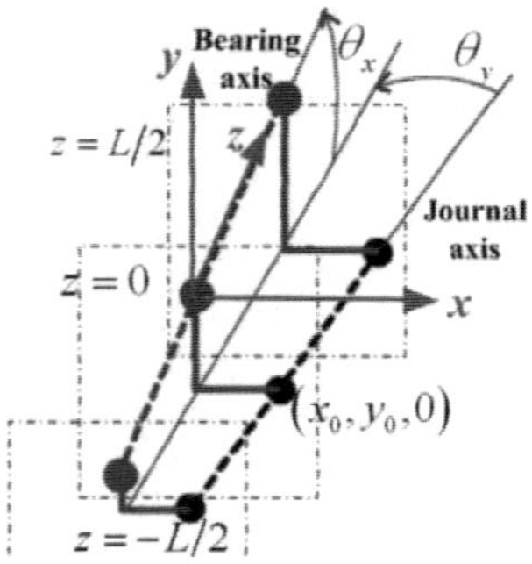

Figure 3 Eccentricities in different axial section of a tilting journal

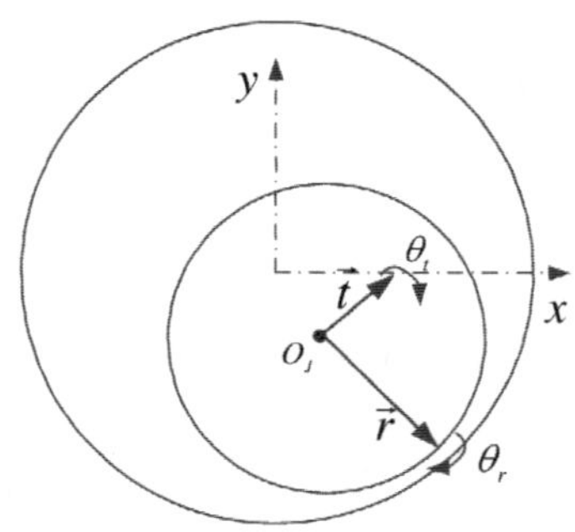

Figure 4 Tilting attitude of journal with two typical rotational axises.

2.4. *Boundary conditions*

Because the top foil deviates from the bump film under negative pressure, the pressure should be larger than the ambient pressure in the downstream area following the minimum film thickness position. Thus, Reynolds boundary condition is applied as given in Eqn.(7), where the position θ^* at zero pressure point is figured out during the numerical process.

$$\bar{p}\big|_{\bar{z}=\pm\frac{L}{2}}=1,\quad \bar{p}\big|_{\theta=0}=\bar{p}\big|_{\theta=2\pi},\quad \frac{\partial\bar{p}}{\partial\theta}\bigg|_{\theta=\theta^{*}}=0 \tag{9}$$

2.5. Bearing Force

In post process, non-dimensional load capacity and torque evaluated by integral on the pressure distribution in the calculating field[2].

$$\bar{F}_x=-\int_{-1}^{1}\int_{0}^{2\pi}(\bar{p}-1)\cos\theta d\theta d\bar{z},\bar{F}_y=-\int_{-1}^{1}\int_{0}^{2\pi}(\bar{p}-1)\sin\theta d\theta d\bar{z},\bar{W}=\sqrt{\bar{F}_x^2+\bar{F}_y^2},\bar{T}=\int_{-1}^{1}\int_{0}^{2\pi}\left(\frac{\bar{h}}{2}\frac{\partial\bar{p}}{\partial\theta}+\frac{\Lambda}{6\bar{h}}\right)d\theta d\bar{z} \tag{10}$$

Reaction torque caused by misalignment is

$$\bar{M}_x=-\int_{0}^{2}\bar{z}\int_{0}^{2\pi}(\bar{p}-1)\cos\theta\cdot d\theta d\bar{z},\bar{M}_y=-\int_{0}^{2}\bar{z}\int_{0}^{2\pi}(\bar{p}-1)\sin\theta\cdot d\theta d\bar{z},\quad \bar{M}=\sqrt{\bar{M}_x^2+\bar{M}_y^2} \tag{11}$$

Recovering dimensional form of the load capacity and equivalent torque are shown below.

$$W=\bar{W}\cdot p_aR\frac{L}{2},\quad T=\bar{T}\cdot p_acR\frac{L}{2},\quad M=\bar{M}\cdot p_acR\frac{L^2}{4} \tag{12}$$

3. Results

To analyze the gas film and the performance of the foil journal bearing with misalignment, two kinds of typical tilting attitude of are defined according to the rotational axis of the journal as shown in Figure 4: a) Radial tilting, rotation around a radial axis along eccentric line, shown in Figure 5b); b) Tangential tilting, journal centers in all sections are located on the eccentric line, shown in Figure 5a).

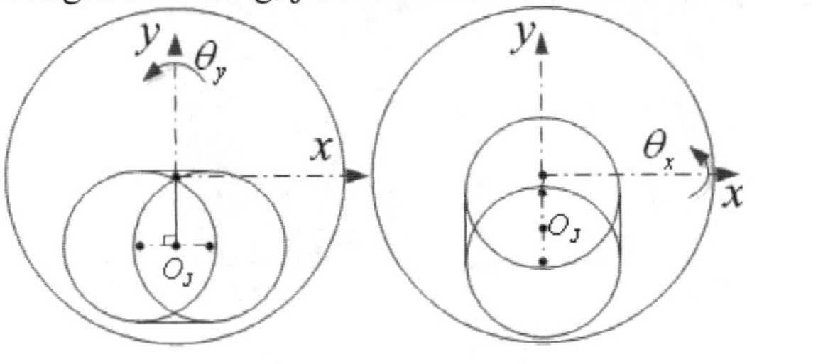

Figure 5 Typical attitudes with misalignment: a) Radial tilting; b) Tangential tilting

Table 1 Data used in the simulation

Foil bearing parameters	
Bearing diameter ($2R$)	38.1×10^{-3}m
Bearing length (L)	38.1×10^{-3}m
Diametral clearance ($2C$)	63.6×10^{-6}m
Rotational speed(Ω)	30000r/min
Pressure(p_a)	101325Pa
Viscosity(μ)	1.932×10^{-5}Pas
Density(ρ_0)	1.29kg/m^3

The foil bearing configuration and lubrication parameters used in this simulation are listed in Table 1. To simplify the input for the simulation, the eccentricity is set only in vertical direction with(x_0, y_0)=(0,e_y) shown in Figure 5.

Figure 6 presents the film thickness distribution and pressure distribution of a typical misalignment with radial tilting journal corresponding to Figure 5a). From Figure 6a), the minimum film thickness positions are distributed along the tilting journal axial direction. The gas film achieves thinnest near two bearing edges. Figure 6b) shows the pressure distribution of the misalignment with radial tilting. Different from the symmetrical pressure distribution with perfect alignment, two pressure peaks emerge along the tilting direction. It is deduced that the two peaks are not at the same height due to the existence of eccentricity.

Figure 7 presents the film thickness distribution and pressure distribution of another typical misalignment with tangential tilting journal corresponding to Figure 5b). From Figure 7a), the film thickness shows a tilting surface along the tilting journal axial direction. The gas film achieves thinnest near an edge of the bearing. Figure 7b) shows the pressure distribution of the misalignment with tangential tilting. Different form the symmetrical pressure distribution with perfect alignment, the pressure peak moves close to an edge of the bearing.

The reaction torque caused by the misalignment varies with the tilting angle. The resultant reaction torque variations are given in Figure 8. It is shown that resultant reaction torques increase with the tilting angles. Compared with tangential tilting, radial tilting usually leads to a larger reaction torque. Small tilting angle leads to a very thin gas film under large eccentricities, the curves with $e_x = 0.6$ and $e_x = 0.8$ contain only several data points in this simulation.

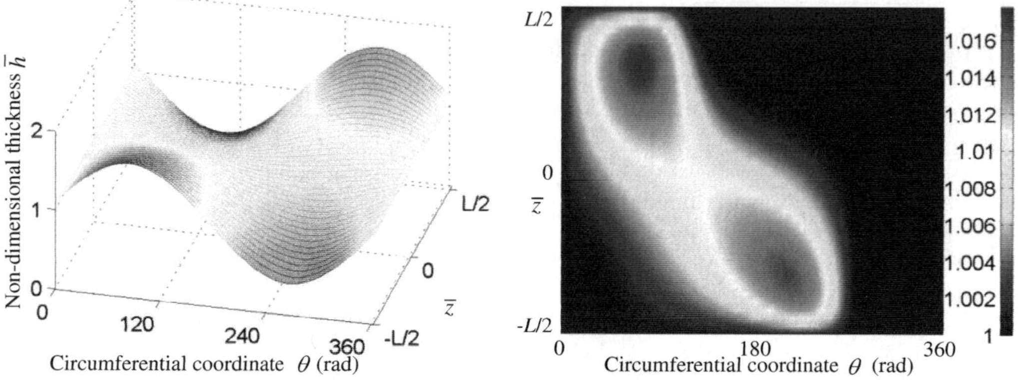

Figure 6 Non-dimensional thickness and pressure distribution ($e_y = -0.1, \theta_y = 1\times10^{-3}$ rad)

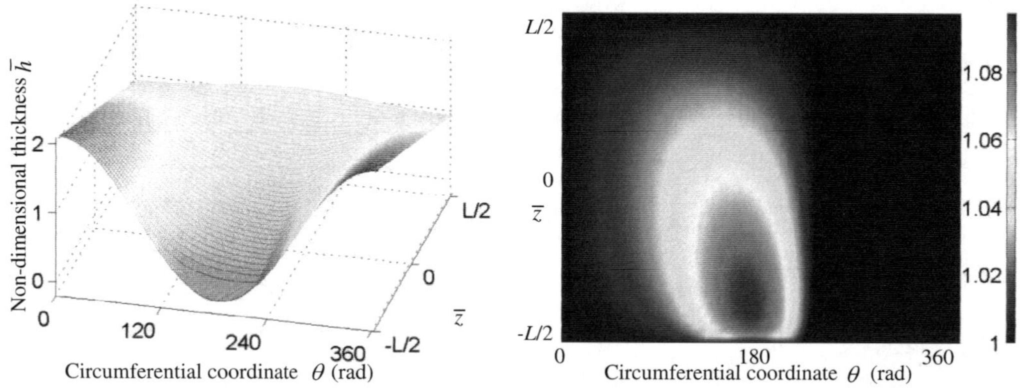

Figure 7 Non-dimensional thickness and pressure distribution ($e_y = -0.5, \theta_x = 9\times10^{-4}$ rad)

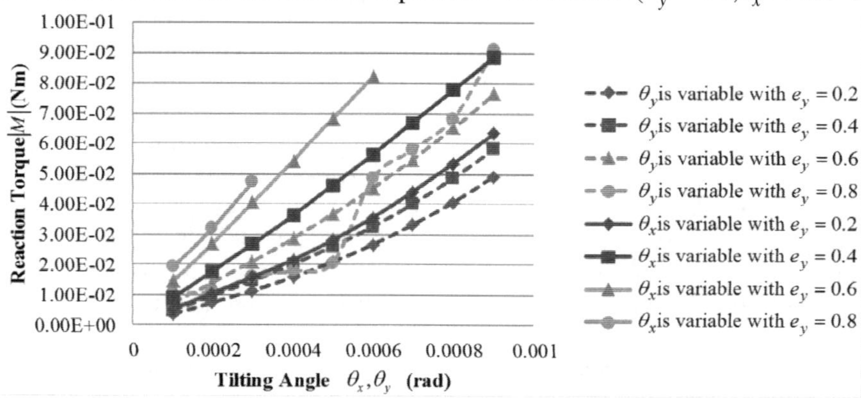

Figure 8 Reaction torque M v.s. tilting angles with different vertical eccentricities

4. Conclusions

Misalignment in a foil journal bearing system makes lubricant performance worse. In this paper, effect of misalignment on gas film thickness is introduced into the lubrication model. Main conclusions are summarized as follow.

a) The two typical misalignment attitudes of the journal were defined as radial tilting and tangential tilting.

b) Two pressure peaks may emerge in some radial tilting attitudes conditions. For tangential tilting attitude, the unique pressure peak moves toward the edge of the bearing.
c) The reaction torques caused by misalignment increase with the tilting angle of the journal. A larger eccentricity leads to a larger reaction torque.

Reference

[1]. Heshmat, H., J.A. Walowit and O. Pinkus, Analysis of Gas Lubricated Compliant Thrust Bearings. Journal of Lubrication Technology, 1983. 4(105): p. 638-646.
[2]. Heshmat, H., J.A. Walowit and O. Pinkus, Bearings, Analysis Of Gas-Lubricated Foil
. Journal of Lubrication Technology, 1983. 4(105): p. 638-646.
[3]. Kim, D. and S. Park, Hybrid air foil bearings with external pressurization, in 2006 ASME International Mechanical Engineering Congress and Exposition, IMECE2006. 2006: Chicago, IL, United States.
[4]. Kim, M.K., Parametric Studies on Dynamic Performance of Hybrid Airfoil Bearing. Journal of Engineering for Gas Turbines and Power, 2008(6).
[5]. Daejong Kim, S.P., Hydrostatic air foil bearings: Analytical and experimental investigation. Tribology International, 2009(3).
[6]. H, K.T., Analysis of Side End Pressurized Bump Type Gas FoilBearings: A Model Anchored to Test Data. 2007, Texas A&M University: Texas.
[7]. Lee, D., Three-Dimensional Thermohydrodynamic Analyses of Rayleigh Step Air Foil Thrust Bearing with Radially Arranged Bump Foils. Tribology Transactions, 2011(3).
[8]. Fangcheng Xu, Z.L.G.Z., Effects of shear stiffness in top foil structure on gas foil bearing performance based on thick plate theory. Proceedings of the Institution of Mechanical Engineers, Part J: Journal of Engineering Tribology, 2013(7).
[9]. M, C. and P.J. P, Effects of Membrane Stresses in the Prediction of Foil Braeing Performanc. Journal of Tribology-Transactions Of The Asme, 1994. 1(116): p. 83-89.
[10]. M, C. and P.J. P, Theoretical Performance of a Hydrostatic Foil Bearing. Journal Of Tribology-Transactions Of The Asme, 1994. 1(116): p. 83-89.
[11]. R, K.C.P. and H. H, Compliant Foil Bearing Structural Stiffness Analysis: Part I-Theoretical Model Including Strip and Variable Bump Foil Geometry. Journal Of Tribology-Transactions Of The Asme, 1992. 2(114): p. 394-400.
[12]. R, K.C.P. and H. Heshmat, Structural Stiffness and Coulomb Damping in
Compliant Foil Journal Bearings: Parametric Studies. Tribology Transactions, 1994. 3(37): p. 455-462.
[13]. R, K.C.P. and H. Heshmat, Structural Stiffness and Coulomb Damping in
Compliant Foil Journal Bearings: Theoretical Considerations. Tribology Transactions, 1994. 3(37): p. 525-533.
[14]. I, I., Analysis of an aerodynamic compliant foil thrust bearing: Method for a rapid design. Journal of Tribology-Transactions Of The Asme, 1999. 4(121): p. 816-822.
[15]. Khonsari, Z.C.P.A., A Thermohydrodynamic Analysis of Foil Journal Bearings. Journal of Tribology, 2006(3).
[16]. Dong-Jin Park, C.K.G.J., Theoretical considerations of static and dynamic characteristics of air foil thrust bearing with tilt and slip flow. Tribology International, 2008(4).
[17]. A., H.S., Gas Foil Bearing Misalignment and Unbalance Effects. 2008, NASA Glenn Research Center: Cleveland, Ohio.
[18]. Howard, S.A., Misalignment in Gas Foil Journal Bearings: An Experimental Study. 2008, NASA Glenn Research Center: Cleveland, Ohio.
[19]. Kai Feng, X.Z.A.Z., Design and structural performance measurements of a novel multi-cantilever foil bearing. Proceedings of the Institution of Mechanical Engineers, Part C: Journal of Mechanical Engineering Science, 2015(10).
[20]. Abdelrasoul M. Gad, On the performance of foil thrust bearing with misaligned bearing runner. Industrial Lubrication & Tribology, 2017(2).

Measuring the likeliness among young adults to green their homes

Tien-Choon Toh[1], Kai-Chen Goh[2], Hui-Hwang Goh[2], Felicia Yan-Yan Yong[1], Cheng-Sim Lim[1] and Chia-Kuang Lee[3]

[1]Universiti Tunku Abdul Rahman, Sungai Long Campus, Kajang, Selangor, Malaysia
[2]Universiti Tun Hussein Onn Malaysia, Parit Raja, Batu Pahat, Johor, Malaysia
[3]Universiti Malaysia Pahang, Gambang, Kuantan, Pahang, Malaysia
E-mail: kaichen@uthm.edu.my

Abstract. It requires a team effort to reduce our overall impact on the planet earth. If each of us would do our parts, we could preserve the environment well and give our future generations a clean and comfortable place to live. Sustainability starts at home. Ways to green our homes have been suggested by people from all walks of life and are easily available as well. However, very little attention has been paid to the extent to which the people would take the initiative to green their homes. With the intention to explore the topic of home greening, this research aims to measure the likeliness among young adults to green their homes. Young adults are targeted because they are future leaders and their actions today will have an effect on the quality of the environment for the next fifty to one hundred years. The data needed are collected using a self-administered structured questionnaire. Eighty-eight young adults living in the Klang Valley of Malaysia have responded to the survey form but only eighty-five sets of it are analysable after detecting outliers. Mean ranking analysis is performed to rank the home greening efforts made by the respondents. The results show that the respondents have highest tendency to use energy-efficient appliances and water-saving devices, fixtures, and technologies for their homes to save the environment. On the contrary, the respondents have lowest tendency to find creative ways to recycle household goods at their homes to save the environment. In general, there is no significant difference in the home greening tendency levels of males and females.

1. Introduction and methods

Every one of us shares the responsibility to preserve and protect the environment as resources from the environment are belonged to the society. We owe our next generations a clean and green environment so they would have a cleaner and brighter future. Sustainability begins at home [1,2]. Guides to green living including green home solutions are readily available and they are often very informative [3,4,5]. Nevertheless, very few researches have been done to investigate the extent to which the people would green their homes. In order to explore the topic of home greening, this research aims to measure the likeliness among young adults to green their homes. Young adults are studied since they will be future policy makers and that whatever they do today will have an impact on the planet and ecosystem for the next fifty to one hundred years.

A questionnaire survey has been conducted in one of the cities in Klang Valley, Malaysia to collect the quantitative data needed in achieving the research objective. Eighty-eight young adults who know about green living and green buildings answer the self-administered structured questionnaire but only eighty-five sets of the survey form are analysable after detecting outliers [6]. A mean ranking analysis, bivariate analysis on means and gender, and the Mann-Whitney U test are done to generate the results.

2. Results and discussion

There are 45 males and 40 females out of the total of 85 respondents. This means that 52.94% of the total number of respondents are male and 47.06% of the total number of respondents are female. All of the responses from these 85 respondents have been analysed statistically. Table 1 shows the ranking of overall respondents' levels of tendency in home greening efforts in Klang Valley, Malaysia. 'Use energy-efficient appliances and water-saving devices, fixtures, and technologies for your home to save the environment' that achieves a mean of 4.08 is ranked first. 'Switch to compact fluorescent light bulbs for your home to reduce electricity use to save the environment' that achieves a mean of 4.01 is ranked second. 'Reduce energy use at your home to save the environment' that achieves a mean of 3.82 is ranked third. 'Reduce water use at your home to save the environment' that achieves a mean of 3.78 is ranked fourth. 'Install motion detectors at your home to switch off lights if there is no one in the room to reduce electricity use to save the environment' that achieves a mean of 3.47 is ranked fifth. 'Select products made from recycled materials or designed to be safer for the environment for your home to save the environment' that achieves a mean of 3.38 is ranked sixth. 'Use reusable products or borrow, rent, or share items used infrequently for your home to save the environment' that achieves a mean of 3.05 is ranked seventh. 'Install a photovoltaic system for your home to generate electricity for household use to save the environment' that achieves a mean of 3.04 is ranked eighth. 'Install a rainwater harvesting system for your home to collect rainwater for household use to save the environment' that achieves a mean of 3.01 is ranked ninth. 'Find creative ways to recycle household goods at your home to save the environment' that achieves a mean of 2.95 is ranked last.

Table 1. Ranking of overall respondents' levels of tendency in home greening efforts in Klang Valley, Malaysia.

Question	Mean	Standard Deviation	Rank
Q10 – How likely is it that you would use energy-efficient appliances and water-saving devices, fixtures, and technologies for your home to save the environment?	4.08	0.790	1
Q09 – How likely is it that you would switch to compact fluorescent light bulbs for your home to reduce electricity use to save the environment?	4.01	0.779	2
Q02 – How likely is it that you would reduce energy use at your home to save the environment?	3.82	0.804	3
Q01 – How likely is it that you would reduce water use at your home to save the environment?	3.78	0.746	4
Q08 – How likely is it that you would install motion detectors at your home to switch off lights if there is no one in the room to reduce electricity use to save the environment?	3.47	1.385	5
Q04 – How likely is it that you would select products made from recycled materials or designed to be safer for the environment for your home to save the environment?	3.38	0.963	6
Q03 – How likely is it that you would use reusable products or borrow, rent, or share items used infrequently for your home to save the environment?	3.05	0.962	7
Q07 – How likely is it that you would install a photovoltaic system for your home to generate electricity for household use to save the environment?	3.04	1.200	8
Q06 – How likely is it that you would install a rainwater harvesting system for your home to collect rainwater for household use to save the environment?	3.01	1.277	9
Q05 – How likely is it that you would find creative ways to recycle household goods at your home to save the environment?	2.95	1.079	10

Table 2 shows the respondents' levels of tendency in home greening efforts in Klang Valley, Malaysia based on gender (male and female). The female respondents have higher tendency than the male respondents to: 'reduce water use at their homes to save the environment', 'reduce energy use at their homes to save the environment', and 'use reusable products or borrow, rent, or share items used infrequently for their homes to save the environment'. But, the male respondents have higher tendency than the female respondents to: 'select products made from recycled materials or designed to be safer for the environment for their homes to save the environment', 'find creative ways to recycle household goods at their homes to save the environment', 'install a rainwater harvesting system for their homes to collect rainwater for household use to save the environment', 'install a photovoltaic system for their homes to generate electricity for household use to save the environment', 'install motion detectors at their homes to switch off lights if there is no one in the room to reduce electricity use to save the environment', 'switch to compact fluorescent light bulbs for their homes to reduce electricity use to save the environment', and 'use energy-efficient appliances and water-saving devices, fixtures, and technologies for their homes to save the environment'. Table 3 shows the result of the Mann-Whitney U test comparing the male and female respondents' levels of tendency in home greening efforts in Klang Valley, Malaysia. The p-values for Q06 and Q07 are each < 0.05, indicating significant results. Thus, except for Q06 and Q07, the null hypothesis is accepted at a 95% confidence level and that the alternative hypothesis is rejected as the (p-values for Q01 to Q05 and Q08 to Q10 are each > 0.05). In general, there is no significant difference in the home greening tendency levels of males and females.

Table 2. Respondents' levels of tendency in home greening efforts in Klang Valley, Malaysia based on gender (male and female).

Question	Gender					
	Male		Female		Total	
	Mean	Standard Deviation	Mean	Standard Deviation	Mean	Standard Deviation
Q01	3.71	0.727	3.85	0.770	3.78	0.746
Q02	3.71	0.895	3.95	0.677	3.82	0.804
Q03	2.93	1.053	3.18	0.844	3.05	0.962
Q04	3.49	1.036	3.25	0.870	3.38	0.963
Q05	2.98	1.158	2.93	0.997	2.95	1.079
Q06	3.33	1.297	2.65	1.167	3.01	1.277
Q07	3.36	1.300	2.68	0.971	3.04	1.200
Q08	3.62	1.403	3.30	1.363	3.47	1.385
Q09	4.07	0.837	3.95	0.714	4.01	0.779
Q10	4.11	0.804	4.05	0.783	4.08	0.790

Table 3. Result of the Mann-Whitney U test comparing the male and female respondents' levels of tendency in home greening efforts in Klang Valley, Malaysia.

Question	Mann-Whitney U	Wilcoxon W	Z	Asymptotic Significance (2-tailed)
Q01	803.500	1838.500	-0.939	0.348
Q02	774.500	1809.500	-1.216	0.224
Q03	741.000	1776.000	-1.473	0.141
Q04	779.000	1599.000	-1.118	0.264
Q05	869.000	1689.000	-0.284	0.776
Q06	634.500	1454.500	-2.394	0.017 *
Q07	618.000	1438.000	-2.555	0.011 *
Q08	768.500	1588.500	-1.195	0.232
Q09	820.000	1640.000	-0.759	0.448
Q10	866.000	1686.000	-0.320	0.749

Grouping variable: gender.
* Significant as p-value < 0.05.

The result of the mean ranking analysis as shown in Table 1 reveals that the respondents are likely to green their homes by changing their lifestyles. It seems that they do not mind spending more money on energy-efficient appliances and water-saving devices, fixtures, and technologies for their homes to save the environment. It is obvious that the young adults studied are aware of the environmental issues surrounding them and committed to reducing the usages of energy and water at their homes to save the environment. Although they may be willing to spend more money on getting the stuffs that are more environmental-friendly, cost remains as a major barrier for the young adults to invest in green building designs that are clearly more expensive than adopting any other green living measures. As we already know, young adults are still students and thus do not have much buying power. Also, they will still be staying with their parents at their parents' houses or perhaps staying at some rented houses at that age so the young adults have no control on the designs of those houses. This can be the reason why 'switch to compact fluorescent light bulbs for your home to reduce electricity use to save the environment' and 'install motion detectors at your home to switch off lights if there is no one in the room to reduce electricity use to save the environment' are ranked higher than 'install a photovoltaic system for your home to generate electricity for household use to save the environment' and 'install a rainwater harvesting system for your home to collect rainwater for household use to save the environment'. The respondents may have concerns over the reliability and durability of products made from recycled materials or designed to be safer for the environment for their homes to save the environment and therefore selecting those products is not as favourable as investing in energy-efficient appliances and water-saving devices, fixtures, and technologies for their homes to save the environment. There is also proof that the young adults use reusable products or borrow, rent, or share items used infrequently for their homes to save the environment. But, it seems difficult for the young adults to find creative ways to recycle household goods at their homes to save the environment. This may be because it is not easy for them to recycle household goods at their homes to save the environment or they may not have good knowledge in this area.

The result of the bivariate analysis on means and gender as shown in Table 2 reflects that the male respondents have more interest than the female respondents in greening their homes even though the female respondents are better off at reducing the usages of water and energy at their homes to save the environment, and using reusable products or borrowing, renting, or sharing items used infrequently for their homes to save the environment seeing that the means of the male respondents for seven out of the total of ten questions, namely, Q04 to Q10 are higher than the means of the female respondents for the same questions.

Nonetheless, the result of the Mann-Whitney U test as shown in Table 3 finds that, in general, there is no significant difference in the home greening tendency levels of males and females. Only two out of the total of ten questions, namely, Q06 and Q07 are perceived differently by the males and females. This signifies that the male respondents have more interest than the female respondents to modify their homes so as to tap natural resources from the environment and exploit them as resources for household use to save the environment.

References

[1] Crabtree, L. (2006). Sustainability Begins at Home? An Ecological Exploration of Sub/Urban Australian Community-Focused Housing Initiatives. *Geoforum*. 37(4), 519-535.

[2] Organo, V., Head, L., and Waitt, G. (2013). Who Does the Work in Sustainable Households? A Time and Gender Analysis in New South Wales, Australia. *Gender, Place and Culture*. 20(5), 559-577.

[3] Jeffery, Y., Barclay, L., and Grosvenor, M. (2008). *Green Living for Dummies®*. Hoboken, New Jersey: Wiley Publishing, Inc.

[4] Bonta, D. and Snyder, S. (2008). *New Green Home Solutions: Renewable Household Energy and Sustainable Living*. Layton, Utah: Gibbs Smith.

[5] Conner, N. (2009). *Living Green: The Missing Manual®*. Sebastopol, California: O'Reilly Media, Inc.

[6] Kumar, M., Abdul Talib, S., and Ramayah, T. (2013). *Business Research Methods*. Shah Alam, Selangor Darul Ehsan, Malaysia: Oxford Fajar/Oxford University Press.

Calculating optimum gear ratios of a two-stage helical reducer with first stage double gear sets

Vu Ngoc Pi[1], Tran Thi Hong[2], Tran Thi Phuong Thao[1], Nguyen Khac Tuan[1], Le Xuan Hung[1], Luu Anh Tung[1]

1 Thai Nguyen University of Technology, Thai Nguyen city, Vietnam
2 Nguyen Tat Thanh University, Ho Chi Minh city, Vietnam.
E-mail: vungocpi@tnut.edu.vn

Abstract. This paper presents a study on the determination of the optimum gear ratios of a two-stage helical reducer with first stage double gear sets. In the study, an optimization problem was built in order to find the optimum gear ratios of the reducer. In the optimization problem, the objective function was the reducer cross section area. Also, the influence of the input factors including the total reducer ratio, the wheel face width coefficient, the allowable contact stress and the output torque were evaluated. In order to investigate the effects of these factors on the optimum gear ratios, a simulation experiment was designed and conducted. In addition, equations for calculating the optimum gear ratios were introduced. By using these equations, the gear ratios can be determined accurately in a simple way.

Keywords. Transmission ratio, optimum reducer design, helical reducer.

1. Introduction

Until now, many studies have been done for the determination of optimal gear ratios of the reducer of a mechanical driven system. This is because the gear ratios powerfully affect the size, the mass and therefore the cost of the reducer.

In previous studies, the optimal gear ratios have been found for two-step reducers [1-4], three-step reducers [4- 8], and four-step reducers [4, 8-10]. Besides, the gear ratios have been determined by one of three basically methods including the graph method, the "practical method" and the model method.

The graph method has used in many studies, such as in [1, 2 and 4]. An example of it is shown in Figure 1. In the figure, the gear ratios of a three-step helical reducer are determined graphically.

The "practical method" was first introduced in [2]. In this method, the gear ratios were determined based on the practical data. For example, the weight of a two-step helical reducer is minimal when the ratio of center distance of the second step to that of the first step is from 1.4 to 1.6 [2]. From that, the optimal gear ratios can be proposed.

The model method is the most common and the best method for calculating the optimal gear ratios. This is because it can give the values of the gear ratios directly and speedily. In the model method, the objective can be the minimal volume of gears [4], the minimal acreage of the cross section of the reducer [5], the minimal length of the reducer [7], the minimal total mass of the reducer [8] or the minimal mass of gears [9, 10].

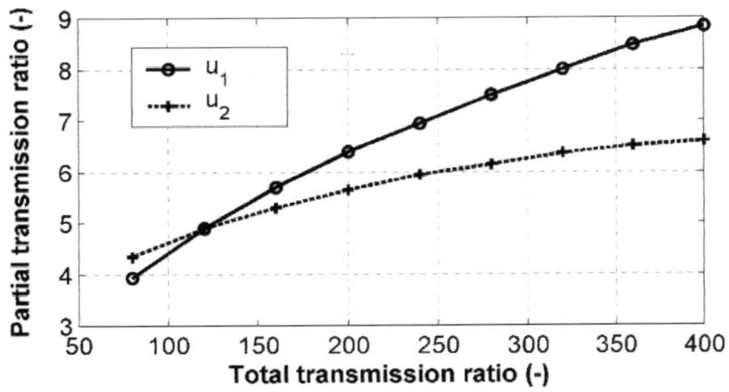

Figure 1. Gear ratios versus total ratio [1].

This paper introduces a study for the determination of the optimum gear ratios of a two-stage helical reducer with second stage double gear sets. In this study, an optimization problem in which the objective is the minimum reducer cross section area was carried out. In addition, the effect of the input factors on the optimum gear ratios was investigated.

2. Optimization problem

Please follow these instructions as carefully as possible so all articles within a conference have the same style to the title page. This paragraph follows a section title so it should not be indented.

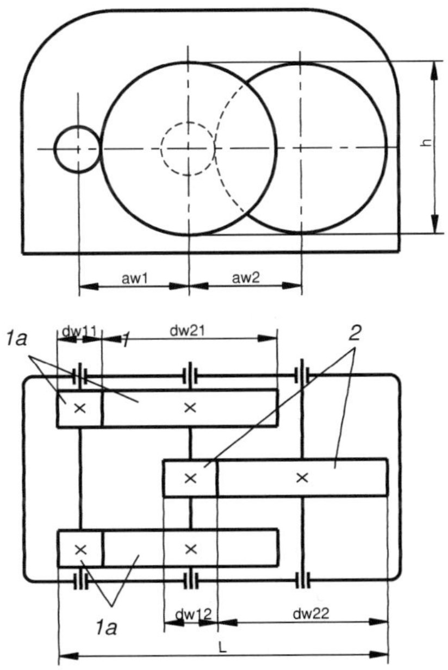

Figure 2. Calculation schema.

For a two stage helical reducer with first stage double gear sets, its cross section area is calculated by the following equation (see figure 2):

$$A = L \cdot h \quad (1)$$

In which, L and h are determined by:

$$L = d_{w11}/2 + a_{w1} + a_{w2} + d_{w22}/2 \quad (2)$$

$$h = \max\left(d_{w21}, d_{w22}\right) \quad (3)$$

In above equations, a_{w1} and a_{w2} are the center distances of the first and the second stage; d_{w11}, d_{w21} and d_{w22} are pitch diameters (mm) of the first and the second stage, respectively. The diameters d_{w11}, d_{w21} and d_{w22} can be calculated as [11]:

IOP Conf. Series: Materials Science and Engineering **542** (2019) 012017 doi:10.1088/1757-899X/542/1/012017

$$d_{w11} = 2 \cdot a_{w1} / (u_1 + 1) \tag{4}$$

$$d_{w21} = 2 \cdot a_{w1} \cdot u_1 \cdot / (u_1 + 1) \tag{5}$$

$$d_{w22} = 2 \cdot a_{w2} \cdot u_2 / (u_2 + 1) \tag{6}$$

In which, u_1, u_2 are the gear ratios of the first and the second stages; $u_1 \cdot u_2 = u_g$ with u_g is the total reducer ratio;

Consequently, the optimization problem can be expressed as

$$\text{minimize } A = L \cdot h \tag{7}$$

With the following constraints

$$1 \le u_1 \le 9 \tag{8}$$

$$1 \le u_2 \le 9$$

From above equations, for solving the optimization problem we need to determine the center distances of the first stage a_{w1} and the second stage a_{w2}.

2.1. Determining the center distance of the first stage

$$a_{w1} = K_a \cdot (u_1 + 1) \cdot \left(\frac{T_{11} \cdot k_{H\beta}}{[\sigma_H]^2 \cdot u_1 \cdot \psi_{ba1}} \right)^{1/3} \tag{9}$$

Where, $K_{H\beta}$ is the contact load ratio for pitting resistance; For the first stage of the reducer $k_{H\beta} = 1.02 \div 1.28$ [11] and we can chose $k_{H\beta} = 1.1$; $[\sigma_H]$ is the allowable contact stress (MPa); In practice, $[\sigma_H] = 350 \ldots 420$ (MPa); k_a is the material coefficient; For steel helical gear we have k_a =43 [11]; ψ_{ba} is the coefficient of wheel face width; for the for the first stage of the reducer $\psi_{ba1} = 0.3 \ldots 0.35$

From the moment equilibrium condition of the reducer we have:

$$T_{out} = 2 \cdot T_{11} \cdot \eta_{hg}^2 \cdot \eta_{be}^3 \cdot u_g \tag{10}$$

In which, η_{hg} is the helical gear transmission efficiency ($\eta_{hg} = 0.96 \ldots 0.98$ [11]); η_{be} is the transmission efficiency of a pair of rolling bearing ($\eta_{be} = 0.99 \ldots 0.995$ [11]). Choosing $\eta_{hg} = 0.97$ and $\eta_{be} = 0.992$ and substituting them into (10) gets:

$$T_{11} = 0.5444 \cdot T_{out} / u_g \tag{11}$$

Substituting (11) and $k_{H\beta} = 1.1$ into (9) we have:

$$a_{w1} = 36.2442 \cdot (u_1 + 1) \cdot \left(\frac{T_{out}}{[\sigma_{H1}]^2 \cdot u_1 \cdot u_g \cdot \psi_{ba1}} \right)^{1/3} \tag{12}$$

2.2. Calculating the center distance of the second stage

The center distance of the second stage a_{w2} can be determined by [11]:

$$a_{w2} = K_a \cdot (u_2 + 1) \cdot \left(\frac{T_{12} \cdot k_{H\beta}}{[\sigma_H]^2 \cdot u_2 \cdot \psi_{ba2}} \right)^{1/3} \tag{13}$$

In addition, for the second stage we have

$$T_{out} = T_{12} \cdot \eta_{hg} \cdot \eta_{be}^2 \cdot u_2 \tag{14}$$

Choosing $\eta_{hg} = 0.97$ and $\eta_{be} = 0.992$ as in section 2.1 and substituting them into (14) gives

$$T_{12} = 1.0476 \cdot T_{out} / u_2 \tag{15}$$

Substituting (15), k_a =43 and $k_{H\beta}=1.1$ (as in section 2.1) into (13) gets:

$$a_{w2}=45.0814\cdot(u_2+1)\cdot\left(\frac{T_{out}}{[\sigma_H]^2\cdot u_2^2\cdot\psi_{ba2}}\right)^{1/3} \tag{16}$$

2.3. Experimental work

For exploring the effect of the input factors on the optimum gear ratios, a simulation experiment was designed and conducted. In this case, a 2-level full factorial design was carefully chosen. Table 1 presents 5 input factors which were selected for the investigation. Consequently, there are $2^5=32$ number of tests for conducting. From equation (7) and (8), a computer program was built for performing the experiment. The various levels of input factors and the results of the output of the program (the optimum gear ratio of the first stage u_1) are shown in Table 2.

Table 1. Input parameters.

Factor	Code	Unit	Low	High
Total reducer ratio	u_g	-	5	30
Coefficient of wheel face width of stage 1	X_{ba1}	-	0.3	0.35
Coefficient of wheel face width of stage 2	X_{ba2}	-	0.35	0.4
Allowable contact stress	AS	MPa	350	420
Output torque	T_{out}	Nmm	10^5	10^7

3. Results and discussions

Figure 3 shows the effect of input factors on the response and the relative strength of the effect. From the figure, it was found that the optimum gear ratio of the first stage u_1 depends strongly on the total reducer ratio u_g. It increases significantly with the increase of the total reducer ratio. In addition, it is effected by the wheel face width coefficients of the first and the second stages (ψ_{ba1} and ψ_{ba2}). Moreover, the optimum gear ratio is not effected by the allowable contact stress of the helical gear set AS, and the output torque T_{out}.

Table 2. Experimental plans and output response.

StdOrder	RunOrder	CenterPt	Blocks	u_g	Xba1	Xba2	AS (MPa)	Tout (Nm)	u_1
25	1	1	1	5	0.3	0.35	420	10000	3.45
17	2	1	1	5	0.3	0.35	350	10000	3.45
29	3	1	1	5	0.3	0.4	420	10000	3.3
6	4	1	1	30	0.3	0.4	350	100	9
15	5	1	1	5	0.35	0.4	420	100	3.47
19	6	1	1	5	0.35	0.35	350	10000	3.63
31	7	1	1	5	0.35	0.4	420	10000	3.47
...									
2	31	1	1	30	0.3	0.35	350	100	9
28	32	1	1	30	0.35	0.35	420	10000	9

The Pareto chart of the standardized effects is shown in figure 4. It was found from this graph, the bars that represent the total reducer ratio (factor A), the coefficients of wheel face width of the first and the second stages (factors B and C) and the interactions between them (AB and AC) cross the reference line. Therefore, these input factors are statistically significant at the 0.05 level with the response model (the gear ratio of the first stage u_1). Also, because the bars which represent the allowable contact stress of the helical gear set (factor D), and the output torque (factor E) do not cross the reference line, these factors (D and E) do not affect the response model.

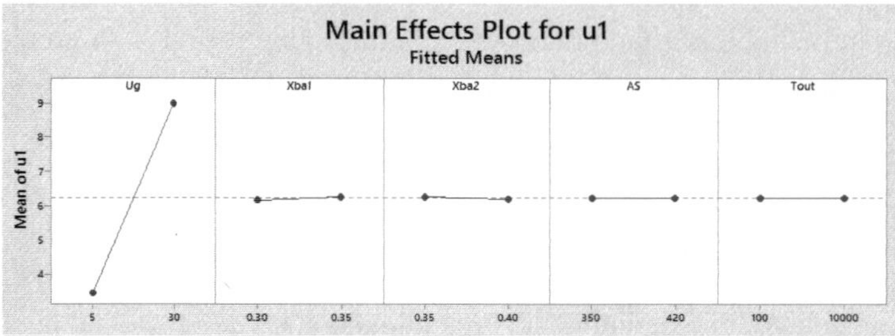

Figure 3. Main effects plot for the optimum gear ratio of the first stage.

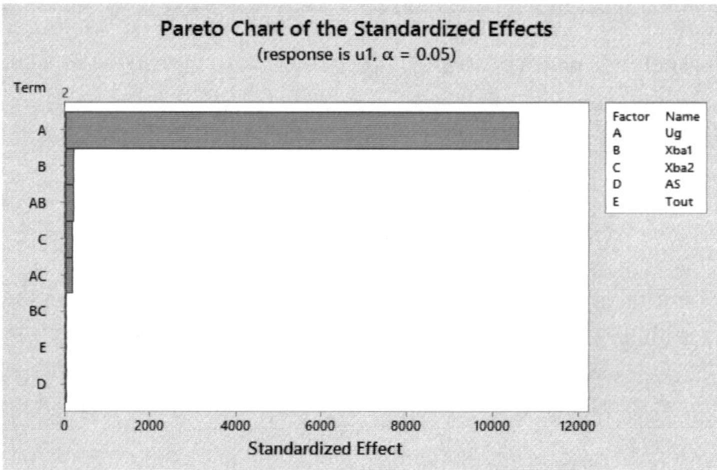

Figure 4. Pareto Chart of the Standardized Effects.

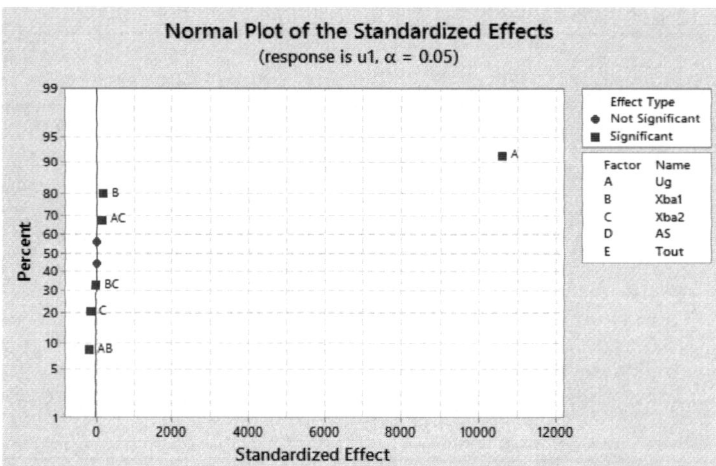

Figure 5. Normal Plot for the optimum gear ratio of the first stage.

Coded Coefficients

Term	Effect	Coef	SE Coef	T-Value	P-Value	VIF
Constant		6.23125	0.00026	23907.22	0.000	
Ug	5.53750	2.76875	0.00026	10622.77	0.000	1.00
Xba1	0.087500	0.043750	0.000261	167.85	0.000	1.00
Xba2	-0.077500	-0.038750	0.000261	-148.67	0.000	1.00
AS	-0.000000	-0.000000	0.000261	-0.00	1.000	1.00
Tout	-0.000000	-0.000000	0.000261	-0.00	1.000	1.00
Ug*Xba1	-0.087500	-0.043750	0.000261	-167.85	0.000	1.00
Ug*Xba2	0.077500	0.038750	0.000261	148.67	0.000	1.00
Xba1*Xba2	-0.002500	-0.001250	0.000261	-4.80	0.000	1.00

Model Summary

S	R-sq	R-sq(adj)	R-sq(pred)
0.0014744	100.00%	100.00%	100.00%

Figure 6. Estimated Effects and Coefficients for the optimum gear ratio of the first stage.

Figure 5 presents the Normal Plot of the standardized effects. This graph is used for describe which effects increase or decrease the response. From the Figure, it was found that the total reducer ratio (factor A) is the most significant factor for the optimum gear ratio. In addition, the total reducer ratio and the coefficient of wheel face width of the first stage (factor B) have a positive standardized effect. If they change from the low level to the high level of the factors, the optimum gear ratio of the first stage increases. Also, the wheel face width coefficient of the second stage (factor C) has a negative standardized effect. The optimum gear ratio decreases when it increases.

Figure 6 presents the estimated effects and coefficients for the optimum gear ratio u_1. From the figure, the total reducer ratio u_g, the wheel face width coefficient of the first and the second stage (ψ_{ba1} and ψ_{ba2}) and their interactions have P-values lower than 0.05. Therefore, these factors are significant to the optimum gear ratio. In addition, the relation between the optimum gear ratio and these significant effect factors can be described as follows:

$$u_1 = 2.1413 + 0.2205 \cdot u_g + 4.95 \cdot \psi_{ba1} - 3.07 \cdot \psi_{ba2} - 0.14 \cdot u_g \cdot \psi_{a1} + 0.124 \cdot u_g \cdot \psi_{ba2} - 2 \cdot \psi_{ba1} \cdot \psi_{ba2} \quad (17)$$

As the adj-R^2 and pred-R^2 are in the high values (figure 6), the above equation fit the data very well.

Equation (17) is used for determining the optimum gear ratio of the first stage u_1. After getting u_1, the gear ratio of the second stage can be found by $u_2 = u_g / u_1$.

4. Conclusions

The minimum cross section area of a two-stage helical reducer with second stage double gear sets can be obtained by determination of optimum gear ratios of the reducer.

Equations for calculating the optimum gear ratios of a two-stage helical reducer with second stage double gear sets were introduced for getting the minimum reducer cross section area.

The optimum gear ratios of the reducer can be calculated accurately and simple as the estimated models are explicit.

Acknowledgements

The work described in this paper was supported by Thai Nguyen University of Technology for a scientific project.

Reference lists

[1] Kudryavtsev V N, Derzhavets Yu A and Glukharev E G 1971 *Design and calculation of gear reducers. Reference guide* (Saint-Petersburg: Mechanical Engineering) p 328

[2] G Milou, G Dobre, F Visa and H Vitila 1996 Optimal Design of Two Step Gear Units, regarding the Main Parameters *VDI Berichte* No 1230 p 227

[3] Vu Ngoc Pi 2001 A new method for optimal calculation of total transmission ratio of two step helical gearboxes, *The National conference on Engineering Mechanics* (Ha Noi) pp 133–136

[4] A N Petrovski, B A Sapiro and N K Saphonova 1987 About optimal problem for multi-step gearboxes (in Russian) (*Vestnik Mashinostroenie) No 10* p 13

[5] V N Pi and N K Tuan 2016 Optimum determination of partial transmission ratios of Three-step helical gearboxes for getting minimum cross section, *Journal of Environmental Science and Engineering A 5* pp 570–573

[6] V N Pi et al 2006 Optimal Calculation of Total Transmission Ratio of Three-Step Helical Gearboxes for Minimum Mass of Gears (In Vietnamese) *Journal of Science and Technology of 6 Engineering Universities* pp 91–93

[7] Vu Ngoc Pi 2008 A new study on optimal calculation of partial transmission ratio of three-step helical reducers, *The 3rd IASME / WSEAS Int. Conf. on Continuum Mechanics* (Cambridge: UK*)* p 23

[8] Romhild I and Linke H 1992 Gezielte Auslegung Von Zahnradgetrieben mit minimaler Masse auf der Basis neuer Berechnungsverfahren *Konstruktion 44* pp 229–236

[9] V N Pi 2008 Optimal determination of partial transmission ratios for four-step helical gearboxes with firt and third step double gear-sets for minimal mass of gears, *Applied Computing Conference ACC' 08* (Istanbul, Turkey) p 27–30

[10] L X Hung, V N Pi and N V Du 2009 Optimal calculation of partial transmission ratios of four-step helical gearboxes with second and fourth-step double gear-sets for minimal mass of gears, *The int. Symp. on Mechanical Engineering (ISME Ho Chi Minh city, Vietnam)* pp 21–23

[11] T Chat and L V Uyen 1998 Design and calculus of Mechanical Transmissions (Hanoi: Educational Republishing House)

IOP Conf. Series: Materials Science and Engineering **542** (2019) 012018 doi:10.1088/1757-899X/542/1/012018

Study on Film Characteristics of Piston-Cylinder Interface of High-Pressure Common Rail Radial Piston Pump with Micro Motion

Bo Qi[1], Yong Zhang[1,*], Guoyou Meng[1], Yao Ding[1], and Ming Ye[2]

[1]School of Mechanical and Automotive Engineering, South China University of Technology, Wushan Road No 381, Tianhe District, Guangzhou, 510641,China
[2]Guangzhou Huagong Motor Vehicle Inspection Technology Co., LTD, Wushan Road No 381, Tianhe District, Guangzhou, 510641,China
Corresponding author: 343241344@qq.com

Abstract. When the high-pressure common rail radial piston pump works, the piston is subjected to film pressure and cam driving force, resulting in some micro-motion such as offset and inclination. A large number of theoretical studies have shown that the micro-motion of piston has an important effect on the film characteristics of piston-cylinder interface. In this paper, according to the theory of hydrodynamic lubrication, Reynolds equation and film thickness equation of piston micro-motion are established, and solve the problem of the lubricant film of the piston-cylinder interface under different working conditions. The results show that the working pressure and cam speed have great influence on the film characteristics of piston-cylinder interface with micro-motion. The research results provide theoretical support for the optimization of the film parameters of the piston-cylinder interface.

1. Introduction

In recent years, with the increasingly stringent emission regulations and the continuous progress of electronic control technology of diesel engine, high-pressure common rail fuel injection system with its remarkable advantages has become the main trend of modern diesel engine technology development [1]. Piston-cylinder interface is the key friction pair of radial piston pump in common rail system, the film characteristics of which determine the working state of the piston-cylinder interface.

In addition to macro reciprocating motion, the piston also has micro motion while it is working on the micro scale. The micro-motion of the piston determines the film thickness of the piston-cylinder interface, the pressure distribution, and the force applied to the piston. Compared with the piston-cylinder interface of axial piston pump, the working pressure and movement of the piston-cylinder interface of radial piston pump are more complicated. Based on the theory of hydrodynamic lubrication, the film thickness equation and Reynolds equation of piston-cylinder interface with micro-motion are established in this paper. By solving the above equations, the film thickness and film pressure taking into account the micro-motion of piston are studied. The research results have certain theoretical and engineering value.

2. Mathematical model

The structural model of the piston-cylinder interface is shown as Fig.1. The centre of the end face of the cylinder is taken as the origin, e as the eccentric distance of the piston-cylinder interface and θ as the eccentric angle of the piston. A is near the piston cavity and B is near the slipper. The film of piston-cylinder interface is expanded along the circumference direction, as shown in Fig.2. x is the direction of the cylinder circumference, y is the

IOP Conf. Series: Materials Science and Engineering **542** (2019) 012018 doi:10.1088/1757-899X/542/1/012018

axial direction of the cylinder, and z is the direction of the piston-cylinder interface film thickness. u, v and w are the speed components of lubricating film in x, y and z directions.

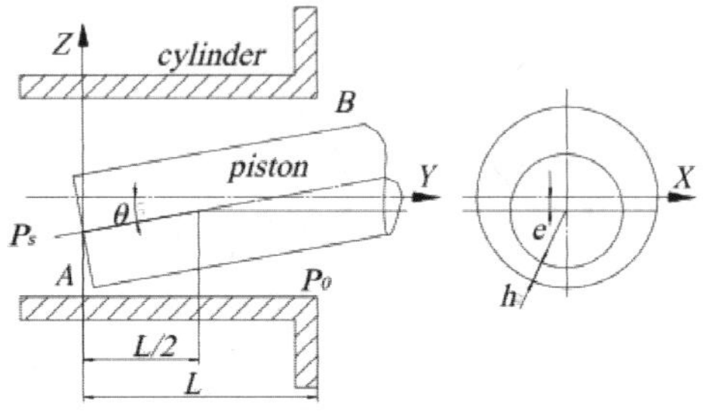

Fig 1. Eccentric model of piston-cylinder interface

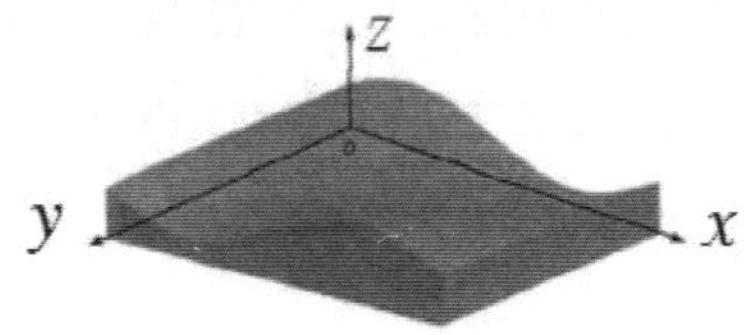

Fig 2. Lubricant film model for piston-cylinder interface

2.1. Reynolds equation of lubrication film

In the study of Reynolds equation of piston-cylinder interface, the following hypothesis is made: Film flow is laminar flow, without turbulence and eddy current. The film has no slip on the piston-cylinder interface, neglecting inertia force and volume force. The film of the piston-cylinder interface is very thin. The pressure variation along the thickness direction of the film is not considered in modeling, but the velocity gradient along the X and Y directions is considered. Based on the above assumptions,under the micro motion conditions, Reynolds equation [2,3] of lubricating film of piston-cylinder interface can be expressed as:

$$\frac{\partial}{\partial x}\left(\frac{\rho h^3}{\mu}\frac{\partial p}{\partial x}\right)+\frac{\partial}{\partial y}\left(\frac{\rho h^3}{\mu}\frac{\partial p}{\partial y}\right)=6V_1\frac{\partial(\rho h)}{\partial y}+12\frac{\partial(\rho h)}{\partial t} \tag{1}$$

Where, h is the film thickness, ρ is the lubricating film density, p is the film pressure, and μ is the film viscosity.

2.2. Film thickness equation

The The piston sub- film thickness expansion equation can be expressed [4] as:

$$h(\theta, y)=C+\left[e+\theta(y-0.5)\right]\cos(\varphi) \tag{2}$$

Where C is the average gap between the piston and the cylinder, φ is the angle between the center line of any two sections in the axial direction of the piston and the x positive half axis.

2.3. Reynolds boundary condition of piston-cylinder interface

The Reynolds boundary of piston-cylinder interface has three kinds: Reynolds boundary, half Sommerfield boundary and all Sommerfield boundary [5,6,7]. When the full Sommerfield is used to solve the problem, the bearing capacity of the lubricant film is small, which is not consistent with the actual situation. The semi Sommerfield boundary is simple and direct, but it can not meet the continuity requirement of film flow. The Reynolds boundary not only satisfies the continuity requirements of film flow, but also avoids negative points [8]. Therefore, the Reynolds boundary is used to solve the Reynolds equation of lubricant film of piston-cylinder interface.

Pressure outlet boundary conditions:

$$P(x,0)=0 \tag{3}$$

Pressure inlet boundary conditions:

$$P(x, L)=P_s \tag{4}$$

Periodic boundary conditions:

$$\begin{aligned} &P(0,\ y)=P(2\pi d,\ y)\\ &\frac{\partial P}{\partial x}(0,\ y)=\frac{\partial P}{\partial x}(2\pi d,\ y)\end{aligned} \tag{5}$$

2.4. Performance parameters of piston-cylinder interface

In the lubricating film of piston-cylinder interface, a micro-surface with a length of *dx* and a width of *dy* is selected. The pressure on the micro-surface is *p*, and the force of micro-surface acting on the piston is *pdxdy*.

The force of the bearing capacity F_x and F_y of the lubricating film in the *x* and *y* directions can be expressed as:

$$\begin{cases} F_x = -\int_0^L \int_0^{2\pi R_p} p(x,y)\cos(x/R_p)dxdy \\ F_y = -\int_0^L \int_0^{2\pi R_p} p(x,y)\sin(x/R_p)dxdy \end{cases} \quad (6)$$

The force of the bearing capacity of the lubricating film on *x* and *y* is:

$$\begin{cases} M_x = -\int_0^L \int_0^{2\pi R_p} yp(x,y)\sin(x/R_p)dxdy \\ M_y = -\int_0^L \int_0^{2\pi R_p} yp(x,y)\cos(x/R_p)dxdy \end{cases} \quad (7)$$

The flow rate of the lubricant film along the coordinate axis y is:

$$Q=\int_0^{2\pi R}\int_0^L vdzdx \quad (8)$$

3. Calculation results and analysis

The generalized Reynolds equation of the piston-cylinder interface is an elliptic equation. The Reynolds boundary and the explicit point iterative method can satisfy the solution requirements, and the numerical solution is solved in MATLAB [9]. The cam speed and the film inlet pressure are the main parameters that affect the characteristics of the piston-cylinder interface sub-film. When the working condition is fixed, the greater the cam pressure angle, the greater the external load the piston bears. And the film characteristics of the piston-cylinder interface becomes the worst. The influence of cam speed and film inlet pressure on the film characteristics of piston-cylinder interface with micro-motion is studied under the max cam pressure angle.

3.1. Variation characteristics of film thickness and pressure of piston-cylinder interface

3.1.1 Variation characteristics of film thickness

Taking the cam speed 500 r/min, the film inlet pressure 40 MPa and the cam pressure angle maximum as example, the film thickness distribution characteristics of piston-cylinder interface are studied, as shown in Fig.3.

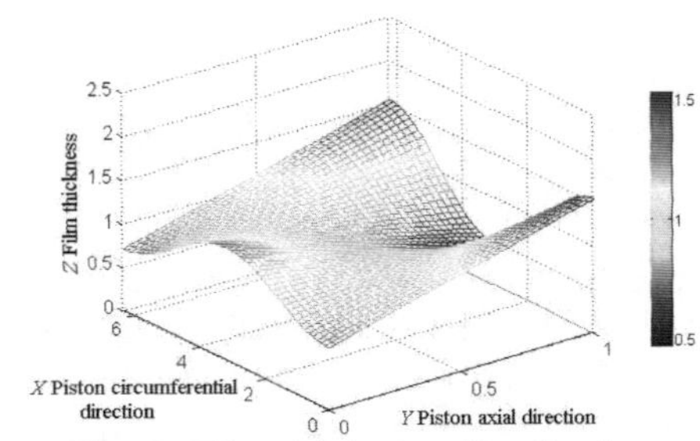

Fig 3. Film thickness distribution

In the case of piston micro-motion, the thickness of the film changes continuously in the axial and circumferential directions. The minimum thickness of the film is at the intersection of the axis corresponding to the expansion angle π and the end face of the film outlet. The thickness of film is closely related to the lubrication characteristics of the piston-cylinder interface. The lubrication characteristics of the thick film are better than those of the thin film. The thickness of film of piston-cylinder interface is different under different working conditions, but the distribution of film thickness is basically the same.

3.1.2 Variation characteristics of film pressure

When the cam speed is 500 r/min, the working pressure is 40 MPa and the cam pressure angle is maximum, the dimensionless value of the eccentricity angle of the piston is 0.55, and the film pressure distribution is compared with the reference [10], as shown in Fig.4.

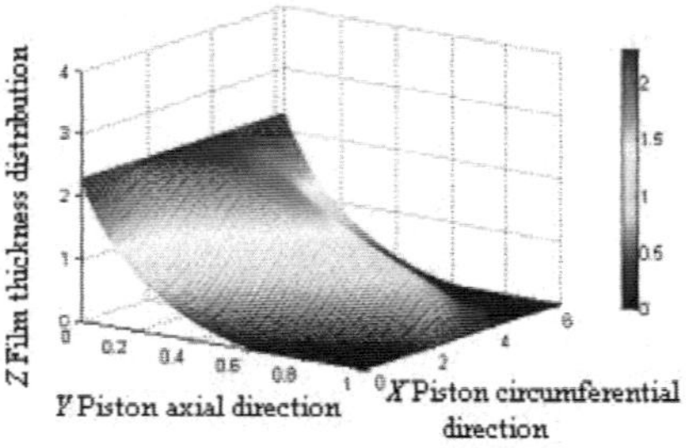

a. The results of this chapter are calculated.

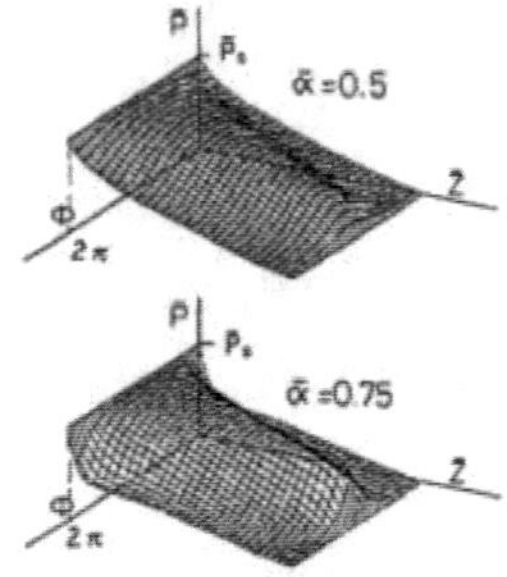

b. Reference [10] calculation results

Fig 4. Comparison of film pressure distributions

In this paper, the eccentric angle is 0.55, and the reference [10] is 0.5. The eccentric angles of the piston in this paper and the reference are not much different. Compared with reference [10], the variation trend of film pressure is basically the same, but because of the different structural parameters and working conditions, there are some differences in the details of film pressure. To some extent, the reference [10] verifies the correctness of the calculation in this paper. At the same sealing position, the smaller the film thickness, the greater the film pressure, and the effect of the film on the extrusion effect becomes more and more obvious. At the axial direction corresponding to the expansion angle, the film pressure gradually decreases from the pressure inlet to the pressure outlet.

3.2. Effect of working pressure on film characteristics of piston-cylinder interface

The film thickness and pressure distribution of piston-cylinder interface under different inlet pressure conditions were studied with the cam speed of 500 r/min and the maximum pressure angle as an example. The effect of different inlet pressure on the minimum film thickness of piston-cylinder interface is shown in Fig.5.

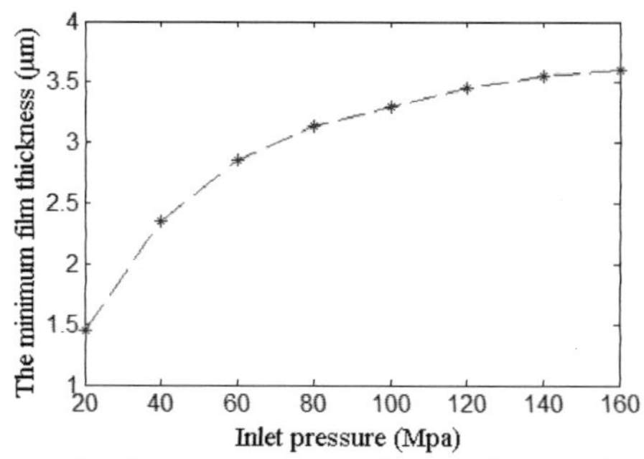

Fig 5. Effect of inlet pressure on the minimum film thickness

The minimum film thickness of the piston-cylinder interface increases with the increase of the working pressure, and the minimum film thickness increases obviously when the film inlet pressure is low. When the film inlet pressure increases to a certain value, the minimum film thickness growth rate slows down. Increasing the inlet pressure of the film will help to increase the minimum film thickness and improve the lubrication characteristics of the piston-cylinder interface.

Under different inlet pressure conditions, the film pressure of the piston-cylinder interface varies along the axis direction corresponding to the expansion angle π, as shown in Fig 6.

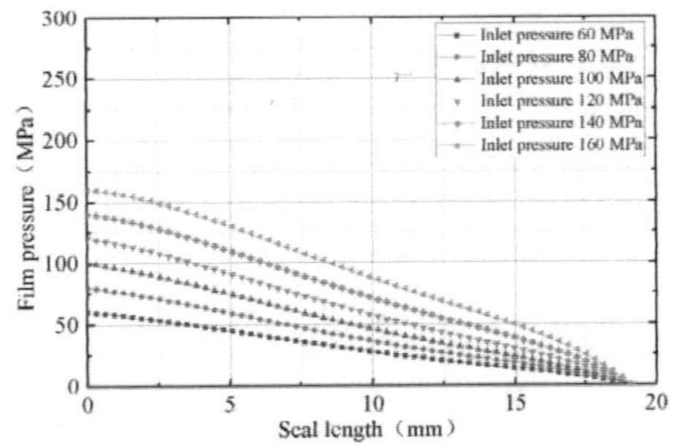

Fig 6. The effect of film inlet pressure on film pressure variation along the expansion angle π corresponding to axial direction

The film pressure gradually decreases along the axis direction corresponding to the expansion angle π, and the inlet pressure is relatively linear when the inlet pressure is low. When the inlet pressure is high, the nonlinear change of the film pressure gradually appears. The greater the inlet pressure, the greater the rate of decline of the film pressure in the axial direction.

3.3. Effect of cam speed on film characteristics of piston-cylinder interface

The effect of the cam speed on the minimum film thickness is shown in Fig.7 when the film inlet pressure is 120 MPa and the cam pressure angle is at its maximum. The blue point in the figure is the minimum film thickness corresponding to different rotation speeds, and the red line is the fitting curve of each point.

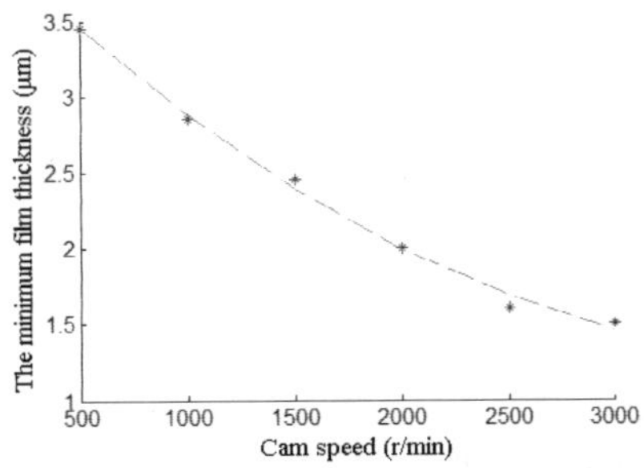

Fig 7. Effect of cam speed on the minimum film thickness

The minimum lubricant film is 3.45 μm at a cam speed of 500 r/min and 1.5 μm at a cam speed of 3000 r/min. With the increase of the cam speed, the minimum film thickness decreases in a quadratic curve, and the decline rate decreases with the increase of the cam speed. With the increase of cam speed, the axial velocity of the piston increases, the movement of the piston becomes more and more complex and the extrusion effect becomes more obvious.

Under different cam speed conditions, the film pressure of the piston-cylinder interface changes along the axis direction corresponding to the expansion angle π, as shown in Fig.8.

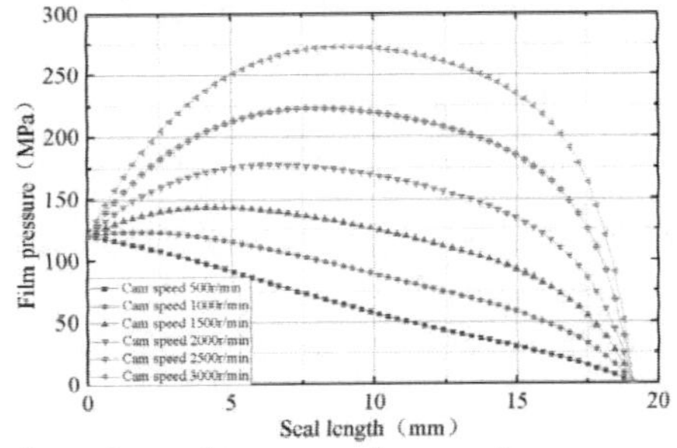

Fig 8. Film pressure variation along the expansion angle π corresponding to axial direction

When the sealing position is fixed, the minimum film thickness decreases gradually with the increase of eccentric angle of piston, resulting in the gradual increase of film pressure. When the cam speed is low, the eccentric angle and offset of the piston are small, and the film pressure changes linearly along the axis direction corresponding to the expansion angle π. As the cam speed increases gradually, the eccentric angle and offset of the piston increase, the film pressure first increases and then decreases, and the larger the rotational speed, the greater the pressure increase. Piston velocity is one of the main factors affecting the lubrication characteristics of the piston-cylinder interface. The greater the cam speed, the greater the reciprocating velocity of the piston, and the larger the offset and the offset angle, the greater the influence of the piston motion on the film pressure.

4. Conclusion

Based on the fluid lubrication theory, this paper establishes a hydrodynamic lubrication model for the piston of the high-pressure common rail radial piston pump that is included in the micro movement of the piston. The film properties of the piston-cylinder interface under different working conditions were discussed by solving the Reynolds equation and the film thickness equation of the piston. The following important conclusions can be drawn from the research:

（1）Due to the different working conditions, the thickness of piston vice film is changing. When the inlet pressure of film is fixed, the eccentric angle of piston increases with the increase of cam speed, and the minimum film thickness decreases in a quadratic curve. When the cam speed is fixed, the eccentric angle of the piston decreases with the increase of the film inlet pressure, and the minimum film thickness increases in a quadratic curve. The film pressure is closely related to the thickness of the film. The sealing position is fixed, the pressure at the position where the film thickness is large is small, and the pressure at the position where the film thickness is small is large.

（2）When the inlet pressure of the film is fixed, the greater the cam speed, the smaller the minimum film thickness, and the lower the film bearing capacity, the worse the lubrication performance of the film. When the cam speed is low, the film pressure changes linearly along the axis direction corresponding to the expansion angle π; when the cam speed is high, the film pressure first increases and then decreases, and the rotational speed increases as the film pressure increases. The speed of the cam is fixed, and the greater the pressure of the film inlet, the greater the minimum film thickness. The lubrication performance of the piston-cylinder interface has been improved to a certain extent due to the enhancement of the bearing capacity of the lubricant film.

References

[1] Qi B, Zhang Y, Guo D Q. Study on seal performance of injector nozzle in high-pressure common rail injection system[J]. 2018 *Journal of the Brazilian Society of Mechanical Sciences and Engineering* , **40**:97-07.

[2] Schlichting H.1979 *Boundary-layer theory*. (New York: Mc Graw-Hill)

[3] Houper L. New results of traction force calculations in elastohydrodynamic contacts [J]. 1985 *Transaction of ASME*, **107**(2):241-48

[4] Uwe wieczorek, Monika Ivantysynova. Computer aided optimization of bearing and sealing and sealing gap in hydrostatic machines the simulation tool CASPAR [J]. 2002 *International Journal of Fluid Power*, **3**(1):7-20

[5] Yang P R.1998 *Numeriacal analysis of fluid lubrication*. (Beijing:National Defend Industry Press)

[6] Meng G. Theoretical and experimental research on the lubrication characteristics in whole cycle of the gap between piston and cylinder in axial piston pumps [D]. 2015 *Beijing Institute of Technoligy*

[7] D.Dawson, C M Taylor. Cavitation in bearings[J]. 1979 *Annual Review of Fluid Mechanics*, **11**:35-65

[8] Krasser, J, Laback, O, Loibennegger B. Anwendung eines elastohydrodrodynamischen verfahrens zur berechnung von kurbeltriebslagern [J]. 1994 *Motortechnische Zeitschrift*, **55**:656-63

[9] Wang Z S, Mu G H, Niu Y L. Reserch on simulation of oil film characteristics of dynamic pressure bearing based on matlab [J]. 2013 *Design & Research*, **40**(2):9-11

[10] Yi F, Masataka S. Mixed lubrication characteristics between the piston and cylinder in hydraulic piston pump-motor [J]. 1995 *Transactions of the ASME*, **117**(1):80-85

Simulation of finding the optimized gravel radius of the arrester bed on truck escape ramp

Qin, Pin-Pin[1], Qin Guang-Yi[2], Mo Ji-Qiang[2], Wu Da[2]
1 Associate professor, Mechanical Engineering of Guangxi University, Nanning, China
2 Post graduate, Mechanical Engineering of Guangxi University, Nanning, China
E-mail: qpinpin@gxu.edu.cn

Abstract. Propose a simulation method of finding the optimized gravel radius of the arrester bed on truck escape ramp in order to obtain the shortest stopping distance of runaway vehicle. A simulation model between gravel and tire is generated and calibrated by PFC2D (Particle Flow Code in two-dimensions) respectively. Utilize the model to simulate the relationship between radius and stopping distance at varied slopes and initial vehicle speeds. Simulation shows that whatever the slopes and the speeds are, the shortest stopping distance corresponds to gravel radius range is between 8.3mm and 15.7mm, that average radius is 12mm. The average radius is basically consistent with the ideal gravel radius 12.7mm, which indicates that the method is effective and could be a substitute for the full-scale test.
Keywords. truck escape ramp; stopping distance; gravel; simulation

1.Introduction

At present, aggregate of the arrester bed on escape ramp is composed of gravel, sand and broken stone etc. The round gravel becomes an ideal aggregate due to its small particle radius and single gradation [1]. According to the results of the full-scale test on truck escape ramp in USA, the most ideal gravel size recommended by American Association of State Highway and Transportation Officials (AASHTO) is 12.7mm, the minimum is 0.63cm, and the maximum is 3.81cm[2]. These full-scale tests provide a reliable basis for the arrester bed design, while also paying huge economic costs [3]. Due to the huge cost and limited by the experimental site, vehicle, personnel, weather and the safety, there is not much domestic full-scale test in China [4]. Because the test conditions are relatively ideal and aggregate will be maintained in time after test, the loose, flat and clean aggregate ensures the good braking effect of the arrester bed. However, as the usage frequency increases, the rolling resistance of the arrester bed become smaller, the braking effect decrease, the longer stopping distance is required, which means the possibility of rushing out of the arrester bed increases. If the aggregate lacks maintenance timely, it is prone to stagnation, which can easily cause wheels uneven subsided, body roll and deviation of vehicle gravity centre, which all may cause roll over[5].

In summary, the existing researches have limits on the braking effect of gravel radius under varied conditions. In practical applications, it is hoped that stopping distance could be calculated and determined by the aggregate state, so as to guide maintenance, maintain good braking effect and avoid accident caused by the shorter stopping distance. Therefore, this paper dedicates to establish a numerical simulation method for the relationship between gravel radius and stopping distance, which could provide a guide for length design of the arrester bed.

IOP Conf. Series: Materials Science and Engineering **542** (2019) 012019 doi:10.1088/1757-899X/542/1/012019

2.Methodology

2.1 Simulation model

The runaway vehicle is assumed to be rigid and brakeless. Its weight is assumed to be equally distributed on each tire. The tires are assumed to be rigid free-rolling wheels. The gravels are assumed to be rigid, spherical, homogeneous, and uniform. The tire and the aggregate are abstracted into a big particle unit and particle combination unit consisting of large amount of aggregates, respectively. The tire-particle unit is generated by PFC2D, which properties were represented by particle radius, particle density, particle normal stiffness, particle tangential stiffness [6-7]. The aggregate-particle unit is generated by PFC2D, which properties were represented by particle radius, particle density, particle normal stiffness, particle tangential stiffness and friction coefficient between particles [6-7].Steps are the following[8]:

Step 1: Establish the wall. Generate a polygon with certain length and wide according to the actual arrester bed scale.

Step 2: Randomly generate a certain radius of particles. Let the particles fill the whole wall, and then compaction particles to achieve stability. Here is the stability condition: the ratio of the average contact force and the average unbalanced force is less than 1%.

Step 3: Initialize settings. It is mainly the setting of gravity acceleration and friction coefficient.

Step 4: Load experiments. Remove the wall and represent the tire with a big particle unit. Impose a downward force on the big particle, which is equivalent to the actual load of the tire.

Step 5: Run the program. Give the tire a running speed until it stops. The running speed is equal to entry speed of runaway trucks. Calculate the distance from the running to the stopping, which is called stopping distance.

2.2 Calibration of model

There are two steps for calibration.

Step1: Referencing to the static tri-axial test results, three groups of numerical simulation were carried out by PFC2D embedded fish language program[9-10]. (a) The relative density of 25%, confining pressure of 69,138kpa; (b) The relative density of 50%, confining pressure of 69,207kpa; (c) The relative density of 75%, confining pressure of 138,207kpa. The results are shown in table 1.

Step2: The parameters of the tire-particle unit and the gravel -particle unit are determined (see table 2 to table 3). Referencing to the data from the reference [4,11], which are results of the full-scale test. The aggregate filled arrester bed on G139 highway section is river gravel with 4.75mm-53mm radius.

2.3Validation of model

2.3.1. Data. The stopping distance is chosen as an indicator of the validation. The validation of model is to compare to full-scale test and see whether simulation accuracy of model is acceptable. Only the verified model can be applied to simulation. The results of predicting and full-scale test are shown in table 4. Test numbers 1 to 4 represent the data from G139 highway section [4].

2.3.2. Analysis. In table 4 the average error between predicting and full-scale test is equal to or below 10.0%, which shows that the predicting accuracy is good. When the vehicle weight is the same, the larger the speed and the longer the stopping distance. The minimum difference between predicting and full-scale test is 0.3m for all the 11 data. Though four predicted stopping distances are longer than the full-scale measured, which means that the application of simulation model tends to be a little risky. The reason is probably the state of aggregates-filled on arrester bed is too ideal for simulation. Usually the ideal aggregates should be spherical, rigid, homogeneous, and uniform.

Table 1. Calibration of model compared to static tri-axial test

Relative Density/%	Confining Pressure/kpa	Peak press/kpa		Peak strain		Volume strain	
		Test	Simulation	Test	Simulation	Test	Simulation
25	69	284	220	0.073	0.03	+0.0032	-0.025
	138	459	410	0.077	0.05	-0.0046	-0.004
50	69	332	310	0.037	0.03	-0.0012	-0.02
	207	558	630	0.066	0.07	+0.002	-0.05
75	138	518	550	0.033	0.06	-0.0037	-0.045
	207	692	760	0.057	0.08	+0	——

Table 2. The parameters of tire-particle unit on G139 site

radius (m)	normal stiffness (N/m)	shear stiffness (N/m)	friction parameters
0.6	2.5e5	2.5e5	0.25

Table 3. The parameters of gravel- particle on G139 site

radius (mm)	density (kg/m^3)	number	Normal stiffness (N/m)	Shear stiffness (N/m)	friction parameters
4.75-53	2630	42000	4.8e6	2.4e7	30

Table 4. Stopping distance compared to full-scale test

Num.	vehicle weight (t)	Entry speed (km/h)	Stopping distance (m)		
			Measured	Predicting	Relative error (%)
1	12.4	40	18.0	16.2	10.0%
2	12.4	50	13.6	12.8	6.0%
3	22.46	45	18.8	17.4	7.0%
4	22.46	48	25.8	27.3	6.0%
5	13.02	27.4	19.4	20.6	6.0%
6	13.02	40.2	32.3	31.9	2.0%
7	13.02	41.8	31.5	33.5	6.0%
8	40	24	9	9.6	6.0%
9	40	28.8	12.2	11.9	2.0%
10	40	28.8	12.4	11.9	4.0%
11	56.5	120	126	135	4.0%

3.Simulation

The length and width of the gravel arrester bed are 80 meters and 0.6 meter respectively. The gravel radius range is divided into four groups with an 7.4mm interval: (a) 8.3mm-15.7mm; (b) 15.8mm-23.2mm; (c) 23.3mm-30.7mm;(d) 30.8mm-38.2mm.

The simulation results of the stopping distance on the gravel arrester bed at various parameters are shown in table 5. For a given grade and entry speed, the stopping distance with radius 8.3mm-15.7mm is less than the other three radius groups. And the influence of the gravel radius on stopping distance shows some certain regularity. For a given grade, the stopping distance increases with the increasing speed and radius. For a given speed, the stopping distance decreases with the increasing grade and decreasing radius. The mechanism is that the smaller the gravel radius, the deeper the tire falling into the aggregate. The contact surface of tire and gravel becomes larger with larger subsidence, so the friction is bigger and the resistance of the vehicle is increased. That's the reason why the stopping distance becomes smaller. So it is found that when the radius scope is between 8.3mm and 15.7mm or average radius is 12mm, the stopping performance is the best. This average radius is very close to the AASHTO recommended ideal 12.7mm. The results of simulation are in good agreement with the experimental results.

Table 5. Results of the stopping distance at varied parameters

Grade	radius（mm）	Entry speed（km/h） 72	90	108
		Stopping distance（m）	Stopping distance（m）	Stopping distance（m）
5%	8.3-15.7	61.43	83.89	94.68
	15.8-23.2	63.27	85.08	95.21
	23.3-30.7	63.42	85.44	95.40
	30.8-38.2	66.18	87.97	96.60
10%	8.3-15.7	45.98	71.43	83.27
	15.8-23.2	47.25	73.17	83.74
	23.3-30.7	47.32	74.16	86.15
	30.8-38.2	47.74	77.16	86.58
15%	8.3-15.7	36.07	48.98	71.80
	15.8-23.2	37.12	50.82	72.94
	23.3-30.7	38.11	52.96	74.74
	30.8-38.2	39.47	53.87	76.82

4.Conclusion

Analyse the relationship between the gravel radius and the stopping distance and the following conclusions are drawn.

a. Aggregate filled on the arrester bed is an important factor affecting the stopping distance of runaway vehicle on the escape ramp. The smaller gravel radius is, the easier tires are to fall into the aggregate. So the greater the amount of subsidence, the greater the resistance, and the shorter the stopping distance required for the same driving vehicle speed.

b. The simulation data shows that the stopping distance is the shortest when the gravel radius is between 8.3mm and 15.7mm. Adjusting the gravel radius is an effective measure to adjust the stopping distance when the terrain is limited.

c. Through numerical simulation of varied gravel radius at different loads, speed and slope, stopping distance can be obtained, which can provide a guide for length design of the arrester bed.

5.References

[1] Zhang Jianjun, Zhang Gaoqiang. *Analysis of incident in emergency escape ramps and legal risk aversion* [J].ICTE,2013: p706-715.

[2] American Association of State Highway and Transportation Officials（AASHTO）. *A Policy on Geometric Design of Highway and Streets （Green Books）* [S]. USA， 2001.

[3] YUAN Zhiying. *The Study on Length of Emergency Escape Ramp in Mountain Road*[D]. Chang'an University, Xi'an,2006.(in Chinese)

[4] Lai Shigui. *The Experimental Study on Stopping Distance of Truck Escape Ramp in Longyan Mountainous Area* [J]. Traffic Engineering, 2016.01.80-88 (in Chinese).

[5] QIAN Yongsheng, ZHOU Bo, CHENG Yonghua, WANG Haiyang. *Mechanism and Experimental Analysis of a Class of Decelerating Downhill Lanes*[J]. Journal of China & Foreign Highway,2007,(04): p5-8. (in Chinese)

[6] *Itasca Consulting Group PFC Theory and Background*[M]. Minneapolis, Minnesota .2014

[7] ZHANG Gaoqiang,SUN Chuanxia,CHENG Xiaohui,etc. *Determining Method of Length of Arrested Bed of Truck Escape Ramp Based on Particle Flow Simulation*[J]. Journal of Highway and Transportation Research and Development, 2011,28(10): p 118-123.

[8] QIN Pin-pin, CHEN Chui-ce, PEI Shi-kang, LI Xin.*A PFC2D model of the interactions between the tire and the aggregate filled arrester bed on escape ramp* [A]. 2017 2nd International Conference on Civil Engineering and Materials Science, ICCEMS 2017[C]. Seoul, Korea,May 26-28 , Pages 243-249, 2017.

[9] Al-Qadi I L and Rivera-Ortiz L A. *Laboratory Testing of River Gravel Used in Arrester Beds* [J]. Journal of Testing and Materials, 1991,19(4): p280-291.

[10] Al-Qadi I L and Rivera-Ortiz L A. *Use of Gravel Properties to Develop Arrester Bed Stopping Model*[J]. Journal of Transportation Engineering, 1991, 117(5): 566-584.

[11] Chen Bo. *Study on the Truck Escape Ramp Design Method of Mountainous Continue, Long and Steep Downgrade* [D]. Southwest Jiao Tong University, 2007 (in Chinese).

Acknowledgments
The authors gratefully acknowledge the support of the Guangxi Natural Science Foundation, Project 2015GXNSFAA139280, in supporting all parts of this study.

IOP Conf. Series: Materials Science and Engineering **542** (2019) 012020 doi:10.1088/1757-899X/542/1/012020

Development and characterization of a MnCu-based high damping alloy plate

ZC JIANG[1], QC TIAN[1,*], ZM REN[1], PG JI[2], JH FENG[2] and FX YIN[2]
1 State Key Laboratory of Advanced Special Steels, Shanghai University, China
2 Research Institute for Equipment Materials, Hebei University of Technology, China
E-mail: tctian@shu.edu.cn

Abstract. From the viewpoint of engineering application, the Mn-Cu based alloy plates are prepared to characterize both the mechanical properties as well as the damping capacity by employing the tensile test and dynamic mechanical analysis (DMA), respectively. The microstructure of MnCu alloy were determined by optical microscopy, while the tensile fracture morphology was observed by using a scanning electron microscopy. It is verified that the mechanical properties are similar to that of A3 steel while the vibration noise across the MnCu alloy plate cut about 5-15dB off during all the experiment frequency range from 20 to 300 Hz when the alloy backing plates are applied under a turbine electric generator. The internal friction characteristic of the alloy is disclosed to be dynamic hysteresis closely related to the vibration frequency, as well as to be static hysteresis that depending on strain amplitudes. With whatever vibration frequencies or strain amplitudes, the damping capacity (tan δ) of the alloy plate still shows satisfactory engineering application value higher than 0.02 at ambient temperature range around room temperature.

1. Introduction

Mn-Cu Alloys have been known for a long time as high damping capacity and high mechanical strength materials. It can absorb vibrational energy significantly to decrease noise and vibration. The damping capacity is contributed by twin boundary damping peak, phase-transformation damping peak, and the antiferromagnetic damping capacity of transformed *fct* phase [1]. M2052 alloy (Mn-20Cu-5Ni-2Fe, at. %) is a representative Mn-Cu based alloy showing superior temperature stability of damping capacity [2,3]. In order to obtain the high damping capacity at room temperature, an aging treatment is necessary in the range of 623K-823K to raise the martensitic phase transformation temperature [4]. The martensitic transformation (*fcc* - *fct*) is preceded by decomposition of original γ-solid solution complying with the spinodal mechanism [5-7]. And the twin damping peak is of dynamic hysteresis associated with the movement of {011} twin boundaries in the fct γ-Mn phase [8].

It is known that the application of the damping alloy is closely related with the temperature, amplitude and frequency, therefore, the damping properties are measured only for specific conditions [9, 10]. The mechanical spectroscopy can be used not only to evaluate the damping properties, but also is a tool for the study of phase transitions. Damping behavior related to different temperatures in Mn-Cu alloys exhibits clearly when low vibration frequencies (0.1-10Hz) are applied in the dynamic mechanical analyzer (DMA) measurement in most researches [11]. There are rare researches concerning higher frequencies. However, in the sense of engineering application, a broad and higher twin-boundary damping peak is expected for the high damping Mn-Cu alloy in higher frequency. In this study, the mechanical properties and the damping characteristics with a series of frequencies were investigated from the viewpoint of practical application.

2. Experimental

The damping alloy (Mn-20Cu-5Ni-2Fe at. %) was prepared by induction-melting the pure metals in argon atmosphere. The ingot was forged and rolled to plate of 20mm thick. After solid solution treatment at 900°C with a staged heating for 2h in vacuum (0.1bar), the plates are oil-quenched, and then are further aged at 435°C in a N_2 atmosphere (0.4Pa) for 6h.

The tensile tests were conducted to test the mechanical properties at the speed of 2 mm / min at 298 K, and the yield stress of the specimens was determined as the flow stress was at 0.2% plastic strain. All specimens for metallographic observation were etched after mechanical polishing by using a mixed solution of alcohol, phosphoric and glycerol with the ratio of 2:1:1. DM6000 model Leica optical microscopy was employed to observe the metallographic microstructure. The fracture morphology of tensile specimens was observed by a scanning electron microscopy (Phenom World Phenom Desktop Scanning Electron Microscope).

A dynamical mechanical analyzer (DMA, model Q800, TA instrument Co. Ltd.) was used to measure the temperature dependent damping behavior in a Dual Cantilever mode. The damping capacity was characterized by tanδ, where δ is the phase lag between stress and strain when the specimen is subjected to cyclic loading. Samples with dimensions of $1\times10\times$ 60mm^3 were spark cut from the plate. After carefully polished, they were heated from -120°C to 350°C at a heating rate of 0.083 K/s. Vibration frequencies of 0.1, 1, 10 and 50Hz, and the strain amplitudes of 5×10^{-5}, 1×10^{-4}, and 2×10^{-4} were applied, respectively. Damping capacity (Tanδ) and storage modulus (E) were measured simultaneously as a function of temperature.

In order to further characterize the damping performance of the plate products, an experiment set was established to measure the vibration noise reduction through the backing plate with a dimension of 800 × 400 × 18 mm, as shown in the Figure 1. The vibration come from a turbine electric generator, and the A3 steel plate and MnCu alloy backing plate with the same dimension were placed between the vibration source and the foundation bed for a better comparison, and the vibration signals of below and above the backing plate were collected by sensors 1, 3 and 2, 4 respectively, and then the damping efficiency of different plates was obtained and shown in the display.

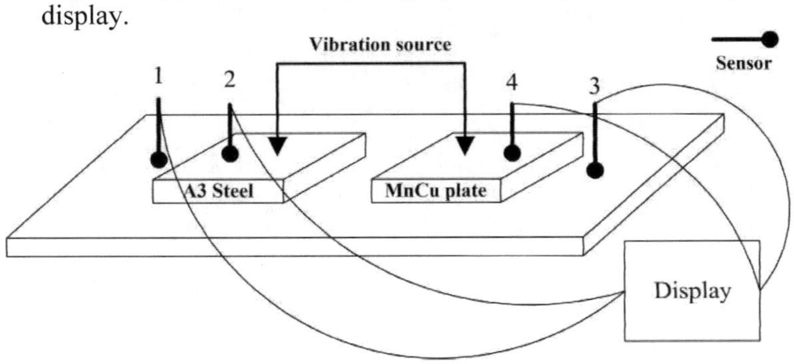

Figure 1. Application of MnCu backing plate under a turbine electric generator, and schematic of the vibration measurement set.

3. Results and discussion

3.1. Mechanical properties

Figure 2 represents the true stress- true strain curve (S - ε) of the MnCu alloy plate, where the yield strength, tensile strength and the elongation can be identified as about 210 MPa, 486 MPa and 37%, respectively, which is similar to that of A3 steel. The hardening index n can be calculated by the curve as about 0.42, which is much higher than that of annealed low carbon steel (about 0.26). It means the Mn-Cu alloy has a uniform strain distribution, and has more excellent formability than annealed low carbon does.

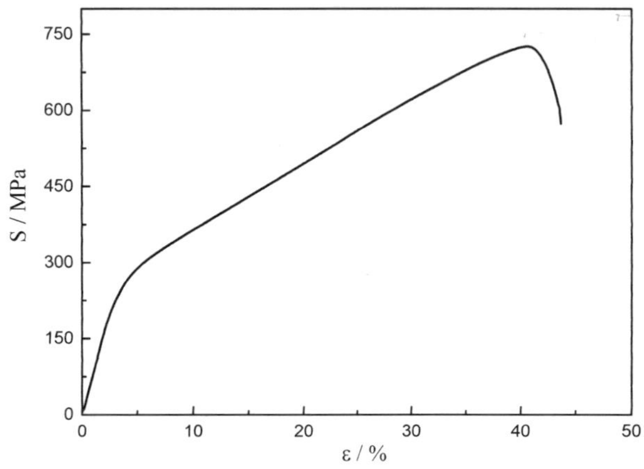

Figure 2. The true stress- true strain curve of MnCu alloy along with rolling direction.

3.2. Fracture morphology

The morphology of the tensile fracture was shown in Figure 3, indicating typical ductile fracture. The circular tensile specimen shows an obvious necking characteristic, and slip bands lying along the circumferential surface can be seen in the necking area, as shown in Fig.3a. Figure 3b shows the morphology of transient break region along the peripheral outside edge where is characterized by the elongated fan-shaped dimples. Figure 3c shows the center of the tensile fracture, it can be seen that there are many different sizes and depths of circular equiaxial dimples.

Figure 3. SEM images of tensile fracture of MnCu alloy.

3.3. Metallographic structure

Figure 4 shows the optical micrographs of the alloy plate at positions of surface, and 1/4 and 1/2 thickness, respectively. It can be seen that the grain size of most grains is about 50μm-100μm, and the average grain size varies from 76μm to 92μm and then to 100μm corresponding to the surface, and 1/4 and 1/2 thickness, respectively. The enlarged microstructure in Fig.4a shows obvious twin microstructure. It is known that there is no twin before phase transformation [12], and the strengthening effect for the nominal friction stress would be only owing to lattice friction. After phase transformation, twin will form and results in matrix strengthening and elevate the yield strength to a certain value. The followed aging treatment can significantly promote twin formation, raise twin density and increase the nominal friction stress remarkably [13].

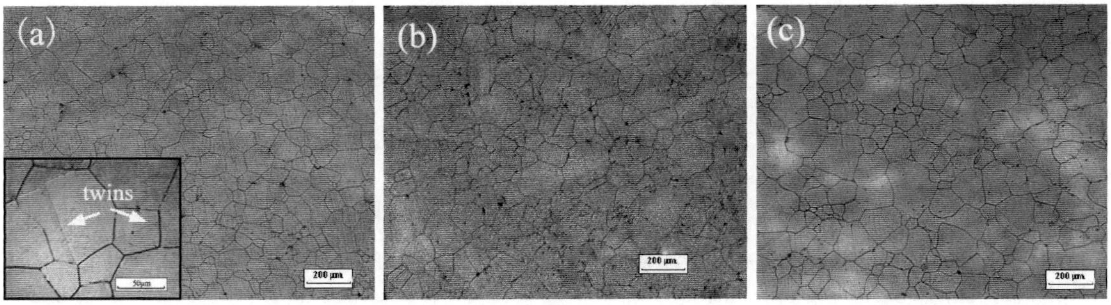

Figure 4. Optical micrographs of MnCu alloy plate at positions of (a) the surface; and (b) 1/4 and (c) 1/2 thickness.

4. Damping capacity of the material

As for Mn-Cu based damping alloys, there exhibited the ''main peak'' and ''sub-peak'' temperature dependent damping behaviour [8]. Figure 5 shows the temperature dependence of storage modulus (E) and the damping capacity (tan δ) of the specimen cut from the MnCu alloy plate under different vibration frequencies at the strain amplitude of 5×10^{-5}. A saddle-shaped two internal friction peaks can be seen for the curve of 0.1Hz. Obviously, the lower temperature peak appeared below 273 K is the twin relaxation peak, which is owing to the formation of martensitic twins and the movement of the {101} twin boundaries in the face-centered tetragonal (fct) phase of Mn-Cu alloys [14], while the higher one is the martensitic transformation peak, which corresponds wells to the soft mode of storage modulus during the phase transformation process from fct phase to face centered cubic (fcc) phase [15]. With the increase of vibration frequency from 0.1Hz to 1Hz and 10Hz, the peak value of *Tanδ* decreases from about 0.045 to about 0.027, meanwhile the higher temperature martensitic transformation peak becomes a sub-peak, whereas the position of soft mode (approximately 85 ° C) remains almost unchanged. However, there are not any "main peak" and "sub-peak" and the damping peak becomes unique with increasing frequency to 50Hz. It might be attributed to the hysteresis of the movement of {101} twin boundaries, which shift to higher temperature under higher vibration frequency, as shown in Figure 5b. As a result, the martensitic transformation peak and twin relaxation peak are coupled to form a wide damping platform and higher damping peaks near room temperature. It should be point out that although the damping capacity decreases significantly with increasing frequency from 0.1 Hz to 10 Hz near room temperature, *Tanδ* is still higher than 0.02 which shows satisfactory engineering application value at ambient temperature range around room temperature. More excitingly, the damping capacity increases near room temperature when the specimen is applied the higher frequency of 50Hz because of coupling mechanism. Therefore, the prepared MnCu alloy plate shows an excellent damping capacity from the viewpoint of practical application.

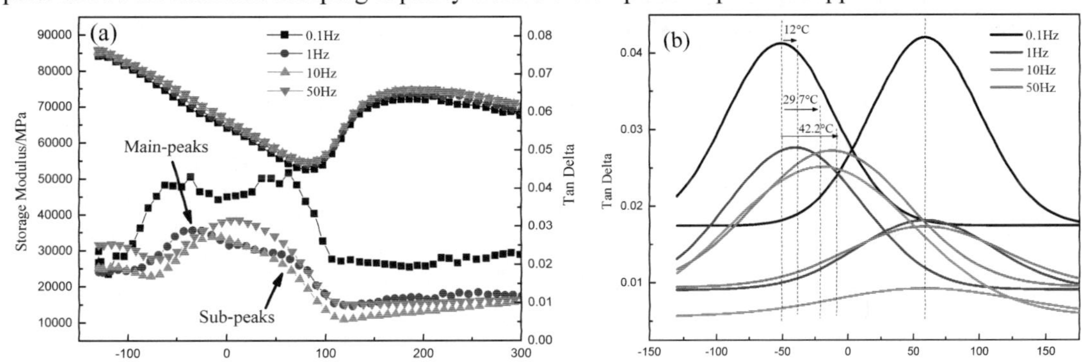

Figure 5. Internal friction and storage modulus (a) and separated peaks of damping peaks (b) for specimen at position of *1/4t* thickness under different frequencies

Figure 6a shows the temperature dependent changes of damping capacity for the specimens at different positions of MnCu alloy plate. It can be seen that the main damping capacity peak decreases gradually from 0.03 to 0.025 corresponding to the specimen from the surface to the center, meanwhile sub-peak also has a small reduction. A sharper minimum peak in Young's modulus for the surface specimen while a broader peak for the 1/4t and 1/2t specimens were observed, indicating a more homogeneous Mn content in the plate surface regions [16]. Figure 6b shows the separated peaks for the specimens at different positions, it can be seen that the peak values of both twin boundary and phase transformation decreases for the specimen from surface to center, the peak position remains unchanged. It was found that the twin density shows grains size dependence, the twin formation can be promoted slightly with large grain size [12, 13]. However, in this experiment the surface position shows a smaller grain size while possesses a better damping capacity than the 1/4t and 1/2t specimens, this phenomenon is properly owing to the relatively uncompleted phase transformation in the alloy plate center .

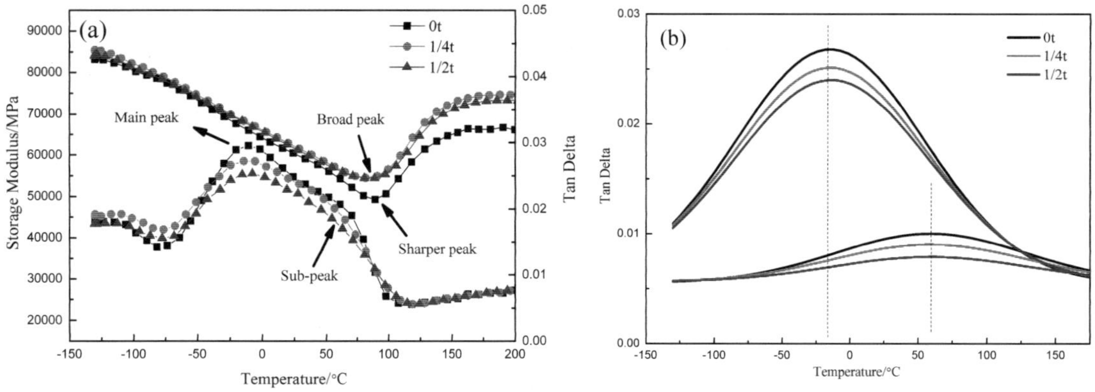

Figure 6. The temperature dependent of damping behavior for the specimens at different thickness positions of MnCu plate under frequency of 10Hz.

Figure 7 shows the temperature dependent damping capacity under different strain amplitudes with the frequency of 10Hz for the MnCu specimen at the plate surface position . It can be seen that the peak damping capacity increases from about 0.037 to 0.047 with the increase of strain amplitude from 5×10^{-5} to 2.0×10^{-4} (Figure 7a), which is determined by the damping mechanism of Mn-Cu based damping alloy. The separated peaks shown in Figure 7b indicate that both twin boundary peak and phase transformation peak increase with increasing stain amplitude. It is known that internal friction of MnCu-based alloys are featured as dynamic hysteresis which strongly depends on vibration frequency, however, this experiment discloses its internal friction is also of static hysteresis, which mechanism might be associated with stress-induced twinning or detwinning of existed twins[17].

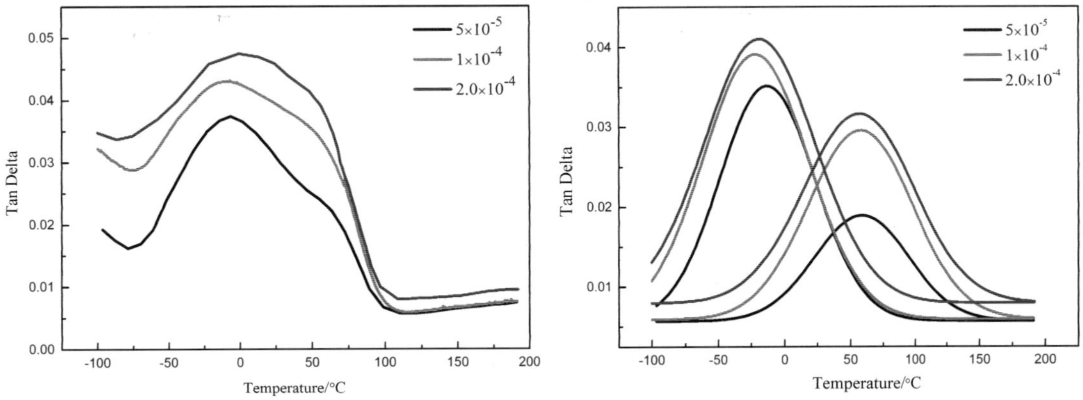

Figure 7. The temperature dependent damping capacity for the MnCu specimen under different amplitudes with the frequency of 10Hz.

Figure 8 represents the vibration measurement results when the backing plates are applied under a turbine electric generator. As shown in the figure, the vibration across the A3 steel plate show small attenuation effect, while that across the MnCu alloy plate shows a significant damping effect of about 5-15dB during all the experiment frequency range from 20 to 300 Hz. Up to now, MnCu based damping alloy has been reported to be used as screws, springs and such small workpiece, it could be expected that large MnCu alloy structural framework would be widely applied in extensive fields including naval ship, aircraft, high-speed rail, automobile and so on.

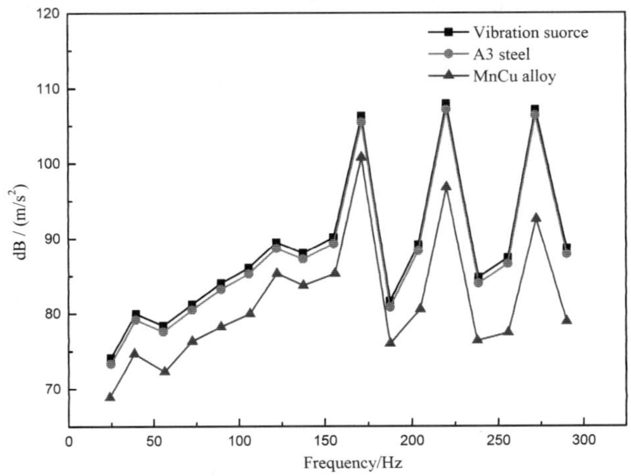

Figure 8. The comparison of vibration and noise reducing capability between A3 steel and MnCu alloy with different frequencies.

5. Conclusions

(1) The MnCu alloy plate were prepared by induction-melting and then forged and rolled to plate of 20mm thick followed by solution treatment and ageing at 435 ° C. The phase transformation temperature corresponding to the position of soft mode occurs approximately at 85 ° C. The yield strength, tensile strength , the elongation and the hardening index *n* can be indentifed as about 210 MPa, 486 MPa, 37%, and 0.42, respectively. The tensile fracture morphology shows a typical characteristic of ductile fracture mechanism.

(2) The internal friction of the alloy is featured as dynamic hysteresis which strongly depends on vibration frequency. With the increase of vibration frequency from 0.1Hz to 1Hz and 10Hz, and then to 50Hz, the internal friction peaks vary from saddle-shaped to the coexistence of a main-peak

and a sub-peak and then to a broaden one, while the peak value of *Tan* δ decreases from about 0.045 to about 0.031

(3) The average grain size varies from 76μm to 92μm and then to 100μm corresponding to the surface, and 1/4 and 1/2 thickness, while the damping capacity decreases gradually from 0.03 to 0.025 corresponding to the specimen from the surface to the center.

(4) This experiment result discloses the internal friction of the alloy is also of static hysteresis. The peak damping capacity increases from about 0.037 to 0.047 with the increase of strain amplitude from 5×10^{-5} to 2.0×10^{-4}.

(5) When the alloy backing plates are applied under a turbine electric generator, the vibration noise across the MnCu alloy shows a significant damping effect of about 5-15dB cut-off during all the experiment frequency range from 20 to 300 Hz.

6. Reference

[1] FX Yin, Nagai K, Watanabe K and Kawahara K 2003 *Mater. Trans.* **44** 1671-4
[2] FX Yin 2003 *Acta Metall. Sin.* **39** 1139-44
[3] Kawahara K Japan Patent No. 2849698
[4] FX Yin, Satoshi I, Takuya S and Nagai K 2006 *Prog. Phys.* **26** 323-30
[5] Vitek J and Warlimont H 2013 *Metal Sci. J.* **10** 7
[6] Markova GV 2004 *Mater. Sci. Eng. A.* **370** 473-76
[7] FX Yin, Ohsawa Y, Sato A and Kawahara K 2000 *Acta Mater.* **48** 1273
[8] Sugimoto K, Mori T and Shiode S 2013 *Metal Sci. J.* **7** 103
[9] FX Yin, Satoshi I, Takuya S and Nagai K. 2006 *Key Eng. Mater.* **319** 67-72
[10] Toru Tanji, Shigenori Moriwaki, Norikatsu Mio, Takayuki Tomaru, Toshikazu Suzuki and Takakazu Shintomi 2003 *J. Alloys Compd.* **355** 207-10
[11] FX. Yin, Takamori S, Ohsawa Y, Sato A and Kawahara K 2001 *INCE* **01** 1906-9
[12] Y Zhong, Takuya S and FX Yin 2008 *Mater. Sci. Eng. A.* **492** 419-27
[13] JZ Yan, N Li, X Fu and Y Zhang 2014 *Mater. Sci. Eng. A.* **618** 205-9
[14] Kato H and Miura S 1995 *Phylosophical Magazine A.* **71** 781-803
[15] FX Yin, Ohsawa Y, Sato A and Kawahara K 1998 *Scripta. Mater.* **38** 1341-46
[16] FX Yin, Satoshi I, Takuya S and Nagai K 2006 *Key Eng. Mater.* **319** 67-72
[17] QC Tian, FX Yin, Sakaguchi T and Nagai K 2006 *Mater. Sci. Eng. A.* **442** 433-438

IOP Conf. Series: Materials Science and Engineering **542** (2019) 012021 doi:10.1088/1757-899X/542/1/012021

Numerical study of the stopping distance for runaway vehicle on truck escape ramp by using the improved equation of car driving

Pinpin QIN[1], Jie WANG[2], Guangyi QIN[2] ,Ming CHENG[2],Qiang MENG[3]

1Associate professor, College of Mechanical Engineering, Guangxi University, Nanning, China
2Post graduate, College of Mechanical Engineering, Guangxi University, Nanning, China
3Post graduate, Automotive college of Jilin University, Jilin, China
E-mail: qpinpin@gxu.edu.cn

Abstract. The stopping distance of the runaway vehicle on the mountain road directly affects the length design of the arrester bed. Since the subsidence depth of runaway vehicle wheel on the arrester bed is neglected in the past researches, the stopping distance estimated by the vehicle driving equation is not accurate. The purpose of this paper is to supplement the influence of subsidence depth on rolling resistance and the influence of mountain highway altitude on air resistance, and improve the estimation accuracy through improved vehicle driving equation. Firstly, the formula for fitting the subsidence depth is established, and the formula for calculating the pushing resistance and compaction resistance is added. Considering the influence of the altitude of the mountain road on the air resistance, the formula for calculating the corrected air resistance and the air lift is added too and the improved vehicle driving equation is generated. Secondly, a simulink model is established based on the improved vehicle driving equation and determine the attester bed entering position and the initial entering speed of the runaway vehicle. Finally, the simulation model is run and analysed. The results show that the error rate of the simulation results is less than 8.0% compared with the full-scale test data. Compared with original equation, the shortening rate of the stopping distance is up to 85.59%.
Keywords. Truck escape ramp; Stopping distance; Simulink

1. Introduction

The stopping distance of the runaway vehicle on the attester bed after entering the truck escape ramp determines the design length of the arrester bed. There are two methods about the design length: the first is direct method which calculates the length of arrester bed directly and the second is indirect method that the length of arrester bed is determined by estimating stopping distance. The direct method is the most classic of the FHWA method proposed by the US Federal Highway Administration [1]. The method assumes that the runaway vehicle that is rushed into the arrester bed performs a uniform deceleration motion, and the wheel that moves forward is only subjected to the rolling resistance and the rolling resistance coefficient is constant, regardless of the influence of the air resistance, and does not consider the influence of the wheel sinking into the attester bed. So the estimated arrester bed length will be too short, and it may be biased to be unsafe if used in engineering[2]. Due to the simple calculation, the FHWA method is still widely used in the length

IOP Conf. Series: Materials Science and Engineering **542** (2019) 012021 doi:10.1088/1757-899X/542/1/012021

design of domestic arrester bed. More scholars have done a lot of research on indirect methods. Based on energy conservation and taking into account air resistance and other factors, STANLEY H T proposed a parking distance calculation method [3], which assumes that the rolling resistance coefficient is constant, and the difference between the actual value and the calculated result is relatively large. Domestic scholar Zhiying Yuan (2006) analysed the driving mechanics characteristics of runaway vehicles on the truck escape ramp and derived the driving equation [4]. However, this equation does not consider the influence of wheel subsidence depth on the stopping distance. Since only the rolling resistance and the slope resistance are considered, the air resistance is neglected, and the rolling resistance coefficient is constant, the estimated stopping distance is short, and the engineering application is unsafe. Yongsheng Qian(2007) based on the vehicle's downhill running resistance, comprehensively considering the air resistance, rolling resistance, slope resistance, inertial resistance and the vehicle's own motion resistance during the operation of the vehicle, calculated the set length of the downhill arrester bed[5]. But the research did not consider the compaction resistance and pushing resistance of the interaction between wheel subsidence and the gravel.

2. Improvement of car driving equation

2.1. *The subsidence depth of wheels on the arrester bed*

When calculating the pushing resistance, the subsidence depth value is required, and the sum of the subsidence depth and the compaction depth is equal to the total subsidence depth [6-7]. Therefore, in order to improve the calculation accuracy of the pushing resistance, in order to obtain a more accurate calculation of the stopping distance, a model of subsidence depth calculation is proposed. The method is to fit the full-scale test data from references [8] by MATLAB, and obtain the fitting formula of the subsidence depth. The formula for subsidence depth is expressed in equation (1):

$$Z = f(\mathrm{G}, \mathrm{v}) = 13.11 + 3.582 \times \sin(0.6571 \times \pi \times \mathrm{G} \times \mathrm{v}) + 0.7563 \times \mathrm{e}^{-(0.5635 \times \mathrm{v})^2} \tag{1}$$

2.2. *Compaction resistance*

After the runaway vehicle enters the brake bed, the interaction between the wheel and the gravel is complicated. From the classic Bekker model, the compaction of the gravel by the tire will produce compaction resistance [9-10], and the concrete calculation formula is shown in equation(2):

$$R_c = P \times w \times k / n + 1 \times \left({3 \times G \times \cos(\mathrm{arctan} i)} \Big/ {2 \times w \times k \times \sqrt{D(3-n)}} \right)^{\left(2n+2 / 2n+1\right)} \tag{2}$$

2.3. *Pushing resistance*

Similar to the compaction resistance, it is known from the classic Bekker model that the tire will have a pushing resistance against the gravel and the concrete calculation formula is expressed in equation(3) [9-10].The modulus and index of deformation in the formula can be measured by static tri-axial test according to references [9-10] and the results are shown here. When the gravel relative density is 13%, 45% and 80%, the deformation modulus is 77.91, 104.44 and 175.24 and the deformation index is1.63, 1.67 and 1.75 respectively.

$$R_b = P \cdot \frac{w}{2} \cdot Z_B{}^2 \cdot \gamma\left(\frac{2N}{\tan\alpha} + 1\right) \cdot \cos^2\left(\tan^{-1}\left(\frac{2}{3}\tan\alpha\right)\right) \cdot 0.45 \cdot \mathrm{g} \tag{3}$$

2.4. *Air lift*

The formula for calculating air lift based on automobile theory[10], and then substituting longitudinal air resistance coefficient, that more accurate and more suitable air lift formula of runaway vehicles in specific mountainous areas could be obtained, such as the following formula.

$$F_z = C_{fz} \cdot b \cdot A \cdot v^2 / 21.15 \tag{4}$$

The windward area of the vehicle can be calculated by calculating formula, but the series of vehicles are too many to measure and calculate completely, so we can use its empirical values, as shown in references [2].In the same way, the air resistance coefficient can also be determined by its empirical values, as shown in references [2].

2.5. Improvement of car driving equation

To add the compaction resistance, the pushing resistance and air lift to the vehicle driving equation, the improved vehicle driving equation is as follows:

$$F_f + F_a + F_i + F_j + R_c + R_b = 0 \tag{5}$$

By inserting the above formula (2) to formula (4) into formula (5), the improved vehicle driving equation can be obtained as is shown in equation (6):

$$\frac{C_{fa} \bullet b \bullet A \bullet v^2}{21.14} + (G \times \cos(\arctan i) - \frac{C_{fz} \bullet b \bullet A \bullet v^2}{21.14}) \times f + G \times \sin(\arctan i) + \frac{(1+\delta_1+\delta_2 \times i_k^2) \times G}{g} \times \frac{dv}{dt}$$
$$+ P \times \frac{w \times k}{n+1} \times \left(\frac{3 \times G \times \cos(\arctan i)}{2 \times w \times k \times \sqrt{D(3-n)}} \right)^{(2n+2/2n+1)} + P \times \frac{w}{2} \times Z_B^2 \times \gamma \left(\frac{2N}{\tan\alpha} + 1 \right) \times \cos^2\left[\tan^{-1}\left(\frac{2}{3}\tan\alpha \right) \right] \times 0.45 \times g + = 0 \tag{6}$$

3. Simulation

3.1. Simulink model of the stopping distance

The acceleration function can be obtained by the formula (6), and then the integral function can be used to obtain the speed function, and then the function is subjected to the definite integral from the initial speed to zero, and get the stopping distance. The acceleration function is shown in equation (7):

$$a = \frac{dv}{dt} = \frac{C_{fa} \bullet b \bullet A \bullet v^2}{21.15} + (G \times \cos(\arctan i) - \frac{C_{fz} \bullet b \bullet A \bullet v^2}{21.15}) \times f + G \times \sin(\arctan i) + P \times \frac{w \times k}{n+1} \times \left(\frac{3 \times G \times \cos(\arctan i)}{2 \times w \times k \times \sqrt{D(3-n)}} \right)^{(2n+2/2n+1)}$$
$$+ P \times \frac{w}{2} \times Z_B^2 \times \gamma \left(\frac{2N}{\tan\alpha} + 1 \right) \times \cos^2(\tan^{-1}(\frac{2}{3}\tan\alpha)) \times 0.45g \times \frac{g}{(1+\delta_1+\delta_2 \times i_k^2) \times G} \tag{7}$$

The stopping distance is shown in equation (8). In order to calculate the stopping distance, a calculation model is established by Matlab/Simulink.

$$s = \int_{v_x}^{0} \left(\int a dt \right) dt \tag{8}$$

3.2. Entrance position of the arrester bed

The formula for calculating thermal failure of runaway vehicle brakes is expressed in equation (9) [5]:

$$T = -400.7 + 110.8 In(10^3 \mathrm{L}) + 132.8 \,\mathrm{In}\, \beta + 42.3 \,\mathrm{In}(\mathrm{v}_i / 3.6) \tag{9}$$

According to the GSRS system developed by the US Federal Highway Administration, for different vehicle types and different slopes, the failure temperature is limited to 260 °C, and the formula (9) is inversely solved by using L as the dependent variable and get the distance from the entrance position to the top of the slope:

$$L = 10^{-3} \times e^{[T+400.7-132.8 Ini - 42.3 In(v_i/3.6)]/110.8} \tag{10}$$

According to incomplete statistics, the position of the runaway vehicle at the top of the slope is the position of the vehicle that is out of control when the brake is in thermal failure, that is to say, the position where the vehicle is out of control and becomes a runaway vehicle and needs to enter the truck escape ramp to perform forced deceleration stopping.

3.3. Initial vehicle velocity

In general, under the same mechanical environment, different initial velocity will get different stopping distance, and with the increase of initial speed, stopping distance will also increase. In this paper, the vehicle velocity prediction model will be used to calculate the speed. The formula of the model is as follows[13]:

$$v_x = 8.82 \times \left(0.01291 \times v_i^2 - 3.28 \times h_x - 3.28 \times f_1 \times L - 0.000033 \times v_m \times L - \frac{0.0499 \times A \times v_n^2 \times g}{G} \right)^{1/2} \quad (11)$$

So far, the thermal failure model of the brake can be used to calculate the L and then replace it into the formula (11). Obviously, the slope can be determined from the exact long slope and the initial vehicle velocity also can be obtained by solving the equation.

3.4. Analysis

The initial conditions of the data from the references [8] were substituted into the simulation model of the stopping distance to calculate the stopping distance, and compared with the results of the original equation from reference [11]. The comparison is shown in table 1. It can be seen from table 1 that under the conditions of the total mass of the two vehicles and the four different initial speeds, the calculated stopping distance of the improved vehicle driving equation is less than 8% compared with the full-scale test data; compared with the original equation, the stopping distance is greatly reduced, and the maximum reduction rate is up to 85.89%.In addition, it is found that the stopping distance calculated by the improved equation is slightly larger than the actual stopping distance, and it is biased safe for practical engineering applications. The stopping distance calculated by the original equation is also much larger than the actual stopping distance. In theory, it is biased safely in practical engineering applications, but it also wastes valuable highway resources. If limited by the topography, there will not be enough space to build a long truck escape ramp.

Table 1. Comparison of the stopping distance (m)

vehicle quality / kg	Initial velocity $/ km \cdot h^{-1}$	Full-scale test data	Improved equation	Error rate(%)	Original equation	Contribution of the improved equation	
						Shortening distance	Shortening rate(%)
24642.46	66.94	36.27	38.52	6.2%	273.08	234.56	85.89%
	73.38	45.42	48.28	6.29%	327.51	279.23	85.26%
10776.37	63.72	39.93	40.82	2.22%	245.03	204.21	83.34%
	75.31	52.73	56.94	7.99%	339.70	282.76	83.24%

4. Conclusion

Using the improved vehicle equation to calculate the stopping distance, the improved equation is shown to be effective by comparing with the results of the full-scale test data and the original car equation. The contribution of the compaction resistance, the pushing resistance and the air lift added into driving equation is not only reflected in shortening the stopping distance but also has high estimation accuracy.

5. Appendix

F_f : Rolling resistance; R_b :Pushing resistance

W :The normal pressure between wheel and contact surface; f :Rolling resistance coefficient; G :The total weight of a runaway vehicle; i :The slope of truck escape ramp

F_z :Air lift force; F_i :Slope resistance

R_c :Compaction resistance;

P :Number of wheels pushing soil on one side

w :Section width of for vehicle wheel

k :Deformation modulus of aggregate

D :Wheel diameter

α :Internal friction angle of gravel(the internal friction angle of the gravel aggregate from 0.8cm to 3.2cm is 35.4°)

T :Temperature of brake during thermal failure

IOP Conf. Series: Materials Science and Engineering **542** (2019) 012021 doi:10.1088/1757-899X/542/1/012021

γ :Weight of gravel per unit volume (and the accumulation density of gravel is $2600\text{kg}\cdot m^{-3}$)

Z_B :Wheel depth pushed by the compaction of gravel(Theoretically, the subsidence depth is equal to compaction depth and subsidence depth is approximately equal to wheel depth when compaction depth is shallow)

N :Local carrying coefficient(and its desirable value is 1); g :Local gravity; Z :Subsidence depth

v :The initial vehicle velocity

F_j :Acceleration resistance

i_k :Transmission ratio; b :Elevation correction coefficient

H :Average elevation; F_a :Air resistance

A :Cross-section windward area of a vehicle

F_z :Air lift force; C_{fz} :Vertical air drag coefficient

β :Slope angle ; i_1 :Highway slope

6. References

v_i :The speed of a runaway vehicle running at the top of a long slope

L :The driving distance of runaway vehicle during thermal failure of the brake

h_x :Altitude difference(and here is the formulation: $h_x = L\times\sin(\arctan(i_1))$)

f_1 :The rolling resistance of mountainous highway

v_m :The average value of v_i and V_x

v_n :The average value of v_i^2 and V_x^2

δ_1 :The influence coefficient of inertia force on the wheels (and its range is 0.03~0.05)

δ_2 :The inertial influence coefficient of a runaway vehicle's engine(and its range is 0.05~0.07)

C_{fa} :Longitudinal air drag coefficient

a :Acceleration; s :Stopping distance

v_x :The initial velocity of the runaway vehicle when it enters the truck escape ramp

[1] American Association of State Highway and Transportation Officials (AASHTO) 2013 GDSH-6-E1 A policy on geometric design of highway and streets (Washington)
[2] ZHANG Gaoqiang et al 2011 Determining method of length of arrested bed of truck escape ramp based on particle flow simulation *(Highway transportation technology)* 118-123
[3] STANLEY H T 1978 A Calculator program to estimate truck coasting speeds for designing gravel arrester beds (Boise,Idaho:Idaho Transportation Department)
[4] YUAN Zhiying 2006 Study on the length of emergency escape ramp of mountain expressway (Xi'an: Chang'an University)
[5] QIAN Yongsheng, ZHOU Bo, CHENG Yonghua 2007 Mechanism and experimental analysis of a class of decelerating downhill driveway (*Chinese and foreign highway*) (04) 5-8
[6] Al-Qadi I L and Rivera-Ortiz L A 1991 Laboratory testing of river gravel used in arrester beds (*Journal of Testing and Materials*) 19(4) 280-291
[7] Al-Qadi I L and Rivera-Ortiz L A 1991 Use of gravel properties to develop arrester bed stopping model (*Journal of Transportation Engineering*) 117(5) 566-584.
[8] Wambold J C, Rivera-Ortiz L A and Wang M C 1998 A field and laboratory study to establish truck escape ramp design methodology (*Final report,Deceleration*) Vol I ;Vol II.
[9] Bekker M G 1969 *Introduction to Terrain-Vehicle Systems* (Michigan: The University of Michigan Press)
[10] ZHUANG Jide 2001 *Computational Automotive ground Mechanics* (Beijing:Mechanical Industry Press) pp 14-101
[11] YU Zhisheng *Automotive Theory* (Beijing: Mechanical Industry Press)
[12] LIANG Ruijuan 2012 The rationality evaluation on the design of emergency escape ramp of mountain expressway (Xi'an: Chang'an University)
[13] QIN Pin-pin, CHEN Chui-ce, PEI Shi-kang, LI Xin .2017 A PFC2D model of the interactions between the tire and the aggregate filled arrester bed on escape ramp (*2017 2nd, Int., Conf., on Civil Engineering and Materials Science, ICCEMS,May 26-28,*Seoul, Korea) pp 243-249

IOP Conf. Series: Materials Science and Engineering **542** (2019) 012022 doi:10.1088/1757-899X/542/1/012022

Advantages of using virtual reality and building information modelling when assessing suitability of various heat sources, including renewable energy sources

B Chudikova[1] and M Faltejsek[2]

1 VSB – Technical University of Ostrava, Faculty of Civil Engineering, Czech Republic
2 VSB – Technical University of Ostrava, Faculty of Civil Engineering, Czech Republic
E-mail: blanka.chudikova@vsb.cz

Abstract. The pace of changes in the construction industry is constantly accelerating. It includes 3D modeling, utilization of drones, 3D printing and virtual or augmented reality. Augmented reality combines something real, such as existing buildings, with expanded 2D or 3D digital information about individual technologies or planned changes, and the actual projection of individual elements into a given (even hidden) space. It can be a key for improving project engineering processes, subsequent maintenance or more efficient renovation procedures. Virtual and augmented realities need 3D building models, i.e. incorporation of building information modelling. Building information modelling (BIM) is a modern and efficient solution process that incorporates the entire building life cycle, based on a realistic model and information. Individual building technical equipment, i.e. including heat sources, to a certain extent influence the internal structure of individual buildings, particularly in the case of a more complex structures. The article describes some of the modern technologies that can be applied to the administration, maintenance and implementation of building technical equipment. Such technologies can also help when deciding about the most suitable heat source for a given building. The article points out the benefits of incorporating these modern technologies.

1. Introduction

The everyday progress in the field of information technologies allow project engineers to utilize modern technologies that make human activities easier or of a higher quality. They can also prevent problems thanks to timely analyses. Modern technologies play a decisive role in the Industry 4.0 revolution. The use of modern technologies is accompanied by many benefits, such as lower cost, lower risks, faster processes, better quality of individual proposals and improved efficiency [1, 2].
In many fields, virtual and augmented realities often represent a welcome aid. Industrial revolution, objective of which is to fully digitalize the construction industry and to expand application of building information modelling in this sector, offers many new opportunities when it comes to applying and utilizing these realities in the work of project engineers. Augmented reality has a great potential to be one of the leading factors of the revolution of the existing user interface. It is because it can offer a unique quality, which provides a direct link between reality and virtual information about this reality. Our entire world can this become a user interface [3].

2. Virtual reality and its degrees

Virtual reality, used and discussed so much these days, allows its users to submerge into artificial worlds and to create visual, auditory and tactile experience, which arouse subjective impressions, using only a computer or a special helmet or googles. One can also focus on the given user interface that provides a combination of virtual reality and reality, thus providing information about small and, at the same time, easily understandable parts. Virtual reality has come to a point where virtual images are often indistinguishable from the real world. It means that basically anything is possible without almost any limitation [3].

Users submerged into virtual reality perceive only virtual impulses, for example, inside of a room with walls that allow for stereoscopic projections or while wearing enclosed virtual reality goggles. The space between reality and virtual reality, which allows us to combine both of them to various degrees, is called mixed reality. One of the degrees of mixed reality is extended reality, where the number of real elements is greater than the number of virtual ones.

The term extended reality is used for creating direct, automatic and operational relations between the physical, real world and electronical, added information that includes digital objects created by a computer. We could say that it is a somehow improved reality, when users can see the real world, however, with added, computer-generated images, objects or information. This view can be generated using special goggles or using a tablet or a smartphone. Added information, which will be displayed as a part of the extended reality, is always supposed to help the given users to conduct their activities, to better imagine certain space with new or hidden elements, or to display relevant information about real elements in the real world [4].

2.1. Usage possibilities of virtual reality in the construction industry

All technological equipment as well as construction details have been getting ever more complicated, which has a great impact on their planning, implementation and operation. Architectural buildings, infrastructure as well as technological support facilities of the buildings are designed using numerous software applications. Various project changes have to be usually incorporated during the implementation process. It does not matter if they are just small changes or relatively extensive interventions. It is no problem to incorporate them back into the given model and to create an easy and transparent as-built documentation of the given construction. However the current or past practices have not been working this way so far and the changes are usually not incorporated back to the drawing documentation in a good way. That causes problems later, for example, during renovations or defect situations, in relation to individual structures, technologies or, more generally, facility management practices.

Virtual reality can be also used in architecture, when engineers can go through proposed constructions and assess if the given buildings are typologically in good order and if individual building areas comply with the given functional as well as operational requirements. They can also assesses aesthetics of corresponding interiors.

Thanks to virtual reality, we can control static and structural proposals, particularly locations and connections of individual elements, since it is easier to locate components that are incorrectly placed when we are fully submerged in the 3D environment. These elements cannot be, so far, corrected directly in the virtual reality environment. The corresponding changes have to be adopted subsequently, using the software application, in which the model was created. Nevertheless, considering the speed of the development in this area, we can expect that even this problem will be solved in a near future and individual element will not be only static, which will allow us to work with them directly in the virtual reality environment.

3. Building information modelling

Building information modelling (BIM) provides a tool for preparing complex construction proposals. Visualization via BIM allows us to determine structural discrepancies and to locate problematic spots already during the pre-investment stage. Problem prediction and problem solving already during the planning stage can save us a lot of money during the following implementation stage and, subsequently, during the given building operation. All changes during the construction implementation

process can be easily incorporated into the model, thus allowing us to simultaneously create the corresponding as-built documentation. After the completion of the construction implementation stage, the model should assist the given building operator during the entire building lifespan period [5, 6].
An integral part of every building is formed by the building technical equipment, which can be created using a realistic 3D model. Collisions with other structures or building installations are inspected in a realistic 3D model. Moreover, the model is amended by information about all technologies, pipelines and valves. The data included in the model show the planned state of the given building and thus make the work of people who subsequently take care of the building easier. The information is complex with regard to all applicable technical systems in the given building, including detailed information about heat sources. It can also involve heating systems, water distribution lines or air-conditioning and ventilation systems. Overviews from all technological systems are available. They include lists of individual elements, material prices or other specifications defined individually in advance. Based on such detailed information, we are able to determine real expenses for, for example, a heating system implementation, and to assess multiple options in advance and subsequently select the most convenient one. If we link a BIM model with a CAFM (computer aided facility management) system, it can be used including attached documents and information related to revisions, inspections and, if applicable, repairs. We could also easily find contacts for individual technicians. Moreover, the model carries information about the contractor who is responsible for correct operation of the given system and about warranties and lifespans of individual elements.

4. Heat sources

Regardless of the fact if it is a new building or renovation of an older one, there always are several heat source options to choose from. Heat sources will differ for local and central heating systems. This article addresses only central heating heat sources. There are several options when it comes to central heating as well. Some of them include:

- Radiators;
- Low-temperature heating (floor, wall, ceiling);
- Hot-air heating.

We choose heat sources based on the given system type. From that perspective, we have the following options available to us:

Electric boilers

Electric energy can easily be regulated and small sources can be used, however, it is also the most demanding energy when it comes to its production. That is why it is the most expenses energy source. Since electric energy in the Czech Republic is mostly made in coal power plants, efficiency of which is relatively small and CO_2 emissions of which are high, there is a great pressure to limit the use of electric energy as the main source for heating buildings. Electric energy is usually proposed only as an auxiliary source to, for example, renewable energy sources.

Gas boilers

From the perspective of comfort and price, gas comes second after electric energy. However, the gas distribution network is not as wide-spread as the electric one. Natural gas heaters are structurally more complex and one must address the issue of where to discharge combustion products and how to ensure operation safety when the flame goes off or when the gas combustion is imperfect. Installation of autonomous sensors, which can detect even very low carbon monoxide concentrations, represents an easy prevention measure against CO leaks.

Solid fuel boilers

Contemporary automatic coal or wood boilers represent yet another heat source that is proposed relatively often these days. And since their service comfort has improved and emissions have been reduced in comparison with the past, solid fuel boilers can sometimes also be a suitable alternative for heating buildings.

Heat pumps

Heat pumps need electric energy for their operation – to drive their compressors. It means that they cannot be used at locations where there is no electric outlet. Heat pumps utilize heat from their surroundings, thus boosting a relatively low operation cost. There are few types of heat pumps, for example: Air-To-Air, Water-To-Air, Water-To-Water. Each type has different efficiency, but it also varies considerably by the purchase price. Their high acquisition prices represent a big disadvantage though.

Solar systems

Solar systems can be designed for water heating purposes and, partially, heating or for the production of electricity - photovoltaic. Their advantage is their wide-spread availability. On the other hand, their big disadvantage is represented by their seasonal and even daily fluctuations. Energy can be, to a certain extent, stored in accumulators. Nevertheless, the relatively high acquisition prices and not a great efficiency of these systems mean that investments in them are economically irretrievable without grants and subsidy programs.

Wind power plants

Small wind power plants can be installed even on flat building roofs. They represent clean energy sources, thus contributing to a reduction of CO_2 emissions. There are several options when it comes to their configuration. One of them is the island system, when surpluses of the produced energy are stored in accumulators for later use.

Considering the current and future legislature, when newly built houses will have to boost almost zero energy consumption, it is clear that renewable energy sources will be widely used in the future. The challenge is to find the best alternative for a given building, while taking into account all its characteristics. This process can be relatively complicated since many factors have to be considered.

5. Choosing suitable heat sources

Several factors have to be considered when selecting a heat source. No qualified decision can be adopted based only on the current energy prices since these prices always change. The total heating cost can be relatively realistically determined, provided good planning practices are put in place. Prior to adopting a decision about a heat source, it is recommended to assess the following items.

Construction preparations

- Financial aspect – gas or electricity connection cost, cost of building a fuel storage space in the case of solid fuel boilers, wellbore prices in the case of heat pumps, etc.
- Time aspect – how long it will take to build the given connections and, if applicable, time needed for the necessary earthworks
- Spatial limitations – necessity of a technical room – it is unavoidable for some heat sources, layout of individual radiators – it is not always possible to install radiators in some rooms; when floor heating is used, a heat source that is compatible with this system has to be selected.

System acquisition price

- The acquisition price should include the price of the actual heat source as well as the price of the entire system. The investment cost should include expenses related to the given heat distribution network. It should take into account the material of the lines, insulation, valves, accumulators (if needed), regulation unit, etc.

Regulation system

- When the selected system can be well regulated, you can potentially save significant money associated with the cost of heating. Regulation can be manual in individual rooms, or automatic, which is a more expensive option; these expenses have to be also included.

System operation

- Operation personnel – will you need any operation personnel to operate the system? Annual or more frequent maintenance, regular cleaning, service, revisions, or will the system be operator and maintenance free? You need to answer also these questions. Should day operation personnel and/or often revisions be needed, you need to expect higher operation cost.
- Operation comfort – will the selected heat source and the entire system be able to achieve the required temperature and to maintain it? There is a risk of room overheating, particularly when a fireplace is additionally used. This has a negative impact on the operation cost [9].

IOP Conf. Series: Materials Science and Engineering **542** (2019) 012022 doi:10.1088/1757-899X/542/1/012022

- Operation hygiene - will the selected heat source and the entire system have an impact on the cleanliness and health conditions in individual rooms? For example, more dust or the possibility to eliminate cold floors.
- Operation expenses – actual cost of the used fuel or energy tariffs.

Lifespan

- The lifespan of the entire system, regulations, the actual source, valves, etc. should be assessed.

Using the above mentioned information, we can arrive to general conclusions about a suitable heat source. Nevertheless, some of the items cannot be properly answered when only a 2D drawing is available. For a 3D model prepared using the building information modelling, all lengths and dimensions of individual pipes, number of branches, elbows, distribution line materials, number and types of valves, etc. would be available. An exact appraisal of individual options can be easily prepared based on such very detailed information.
For proposals of renewable energy sources, you also need to consider the particular situation of a given building. It cannot be unambiguously said that a solar system for preparation hot water is suitable for every building since some building roofs can be shaded most of the day due to the surrounding buildings. In that case, the system would lose precious solar yields and its efficiency would decrease. This situation can be prevented by preparing a 3D model of the given building, including the surrounding buildings, and a software simulation - ArchiCad is used here, which clearly shows how shades will move throughout the entire day. Based in this simulation, we can assess if it is suitable to even propose a solar system. If it is, we can see where individual solar collectors should be ideally installed in order to gain the biggest yield possible [7].

Figure 1: Simulation of shadows during the day in Archicad software, evaluation of suitable location of solar collectors for prepare hot water

When it comes to a small wind power plant proposal, the situation is similar. We cannot objectively assess the suitability or unsuitability of a wind power plant without knowing the layout of the surrounding buildings and appropriately accurate assessment of the wind flows. When simulation methods are utilized (in this case we use a 3D model created using BIM, which was exported to Ansys), we can professionally determine a suitable location for the given turbine installation based on the simulation results.

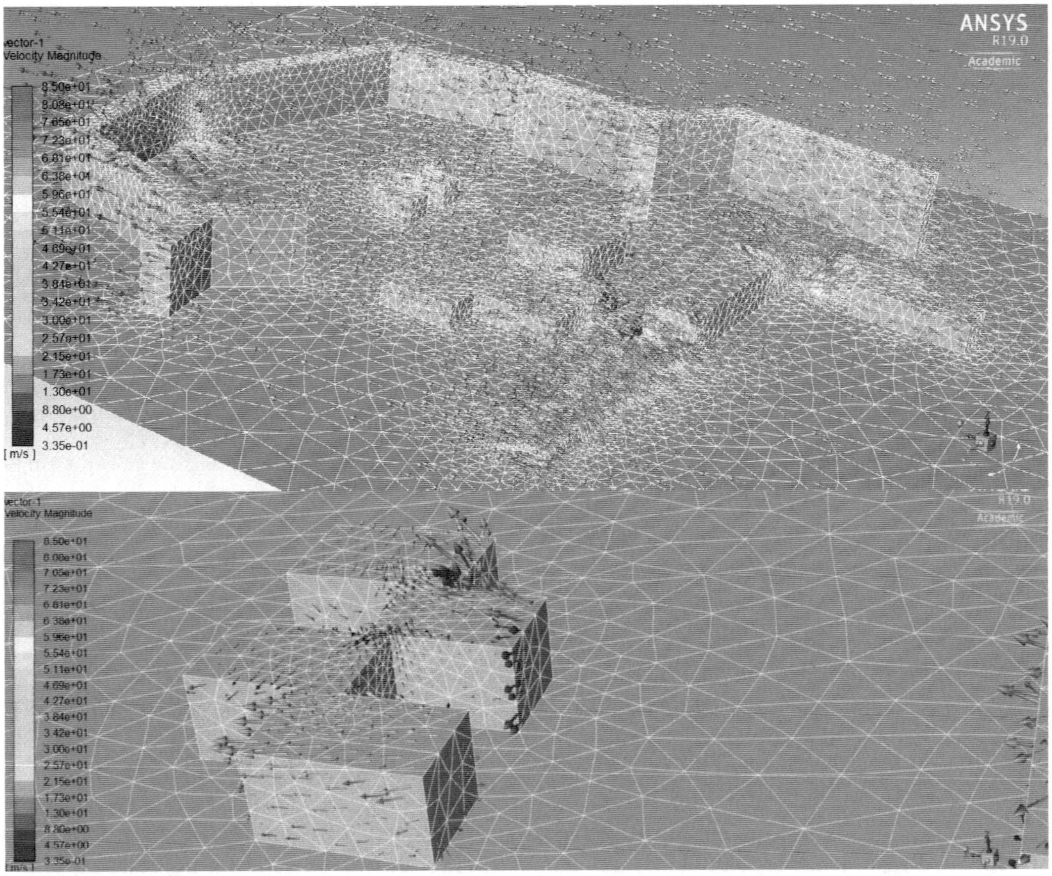

Figure 2: Wind simulation in Ansys program - wind flow overall situation including surrounding buildings and wind movement on the roof of a particular building

Incorporation of virtual reality into project engineering and implementation of technological systems can improve orientation in the project intention, which can be quite difficult in more complex technological systems. In virtual reality, we can walk about an extensive and complex machine room, we can inspect accuracy of all connections and we can also carefully walk through all parts of the given system, while having all relevant information about the given elements available to us. Moreover, we can also assess suitability of individual heating systems for the given space.

IOP Conf. Series: Materials Science and Engineering **542** (2019) 012022 doi:10.1088/1757-899X/542/1/012022

Figure 3: Technological Background of an Experimental House - created 3D model and then showing how it might look like a virtual reality, including element information

Extended reality has also its place. It can be used as an analysis of discrepancies in, for example, large technological facilities. When inspecting an implemented machine room, you can, for example, control the planned project plan in relation to the already implemented parts on your tablet. Should there be any differences or missing components, you can thus prevent huge problems and financial damages, which could occur upon launching the system.

Heat sources have to be also designed pursuant to the particular energy demands of the given building, which is a very important and monitored parameter for new as well as existing buildings, which go through a renovation process. Considering the requirements for the lowest possible energy needs of buildings, extended reality has a potential to develop further, with the expectation that emphases will be put on incorporating detailed information directly into the 3D model of the given building [8]. The corresponding model could include information about the heat transfer coefficient of individual structures – the existing ones and those required by the given standard, or the values of individual linear and point heat transfer coefficients. Using a tablet, experts would then be able to immediately see the state of the given building and to modify their heat source selection or to propose structural modification and, subsequently, a heat source.

Figure 4: View detailed information about the existing building

6. Conclusion

The heat source selection process should particularly take into account whatever is available at the given location. Based on the information stated in the article, you can assess the theoretical benefits and negatives of individual heat sources as well as entire systems. Theoretical considerations should be followed by practical simulations and analyses. This particularly applies to proposals of renewable energy sources, when the investment cost for acquiring such systems is relatively high. The investment return period depends on the achieved profits. Incorporation of modern technologies into the decision making process related to suitable heat sources and thus into the entire building proposal, planning and implementation process, can bring many benefits and opportunities for making these procedures more time-efficient, more exact and, most of all, more profitable. It is necessary to admit that only 2D drawings will not soon be sufficient and that the ongoing technological development allows for making project engineering practices more efficient. Project engineers and technician should thus learn how to utilize this potential for their own benefit as well as for the benefit of investors. Should it be the application of the building information modelling, or virtual or extended reality, all of these modern technologies can accelerate and simplify construction processes and all of their parts.

7. References

[1] VERBA, N., CHAO, K.-M., LEWANDOWSKI, J., SHAH, N., JAMES, A., TIAN, F. Modeling industry 4.0 based fog computing environments for application analysis and deployment. *In: Future Generation Computer Systems, vol. 91, pp 48-60, 2019.DOI: 10.1016/j.future.2018.08.043.*

[2] LU, Y. Industry 4.0: A survey on technologies, applications and open research issues. *In: Journal of Industrial Information Integration, vol. 6, pp 1-10, 2017. DOI: 10.1016/j.jii.2017.04.005.*

[3] AUKSTAKALNIS, Steve. Practical Augmented Reality: A Guide to the Technologies, Applications, and Human Factors for AR and VR (Usability). Indiana: Addison-Wesley, 2016. ISBN 978-0134094236.

[4] BENFORD, S., GREENHALGH, C., REYNARD, G., BROWN, C., KOLEVA, B. Understanding and Constructing Shared Spaces with Mixed-Reality Boundaries. *In: ACM Transactions on Computer-Human Interaction, vol. 5, pp 185-223, 1998. DOI: 10.1145/292834.292836.*

[5] BONENBERG, W., WEI, X., ZHOU, M. BIM in prefabrication and modular building. *In: Advances in Intelligent Systems and Computing, vol. 788, pp 100-110, 2019. DOI: 10.1007/978-3-319-94199-8_10.*

[6] EASTMAN, Ch., TEICHOLZ, P., SACKS, R., LISTON, K. BIM Handbook, A Guide to Building Information Modeling for Owners, Managers, Designers, Engineers, and Contractors. New Jersey: John Wiley & Sons, Inc, 2008. ISBN: 978-0-470-18528-5.

[7] SALKUTI, S. R. Day-ahead thermal and renewable power generation scheduling considering uncertainty. *In: Renewable Energy, vol. 131, pp 956-965, 2019. DOI: 10.1016/j.renene.2018.07.106.*

[8] ZUAZUA-ROS, A., MARTÍN-GÓMEZ, C., IBANEZ-PUY, E., VIDAURRE-ARBIZU, M., GELBSTEIN, Y. Investigation of the thermoelectric potential for heating, cooling and ventilation in buildings: Characterization options and applications. *In: Renewable Energy, vol. 131, pp 229-239, 2019. DOI: 10.1016/j.renene.2018.07.027.*

[9] VANUS, J., MARTINEK, R., BILIK, P., ZIDEK, J., SKOTNICOVA, I. Evaluation of thermal comfort of the internal environment in smart home using objective and subjective factors. *In: 17th International Scientific Conference on Electric Power Engineering, EPE 2016; Prague; Czech Republic, article number 7521768, 2016. DOI: 10.1109/EPE.2016.7521768.*

Acknowledgments

This article was created with the support of the Student Grant Competition of VŠB-Technical University Ostrava, registration number of the project SGS SP2018/66.

Simulation of heavy-duty engine based on flexible multi-body dynamics

Wen Li[1, 2, 3], Lei Chen[1, 2], Jintao Liu[1, 2], Zhenxing Liu and Baotang Zhuang[1, 2]

[1] Beijing Institute of Control Engineering, Beijing, 100190, China
[2] Beijing Engineering Research Center of Efficient and Green Aerospace Propulsion Technology, Beijing 100094, China

[3] E-mail: gradylee@126.com

Abstract. For heavy-duty diesel engine, the structural flexibility of the crankshaft and the engine block has an important influence on the dynamic characteristics of the engine. The deformation and modal resonance generated at work must be considered. Therefore, the multi-body dynamic analysis of the crankshaft and the body structure is based on the flexible body model of the crankshaft and the body. In this paper diesel engine dynamics behaviour is analysed based on flexible multi-body method. The dynamic substructure condensation method is used to reduce the he shaft and the bearing block, obtain the flexible body model of the crankshaft and the body, reduce the computational solution scale, and improve the computational efficiency.

1. Introduction

With the continuous strengthening of engine, its speed, load and power density are gradually improved. The structural flexibility of the crankshaft and the engine block has an important influence on the dynamic characteristics of the engine [1]. The deformation and modal resonance generated at work must be considered in the analysis. Therefore, the multi-body dynamic analysis of the crankshaft and the body structure is based on the flexible body model of the crankshaft and the body. The research object of this paper is a six-cylinder v-type diesel engine. The flexibility of the crankshaft, the body and the connecting pairs is the main non-linear factor that needs to be considered [2].

Firstly, the finite element model of the crankshaft and body was established, and the principal degree of freedom node of the above model was defined, then the dynamic substructure condensation model of the crankshaft was obtained by calculation. After importing the quality matrix, stiffness matrix and geometry of the condensation, to the software of AVL EXCITE [3], the system parameters are defined in the POWER UNIT module, including the body element and the connection UNIT, the mechanical coupling relation, the engine structure parameters. According to the analysis needs, select the appropriate kinematics model, which is to select the different connection subunit models in the software. Considering the lubrication of oil film, the bearing adopts elastic hydrodynamic lubrication model (EHD2) to define the oil supply characteristics, oil tank and lubricating oil type of the bearing. The burst pressure curve of each cylinder is imported into the software, which is applied to the engine dynamics model as the load. The simulation control parameters are set in the software, including simulation iteration step, iteration precision, result storage information and integral method. After completing the calculation, the desired results are extracted and the results are processed [4].

2. Dynamic substructure condensation

IOP Conf. Series: Materials Science and Engineering **542** (2019) 012023 doi:10.1088/1757-899X/542/1/012023

The mesh model of the body was established by HYPERMESH finite element pre-processing software, the condensation process is accomplished by ABAQUS finite element software.

2.1. Substructure condensation model of crankshaft

The substructure condensation model of crankshaft is condensed from the finite element model (FEM) of the crankshaft assembly, which including crankshaft, flywheel and balance weight, as shown in Figure 1. The FEM of crankshaft assembly is composed of hexahedral mesh, in order to facilitate the definition of the connecting between main bearing and connecting rod large end bearing.

6 layers of hexahedral elements are arranged along the length direction of the main journal, in which the size of the middle 4 layer element equal to 1/4 size of main bearing width. 10 layers hexahedral elements are arranged along the length direction of the crank pin, in which the thickness of the middle 8 layer grid is the 1/8 size of the width of the bearing of connecting rod. For both main journal and crank pin, 48 grid elements are arranged in the circumferential direction. At the same time, it is necessary to ensure that there are nodes in the center of each layer of the main journal and crank pin length direction [5], so as to facilitate the binding of RBE2. The model has 25,914 units and 32,768 nodes.

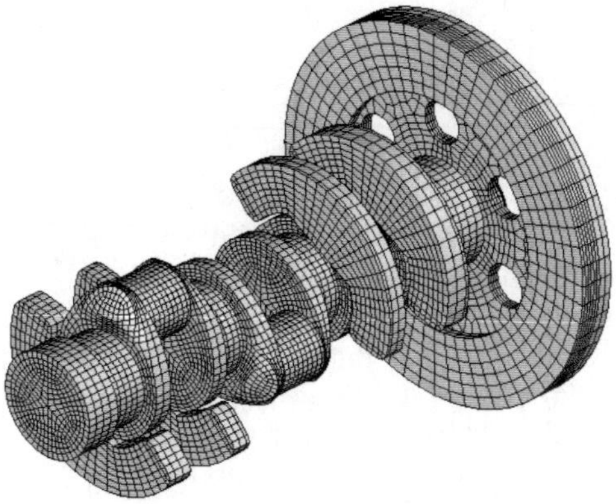

Figure 1. FEM of crankshaft

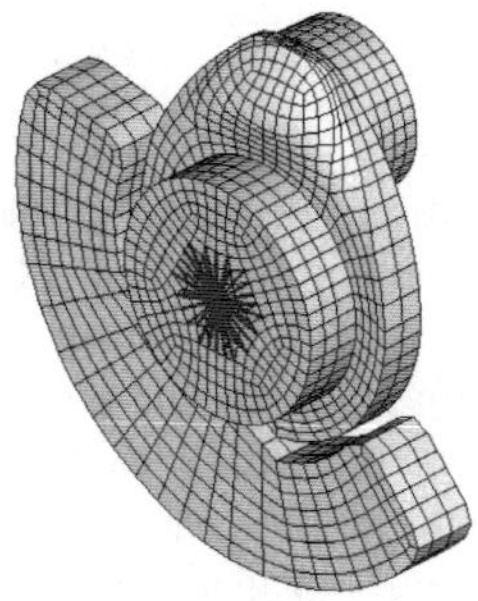

Figure 2. The RBE2 connection at the centre of the main journal

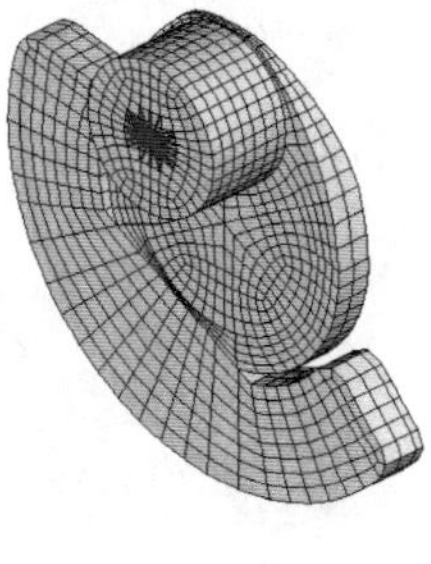

Figure 3. The RBE2 connection at the centre of connecting rod shaft

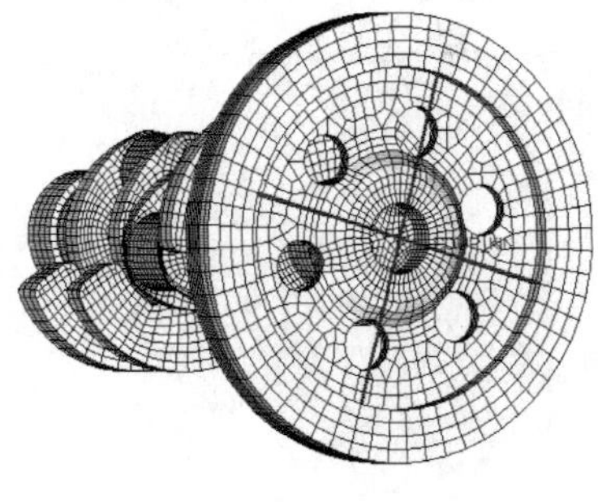

Figure 4. The RBE2 connection at the centre of the flywheel

5 nodes at the center of main journal section are defined as master nodes, while 2 nodes at the center of crank pin section are defined as master nodes. The six degrees of freedom (DOF) between the master node at the center of the section and the slave nodes within the radius of 1/2 of the surrounding section are coupled by the DOF coupling element RBE2, as shown in Figure 2 and Figure 3, in which the master nodes are the main DOF nodes and the slave nodes are the subordinate nodes. Master nodes are also defined on the flywheel, inertia body and shock absorber body, which are shown in figure 4. The master node on the flywheel is used to load torque, and the master nodes on the inertial body and the shock absorber body are used to characterize the dynamic behavior of the torsional damper.

IOP Conf. Series: Materials Science and Engineering **542** (2019) 012023 doi:10.1088/1757-899X/542/1/012023

It is necessary to ensure that the addition of RBE2 connections does not affect the stiffness and mode of the original model, as they directly control the torsional vibration and fillet stress of the shafting. In order to ensure the accuracy of the calculation results, the model compliance must be verified. The crankshaft finite element model and the original model must have the same dynamic characteristics. Table 1 introduces the modal frequencies of the crankshaft model before and after the RBE2 coupling connection. The result shows that the addition of RBE2 has little effect for the crankshaft modal, therefore the modified model has the same dynamic characteristics as the original model.

Table 1. The modal analysis result of the crankshaft before and after set contrast RBE2

Modal order	Modal frequency（HZ）（nonRBE2）	Modal frequency（HZ）（RBE2）	Error（%）
7	95.418	97.91	2.61
8	100.76	103.8	3.01
9	218.16	223.62	2.50
10	244.71	246.41	0.69
11	270.82	277.25	2.37
12	386.66	394.31	1.97
13	436.26	448.42	2.78
14	549.29	555.83	1.19

2.2. Engine Block Dynamic Substructure Condensation

The engine block include the body, the oil pan, the main bearing cover and other parts, and the cylinder block and the main bearing are also installed on the body, as shown in figure 5. The cylinder sleeve, bearing and bearing backing by 8 node hexahedron unit, and the rest of the 10 node tetrahedral two unit. The width direction of the bearing shell divided into 4 equal parts, and the main journal consistent. The total number of nodes in the finite element model is 341391, and the number of elements is 188492. 8 nodes hexahedron element is 20352, and 10 node tetrahedron element is 168140.

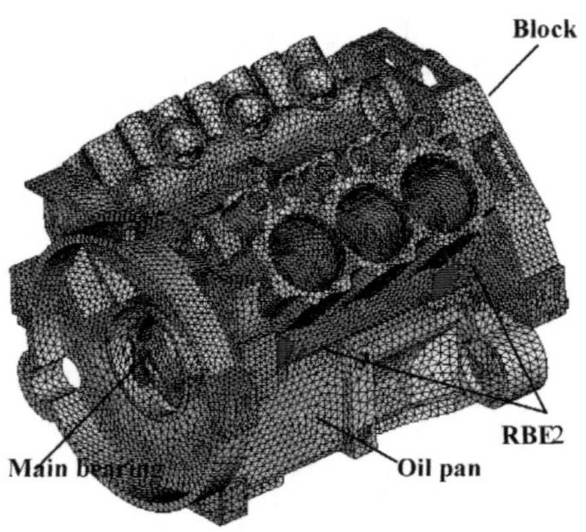

Figure 5. Mesh model of engine Block

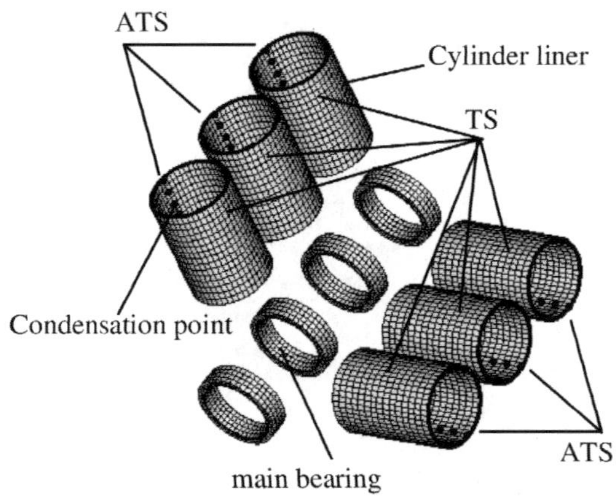

Figure 6. Engine model after condensation

2.3. Substructure condensation model of cylinder liner

In the condensation process of the cylinder liner, 10 main degrees of freedom nodes are defined on the TS and ATS sides of each cylinder sleeve to establish a coupling connection with the mass point at the small end of the connecting rod, as shown in Figure 6.

3. The establishment of dynamic model of the engine

3.1. The condensation dynamic model of the engine

The resulting sub-structure condensation file is imported into AVL EXCITE, and the condensation dynamic model of the whole engine is obtained as shown in Figure 7.

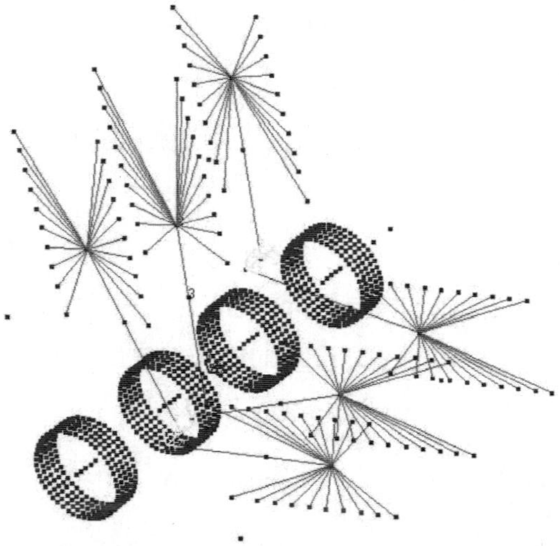

Figure 7. The condensation dynamic model of the whole engine

3.2. The flexible multi-body dynamic model of engine

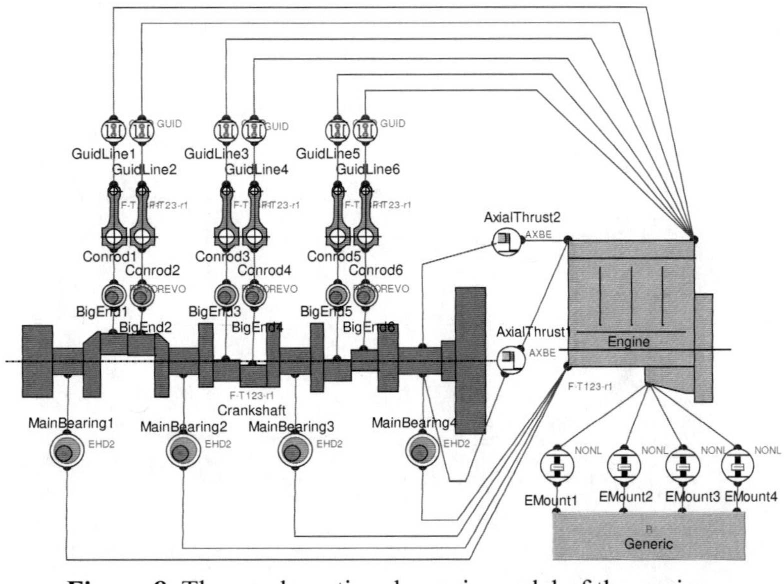

Figure 8. The condensation dynamic model of the engine

Through the POWER UNIT module in AVL EXCITE, a flexible multi-body dynamic model of a six-cylinder high-power diesel engine is established. The model fully considers the major components of the elastomer nonlinear link element interaction between deformation and elastomer. The connections between the various moving parts are defined in the POWER UNIT module of AVL EXCITE, as shown in Figure 8.

The component parameters and global parameters of the model are analyzed, and the analysis step parameters are defined. Each working condition is defined as three cycle simulation calculations, and the simulation results of the second cycle are extracted as the analysis data. In the analysis process, the iteration step value is 0.0625 degree, the minimum iteration step is 0.015625 degree, and the maximum iteration step is 0.5 degree.

3.3. Connection definition of main bearing

An elastic hydrodynamic lubrication bearing model (EHD2) is used to connect the main bearing pairs [6]. The lubrication model can describe the axle neck and bearing elastic deformation, bearing clearance, axis of mass conservation (non-cavitation model), the supply of the lubricating oil (oil tank, oil hole and line etc.) and other factors, is a more accurate to obtain the bearing pressure distribution model. The parameters of main bearing are shown in table 2 [7].

Table 2. Parameters of Main bearing EHD2 model

Parameter	Value	Parameter	Value
Main bearing width	30mm	Oil type	SAE5W-30
Main bearing diameter	108mm	Oil supply pressure	4.6 bar
Radial clearance	0.025mm	Cavitation pressure	0.6 bar
Modulus of elasticity	0.008	The coefficient of friction	0.05
The root-mean-square root of the peak surface of crankshaft	0.4	The average height of the main shaft neck	0.3
The mean-square root of the peak roughness of the bearing surface	0.8	The average peak point height of the bearing	0.5
Oil temperature	100 ℃	Oil groove position	315°~135°

4. Flexible multi-body dynamic Simulation

The flexible multi-body dynamics model of journal bearing is driven by the explosion pressure curve of the cylinder. The burst pressure curve in this paper is obtained through the actual engine cylinder pressure measurement experiment [8], which is shown in figure 9.

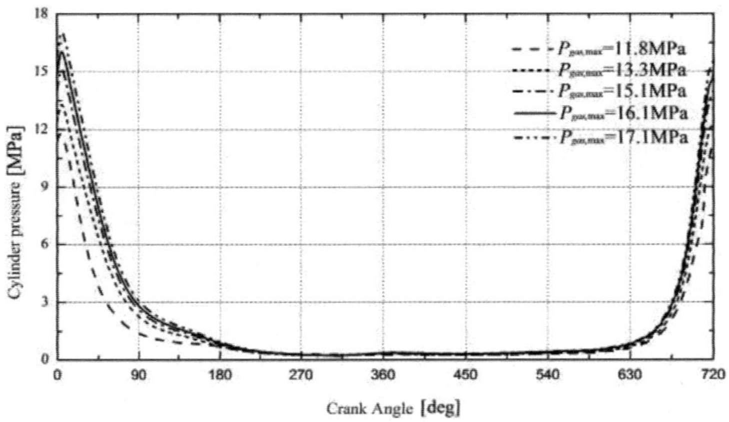

Figure 9. The explosion pressure curve of the cylinder of engine.

The outburst pressure curve of the pressure Pgas,max=11.8MPa was obtained through the test experiment of the pressure test of the prototype cylinder under the rated speed of 2200rpm. The peak burst pressure Pgas,max was 13.3MPa, 15.1MPa, 16.1MPa and 17.1MPa. The pressure curve was obtained through the GT Power calculation on the basis of the burst pressure curve of the experimental test and the parameters of the known engine oil injection volume and the intake and exhaust Angle.

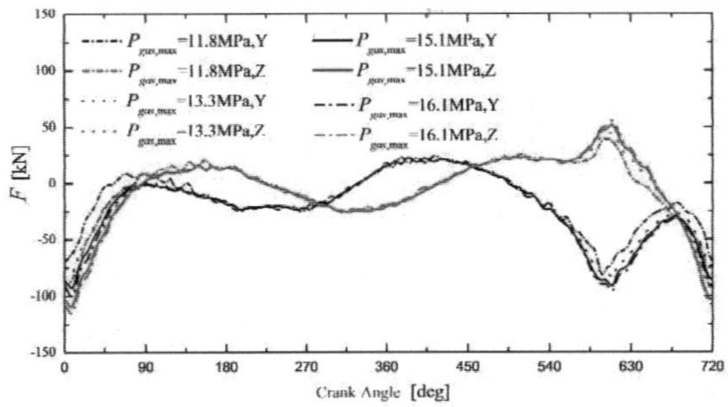

Figure 10. The response situation of the bearing force of the 1st bearing is given by different outburst pressure curve

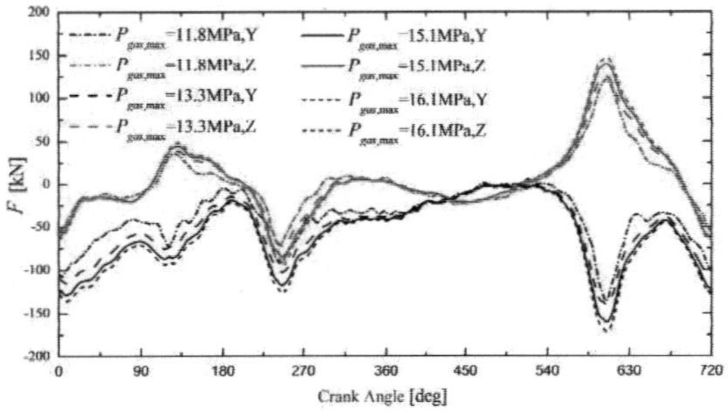

Figure 11. The response situation of the bearing force of the 2nd bearing under different outburst pressure curve

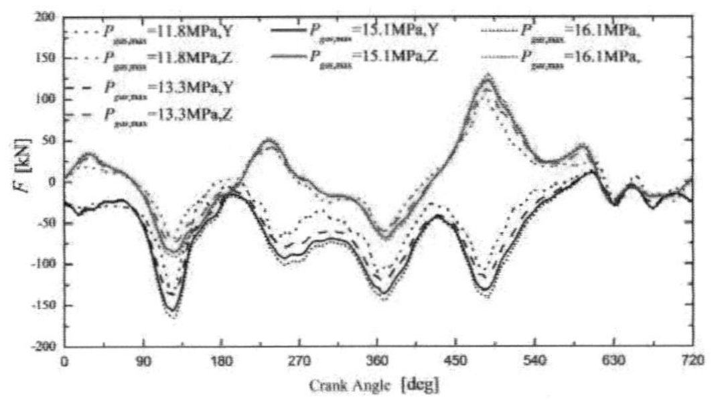

Figure 12. The response situation of the bearing force of the 3rd bearing under different outburst pressure curve

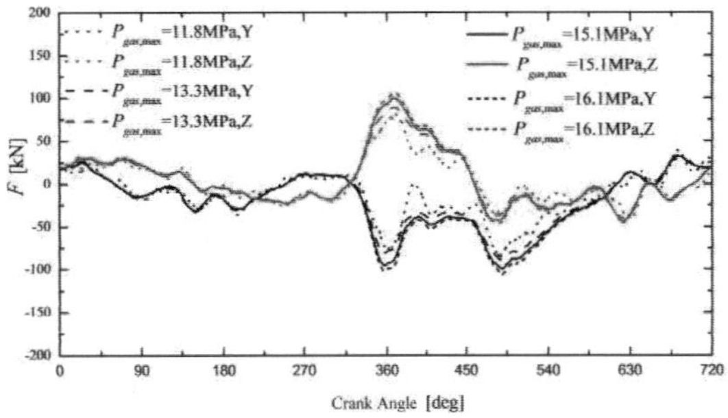

Figure 13. The response situation of the bearing force of the 4th bearing under different outburst pressure curve

5. Conclusions

The structural flexibility of the shaft and the bearing block has an important influence on the dynamic characteristics of the engine. The following conclusions can be obtained through the above analysis:

- The dynamic substructure condensation method is useful to reduce the computational solution scale, and improve the computational efficiency.
- The main bearing force of the two sides of the body is relatively small, while the main bearing force of the middle partition is larger
- The main bearing is mainly affected by adjacent explosion pressure of the cylinder, not adjacent to each cylinder explosion pressure on the main bearing force effect is small.
- With the increase of engine outbreak of peak stress, peak bearing force of the bearing is also increasing, of which the second and third main bearing of the bearing force was bigger.

Acknowledgments
In this paper, the research was sponsored by the National Natural Science Foundation of China (Project No. 51305025).

References

[1] Zhao EH 2017 Numerical and Experimental Studies on Tribological Behaviors of Cu-Based Friction Pairs from Hydrodynamic to Boundary Lubrication *Tribology Transations* 2 **4** 347-356

[2] Sander DE 2016 Simulation of journal bearing friction in severe mixed lubrication – Validation and effect of surface smoothing due to running-in **96** 173~183
[3] AVL Workspace. Excite-Release notes, AVL User Manuals, 2003
[4] Li G and Gu F 2017 An Improved Lubrication Model between Piston Rings and Cylinder Liners with Consideration of Liner Dynamic Deformations *Energies* **10**(12): 2122
[5] Hwang Sheng Jiew, Chen Jer-Si and Jiang Yaqun 2000 Engine Cranktrain System Simulation and Validation *International Adams User Conference*
[6] SeokJooPark 2000 Electrohydrodynamic Flow and Particle Transport Mechanism in Electrostatic Precipitators with Cavity Walls *Aerosol Science &Technology* **33**(3): 205~221
[7] Kumar A, Goenka P K, Booker J F. Modal analysis of elastohydrodynamic lubrication: a connecting rod application [J]. ASME Journal of Tribology, 1990, 112: 524~534
[8] Sheng Jiew Hwang, Jer-Si Chen, Yaqun Jiang. Engine Cranktrain System Simulation and Validation [C]. International Adams User Conference 2000

IOP Conf. Series: Materials Science and Engineering **542** (2019) 012024 doi:10.1088/1757-899X/542/1/012024

Simulation Research on Characteristics of Paint Deposition on Spherical Surface

Yan Chen[1], Jun Hu, Wenzhuo Chen, Gang Zhang, Shuang Zhou and Bo Li

1 Petroleum Department, Army Logistics University, Chongqing, China
E-mail: yansohucom@sohu.com

Abstract. In order to describe the gas-liquid two-phase coupling flow process of paint deposition on spherical surface, the Euler-Lagrangian method is adopted to establish the model of paint deposition on spherical surface, including continuous phase model, discrete phase model and impingement and sticking model. By comparing and analyzing the simulation results of flat and spherical surface in *XZ*-plane, the uniformity of the paint film on spherical surface is better than that on flat. As the spherical surface diameter increases, the distribution range of the paint film expands, the film thickness decreases, and the deposition rate improves. The characteristics of paint deposition on spherical surface are verified by experiments.

1. Introduction

Robotic painting is a mechanized and automatic painting technology[1-3], where the spray gun is usually under the predetermined path provided through off-line programming. Fully understanding the characteristics of paint deposition process on the workpiece to be painted is important for obtaining the path through off-line programming. Air spray gun is one type of the common spray guns, and the characteristics of the paint deposition process of using it to paint flat workpiece have been studied by many researchers. But there are few studies on the paint deposition process of painting spherical surface, which is also a common curved surface in industry.

Paint deposition process is a gas-liquid two-phase transient flow phenomenon, which is usually simulated by applying CFD (computational fluid dynamics)[6-7]. The Euler-Lagrangian method is a commonly-used method for studying paint deposition, which treats the air as a continuous phase obtained by solving the time-average N-S equation, and treats the atomized paint droplets as a discrete phase obtained by capturing the motion of paint droplets. And the Euler-Lagrangian method is widely adopted to study paint deposition by most researchers, such as Flynn[8], Ye[9], Chen[10] and Hilton[11].

Therefore, in this paper, the Euler-Lagrangian method is used to establish a paint deposition model to study the characteristics of paint deposition on spherical surface. Simulation cases on flat and spherical surface with different diameters are simulated. The characteristics of paint deposition on

IOP Conf. Series: Materials Science and Engineering **542** (2019) 012024 doi:10.1088/1757-899X/542/1/012024

spherical surface are contrastively analyzed. And Simulation results are verified by experiments.

2. Paint deposition model

The spray flow field is a coupling flow field of the continuous phase and the discrete phase, which is described by adopting the Euler-Lagrangian method.

2.1 Continuous phase model

During the spraying process, the ambient temperature remains constant. Therefore, the gas phase flow mainly considers the mass conservation equation and the momentum conservation equation.

The mass conservation equation is

$$\frac{\partial \rho}{\partial t}+\frac{\partial}{\partial x_i}(\rho u_i)=0 \tag{1}$$

in which ρ is the air density and u_i is the air velocity in the *i* direction.

The momentum conservation equation is

$$\frac{\partial}{\partial t}(\rho u_i)+\frac{\partial}{\partial x_j}(\rho u_i u_j)=-\frac{\partial p}{\partial x_i}+\frac{\partial}{\partial x_j}\left(\mu\frac{\partial u_i}{\partial x_j}-\rho u_i^{'} u_j^{'}\right) \tag{2}$$

where *p* is the air pressure, μ is the air viscosity, and $u_i^{'}$ is the pulsating speed in the *i* direction.

Besides, Realizable $k-\varepsilon$ turbulence model is applied, which can avoid the situation that the positive stress is negative due to the extremely high time-average strain rate.

2.2 Discrete phase model

The paint droplets are mainly affected by gravity and air impact force in the spray flow field. In the Cartesian coordinate system, the force balance equation of the paint droplet in the *x* direction is

$$\frac{du_p}{dt}=F_D(u-u_p)+\frac{g_x(\rho_{p-}\rho)}{\rho_p} \tag{3}$$

in which ρ_p is the paint density, u_p is the velocity of the paint droplet and $F_D(u-u_p)$ indicates the unit mass drag of the paint droplet.

2.3 Paint impingement and sticking model

When spraying, the discrete paint droplets impinge on the workpiece surface and stick to form the paint film, which is described by the mass conservation and the momentum conservation .

The mass conservation shows as follows

$$\frac{\partial h}{\partial t}+\nabla_s\cdot\left[h\vec{V_l}\right]=\frac{\dot{m}_s}{\rho_l} \tag{4}$$

in which *h* is the height of paint film, $\vec{V_l}$ is the flow velocity of the newly formed paint liquid film and $\dot{m}_s$ is the paint mass source of unit area, which is given by $\dot{m}_s=\alpha_d\rho_d V_{dn}A$.

The momentum conservation is

IOP Conf. Series: Materials Science and Engineering **542** (2019) 012024 doi:10.1088/1757-899X/542/1/012024

$$\frac{\partial h\vec{V}_l}{\partial t}+\nabla_s\cdot\left[h\vec{V}_l\vec{V}_l\right]=\frac{h\nabla_s P_L}{\rho_l}+(\vec{g}_\tau)h+\frac{3}{2\rho_l}\vec{\tau}_{fs}-\frac{3\nu_l}{h}\vec{V}_l \tag{5}$$

Where the transient term and the convection term are shown on the left successively. On the right, the first term is the combined effect of the gas flow pressure, the normal component of gravity along the surface and the surface tension of the liquid paint film; the others are the tangential component of gravity along the surface, the viscous shear stress at the intersection interface between the gas phase and the liquid phase and the viscous stress of the liquid film, respectively.

3 Simulation cases setup

The geometric model of spray nozzle adopted in simulation is shown in Figure 1. The paint hole with a diameter of 1.1mm lies in the center of the spray nozzle. Concentric with the paint hole is the central air hole—a ring with an outer diameter of 2mm and an inner diameter of 1.6mm. And there are two identical assistant air holes with a diameter of 0.5mm on each side of the central air hole. On the horn mouth of the spray nozzle are two shaping air holes respectively, and the one closer to the paint inlet hole has a diameter of 0.6mm while the further 0.8mm.

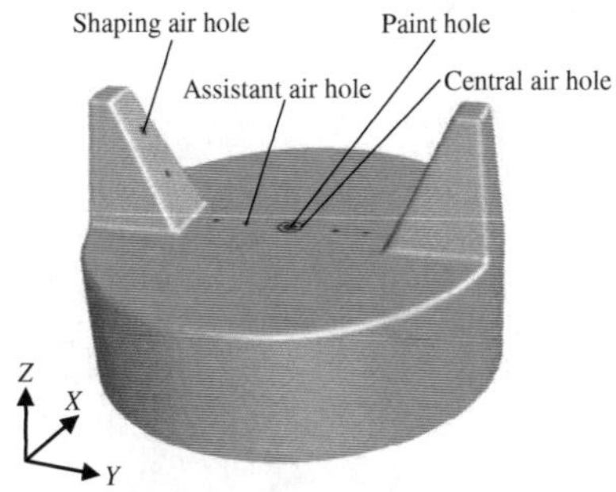

Figure 1. Geometric model of spray nozzle

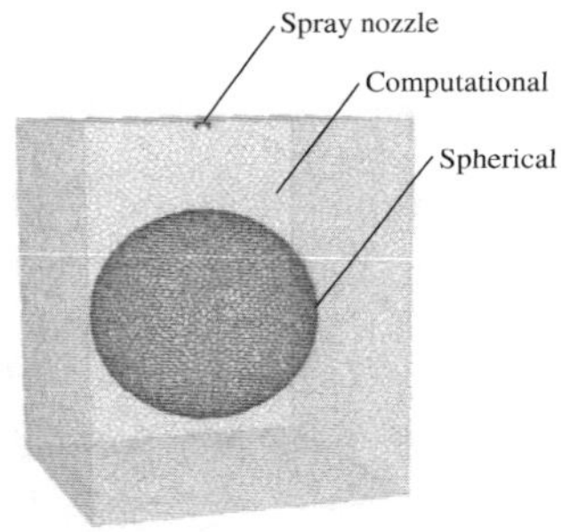

Figure 2. Mesh of computational domain

The simulation cases are conducted by spraying spherical surfaces with diameters of 380mm, 440mm and 550mm. The spraying distance is set at 180mm from the paint hole to the workpiece surface, and the axis of the spray nozzle is perpendicular to the workpiece. The computational domain is meshed by polyhedral mesh method as shown in Figure 2. And a comparative case is carried out by spraying a 400 mm×400 mm flat plate.

The central air hole, the assistant air hole and the shaping air hole are set as pressure inlets with 1.5atm, 1.5atm and 0.8atm, respectively, of which the hydraulic diameters are set as 0.4mm, 0.4mm and 0.6mm, and the turbulent intensities are kept the same as 5%. The liquid phase consists of 71 discrete particle parcels, which are distributed in the circular region of the paint hole by adopting the Rosin-Rammler distribution. The density of the liquid phase is 1200kg/m^3, and the viscosity is 0.0979kg/(m·s). According to the DPM model in ANSYS Fluent, the workpiece is set as the wall-film model. The time step is $\Delta t = 1\times10^{-4}$ s, and the spraying time is 0.5s.

4. Simulation results and analysis

Figure 3 shows the contours of the air velocity in *XZ*-plane in two spray flow fields simulated on flat and spherical surface. To obtain appropriate contour shapes, the variation range of the air velocity is limited to 0-20m/s, and the *XZ*-plane is chosen to illustrate. According to the change of the air velocity,

IOP Conf. Series: Materials Science and Engineering **542** (2019) 012024 doi:10.1088/1757-899X/542/1/012024

the spray flow field is approximately divided into two parts: the diffusion zone and the deposition zone. The former refers to the area where the liquid paint is discretized and diffuse, while the latter is the area in which a large number of paint droplets deposit on the workpiece.

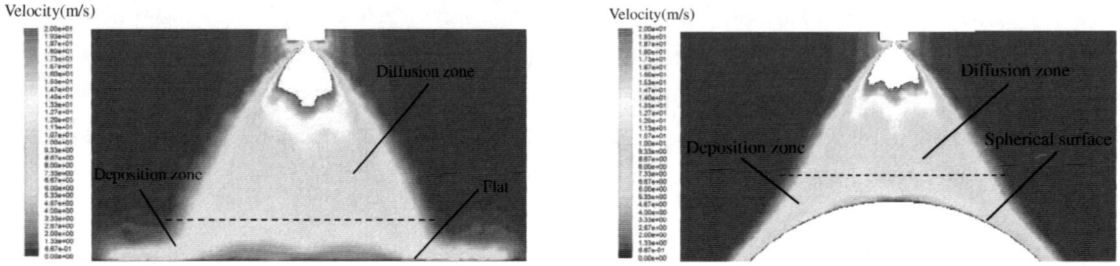

(a) Change of air velocity above flat **(b)** Change of air velocity above spherical surface

Figure 3. Change of air velocity above different workpieces

Figure 4 shows the contours of the paint film thickness on flat and spherical surface. 24.5μm is the maximum value of paint film thickness on spherical surface, which is much smaller than 29.7μm on flat, and the elliptical rings of paint film on spherical surface are more obvious than that on flat, which indicates that the uniformity of paint film on spherical surface is better than that on flat.

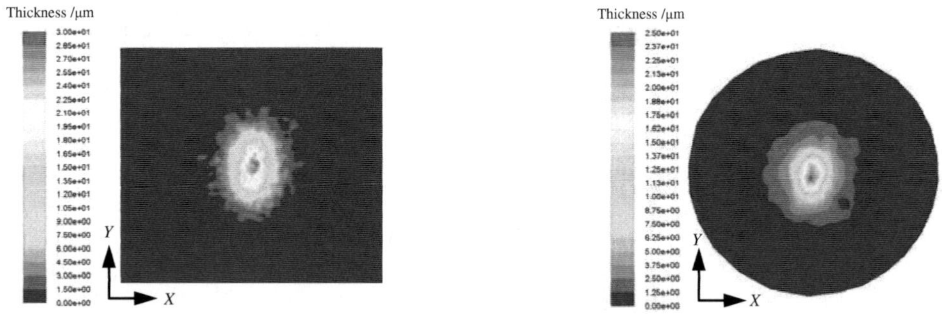

(a) Thickness distribution on flat **(b)** Thickness distribution on spherical surface

Figure 4. Spatial distribution of paint film thickness on different surfaces

Figure 5 shows the comparison of the thickness distribution of paint film on different workpieces in *XZ*-plane between simulation and experiment. The simulations and experiments of painting the three spherical surfaces with different diameters and the flat were carried out under the same painting parameters. In Figure 5, the curves are obtained by fitting the thickness data of simulation cases, while the dots represent the points on the paint films at an interval of 5mm, and their data are the averages of the film thicknesses measured three times by a thickness gauge at every point.

As is shown in Figure 5, the larger the spherical surface diameter is, the more similar the thickness distribution on is to that on flat. Besides, when the diameter increases, the distribution range of the paint film enlarges and the height of the paint film increases, which indicates that the deposition rate improves.

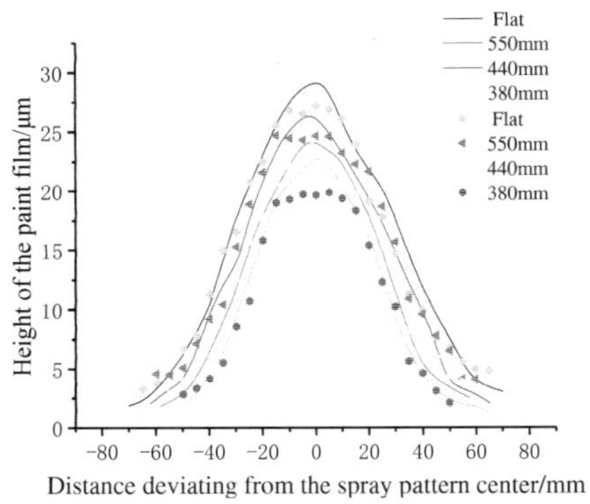

Figure 5. Thickness distribution on different workpieces in *XZ*-plane
(curve: simulation, dot: experiment)

5. Conclusions

According to the contrastive analysis of simulation cases and the verification of experiments, the characteristics of paint deposition on spherical surface are revealed, and two conclusions are obtained:

(1) Comparing with flat, spherical surface has a great influence on paint deposition, which leads to smaller paint film thickness, more obvious the elliptical rings and better uniformity.

(2) As the spherical surface diameter increases, the distribution range of the paint film expands, the film thickness decreases, and the deposition rate improves.

Acknowledgments

This project is supported by National Natural Science Foundation of China (Grant No. 51475469).

Reference

[1] Chen Y, Chen W, Li B, Zhang G and Zhang W 2017 *Industrial Robot: an International Journal* **44** pp 629-38.

[2] Lin Y, Yi Z, Li Y, Ji Z, Liu Y, Pei F and Li B 2018 *Robot* **40** pp 115-128.

[3] Wu C and Lu J 2017 *The 2017 IEEE International Conference on Applied System Innovation (Sapporo).*

[4] Chen Y, Chen W, Chen K and Li B 2016 *Journal of Robotics* **2016** pp 1-12.

[5] Chen Y, Chen W, He S, Pan H, Li B and Zhang W 2017 *China Surface Engineering* **30** pp 122-131.

[6] Rahimi M J, Sitaraman H, Humbird D and Stickel J J 2018 *Chemical Engineering Research and Design* **139** pp 283-295.

[7] Hu X, Qian S, Yamauchi K and Okochi H *2017 ASME 2017 Fluids Engineering Division Summer Meeting(Waikoloa).*

[8] Flynn M R and Sills E D 2000 *Annals of Occupatioknal Hygiene* **44** pp 191-202.

[9] Ye Q and Pulli K 2017 *Journal of Coatings Technology and Research* **14** pp 467-476.

[10] Chen Y, He S, Zhang G, Chen W and Jiang J 2016 *Proceedings of the 4th International Conference on Machinery, Materials and Computing Technology (Hangzhou).*

[11] Hilton J E, Ying D Y and Cleary P W 2013 *Chemical Engineering Science* **99** pp 141-160.

IOP Conf. Series: Materials Science and Engineering **542** (2019) 012025 doi:10.1088/1757-899X/542/1/012025

A non-Hertzian contact analysis for the ball and raceway of a deep groove ball bearing

XI qiang[1, 2*], PI yangjun[1, 3] and HU yumei[1, 4]

1 Chongqing University, 174 shazheng street, shapingba district, Chongqing, 400044, China
2 cquxiqiang@163.com
3 cqpp@cqu.edu.cn
4 cdrhym@163.com
* The corresponding author
Abstract. Aiming at the premature damage of deep groove ball bearings, considering axial load, elastic-plastic constitutive, coordinated contact and bearing clearance and so on, a non-Hertzian elasto-plastic contact analysis for the bearing ball and the raceway is performed by trans-scale three-dimensional finite element simulation. The results shown that the failure mechanism of the bearing is extremely similar to the L-P fatigue life theory. The effects of load magnitude and direction on contact state were further studied. It indicates that a certain angle of load is beneficial to reduce the maximum internal stress of the bearing, but the angle should not be too large. Besides, increasing the load magnitude results in an increase in the stress value and the depth of maximum stress, but when the load is too heavy, the area that touches the contact surface will be damaged firstly and quickly.

1. Introduction

Gearboxes are widely used in the transmission system of automobiles, and excessive noise and premature breakdowns of the automotive gearboxes are mainly caused by the failure of rolling-element bearings. A typical failure of ball and raceway of a rolling-element bearing in a twin countershaft transmission is presented Fig.1, in which obvious damages are observed.

The failure of transmission bearings is often analyzed by means of fault diagnosis and signal processing. These studies help to identify and distinguish whether the bearings is working properly, but do not fundamentally reveal the mechanism of transmission bearing failure. The origins of failure in rolling bearings are mainly due to the contact of roller and raceway, resulting in such commonly known damages as pitting, spalling, crushing, peeling, scuffing[1, 2]. Therefore, the study of the real contact situation between roller and raceway is of great importance to the failure prediction of rolling bearings. The contact characteristics between roller and raceway of bearing have been extensively studied by researchers [3,4,5].

The classical Hertz contact theory is extensively adopted in the counter-conforming contact case. There are two main assumptions in the Hertz contact theory, namely the elastic half-space assumption and the linear elastic constitutive assumption. However, when it comes to the contact problems of roller and raceway as shown in Fig.2, the contact between the ball and inner raceway in the axial symmetrical section, and the contacts between the ball and outer raceway both in the circumferential and axial-symmetrical sections are all conformal contact.

IOP Conf. Series: Materials Science and Engineering **542** (2019) 012025 doi:10.1088/1757-899X/542/1/012025

Figure 1. A failure bearing

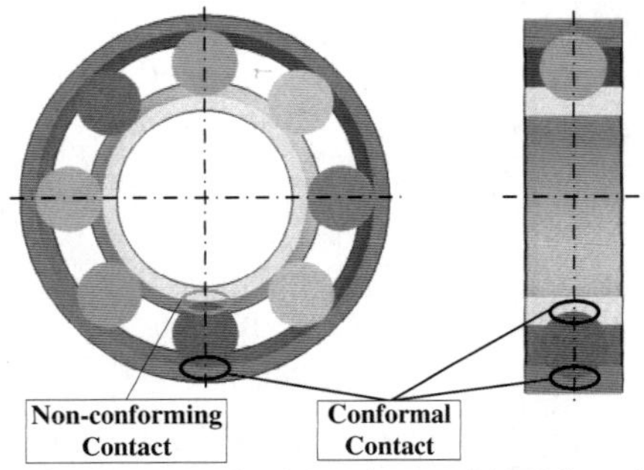

Figure 2. Symmetrical sections of ball bearing

Specifically, in the axial symmetrical section, both inner raceway and outer raceway in contact with ball show a high geometrical adaptation, which no longer meets the Hertzian elastic half-space assumption. Moreover, the real material constitutive model of bearing steel is elasto-plastic [6]. When the load exceeds the elastic limit and no longer meets the Hertzian linear elastic assumption, it is hard to describe the contact characteristics using elastic constitutive model and the existing theoretical models. It is worthwhile to emphasize that these two non-Hertzian contact factors are significantly important in the contact of ball and raceway of deep groove ball bearings. Goodman and Keer [7] analyzed the contact of a frictionless sphere in a conforming cavity, they found the contact stiffness in compression deviates more than 25% as predicted by the Hertz theory. Kang et al [8] also found the contact between balls and raceway of a deep groove ball bearing no longer meets the elastic half-space assumption. They also revised the Hertzian contact equation which determines the relationship of load and deflection proposed by Jones-Harris, and the modified equation is more consistent with the experimental results. Zhupanska[9] analyzed the problem of normal contact of two identical elastic spheres and equationted the corresponding mixed boundary-value problem at the surface of the sphere, thereby relaxing the assumption of the Hertz theory about small size of the contact area compared to the sizes of the contacting bodies. Comparison of the results with the Hertz theory shows that the latter predicts contact stresses with high accuracy even for relatively large contact areas. The frictional contact problem of a mixed ceramic deep groove ball bearing based on ABAQUS and experiments is analyzed in [10]. The experiments find the maximum contact stress of the ball and raceway exceeds the yield limit, and plastic deformation occurs at the contact area. By comparing the experimental with the simulating results, they find that if nonlinear property of bearing material is not considered, the mutual approach of the ball and raceway would produce large errors.

At present, it is hard to give an accurate theoretical solution considering the conformal contact in the plastic zone to the contact between the ball and raceway of a deep groove ball bearing. An extremely effective means for solving such problem would be FEM. The FEM divides the solution area into a series of units, the denser the unit division, the more accurate the solution. During a two-dimensional contact analysis between the ball and the rigid plate, R L. Jackson [11] found that when the meshed contact area is controlled to ensure that at least 30 contact elements are always in contact, the maximum contact radius error of 3.3%. However, it is very difficult for three-dimensional contact simulation model to mesh so fine. The trans-scale finite element method enables micron-scale meshing. The main idea of the trans-scale finite element method is perform a coarse mesh finite element analysis first, then extract the contact area for micron-scale modeling, and finally load the node displacement in the coarse mesh model result as a boundary condition into the micron-scale model. The modeling and analysis process of the trans-scale finite element analysis method is described in detail in [12].

In this paper, a non-Hertz contact between the ball and raceway is analyzed by the trans-scale finite element method under actual load boundary conditions. Then, the effect of load magnitude and direction on the contact state is compared. This study will be benefit to the design, wear, lubrication and contact fatigue researches of rolling-element bearings.

2. Method validation

Taking the contact between two hemispheres as an example, the accuracy of the trans-scale simulation method is verified by the three-dimensional Hertz contact theory.

IOP Conf. Series: Materials Science and Engineering **542** (2019) 012025 doi:10.1088/1757-899X/542/1/012025

2.1. Obtain theoretical solution

According to the Hertz contact theory, for the two hemisphere contact problems:

Contact radius a:

$$a = \sqrt[3]{\frac{3}{4}\frac{RQ(1-\mu^2)}{E}} \tag{1}$$

Elastic approaching quantity δ:

$$\delta = \sqrt[3]{\frac{9Q^2(1-\mu^2)^2}{2RE^2}} \tag{2}$$

Maximum contact stress P_{max}:

$$P_{max} = \sqrt[3]{\frac{6QE^2}{\pi^3 R^2 1-\mu^2)^2}} \tag{3}$$

Where R is the radius of the sphere, Q is the compressive load, E and μ is the elastic modulus and Poisson's ratio.The sphere studied here has a spherical radius of 12.5 mm and is made of bearing steel (considering it as an ideal linear elastic material) with a density of 7.85e3Kg/m3, an elastic modulus of 210 GPa and a Poisson's ratio of 0.3. The load of the two hemisphere contact models is set to a radial load of 1000N. The theoretical solution of the contact radius, elastic approaching value and maximum contact stress of the two hemisphere contact model are obtained and shown in table 1.

Table 1. Theoretical solution of two hemisphere contact model

load	contact radius	elastic approaching quantity	maximum contact stress
1000N	0.273mm	23.8μm	6413 MPa

2.2. Obtain numerical solution and comparative analysis

Firstly, a coarse mesh model is solved for the two hemisphere contacts. The spheres were discrete in all hexahedral mesh with an average grid size of about 0.1 mm. The bottom surface of the bottom hemisphere in the three-dimensional contact finite element model is fully constrained, the upper hemisphere constrains all rotation and the translational degrees of freedom in all directions except radial direction (Y-axis). And applies a load of 1000N to the upper surface of the upper hemisphere along the negative direction of the Y-axis. Contact is established between the surfaces of two spheres, regardless of friction, and the coefficient of friction is zero. The material was set to an ideal linear elastic material. The LS-DYNA software is used to solve the coarse mesh model. Based on the simulation results, the micro-scale model meshing area is determined and then solved. The partial model extracted from the coarse mesh model is refined to ensure that the mesh size of the contact area is about 9 μm. So, there are more than 30 units to describe the radius of the contact circle, which can ensure that the number of units participating in the contact is always not less than 30, thereby ensuring the accuracy of the solution.

The results of micro-scale model shows that the maximum contact pressure of the two hemispheres is in the center of the circle, which is about 6067 MPa. The distance between the outer edge nodes of the contact circle is extracted as the diameter of the contact circle. The distance between two nodes that symmetrical about the center of the contact area was measured, and the distance between the two points was 0.562083 mm. That is to say, the contact radius is about 0.281 mm. The comparison between the numerical solution and the theoretical solution of two hemisphere contact is shown in table 2.

Table 2. Comparison between numerical and theoretical solution

	maximum contact stress(MPa)	contact radius (mm)	elastic approaching value(μm)
theoretical solution	6413	0.273	23.8
numerical solution	6067	0.281	25.3
error	5.4%	2.9%	6.3%

It can be seen from table 2 that the error of maximum contact stress, contact radius and elastic approaching value are all within 7%. Therefore, the calculation accuracy of the trans-scale finite element simulation method fully meets general engineering requirements. In addition, according to [12], the trans-scale finite element simulation method has nothing to do with the contact object itself. Therefore, the method can be applied to solve the contact between bearing ball and raceway.

3. 3D non-Hertzian contact of ball and raceway

The analysis of the three-dimensional non-Hertzian contact analysis between the ball and outer raceway is combined with the engineering application and realistic load boundary conditions are considered. Therefore, this paper determines the external load of a bearing and analyzes the load distribution over balls based on a twin countershaft transmission before performing the 3D conformal FEA.

3.1 Calculation of bearing loads

The twin countershaft transmission is quite common in the front-engine front-wheel driving car. The input shaft of the transmission studied here is supported by two deep groove ball bearings on each side, while each of the two countershafts is supported by a deep groove ball bearing on the left side and a cylindrical roller bearing on the right side. The right side of the transmission is close to the engine.

Transmission working at the first and reverse gears, corresponding to the two limit-load conditions of maximum torque, is selected in this analysis. The maximum output torque of the matched engine is 190Nm. Thereby the external loads of the supporting bearings at the operating conditions of first gear and reverse gear are obtained [13]. The result figures out that the maximum loads of deep groove ball bearings appears at the the bearing at the right end of the input shaft, which is consistent with the experimental location of the bearing failure in advance, and the radial force is about 12330N, corresponding to an axial force of about 4678N.

3.2 Load distribution over balls

Load distribution is one of the most important operating characteristics for the ball bearings, and can affect the running accuracy and load carrying capacity. When only the radial external load is considered, Stribeck empirical equation is always used to determine the relationship of the maximum contact force and the radial external load. When the external loads in both radial and axial directions are taken into account simultaneously, the radial and axial integral calculus proposed by Sjövall should be used to get the load distribution over balls. However, the method is quite complex, and it can't consider the influence of clearance which is of critical importance. It should be noted that the assumption of Sjövall is $\gamma \geq \alpha'$, where $\gamma = \arctan\frac{Fa}{Fr}$ and α' is the contact angle, in which Fa, Fr is the axial and radial external load respectively. When the assumption is not satisfied, the ball bearing will be unstable, and it is hard to analyze the load distribution by theoretical solution. This paper analyzes the load distribution over the balls in a deep groove ball bearing, in which both the axial and radial external loads are taken into consideration, as well as many other factors, such as the clearance, effect of cage-aligning, the gravitational influence of bearing components, are counted in for a more accurate and realistic results. Because of the complexity of the problem, FEA is applied in this paper to determine the load distribution over balls, which produces a more accurate boundary condition for the non-Hertzian contact between ball and outer raceway.

The finite element model (FEM) is established as shown in Fig.3 basing on the structural parameters of the most dangerous bearing, which consists of about 380,000 hexahedral elements.

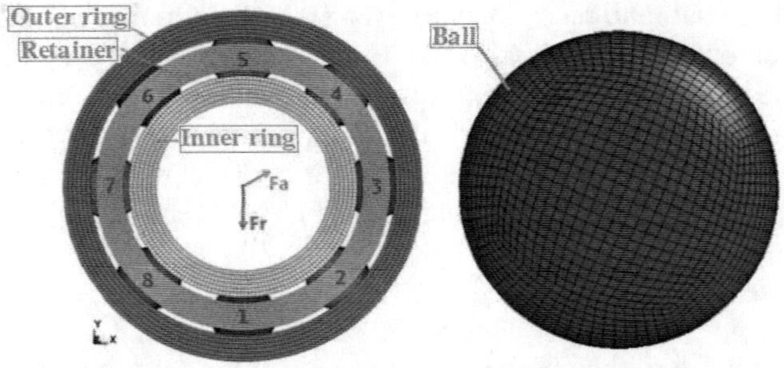

Figure 3. Finite element model of a bearing

As the balls will contact with the retainer and raceway in realistic operating conditions, we make the definition of contacts respectively, totally 24 contact pairs created. Because retainer may contact with the raceway, two more contact pairs are defined. To simulate the interference fit of the outer raceway with the bearing housing, fixed boundary condition is applied to the nodes on the outer surface of the outer raceway. Radial and axial external forces are exerted to the nodes on the inner surface of

IOP Conf. Series: Materials Science and Engineering **542** (2019) 012025 doi:10.1088/1757-899X/542/1/012025

the inner raceway, and gravitational acceleration (g) is applied to all the bearing components in the opposite of Y-direction. The material of the balls and raceways are bearing steel, with an elastic modulus of 207Gpa, a density of 7830kg/m^3 and a Poisson's ratio of 0.3. The material of the retainer is PA66 nylon, with an elastic modulus of 8.3Gpa, a density of 1145kg/m^3 and a Poisson's ratio of 0.28.The load distribution of radial and axial contact forces over the balls are shown in table 3.

Table3. Load distribution of balls (N)

	Ball1	Ball2	Ball3	Ball4	Ball5	Ball6	Ball7	Ball8
Radial	7615.11	4449.48	1018.70	608.62	672.47	608.41	1049.46	4384.27
Axial	385.33	747.85	547.58	532.68	641.45	536.33	555.82	731.83
Total	7624.85	4511.89	1156.54	808.81	929.13	811.06	1187.56	4444.93

It is clear that the maximum total contact force of the bearing occurs at ball 1, with a radial contact force of 7615.11N and an axial contact force of 385.33N. Over the entire load distribution, it exhibits a trend of symmetrical distribution, with the axis of symmetry passing through the sphere centers of ball 1 and ball 5.

3.3 3D non-Hertzian contact

Taking ball 1 as the analysis object, the three-dimensional non-Hertzian contact analysis of the ball and the outer raceway is carried out according to the trans-scale simulation method in literature [12]. The simulation results obtained from the previous ball load distribution analysis are used as the results of the coarse mesh model, and the determined micro-scale model meshing area is judged accordingly. As shown in Fig.4, the inside of the white frame is the micro-sized model meshing area.

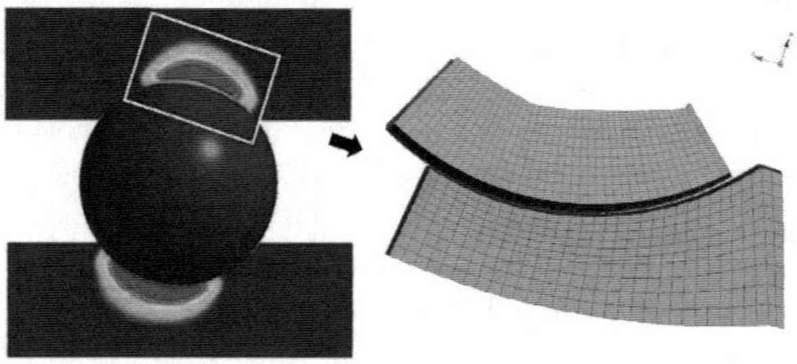

Figure 4. Micro-sized model meshing area of ball and raceway contact

The micro-scale model is then solved. The partial model extracted from the coarse mesh model is refined to ensure that the mesh size of the contact area is about 20 μm. So, the ratio of the short half axis of the contact ellipse to the unit size is about 18, which can ensure that the number of units participating in contact is always more than 30, thereby ensuring the accuracy of the solution. The load of the micro-size model is consistent with the coarse model. All degrees of freedom on the outer surface of the outer raceway are constrained. Contact is established between the ball and outer raceway, and the dynamic and static friction coefficients are both set to 0.002. This paper selects an elasto-plastic constitutive model for the material of the ball and raceway, with MAT024-PIECEWISE-LINEAR-PLASTICITY in the commercial finite element software LS-DYNA to describe. Since the contact analysis of ball and raceway under radial and axial loads belongs to quasi-static analysis, it is unnecessary to consider the effect of strain rate on the plasticity of the material. In addition, a static compression experiment of the bearing material is carried out, and the true stress-strain curve of the material is obtained, and then loads into the model. The yield strength of the material is about 1684Mpa.

4. Results and discussion

4.1. Results

First, as shown in Fig.5 and Fig.6, the contact pressure distribution on the contact surface between the ball and raceway is substantially elliptical (not completely elliptical for axial force existing). And the contact pressure on surface is gradually increased from the outside to the inside, and the maximum value appears at the center of the contact area. The maximum contact pressure on surface of the ball is about 2984 MPa, and the maximum contact pressure on raceway surface is about 2759 MPa.

Figure 5. Contact pressure distribution of the contact surface of ball

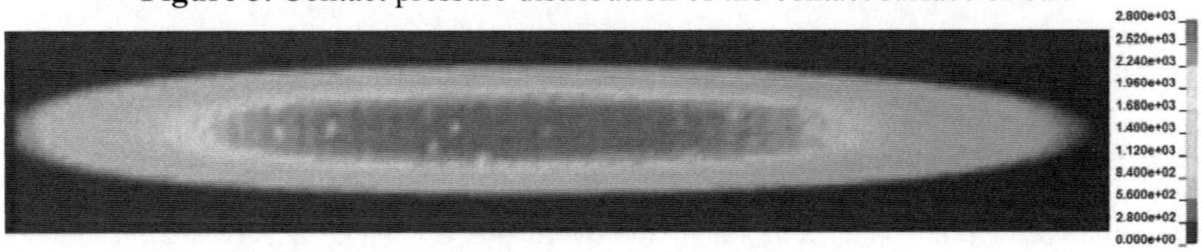

Figure 6. Contact pressure distribution of the contact surface of raceway

In addition, the premature damage of bearing is also closely related to the V-M stress distribution of the contact area. In order to obtain the V-M stress distribution of the contact area, the micro-size model is cut along the axial and circumferential symmetrical middle planes of the bearing. The V-M stress distribution in the cutting plane of the micro-scale model is shown in Fig.7 and Fig.8.

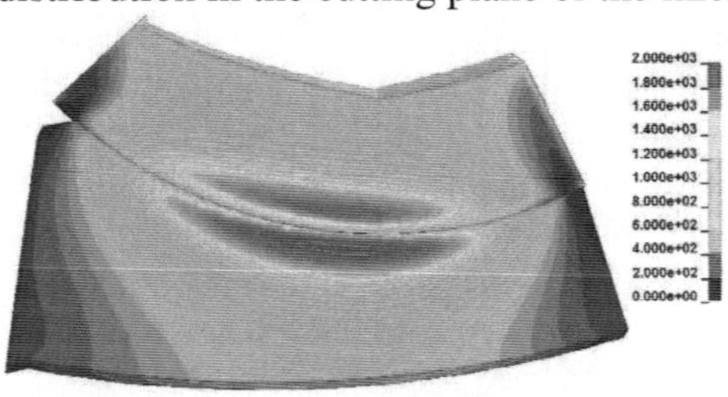

Figure 7. Axial sectional view

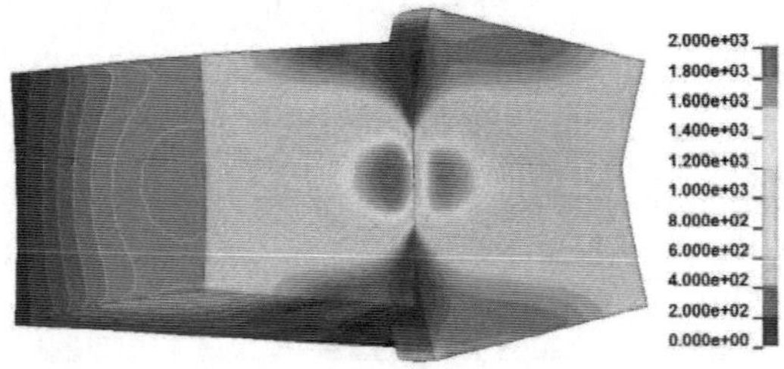

Figure 8. Circumferential sectional view

It can be clearly shown from the two cross-sectional views that the distribution of effective stress distribution in the axial symmetrical section with higher geometrical adaptation is also significantly different from the circumferential symmetrical section with lower geometrical adaptation. The V-M stress distribution is semi-elliptical in the circumferential symmetrical section, and is quite similar with the 2D Hertzian results. However, within the axial symmetrical section, with higher geometrical adaptation, the effective stress distribution forms an elongated strip, distinguishing from the Hertzian distribution. Besides, the maximum V-M stress does not appear on the contact surface, but below the contact surface.

In addition, the simulation results show that the maximum V-M stress of the ball and the raceway is about 1752 MPa and 1968 MPa. However, the yield strength of the bearing material is 1684 MPa. That is to say, the V-M stress of the bearing analyzed in this paper exceeds the yield strength of the material under the limit load condition. Therefore, we extract the variation curve of the maximum contact pressure, maximum equivalent stress and maximum shear stress of the ball and the outer raceway during the contact process as Fig.9 shown.

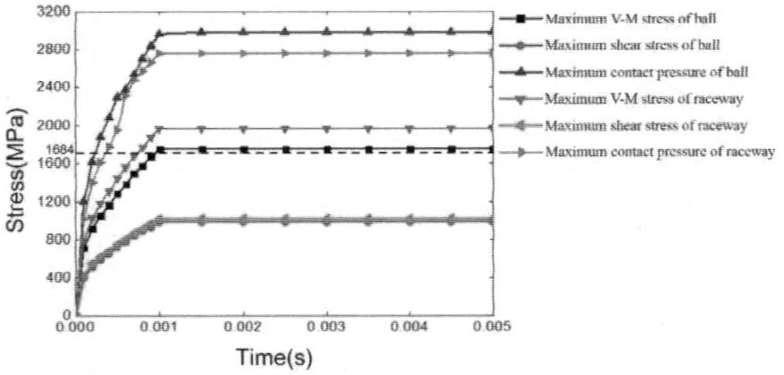

Figure 9. Contact characteristics change process

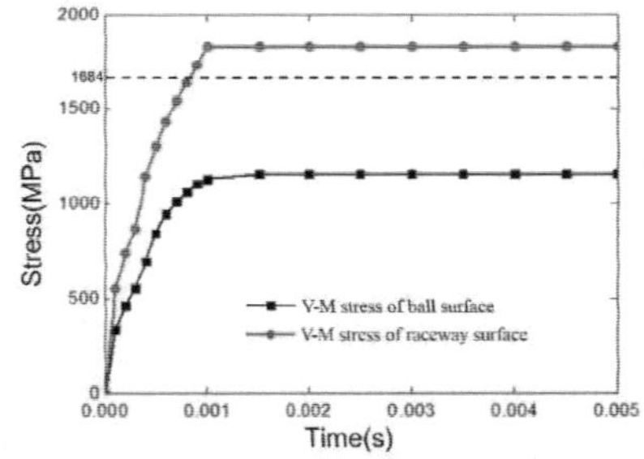

Figure 10. Maximum V-M stress on surface

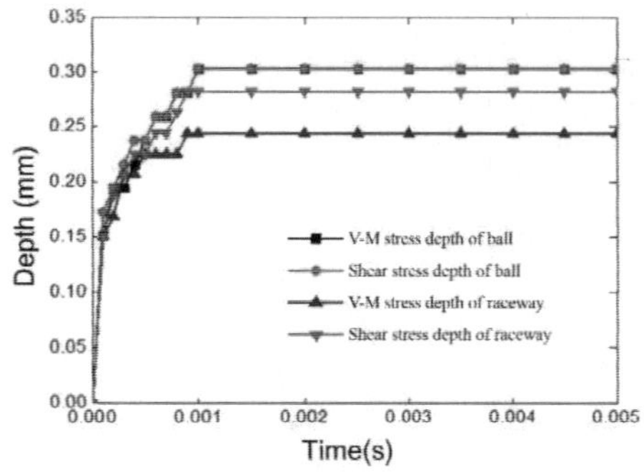

Figure 11. Stress depth change process

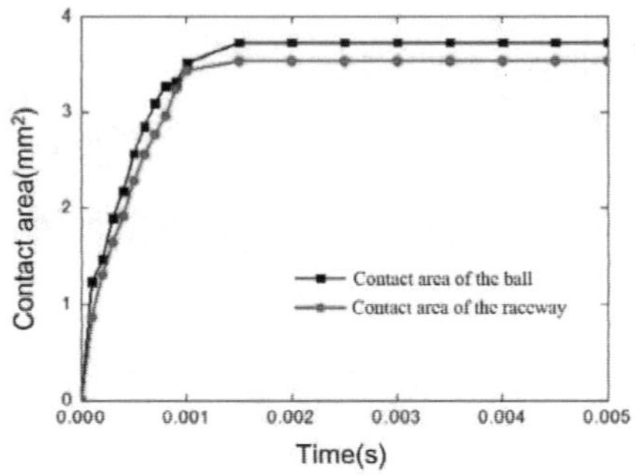

Figure 12. Contact area change process

It can be seen from the figure that the maximum shear stress values of the balls and the outer raceway are not much different at each moment, and the former is always slightly smaller than the latter. When the load is stabilized, the maximum shear stress of the ball and the outer raceway is stable at about 1000 MPa. However, the bearing steel is a plastic material with a tensile strength of about 1600 MPa. The shear stress limit is generally considered to be 0.5 to 0.7 times the tensile strength, that is, 800 to 1120 MPa. That is to say, both ball and raceway are highly likely to be damaged. The maximum V-M stress of the ball slightly exceeds the yield strength of the material at the time of loading, and finally stabilizes at about 1752 MPa, exceeding the yield strength of the material by 4%; The maximum V-M stress of the outer raceway reached the yield strength of the material at the time of loading too, and finally stabilized at around 1968 MPa, exceeding the yield strength of the material by 17%. The maximum contact pressure of the ball and the outer raceway are both quickly exceed the yield limit of the material, but the V-M stress of the ball contact surface ball never reaches the material yield limit, and is stable at about 1157 MPa; The V-M stress value of the outer raceway contact surface has reached the yield limit and is stable at about 1832 MPa. The stress variation curve of the unit with the largest V-M stress on the contact surface is shown in Fig.10. The maximum V-M stress depth, the maximum shear stress depth and the contact area of the ball and the outer raceway are shown in Fig.11 and Fig.12.

It can be seen that, as the load increases, the maximum V-M stress and the maximum shear stress depth of the ball gradually increase, and finally reach stability after loading is completed, and the stable value is 0.303 mm; The maximum V-M stress and the maximum shear stress depth of the outer raceway is also gradually increasing, the final stability is 0.245mm and 0.282mm. Besides, the contact area of the ball is always slightly larger than the outer raceway, and the stress depth of the ball is always greater than the outer raceway. Therefore, the bearing volume of the ball will always be larger than the outer raceway, which is an important reason why the V-M stress of the outer raceway is higher than the ball. In addition, the maximum shear stress and the maximum V-M stress of the ball appear at the same depth at each moment. However, the maximum shear stress depth of the outer raceway begins to exceed the maximum equivalent stress depth when the load is loaded halfway.

In summary, the bearing volume of the ball is larger than outer raceway, the maximum V-M stress value of the ball is smaller than outer raceway, and the maximum shear stress value of the ball is slightly smaller than outer raceway. Therefore, the bearing analyzed in this paper will yield at the subsurface of outer raceway first. Meanwhile, the V-M stress on the outer raceway surface exceeds the yield strength of the material too. So, both the surface and the subsurface of outer raceway will initiate cracks. Subsequently, for the continuously rolling effect of the ball and raceway, crack that initiated at the subsurface of raceway extends to the surface, and connects with the crack initiated at the surface. As we can see in Fig.1, the inner raceway and the ball are also damaged. The reason is that parts of the lubricating oil immerses into these cracks. Under the hydrodynamic pressure and alternating loads of bearing operating, parts of the material are removed from the original base of ball and raceway, which results in pitting and spalling. Furthermore, these removed materials will act as abrasives, with the bearing operating, piercing the film and scratching the ball and raceway, which significantly reduce the lubricating properties of the grease. Meanwhile, due to the large pieces of peeling, the running accuracy of the bearing decreases, and additional impact is generated, which accelerates bearing failure.

Obviously, the damage mechanism of the bearing analyzed in this paper is very close to the fatigue life theory of Lundberg and Palmgren [14]. According to the subsurface initiated crack mechanism, the crack will first occur at the core of maximum shear stress under alternating loading, and gradually extends to the surface, and connects with the surface initiated pitting, which results in pitting and spalling. Besides, the survival probability of the load carrying volume is determined jointly by the stress cycles, Weibull life distribution, the peak of the maximum shear stress and the contact depth of the peak maximum shear stress, whereas the load carrying volume is related to the sizes of the contact area.

4.2 Further discussion

The magnitude and direction of the load have a significant impact on the static contact state of the ball and raceway. Therefore, this paper analyzes the effect of load magnitude and direction on the non-Hertzian contact of bearing ball and raceway. The load vector in the previous article has an angle of about 2.9 with the radial direction of the ball. Keep the load size constant, set the load angle to 0, 2.9°, 5°, 10°, 15°, 20°, and extract the maximum equivalent stress, maximum shear stress and stress depth of the outer race as Fig.13 and Fig.14 shows.

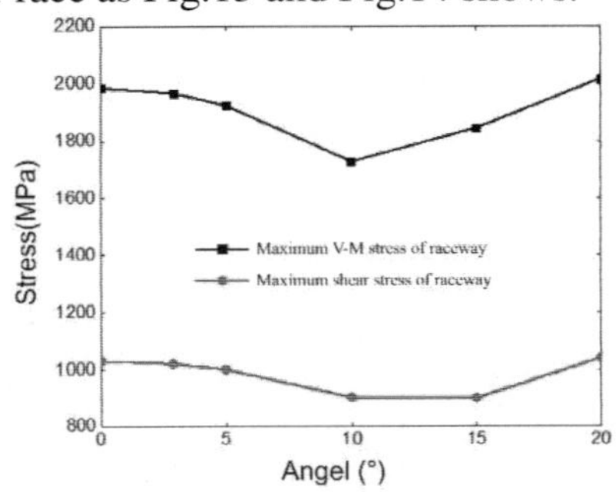

Figure 13. Effect on maximum stress

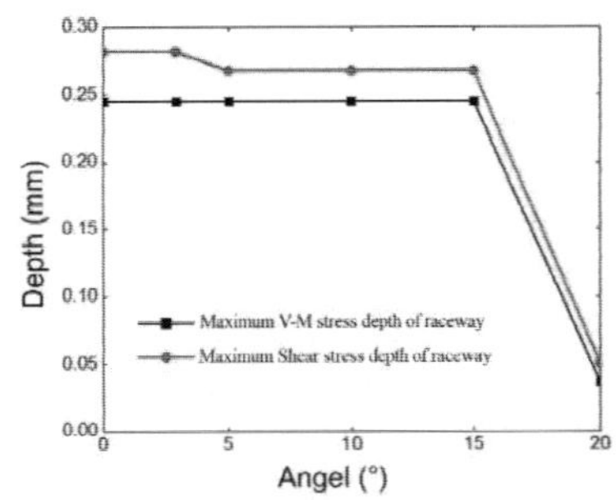

Figure 14. Effect on the maximum stress depth

It can be seen that, as the angle increases, the maximum V-M stress and the maximum shear stress decrease first and then increase. It reaches the minimum when the angle is 10°, and decreases by 7.1% and 6.9% compare to the value while the angle is 0°. The maximum V-M stress and the maximum shear stress depth do not change much as the angle increases, but suddenly decrease when the angle reaches 20°. It indicates that the deep groove ball bearing can withstand a certain degree of axial load, and a certain angle of load is beneficial to reduce the maximum stress value of the bearing, but the angle should not be too large. The reason might be that, due to the geometric coordination, the load with a certain angle can make the load on the ball more transmitted to the raceway through the periphery of the contact area instead of being excessively concentrated in the contact center, thereby reducing the stress value. When the angle exceeds a certain limit, the contact center is excessively deviated due to the excessive declination, so that the contact area is likely to reach the shoulder of the raceway, which resulting in a decrease in the bearing volume, thereby resulting in an increase in the stress value.

Then, keep the angle constant and solve the model with radial loads of 5000N, 7615.11N, 10000N, 15000N, 20000N. The changes in maximum V-M stress, maximum shear stress, and stress depth are shown in Fig.15 and Fig.16.

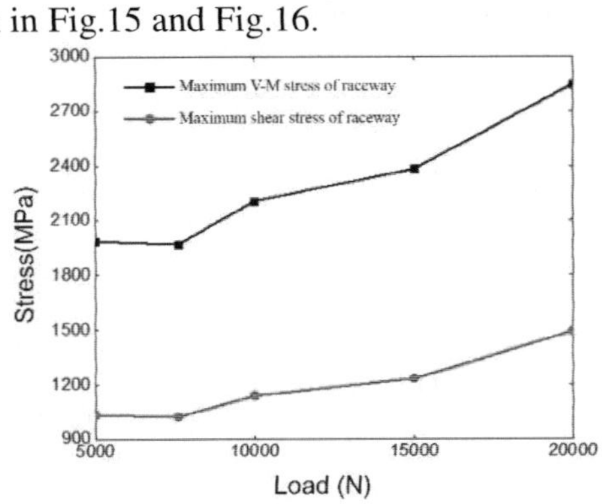

Figure 15. Effect on maximum stress

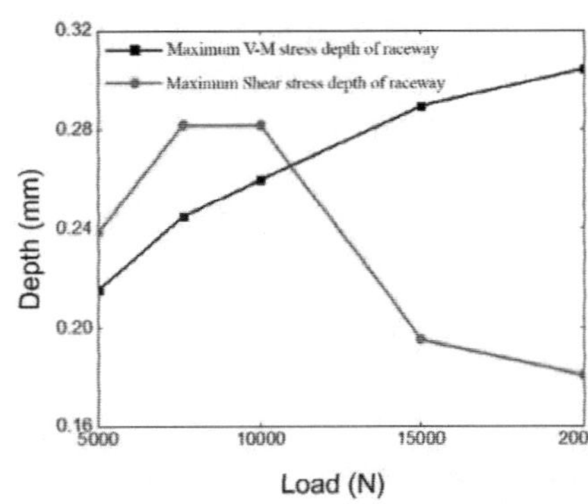

Figure 16. Effect on maximum stress depth

It can be seen that, as the load increases, the maximum V-M stress, the maximum shear stress, and the maximum V-M stress depth increase gradually, the maximum shear stress depth increases first, then flattens, and finally decreases. This indicates that increasing the load value results in an increase in the stress value and yield depth. However, when the load is too heavy, the area that touches the contact surface will be damaged firstly and quickly. The drop in the maximum shear stress depth in Fig.16 maybe due to this.

5. Conclusion

This paper analyzed the contact between ball and raceway of a deep groove ball bearing of a two-shaft transmission by trans-scale finite element method. The non-Hertzian contact between the ball and outer raceway under extreme working conditions is obtained. Based on this, the bearing damage mechanism and process is analyzed. The result indicates that the damage mechanism of the bearing is very similar to the failure mechanism predicted by L-P fatigue life theory.

This paper analyzes the effect of load magnitude and direction on the non-Hertzian contact of bearing ball and raceway. The result shows that a certain angle (about 10) of load is beneficial to reduce the maximum internal stress of the bearing, but the angle should not be too large. Besides, increasing the load magnitude results in an increase in the stress value and the depth of maximum stress, but when the load is too heavy, the area that touches the contact surface will be damaged firstly and quickly.

6. References

[1] M Littmann, W. E. The mechanism of contact fatigue. Interdisciplinary approach to the lubrication of concentrated contacts, (Ed. P. M. Ku) Proceedings of a Symposium, Troy, NY, July 1969;NASA Special Report, SP-237, 1970; PP. 309-378(NASA, Washington DC).

[2] Olver AV, The mechanism of rolling contact fatigue: an update, Proceedings of the Institution of Mechanical Engineers, Part J: Journal of Engineering Tribology. 2005; 219:313-30.

[3] Srečko Glodež, Rok Potočnik, Jože Flašker. Computational model for calculation of static capacity and lifetime of large slewing bearing's raceway. Mechanism and Machine Theory. 2012; 47:16-30.

[4] Albert V.Korolev, Andrey A.Korolev, Radoslav Krehel̆. Character of distribution of the load between the balls in the ball bearings under the action combined of external load. Mechanism and Machine Theory. 2014; 81:54-61.

[5] EI-Thalji I, Jantunen E. A descriptive model of wear evolution in rolling bearings. Engineering Failure Analysis. 2014; 45:204-224.

[6] Linares Arregui I, Alfredsson B. Elastic-plastic characterization of a high strength bainitic roller bearing steel-experiments and modeling. International Journal of Mechanical Sciences. 2010; 52:1254-1268.

[7] L.E.Goodman, L.M.Keer. The contact stress problem for an elastic sphere indenting an elastic cavity. International Journal of Solids and Structures. 1965; 407-415.

[8] Kang Y, Shen PC, et al. A modification of the Jones-Harris method for deep-groove ball bearings. Tribology. 2006; 39:1413-1420.

[9] Zhupanska OI. Contact problem for elastic spheres: Applicability of the Hertz theory to non-small contact areas. International Journal of Engineering Science. 2011; 49:576-588.

[10] Xu Q, Ren CZ, Chen JJ. Three dimension FEM analysis of interface and contact stress of hybrid ceramic ball bearing. Machine tool & Hydraulics. 2003; No2:203-204,206.

[11] Robert L.Jackson, Itzhak Green. A Finite Element Study of Elasto-Plastic Hemispherical Contact Against a Rigid Flat [J]. ASME .Journal of Tribology.2005, 127(2):343-354.

[12] Xie Hao. Research on Local Model-Based Contact Simulation of Ball Bearings [D]. Chongqing University, 2016.

[13] Lan Jiangming. A non-Hertzian contact analysis for the ball and raceway of a transmission bearing [D]. Chongqing University, 2015.

[14] G. Lundberg, A. Palmgren. Dynamic capacity of rolling bearings. Acta Polytechnica Mechanical Engineering Sciences. 7(3):1-7, 1947.

IOP Conf. Series: Materials Science and Engineering **542** (2019) 012026 doi:10.1088/1757-899X/542/1/012026

Simulation of nozzle position optimization based on single-phase flow

Guang Chen[1], Yuancong Zhou[1], Xuelin Qian[1],Wenmin Wu[2], Yifan Zhao[2],Yumei Hu[1]*

1 State Key Laboratory of Mechanical Transmission Chongqing University
Chongqing 400044, China
2 Chinese helicopter Desigh and Research Institude 333001,China
E-mail: 617799949@qq.com

Abstract: The numerical simulation on the lubrication of the complex gearbox was studied and an optimization for its injection lubrication system was carried out. Firstly, a pair of high speed gear test bench was built and related experiments were completed. The experimental results verified the correctness of the numerical simulation modeling method. Then the numerical model of the injection lubrication of the complex gear box was established, the simulation analysis was carried out. The results show that the overlong injection distance, the angle of the end face and the arrangement of the nozzle in the out of mesh side are not good for lubrication. And there is a trace in the center plane of the two meshing gear, along which the ejected oil will be under the least impact of the high speed airflow that driven by the rotating gears. When the nozzle axis is in the direction of the trace, the gear lubrication condition is the best.

1. Introduction

The gear transmission is gradually developing towards high speed, heavy load, compact and long-time operation. The harsh working environment of the gears causes a large amount of heat to be generated due to sliding friction between the tooth surfaces. If the heat cannot be dissipated in a time, the surface temperature of the gear teeth will rise sharply, the lubricant film will break, and the gear surface will be damaged by glueing, which will lead to transmission failure [1-4]. Therefore, a certain amount of lubricating oil supply is a prerequisite for high speed gear lubrication and timely heat dissipation.

High-speed gear transmission mostly adopts injection method to achieve tooth surface lubrication and cooling. On the one hand, the movement of the oil is greatly affected by the air flow driven by the rotating gear near the tooth surface [5]. Predecessors have conducted a lot of research on this phenomenon. Literature [6-7] studied the effect of a rotating single gear on the oil jet under different speeds and pressures through experimental methods. The results showed that the air flow field caused by the high speed rotation of the gear could make the oil jet collapse. At the same time, the oil in the meshing area that actually participates in lubrication will decrease as the gear speed increases. Literature [8] studied the impact depth of oil jets with different transmission ratios through the combination of experiment and simulation, and reached the relevant conclusions. In literature [9], the VOF model was used to simulate the process of the oil jet impinging on the tooth surface to evaluate the resistance torque of the oil jet lubrication. On the other hand, the structure characteristics and the azimuth parameters of the nozzle will directly affect the movement of the oil jet, which will eventually affect the lubrication and cooling of the gear. The literature [10-15] shows that the internal geometry of the nozzle will affect the nozzle mass flow rate, particle size, spray macroscopic characteristics, and thus affect the cooling performance. The influence of injection distance and spray angle on the motion of lubricating oil is studied in literature [5,16], and valuable laws are obtained.

Taking the single stage gear as the research object, a pair of high speed spur gear transmission test rig using jet lubrication has been built. The related experiments were conducted. The experimental results provide a reference for the subsequent validation of the correctness of the numerical simulation of jet lubrication, and the validated numerical simulation modeling technique is applied to the jet lubrication simulation of the complex gear box. The lubrication condition under its working state is predicted, and a calculation method base on single-phase flow is proposed to find the optimal injection angle of the nozzle arrangement. After improving the lubrication effect of the meshing gear, the reliability and practicability of the single-phase flow calculation method are proved. Finally, this method is applied to the optimization of the entire complex gearbox to solve practical engineering problems.

2. Injection lubrication experiment

In order to verify the correctness of the numerical modelling of the injection lubrication, a high-speed gear transmission test bench using injection lubrication was built. As shown in Fig. 1. The test bench mainly includes four parts: driving and transmission system, nozzle position control system, oil supply system and experimental base. The left side of the picture is the assembly of the test bench, and the right side is the high-speed camera for data acquisition.

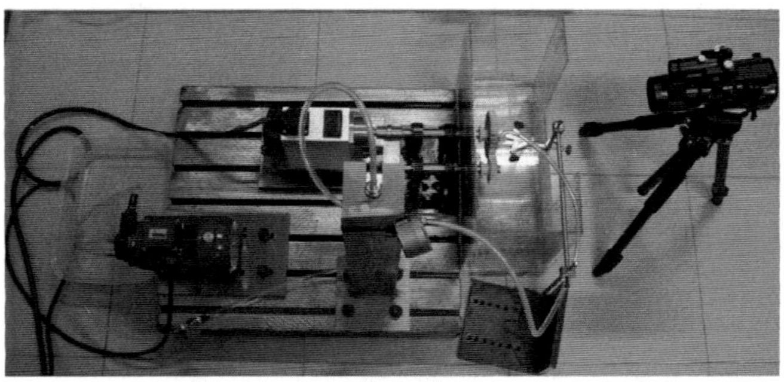

Fig.1 Experimental rig for spray lubrication

The exact position of the nozzle in space is determined by the three variables: jet distance, jet angle and end face angle. The three variables are defined as follows:

1) Injection distance: The injection is defined as the distance from the nozzle outlet to the intersection of the two gears' pitch circle, as L shown in Figure 2.

IOP Conf. Series: Materials Science and Engineering **542** (2019) 012026 doi:10.1088/1757-899X/542/1/012026

2) Injection angle: The injection angle is defined as the angle between the central axis of the nozzle and the common tangent of the two gears' pitch circle, as angle α shown in Figure 2.

3) End face angle: The end face angle is defined as the angle between the central axis of the nozzle and the gear face, as the angle β shown in Figure 2.

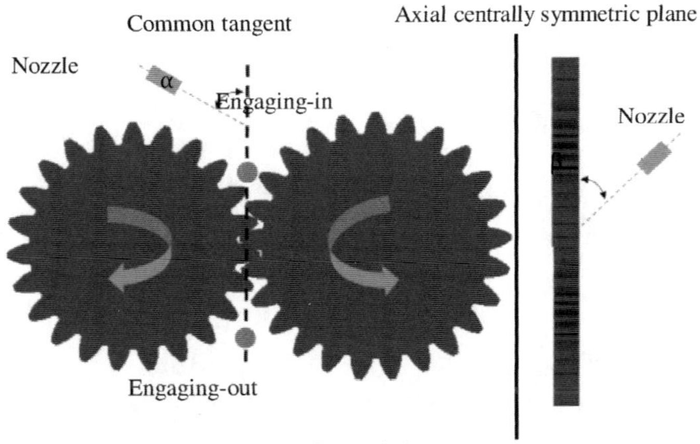

Fig.2 Parameters of nozzle's arrangement

Table 1 Experimental conditions of spray lubrication

Experimental condition	Flow rate ($L\cdot min^{-1}$)	Injection method	Injection distance (mm)	Injection angle (°)	End face angle (°)	Rotational speed ($r\cdot min^{-1}$)
①	0.3	Engaging-in	30	0	0	13371
②	0.3	Engaging-in	40	0	0	13371
③	0.3	Engaging-in	50	0	0	13371
④	0.3	Engaging-in	60	0	0	13371

The high speed cameras are mainly composed of 0~6 times lens, CCD, three tripod, main engine, optical fiber and other components. During the experiment, experimental data was collected with a shooting frequency of 250 frames per second. Since water and lubricating oils are very similar in many properties, the amount of oil needed in the experiment is very large. For environmental considerations and other factors, water is used instead of lubricating oil in experiments.

The high speed camera is used to collect and observe the jet flow's movement in the box as shown in Figure 3. As can be seen from the figure, due to the windage resistance effect caused by the rotation of the high speed gear, the liquid jet from the nozzle tends to deflect toward the bull gear side, which indicates that the pinion gear has a stronger windage resistance effect than the large gear. In addition, the deflection of the jet to the side of the large gear increases with the increase of the jet distance. The deflection of the oil trajectory is quite obvious when the injection distance is 60mm. It can be predicted that when the injection distance continues to increase to a certain value, the jet will deflect too far toward the bull gear to enter the meshing zone, and the amount of jet liquid in the meshing zone will be greatly reduced. Therefore, in the design process of the high-speed gear transmission injection lubrication system, the influence of the injection distance must be considered, and the injection distance should not be too large.

IOP Conf. Series: Materials Science and Engineering **542** (2019) 012026 doi:10.1088/1757-899X/542/1/012026

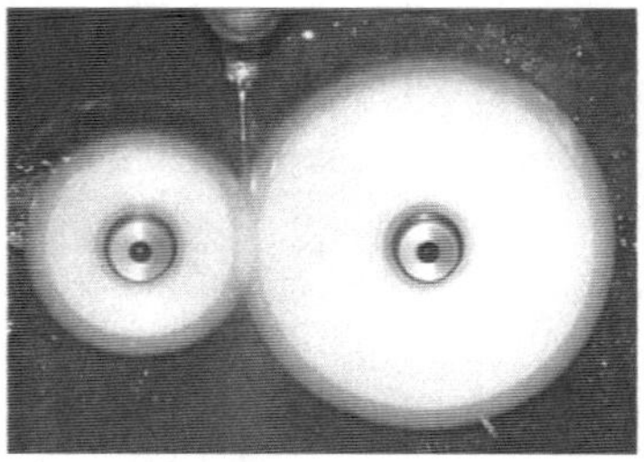
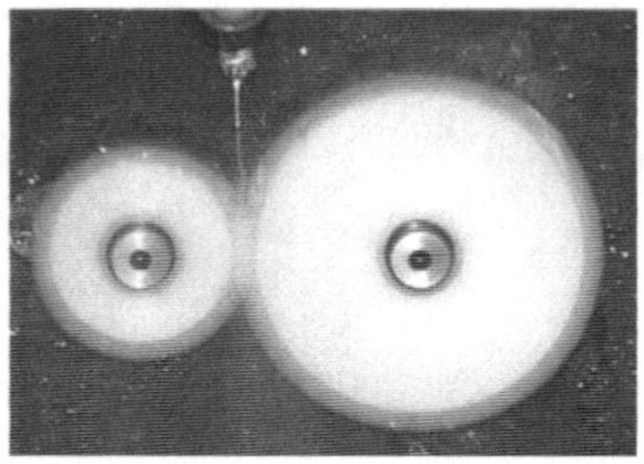

(a) L=30mm (b) L=40mm

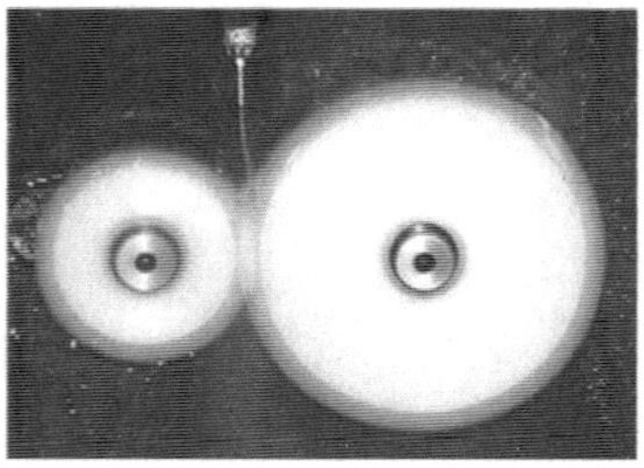
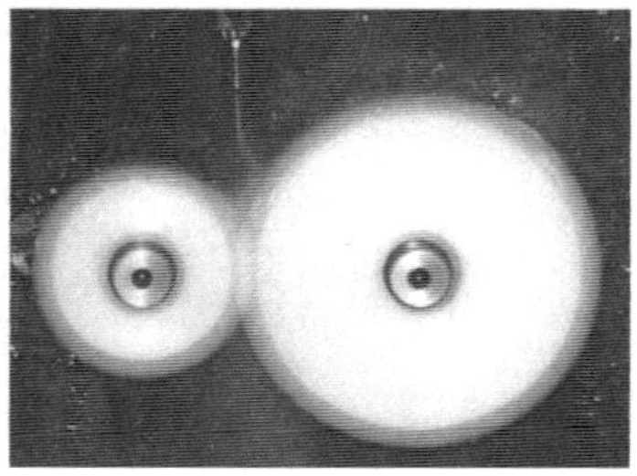

(c) L=30mm (d) L=40mm

Fig.3 Jet flow at different spray distance

3. Injection lubrication simulation technology

In order to verify the correctness of the numerical modeling of the gear jet lubrication, a three-dimensional model of a pair of gear and box is built with the same size as the experimental size. The mesh model is shown in Figure 4.

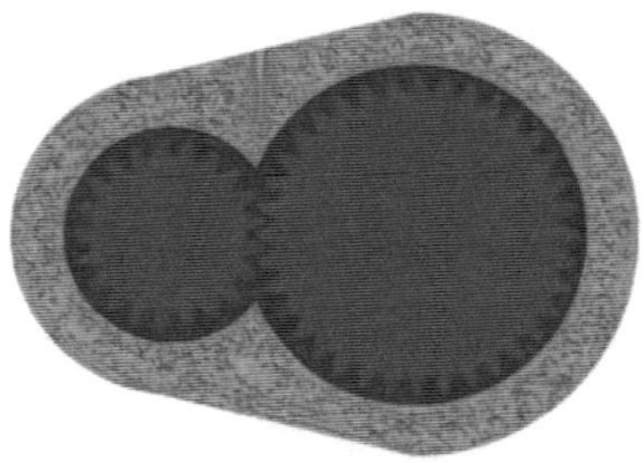

Fig.4 Three-dimensional numeric model

Because the nozzles are arranged on the engaging-in side, the movement of the jet on the this side is a focus of attention. In order to facilitate the observation and analysis, the motion trajectories of the fluid are locally enlarged and divided into three motion paths A, B, and C as shown in Fig. 5 . As we can see, the trajectory of the fluid in simulations and experiments agree well .

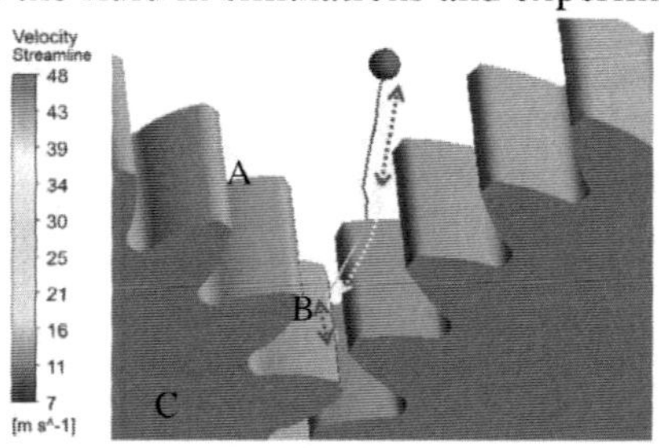

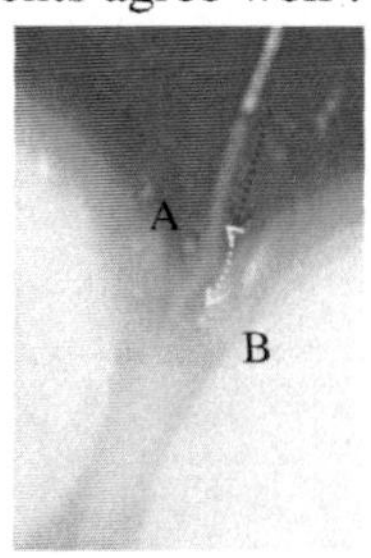

Fig.5 Comparison of the moving trail of jet flow between simulation and experiment

IOP Conf. Series: Materials Science and Engineering **542** (2019) 012026 doi:10.1088/1757-899X/542/1/012026

Path A: the liquid moves along a straight line. The path is relatively far from the gear, and the strength of the air is weak. The momentum of the liquid itself is big enough to overcome the influence of the airflow's effect.

Path B: The motion trajectory of the liquid appears to be deflected. The path now is much closer to the gear and in the region where the airflow driven by the bull gear and pinion gear intensely interact with each other. The limited momentum of the liquid is insufficient to overcome the resistant effect of the airflow. At the same time, the airflow driven by the pinion has greater strength than the one driven by the bull gear, which eventually causes the liquid to deflect toward the large gear side;

Path C: The liquid reaches the gear tooth surface for lubrication. The condition of the gear lubrication and the heat dissipation performance mainly depend on the amount of liquid that reaches this region.

4. Numerical simulation of jet lubrication for complex gear box

Taking a high-speed gear transmission system as the research object, a three-dimensional model was established. The model mainly includes the box and its five pairs of meshing gears on the 01 to 06 axis, and nozzles name A, B, D, E, and G. Five nozzles lubricate the 01-02, 02-03, 03-04, 04-05, and 01-06 gears in turn. The gear speed of the 05-axis gear is the highest, which is 27045r/min; the gear speed of the 03 and 06 gear is the lowest, which is 9280r/min. The rotation direction and nozzle distribution are shown in Fig. 6.

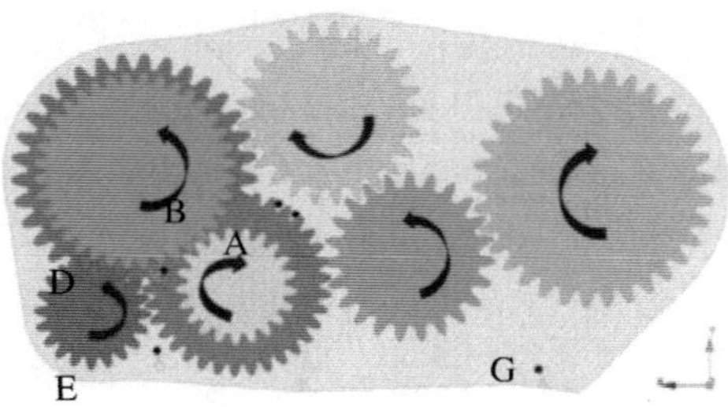

Fig.6 Rotational directions of the gears and distribution of the nozzles in a high-speed gear transmission system

The verified numerical simulation method mentioned has been applied to the simulation of the complex gear box model. The simulation calculation domain is a gas-liquid two-phase fluid domain in a complex gear box. The nozzle orientation information is as follow:

1) Injection distance: Nozzle G has the largest injection distance about 90mm, and the other nozzles have a injection distance of 31~53mm.

2) End face angle: Nozzle D has the largest end face angle about 50 degrees. Nozzle E has the end face angle about 48 degrees, and the end face angle of other nozzles are about 20 degrees.

3) Injection angle: The injection angle of nozzles are between 1 degrees ~9 degrees, nozzle D has the largest injection angle.

4) Nozzle arrangement: Nozzle A is arranged on the engaging-out side, other nozzles are arranged on the engaging-in side.

Three typical nozzles A, D and G are analyzed respectively (nozzle A is located on the engaging-out side, nozzle D has the largest end face angle and nozzle G has the largest injection distance). The oil flow velocity streamline of the three nozzles is shown in Figure 7, respectively. The red solid arrow is the axis of the nozzle, that is, the theoretical injection direction of the nozzles.

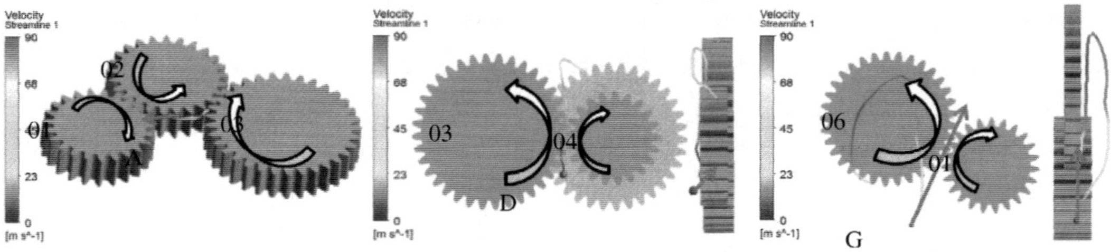

Fig.7 Velocity streamline of oil starting from nozzle A、 D and G

As can be seen from the figure, none of the streamlines has entered the corresponding meshing zone along the theoretical injection direction. The theoretical injection direction of nozzle A is directing to the 01-02 gear meshing area, but its oil is brought into the 02-03 gear meshing area under the reversed action of the airflow driven by the gear0q and 02 gear. The oil velocity streamline ejected from nozzle D is initially biased toward the 03 gear, moving above the meshing area of the 03-04 gear, and finally moving with the direction of the 04 gear under the combined influence of the 04-shaft twin gear. The fluid velocity streamline ejected from the nozzle G is deflected directly towards the gear06.

Combining the information of each nozzle, the reasons are as follows:

1) Injection angle: Excessive injection distance will increase the difficulty of lubricating oil entering the meshing zone, especially for nozzle G. The actual design value should not be too large (this conclusion is consistent with the experimental conclusion)

2) End face angle: Nozzle A, B, D and E all have a certain end angle, D and E have end face angles much larger than the rest. The angle of end face will increase the difficulty of lubricating oil entering the meshing area.

3) The nozzle position relative to the gear rotational speed: The nozzle A is arranged on the engaging-in side and the lubricating oil flows backwards and cannot enter the meshing area directly.

5. Study on nozzle injection angle

According to previous studies, the injection angle has a great influence on the amount of the oil that can actually enter the gear meshing zone [16], which directly affects the lubrication condition of the gear. So what is the certain value of the injection angle to get the best fuel injection effect?

Taking the 03-04 gear as an example, in order to arrange the nozzle properly and make the gear lubrication best, we need to know the airflow distribution in the box first. At the same time, the presence of the end face angle of the nozzle will make the oil hard to get into the meshing zone, so the angle of the end face is set to 0 degrees. In addition, in order to explore the influence of the physical structure of the nozzle on the calculation results, three models were created. One has no nozzle body; one has a 0 injection angle, and the last one has a injection angle that is 10°. The model is shown in Figure 8.

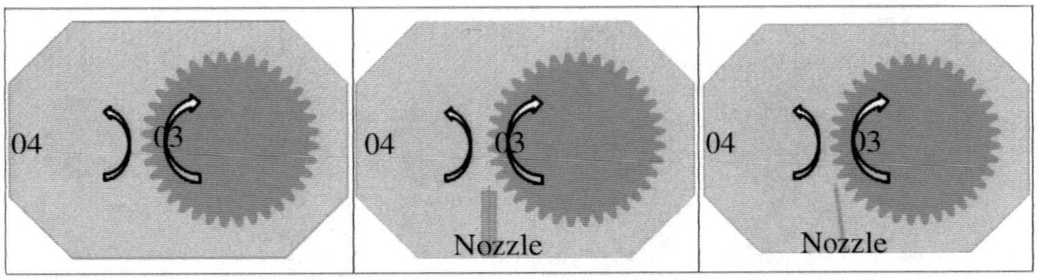

(a) No nozzle (b) Injection angle = 0° (c) Injection angle = 10°

Fig.8 Geometric model with no nozzle entity and different spray angles

The air velocity vector distribution diagram in the center plane is shown in Figure 9. It is found that there is a trace with the weakest air motion on the gear engaging-in side. Points A and B in the figure are located on the trace. Through data collection, it is found that the location of the trace in the three models basically coincides, which indicates that the solid structure of the nozzle does not affect the location of the AB trace. Based on this, it is inferred that if the oil is injected into the meshing area along the direction of the AB track, the impact of the airflow on the motion of the oil will be minimum, and the gear has the best lubrication condition

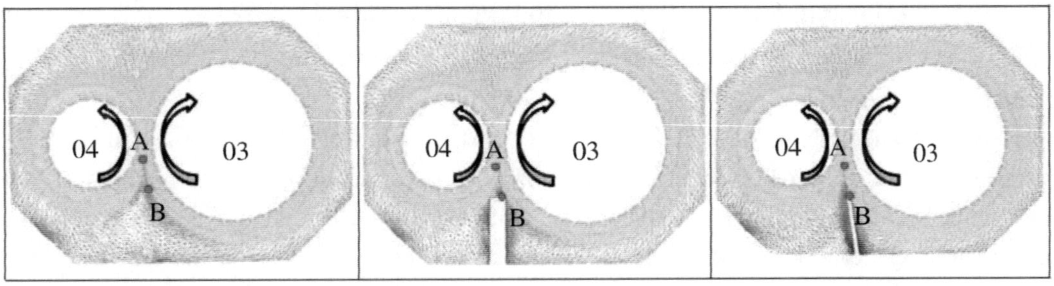

(a)

No nozzle (b) Injection angle = 0° (c) Injection angle = 10°

Fig.9 Velocity vector of airflow in the center plane of meshed gears under the circumstance of no nozzle entity and different injection angles

In order to verify the above conjecture, the nozzles of three pairs of meshing gears are arranged as follows: (1) the nozzle axis is arranged along the direction of AB trace.(2) the nozzle injection angle is set to 0° .(3)the nozzle injection angle is set to 5° . The oil volume fraction distribution on the gear surface is shown in Fig10-11.

The oil volume fraction on the tooth surface is used to evaluate the lubrication of gears. The oil volume fraction is defined as the volume of the oil in unit grid space. So the oil volume fraction on the tooth surface can reflect the amount of oil in the gear meshing position, which also reflects the gears meshing area lubrication condition.

The left part of the Fig 10 shows the oil volume fraction distribution on the gear surface, the right part shows the velocity streamline of the ejected oil. Under this condition, the oil can be directly injected into the mesh area after leaving the nozzle. At the same time, comparing figure 10 and figure 11, it is found that when the nozzle axis is arranged along the trace, the distribution of oil volume fraction on the tooth surface in the meshing area is obviously superior to those of the other two, and the gear lubrication of it is the best, which proves the correctness of the above conjecture. This single phase flow method can be extended to obtain the optimal location of the nozzle for any spur gear pair.

IOP Conf. Series: Materials Science and Engineering **542** (2019) 012026 doi:10.1088/1757-899X/542/1/012026

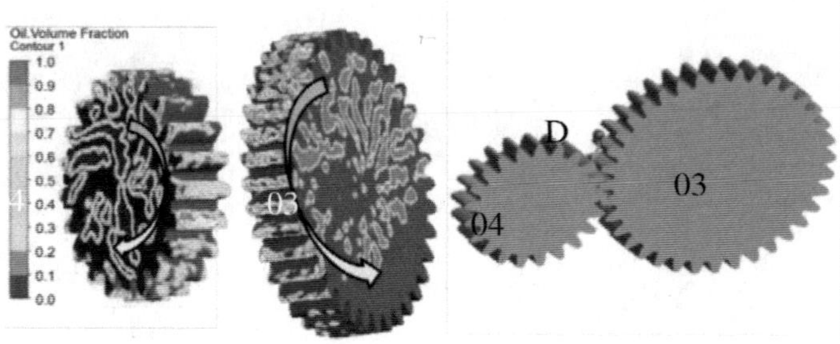

Fig.10 Axis of the nozzle is in the direction where the motion of air is weakest

(a) Injection angle = 0° (b) Injection angle = 5°

Fig.11 Axis of the nozzle is in the direction where the motion of air is not weakest

6. Conclusion

In this paper, the influence of nozzle injection distance on the motion of oil jet is obtained through high-speed gear jet lubrication experiment, and the experimental results verify the correctness of the numerical simulation model of high-speed gear under jet lubrication.

In a pair of meshing gears, the wind resistance effect caused by pinion gear is stronger than that caused by big gear, which results in that the oil trajectory tends to deflect to the side of big gear when the nozzle is arranged at the engaging-in side, the injection angle and the end angle are all 0 degrees, and the deflection degree increases with the increase of injection distance.

The excessive injection distance directly reduces the amount of oil in the meshing area, so the value should not be too large. The existence of the end angle will increase the difficulty of oil entering the meshing area, so it should be as small as possible or even 0 in design.

When the nozzle is located at the engaging-out side, most of the oil is affected by the cross wind resistance effect, and can not enter the meshing area because of the direct reverse flow. Therefore, in practical application, as far as possible, avoid the oil supply on the gear's engaging-out side.

In addition, this paper explores the optimum injection angle by solving single-phase flow, and obtains the optimum layout orientation of the nozzle in the central plane.

References

[1] Handbook A S M. Friction, lubrication, and wear technology [J]. *ASM International*, 1992, 18: 362-369.

[2] Hamel M, Addali A, Mba D. Investigation of the influence of oil film thickness on helical gear

defect detection using acoustic emission [J]. *Applied Acoustics*, 2014, 79: 42-46.

[3] Dowson D. Elastohydrodynamic and micro- elastohydrodynamic lubrication [J]. *Wear*, 1995, 190(2): 125-138.

[4] Castro J, Seabra J. Global and local analysis of gear scuffing tests using a mixed film lubrication model [J]. *Tribology International*, 2008, 41(4): 244-255.

[5] Wang Y, Niu W, Wei S, et al. Influence of spin flow on lubricating oil jet-design method of oil spray parameters to high speed spur gears [J]. *Tribology International*, 2015, 92: 290-300.

[6] Akin L S, Mross J J, Townsend D P. Study of lubricant jet flow phenomena in spur gears [J]. *Journal of Tribology*, 1975, 97(2): 283-288.

[7] Massini D, Fondelli T, Facchini B, et al. High speed visualizations of oil jet lubrication for aero-engine gearboxes [J]. *Energy Procedia*, 2016, 101: 1248-1255.

[8] Townsend D P, Akin L S. Study of lubricant jet flow phenomena in spur gears-out of mesh condition [J]. *Journal of Mechanical Design*, 1978, 100(1): 61-68.

[9] Fondelli T, Andreini A, Da Soghe R, et al. Numerical simulation of oil jet lubrication for high speed gears [J]. *International Journal of Aerospace Engineering*, 2015, 2015.

[10] Wang Yanzhong. NIU Wentao, Tang wen, et al. Research on deflection of nozzle spray direction [J]. *Journal of Aerospace Power*, 2012, 27(7):1665-1670. (in Chinese)

[11] Blessing M, König G, KrÜger C, et al. Analysis of flow and cavitation phenomena in diesel injection nozzles and its effects on spray and mixture formation [C]// *Fuel Injection Systems 2003: IMechE Conference Transactions*, 2003. 2003, 2:21.

[12] Payri F, BermÚdez V, Payri R, et al. The influence of cavitation on the internal flow and the spray characteristics in diesel injection nozzles [J]. *Fuel*, 2004, 83(4): 419-431.

[13] Zhang A, Montanaro A, Allocca L, et al. Measurement of diesel spray formation and combustion upon different nozzle geometry using hybrid imaging technique [J]. *SAE International Journal of Engines*, 2014, 7(2014-01-1410): 1034-1043.

[14] Payri R, Viera J P, Gopalakrishnan, et al. The effect of nozzle geometry over internal flow and spray formation for three different fuels [J]. *Fuel*, 2016, 183: 20-33.

[15] He Z, Guo G, Tao X, et al. Study of the effect of nozzle hole shape on internal flow and spray characteristics [J]. *International Communications in Heat and Mass Transfer*, 2016, 71:1-8.

[16] Wang Yanzhong. NIU Wentao, Tang Wen, et al. Influence of spray orientation parameters on spray lubrication process of aero spur gears [J]. *Journal of Aerospace Power*, 2015,30(7):1605-1610. (in Chinese)

[17] Anderson J D, Wendt J. Computational fluid dynamics [M]. *New York: McGraw-Hill*, 1995.

Study and comparison on mechanical properties of various polymers reinforced with ground tires rubber (GTR)

M Marín-Genescà[1], J García-Amorós[2], J Bordes-Arroyo; R Mujal-Rosas[1], Ll Massagués Vidal[2]

[1]Mechanical Engineering Department, ETSEQ-URV, Països Catalans, 26, Tarragona – 43007 Spain
[2]Electrical Engineering Department, ETSE-URV, Països Catalans, 26, Tarragona – 43007 Spain
[3]Electrical Engineering Department, EET-UPC, Colom,1 Terrassa - 08222 Spain
E-mail: marc.marin@urv.cat

Abstract. The recycling of worn tires is a big challenge nowadays and it is necessary for environmental and economic reasons: mass production of tires, as well as the difficult storage or elimination is a real environmental problem. Various methods for recycling tires are currently used, such as mechanical grinding, which puts vulcanized rubber, steel and fibers apart. The rubber may be used in several industrial applications as flooring, insulations, footwear, etc. The aim of this paper focuses on finding a new application for the old used tires (GTR). Tires dust and 7 different polymers have been mixed, and we have checked the maximum accepted values of GTR that can be admitted while keeping mechanical properties within acceptable values, as well as initial polymer microstructure. This would allow including GTR in industrial applications of recycled polymers. The recycled tire dust which results from the industrial milling processes has been divided by categories according to the size of the particles. This has been mixed with different polymers, in different GTR concentrations in order to establish its conduct through mechanical test. With the present study it is intended to give a second outlet to the tires out of use, demonstrating the feasibility of the mechanical properties analyzed.

1. Introduction

The important issue of the accumulation of used tires [1-3] has driven the efforts of the international scientific community to seek solutions for recovery and reuse. Many plastic materials include elastomers to improve its toughness. In general, a thermoplastic or thermosetting polymer acts as a matrix and the elastomer acts as a dispersed phase [4-6]. Moreover, as in other two-phase polymer blends [7-8], the interfacial compatibility between the components is important for achieving the desired properties. In the case of recycled elastomers, expected compatibility is low. One way to increase the compatibility between components is to reduce the degree of cross-linked GTR by devulcanization [9-11]. Significant changes in properties are observed when we change the size of the reinforcement particles [12]. The size of particles is restricted because of the pulverization technical procedures, so we have chosen a simple and cheap way to obtain the classification in the three desired particle sizes (<200μm, 200μm-500μm and >500μm) such as screening, although this method has the disadvantage of using only a part of the initial amount of GTR. It is, therefore, to determine what percentage of GTR can be added to different polymers matrix (PVC, EVA, HDPE, PP, PA, ABS and PS), keeping mechanical properties, as well as the polymer initial microstructure [13-14] within an acceptable range of mechanical values. This would allow to add GTR to various industrial processes. To this end, we have analyzed various concentrations of polymer/GTR (from 0% to 70% of GTR), with the three indicated particle sizes. Finally, we will give some applications that could have these compounds analyzed, there will be a valuation of the residual GTR for applications that may be interesting, and thus give out a waste like the GTR, which today represents a very important problem in developed societies.

2. Methodology

2.1 Materials

The polyvinyl chloride (PVC) with a fluidity index of 1.35 g/min and a density of 1.225 kg/m^3. High density polyetilene (HDPE) recycled, with a melt flow index of 1.35 g/min and density of 960 kg/m^3 was used in this study. The EVA (Etylene Vinyl Acetate) copolymer, with 18% of vinyl acetate and 82% ethylene, melt flow rate of 0.2 g/min and a density of 937 kg/m^3. Polypropylene (PP) with a melting

IOP Conf. Series: Materials Science and Engineering **542** (2019) 012027 doi:10.1088/1757-899X/542/1/012027

temperature of 165 ºC, a flow rate of 0.55 g/min, and a density of 902 kg/m^3. ABS it is made up of 30% acrylonitrile, 20% butadiene, and 50% styrene, its melting temperature is 230 ºC and its density is 1050 kg/m^3. Polyamide 6 (PA), also known as Nylon 6, its melting temperature is 220 ºC, its melt flow rate is 1.55 g/min and its density is 1130 kg/m^3. Polystyrene, PS, (styrene-butadiene-styrene), its melting temperature is 180°C, its melting index is 1.45 g/min and its density is 1050 kg / m^3. On the other hand the old used tire (GTR), with a particle size less than 700 µm, was verified by TGA analysis that carbon black content was about 35%. The original GTR was separated by sieving into three categories of size particles: <200 µm, 200-500 µm and >500 µm.

2.2 Preparation of the compound and mechanical analysis

The recycled tire powder was dried in an oven at 100ºC for 24h. Five samples of Polymer/GTR compound, varying the composition (5%, 10%, 20%, 40%, 50% and 70% of GTR), were prepared for each particle size. The mixing process was done with a *Brabender* mixer machine, at different temperatures. Polymer/GTR laminates were obtained by using a hot plates press at constant pressure 200 bar and different temperatures for 10 min depending on the polymer to be treated. Samples for testing were properly set up according to the specifications of ASTM-D-638 type V. A sample of the pure polymer was also prepared with the same requirements in order to obtain comparable results. Stress–strain tests were carried out with an *Instron 3366–10kN* universal machine, as the specifications of ASTM-D-638 type V. The test speed was 20 mm/min. The test temperature was 23±2 ºC, and the relative humidity was of 50±5%. The study of mechanical properties, according to the GTR concentration in the matrix and the particle sizes include the Young's modulus, tensile stress, elongation at break and toughness. Five specimens for testing (Fig 1) were used.

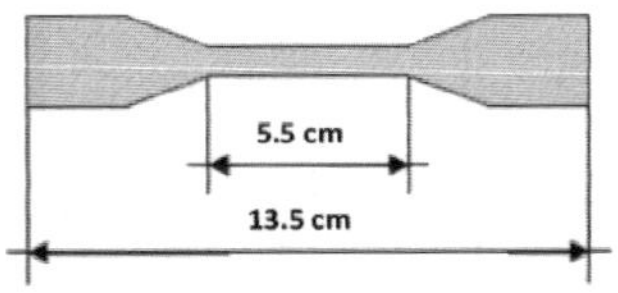

Figure 1. Obtained samples for testing.

3. Results and discussion

3.1 Mechanical Properties

Figure 2 and 3 shows the results by testing the stress-strain mechanical properties of different polymers analyzed. These properties were analyzed with different GTR concentrations and the three categories of sizes of reinforcing particles in the Polymer matrix [15-19]. Figure 2-a shows the Young's modulus of the composite, the rigidity for small GTR concentrations (>10%) decreases significantly regarding the original PVC, from a value of 2800 MPa (5% GTR) to 1627 MPa (70% GTR) for small particle sizes (<200 µm). We can also see how rigidity decreases with the other particle sizes, from original Young's modulus value of 2461 MPa (5% GTR) to 1120 MPa (70% GTR) for particle size of 200-500µm, and to 921 MPa for particle size of >500µm for the same GTR concentration. This is due to the fact that large particles have higher probability of cracks, and they also crowd during the melting process. When the content of GTR continues to increase, interfacial adhesion is weakened, and it causes a diminution of stiffness in all the cases. Thus, for concentrations of 40% and 50% GTR, values are 2005 MPa and 1900 MPa (28.4% and 32.14% lower than those of the EVA/GTR-5% compound for <200µm particles), reaching slightly higher values than those of pure PVC for 70% GTR concentrations, no matter the size of the particles is (1627MPa, 1120MPa and 921MPa, for the three particle sizes 200µm, 200-500µm and >500µm, respectively). In the HDPE composites with GTR, can check that the decrease is so high for the 5% GTR compounds (figure 2-b), for HDPE/GTR-5%, we have this values 33.84% for 200µm size of GTR particles, 28.04% for 200-500µm of size, and 23.05% >500µm, this probes that the decrease is 18% and 32% respectively in function of the particle size, is really proved that the particle size is a very important fact for the mechanical behavior [20-21].

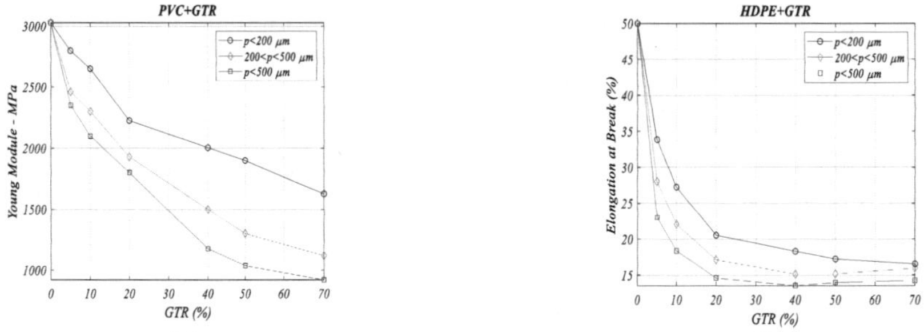

(a) (b)

Figure 2. a) Young's modulus (MPa) for different concentrations of PVC/GTR and particles sizes. b) Elongation at break (%) for different concentrations of HDPE/GTR and particles sizes.

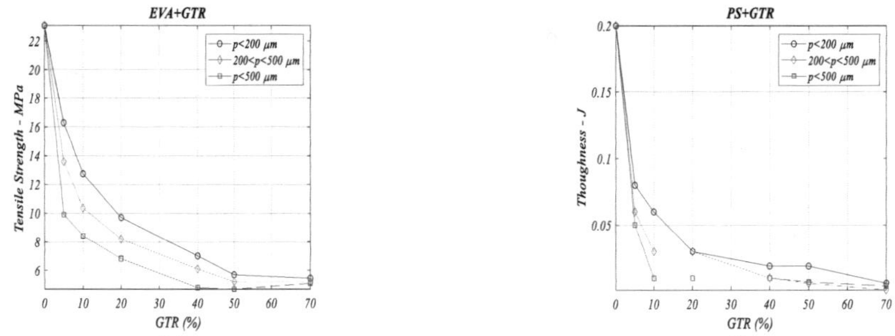

(a) (b)

Figure 3. a) Tensile Stress (MPa) for different concentrations of EVA/GTR and particles sizes. b) Thoughness (J) for different concentrations of PS/GTR and particles sizes.

Figure 3-a shows the tensile stress, which differs from the stiffness, showing a decrease since low GTR concentrations. Thus, for 5-10% GTR concentrations, the decrease of tensile stress compared to pure EVA are quite significant, from 23MPa to 16.2MPa and to 12.7MPa (28% and 44% respectively for particles size <200μm). These decreases are more significant for larger particle sizes, reaching 13.6MPa (41%) and 10.4MPa (54%) respectively for particles sizes between 200μm-500μm, and 9.9MPa (57%) and 8.4MPa (63%) respectively for >500μm particles. From 20% GTR, values fall sharply regardless of particle size or GTR concentration in the matrix. Values are 3.5 - 4 times lower than the original EVA. It shows poor compatibility between components when the percentage of reinforcement is increased. We can see similar trends in the figure 3b) for a PS composites analyzed. The optimum particle size is the p<200μm. The addition of GTR decrease the mechanical properties analyzed [22].

As we can see in the figures 2 a) and b) and 3 a) and b) analyzed, can assure that the size particle of p<200μm is the optimum particle size analyze. With this condition are analyzed the 7 polymers mixed with GTR generated in the next figures 4 a) and b) and 5 a) and b). The addition of old used tires particles in all compounds produces a decrease in the four mechanical properties analyzed in general like young modulus (Fig 4-a) and tensile strength (Fig 4-b) elongation at break (Fig. 5-a) and toughness (Fig. 5-b).

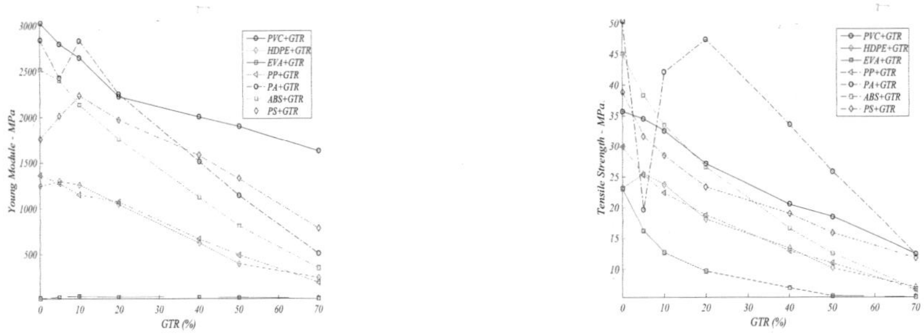

(a) (b)

Figure 4. a) Young Module (MPa), b) Tensile Strength (MPa), for seven polymers/GTR concentrations and particle size <200μm.

For the Young Module analysis, we can see how the PVC and EVA compounds has better behavior with the addition of the GTR percent in the composite, and in the Tensile Strength analyzed we can see how the PA and PVC compounds has better behaviors with the addition of quantities of GTR (near 70%).

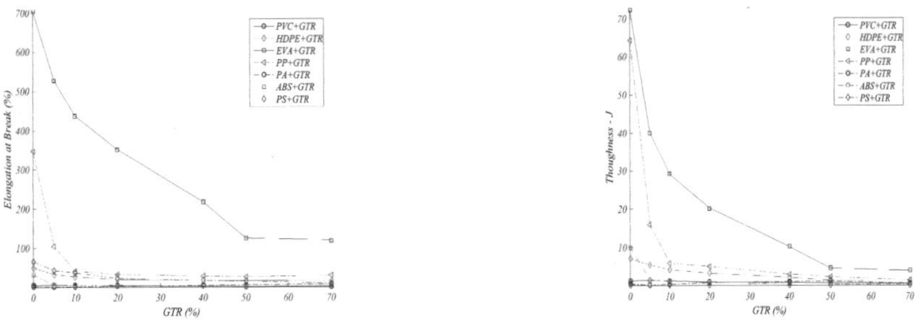

(a) (b)

Figure 5. a) Elongation at Break (%), b) Toughness (J), for seven polymers/GTR concentrations and particle size <200μm.

For particle sizes <200μm elongation at break of the GTR compounds [23], the optimum behavior is for EVA compounds, that goes from 704% to 528%, 437% and 351%, which means decreases of 29%, 43% and 50% for GTR concentrations of 5%, 10% and 20%. From 40% concentration, the values are always lower than 3.5 times those of pure EVA. The reduction of deformation capacity of the elastomer influences the decrease in elongation and, subsequently, the decrease of hardness, and we have similar behaviors of the Toughness property for the composites with EVA/GTR. Regarding toughness, these drops are even greater. Actually, regarding the compound with the smallest particle size (<200μm), the toughness goes from 72.3J for pure EVA to 40J or 29.2J for 5% and 10% GTR concentrations. It will decrease even more for larger particle sizes. For concentrations above 20%, considerations on the particles sizes or the percentage of GTR are not important at all, since even 10J of energy are not reached in any case, and it means that these values are seven times lower than the original EVA. The decrease in elongation at break is related to the imperfect interfacial adhesion between components. The rest of polymers are so low values for the elongation at break and toughness properties. The incidence of poor adhesion between phases is a particularly important result [23].

4. Conclusions

Firstly, it can be clearly seen how the particle size decisively influences the properties analyzed, it is observed as GTR particle of lower sizes (<200μm) have better mechanical behavior in the figures 2 a) and b) and 3 a) and b), so, the mechanical properties analyzed have a better behavior with a GTR minimum particle size (<200μm), this is due to the imperfect interfacial adhesion between components is better for a low size of particle than a highest size of particle (p>500μm). Stress-strain tests show how, by adding the GTR in the matrix up to 10%, and especially for small particle sizes, the Young's modulus of the compound increases, although other mechanical properties decrease. With the EVA,

IOP Conf. Series: Materials Science and Engineering **542** (2019) 012027 doi:10.1088/1757-899X/542/1/012027

PVC and HDPE composites. This behavior may be due to the fact that reinforcement-matrix is correct for these formulations, and therefore, some mechanical properties such as stiffness improve. However, for GTR concentrations above 10%, with any particles size, all mechanical properties decrease, and the larger the size of the particle is, the more significant the decrease is. The results obtained from the analysis of these compounds, only obtained from recycled polymers, show that that at 10-20% GTR concentration is the limit concentration value for keeping acceptable values of mechanical properties of the compound, considering exclusively the GTR particle size as a variable, and without any previous pre-treatment considered. Other methods such as prior devulcanization of GTR should be tested, to see if values could be increased up to 20-25% of GTR. This would allow its use in various fields of industry, as for example in pipes for waste in construction works and conduits for electrical cables, as well as for waste deposits in construction works, through these applications we could introduce recycled GTR in industrial applications, and so we could give also an exits to quantities of GTR in disuse that nowadays represent a problem both its storage and its recycling.

5. References

[1] European Tyre Recycling Association (ETRA). Available at: www.etra-eu.org.

[2] Used Tyre Working Group (UTWG). *Tyre Recycling*; Department of Trade and Industry: London, UK. Available at: www.tyredisposal.co.uk, November, 2006.

[3] Liu HS, Richard CP, Mead JL and Stacer RG. *Development of Novel Applications for Using Recycled Rubber in Thermoplastics*; Technical Research Program, Chelsea Center for Recycling and Economic Development, University of Massachusetts: Lowell, 2000.

[4] Figovslq O, Beilin D, Blank N, Potapo J and Chernyshe V. *Development of polymer concrete with polybutadiene matrix. Cem Concr Compos* 1996, 18, 437-444.

[5] Hernandez-Olivares F, Barluenga G, Bollatib M and Witoszekc B. *Static and dynamic behaviour of recycled tyre rubber-filled concrete. Cem Concr Compos* 2002, 32, 1587-96.

[6] Goncharuk GP, Knunyants MI, Kryuchkov AN and Obolonkova ES. *Effect of the specific surface area and the shape of rubber crumb on the mechanical properties of rubber-filled plastics. J Polym Sci Part B: Polym Chem* 1998, 40, 166-169.

[7] Dierkes WK. *Rubber recycling.* In Recent research developments in macromolecules, Pandalai SG, Ed.; Trivandrum: Research Signpost, 2003, vol. 7, pp. 265–292.

[8] Radeshkumar C and Karger-Kocsis J. *Thermoplastic dynamic vulcanisates containing LDPE, rubber, and thermochemically reclaimed ground tyre rubber. Plast Rubber Compos* 2002, 31,99-105.

[9] Yehia A, Mull MA, Ismail MN, Hefny YA and Abdel-Bary EM. *Effect of chemically modified waste rubber powder as a filler in natural rubber vulcanizates. J Appl Polym Sci* 2004, 93, 30-36.

[10] Colom X, Andreu-Mateu F, Cañavate FJ, Mujal R and Carrillo F. *Study of the influence of IPPD on thermo-oxidation process of elastomeric hose. J Appli Polym Sci* 2009, 5, 2011-2018.

[11] Cepeda-Jimenez CM, Pastor-Blas MM, Ferrándiz-Gómez TP and Martín-Martinez JM. *Surface Characterization of vulcanized Rubber treated with sulphuric acid and its adhesion to polyurethane adhesive. J Adhesion* 2000, 73, 135-160.

[12] Mujal R, Marín-Genescà M, Orrit J, Rahhali A, Colom X, *Dielectric, mechanical, and thermal characterization of high-density polyethylene composites with ground tire rubber.* Journal of thermoplastic composite materials, 2012, vol. 25, núm. 5, p. 537-559

[13] Nakason C, Kaesaman A and Supasanthitikul P. *The grafting of maleic anhydride onto natural rubber. Polym Test* 2004, 23, 35-41.

[14] La Mantia FP, Lo Verso S and Tzankova Dintcheva N. *EVA Copolymer Based Nanocomposites. Macro Mat Eng* 2002, 287, 12, 909-914.

[15] Kim JI, Ryu SH and Chang YW. *Mechanical and dynamic mechanical properties of waste rubber powder/HDPE composite. J Appl Polym Sci* 2000, 77, 2595-2602.

[16] Colom X, Cañavate J, Carrillo F, Velasco JI, Pages P, Mujal R and Nogues F. *Structural and mechanical studies on modified reused tyres composites. Eur Polym J* 2006, 42, 2369-2378.

[17] Ganesh B and Unnikrishnan GJ. *Cure characteristics, morphology, mechanical properties, and aging characteristics of silicone rubber/ethylene vinyl acetate blends. Appl Polym Sci* 2006, 99, 3, 1069-1082.

[18] Tang LW, Tam KC, Yue CY, Hu X, Lam YC and Li L. *Rheological and mechanical properties of compatibilized polystyrene/ethylene vinyl acetate blends. J Appl Polym Sci* 2004, 94, 5, 2071-2082.

[19] Shih-Kai C, Po-Tsun C, Cheng-Chien W and Chuh-Yung C. *Mechanical properties of ethylene-vinyl acetate/ polystyrene blends studied by in situ polymerization.* Journal of Applied Polymer Science. *J Appl Polym Sci* 2003, 88, 3, 699-705.
[20] Mujal R, Orrit-Prat J, Ramis-Juan X, Marin-Genesca M, Ahmed Rahhali. *Dielectric, Thermal, and Mechanical Properties of Acrylonitrile Butadiene Styrene Reinforced with Used Tires.* Advances in Polymer Technology 32:E399-E415. Wiley 2013. 0730-6679
[21] Mujal R, Orrit-Prat J, Ramis-Juan X, Marin-Genesca M. *Electrical application of polyamide reinforced with old tire rubber (ground tire rubber): Dielectric, thermal, mechanical and structural properties. Journal of thermoplastic composite materials.* 27(9):1209-1231. 2014. 0892-7057
[22] El-Nashar DE, Eid MAM, Abou Aiad TH and Abd-El-Messieh SL. *Electrical and Mechanical Investigations on Polyvinyl Chloride Filled with HAF Black. J Reinforc Plast Compos* 2009, 28, 14, 1763-1773.
[23] Saad ALG, Aziz HA and Dimitry OIH. *Studies of Electrical and Mechanical Properties of Poly(vinyl chloride) Mixed with Electrically Conductive Additives. J Appl Polym Sci* 2004, 91, 1590-1598.

Acknoledgements
This work was funded by the Ministerio de Economía y Competitividad de España ENE2015-64117-C5-3-R (MINECO/FEDER).

IOP Conf. Series: Materials Science and Engineering **542** (2019) 012028 doi:10.1088/1757-899X/542/1/012028

Study on improvement of design method for airport flexible pavement in Vietnam

Huy Khang Pham[1] and Dinh Chung Nguyen[2]

1 Professor, University of Transport and Communications, Hanoi, Vietnam
2 Airport Design and Construction Consultancy One Member Limited Liability Company ADCC, Hanoi, Vietnam
E-mail: huykhangpham.utc@gmail.com

Abstract. Presently, there are 22 airports in operation comprising of 9 international airports and 13 domestic airports in Vietnam. According to the master plan of airports national wide, there are 124 airports and aerodromes both civil airports and military airfield. Design of airport flexible pavement is an important step during design the airport runway and taxi. Currently airport flexible pavement in Vietnam designs based on Vietnamese national standards, American standards and Russian standards. However, these standards have some issues such as design load, effect depth of the aircraft load, relative deflection caused by the aircraft and critical relative deflection. Therefore, this research points out improper issues that is not compatible with current practices of increase in airplane total weight. It also clarifies unfavorable effects of existing flexible pavement design method to the performance of runway pavement in some airports in Vietnam. Some technical measures are proposed for improvement of the current flexible pavement design standards and specifications are also recommended.

1. Introduction

Vietnam has 22 airports in operation comprising of nine international airports and 13 domestic airports. According to the master plan of airports national wide, there are 124 airports and aerodromes both civil airports and military airfields; of which, 28 dual-function airport for both civil and military, 96 military airports. The number of military airports need for new construction is 30; 76 airports and military airfields should be upgraded or rehabilitated in the coming time.

Regarding to airport pavement structures in operation, the sharing of asphalt concrete, composition structure of asphalt concrete and cement concrete, and cement concrete are 50%, 14% and 36%, respectively. Applying asphalt concrete pavement is a common trend due to its advantages such as: high strength, good riding comfort, low noise, no dust, good skid resistance, fast construction progress, as well as reasonable cost. Moreover, it is also quite convenient for maintenance, rehabilitation or upgrading. Therefore, asphalt pavement has become more proper application especially in case of airports in operation.

National standard coded TCVN 10907:2015 is the current standard for design of airport pavement. The standard is originated from Russian standard SNIP with less localization to actual conditions of airport operation in Vietnam. Therefore, there exists shortcomings in the standard as the result of less attention and limited investment on standardization. In some airports, there have occurred some damages on new repaired pavements by asphalt concrete or polymer modified asphalt such as rutting, deformation that affect safety operation of airplanes. The main causes of the damages have been identified that consist of design method, the effect of aircraft load, operation conditions, quality of construction materials, construction quality as well as maintenance performance [1]. Vietnam is a member of the International

Civil Aviation Organization (ICAO). The current aircraft fleet are mainly from Boeing and Airbus. Design of pavement structures of runways, taxiways, and aprons must comply with ICAO regulations and meet the requirements for the operation of the aircraft used by international airlines. Therefore, the research on the method of flexible airport pavement design in order to contribute for the improvement, supplementation to current design standards as well as propose new design method to meet the condition or operation of heavy commercial aircrafts with high frequencies would be in necessary in Vietnam. Therefore, in order to contribute for the improvement, supplementation to current design standards as well as propose some technical measures meet the condition or operation of heavy commercial aircrafts with high frequencies would be in necessary in Vietnam.

2. Some shortcomings of Vietnam standard

Vietnam standard TCVN 10907:2015 is compiled from СНИП standard from Russia with supplementation of provisions of the survey based on Advisory Circular - AC No 150 / 5320-6 [2]. When applying these standards during the design of airport pavements for heavy aircrafts with high operation frequency, the shortcomings are understood as follows.

2.1. Design load

Design load is requested to convert to design aircraft. Maximum weight of 85 tons tangdem, 1.5 MPa tire pressure. Actual operation at airports with the B787-8 / 9, B777-300 / 200, B747-400 / 800, A350-900, A340, A330, A321,... aircraft load and tire pressure are higher than the standard values (Figure 1). The load on the main gear of the aircraft in operation is much greater than the standard value. In pavement structural analysis, the total pavement thickness usually at least two times larger than the equivalent wheel path diameter, beyond the range in analysis nomogram.

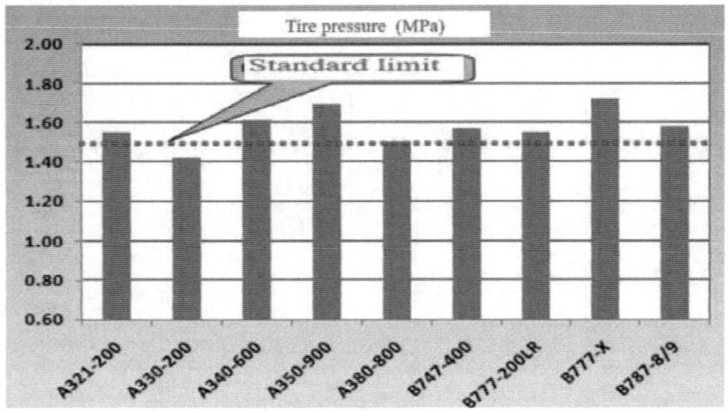

Figure 1. Comparison of air pressure of aircraft in operation with standard

2.2. The effect depth of the aircraft load

The effect depth of the aircraft load is the thickness of the compressed soil layer as prescribed in Tables 6, 7 in TCVN 10907: 2015 varies from 5 meters to 6 meters depends on the load on one wheel of the main axle and consideration of soil property in the ground. For aircrafts with operation weight higher than the specified load in the standard, actual depth would be more than 6 meters depending on the load and bearing capacity of the ground. The depth needs being determined in order to consider for proper measures of treatment if necessary for improvement of the load bearing capacity.

2.3. Calculation of the relative deflection caused by the aircraft λ_d

The relative deflection caused by the aircraft's load depends on the tire pressure Pa and the elastic equivalent modulus Eed (overall elastic modulus) of the pavement and subgrade. Actual structural analysis for new aircraft with modulus of subgrade less than 45MPa, the total thickness of pavement layers is usually two times higher than the equivalent diameter of the aircraft wheel path that is out of range of nomograph that helps to identify coefficient Ψ_k [1]. Therefore, it is impossible to identify Eed and relative deflection by aircraft load λd.

2.4. Calculation of the critical relative deflection λ_u
According to TCVN 10907:2015, critical relative deflection of pavement is determined based on the take-off frequency of aircraft per day (no more than 100 takeoffs per day), tire pressure (less than or equal 1.5 Mpa) and subgrade condition following nomograph 8G [1].

3. Proposal of technical measures to improvement the design of airport flexible pavement

3.1. Improvement of the Vietnam standard
In order to meet the actural conditions of airport operation, when applying TCVN 10907: 2015, it is necessary to study and supplement the following issues.

3.1.1. Determination of ground depth compressed under the effect of aircraft load
Application of finite element method to calculate the depth of effect of the load of the types of aircraft in operation. Condition determined by the finite element method is deformation or vertical displacement under the effect of the load U ≈ 0 (ε ≈ 0) or the load caused by the aircraft that is less than 5% of the load on the structural and ground layers itself. Based on the strain contour, displacement at the position of the displacement or deformation node to zero to determine the depth of effect of the load on the ground.
Analysis for Airbus, Boeing aircrafts with subrade loading capacity in elastic modulus varies from 28MPa to 100MPa. Calculated results are presented as nomographs in Figure 2 and Figure 3.

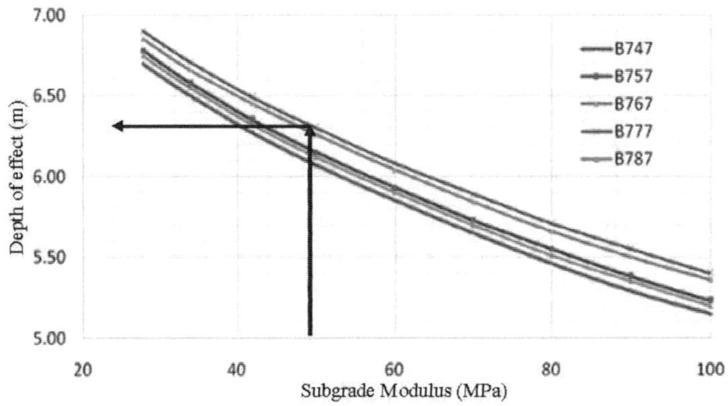

Figure 2. The depth of effect on the ground of the Boeing aircraft

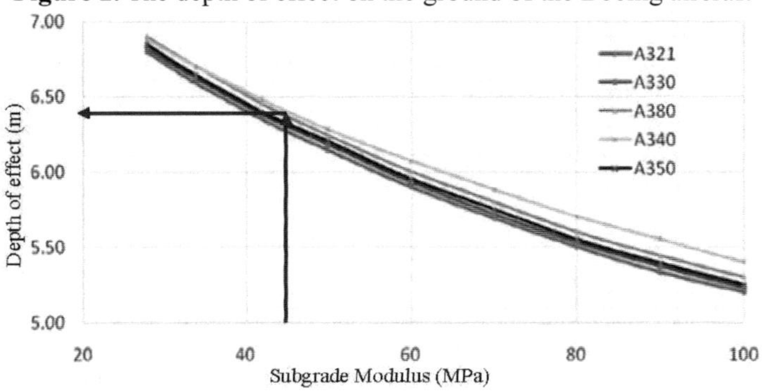

Figure 3. The depth of effect on the ground of the Airbus aircraft

IOP Conf. Series: Materials Science and Engineering **542** (2019) 012028 doi:10.1088/1757-899X/542/1/012028

The depth of effect of the payload of Boeing and Airbus aircrafts depends on the subgrade strength in elastic modulus or CBR value and the aircraft load; with subgrade elastic modulus of equal or less than 60MPa, the depth of effect of the type of aircraft is usually deeper than 6m. Nomograph in Figures 2 and 3 only applicable for new airport pavements.

3.1.2. Calculate relative deflection when design load out of range of nomograph

When calculating the type of aircrafts with a heavy load and low load bearing capacity of subgrade, the total thickness of the large structure, $t_{tot}/D_e \geq 2$ using the empirical formula (1) proposed by Bacburg (Russia) as follows [3]:

$$\psi_k = \frac{1{,}05E/E_{mt}}{(1-E/E_{mt})/\sqrt{1+4(t_{tot}/D_e)^2(E/E_{mt})^{-2/3}}+E/E_{mt}} \quad (1)$$

Where:

$E_{m1}, E_{m2}, \ldots E_{mt}$ is the elastic modulus of material layers (MPa);
$t_1, t_2, \ldots t_n$ is the thickness of structure layers (cm);
t_{tot} is the total thickness of the pavement structure (cm);
E is the elastic modulus of the existing ground (MPa);
D_e is the diameter of the circle (m) of the equivalent axle of the equivalent axle load.

3.1.3. Calculate the critical relative deflection when design load out of range of nomograph

The critical relative deflection of the pavement λ_u depends on the type of soil, pressure on tires and number of times the conversion effect of N_r of aircraft load. From the nomograph, the relation between the deflection λ_u and the pressure on tires, frequency of aircraft operation is nonlinear. To make extrapolation for the case that design load out of range of the nomograph (in case Pa> 1.5Mpa, Nr> 100 passes / day), digitize the nomograph 8G in Excel, construct a regression function that determines the critical relative deflection following exponential rule on the Matlab software. The results of the regression functions are as follows:

- Clayey, clay, sandy (including laterite soil), determined by the formula:

$$\lambda_u = \frac{\left[3{,}165\left(-0{,}8395p_a^2+3{,}208p_a+0{,}3722\right)\right]}{N_r^{0{,}1313}} \times 10^{-3} \quad (2)$$

- Dust sandy determined by the formula:

$$\lambda_u = \frac{\left[3{,}163\left(-0{,}9266p_a^2+3{,}133p_a+0{,}0255\right)\right]}{N_r^{0{,}1023}} \times 10^{-3} \quad (3)$$

- Medium and large grain sand, determined by the formula:

$$\lambda_u = \frac{\left[1{,}889\left(-0{,}0702p_a^2+1{,}793p_a+0{,}7897\right)\right]}{N_r^{0{,}1125}} \times 10^{-3} \quad (4)$$

The error calculated by regression formula and nomograph $\Delta\lambda_u < 1.79\%$. According to the Russian Airport Design Guide [4], it is acceptable for the error of less than 10% in case of relative deflection, and error of less than 5% in case of flexural analysis for asphalt layers. Therefore, the error between the regression and nomograph of 1.79% is acceptable.

3.2. Modification and application of AC 150 / 5320-6 (2016) in Vietnam

Vietnam is a member of the International Civil Aviation Organization (ICAO). The current aircraft fleet are mainly from Boeing and Airbus. Design of pavement structures of runways, taxiways, and aprons must comply with ICAO regulations and meet the requirements for the operation of the aircraft used by international airlines. Therefore, it would be in necessary to improve and supplement to current design standards as well as propose some calculation methods to meet the condition or operation of heavy commercial aircrafts with high frequencies in Vietnam.

3.3. Concerned issues for application in Vietnam

This method is built based on natural conditions, materials, construction technologies of America. To apply in Vietnam, there should be researches, evaluation and adjustment for customization due to local conditions of geographical, natural conditions, materials, construction technology;
Detailed regulations on materials, construction specification and acceptance of structural layers issued by FAA in standard AC 150 / 5370-10 (2014), standard construction and acceptance of the airport [4]. When applying design standards AC 150 / 5320-6F, it is necessary to compile AC150 / 5370-10 for consistent application in Vietnam as a basis for clearly defining construction material standards, test methods, construction technical guidelines and construction acceptance in construction of the flexible pavement of the airports. As the fact that Vietnam does not have separated standards in materials testing, construction and acceptance for airport pavement; standards of road pavements are referred.

4. Conclusions

In the short term, parallel use of Vietnam standard (TCVN 10907:2015) with some supplementation reasonable results from studies, and AC 150/5320-6 for comparison, verification and selection of thicker structure for more safety. After completion researches, analysis, and evaluation, new Vietnam standard should be promulgated for unified application.
In the long term, it is recommended to apply the standard AC 150 / 5320-6 to design pavement structures for airports and airfields in Vietnam. This is the standard recommended by the International Civil Aviation Organization (ICAO) with practical application in many countries.
The Civil Aviation Administration of Vietnam is recommended to study and promulgate AC 150/5320-6 as in-house standard to apply with additional researches, adjustment taking into consideration of local conditions. Promulgation national standard for airport pavement design and its quality evaluation is also recommended.
Material investigation, design of asphalt mix that proper for conditions of high temperature of 600C and aircraft operation frequency of more than 100 operations/day at high air tire pressure up to 1.75MP for new construction of airports in Vietnam in the future.
Research, compile the set of standards on testing, construction and acceptance of airport pavement according to AC150 / 5370-10 (2014) for completion.

Acknowledgments

Authors wishing to acknowledge to all colleagues in University of Transport and Communications, Hanoi, Vietnam for their supporting and cooperation to complete this research work.

References

[1]. Vietnam standard TCVN 10907:2015, Civil Airport - Airport Pavment - Design specifications.
[2]. U.S. Department of Transportation (2016), "Airport Pavement Design and Evaluation", Advisory Circular - AC No 150/5320-6F, Ferderal Aviation Administration.
[3]. В.А. Кульчицкий, В.А. Макагонов (2012), АЭРОДРОИМНЫЕ ПОКРЫТИЯ СОВРЕМЕННЫЙ ВЗГЛЯД, Физматлпт, Москва.
[4]. Глушков.Г.И (1994), ЖЕСТКИЕ ПОКРЫТИЯ АЭРОДРОМОВ И АВТОМОБИЛЬНЫХ ДОРОГ, Транспорт, Москва.

IOP Conf. Series: Materials Science and Engineering **542** (2019) 012029 doi:10.1088/1757-899X/542/1/012029

Fabrication of Ni60+25%WC reinforced steel matrix surface composites by induction cladding

Liu Heping[12]*, **Sun Fenger**[1], **Cheng Shaolei**[1], **Gao Yibo**[1], **Liu Langlang**[1]

1School of Materials Science and Engineering, North University of China, Taiyuan 030051, China
2School of Materials Science and Engineering, Taiyuan University of Technology, Taiyuan 030024, China
E-Mail: peace666@126.com

Abstract: Failure of mechanical parts usually occur on the surface or start from the surface. such as wear, oxidation corrosion and fatigue. This work uses high frequency induction cladding technology to improve the overall performance of stainless steel surface. A layer of WC and stainless steel is coated on the surface of the stainless steel. The effects of different WC material contents and different induction heating current intensities on the cladding were analyzed. After the preparation of the sample, the morphology and properties of the cladding layer were analyzed by SEM, XRD and other test methods. The results show that when the content of WC is 2%-3%, the heating time is 90s and the heating current is 860A-890A. The surface of the cladding layer is most densely formed and there is no obvious over-burning phenomenon and pore generation.

1. Introduction

Induction cladding technology is developed on the basis of induction heating technology. Induction cladding is the pre-preparation of an alloy layer on the substrate of a work piece. The method uses high frequency induction equipment to heat. This cladding technique allows the preformed alloy layer to melt and form a metallurgical bond with the substrate. The cladding layer prepared by the high frequency cladding method is called a high frequency cladding layer. There are two main pre-fabrication methods for the induction cladding alloy layer: a pre-formed powder method and a pre-formed powder block method. High frequency induction cladding is a relatively economical high density energy surface modification technology. This technology adjusts the operating frequency of the device to allow heat to be concentrated in the area where heating is required. Thereby, the

* Corresponding author.Tel.: +86-0351-3557519
E-mail address: peace666@126.com

workpiece and the substrate can be metallurgically bonded. The heat of induction heating is concentrated on the surface layer, and the heat influence on the substrate is small. The induction cladding technology achieves two requirements of small thermal influence and strong bonding force between the substrate and the cladding layer[1-3].

The composite material obtains higher specific stiffness and specific strength, better heat resistance and lower coefficient of thermal expansion. At the same time, the particle reinforced metal matrix composite has good shape, toughness and workability. Particle reinforcement technology is mainly used under working conditions such as wear resistance, heat resistance and corrosion resistance. The ceramic particle reinforced metal-based surface composite material has the characteristics of good metal plastic toughness[4]. The surface cladding layer will also have the advantages of high hardness of ceramic particles and good wear resistance. The surface composite exhibits superior performance unmatched by single metal or ceramic particles. The Ni60+25%WC particles are used as reinforcements for the cladding layer. This is because the high melting point, high hardness, good wear resistance and stable performance[5,6]. WC has good wettability with steel, and the wetting angle between the two is almost zero.

In this work, the mixed Ni60+25%WC/316 stainless steel composite powders was prepared on the surface of 316L stainless steel by induction cladding. The purpose is to improve the mechanical properties of the stainless steel surface such as hardness and friction. Extend the service life of 316L stainless steel. By observing the macroscopic and microscopic morphology of the ball milled powder and the coated coating, the effect of Ni60+25%WC content and induction cladding current on the cladding layer was analyzed. It has certain reference significance for the preparation of stainless steel surface cladding layer.

2. Experiment

The matrix material in this work is 316L stainless steel. The cladding material is WC powder and 316 stainless steel powder. Ni60+25%WC powders with a mass fraction of 1%, 2%, 3%, and 4% were placed in the ball mill jar. The powder was uniformly mixed by ball milling for 1.5 h at a ball to material ratio of 7:1. The base 316L stainless steel was cut into a metal piece with a size of 25 mm × 25 mm × 4 mm. The 316L surface was pretreated with a 600-1500 mesh corundum gauze. The 316L stainless steel was then ultrasonically cleaned with a 98% alcohol solution. A rosin and turpentine mixture in a ratio of 1:3 was used as a binder. The powder of different ratios after the obtained ball mill was mixed and coated on the surface of 316L stainless steel by induction cladding.

3. Results and Discussion

3.1 Effect of different proportions of Ni60+25%WC and 316 stainless steel mixed powder on ball milling

The SEM photographs at different contents of the Ni60+25%WC/316 stainless steel composite powders after milling with content of 1%, 2%, 3% and 4% are in Figure1. As can be seen from the Figure 1, the mixed powder having a Ni60+25%WC content of 2% showed that the ball-milled particles were most uniform. At the same time, the ball dispersion showed the most uniform dispersion of Ni60+25%WC and 316 stainless steel powder. Moreover, there is no obvious bonding phenomenon on the surface of the 316 stainless steel powder. When the Ni60+25%WC content is 1% and 3%, the surface of the 316L stainless steel powder has obvious surface adhesion. The mixed powder particles are irregular in shape. When the content of Ni60+25%WC reaches 4%. The surface of 316L stainless steel powder is seriously damaged. A large amount of bonding occurs in the mixed Ni60+25%WC/316 stainless steel composite powder.

Figure1different contents of the Ni60+25%WC/316 stainless steel composite powders after ball milling

(a) 1%, (b) 2%, (c) 3%, (d) 4%

3.2 Effect of different proportions of Ni60+25%WC mixed powder on the cladding layer

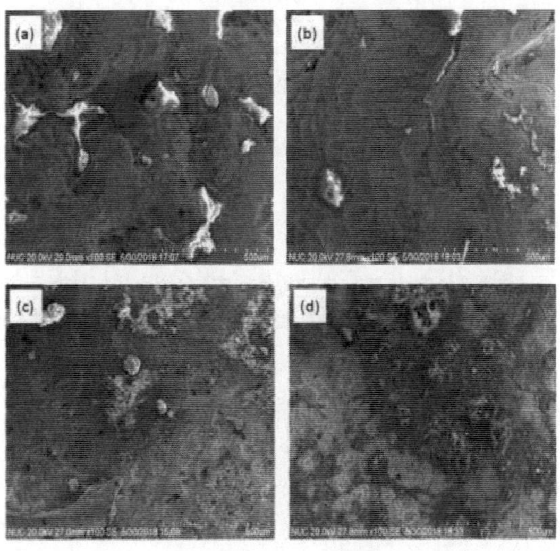

Figure 2 The SEM topography of coating microstructure with different Ni60+25%WC content

(a) 1%, (b) 2%, (c) 3%, (d) 4%

Figures 2 is SEM images of the cladding layer with WC content of 1%, 2%, 3%, and 4%, respectively. It can be seen from the picture that white particles appear in the cladding layers of different Ni60+25%WC content. These white particles are all melted precipitate phases of Ni60+25%WC. The picture shows the different Ni60+25%WC content in the cladding layer. SEM images of 1% and 2%

show that the surface of the cladding layer is relatively flat. There are no obvious holes and cracks on the surface of the cladding layer. And when the content of Ni60+25%WC is higher than 2%, porosity begins to appear on the surface of the cladding layer. The surface of the cladding layer becomes rough. It can be seen from the picture that the Ni60+25%WC content increases from 1% to 4%. The large particles originally formed on the surface of the cladding layer are gradually reduced and dispersed. At the same time they are accompanied by an increase in surface porosity.

In summary, with the increase of Ni60+25%WC content, it will reduce the deposition of WC on the surface of the cladding layer and increase the surface strength of the cladding layer. At the same time, however, it causes thermal agglomeration of Ni60+25%WC particles in the cladding layer. This will create a large number of pores on the surface of the cladding layer. Therefore, under the premise of combining SEM images, the WC content is most suitable at 2%-3%.

3.3 Effect of different induction heating current on cladding layer

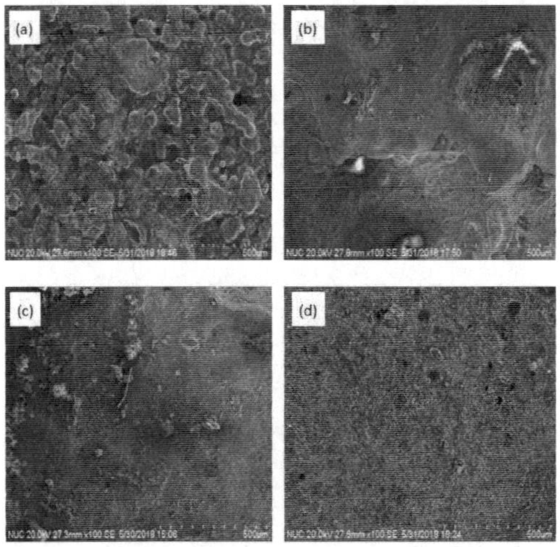

Figure 3 Topography of the cladding layer with different induction heating currents
(a) 830A, (b) 860A, (c) 890A, (d) 920A

Figure 3 shows the topography of the cladding layer with different induction heating currents of 830A, 860A, 890A, and 920A, respectively. When the induction heating current is 830A,a large number of grooves and holes appear on the surface of the cladding layer. The induced current is set to 860A-890A. It can be seen from the SEM image that the surface of the cladding layer is relatively flat. At the same time, there are no obvious defects and cracks on the surface. When heating current reaches 930A, the surface roughness of the cladding layer is increased. The surface produces a large amount of white particles[7,8].

In summary, the induction heating current is higher in a certain range. The resulted cladding layer will be denser and the surface strength will be better. It performs best in the range of induced currents of 860A-890A. When the induction heating current exceeds 890A, over burning may occur. This leads to a decrease in the strength of the cladding layer.. At the same time, it will cause damage to the substrate.

4 Conclusions

In this work, high frequency induction cladding technology is used to clad a Ni60+25%WC and stainless steel composite coating on the surface of stainless steel, in order to improve the comprehensive properties of stainless steel surface. With the increase of Ni60+25%WC content, the deposition of Ni60+25%WC on the cladding layer will be reduced, and the surface strength of the cladding layer will be increased. But at the same time, it will cause the thermal agglomeration of Ni60+25%WC particles in the cladding layer, resulting in a large number of surface pores. Therefore, in combination with the SEM image analysis, the WC content is most suitable for 2%-3%. The higher the induction heating current in a certain range, the denser the surface of the cladding layer and the better the surface strength. When the current is in the range of 860A-890A, the coating performs best. When the induction heating current exceeds 890 A, over burning may occur, which will lead to the decrease of the strength of the cladding layer and the damage to the matrix.

Acknowledgments

This research work was financially funded by the Science and Technology Innovation Project of Shanxi Province (No.2016156), the Postdoctoral Science Foundation of China (No.2016M590214), the Key Research and Development Program of Shanxi Province(No.201803D121028), and the Natural Science Foundation of Shanxi Province, China (No. 2015011036).

References

[1] Chen X, Qin X, Gao K, et al. 2018 Microstructural Evolution of NiCrBSi Coatings Fabricated by Stationary Local Induction Cladding. Journal of Materials Engineering & Performance, p1-11.

[2] Yin J, Wang D Z, Meng L, et al. 2017 High-temperature slide wear of Ni-Cr-Si metal silicide based composite coatings on copper substrate by laser-induction hybrid cladding. Surface & Coatings Technology, p 325.

[3] Yu H L, Zhang W, Wang H M, et al. 2017 In-situ synthesis of TiC/Ti composite coating by high frequency induction cladding. Journal of Alloys & Compounds, p 244-255.

[4] Pines M L, Bruck H A. 2006Pressureless sintering of particle-reinforced metal–ceramic composites for functionally graded materials: Part I. Porosity reduction models. Acta Materialia, , p1457-1465.

[5] Sui Y, Han L, Jiang Y, et al. 2018 Effects of Ni60WC25 powder content on the microstructure and wear properties of WC p, reinforced surface metal matrix composites. Transactions of the Indian Institute of Metals,p1-8.

[6] Liu S L, Zheng X P. 2009 Microstructure and properties of AC-HVAF sprayed Ni60/WC composite coating. Journal of Alloys & Compounds, p254-258.

[7] Hong S I, Lee K W. 2005 Stress-induced reorientation of hydrides and mechanical properties of Zircaloy4 cladding tubes. Journal of Nuclear Materials, p203-208.

[8] Pinkerton A J, Li L. 2003The effect of laser pulse width on multiple-layer 316L steel clad microstructure and surface finish. Applied Surface Science, p411-416.

IOP Conf. Series: Materials Science and Engineering **542** (2019) 012030 doi:10.1088/1757-899X/542/1/012030

Orthogonal Experimental Study on Two Side-direction Burrs of Grind-hardening Workpiece Based on Transverse Feed

Shunxing Gao, Judong Liu, Jiawei Jin, Zhilong Xu and Xiaofan Yang

College of Mechanical and Energy Engineering, Jimei University, Xiamen 361021, China
Email: 785100937@qq.com, liujd@jmu.edu.cn

Abstract: Based on the orthogonal experiment, the influences of grinding depth (a_p), the workpiece feeding rate (v_w) and the transverse regrinding value (C_r) on two side-direction burr of 40Cr steel large-plane grind-hardening workpiece were investigated. The results show that, under the coupling of heat and force in the grind-hardening process, the surface metal of the workpiece metal flows toward the unconstrained direction on both sides, resulting in the formation of the burr. The grinding depth and the workpiece feeding rate become leading factors affecting the size of the burr. Moreover, the significance of three factors to the size of the burrs is as follows: the grinding depth (a_p) > the workpiece feeding rate (v_w) > the transverse regrinding value (C_r). The result showed that the maximum size of the first-pass and the second-pass burr would increase with the increasing of the grinding depth and the transverse regrinding value in grind-hardening processes, but it decreases with the workpiece feeding rate. The optimal process parameters are a_p=0.1mm,v_w=0.8m/min, and C_r=1mm under the experiment conditions.

1. Introduction

Hardening layer with a certain thickness will be formed on the workpiece surface when the grinding depth reached to a certain value in the dry grinding process. This process could be called grind hardening. It realizes the integration of grinding and surface quenching, and has broad application prospects in the field of metal material grinding and surface modification [1]. At present, scholars mainly study on plunge grinding [2-11]. Nevertheless, the research of the burrs on both sides of the workpiece in the grind-hardening process based on the transverse feed is concerned less. When the width of the workpiece is wider than the width of the grinding wheel, the workpiece needs to be longitudinal reciprocating motion. At the same time, the grinding wheel needs to be transverse intermittent feed motion. Transverse feed grinding simplified schematic diagram is shown in Fig.1. As shown in the figure, a regrind zone is the overlapping portion of between the first grinding process and the second grinding process. Combined with the previous researches on the deep microstructure and hardness of the hardening layer and the depth of the hardened layer and its uniformity [8], this paper will analyze the effect of grinding depth, workpiece feeding rate and transverse regrinding value on the size of the burrs on both sides of the surface grind-hardening workpiece. Then the size of the burrs on both sides of the workpiece is optimized based on the orthogonal experiment.

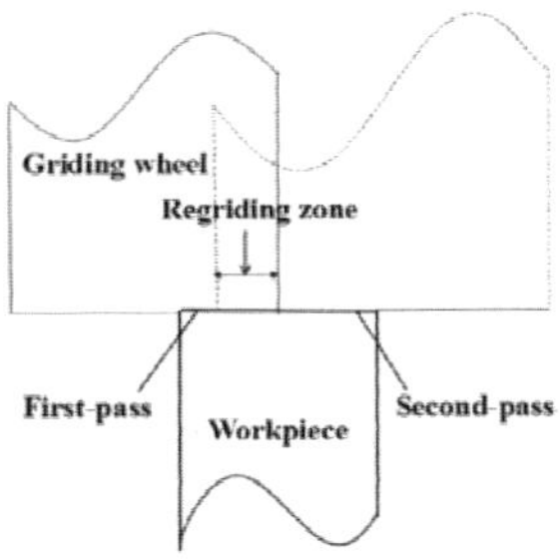

Figure 1. Transverse feed grinding

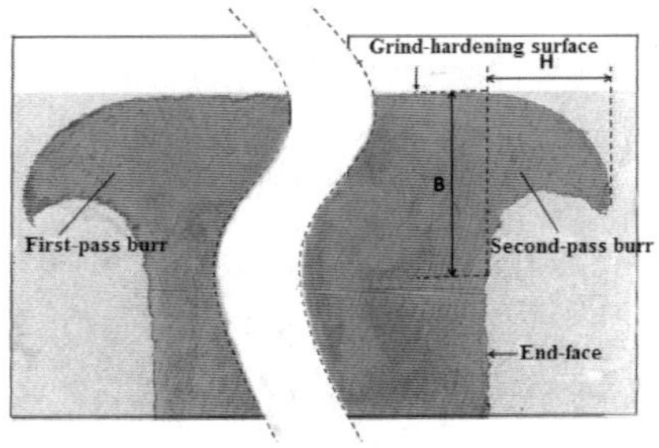

Figure 2. The size of two side-direction burrs

2. Experiment of grind hardening

Experiment of grind-hardening process is based on 40Cr specimens. A two-pass and two-way reciprocating plunge grinding experiment is designed to analyze the effect of the transverse regrinding value on the size of the burrs on both sides of the surface grind-hardening workpiece. In addition, the grind-hardening process is shown in Fig. 1.The area of experimental workpiece materials is 50 × 10 × 20 mm. In the grind-hardening experiment, the first-pass width of grind-hardening is 5 mm and the second-pass width of grind-hardening are 5 mm, 6 mm, 7 mm, and 8 mm, respectively, which are corresponding to the transverse regrinding value of 0mm, 1mm, 2mm and 3mm. Then process conditions of the experiment are shown in Table 1. After grind-hardening process, the wire electric discharge machine is used to evenly divide the specimen into four equal parts along the longitudinal direction of the workpiece. Then the size of each part is 5×10×5mm. It is made into four metallographic specimens after being inlaid and polished. VK-X100K microscope is used in the size and the surface morphology of the burrs on both sides of the surface grind-hardening workpiece. The size of burrs is shown in Fig. 2, which H is the burr height and B is the burr root depth. The left of the part is first-pass burr zone and the right of the part is second-pass burr zone. Since the maximum burr size play a leading role in the burr removal, the maximum burr size is used as a crucial index.

Table 1. Process conditions of the experiment

Machine tool	Grinding wheel	v_s/(m/s)	a_p/(mm)	v_w/(m/min)	C_r/(mm)	Grinding mode
M7130 surface grinder	P340×40×127 WA60L6V	25.6	0.1,0.2, 0.3, 0.4	0.2, 0.4, 0.6, 0.8	0,1, 2,3	first-pass down-grinding + second-pass up-grinding

3. Experiment and analysis of grind hardening

The range of the grinding depth for the maximum height of the first-pass burr is 0.578, the range of the workpiece feeding rate for the maximum height of the first-pass burr is 0.218 and the range of the transverse regrinding value for the maximum height of the first-pass burr is 0.139. The results show that the influence of each factor on the maximum height of the first-pass burr is the grinding depth (a_p) > the workpiece feeding rate (v_w) > the transverse regrinding value (C_r). The influence of the grinding depth, the workpiece feeding rate and the transverse regrinding value on the maximum height of the first-pass burr is shown in Fig. 3. Moreover, the optimal process parameters for the maximum height size of the first-pass burr of the 40Cr steel are a_p=0.1mm,v_w=0.8m/min, and C_r=0mm under the experiment conditions.

The variance analysis of the maximum height of the first-pass burr is shown in Table 2. In the case of significant levels, α=0.1, α=0.05 and α=0.01, the F-values of the grinding depth, the workpiece feeding rate and the transverse regrinding value are larger than the critical value. Thus, the grinding depth and the workpiece feeding rate highly become leading factors affecting the maximum height of the first-pass burr, and the transverse regrinding value becomes a leading less factor in the grind-hardening process.

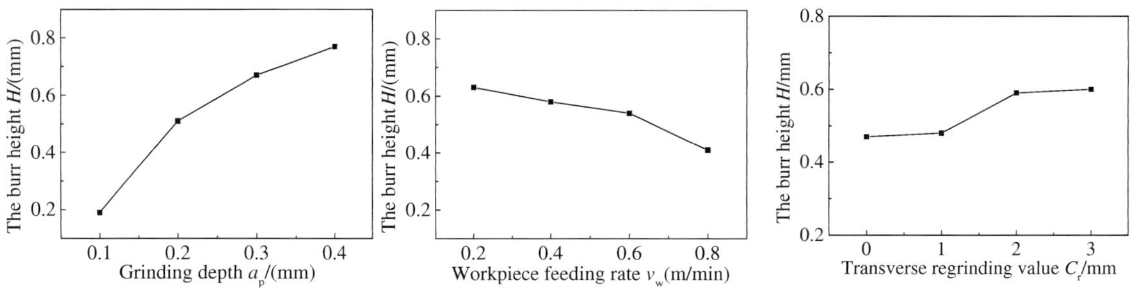

Figure 3. Effect of the grinding depth (a_p), the workpiece feeding rate (v_w) and the transverse regrinding value (C_r) on the maximum height of the first-pass burr

Table 2. Variance analysis of the maximum height of the first-pass burr

Source of variance	Sum of squares, *S*	Freedom, *f*	*MS*	*F*	Criticality value	Significance
a_p	0.772	3	0.257	49.941	$F_{0.90}(3,6)=3.29$	* * *
v_w	0.107	3	0.036	6.901	$F_{0.95}(3,6)=4.76$	* *
C_r	0.061	3	0.020	3.940	$F_{0.99}(3,6)=9.78$	*
e	0.031	6	0.005			
sum	0.97	15				

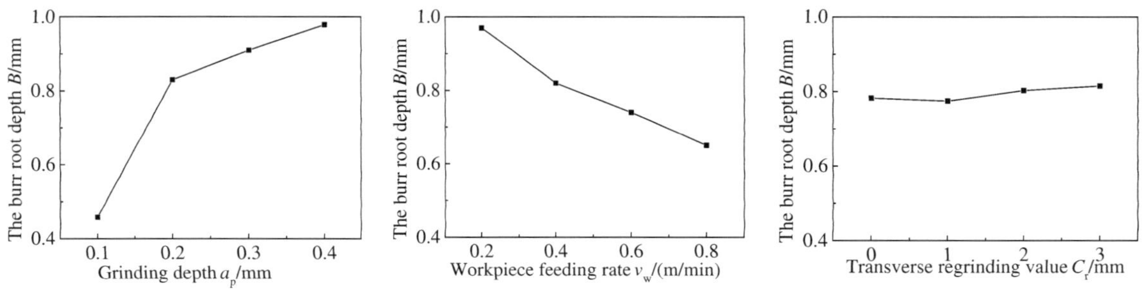

Figure 4. Effect of the grinding depth (a_p), the workpiece feeding rate (v_w) and the transverse regrinding value (C_r) on the maximum root depth of the first-pass burr

The range of the grinding depth for the maximum root depth of the first-pass burr is 0.521, the range of the workpiece feeding rate for the maximum root depth of the first-pass burr is 0.313 and the range of the transverse regrinding value for the maximum root depth of the first-pass burr is 0.041. The results show that the influence of each factor on the maximum root depth of the first-pass burr is the grinding depth (a_p) > the workpiece feeding rate (v_w) > the transverse regrinding value (C_r). The influence of the grinding depth, the workpiece feeding rate and the transverse regrinding value on the maximum root depth of the first-pass burr is shown in Fig. 4. Moreover, the optimal process parameters for the maximum root depth of the first-pass burr of the 40Cr steel are a_p=0.1mm,v_w=0.8m/min, and C_r=0mm under the experiment conditions.

The variance analysis of the maximum root depth of the first-pass burr is shown in Table 3. In the case of significant levels, α=0.1, α=0.05 and α=0.01, the F-values of the grinding depth, the workpiece feeding rate and the transverse regrinding value are larger than the critical value. Thus, the grinding depth and the workpiece feeding rate highly become leading factors affecting the maximum root depth of the first-pass burr, and the transverse regrinding value becomes no obvious in the grind-hardening process.

Table 3. Variance analysis of the maximum root depth of the first-pass burr

Source of variance	Sum of squares, *S*	Freedom, *f*	*MS*	*F*	Criticality value	Significance
a_p	0.646	3	0.215	293.246	$F_{0.90}(3,6)=3.29$	* * *
v_w	0.210	3	0.070	95.202	$F_{0.95}(3,6)=4.76$	* * *
C_r	0.004	3	0.001	1.907	$F_{0.99}(3,6)=9.78$	
e	0.004	6	0.001			
sum	0.865	15				

The range of the grinding depth for the maximum height of the second-pass burr is 0.577, the range of the workpiece feeding rate for the maximum height of the second-pass burr is 0.188 and the range of the transverse regrinding value for the maximum height of the second-pass burr is 0.112. The results show that the influence of each factor on the maximum height of the second-pass burr is the grinding depth (a_p) > the workpiece feeding rate (v_w) > the transverse regrinding value (C_r). The influence of the grinding depth, the workpiece feeding rate and the transverse regrinding value on the maximum height of the second-pass burr is shown in Fig. 5. Moreover, the optimal process parameters for the maximum height of the second-pass burr of the 40Cr steel are a_p=0.1mm,v_w=0.8m/min, and C_r=0mm under the experiment conditions.

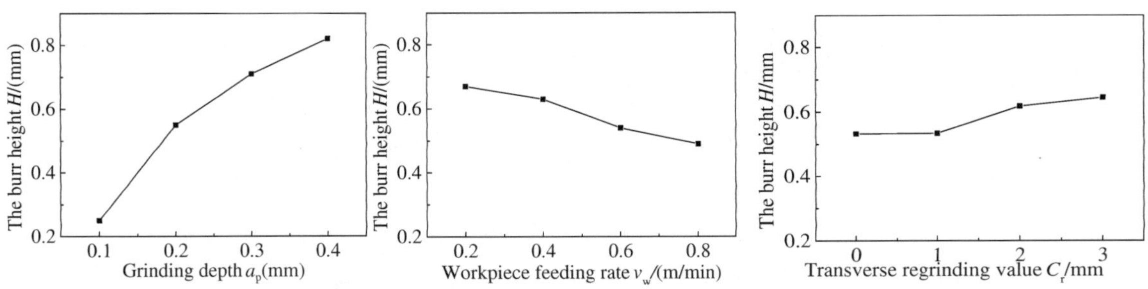

Figure 5. Effect of the grinding depth (a_p), the workpiece feeding rate (v_w) and the transverse regrinding value (C_r) on the maximum height of the second-pass burr

The variance analysis of the maximum height of the second-pass burr is shown in Table 4. In the case of significant levels, α=0.1, α=0.05 and α=0.01, the F-values of the grinding depth, the workpiece feeding rate and the transverse regrinding value are larger than the critical value. Thus, the grinding depth and the workpiece feeding rate highly become leading factors affecting the maximum height of the second-pass burr, and the transverse regrinding value becomes a leading factor in the grind-hardening process.

Table 4. Variance analysis of the maximum height of the second-pass burr

Source of variance	Sum of squares, S	Freedom, f	MS	F	Criticality value	Significance
a_p	0.753	3	0.251	91.141	$F_{0.90}(3,6)=3.29$	* * *
v_w	0.088	3	0.029	10.661	$F_{0.95}(3,6)=4.76$	* * *
C_r	0.040	3	0.013	4.842	$F_{0.99}(3,6)=9.78$	* *
e	0.017	6	0.003			
sum	0.898	15				

The range of the grinding depth for the maximum root depth of the second-pass burr is 0.588, the range of the workpiece feeding rate for the maximum root depth of the second-pass burr is 0.239 and the range of the transverse regrinding value for the maximum root depth of the second-pass burr is 0.072. The results show that the influence of each factor on the maximum root depth of the second-pass burr is the grinding depth (a_p) > the workpiece feeding rate (v_w) > the transverse regrinding value (C_r). The influence of the grinding depth, the workpiece feeding rate and the transverse regrinding value on the maximum root depth of the second-pass burr is shown in Fig. 6. Moreover, the optimal process parameters for the maximum root depth of the second-pass burr of the 40Cr steel are a_p=0.1mm,v_w=0.8m/min, and C_r=1mm under the experiment conditions.

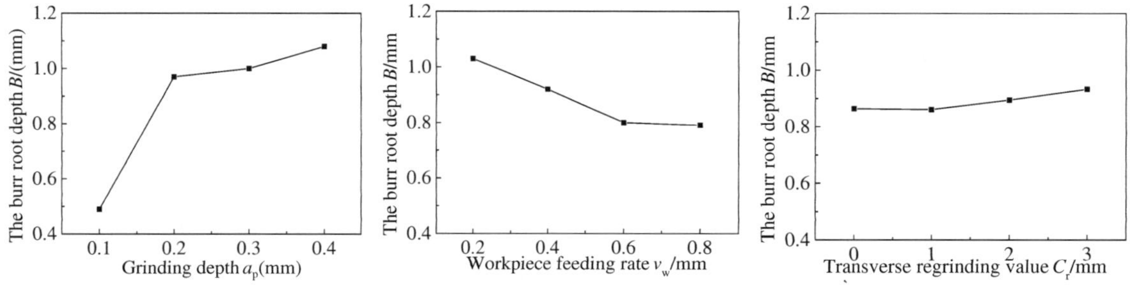

Figure 6. Effect of the grinding depth (a_p), the workpiece feeding rate (v_w) and the transverse regrinding value (C_r) on the maximum root depth of the second-pass burr

Table 5. Variance analysis of the maximum root depth of the second-pass burr

Source of variance	Sum of squares, S	Freedom, f	MS	F	Criticality value	Significance
a_p	0.863	3	0.288	264.109	$F_{0.90}(3,6)=3.29$	* * *
v_w	0.148	3	0.049	45.365	$F_{0.95}(3,6)=4.76$	* * *
C_r	0.014	3	0.005	4.168	$F_{0.99}(3,6)=9.78$	*
e	0.007	6	0.001			
sum	1.031	15				

The variance analysis of the maximum root depth of the second-pass burr is shown in Table 5. In the case of significant levels, α=0.1, α=0.05 and α=0.01, the F-values of the grinding depth, the workpiece feeding rate and the transverse regrinding value are larger than the critical value. Thus, the grinding depth and the workpiece feeding rate highly become leading factors affecting the maximum root depth of the second-pass burr, and the transverse regrinding value becomes a leading factor in the grind-hardening process.

4. Method of reducing the burr size

According to our previous research results, the softened band between first-pass grind-hardening process and second-pass grind-hardening process would increase with the increasing of the workpiece feeding rate or with the decreasing of the grinding depth, but it increases with the increasing of transverse regrinding value. When the transverse regrinding value (C_r) is 1mm, the softened band between first-pass grind-hardening process and second-pass grind-hardening process is quite narrow, which means the grind-hardening quality is obvious highly [8]. Therefore, under the premise of reducing the burr size and improving uniformity of grind-hardened layer, the optimal process parameters are a_p=0.1mm,v_w=0.8m/min, and C_r=1mm under the experiment conditions.

5. Conclusions

1. The surface metal of the grind-hardening workpiece flows to the unconstrained direction on both sides under the coupling of heating and force in the grind-hardening process. Therefore, the burrs on both sides formed. Then the size of the burrs on both sides of the surface grind-hardening workpiece gradually increases with the decreasing of workpiece feeding rate and with the increasing of the grinding depth and the transverse regrinding value in the grind-hardening process.
2. The grinding depth and the workpiece feeding rate become leading factors affecting the size of the burr, and the transverse regrinding value becomes not a leading factor in the grind-hardening process relatively. Moreover, the significant factors of the size of the burrs are as follows: the grinding depth (a_p)> the workpiece feeding rate (v_w) > the transverse regrinding value (C_r).
3. Under the premise of ensuring the grind-hardening quality of the specimen, the optimal process parameters are a_p=0.1mm,v_w=0.8m/min, and C_r=1mm under the experiment conditions.

Acknowledgments

This work was financially supported by the National Natural Science Foundation of China (grant no.51676085) and the Natural Science Foundation of Fujian Province (grant no.2016J01246, 2018J01489).

References

[1] Brinksmeier E and Brockhoff T 1996 *Annals of the CIRP*, vol **45** (1): 283-286.
[2] Brockhoff T, 1999 *CIRP Annals-Manufacturing Technology*,vol **48**(1):255-260.
[3] Brinksmeier, E and T. Brockhoff 1999 *Metallurgia Italiana(Italy)* vol **91.4**: 19-23.
[4] Liu L F, Zhuang J Z and Huang S W 2010 *Key Engineering Materials*, vol **455**:580-584.
[5] Liu J D, et al 2011 *Advanced Materials Research* vol **211**-212(211-212):36-39.
[6] Liu J D, et al 2012 *Applied Mechanics and Materials* vol **217**-219:1869-1873.
[7] Huang X, et al 2013 *The International Journal of Advanced Manufacturing Technology*, vol **66**(9-12): 1555-1561.
[8] Jin J W, Liu J D, et al 2015 *Hot Working Technology*. vol **(22)**:173-175. (In Chinese)
[9] Huang X, et al 2016 *Journal of Mechanical Science and Technology*, vol **30**(8): 3819-3827.
[10] Guo Y, et al 2017 *The International Journal of Advanced Manufacturing Technology*, vol **89**(5-8): 2001-2010.
[11] Xiu S, et al 2018 *The International Journal of Advanced Manufacturing Technology*, vol 1-15.

IOP Conf. Series: Materials Science and Engineering **542** (2019) 012031 doi:10.1088/1757-899X/542/1/012031

Anthocyanins extraction from Purple Sweet Potato (*Ipomoea batatas* (L.) Lam): The effect of pH values on natural color

Tri Nhut Pham[1,2], Tran Quoc Toan[3], Tri Duc Lam[1], Hieu Vu-Quang[1], Dai-Viet N. Vo[1,4], Tran Anh Vy[5], Le Minh Bui[1,*]

[1]NTT Hi-Tech Institute, Nguyen Tat Thanh University, Ho Chi Minh City, Vietnam
[2]Faculty of Chemical Engineering and Food Technology, Nguyen Tat Thanh University, Ho Chi Minh City, Vietnam
[3]Institute of Natural Products Chemistry, Vietnam Academy of Science and Technology, Ha Noi, Vietnam
[4]Faculty of Chemical and Natural Resources Engineering, Universiti Malaysia Pahang, Malaysia
[5]Department of Chemical and Biochemical Engineering, Gachon University, South Korea
*Email: labasm2013@gmail.com

Abstract. Anthocyanins are widely used in the food industry as safe natural colorants thanks to their benefits, attractive colors and safety for consumption. The first objective of this study was to evaluate anthocyanins extraction from Vietnam purple sweet potato (Ipomoea batatas (L.) Lam) for use as food coloring. Different extraction temperatures of solvent ethanol (40 - 70%), duration of extraction (40 - 80 min), temperature extraction (30 - 70°C) and liquid-solid ratios (4:1 - 8:1 mL/g) were selected in order to extract purple sweet potato. Second, this study examines the anthocyanin color behavior at various pH levels in aqueous solutions levels ranging from 1.0 to 14.0. At low pH (acidic conditions), anthocyanins are stable and gives a red color. Meanwhile, increasing the pH value of anthocyanin will change the color from red to pink, violet, blue, green and yellow. This variation suggested that anthocyanin could be utilized as a possible pH color indicator in commercial packaging industry and agriculture.

1. Introduction

Natural products chemistry plays a pivotal role in the physical and biological sciences. Developing new avenues by presenting novel applications, broad perspective, valuable inputs, and a new vision for the future is described [1-8]. Since color and texture of food are often perceptually associated with freshness, ripeness or safety, coloring alternatives from natural sources to synthetic dyes have been an interest of research [9,10]. Among many natural colorants, anthocyanins are commonly used due to their health benefits, attractive colors and consumption safety. Structures of anthocyanins are polar consisting aromatic rings bound with hydroxyl, carboxyl, methoxyl and glycolyl groups. In addition, anthocyanins also possess other various beneficial properties, including antioxidising activity and mitigation of risk of circulatory diseases [11,12]. Multiple physiological functions including antimutagenicity, antihyperglycemic effect and free radical-scavenging activity are exhibited from natural purple sweet potato color. Back in time, purple sweet potato (Ipomoea batatas (L.) *Lam*) was derived from Japan and was considered as a safe and stable source of purple color thanks to its high volume of anthocyanin [13,14]. In the past few years, owing to the growing demand pressed by rapid development in the food industry, purple sweet potato varieties have been cultivated in many countries. This is in line with the current trend of countries around the world to study the use of natural colors used in food, because they are fairly safe for users.

One characteristic that has made anthocyanins utilizable in the food industry is their capability to change color. There are many factors that could alter color stability of anthocyanin including pH, storage temperature, chemical structure, concentration, light, oxygen, solvents the presence of enzymes, flavonoids, proteins and metallic ions. In food applications, pH also greatly impacts microorganisms in terms of denaturization, gelification, enzymatic activities, growth and mortality, muscle foods in terms of water-holding capability, bacterial spores and other chemical reactions such as the Maillard reaction. Therefore, it is crucial to control pH to produce safe and high-quality food products [15,16]. From the current trend, it can be anticipated that the use of natural food colorants relative to the synthetic ones will keep increasing. Therefore, the goal of the present study is two-fold. First, we attempt to investigate extraction conditions that affect anthocyanin extraction from Vietnamese purple sweet potato (*Ipomoea batatas* (L.) Lam) including ethanol solvent, duration of extraction, extraction temperature and liquid-solid ratios. Second, we also aim to determine the color behavior of anthocyanin towards various pH levels.

2. Material and Methods

2.1. Extraction of anthocyanins from Vietnam purple sweet potato

Purple sweet potato (*Ipomoea batatas* (L.) Lam) were grown in Lam Dong, Vietnam. The harvested potatoes were washed, cut into cubes (3-5mm), dried by microwave 400W for 30 min, pulverized, then the collected powder was stored in dark black botte in room temperature for further use. All chemical and solvent are purchased from Sigma Aldrich.

For extraction parameter study, 15g of purple sweet potato powder was placed in the two neck round bottom flask and was extracted by H_2O: Ethanol solutions, liquid/solid ratio was 4:1 - 7:1 (mL/g). The extraction temperature was adjusted to about 40 - 70 (^oC) and time was 40 - 70 (min). Then, centrifugation took place at 4000 rpm for 15 min by high speed centrifuge Model LACE16 (COLO lab expert, China). The supernatant was collected and the extract, after being filtered with filter paper, was transferred into plastic bottle to estimate anthocyanin content. The pH scanning of supernatant ranges from λvis_{max} to 700 nm were recognized for anthocyanin content by UV-Visible Spectrophotometer (Evolution 60S) calibrated with distilled water as the blank. For determination of total monomeric anthocyanin content, the pH-differential method is used [17].

2.2. Anthocyanin extract as pH indicator application

To develop a natural liquid pH indicator from PSP, different test tubes containing a specified volume of anthocyanin extract was prepared. Following that, HCl 1N or NaOH 1N was added into tubes accordingly to produce samples with pH ranging from 1 to 14. Before visual inspection of color, the contents in each tube were thoroughly mixed. [18].

3. Result and Discussion

3.1. Determination of anthocyanin content

Effects of four individual factors on the anthocyanin content were summarized in Table 1. The rapid increase in anthocyanin content with the increase of the ethanol concentration is recognized in Table 1A. Besides, it is shown that the content and quality of anthocyanin pigments strongly depend on the solvent concentration. When rising ethanol concentration from 40% to 70%, the anthocyanin content increased and reached the highest point with EtOH60 at 200,79 mg/L. Ethanol concentration strongly influences the composition of the scavengers from the material, changes the translation properties so it strongly influences the other parameters of the color solution, which contains anthocyanin content. The anthocyanin content in the fluid as well as the effect of color separation from the material was increasing with temperature and reached the highest temperature of 60 - 65^oC, as showed in Table 1B. However,

IOP Conf. Series: Materials Science and Engineering **542** (2019) 012031 doi:10.1088/1757-899X/542/1/012031

when the temperature increased to 70°C, the change in anthocyanin content became subtle. The temperature is so low that only small amounts of anthocyanin are present in sweet potato powder, and the high temperature makes the anthocyanin come out more. By increasing the liquid-solid ratio from 4:1 to 6:1 mL/g, the extracted anthocyanin content increased from 170,13 mg/L to 198,48 mg/L (Table 1C). However, the anthocyanin content decreased sharply after increasing the ratio from 6:1 to 7:1 mL/g. The liquid/solid ratio reaches a certain value (depending on the characteristics of the material), the cell, owing to its efficient water absorption, swells to the maximum and bursts out simultaneously, releasing the color within the vacuole. From Table 1D, the level of anthocyanin in the fluid tends to decrease with different extraction times. When extracting for 40 - 70 minutes, the anthocyanin content is relatively stable, but when the time is increased over 60 minutes, it decreases markedly due to anthocyanin decomposition caused by the long exposure to high temperature.

Table 1. Fixed and variable factors in the single factor investigation of each variable

A \ Experiment	A11	A12	A13	A14
Solvent Concentration	Ethanol 40%	Ethanol 50%	Ethanol 60%	Ethanol 70%
Absorption	0.7371	0.7679	0.8016	0.7854
Total anthocyanin content (mg/L)	184.63	192.35	200.79	196.73
B \ Experiment	**B11**	**B12**	**B13**	**B14**
Temperature (°C)	40	50	60	70
Absorption	0.7657	0.8308	0.8477	0.8506
Total anthocyanin content (mg/L)	191.82	208.15	212.33	213.06
C \ Experiment	**C11**	**C12**	**C13**	**C14**
Liquid/solid ratio (mL/g)	4:1	5:1	6:1	7:1
Absorption	0.6792	0.7926	0.7804	0.6535
Total anthocyanin content (mg/L)	170.13	198.53	195.48	163.57
D \ Experiment	**D11**	**D12**	**D13**	**D14**
Time (min)	40	50	60	70
Absorption	0.8497	0.8552	0.7977	0.7608
Total anthocyanin content (mg/L)	212.83	214.21	199.81	190.57

3.2. The effect of pH values on anthocyanin color

Changing the pH of the anthocyanin pigment significantly affects the tone of the colorant. To be specific, basic colors of anthocyanins including blue, purple, red and orange are directly associated with quantities of hydroxyl groups and are indirectly related to quantities of methoxyl groups [19]. In the liquid indicator pH scale development, the anthocyanin extract showed different colors under different pH conditions (Figure 1). Therefore, it can be used as a natural pH indicator. Visually, the color varies from red in acidic solution to purplish to blue to green in mildly alkaline solution to yellow in a high alkaline solution. These anthocyanin-exclusive changes of color range are similar to red cabbage purple, corn flowers blue, and poppies red. This is contrast with some commercialized indicator, such as phenolphthalein, whose color change only occurred from colorless or pink in acid base titration. It is because phenolphthalein undergoes a structural rearrangement of one-proton removal from one of its phenol group as the pH rises. Anthocyanin extract, on the other hand, was able to show multiple color changes due to more than one proton donation or acceptance group stability over a fairly wide pH range when they contain two or more acyl groups [20].

IOP Conf. Series: Materials Science and Engineering **542** (2019) 012031 doi:10.1088/1757-899X/542/1/012031

At the equilibrium state, anthocyanins are characterized by five different equilibrium structures, including red flavylium cation, colorless carbinol pseudo base, purple quinoidal base, blue quinoidal base and and yellowish chalcone, and ratio of such structural anthocyanins, determined by pH value, results in color change. Among those structures, flavylium form could not be recovered or formed by re-acidification.. At acidic medium of pH = 1 - 3, red flavylium cation (2-phenylchromenylium cation) is predominant in the solution. . However, increasing the pH leads to a decrease in in concentration of the flavylium cation due to the hydration which in turn reduces the color intensity and produces the colorless carbinol pseudobase (hemiacetal or chromenol). In the hydration process, the anthocyanidin skeleton of the conjugated 2-benzopyrilium system is disrupted at the 2 position by a nucleophilic attack of water (Figure 2). As the pH shifts higher, flavylium cation loses proton rapidly. At pH < 7, purple quinoidal anhydrobase is dominant in the equilibrium. at pH < 8, the color turned to the deep blue characterized by ionized anhydrobase. When pH exceeds 8, the central pyran ring opened, allowing the formation of carbinol form and the light yellow chalcone. At any point, the color could be altered by adjusting the pH of the solution to acidic or alkaline. In general, it is the transformation of ions that cause color change and pH adjustment could be utilized to make such changes.

Table 2. Results of pH application

pH	**1**	**2**	**3**	**4**	**5**	**6**	**7**
Color	Red	Dark pink	Pinkish red	Pinkish red	Lavender	Faint pink	Faint violet

pH	**8**	**9**	**10**	**11**	**12**	**13**	**14**
Color	Faint blue	Dark blue	Dark green	Faint green	Green	Yellowish green	Yellow

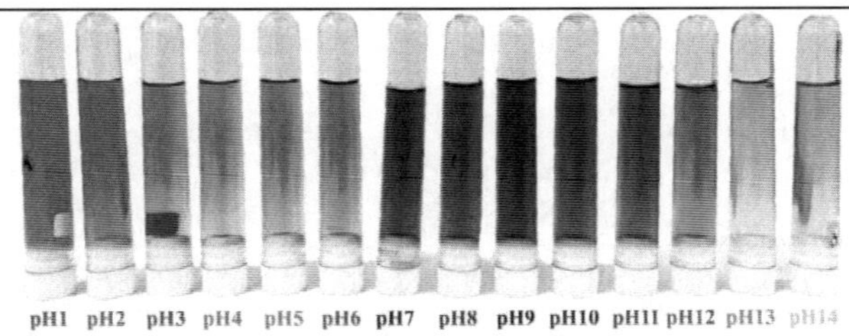

Figure 1. The color change of the anthocyanin corresponds to the different pH

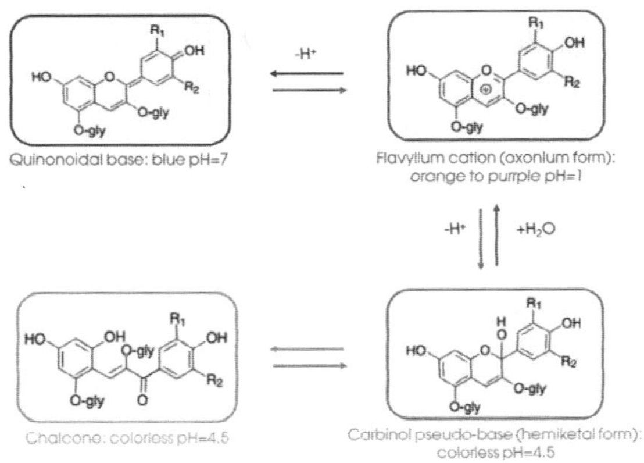

Figure 2. Predominant structural forms of anthocyanins present at different pH levels

4. Conclusion

We have successfully extracted anthocyanins from Vietnamese purple sweet potato (Ipomoea batatas (L.) Lam) and have obtained the set of condition variables including solvent ethanol of 60%, temperature of 70°C, liquid-solid ratio of 5:1 (mL/g), and time of 50 (min), whose corresponding anthocyanin content was the highest at 214.21 mg/L. In addition, the study on purified anthocyanins as natural pH indicator revealed that the color of anthocyanin pigments was drastically altered with adjustment in pH value of the medium. To be specific, anthocyanins are red at low pH (acidic conditions) and higher pH value will alter the color from pink to violet, then blue, green and finally, yellow. These results suggest that anthocyanin extracts from Vietnam purple sweet potato (Ipomoea batatas (L.) Lam) represent a feasible solution for pigmentation for the food colorant market.

References

[1] Vo T-S, Ngo D-H, Bach L G, Ngo D-N and Kim S-K 2017 *Process Biochemistry* **54** 188–94
[2] Mai H C, Le T T T, Diep T T, Le T H N, Nguyen D T and Bach L G 2018 *Asian Journal of Chemistry* **30** 293–7
[3] Thinh P V, Trinh N D, Thuong N T, Ho V T T, Quynh B T P and Bach L G 2018 *Solid State Phenomena* **279** 230–4
[4] Nhan N P T, Hien T T, Nhan L T H, Anh P N Q, Huy L T, Nguyen T C T, Nguyen D T and Bach L G 2018 *Solid State Phenomena* **279** 235–9
[5] Hien T T, Nhan N P T, Trinh N D, Ho V T T and Bach L G 2018 *Solid State Phenomena* **279** 217–21
[6] Minh N P and Giang B L 2018 *Journal of Global Pharma Technology* **0**
[7] Phan A N Q, Bach L G, Nguyen T D and Le N T H 2019 *Journal of Nanoscience and Nanotechnology* **19** 974–8
[8] Hien Tran T, Huu Hao Nguyen H, Nguyễn Q, Tan H, Thi Hong Nhan L, Nguyen D H, Lam T, Trung Do S and Nguyen T 2018 *Processes* **6** 206
[9] Riaz G and Chopra R 2018 *Biomedicine & Pharmacotherapy* **102** 575–86
[10] Ongkowijoyo P, Luna-Vital D A and Gonzalez de Mejia E 2018 *Food Chemistry* **250** 113–26
[11] Khoo H E, Azlan A, Tang S T and Lim S M 2017 *Food & Nutrition Research* **61** 1361779
[12] Braga A R C, Murador D C, de Souza Mesquita L M and de Rosso V V 2018 *Journal of Food Composition and Analysis* **68** 31–40
[13] Yamakawa O and Yoshimoto M 2002 *Acta Horticulturae* 179–85
[14] Steed L E and Truong V-D 2008 *Journal of Food Science* **73** S215–21
[15] Roobha J J, Saravanakumar M, Aravindhan K M and Devi P S 2011 *1* **1**
[16] Kalthum Ib U, Idayu Muha I and Mohd Salle R 2011 *Journal of Applied Sciences* **11** 2406–10
[17] Wrolstad R E, Durst R W and Lee J 2005 *Trends in Food Science & Technology* **16** 423–8
[18] Mohd P, Khan A and Farooqui M 2011 *Journal of Advanced Scientific Research* **2** 20–7
[19] He F, Mu L, Yan G-L, Liang N-N, Pan Q-H, Wang J, Reeves M J and Duan C-Q 2010 *Molecules* **15** 9057–91
[20] Bondre S, Patil P, Kulkarni A and Pillai M M 2012

Extraction of anthocyanins from Butterfly pea (Clitoria ternatea *L. Flowers*) in Southern Vietnam: Response surface modeling for optimization of the operation conditions

Tri Nhut Pham[1,2], Duy Chinh Nguyen[1], Tri Duc Lam[1], Pham Van Thinh[3], Xuan Tien Le[4], Dai Viet Vo Nguyen[5], Quang Hieu Vu[1,*], Trinh Duy Nguyen[1,6], Long Giang Bach[1,7,*]

[1]NTT Hi-Tech Institute, Nguyen Tat Thanh University, Ho Chi Minh City, Vietnam
[2]Faculty of Chemical Engineering and Food Technology, Nguyen Tat Thanh University, Ho Chi Minh City, Vietnam
[3]Dong Nai Technology University, Bien Hoa City, Dong Nai Province, Viet Nam
[4]Department of Chemical Engineering, HCMC University of Technology, VNU-HCM, Ho Chi Minh City, Vietnam
[5]Faculty of Chemical and Natural Resources Engineering, Universiti Malaysia Pahang, Malaysia
[6]Department of Imaging System Engineering, Pukyong National University, Busan, South Korea
[7]Department of Chemical Engineering, Pukyong National University, Busan, South Korea
Email:blgiang@ntt.edu.vn

Abstract. In this paper, the Response Surface Methodology (RSM), in conjunction with Central Composite Design (CCD), was used to optimize the extraction of anthocyanins from Butterfly pea (Clitoria ternatea L.Flowers) cultivated in Southern Vietnam. The effect of extraction temperatures of solvent ethanol (50-70 °C), duration of extraction (40-50 min) and solid-liquid ratios (20:1–30:1) was measured as independent variables on the total extraction anthocyanins in the response function. The highest anthocyanin content of 132.756 mg/L of butterfly pea anthocyanin was collected at the solid liquid ratio of 23:1, extraction time of 46 min, and temperature 60.6°C. Butterfly pea anthocyanins yield detailed significant correlation with high F values, low P values (<0.0001), and desirable determination coefficient ($R^2 = 0.9994$).

Keywords: Response Surface Methodology, Central Composite Design, Optimizing conditions, Anthocyanins extraction, Butterfly pea

1. Introduction

Clitoria ternatea, commonly known as butterfly pea or blue pea, is found throughout the tropical regions of Vietnam. The plant bears white or blue flowers and grows in wild and also in gardens. It is well known as tropical perennial climber herb from family Fabaceae with slender downy stem [1,2]. Thanks to its anthocyanin compound, the flower could be used as a natural colorant. Anthocyanins in the flower are made up of basic classes and this determines the color. To be specific, pelargonidin is responsible for orange-red color, cyanidin for red hues and delphinidin for lilac to blue hues. Anthocyanins are also widely used as a natural supplements and have been shown to have antiinflammatory activity, anti-oxidant, anticarcinogenic and anti-microbial including anti-candida activities [3-5].

Newly discovered health benefits and applications in the food processing industry has rapidly pushed the demands for anthocyanin. In studies utilizing the single investigation approach, variable was studied individually, therefore failing to control the simultaneous effects of other parameter. As such, number of experimental attempt will be numerous and materials needed are increased. The Response Surface Methodology (RSM) could overcome such limitations by adopting a multivariate system that fits experimental data in a statistical model by a response function [6-10].
Given the huge potential of anthocyanin and the abundance of the substance in Butterfly pea, this study aim to facilitate anthocyanin exploitation in nutraceutical/food industry. In this study, we investigate the total extracted anthocyanins from Butterfly pea (Clitoria ternatea *L.Flowers*) and its experimental parameters. The effect of extraction temperatures of solvent ethanol, duration of extraction, solid-liquid ratios and their interactions was evaluated using a Central Composite Design (CCD) combined with RSM.

2. Material and methods

2.1. Extraction of totals anthocyanins from Butterfly pea

Dried petals of butterfly pea were ground using a commercial grinder and were weighed to 10g, put in the two neck round bottom flask and was extracted by Ethanol 50° solutions. Ethanol (C_2H_5OH) is purchased from Sigma Aldrich (US). Monomeric anthocyanin was calculated as cyanidin-3-glycoside using a pH differential method [11]
First of all, each experimental condition will be tested individually to determine a set of optimal condition. For extraction parameter study, 10g of butterfly pea powder was placed in the two neck round bottom flask as and was extracted by ethanol with concentration at 50. The liquid/solid ratio in this experiment ranges from 10:1 to 30:1 (mL/g) and. The extraction temperature is adjusted from 40 to 80 (°C) and time varies from 15 to 75 (min). Then, centrifugation took place at 4000 rpm for 15 min by high speed centrifuge Model LACE16 (from COLO lab expert). The supernatant was collected and the extract, after being filtered with filter paper, was transferred into plastic bottle for yield estimation. The pH scanning of supernatant ranges from 400 nm to 700 nm.

2.2. Statistical analysis

Response surface analysis conducted by Design-Expert® Version 11 software was employed to determine the optimum result of response affected by three factors (solid/liquid ratio, extraction temperature and extraction time). The analysis results is expressed in the form a statistical model in which anthocyanin content is formulated as a function relating to considered factors. To verify the optimum process, 5 replications of the extraction process at optimum conditions were conducted. The resulting values was then compared to the volume predicted by the model.

Table 1. Independent variables matrix and their encoded levels for RSM model.

Code	Name	Units	Levels				
			-α	-1	0	+1	+α
A	Water to raw material ratio	mL/g	16.59	20	25	30	33.40
B	Temperature	°C	51.591	55	60	65	68.409
C	Extraction time	Min	36.59	40	45	45	53.41

IOP Conf. Series: Materials Science and Engineering **542** (2019) 012032 doi:10.1088/1757-899X/542/1/012032

3. Result and discussion

The results of 20 experimental runs with the RSM model were shown in Table 2. The results implied that the yield of anthocyanin would be severely altered as survey parameters change (ratio, temperature, time). In the table 2, No.1~14 were the factorial experiments, and No.15~20 were the central experiments. The resulting anthocyanin content ranges from 181.826 mg/L to 212.585 mg/L.

Table 2 *The matrix of observed and predicted values for RSM model*

No	Independent factors			Y (mg/L)		No	Independent factors			Y (mg/L)	
	X_1	X_2	X_3	Actual	Predicted		X_1	X_2	X_3	Actual	Predicted
1	20	55	40	118.02	113.39	11	25	51.591	45	121.86	127.40
2	30	55	40	120.80	113.94	12	25	68.409	45	122.8	127.95
3	20	65	40	130.332	117.94	13	25	60	36.59	127.31	128.75
4	30	65	40	119.75	119.68	14	25	60	53.41	128.7	130.31
5	20	55	50	125.86	120.74	15	25	60	45	132.26	132.25
6	30	55	50	124.78	121.96	16	25	60	45	132.22	132.25
7	20	65	50	126.99	122.85	17	25	60	45	132.21	132.25
8	30	65	50	113.81	123.83	18	25	60	45	132.22	132.25
9	16.59	60	45	123.77	124.70	19	25	60	45	132.38	132.25
10	33.40	60	45	113.86	125.83	20	25	60	45	132.25	132.25

Table 3 ANOVA data for removal models

Source	Sum of Squares	Degree of freedom	Mean Square	F-value	Prob. > F		Comment
Model	733.66	9	81.52	10544.99	< 0.0001	significant	SD = 0.2024
A-A	117.91	1	117.91	15252.80	< 0.0001		Mean = 125.61
B-B	0.9505	1	0.9505	122.95	0.0025		CV(%) = 0.1611
C-C	2.20	1	2.20	284.00	< 0.0001		R^2 = 0.9994
AB	90.20	1	90.20	11667.55	< 0.0001		AP = 130.12%
AC	7.72	1	7.72	998.45	< 0.0001		
BC	52.54	1	52.54	6796.66	< 0.0001		
A^2	321.95	1	321.95	41646.26	< 0.0001		
B^2	174.90	1	174.90	22625.25	< 0.0001		
C^2	31.45	1	31.45	4068.76	< 0.0001		
Residual	0.0773	10	0.0077				
Lack of Fit	0.0578	5	0.0116	2.97	0.1290	not significant	
Pure Error	0.0195	5	0.0039				

The difference between actual values and predicted values was not substantial indicating that the results of the experimental experiments have high accuracy. As indicated by Table 3, determination coefficient $R^2 = 0.9994$ shows a high goodness-of-fit of the predicted values to the real data. Overall, the model is significant as demonstrated by the F-value of 10544.99. The probability of the model to obtain such value owing to noise is minimal (0.016%). All model terms are also statistically significant as indicated by low p-values. Insignificance of the lack of fit shows that the model fits well with the data, thus no further specification of the model is required. After estimating model coefficients, the following model is obtained: $Y = 132.26 - 2.84X_1 + 0.2199X_2 + 0.357X_3 - 3.18X_1X_2 - 0.8073X_2X_3 - 2.64X_1X_3 - 4.74X_1^2 - 3.5X_2^2 - 1.49X_3^2$ (3). The model is fit with the response variable and independent variables. Therefore, anthocyanin content of 132.756 mg/L was extracted by the optimum parameters as $X_1 = 23{:}1$ (mL/g), $X_2 = 60.6$ (°C) and $X_3 = 46$ (min), which was completed by DX11 software.

The overall significance of the model is implied by The Model F-value of 1953.19, suggesting that there is only a 0.01% chance that an F-value this large could occur due to noise. In addition, all model terms are statistically significant suggesting that model reduction is unnecessary. The Lack of Fit F-value of 20.01 implies the Lack of Fit is significant. There is only a 0.25% chance that a Lack of Fit F-value this large could occur due to noise.

Based on the model estimated by the software, three response surface plots of the anthocyanin yields, corresponding to different combinations of X_1, X_2 and X_3 were plotted as in Fig.2. The contour plots represented impact intensity of the interaction variables on the yield. According to the model results, all three interaction terms representing combinations of three factors are significant ($P<0.05$). From the response surface analysis, optimum process of anthocyanin extraction are identified with the following parameters: temperature 60.6 (°C), time 46 (min), and liquid-solid ratio 23:1 (mL/g). The maximum predicted anthocyanin content was 132.756 mg/L.

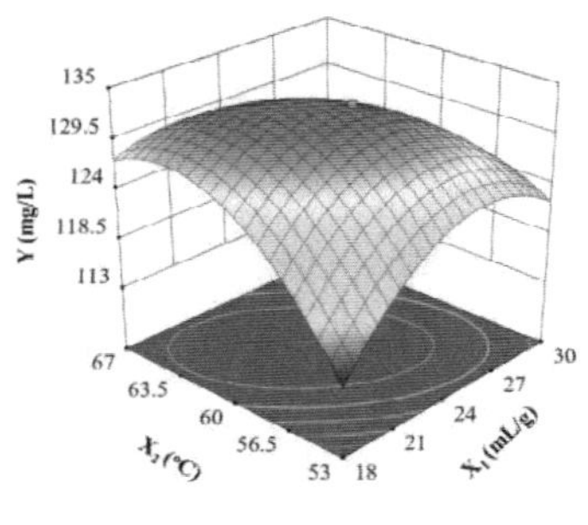

Figure 1. The influence and interaction of Y with X_2 and X_1.

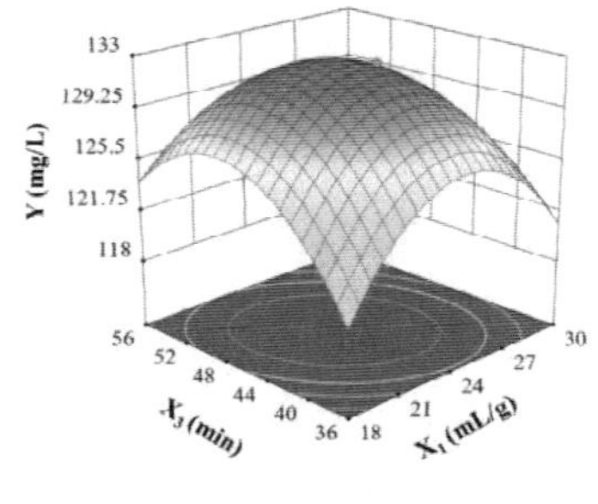

Figure 2. The influence and interaction of Y with X_1 and X_3.

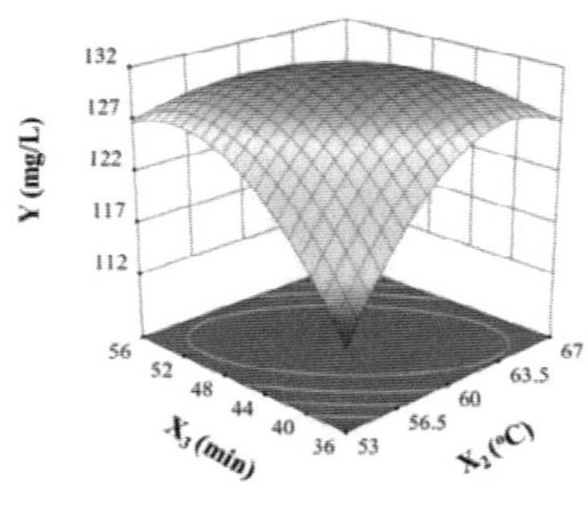

Figure 3. The influence and interaction of Y with X_2 and X_3.

4. Conclusion

We have successfully extracted and optimized of the anthocyanin extraction process from the Butterfly pea (Clitoria ternatea L.Flowers) in Southern Vietnam. Response surface methodology (RSM), based on a CCD design, was used. Three-dimensional and contour-response surface plots clearly suggested the presence of impact of the different levels of liquid/solid ratio, extraction time, and reaction temperature on the total anthocyanin content. Following the single factor assays, we conducted 20 experiments planned by RSM to optimize of the extraction process of anthocyanin content. The results of surface response methodology revealed that the optimum set of experimental conditions to achieve the highest yield includes temperature 60.6 (°C), time 6 (min), and liquid-solid ratio 23:1 (mL/g), corresponding to the highest anthocyanin content of 132.756 mg/L. Therefore, through the use of RSM, significant factors impacting the extraction yield has been identified, proving a framework for further investigation

Acknowledgments
This work was supported by grants from Nguyen Tat Thanh University, Ho Chi Minh City, Viet Nam.

References

[1] Clausen H D, Holbeck H B and Reddersen J 2001 Factors influencing abundance of butterflies and burnet moths in the uncultivated habitats of an organic farm in Denmark *Biolo. Conservation* **98** 167-78
[2] Mehmood A, Ishaq M, Zhao L, Yaqoob S, Safdar B, Nadeem M, Munird M and Wang C 2019 Impact of ultrasound and conventional extraction techniques on bioactive compounds and biological activities of blue butterfly pea flower (Clitoria ternatea L.) *Ultrasonics Sonochemistry* **51** 12-9
[3] Ongkowijoyo P, Luna-Vital D A, Gonzalez de Mejia E 2018 Extraction techniques and analysis of anthocyanins from food sources by mass spectrometry: An update *Food Chemistry* **250** 113-26
[4] Xie L, Su H, Sun C, Zheng X and Chen W 2018 Recent advances in understanding the anti-obesity activity of anthocyanins and their biosynthesis in microorganisms *Trends in F. Sci. & Tech* **72** 13-24
[5] Rupasinghe H P V, Arumuggam N, Amararathna M, De Silva A B K H 2018 The potential health benefits of haskap (Lonicera caerulea L.): Role of cyanidin-3-O-glucoside *J. of Func. F.* **44** 24-39
[6] Tran T H, Nguyen P T N, Nguyen D T, Ho V T T and Bach L G 2018 Optimizing the Pomelo Oils Extraction Process by Microwave-Assisted Hydro-Distillation Using Soft Computing Approaches *Solid State Phenomena*, **279** 217-21
[7] Nguyen P T N, Tran T H, Le T H N, Pham N Q A, Huy L T, Nguyen T C T, Nguyen D T and Bach L G 2018 Application of Response Surface Methodology to Optimize the Process of Saponification Reaction from Coconut Oil in Ben Tre-Vietnam *Solid State Phenomena* **279** 235-39
[8] Tran T V, Bui Q T P, Nguyen T D, Ho V T T and Bach L G 2017 Application of response surface methodology to optimize the fabrication of $ZnCl_2$-activated carbon from sugarcane bagasse for the removal of Cu^{2+} *Water Sci. and Tech.* **75** 2047-55
[9] Tran V T, Bui T P Q, Nguyen T D and Bach L G 2017 Response surface methodology approach for optimization of Cu^{2+}, Ni^{2+} and Pb^{2+} adsorption using KOH-activated carbon from banana peel *Surf. and Interf.* **6** 209-17
[10] Tran T V, Bui Q T P, Nguyen T D, Le N T H and Bach L G 2017 A comparative study on the removal efficiency of metal ions (Cu^{2+}, Ni^{2+}, and Pb^{2+}) using sugarcane bagasse-derived $ZnCl_2$-activated carbon by the response surface methodology *Ads. Sci. and Tech.* **35** 72-85
[11] Anthika B, Kusumocahyo S P and Sutanto H 2015 Ultrasonic Approach in Clitoria ternatea (Butterfly Pea) Extraction in Water and Extract Sterilization by Ultrafiltration for Eye Drop Active Ingredient *Procedia Chemistry* **16** 237-44

IOP Conf. Series: Materials Science and Engineering **542** (2019) 012033 doi:10.1088/1757-899X/542/1/012033

Effects of various solvent concentration, liquid-solid ratio, temperatures and time values on the extraction yield of anthocyanin from Vietnam Hibiscus sabdariffa L. (*Roselle*)

Tri Nhut Pham[1,2], Thuong Nhan Phu Nguyen[1], Tri Duc Lam[1], Thien Hien Tran[1], Duy Chinh Nguyen[1], Dai-Viet N. Vo[3], Xuan Tien Le[4], Sy Trung Do[5], Long Giang Bach[1,2,*]

[1]NTT Hi-Tech Institute, Nguyen Tat Thanh University, Ho Chi Minh City, Vietnam
[2]Faculty of Chemical Engineering and Food Technology, Nguyen Tat Thanh University, Ho Chi Minh City, Vietnam
[3]Faculty of Chemical and Natural Resources Engineering, Universiti Malaysia Pahang, Malaysia
[4]Department of Chemical Engineering, HCMC University of Technology, VNU-HCM, Ho Chi Minh City, Vietnam
[5]Institute of Chemistry, Vietnam Academy of Science and Techology, Hanoi City, Vietnam
*Email: blgiang@ntt.edu.vn , labasm2013@gmail.com

Abstract. Nowadays, the trends in human health and fossil fuel dependent markets present opportunities for agricultural crops as renewable sources in several substitutes of synthetic components in clothing, fuels, and food. This research focused on anthocyanins along with protocatechuic acid and quercetin, which have been recognized as bioactive compounds in Hibiscus sabdariffa L. aqueous extracts. Laboratory extraction study was conducted on various solvent concentration, liquid-solid ratio, temperatures and time values on the extraction yield of anthocyanin from Vietnam Hibiscus sabdariffa L. (Roselle). On theother hand, total anthocyanins wereinvestigated in Hibiscus sabdariffa L. extract as μg/mL cyanidin-3-glucoside (cyd-3-glu). Extraction temperature of 60◦C using solvent concentration 50%, extraction time of 30 min at liquid-solid ratio of 8:1 showed the highest anthocyanin recovery at 180,821 mg/L. Overall, Vietnam Hibiscus sabdariffa L. (Roselle) are responsive to anthocyanin extraction, making it an appropriate substrate for the development of industrial colorants and dyes.

1. Introduction

Nowadays, the application of natural ingredients become more and more popular in various fields such as global food, beverages, cosmetic and pharmaceutical industries [1-3]. Generally, food pigments know dangerous degeneration during processing, and recovery of the missing color is a method to preserve the freshness perspective of foods [4,5]. In food colors of natural origin, anthocyanins are the most common color family. Anthocyanin is a water-soluble phenolic compound that functions as a color from red to dark green to plants that have long been used as a natural coloring ingredient for food safety. Anthocyanin colorants can be extracted from a variety of materials. Anthocyanins, apart from being recognized as natural plant pigments from the flavonoid family, possess valuable pharmacological properties such as anti-oxidative, anti-inflammatory, and antineurodegenerative effects [6,7].

IOP Conf. Series: Materials Science and Engineering **542** (2019) 012033 doi:10.1088/1757-899X/542/1/012033

Hibiscus Sabdariffa is one of the Malvaceae family of the Hibiscus genus, also identified as Roselle. This plant can tolerate with short terms of drought and can be grown throughout the subtropics and tropics during rainy and hot seasons. The different parts in a Roselle plant including calyces, flower, and petals are wealthy in anthocyanins which one of the chemical components can be located naturally in the plant [8,9]. The various extraction conditions affect on extraction yields of anthocyanins from plant tissues [10,11]. By using solvent extraction processes, isolation of anthocyanin pigments from plants is completed. Anthocyanins belong to polar molecules group and more dissolved in polar solvents, however, extraction conditions also play an important role in their overall solubility. The previous studies on extracting anthocyanins from vegetables and fruits such as purple corn, red and black currants, grapes and so on have presented that ethanol extraction is appropriate. The extraction requirements such as incubation temperature and time, solvent concentration, and solid-liquid ratio are significantly critical in the concentration and stability of anthocyanins. The objective of this work was to find a suitable and potential alternative to replace cochineal as the organic dye for red color, which focuses on the extraction of Roselle. Therefore, an examination of the efficient extraction of anthocyanins from Hibiscus sabdariffa L. (*Roselle*) grown in Viet Nam requires evaluation under various solvent type, solid loading, incubation temperature and time valuesThe property parameters analyzed included the color of the extracts by applying the pH-differential method. The results of the project hope to bring out the potential for the application of natural pigments in the food industry and improve the value of using raw materials in Vietnam.

2. Material and methods

2.1 Sample repairation

Hibiscus sabdariffa L. dried calyces were used, provided by farmers of Lam Dong Province, Viet Nam. Dried petals of roselle were ground using a commercial grinder and were weighed to 10g, put in the two neck round bottom flask and was extracted by Ethanol 50° solutions. The various particle diameter ground samples were kept in polyethylene vacuum bags at room temperature until extraction was carried out.

2.2 Extraction of anthocyanin

(A)

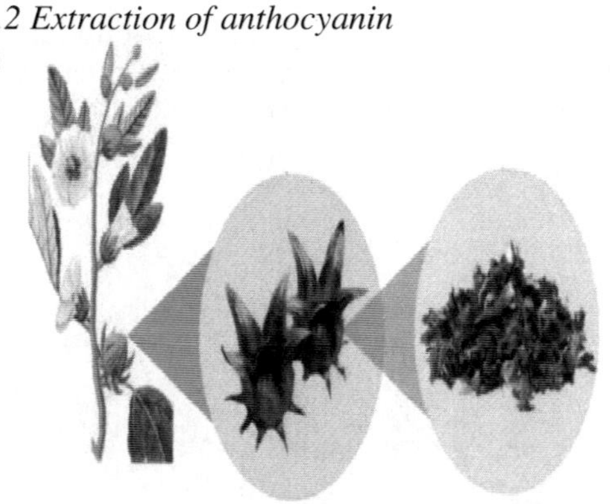

(B)

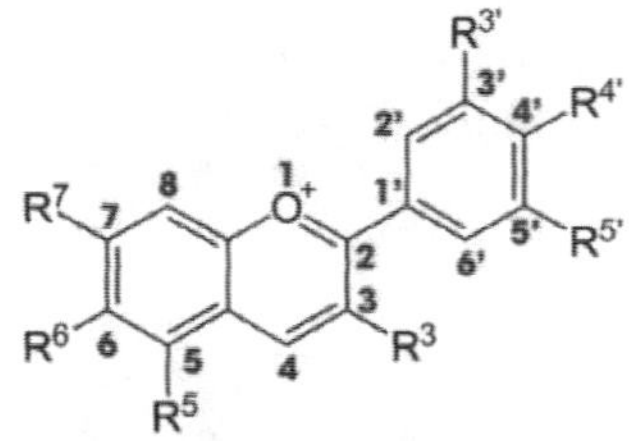

Anthocyanidin	R_3'	R_4'	R_5'	R_3	R_5	R_6	R_7
Cyanidin	-OH	-OH	-H	-OH	-OH	-H	-OH
Peonidin	$-OCH_3$	-OH	-H	-OH	-OH	-H	-OH
Petunidin	-OH	-OH	$-OCH_3$	-OH	-OH	-H	-OH
Malvidin	$-OCH_3$	-OH	$-OCH_3$	-OH	-OH	-H	-OH
Delphinidin	-OH	-OH	-OH	-OH	-OH	-H	-OH
Pelargonidin	-H	-OH	-H	-OH	-OH	-H	-OH

Figure 1. Extraction of the anthocyanin (A) Hibiscus sabdariffa L dried. (B) The structure of anthocyanin.

First of all, each experimental condition will be tested individually to determine the optimum condition. For extraction parameter study, 10g of roselle powder was placed in the two neck round bottom flask as and was extracted by ethanol with concentration at 50. The liquid/solid ratio in this experiment ranges from 2:1 to 10:1 (mL/g). The extraction temperature is adjusted from 40 to 80 (°C) and time varies from 15 to 35 (min). Then, centrifugation took place at 4500 rpm for 15 min by high speed centrifuge Model LACE16 (from COLO lab expert). The supernatant was collected and the extract, after being filtered with filter paper, was transferred into plastic bottle to estimate anthocyanin content. The pH scanning of supernatant ranges from λvis_{max} to 700 nm were recognized for anthocyanin content. The pH-differential method uses to determine the total anthocyanin level [12]. The maximum absorption is 519nm (Figure 1).

IOP Conf. Series: Materials Science and Engineering **542** (2019) 012033 doi:10.1088/1757-899X/542/1/012033

2.3. Total Anthocyanin Content Measurement.

The total anthocyanin content was determined according to the spectrophotometric pH differential method. Samples were diluted separately with 0.025M potassium chloride buffer (pH 1) and 0.4M sodiumacetate buffer (pH 4.5). Absorbance of the mixture was measured at 519 (λvis-max) and 700 nm using a UV-Vis spectrophotometer. Absorbance was calculated as $A = [(A511 - A700)$ pH 1.0 $- (A511 - A700)$ pH 4.5. The total anthocyanin content was calculated as cyanidin-3-glucoside equivalents as in the following equation, anthocyanin content (mg/100 g)

$$= \frac{A \times \mathrm{MW} \times \mathrm{DF} \times \mathrm{V} \times 100}{a \times l \times m}$$

With A is the absorbance, MW is the molecular weight of cyanidin-3-glucosode (449.2g/mol), DF is the) dilution factor, V is the solvent volume (ml), a is the molar absorptivity (26900), l is the cell path length (1cm), and m is the freeze-dried sample weight.

3. Result and discussion

c

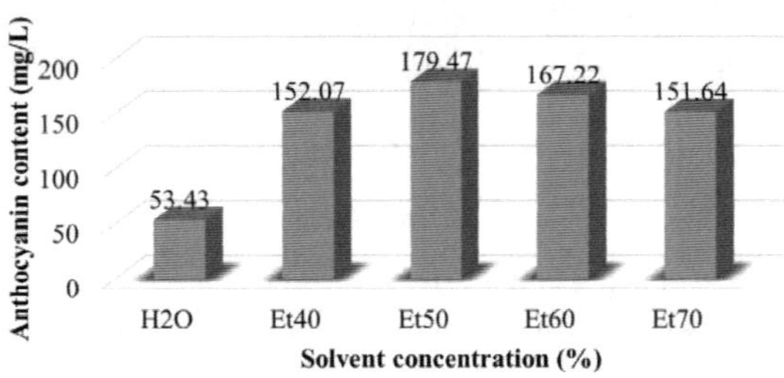

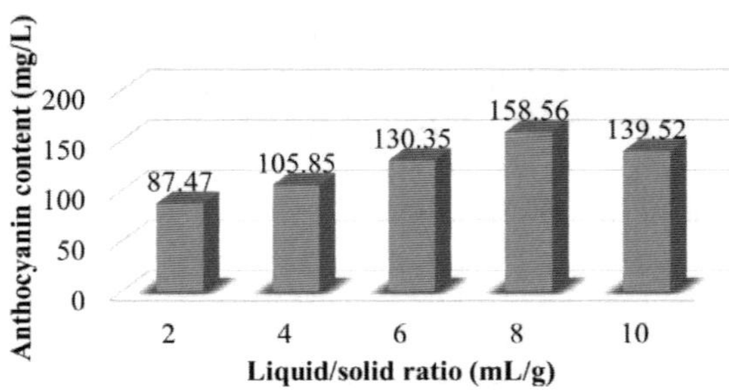

Figure 2. The effects of various solvent concentration on extraction yield of anthocyanin from Vietnam Hibiscus sabdariffa L. (*Roselle*)

Figure 3. The effects of various liquid-solid ratio on extraction yield of anthocyanins from Vietnam Hibiscus sabdariffa L. (*Roselle*)

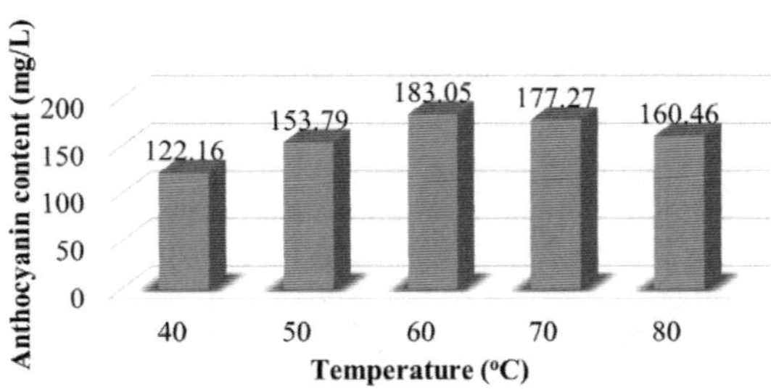

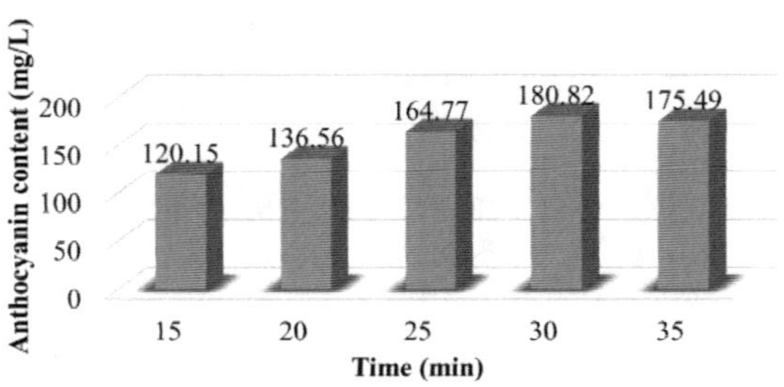

Figure 4. The effects of various temperature on extraction yield of anthocyanin from Vietnam Hibiscus sabdariffa L. (*Roselle*)

Figure 5. The effects of various time on extraction yield of anthocyanins from Vietnam Hibiscus sabdariffa L. (*Roselle*)

3.1. Effect of solvent concentration

As extraction anthocyanins from red onions may have a promising to be applied as eco-friendly dyestuffs and nutraceuticals. In term of stability, the factors affecting their applications are addressed in this study.

Z determining the solvent influence on anthocyanins extracted from Vietnam Hibiscus sabdariffa L. (*Roselle*), the following solvent systems were used water, ethanol/water (40/60), ethanol/water (50/50), ethanol/water (60/40), ethanol/water (70/30). the pH differential spectrophotometric method is applied in order to evaluation of the anthocyanin concentration. As shown in Figure 2, regarding the extraction yield of anthocyanin, 50% ethanol solution demonstrated more efficient than the other systems. The results presnted that the anthocyanin content in the color solution tended to increase with the alcohol content and reached the maximum at EtOH 50, then gradually decreased to the EtOH 70 concentration. With EtOH 50, the anthocyanin content in the color solution was 1.2 times higher (179.47 mg/L) than with EtOH 70 (133.76 mg/L). The best solvent system is 50% ethanol solution which also minimizes the pigment decomposition favoring the extraction of anthocyanin in their original form. EtOH is more selective, so it only dissolves certain compounds, so limiting the solubility of some starch components, the sugar contained in the roselle, makes it easier for the anthocyanin to come out, while also saving more solvent. In addition, increasing the concentration of EtOH will increase the selectivity of the extraction solvent, thus attracting more anthocyanins and reducing other impurities [13]. Using EtOH 50 due to the alcohol content is enough to entice a large amount of anthocyanin out simultaneously while helping to save solvent.

3.2. Effect of liquid-solid ratio

Figure 4 demonstrated the effect of liquid-solid ratio on anthocyanin content. By increasing the liquid-solid ratio from 2:1 to 8:1, the extracted anthocyanin content increased from 87.47 mg/L to 158.56 mg/L. However, from 8:1 to 10:1 of liquid-solid ratio, the anthocyanin content decreased sharply. This can be explained as follows. When the ratio of materials/solvent is insufficient to fill up the material, the hypertonic environment could not be created, detaining the color in the vacuole of the material. However, when the solvent/material ratio reaches a certain value (depending on the characteristics of the material), the cell, owing to its capability to rapidly absorb water, swells to the maximum and bursts out simultaneously, releasing the color within the vacuole [14,15]. Therefore, based on the graph we choose the liquid/solid ratio of 8/1.

3.3. Effect of temperature

As showed in Figure 4, the anthocyanin content in the fluid as well as the effect of color separation from the material increased with increasing temperature and reached the highest level at the temperature of 60°C. There are significant differences from 30 to 60°C but did not observe between 60 and 70 °C. Therefore, in practice experiment, an extraction temperature of 60 °C could be applied, according to the effect of the good extraction yield of anthocyanins from Vietnam Hibiscus sabdariffa L. (*Roselle*). However, from 60 to 70°C, the anthocyanin content in the extract did not change significantly and rising temperature past 70°C caused the content to diminish due to decomposition. Long exposure at high temperatures would be harmful to the wanted compound as it would undergo the process of oxidation and polymerization [16]. The appropriate temperature will allow the anthocyanin to come out of the solvent. Excessive temperature causes only a small amount of anthocyanin in the roselle to come out. High temperatures cause anthocyanin to come out more but also a large amount of anthocyanin has decomposed [17]. Therefore, with low temperature as 40°C can only attract small amount of anthocyanin so the anthocyanin content in the color is low. At 60-70°C proper temperature makes the anthocyanin diffusion volume relatively large and does not decompose much due to the effect of temperature. At 80°C the temperature is so high that large quantities of anthocyanin are released but also decompose many colors. So, we choose the temperature of 60°C as the optimal temperature for obtaining sugar content as well as anthocyanin.

3.4. Effect of time

From Figure 5, the level of anthocyanin in the fluid tends to decrease with prolonged extraction time. Anthocyanin content was highest at 30 minutes, but as the time increased past 35 minutes, the content decreases markedly due to anthocyanin decomposition caused by the long exposure to high temperature [18]. When the extraction temperature was increased, the anthocyanin yield can be raised [19].Therefore, the appropriate time to extract anthocyanin is 30 minutes. As the extraction capability and anthocyanin stability depend mostly on the mixed effects caused by solvent concentration liquid-solid ratio, temperatures and time values. Herein, concentration of ethanol 50%, liquid/solid ratio at 8:1 (mL/g), temperature 60° C and reaction time 30 (min) was considered to be optimal for anthocyanins extraction from Vietnam Hibiscus sabdariffa L. (*Roselle*)

4. Conclusion

In this paper, the extraction of Hibiscus sabdariffa L. anthocyanin, which is increasingly used as a natural and safe coloring agent was investigated. After examining different liquids to support in extraction, the results have shown that the extraction of anthocyanin was higher with the use of non-toxic media composed of water/ethanol mixtures extracts of 50:50 (179.47 mg/L) than with water (53.43 mg/L). An increase in extraction time created with a greater quality of extracts. The solution extracted using 50% ethanol for 30 min, at 60°C was suitable for receiving colorant extracts with high quantities of anthocyanins from Hibiscus sabdariffa L. This research recommends that anthocyanin extraction from Vietnamese *Hibiscus sabdariffa* L. is a usable process, reasonable price and relatively quick, presenting a final extracted dye solution of such a high concentration, making it a hopeful substrate for the development of industrial dyes and colorants.

Acknowledgments: This study was funded by the Nguyen Tat Thanh University, Ho Chi Minh City, Vietnam.

References

[1] Phan Q , Bach G , Nguyen T and Le N 2019 *J. Nanosci. Nanotechnol.* **19** 974-78
[2] Tran H, Nguyen N, Nguyen T, Ho V and Bach G 2018 *Solid State Phenom.* **279** 217-21
[3] Nguyen N, Tran H, Le N, Phan Q, Le H, Nguyen T, Nguyen D and Bach G 2018 *Solid State Phenom.* **279** 235-39
[4] Aryee A, Agyei D and Akanbi T 2018 *Curr. Opin. Food Sci.* **19** 113-19
[5] Braga A, Murador D, Mesquita L and Rosso V 2018 *J. Food Compost. Anal.* **68** 31-40
[6] Zha J and Koffas M 2017 *Synth. Syst. Biotechnol.* **2** 259-66
[7] Gowd V, Jia Z and Chen W 2017 *Trends Food Sci. Technol.* **68** 1-13
[8] Riaz G and Chopra R 2018 *Biomed. Pharmacother.* **102** 575-86
[9] Rocha I, Bonnlaender B, Sievers H, Pischel I and Heinrich M 2014 *Food Chem.* **165** 424-43
[10] Ongkowijoyo P, Luna-Vital D and Mejia E 2018 *Food Chem.* **250** 113-26.
[11] Setford P, Jeffery D, Grbin P and Muhlack R 2017 *Trends Food Sci. Technol.* **69** 106-17
[12] J Lee, R W Durst and R E Wrolstad 2005 *J. AOAC Int.* **88** 1269-78
[13] Ping D, Zheng X, Zhou Y, Li Y, Li S and Li H 2017 *Food Chem.* **217** 552–59
[14] Koksal E, Bursal E, Dikici E, Tozoglu F and Gulcin I 2011 *Food Chem.* **80** 217-22
[15] Maran J, Sivakumar V, Thirugnanasambandham B and Sridhar R 2015 *J. Food Sci. Technol.* **52** 3617-26
[16] Yang L, Cao Y, Jiang J, Lin Q and Zhu L 2010 *J. Sep. Sci.* **33** 1349–55
[17] Anuar N, Adnan A, Saat N, Aziz N and Taha R 2013 *Sci. World. J.*
[18] Galhiane M, Rissato S, Chierice G, Almeida M and Silva L 2006 *Talanta.* **70** 286–92
[19] Zou T, Wang M, Gan R and Ling W 2011 *Int. J. Mol. Sci.* **12** 3006-17

IOP Conf. Series: Materials Science and Engineering **542** (2019) 012034 doi:10.1088/1757-899X/542/1/012034

Application of Augmented Reality in Satellite Cable Network Assembly

Wei Zhang[1], Changyu Chen[1], Huajun Chen[1], Gang Sun[1], and Shuping Zhao[1]

1 Beijing Institute of Spacecraft Environment Engineering, Beijing Engineering Research Center of the Intelligent Assembly Technology and Equipment for Aerospace Product, Beijing, China
E-mail: zhangwei308308@163.com

Abstract. This paper reviews the traditional cable installation in assembly and integration process of satellite. An effective and accurate cable network assembly information transmission method from designers to operators is not formed, in the process of laying complex cables, operators need to frequently confirm cable laying status back and forth between satellite and computer stations, this can easily lead to cable routing error or cable fixing point selection error. In order to improve efficiency and quality of satellite cable installation, a cable assembly system based on AR technology is designed. The successful application of this system provides a new solution for satellite cable laying and marks a new step in the digital assembly capacity.

1. Introduction

Cable network is widely used in final assembly of all platforms of satellite. Its main function is to transmit power and signal between various electronic devices[1]. Satellite cable network consists of many single cables; a single cable includes several branches, with cable electrical connectors at the end of the cable branches or directly connected to the instruments. Each cable and connector has a unique number to ensure that the cable is laid without errors. Cable brackets are used to fix the cable and are installed on the satellite panel before the cable is laid, cable network assembly not only needs to ensure that the cable is tied to the correct cable brackets and routed through correct holes in the deck according to the design requirements, but also needs to ensure that the free length of the cable end meets the requirements of subsequent cable plugging. Due to the huge difference in cable direction between different satellites, the narrow operating space and complicated operating conditions, the cable network assembly is heavily dependent on manual operations. In order to complete the laying, the operator needs to obtain the required information from the drawing or 3D model, lay the cable in the satellite cabin, then adjust and tie the cable.

At the satellite assembly plant, a fixed Kanban system is currently used to display the information required for cable assembly, and the worker operates according to the drawings or the 3D model. The workers perform the operation in the cabin with the memory of the information required for assembly, and needs to constantly switch the viewing angle between the display screen and the satellite due to the poor openness of the satellite cabin. The disadvantage of this type of method is that the relevant design information is manually matched to the real physical environment, and the identification of cable brackets and deck holes which are densely distributed is inefficient and error-prone. Work efficiency is reduced significantly because workers have to bear a lot of additional cognitive burden to confirm the correctness of cable assembly.

IOP Conf. Series: Materials Science and Engineering **542** (2019) 012034 doi:10.1088/1757-899X/542/1/012034

Cable Installation which is flexible assembly is complicated and error-prone, in recent years, virtual assembly technology is widely used[2-6]. In order to improve efficiency and quality of satellite cable installation, a cable assembly system based on AR technology is designed. This method automatically matches the cable laying information with the real physical environment, and guides the workers to complete the cable assembly. In the field of complex mechanical assembly, enterprises have researched the assembly guidance method based on augmented reality, but cannot meet the needs of satellite single-piece production and unable to systematically solve the problem of automatic conversion of cable laying models to visual augmented reality interactive information.

2. Cable Installation in Assembly and Integration Process of Satellite

Satellite comprises several modules which will be mated in subsequent stages in AIT process of satellite, and the cables are installed before equipment in a single module. The cable installation step in AIT process of satellite is showed in Figure 1, cable brackets which contain many nylon bases and several connecter brackets are installed on satellite with bolts before cable laying, cable fixing is performed according to relevant standards after cable laying; the cables are divided into high frequency cables and low frequency cables, and the low frequency cables are divided into signal cables and power cables, power cables are usually laid outside the low frequency cables. Currently, nylon base bundled with nylon tie is the main harness support to fix cable for domestic satellite, as is showed in Figure 2.

Figure 1 Process of Cable Installation.

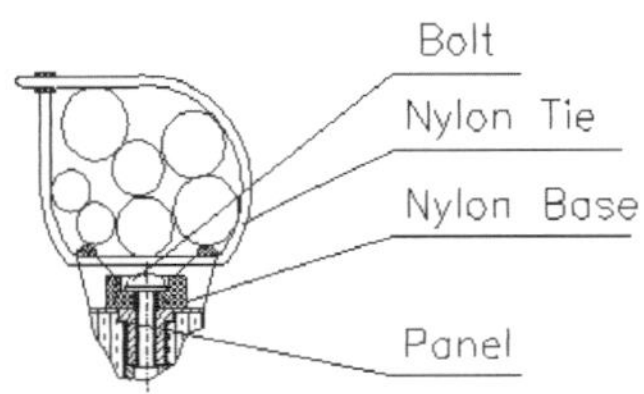

Figure 2 Diagram of Cable Assembly on Nylon Base.

As the satellite cable design changes from the traditional 2D mode to the 3D digital mode[7-8], the satellite cable assembly also enters the 3D assembly mode. Cable Installation is flexible assembly, the design input is inconvenient to use, and an effective and accurate cable network assembly information transmission method from designers to operators is not formed, so it leads to cable routing error and cable fixing point selection error, etc. The current 3D model based on desktop computer has poor human-computer interaction, In the process of laying complex cables, operators need to frequently confirm cable laying status back and forth between satellite and computer stations, and therefore, new technology is needed to improve cable assembly efficiency.

3. Augmented Reality

Compared with Virtual Reality technology, Augmented Reality technology can not only see virtual scenes, but also superimposes virtual scenes in real scenes by calculating operator position and perspective in real time. The Microsoft HoloLens[9] was released as a development edition in 2016 and is now available as a consumer version. The Google Glass was marketed as an AR device, however it is merely a transparent display lacking many necessary features, such as spatial mapping and a usable display, to interact with the real world and provide true AR capabilities. Another AR HMD, the Daqri Smart Helmet, is designed for industrial use but is currently still in development. Since the Google Glass is not capable of running AR assembly applications and the Daqri Smart Helmet is not yet released, the Microsoft HoloLens is the ideal choice for investigating an AR assembly application on a commercially available device.

IOP Conf. Series: Materials Science and Engineering **542** (2019) 012034 doi:10.1088/1757-899X/542/1/012034

The decision to use the HoloLens to investigate an AR assembly application on a HMD is strengthened by its' state of the art capabilities. Unlike AR HMDs in the past, the Microsoft HoloLens is a completely self-contained HMD, i.e, it does not require the HMD to be tethered to a separate computing device. The HoloLens features four Intel Atom x5- Z8100 1.04 GHz Intel Airmont Logical Processors, a HPU/GPU Holographic Processing Unit, 64 GB Flash, 2 GB RAM and 2-3 hours of active battery life that allows standalone operation of this device. All of this processing power is used to run 2 HD 16:9 light engines that project light through holographic lenses leading to a total resolution of 2.3 million light points. High resolution spatially located 3D content is generated by this system. The HoloLens also consists of an Inertial Measurement Unit (IMU), 4 environment-processing cameras, an RGB camera, and 1 depth camera to map its surroundings and allow interaction between the real and virtual world while tracking the device's position. Other features include 4 microphones, gaze tracking, gesture input, spatial sound and voice support[10]. The HoloLens is shown in Figure 3 below.

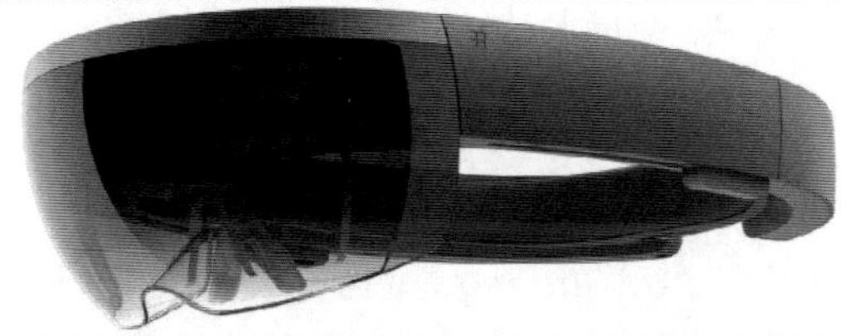

Figure 3 The Microsoft HoloLens.

4. The Process of Application of Augmented Reality in Satellite Cable Assembly

A cable assembly system based on AR technology is designed to improve efficiency and quality of satellite cable installation, the flow chart of the method is showed in Figure 4. Compared with the traditional method, this method automatically matches the cable laying information with the real physical environment, and guides the workers to complete the cable assembly.

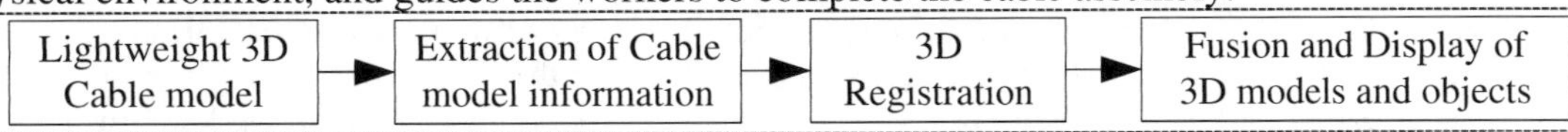

Figure 4 The Process of Application of Augmented Reality in Satellite Cable Assembly.

4.1. Lightweight 3D Cable model

The 3D cable model which is designed using CAD modeling tools such as CATIA and PROE is large, and the required computer performance is relatively high, so the 3D design model which is need to be lightweight is not suitable for application in mobile terminals. The 3D design cable model contains much modeling data, such as deck structure, devices, and electrical connectors. In order to improve the subsequent display rendering effect, the unnecessary model information should be deleted first, and then compress the 3D cable model by format conversion. The most important indicator that affects Unity rendering efficiency or whether the picture is stuck is the Tris and Verts in the states panel, Tris are the total number of triangular faces rendered within the camera's field of view, and Verts are the total number of vertices rendered within the camera's field of view. The lightweight 3D model has the advantages of small file size, easy operation and smooth display.

4.2. Extraction of Cable model information

While laying the satellite cable, the operator need to know the branch relationship of the cable, the length of the branch, the electrical connector code of each branch and the layout position of each branch in the cabin. Because much modelling data information, such as device and electrical connector information, is deleted in the process of 3D Cable model weight reduction, so it is necessary to extract cable model information in design 3D model through secondary development of software. The electrical connector code and location information of cable free end are very important cable model information in laying the cable, and the electrical connector code is added to the corresponding position of the lightweight 3D model.

4.3. 3D Registration

The lightweight 3D model has the advantages of small file size, easy operation and smooth display. The three-dimensional registration technology mainly realizes the correct fusion of virtual cables and real assembly scenes through the conversion between coordinate systems. Accurate 3D registration of virtual and real scenes is a key technical problem to be solved in the real-time fusion of cables, to accurately register the virtual assembly object in the corresponding position in the real environment, the conversion relationship between the virtual space and the real space coordinate system must be established in the cable virtual fusion assembly system. Registration error reaches a certain value will result in erroneous results. In this cable assembly system based on AR technology, two-dimensional codes which can be recognized very quickly are used in 3D registration, and 2D QR codes are pasted respectively on satellite several modules deck structure.

4.4. Fusion and Display of 3D models and objects

The cable assembly system based on AR technology was developed using Unity3D, Vuforia and the Microsoft HoloLens. Unity3D is an excellent development tool for the Microsoft HoloLens. To achieve higher precision, the embedded Vuforia system was used to perform marker based tracking. 2D QR codes which are pasted respectively on satellite several modules deck structure are used to define the separate locations of the assembly workspace. After these locations are defined, the HoloLens can properly track those positions as the user moves around in the assembly area using its IMU and environment processing cameras. An assembly application with a user friendly UI has been built to improve the user experience during cable laying, it has 3 interaction modes, such as menu, 2D QR code and voice. The electrical connector code display orientation is set to always face to user, and the different types of cables are displayed in different colors.

5. The Typical Case of Application of Augmented Reality in Satellite Cable Installation

Cable assembly system based on AR technology is tested during the AIT process of a remote sensing satellite in the paper. Remote sensing satellite comprises service module and payload module, 2 lightweight 3D cable models were built and all cables were classified into 2 modules according to the actual laying process. Lightweight 3D cable model of the service module is showed in Figure 5, it contains 134 cables which were installed first in AIT process of satellite. The outer panel structure, all devices and electrical connectors are deleted, after format conversion, the 3D cable model of service module is compressed to 27.1 MB, and the Tris are 4.1 million, Verts are 2.1 million. Because only the cable to be laid is needed to display in actual use, The electrical connector code of a cable is extracted and added to the corresponding position of the lightweight 3D model which is showed in Figure 6.
As is showed in Figure 7, the 2D QR code which represents service module of the remote sensing satellite is printed on paper and pasted on the satellite deck structure where is easy to determine the location. The operator must first wear Microsoft HoloLens before installing the cable and use the gesture to enter the corresponding module of the remote sensing satellite, secondly, scan 2D QR code in the satellite deck structure to complete 3D registration, and then select cable to be laid from the UI interface. The process of this system is very simple, and the cable assembly information is clear at a glance, thereby greatly improving the visibility of the 3D model and guiding principle of the laying operation. Figure 8 and Figure 9 show a typical case of application in a satellite cable installation, with the help of the cable assembly system based on AR technology, the time of laying 134 cables in one module of a remote sensing satellite can be reduced from 7 days to 5 days, and the cable laying accuracy can reach up to 100%.

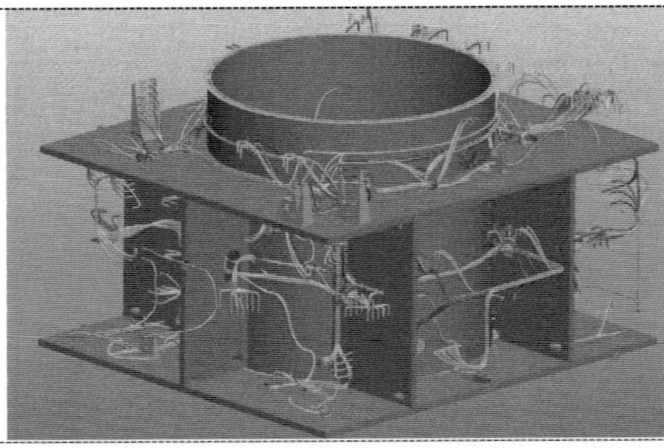

Figure 5 Lightweight 3D Cable Model of the Service Module.

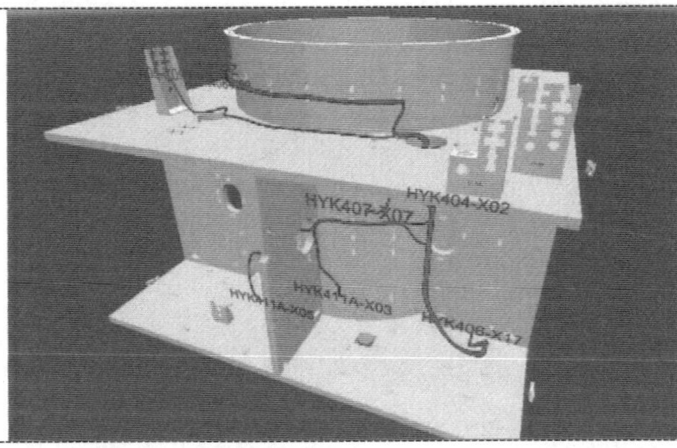

Figure 6 Electrical Connector Code Extraction and Display.

Figure 7. The 2D QR Code Pasted on The Satellite Deck.

Figure 8. Laying Cables in One Module of a Remote Sensing Satellite

Figure 9. Fusion and Display of 3D models and objects.

IOP Conf. Series: Materials Science and Engineering **542** (2019) 012034 doi:10.1088/1757-899X/542/1/012034

6. Conclusion

As the satellite cable design changes from the traditional 2D mode to the 3D digital mode, the satellite cable assembly also enters the 3D assembly mode. Cable Installation is flexible assembly, in order to improve efficiency and quality of satellite cable installation, a cable assembly system based on AR technology is designed. The successful application of this system provides a new solution for satellite cable laying and marks a new step in the digital assembly capacity. Augmented Reality technology is particularly suitable for cable laying applications; subsequent exploration will be carried out in other AIT process of satellite such as pipe valve installation.

7. References

[1] Wang Yingyan,Huang Yeping.The low frequency electric cable network design for satellite,Spacecraft Environment Engineering,2013,30(2)

[2] WANG Zhibin,LIU Jianhua, LIU Jiashun,NING Ruxin. A Multi-branch Mass-spring Model for Virtual Assembly of Cable Harness, Journal of Mechanical Engineering, 2014,50(3)

[3] Meng Guangwei,Wang Meng,Gu Haoqing,Li Lin,Feng Kangjun.3-D Bunched Cable Net Design And Virtual Assembly Technology On Missiles,Electronic Test,2016,(3)

[4] ZHANG Qiuyue,AN Luling. Application of Virtual Reality and Augment Reality in Aircraft Assembly, Aeronautical Manufacturing Technology,2017,530(11)

[5] WANG Falin,GUO Yu,ZHA Shanshan.Assembly System Construction and Its Key Techniques of Cable Harness under Virtual and Real Scene Fusion for Complex Mechatronic Products,JOURNAL OF GRAPHICS,2018, 39 (2)

[6] G Evans,et al. Evaluating the Microsoft HoloLens through an augmented reality assembly application, Proc.of SPIE Vol.10197

[7] Fan Kai Huang,Yeping Liu,Zheng Zhang,Guoqiang. Collaborative design for satellite tridimensional layout and cables based on Pro/E software,Spacecraft Environment Engineering,2015,32(4)

[8] Zhao Jinlong,Li Shu,Ma Yancheng,Chen Shuai.Process Optimizing Method Exploration of Cable Laying for Large Manned Spacecraft,Aerospace Manufacturing Technology,2018,(03)

[9] https://www.microsoft.com/zh-cn/hololens

[10] Hou Ying,Xu Weiwei. A Survey of Augmented Reality Technology,Computer Measurement & Control,2017,25(2)

IOP Conf. Series: Materials Science and Engineering **542** (2019) 012035 doi:10.1088/1757-899X/542/1/012035

Development of a 3-y-old Pediatric Cervical Spine Finite Element Model

Na Li[1], Wei Wei[2],Siwen Wu[3],Xian ping Du[2], Yin Liu[1]and Pengfei Rong[1,*]

[1] Xiangya 3rd hospital, Central South University,138th Tongzipo Rd, Changsha 410013, China
[2] State Key Laboratory of Advanced Design and Manufacturing for Vehicle Body, Hunan University,1st Lushan South Rd, Changsha 410082, China
[3]Xiangya 2rd hospital, Central South University,139th Renmin Rd, Changsha 410013, China

*Corresponding author: Rongpengfei66@163.com

Abstract
A 3-year-old pediatric cervical spine finite element (FE) model with detailed anatomical and material properties was developed and validated against cadaver tests under both quasi-static loadings. First, bone geometry was reconstructed based on high-resolution computed tomography (CT) scans, and elastic-plastic material was defined to simulate the cortical and cancellous bones. To simulate various ligament tears during dynamic tensile, ligament failure was defined using force versus displacement curves, which had a sigmoidal shape governed by three control points. To better represent the complicated structure of the disc, nucleus pulposus, annulus fibrosus substrate and four pairs of reinforced fiber lamina, intervertebral discs were defined using composite materials combined with viscoelastic material, hill foam material and four pairs of reinforced fiber lamina, respectively. This FE model could be utilized in prediction of cervical spine fracture, ligament and disc tear underlying pediatric cervical injuries.

1. Introduction

As the number of people using modern transportation continues to increase, the number of traffic accidents also increases each year, and it is widely accepted that traffic accidents are a major cause of human fatalities[1]. Recently, considerable attention has been paid to the pediatric cervical spine due to its increased vulnerability to injury. It has been shown that upper spinal injuries caused by sudden deceleration resulting from vehicle crash or direct impact in the case of pedestrians frequently occur in the 0-6 year age group[18]. Three years of age is an important growth period for the pediatric cervical spine. The fusion of the posterior synchondrosis first occurs at this age, and the anterior and posterior ossification centers join at this age as well[2]. These unique anatomical characteristics might cause common pediatric upper cervical injuries in 3-year-old children, including occipitoaxial joint dislocation, atlantoaxial joint dislocation, atlantoaxial rotatory subluxation, and spinal injury without radiological abnormalities[8]. To predict the cervical spine injury in 3-year-old children, it is necessary to develop a higher anatomical and mechanical biofidelity finite element (FE) model.

Experimental data obtained from human cadavers available for pediatric cervical spine studies are insufficient in this regard. Over the past three decades, while numerous adult FE models have been developed, only a handful of pediatric FE models have become available[4]. Applying the scaling method proposed by Melvin[15], then a series of pediatric cervical segmental FE models (1 year old, 3 years old, and 6 years old) were developed by scaling an adult cervical spine FE model. Although these models can simulate the pediatric facet joint and nucleus morphology in a satisfactory manner, the material properties of the annulus fibers and other unique morphological features of the pediatric cervical spine were inaccurate because of the limitations of the medical imaging techniques available at that time. Mizuno et al. [16] developed a 3-year-old whole-body human FE model in which both the geometry and material properties were linearly scaled from an adult model (THUMS AM50, Toyota, Japan). The authors validated this model against a 3-year-old dummy test, but it had the same limitations as Kumaransan's FE models[11].

In this study, the objective was to develop a 3-year-old cervical spine FE model with nonlinear material properties. Because it involves all of these methods, our proposed FE model might predict the cervical spine fracture, ligament and disc tear underlying pediatric cervical injury, which would present injury response in traffic accident and the designment of better for 3-year-old children protection system in vehicle.

2. Materials and methods

The images of vertebrae shown in Figure 1a were obtained from a 3-year-old boy (the cause of death was aplastic anemia and infection) with a height (93 cm) and weight (16 kg) similar to average 3-year-old pediatric. The images were obtained using CT scans (Brilliance CT, Philips Medical Systems, Best, The Netherlands) with a 1-mm slice thickness, and the screen resolution of 512×512 pixels. All procedures followed guidelines that were approved by the Central South University Xiangya 3rd Hospital Ethics Committee. Geometric data for the osseous structures were 3-D reconstructed using MIMICS (Version 12, Materialise Inc., Leuven, Belgium), as shown in Figure 1b. The geometries of the ligaments, cartilage, and intervertebral discs not visible in the CT images were filled in between the bony segments based on pediatric cervical anatomy and pediatric spinal models[9]. We adopted a multi-block approach to efficiently meshes. Hypermesh 10.0 (Altair, Troy, MI) was utilized to generate the meshes, as shown in Figure 1c. All calculations were conducted using LS-DYNA version 971 solver (LSTC, Livermore, CA).

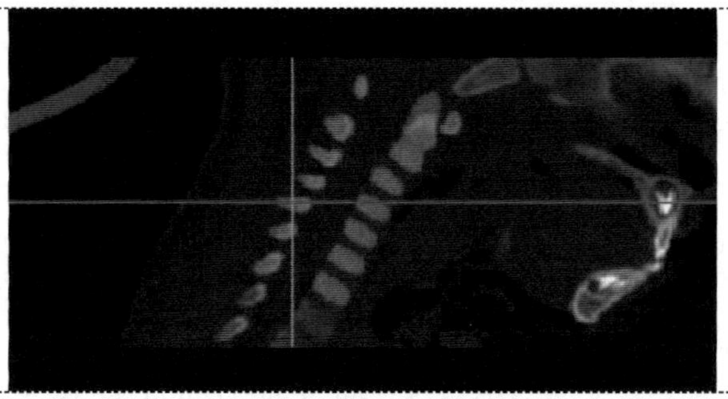

Figure 1a. Images of vertebrae obtained via CT.

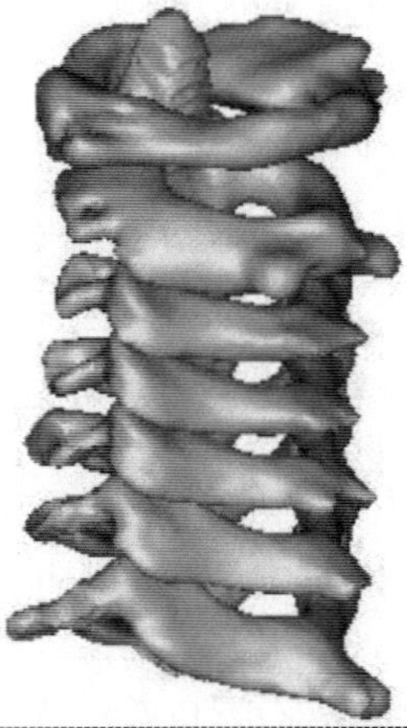

Figure 1b. Geometry of the 3-year-old pediatric FE model reconstructed in 3-D using MIMICS.

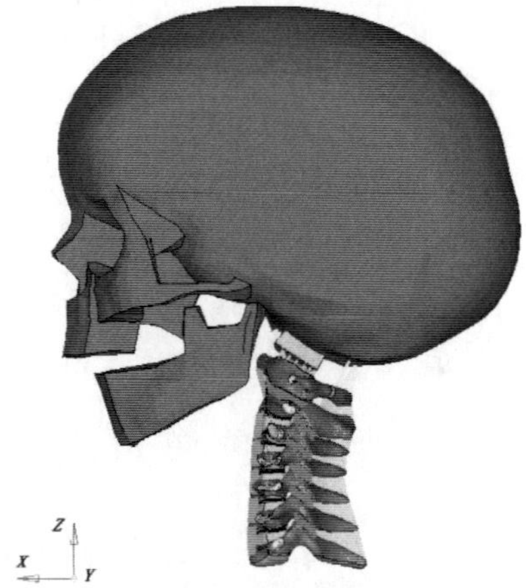

Figure 1c. The entire 3-year-old pediatric FE model including the head.

The material properties for the 3-year-old cervical spine model are showed in Table 1. According to Gilsanz et al.[7], the scale factor according the adult material properties of the cortical and cancellous bones was 0.805. The material parameters of cortical and cancellous bones for FE modeling of the adult spine reported by Panzer et al[17]. were utilized to derive the cortical and cancellous bones for modeling the 3-year-old pediatric cervical spine.

Table 1. Material properties parameters for the 3-year-old pediatric cervical spine FE model.

Tissue name	Element type	Material model	Material parameter	Scale factor
Cortical bone	Shell	Power-law plasticity	ρ=1.61 g/cm3, E=13.52 GPa, μ=0.3, K=354.8 MPa, N=0.2772	0.805 [7]
Cancellous bone	Hexahedron	Power-law plasticity	ρ=0.87 g/cm3, E=234 MPa, μ=0.3 K=5.7 MPa, N=0.2741	0.805 [7]
Endplate	Shell	Power-law plasticity	ρ=1.61 g/cm3, E=4.48 GPa, μ=0.3, K=153.2 MPa, N=0.2772	0.805 [7]
Annulus ground substance	Hexahedron	Hill Foam	m=3, n=2, C1=1.5341 MPa, b1=1, C2=-1.652 MPa, b2=2, C3=0.624 MPa, b3=3	0.705 [15]
Annulus fibers	Shell	Fabric	Stress versus strain curve	0.705 [15]

Nucleus	Hexahedron	General viscoelastic	G0=1.72 GPa G1=0.5930 kPa, β1=0.001477 1/s, G2=0.6763 kPa, β2=0.061524 1/s, G3=0.9516 kPa, β3=1.017893 1/s, G4=2.0384 kPa, β4=13.20041 1/s,	1
Growth plate	Hexahedron	Isotropic elastic	ρ=1.36 g/cm3, E=25 MPa, μ=0.4	1
Endplate cartilage	Hexahedron	Isotropic elastic	ρ=1.36 g/cm3, E=21.25 MPa, μ=0.4	0.85 [15]
Transverse cartilage	Hexahedron	Isotropic elastic	ρ=1.36 g/cm3, E=21.25 MPa, μ=0.4	0.85 [15]
Vertebral cartilage	Hexahedron	Isotropic elastic	ρ=1.36 g/cm3, E=21.25 MPa, μ=0.4	0.85 [15]
Facet cartilage	Hexahedron	Isotropic elastic	ρ=1.36 g/cm3, E=8.13 MPa, μ=0.4	0.782 15]
Ligaments	Beam spring	Non-linear	Force versus displacement curves	0.744 [15]
Dimensional scale factor G_s				0.637 [27]

E, Young's modulus; μ, Poisson's ratio; ρ, density, K, strength coefficient; N, hardening exponent; m, n, Ci, bi, material constant; Gi, viscoelastic modulus; βi, viscoelastic exponent.

Based on the 3-year-old cartilage scale factor of 0.85 [7] and the adult cervical spine study reported by Yamada et al. [21], the Young's modulus values for the pediatric transverse process cartilage, vertebra cartilage and facet joint cartilage were calculated.The disc of the 3-year-old cervix was composed of the annulus fibrosus, the nucleus pulposus, the endplate cartilage and the growth plate to represent the anatomy and material properties of the annulus fibrosus. The substrate of the annulus fibrosus was simulated using *MAT_HILL_FOAM material in LS-DYNA to model non-linear mechanical behavior. Adult annulus fibrosus substrate stress versus stretch curves were measured by Fujita et al. [6] with axial tensile and compression experiments. Yoganandan et al. [26] reported a scale factor for the 3-year-old pediatric cervical spine of 0.705. As shown in Figure 2a, with engineering stress equation (1) as the objective function, we first obtained the material parameters C_j and B_j, which are listed in Table 1. We then referred to a study that used FE modeling of the nucleus by Fice[5] and simulated the nucleus using a viscoelastic material. Combining this with the scale factor reported by Yoganandan et al. [26], we obtained the material parameters shown in Table 1.

$$S_1 = \sum_{j=1}^{m} C_j(\lambda_1^{b_j-1} - \lambda_1^{\frac{-nb_j}{2n+1}-1}) \tag{1}$$

Here, S_1 represents engineering stress, $m=3$, and $n=2$.

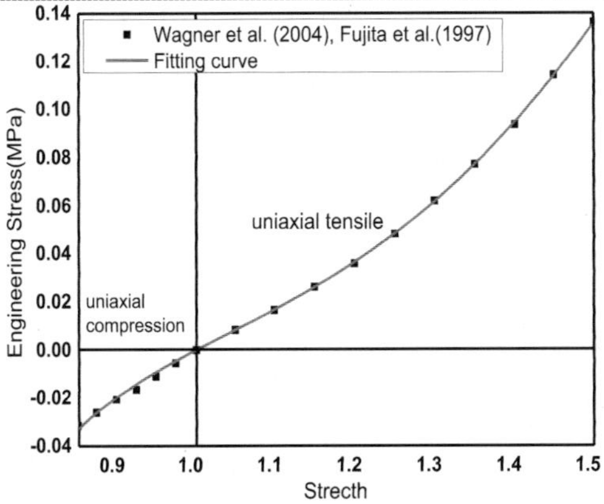

Figure 2a. Stress versus stretch curve of the annulus fibrosus substrate of the 3-year-old pediatric cervical spine, with the stretch being equal to the engineering strain plus one.

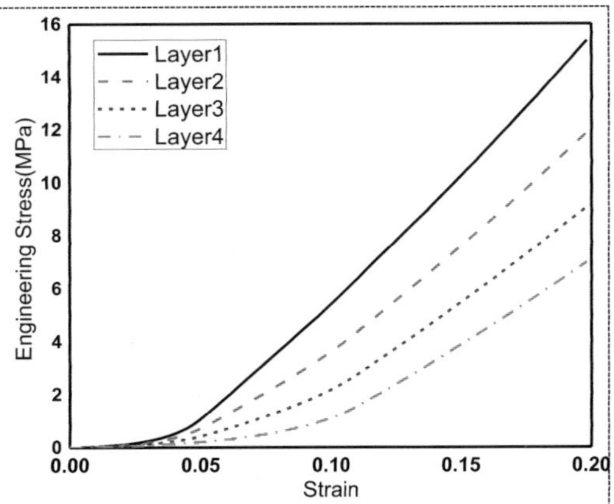

Figure 2b. Engineering stress versus strain curves of the four layers of pediatric AF fiber lamina [19].

The four pairs of reinforced fiber lamina were embedded in the substrate of the annulus fibrosus. The direction of the fibers in each lamina typically vary from +/- 25° in the outer layers to +/- 45° in the inner layers, as measured in the transverse plane of the disc , as shown in the Figure 3a. Using axial tensile experiments on the outer and inner layers of the adult disc AF fiber lamina and the scale factor reported by Yoganandan et al. [26], engineering stress versus strain curves for four layers of pediatric AF fiber lamina were obtained using the linear interpolation method, as shown in Figure 2b.

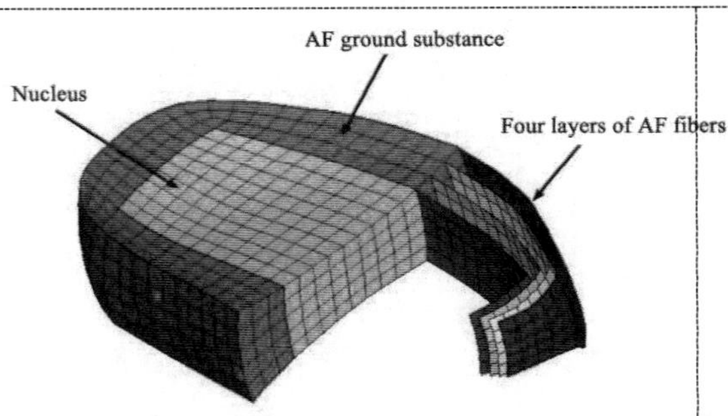

Figure 3a. Disc and fiber orientations of the complex structure simplified into four layers.

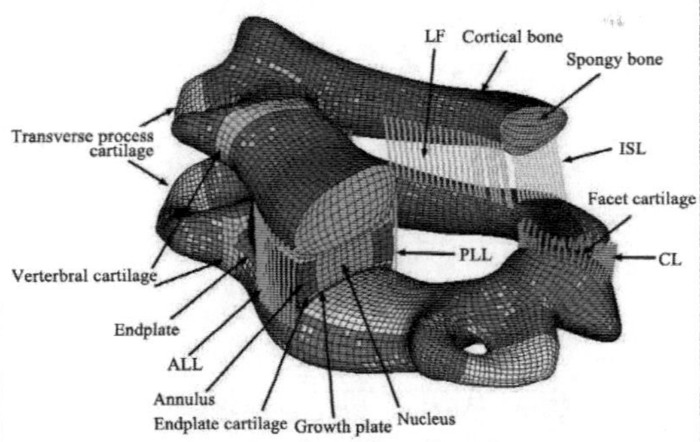

Figure 3b. Sectioned isometric view of the C4-C5 segment model (ALL-anterior longitudinal ligament, PLL-posterior longitudinal ligament, CL-capsular ligament, LF-ligamentum flavum, ISL-interspinous ligament).

Yoganandan et al. 26] reported an adult disc failure force of 571 N. Using the scale factor from equation (2), the disc failure force of a 3-year-old pediatric cervical spine was calculated to be 163 N. In this equation, $F_{\max}$ represents the adult disc failure force, the disc material scale factor α is 0.705, and the disc transverse area scale factor $\alpha_A^{'}$ is the squared value of the geometric scale factor (G_S), which is 0.637 [13]. DeWit and Cronin [3] further defined disc failure stress as the ratio of failure force to transverse area. Thus, the failure stress of the endplate and growth plate was 13.8 Mpa.

$$f = F_{\max} \times \alpha \times \alpha_A^{'} \quad (2)$$

The ligaments were also simulated by force versus displacement curves, which had a sigmoidal shape governed by three control points, as shown in Figure 4a. The failure strain and force for adult cervical spine ligaments in C2-C7 and C0-C2, which was also utilized in Dong's 10-year old pediatric cervical spine FE modeling[4], either reported by Yoganandan et al. [26]. In our study, equation (3) was developed by improving upon an equation reported in Dong's study [4], in which the failure stress and failure strain of 3-year-old pediatric cervical spine ligaments were calculated from the adult data. The force versus displacement curve for 3-year-old pediatric cervical spine ligaments was then obtained using logistic regression curve fitting, as shown in Figure 4b.

$$\left.\begin{aligned} d_i &= \frac{1}{\alpha \times G_S} \times \varepsilon_{\max} \left(\frac{\varepsilon_i}{\varepsilon_{\max}}\right) \times l_p \\ f_i &= \frac{G_S \times F_{\max} \times \left(\frac{F_i}{F_{\max}}\right)}{N_i} \end{aligned}\right\} \quad (3)$$

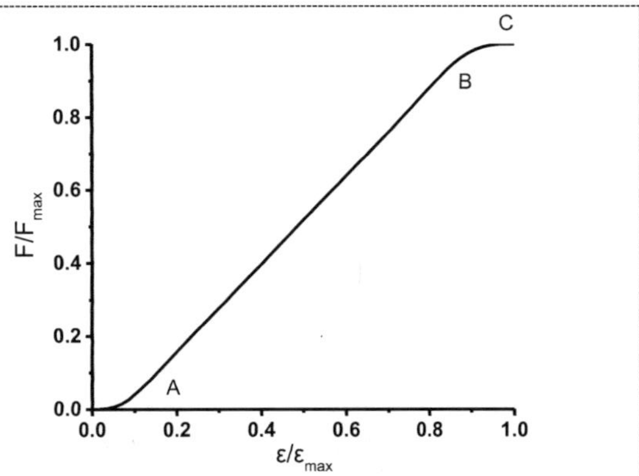

Figure 4a. Force versus displacement curves of the ligaments, which have a sigmoidal shape governed by the three control points A, B and C. The zero to A point was the toe region, the A to B point was the linear region, and the B to C point was the sub-traumatic region. Detailed values for A, B, and C are presented in Table 2.

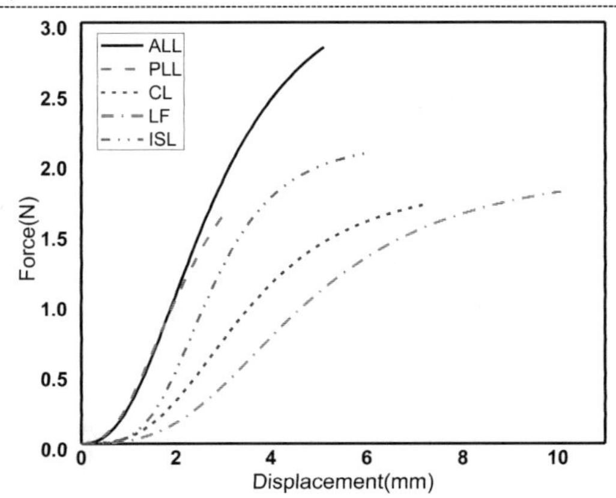

Figure 4b. Force versus displacement curves for 3-year-old pediatric cervical spine ligaments.

In equation (3), $i=1$, 2, and 3 represent three control points; $\varepsilon_3 = \varepsilon_{\max}$; $F_3 = F_{\max}$; d_i represents the deflection of the ligament at the three control points; l_P is the length of the ligament; F_i is the force acting on the ligament at the three control points; α is the scale factor for the material properties of the ligaments, which is 0.744 ; and G_S, the dimensional geometrical scale factor, is 0.637 [26].

Because ligaments are sensitive to strain rate, the visco-elastic material parameters for 3-year-old pediatric cervical spine ligaments had to be included in the FE model. Yoganandan et al. [25] reported ligament tensile responses at different strain rates.

In summary, the 3-year-old cervical spine FE model has 278,492 elements and 243,901 nodes, including 1,384 1-D beam elements, 84,066 2-D shell elements, 192,940 3-D hexahedral elements, and 102 triangular prism elements, in which the element size ranges from 0.5 to 1.0 mm. The head was simulated as a rigid body element.

3. Results

Limited experimental data have been reported for the pediatric cervical spine. Luck et al. [13] utilized cadaver specimens ranging from 20 weeks to 18 years old in quasi-static experiments to obtain axial tensile stiffness, failure force and failure displacement. Because the specimen ages were distributed from neonate to 22 months and 72 months to 216 months, precise data for 36 months were lacking. Therefore, we used the non-linear power-law function equation (4) to represent the relationship between age and C4-C5 quasi-static tensile stiffness for all of the pediatric age groups, as shown in Figure 5a.

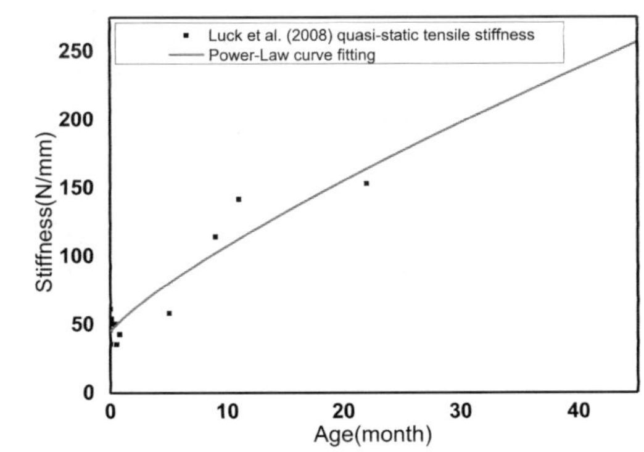

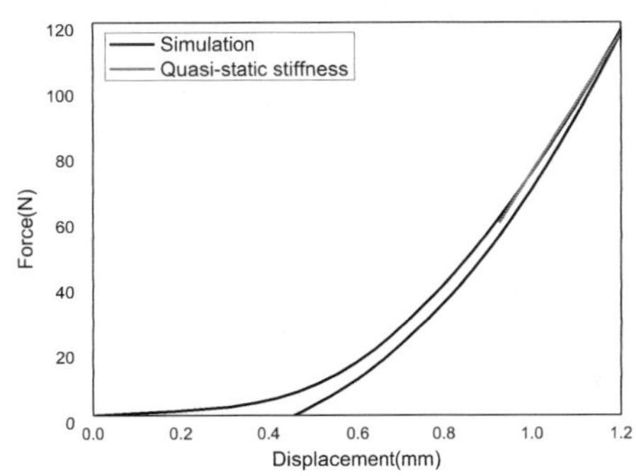

Figure 5a. Age versus C4-C5 quasi-static tensile stiffness curve for children across all age ranges.	Figure 5b. Simulated force versus displacement curve for the C4-C5 cervical segment and quasi-static tensile stiffness obtained via curve fitting.

$$Y = A \bullet X^{B} + C \tag{4}$$

Here, Y is the quasi-static tensile stiffness, X is the age (in months), and the undetermined parameters A, B and C were determined by curve fitting. After curve fitting, the correlation coefficients between the curve and the experimental data were R2=0.87836, A=10.063, B=0.802, and C=45.34. Based on this curve, the 3-year-old pediatric quasi-static tensile stiffness was found to be 223.5 N/mm.

In the quasi-static tensile experiment reported by Luck et al. [13], the C4 upper surface was constrained to six degrees of freedom, whereas the loading was applied to the lower surface of C5. The loading rate was 25.7 N/s, and the load was set to 120 N under quasi-static loading conditions. The force versus displacement curve was compared to the experimental data.

In our quasi-static validation, quasi-static tensile stiffness for C4-C5 segment motion was determined using linear logical regression analysis under loading of 50% to 100%. As shown in Figure 5b, the correlation coefficient between the curve and the experimental data was R2=0.998, and the stiffness was 211.8 N/mm. Compared to the experimental quasi-static tensile stiffness shown in Figure 5a, the deviation was 5.5%.

4. Discussion

The pediatric cervical spine is significantly different from the adult spine with regard to its anatomical and material properties[27]. In this study, the C2-C7 cervical segmental response was first validated under quasi-static loading, then compared with both segmental and global experimental data. However, muscles were not included in this model because accurate muscle force is dependent on physiological cross-section area (PCSA), muscle activation level and muscle fiber length [2], which were unavailable for the 3-year-old pediatric cervical spine [11].

The validation results indicated that the quasi-static stiffness is well correlated with experimental values. Which were within the range of baseline responses or experimental corridors, indicating that all of the segmental FE models have good biofidelity.

In the quasi-static validations, the simulated ROM values were in the range of the experimental corridor, and the deviations relative to the experimental reference value were 14.1%, 7.4% and 1.7%. In contrast, under quasi-static axial rotation loading, the simulated ROM value exceeded the maximum value of the experimental corridor, and the maximum deviation relative to the maximum experimental value was 7.0%. This result indicated that this model could represent 3-year-old C4-C5 cervical extension-flexion and lateral bending responses under 0.905 Nm and 0.517 Nm quasi-static loading.

5. Conclusions

In this study, a 3-year-old cervical spine FE model was developed that includes C0-C7 cervical vertebra, intervertebral discs, cartilage and ligaments. The elastic-plastic material used was defined to simulate the cortical and cancellous bones, and the nucleus pulposus and annulus fibrosus were defined using composite materials combined with viscoelastic material, hill foam material and four pairs of reinforced fiber lamina.The model could be further improved by performing additional experiments involving 3-year-old cervical spine muscles to obtain more accurate material parameters and geometric data.

6. References

[1] Basu S. 2012. Spinal Injuries in Children. Med Monatsschr 3: 441-444.

[2] Christophy M, Senan NAF, Lotz JC, O' Reilly OM. 2012. A Musculoskeletal Model for the Lumbar Spine. Biomech. Model. Mechanobiol. 11: 19-34.

[3] DeWit JA, Cronin DS. 2012. Cervical Spine Segment Finite Element Model for Traumatic Injury Prediction. J. Mech. Behav. Biomed. 10: 138-150.

[4] Dong L, Li G, Mao H, Marek S, Yang KH. 2013. Development and Validation of a 10-Year-Old Child Ligamentous Cervical Spine Finite Element Model. Ann. Biomed. Eng. 41: 2538-2552.

[5] Fice JB, Cronin DS, Panzer MB. 2011. Cervical Spine Model to Predict Capsular Ligament Response in Rear Impact. Ann. Biomed. Eng. 39: 2152-2162.

[6] Fujita Y, Duncan NA, Lotz JC. 1997. Radial Tensile Properties of the Lumbar Annulus Fibrosus are Site and Degeneration Dependent. J. Orthop. Res. 15: 814-819.

[7] Gilsanz V, Perez FJ, Campbell PP, Dorey FJ, Lee DC, Wren TA. 2009. Quantitative CT Reference Values for Vertebral Trabecular Bone Density in Children and Young Adults 1. Radiology 250: 222-227.

[8] Hamilton MG, Myles ST. 1992. Pediatric spinal injury: review of 174 hospital admissions. Journal of Neurosurgery 77: 700-704.
[9] Holzapfel GA, Schulze-Bauer CAJ, Feigl G, Regitnig P. 2005. Single Lamellar Mechanics of the Human Lumbar Anulus Fibrosus. Biomech. Model. Mechanobiol. 3: 125-140.
[10] Irwin A, Mertz HZ. 1997. Biomechanical Basis for the CRABI and Hybrid III Child Dummies. Pages 261-274. Child Occupant Protection 2nd Symposium Proceedings, SAE Technical Paper 973317. Orlando, FL: SAE Conference Proceedings P-316.
[11] Kumaresan S, Yoganandan N, Pintar FA, Maiman DJ, Kuppa S. 2000. Biomechanical study of pediatric human cervical spine: a finite element approach. J Biomech Eng 122: 60-71.
[12] Lavallee AV5, Ching RP, Nuckley DJ. 2013. Developmental Biomechanics of Neck Musculature. J. Biomech. 46: 527-534.
[13] Luck JF, Nightingale RW, Loyd AM, Prange MT, Dibb AT, Song Y, Fronheiser L, Myers BS. 2008. Tensile Mechanical Properties of the Perinatal And Pediatric PMHS Osteoligamentous Cervical Spine. Stapp Car Crash J. 52: 107.
[14] Mccall T, Fassett D, Brockmeyer D. 2006. Cervical spine trauma in children: a review. Neurosurgical Focus 20: E5.
[15] Melvin JW. 1995. Injury Assessment Reference Values for the Crabi 6-Month Infant Dummy in a Rear-Facing Infant Restraint with Airbag Deployment. SAE Technical Paper. Report no. 0148-7191.
[16] Mizuno K, Iwata K, Deguchi T, Ikami T, Kubota M. 2005. Development of a Three-Year-Old Child FE Model. Traffic Inj. Prev. 6: 361-371.
[17] Panzer MB, Fice JB, Cronin DS. 2011. Cervical Spine Response in Frontal Crash. Med. Eng. Phys. 33: 1147-1159.
[18] Ruge JR, Sinson GP, McLone DG, Cerullo LJ. 1988. Pediatric Spinal Injury: The Very Young. J. Neurosurg. 68: 25-30.
[19] Wagner DR, Lotz JC. 2004. Theoretical Model and Experimental Results for the Nonlinear Elastic Behavior of Human Annulus Fibrosus. J. Orthop. Res. 22: 901-909.
[20] Xu J, Li Y, Lu G, Zhou W. 2009. Reconstruction model of vehicle impact speed in pedestrian – vehicle accident. International Journal of Impact Engineering 36: 783-788.
[21] Yamada H, Evans FG. 1970. Strength of Biological Materials. Baltimore: Williams & Wilkins.
[22] Yoganandan N, Kumaresan S, Pintar FA. 2000a. Geometric and Mechanical Properties of Human Cervical Spine Ligaments. J. Biomech. Eng. 122: 623-629.
[23] Yoganandan N, Kumaresan S, Pintar FA 2001. Biomechanics of the Cervical Spine Part 2. Cervical Spine Soft Tissue Responses and Biomechanical Modeling. Clin. Biomech. 16: 1-27.
[24] Yoganandan N, Nahum AM, Melvin JW. 2015. Accidental Injury: Biomechanics and Prevention. New York, NY: Springer.
[25] Yoganandan N, Pintar F, Butler J, Reinartz J, Sances AJ, Larson SJ. 1989. Dynamic Response of Human Cervical Spine Ligaments. Spine 14: 1102-1110.
[26] Yoganandan NA, Kumaresan S, Pintar FA. 1998. Pediatric Cervical Spine Biomechanics Using Finite Element Models. Pages 349-363. Proc. International Research Council on the Biomechanics of Injury Conference: International Research Council on Biomechanics of Injury.
[27] Yoganandan NA, Pintar FA, Kumaresan S, Gennarelli TA, Sun E, Kuppa S, Maltese MR, Eppinger RH. 2000b. Pediatric and Small Female Neck Injury Scale Factors and Tolerance Based on Human Spine Biomechanical Characteristics. Pages 345-359. Proc. International Research Council on the Biomechanics of Injury Conference: International Research Council on Biomechanics of Injury.

IOP Conf. Series: Materials Science and Engineering **542** (2019) 012036 doi:10.1088/1757-899X/542/1/012036

Saponin, Polyphenol, Flavonoid content and α-glucosidase Inhibitory Activity, Antioxidant Potential of Launaea sarmentosa Leaves grown in Ben Tre province, Vietnam

Le Thi Them[1,2,3], Phu Tuong Nguyen Dung[4], Pham Thi Nhat Trinh[5], Quach Tong Hung[1], Le Nguyen Tuong Vi[1], Nguyen Trong Tuan[6], Tri Duc Lam[7], Van Thuy Nguyen[7], Le Tien Dung[1,3,*]

[1]Institute of Applied Materials Science, Vietnam Academy of Science and Technology, Ho Chi Minh city, Vietnam
[2]Faculty of Chemical Engineering and Food Technology, Nguyen Tat Thanh University, Ho Chi Minh City, Vietnam
[3]Graduate University of Science and Technology, Vietnam Academy of Science and Technology, Ha Noi, Vietnam
[4]Vietnam Sarmentosa Joint Stock Company, Ben Tre Province, Vietnam
[5]Faculty of Natural Sciences, Tien Giang University, Tien Giang Province, Vietnam
[6]College of Natural Sciences, Can Tho Univesity, Can Tho City, Vietnam
[7]NTT Hi-Tech Institute, Nguyen Tat Thanh University, Ho Chi Minh City, Vietnam
*Email: inpcdung@yahoo.com, labasm2013@gmail.com

Abstract. Launaea sarmentosa (Willd) Schultz-Bip.ex Kuntze, a creeping herb, belongs to family Asteraceae. It is distributed in sandy coasts of India, Sri Lanka, Malaysia, East Africa and Vietnam. In Vietnamese folk medicine, whole plant possess tonic, soporific, diuretic, laxative properties and the roots are used to benefit milk for mothers after childbirth, instead of *Lactuca indica*. Up to now, the phytochemicals and bioactivities of this plant are not yet been examined. The present study was designed to investigate the saponin, polyphenol and flavonoid contents as well as α-glucosidase inhibitory and antioxidant activities of the leaves of L. sarmentosa. Preliminary phytochemical screening of this plant were found to give positive reactions for flavonoid, saponin, alkaloid, tannin, and steroid compounds. The result of quantitative determination showed that the leaves contains high amount of polyphenol compounds (290.90 mg GAEs/g extract), lower amount of flavonoids (85.47 mg QEs/g extract) and the saponin content was found to be 10.8 % (per dry weight). As expected, the ethanol extract had a antioxidant potential with IC_{50} value of 24.15 μg/mL in DPPH experiment, compared to that of vitamin C. Meanwhile, the α-glucosidase inhibitory assay showed inhibited enzyme activity with IC_{50} value of 67.09 μg/ml.

1. Introduction

It is believed that medicinal plants play an important role in the health care system in Vietnam, based on their secondary metabolites, organic compounds synthesized by plants [1-4]. Due to these metabolites, the extracts or plants have been considered as a promising materials for the development new pharmaceuticals. Recently, polyphenol, flavonoid and saponin in plants are reported to possess many beneficial bioactivities, such as antioxidant, anti-inflammatory, anticancer, antihyperglycemic and antidiabetic activities [5]. *Launaea sarmentosa* (Willd) Schultz-Bip.ex Kuntze, a creeping herb,

IOP Conf. Series: Materials Science and Engineering **542** (2019) 012036 doi:10.1088/1757-899X/542/1/012036

belongs to family Asteraceae. It is distributed in sandy coasts of India, Sri Lanka, Malaysia, East Africa and Vietnam. According to Vietnamese folk medicine, the leaves is consumed as vegetable, whole plant possess tonic, soporific, diuretic, laxative properties and the roots are used to benefit milk for mothers after childbirth, instead of *Lactuca indica* [6]. Up to now, the phytochemical quantification and medicinal values of this plant are not yet been examined. As a part of our continuing studies on phytochemical and bioactivities of medicinal plants in Vietnam, this paper inform the phytochemical screening and to evaluate total polyphenol, flavonoid and saponin contents, as well as α-glucosidase inhibitory, free radical scavenger activities of the *L. sarmentosa* leaves [7-9].

2. Experimental

2.1. Chemicals and reagents

All reagents were of analytical grade and purchased from Merck (Darmstadt, Germany). Distilled and deionized water were obtained from Department of Pharmaceutical Biochemistry (Institute of Applied Materials Science, VAST, Vietnam). Others solvents and reagents were of analytical grade.

2.2. Plant Material

Leaves of *Launaea sarmentosa* were collected in the farm of Vietnam Sarmentosa Joint Stock Company, Ben Tre Province, Vietnam at August, 2018. Plant sample was authenticated by Southern Institute of Ecology, VAST, Vietnam, and voucher specimen (LSL1082018) was deposited at Department of Pharmaceutical Biochemistry, Institute of Material Science. The leaves sample were washed, afterthat cool dried and ground to powder.

2.3. Phytochemical screening

Phytochemical screening was carried out using standard methods to detect the plant secondary metabolites including of alkaloid, tanin, flavonoid, saponin, steroid, phenolic compounds etc. [10]

2.4. Total polyphenol content

Total polyphenol content was determined using Folin-Ciocalteu (FC) reagent [11]. The ethanol extract (10 mg) was dissolved in methanol (2 mL). Solution of 10% Folin-Ciocalteu reagent was prepared by adding Folin-Ciocalteu reagent (10 mL) in water (90 mL). Then, 5% Na_2CO_3 (3 g) was prepared by dissolving Na_2CO_3 (3 g) in water (50 mL). The plant extract (200 μL) was taken in a test tube and added 10% Folin-Ciocalteu reagent (1.5 mL). Then all the test tube was kept in a dark at room temperature for 5 min. Finally, 5% Na_2CO_3 (1.5 mL) was added to the solution and mixed well. The tube was kept again in the dark for 2 h. The absorbance was measured at 760 nm through UV-spectrophotometer, with gallic acid as a standard. The total polyphenol content were expressed as mg gallic acid equivalents (GAE)/ g extract.

2.5. Total flavonoid content

The total flavonoid content (mg/mL) was determined using aluminum chloride ($AlCl_3$) method [12]. The assay mixture consisting of 0.5 mL of the ethanol extract, 0.5 mL distilled water, and 0.3 mL of 5% $NaNO_2$ was stand at 25°C. This was followed by addition of 0.3 mL of 10% $AlCl_3$ immediately. Two milliliters of 1 M NaOH was then added to the reaction mixture, and the absorbance was measured at 510 nm. Quercetin was used as a standard. Total flavonoids content is expressed as quercetin equivalence (QE) mg/g extract.

2.6. Total saponin content

The determination of total saponin was done by slightly modified methods [13]. Briefly, 1 g of powdered leaf was extracted with 100 ml of 20% aqueous ethanol for half hour in an ultrasonic bath (Power sonic, Korea) equipped with digital sonication power, time and temperature controller with a useful volume of 10 L. The mixture was filtered by using Whatman filter paper no 1 and the residue again extracted with another 100 ml of 20% aqueous ethanol in same procedure. The total extract were concentrated under reduced pressure at 40°C to give 50 ml approximately. The residue was extracted three times with 30 ml diethyl ether. Next, the aqueous layer was successfully extracted with 30 ml n-butanol. The n-butanol fraction was washed with 10 ml of 5% aqueous sodium chloride. Finally, the below layer was evaporated and dried until constant weight. Saponin content was calculated and express as a percentage against dried weight of leaves.

2.7. Antioxidant Assay

DPPH scavenging expreiment was carried out as previously described [14]. The experiment was repeated three times. IC_{50} values denote the concentration of sample, which is required to scavenge 50% of DPPH free radicals. Vitamin C was used as standard compound. 1 ml of 0.1 mM DPPH prepared in ethanol was mixed with 1 ml of the leaf extracts ranging from 10 to 150 µg/mL. The reaction mixture was vortexed thoroughly and left in the dark at room temperature for 30 minutes. The absorbance was measured on a spectrophotometer at 517 nm. The scavenging ability of the plant extracts was calculated using the equation:

$$\text{DPPH scavenging activity (\%)} = \frac{Abs_{control} - Abs_{sample}}{Abs_{control}} x100$$

where $Abs_{control}$ is the absorbance of DPPH + ethanol; Abs_{sample} is the absorbance of DPPH radical + sample (i.e. extract or ascorbic acid)

2.8. α-Glucosidase inhibitory activity

The α-glucosidase inhibitory activity was determined according to the described method [15]. The principle of the method is that the enzyme *α*-glucosidase hydrolyzes the substrate *p*-nitrophenyl-*α*-D-glucopyranosid (pNPG) to give *α*-D-glucose and *p*-nitrophenol (pNP). The α-glucosidase inhibitory activity was calculated using the following equation:

$$\text{\% Inhibition} = \frac{A_{control} - A_{sample}}{A_{control}} x100$$

$Abs_{control}$ is the absorbance without plant extract; Abs_{sample} is the absorbance with sample

3. Result and Discussion

Preliminary phytochemicals screening (Table 1) showed that *Launaea sarmentosa* leaves were found to be present alkaloid, flavonoid, triterpenoid, saponin, and only tannin was absent. As we known, the plant secondary metabolites are proved to possess interesting bioactivitie such as anti-inflamatoty, antidiabetic, antibacterial, antioxidant activities, wound healing properties, protection of the skin, health promotion and disease prevention etc. Nowaday, phytochemicals are increasingly attracted to sciencetist for the isolation and structural determination, because they are not only used directly as drugs for treatment of disease but also as modeling compounds for development of new drugs with less toxic and side efects to humans. Previous authors already reported that flavonoid and polyphenol exhibited many bioactivities such as antioxidant, anti-inflammatory, antimicrobial, anticancer, α-glucosidase inhibition and anti-alergic activities [16,17]. Saponin are other type of phytoconstituents which are responsible for tonic and antimicrobial activities of plant [18]. Therefore, the appearance of different phytochemicals in *Launaea sarmentosa* may be responsible for its tonic effect, laxative, antibacterial activities which were used in Vietnam traditional medicine.

Table 1. Phytochemical screening of ethanol extracts

Phytochemical tests	***Launaea sarmentosa* extract**
Flavonoids	+the
Alkaloids	+
Triterpenoids	+
Tannin	-

Saponin	+

Note: (+) indicates present; (-) indicates absent

The results of total flavonoid, polyphenol and saponin contents (Table 2) showed that *Launaea sarmentosa* leaves contained polyphenol (290.9 mg GAEs/g of extract), flavonoid (85.47 mg QEs/g) and saponin (10.8 %). The α-glucosidase inhibitory and antioxidant activities test (Table 3) showed the IC_{50} values of 67.09 mg/ml and 24.15 μg/mL, respectively. The α-glucosidase inhibition may be responsible for the antihyperglycemic effects in plasma blood. Enzyme α-glucosidase is one of the enzyme responsible for carbohydrates destruction to form smaller sugars, such as glucose and fructose. Thus, they will slowly digest glucide and delay glucose absorption in body, leading blood glucose levels following meals increase slowly. The high inhibition activity of *Launaea sarmentosa* leaves on α-glucosidase might be contributed by the high content of polyphenol and flavonoid compounds. Diabetes drugs with *α*-glucosidase inhibitors such as acarbose and voglibose, usually caused serious side effects. Therefore, it is necessary to discover and develop for alternatives with *α*-glucosidase inhibitory activity and without side effects to the body. In recent years, projects were deeply studied to discover potent *α*-glucosidase inhibitors from plant natural products have received great attention due on their highly abundant metabolites in nature and potential biological activities, among them *Launaea sarmentosa* was very promising, based on its alfa-glucosidase inhibitory activity.

Table 2. Selected phytochemical contents of *Launaea sarmentosa* leaves

Items	Yield
Total polyphenol content (mg GAEs/g)	290.90
Total flavonoid content (mg QEs/g)	85.47
Total saponin (% per dry weight)	10.80

It was showed that the extract contains high amount of polyphenol (290.9 mg GAEs/g of extract), but low amount of flavonoid (85.47 mg QEs/g extract). This may be due to most of the polyphenol in the extract are non-flavonoid. The study also showed that the extract exhibited highly antioxidant activity through IC_{50} value of 24.15 μg/mL (in DPPH test) and high amount of polyphenol content may be the cause for their antioxidant ability. Because antioxidants (polyphenol and flavonoid) are able to stabilize or scavenging free radicals, so that they can not cause oxidative damage to the cellular structures. Free radicals and oxidants play a dual role as both toxic and beneficial compounds, since they can be either harmful or helpful to the body. When the balance between antioxidants and free radicals were shifted toward generating radicals, called oxidative stress and causing most of the degenerative diseases including atherosclerosis, cancer, inflammatory joint disease, asthma, diabetes, senile dementia, etc. The human body has several methods to rebalance through generating endogenous or exogenous antioxidants such as vitamin C, vitamin E, coenzyme Q10, catalase superoxide dismutase and polyphenol or flavonoid supplements.

Our results were in good agreement with previous study that polyphenol compounds are potential antioxidants. The alfa-glucosidase inhibitory and antioxidant properties of *Launaea sarmentosa* leaves are meaningful to reduce oxidative damage associated with diabetic complications [19]. An antidiabetic herb with potent antioxidant activity may be prevented oxidative stress, which is considered supporting the relationship between chronic hyperglycemia and diabetes-associated complications.

Morever, total saponin content of *Launaea sarmentosa* leaves were found to be 10.80 %. The bioactivities of saponin are normally assigned to their amphipathic properties, consisting of a hydrophobic triterpene/steroid skeleton and a hydrophilic carbohydrate molecules. It is known that potential biological activities of saponins are dependent on other aspects of their structure. Some saponins are also important pharmaceuticals, and the underexplored biodiversity of plant saponins is likely to prove to be a vital resource for future drug discovery. Saponin compounds is believed to protect living body against hypercholesterolemia and antibiotic properties. Saponins were also responsible for central nervous system activities and being used to create tonics for treating depression, neurasthenia, etc. So, this result can be used to explain the ability of making people more strong when using *Launaea sarmentosa* in daily meals. Hence, *Launaea sarmentosa* may be a potential source of natural products for the development drug for treatment type 2 diabetes, stress oxidative and tonic property.

Table 3: *In vitro* α-glucosidase inhibitory and antioxidant activity of extract

Sample	DPPH (IC_{50})	α-glucosidase inhibitory activity (IC_{50})
Launaea sarmentosa leaves	24.15 (μg/mL)	67.09 (μg/mL)
Acarbose		138.2 (μg/mL)
Vitamin C	6.45 (μg/mL)	

Acknowledgments
This paper was partially supported by the Nguyen Tat Thanh University, Ho Chi Minh City, Vietnam

References

[1] Tran T, Nguyen H, Nguyen D, Nguyen T, Tan H, Nhan L, Nguyen D, Tran L, Do S and Nguyen T 2018 *Processes* **6** 206

[2] Vo T-S, Ngo D-H, Bach L G, Ngo D-N and Kim S-K 2017 *Process Biochemistry* **54** 188–94

[3] Ryu B, Kim S-Y, Vo T-S, Kim W-S, Kim D G and Kim S-K 2017 *RSC Advances* **7** 24561–8

[4] Santhi R., Lakshmi G., Priyadharshini A.M. and Anandaraj L. International Research Journal of Pharmacy, 2011, **2 (1),** 131-135.

[5] Hoensch HP, Oertel R. *Clinical Nutrition Experimental*, 2015, **3**, 8–14.

[6] Loi, D. T. Medicinal plants and remedy of Vietnam. *Publisher of Medicine*, Hanoi, Vietnam, 2004, 816-818.

[7] Tri D.M., Minh P.N., Phat N.T.,Chi M.T., Phong M.T., *Phytochem Lett.*, 2016, **18**,10-13.

[8] Trinh P.T.N., Tri M.D., Hien D.C., AnN.H., Minh P.N., An P.N., Dung L.T., *Nat Prod Res.,* 2016, **30(7**), 761-767.

[9] Trinh P.TN., Luan N.Q., Tri M.D., Khanh V.D., An N.H., Minh P.N., An P.N., Thuy N.T.L., Phung N.K.P., Dung L.T., *Nat Prod Res.*, 2017, **31(15),** 1733-1738.

[10] Harborne JB. Phytochemical methods - a guide to modern techniques of plant analysis. *2nd ed.* London: Chapman and Hall, 1984. pp. 4–16.

[11] Mallick EP., Singh MB. Plant Enzymology and Histoenzymology, *1st ed.* New Delhi, Kalyan Publisher; 1980, 280.

[12] Kariyone T., Hashimoto Y., Kimura M. *J Pharm Soc.* 1953;**73**, 253–256.

[13] Obadoni BO., Ochuko PO. *Glob J Pure Appl Sci*, 2001; **86**, 2003-2008.

[14] Chai TT., Wong FC. *J Med Plants Res.* 2012;**6,** 1730–1735.

[15] Xiao H., Liu B., Mo H., Liang G. *Food Res Int.* 2015;**76**, 605–610.

[16] Kokate KC. *Practical pharmacognosy 4th ed.* Delhi: Vallabh Prakashan; 1997, p. 218.

[17] Ayoola GA., Coker HAB., Adesegun SA., Adepoju-Bello AA., Obaweya K., Ezennia EC., *Trop J Pharm Res. 2008*; **7**, 1019–1024.

[18] Anyasor GN., Ogunwenmo KO., Oyelana OA., Akpofunure BE. *Afr J Biotechnol.* 2010; **9(31**), 4880–4884.

[19] Hill MF. *Curr Cardiol Rev.* 2008; **4(4),** 259-268.

IOP Conf. Series: Materials Science and Engineering **542** (2019) 012037 doi:10.1088/1757-899X/542/1/012037

Preparation and Determination of Total Anthocyanins extraction from the Skin of Vigna cylindrica Skeels (Dolichos catjang Burm. f)

Tri Nhut Pham[1,2], Dai-Viet N. Vo[3], Tri Duc Lam[1], Tran Anh Vy[4], Tran Quoc Toan[5],
Hieu Vu-Quang[1], Le Thi Hong Nhan[6], Phan Nguyen Quynh Anh[6], Le Minh Bui[1,*]

[1]NTT Hi-Tech Institute, Nguyen Tat Thanh University, Ho Chi Minh City, Vietnam
[2]Faculty of Chemical Engineering and Food Technology, Nguyen Tat Thanh University, Ho Chi Minh City, Vietnam
[3]Faculty of Chemical and Natural Resources Engineering, Universiti Malaysia Pahang, Malaysia
[4]Department of Chemical and Biochemical Engineering, Gachon University, South Korea
[5]Institute of Natural Products Chemistry, Vietnam Academy of Science and Technology, Ha Noi, Vietnam
[6]Department of Chemical Engineering, HCMC University of Technology, VNU-HCM, Ho Chi Minh City, Vietnam
*Email: labams2013@gmail.com

Abstract. Because of antioxidative potential, anthocyanins become more and more popular in various fields including food products and pharmaceutical industries. The extraction technique has a great effect on the quality of natural color. Seed coat color is an essential part of customer choices for cowpea (Vigna unguiculata L. Walp). The experimental study was conducted to determine the effect of different extracting parameters such as duration of extraction, different extraction of temperature (20 - 90oC), duration of extraction (30 - 90 min) and solid-liquid ratios (2:1 - 10:1 mL/g) and solvent concentration (H2O : Ethanol) were selected in order to extract anthocyanin from Vigna cylindrica Skeels. The highest anthocyanin content of 205.58 mg/L was collected at 50% enathol, the solid/liquid ratio 6:1 (mL/g), extraction time 60 (min), and temperature 60 (oC). The results showed that different extraction media can significantly affected the anthocyanins content of the valuable source Vigna cylindrica Skeels extracts.

1. Introduction

The common bean (*Phaseolus vulgaris* L.) is one the most popular beans worldwide as a food ingredient. It is most commonly called as "cow pea", is medicinal plant belonging to the family Fabaceae or Leguminosae, having lot of pharmacological properties. In the developing countries, especially Vietnam, the cowpea plays an important source of vitamin, minerals, antioxidants, proteins and so on [1-2]. The coloration of the bean skin from red or purple to even black is a consequence of anthocyanin biosynthesis. The previous studies showed that the amounts of flavonols in cowpea was high including anthocyanins, quercetin, and procyanindins. In addition, the seed coat color is the main determinant of flavonoid formation [3-5].

In the natural food colorants, anthocyanins are the most popular family name was found in not only higher trees but also some fruits, nut, vegetables and so on. Anthocyanins maintain valuable pharmacological characteristics such as anti-inflammatory, antineurodegenerative and anti-oxidative. Nowadays, anthocyanins have advantageous health impacts because of their antioxidant activities. Therefore, it has application as anticancer agents and against cardiovascular diseases. Besides health advantages, anthocyanin pigments have a high trade in the commercial business such as textile industries and natural dyes in the food [6-8]

In term of extraction techniques, the research should maximize the recovery of anthocyanin and minimize the number of supplements and reduce the degradation of the natural state of colorants. Having the different process parameters impact on the extraction of the bioactive compound such as temperature, solvent composition, extraction time and solid to liquid ratio [9-14]. Aside from population studies, previous studies have highlighted the role of anthocyanins in different fields. However, relatively little is explored about the combination of different factors and measurement of optimal conditions in order to obtain a yield of anthocyanins. From a practical viewpoint, the study using random concentrations of solvents, extraction time, liquid-solid ratio and temperatures has been carried to extract anthocyanins from the skin of Vigna cylindrica Skeels (Dolichos catjang Burm. f) of some popular in Vietnam region varieties

2. Material and methods

2.1. Sample preparation

The cowpea samples were collected from Go Vap market (Ho Chi Minh City, Vietnam). Following with washing, soaking in water for about 30 minutes. Next, seeds were peeled and dried to constant volume and store in polyethylene vacuum bags at room temperature until extraction was taken. Ethanol (C2H5OH) is obtained from Sigma Aldrich (US). Distilled water was applied for the preparation of all solutions.

2.2. Anthocyanins extraction from Vigna cylindrica Skeels process

In order to extract anthocyanins, 10g CSC was put in the two neck round bottom flask as and was extracted by ethanol with concentration was investigated at conditions as concentrations of solvents (H_2O: Ethanol), the liquid/solid ratio: 2:1 - 10:1 (mL/g). The extraction temperature is adjusted to about 20–90 (^oC) and 30–90 (min). Then, centrifugation took place at 4000 rpm for 15 min by high speed centrifuge Model LACE16 (from COLO lab expert). The supernatant was collected and the extract, after being filtered with filter paper. The extraction process were done in dark color bottles with screwed caps. Monomeric anthocyanin was calculated as cyanidin-3-glycoside using a pH differential method [15]

3. Result and discussion

Effect of solvent concentration with extraction: Solvent concentration plays an important role in obtaining high extraction efficiency of anthocyanins. The choice of extracting medium is considered a vital factor for the extraction of natural colorants, which is the first crucial step for parameters optimization. Anthocyanins is the compound, which easy solute in water, so H_2O will take anthocyanin out of cell. Though, since H_2O is soluble sugar, starch during extraction should impede many anthocyanin escape processes. EtOH is more selective and therefore only dissolves certain compounds so limiting the solubility of some starch constituents, the sugar contained in pea pods, makes it easier for anthocyanins to come out, while also saving more solvent. Thus, ethanol/water is the common solvent used in the extraction of polar organic compounds. Therefore, the ethanol/water solution is the standard solvent applied in the extraction of polar organic compounds. Moreover, different ethanol concentrations were prepared as the solvent in order to examine the influence of ethanol concentration on the anthocyanin yield, and the results are shown in Figure 2A. The increasing ratio of ethanol to the raw material between 0 % and 100% could grow anthocyanins extraction. The anthocyanin yield was compared with the percentage of ethanol when the concentration from 25% (158.91 mg/L) to 50% (205.55 mg/L). Hence, in term of the cost and yield, 50 % ethanol in water was determined for the following experiments.

Effect of liquid-solid ratio: Liquid-solid ratio acts outstanding function in getting high extraction efficiency of anthocyanins in the method of standard extraction. In general, a larger solvent volume can dissolve components more effectively. In this research, the efficient liquid-to-solid ratio was estimated. Figure 2B presents the effect of liquid-to-solid ratio on the extraction yield of anthocyanins. TThe extraction with various different ratios of solvent to the skin of Vigna cylindrica Skeels including 2:1, 4:1, 6:1, 8:1, 10:1 mL/g were handled in a water bath at a temperature of 60°C for 60 min.. The conclusions from this research revealed that the more the ratio of fresh calyxes to water the more the red color intensity perceived. Below the fixed conditions of different factors, it could be recognized that the extraction efficiency positively rose with the liquid-to-solid ratio, especially when the ratio raised from 2:1 (150.96 mg/L) to 6:1 (205.51 mg/L). However, the anthocyanin content decreased sharply after increasing that ratio from 8:1 to 10:1. This can be explained as when the solvent/material ratio reaches a certain value (depending on the characteristics of the material), the cell then swells to the maximum and bursts out simultaneously, releasing the color within the vacuole. Based on the graph, the ratio 6:1 with an extraction content of 205.51 mg/L was selected to produce the skin of Vigna cylindrica Skeels extract to be concentrated

*Effect of temperature***:** Besides solvent concentration and liquid-solid ratio, temperature extraction is also an essential factor in the extraction of heat compounds. In this research, extraction was taken at various temperatures (30-90 °C) While other extraction parameters did not change. Particularly, other conditions were fixed at 50% ethanol, 6:1 liquid-to-solid ratio, 60 min extraction time. The influence of temperature on the extraction yield of anthocyanins is presented in Figure 2C. When the temperature raised from 20 to 60 °C, the yield of anthocyanins rise from 118.08 to 205.60 mg/L, following with falling at the temperature from 60 to 90 °C. Too low temperatures make only small amounts of anthocyanins available in the black bean shell, and high temperatures cause anthocyanins to come out more, but large amounts of anthocyanins are also decomposed. Therefore, with a low temperature of 20°C, only a small amount of anthocyanin can be attracted. At 60°C the appropriate temperature for anthocyanin diffusion is relatively large and does not decompose much due to the effect of temperature. Thus, the optimal temperature for obtaining sugar content as well as anthocyanin is 60°C.

Effect of time: Time duration plays vital functions in getting high extraction efficiency of anthocyanins as well. After the yield of anthocyanins extraction rises with time, the establishment of equality for the objective components in and out of plant cells. Furthermore, the impact of time on the extraction yield, different time points were examined from 20 to 100 min. Figure 2D presents the extraction results taken under various time duration. Anthocyanins content was highest at 60 minutes and reached 205.57 mg/L, but when the time is increased over 80 minutes, it decreases markedly. Because of its high temperature at long periods of time, the anthocyanin is decomposed. This suggests that extraction time does not increase the anthocyanin content. The appropriate time to extract anthocyanin is 60 minutes.

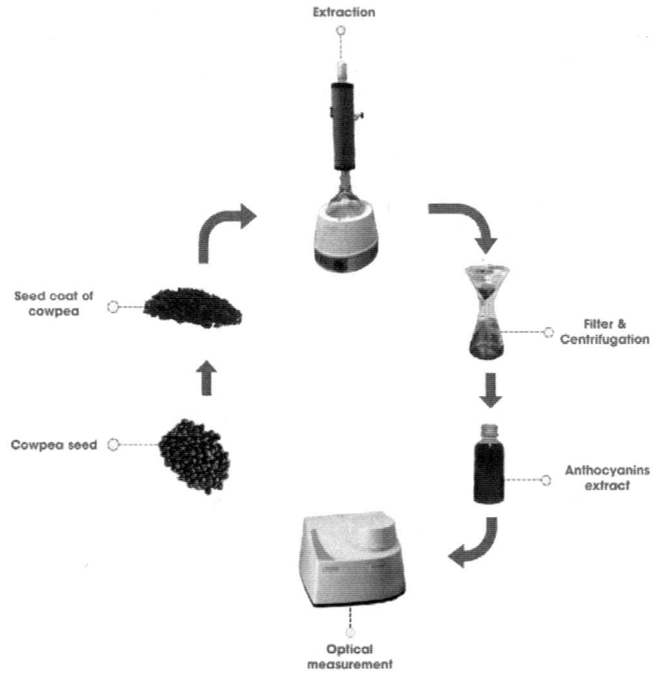

Figure 1. The process of the total anthocyanins extraction from the skin of Vigna cylindrica Skeels (Dolichos catjang Burm. f)

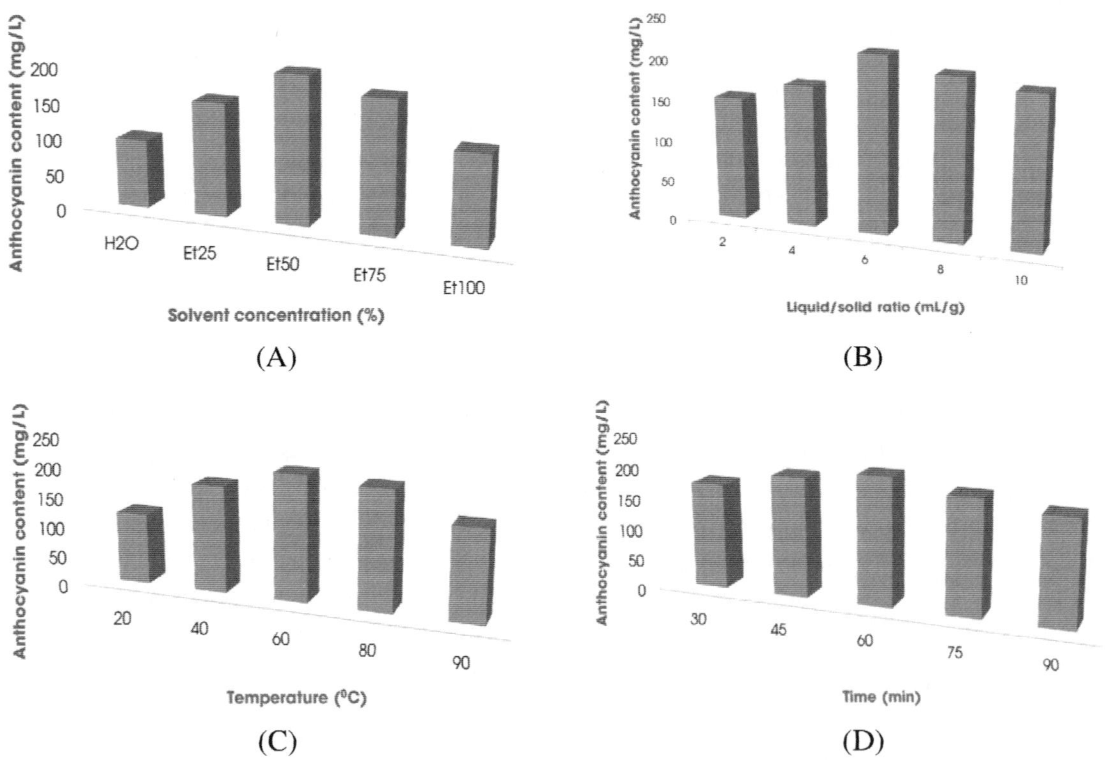

Figure 2. Effects of (A) solven ethanol concentration, (B) liquid/solid ratio, (C) temperature and (D) time factor on the content of anthocyanin

4. Conclusion

In this studies, Though extraction of anthocyanins from the skin of Vigna cylindrica Skeels (Dolichos catjang Burm. f), we genetically analyzed the temperature, extraction time, and liquid/solid ratio had significant impacts on anthocyanin extraction. Various extracting media were applied and the mixture of 50% (v/v) ethanol resulted in maximum anthocyanin content (205.60 mg/L). The increasing of time, regardless of the other variables, led to a reduction in the total anthocyanin content. When antagonize with other verve of natureal anthocyanins, the Vigna cylindrica Skeels is rich in anthocyanin and it can be used as a natural color resources of anthocyanin for food and pharmaceutical industries.

Acknowledgments: This work was supported by grants from Nguyen Tat Thanh University, Ho Chi Minh City, Viet Nam.

References

[1] Zargar S M, Mahajan R, Nazir M, Nagar P, Kim S T, Rai V, Masi A, Ahmad S M, Shah R A, Ganai N A, Agrawal G K and Rakwal R 2017 *Journal of Proteomics* **169** 239–48
[2] Chekanai V, Chikowo R and Vanlauwe B 2018 *Agriculture, Ecosystems & Environment* **266** 167–73
[3] Awika J M and Duodu K G 2017 *Journal of Functional Foods* **38** 686–97
[4] Nderitu A M, Dykes L, Awika J M, Minnaar A and Duodu K G 2013 *Food Chemistry* **141** 1763–71
[5] Nassourou M A, Njintang Y N, Noubissié T J-B, Nguimbou R M and Bell J M 2016 *The Crop Journal* **4** 391–7
[6] Braga A R C, Murador D C, de Souza Mesquita L M and de Rosso V V 2018 *Journal of Food Composition and Analysis* **68** 31–40
[7] Zhu F 2018 *Food Research International* **109** 232–49
[8] Ongkowijoyo P, Luna-Vital D A and Gonzalez de Mejia E 2018 *Food Chemistry* **250** 113–26
[9] Tabani H, Nojavan S, Alexovič M and Sabo J 2018 *Journal of Pharmaceutical and Biomedical Analysis* **160** 244–67
[10] Chen S, Shang H, Yang J, Li R and Wu H 2018 *Industrial Crops and Products* **121** 18–25
[11] Das P R and Eun J-B 2018 *Food Chemistry* **253** 22–9
[12] Hien T T, Nhan N P T, Trinh N D, Ho V T T and Bach L G 2018 *Solid State Phenomena* **279** 217–21
[13] Nhan N P T, Hien T T, Nhan L T H, Anh P N Q, Huy L T, Nguyen T C T, Nguyen D T and Bach L G 2018 *Solid State Phenomena* **279** 235–9
[14] Tran T H, Nguyen H H H, Nguyen D C, Nguyen T Q, Tan H, Nhan L T H, Nguyen D H, Tran L D, Do S T and Nguyen T D 2018 *Processes* **6** 206
[15] Lee J, Durst R and Wrolstad R 2005 *Journal of AOAC International* **88** 1269–78

IOP Conf. Series: Materials Science and Engineering **542** (2019) 012038 doi:10.1088/1757-899X/542/1/012038

New direction in research on extraction of *Citrus aurantifolia* (Lemon fruit) essential oil grown in Mekong Delta - Vietnam via microwave-assisted hydrodistillation

Thien Hien Tran[1,2], Van Tien Nguyen[2], Tan Phat Dao[1,2], Tri Duc Lam[1], Tran Quoc Toan[3], Trinh Duy Nguyen[1], Dai-Viet N. Vo[4], Tran Anh Vy[5], Le Minh Bui[1]

[1]NTT Hi-Tech Institute, Nguyen Tat Thanh University, Ho Chi Minh City, Vietnam
[2]Faculty of Chemical Engineering and Food Technology, Nguyen Tat Thanh University, Ho Chi Minh City, Vietnam
[3]Institute of Natural Products Chemistry, Vietnam Academy of Science and Technology, Ha Noi, Vietnam
[4]Faculty of Chemical and Natural Resources Engineering, Universiti Malaysia Pahang, Malaysia
[5]Department of Chemical and Biochemical Engineering, Gachon University, South Korea
*Corresponding author: labams2013@gmail.com

Abstract. A new research approach has been applied to the extraction of lemon essential oils to increase the extraction efficiency and improve the quality of isolated essential oils. In this study, the combination of microwave extraction method and response surface method is employed to investigate factors influencing lemon peel essential oil extraction process; including size of the raw material particles, raw material to water ratio, extraction time, and microwave power. The results showed that lemon peel of sizes 1 to 2 mm, material: solvent ratio 1:3 g/mL, extraction time of 60 min and microwave power of 450 W gave the highest yield of essential oil (2.4 %). The GC-MS results showed the domination of D-Limonene, the main ingredient of citrus essential oils, in higher concentrations than those reported by previous studies.

1. Introduction

The importance of the genus Citrus has been long recognized both commercially and scientifically. Among a wide variety of species belonging to the genus Citrus, *Citrus aurantifolia* (lemon fruit) offers a widespread coverage on many aspects including medicine, cultivation, industrial use and chemistry [1-4]. In addition to juice production, lemons could be used for the extraction of essential oils from its peel. The oil is a crucial and widely-used flavorant in medicines, cosmetics, beverages, household products, fragrances, and food. The composition of lemon essential oil is complex and its components are classified into three groups: oxygenates, terpenes and sesquiterpenes. The main compounds in oils are limonene, γ-terpinene, citral, linalool and β-caryophyllene.
The conventional methods of essential oil extraction from plant tissues include hydrodistillation, expression, simultaneous distillation extraction and solvent extraction [5-7]. However, these traditional

techniques have several disadvantages regarding the extraction time, quantity of solvent, and oil recovery, rendering them particularly inefficient. The microwave processing technology has been applied in many areas of food processing such as pasteurization, cooking, reheating, drying, baking and thawing. This technique has gained great interest as a tool for synthesis and process development [8-11]. For isolation of target ingredients, particularly extraction of essential oil constituents, microwave heating has been widely used. In comparison to traditional extraction methods, microwave assisted extraction has many benefits such as higher yields of the isolated products, improved extract quality, shorter extraction time and reduced solvent consumption [11-16]. The aim of the present study was to investigate the applicability of microwave-assisted hydrodistillation for the extraction of essential oil from Vietnam *Citrus aurantifolia* (lemon fruit). Optimum values for the process variables, including the size of lemon peels, raw material - water ratio, extraction time, and microwave power were investigated to maximize essential oil yield.

2. Materials and Methods

2.1 Plant sample preparation

Citrus aurantifolia (lemon fruit) was collected from Tien Giang Province, Vietnam. The fruits were washed several times with water to remove impurities and dried. After that, the peels were separated from the sarcocarp. Prior to extraction, raw materials were kept in a non-hygroscopic bag and stored in a cooler (LC-1416B, Alaska, Vietnam) at temperatures below 10°C.

2.2 Extraction method

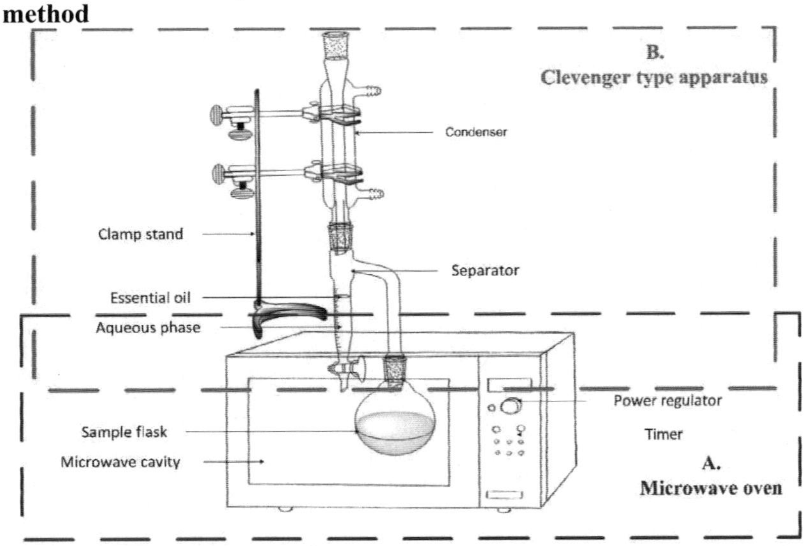

Figure 1. Schematics of microwave-assisted extraction of lemon essential oil

This method was conducted to individually investigate the factors affecting the yield obtained by the extraction process of lemon oils. These factors include the size of lemon peels (1-2 mm, 2-4 mm, and larger than 4 mm), raw material: water ratio (1:1, 1:2, 1:3 and 1:4 g/mL), extraction time (30, 45, 60, 74 and 90 min), microwave power (150, 300, 450 and 600 W). The yield of orange oil obtained (%) is calculated by quotient of the volume of attained oil (V, mL) and the amount of orange peel originally used for the experiment (m, g) were presented as in equation:

$$Y = \frac{V}{m} \times 100$$

Gas Chromatography-Mass Spectrometry (GC-MS) (GC Agilent 6890 N, MS 5973 inert) was used to analyze the components of essential oils. 25 μL sample of essential oil in 1.0 mL *n*-hexane and dehydrated with Na_2SO_4. . The used column was HP5-MS column with head column pressure of 9.3 psi. GC-MS were obtained under the following conditions: carrier gas He; flow rate 1.0 mL/min; split 1:100; injection

IOP Conf. Series: Materials Science and Engineering **542** (2019) 012038 doi:10.1088/1757-899X/542/1/012038

volume 1.0 μL; injection temperature 250°C; oven temperature progress included an initial hold at 50°C for 2 min, a rise to 80°C at 2°C/min, a rise to 150°C at 5°C/min, a rise to 200°C at 10°C/min and a rise to 300°C at 20°C/min for 5 min.

3. Results and Discussion

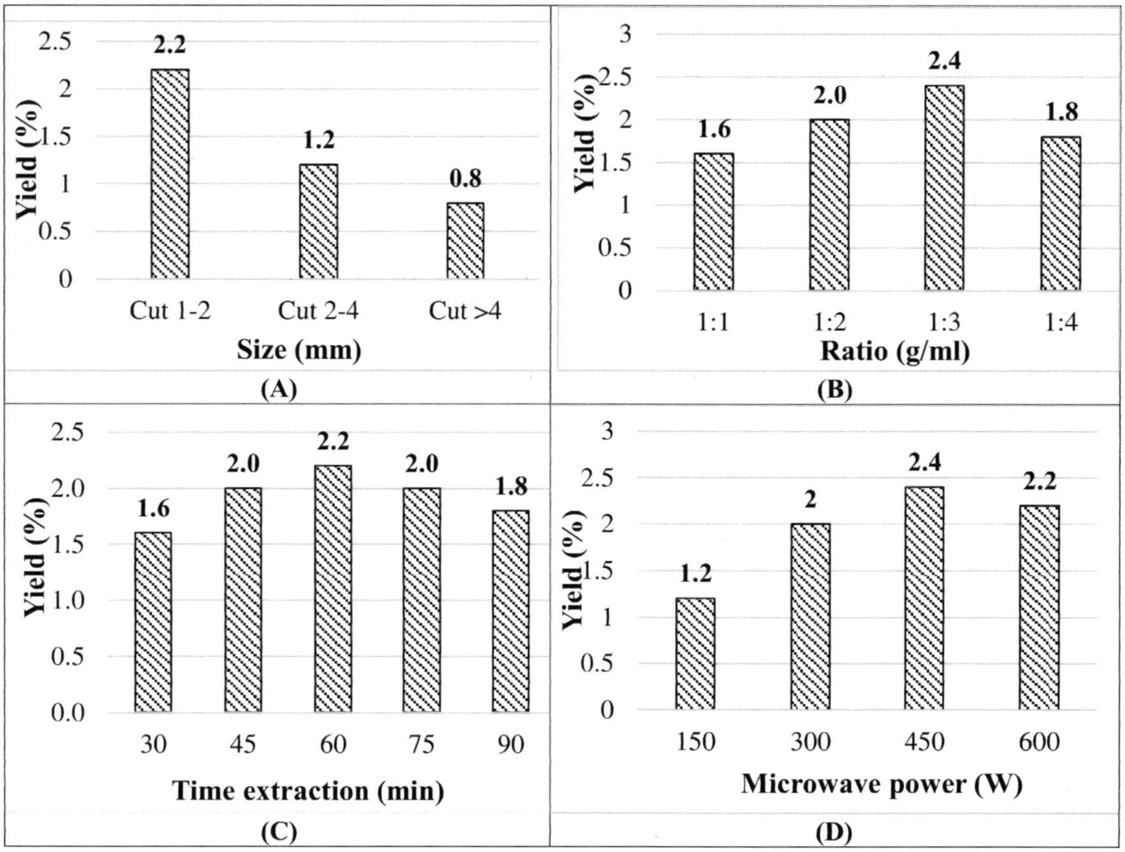

Figure 2. Factors affecting the yield of lemon oil extraction process by Single factor investigation method. The effect of the raw material particle size (A), raw material to water ratio (B), extraction time (C), the applied microwave power (D) on the yield of essential oils.

The factors that affect the yield of lemon oils in the extraction process are shown in Figure 2. Fig 1A illustrated that the yield of lemon essential oil gradually increased as the material size decreases. When the lemon peel size changes from particles larger than 4 mm to those of 1-2 mm, the essential oil yield increases from 0.8% to 2.2%. The reason for this is that by material cutting, the contact surface between water and material is extended, facilitating water diffusion into the essential oil bags, quickly pushing the oil outside under the effect of microwave. However, if the material is cut into very small pieces, more essential oil will be lost by the cutting process, finally resulting in reduced oil yield. Therefore, the best size of lemon peels was 1-2 mm.

In the second survey, the factor of raw material to water ratio was investigated. According to Figure 1B, by using more water, the diffusion of oil into the water is enhanced. The increase of yield of the extracted oil from 1.6% to 2.4% when increasing the raw material to water ratio from 1:1 to 1:3 was obtained. It is because the water could easily be absorbed by the material, promoting the extraction of soluble components which in turn causes higher oil yield. However, at ratio of 1:4, the essential oil content

decreased to 1.8% since excessive water will dissolve or emulsify the essential oil. So, the ratio of raw material to water of 1:3 g/mL gave the best results (yield of 2.4%).
Similarly, Figure 2C and 2D demonstrated the effects of extraction time and microwave power on the yield of extracted lemon oils. When the extraction time prolonged, the efficiency of the extracted oil gradually increases from 1.6% at 30 min to 2.4% at 60 min. However, yield decreased to 1.8% at 90 min because of denaturation of certain substances due to prolonged exposure to high temperature. Figure 2D demonstrated that the higher microwave power leads to better yield of the oil but only to a certain limit of power level. It is because higher temperatures, caused by increased microwave power, incite the movement of the constituents, subsequently raising the velocity of chaotic movement of components in the mixture. Therefore, the extraction time and the microwave power in the survey were chosen at 60 min and 450 W, respectively.
From Figure 3 it can be seen that D-Limonene was detected as the major constituent of the essential oil (71.898%), as expected. This substance is important as it possess numerous biological activities. D-Limonene content in present study is significantly higher than it is reported in previous studies [17,18]. Other identified components were *o*-Cymene, L-beta- pinene, and beta-bisabolene with corresponding levels of 13.765%, 5.221%, and 1.617%, respectively.

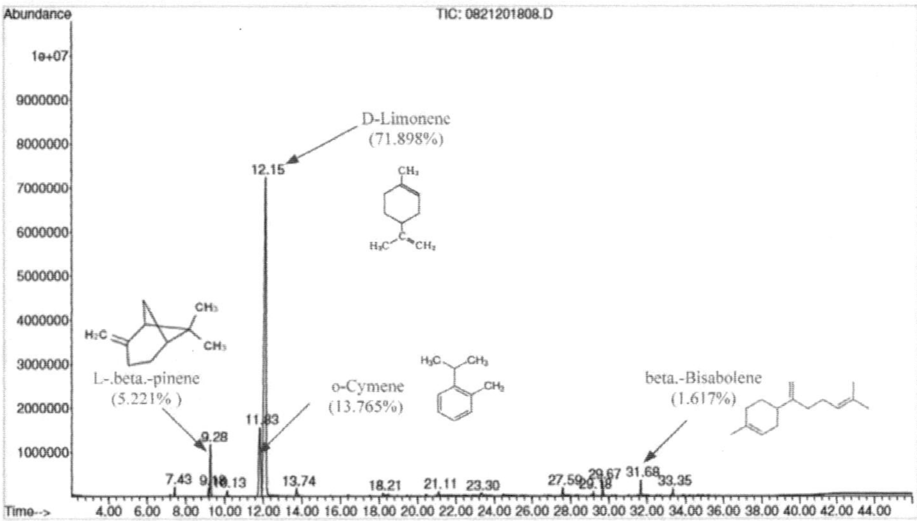

Figure 3. Chromatograph of lemon essential oil obtained by GC-MS with marked peaks of the main components and their chemical structures

4. Conclusion

To conclude, the parameters that affect the extraction of the lemon peel essential oil using microwaves during extraction were optimized. The obtained yield of lemon oil of 2.4% was obtained by use of lemon peel with particle sizes from 1 to 2 mm, water-to-material ratio 3:1 mL/g, at the extraction time of 60 min and microwave power of 450 W. The results of GC-MS analyses also revealed that the content of D-Limonene was the highest in samples produced using described parameters. The results showed that D-Limonene, a main compound of lemon peels essential oil can be effectively isolated by microwave technology what could be of great importance for food industry as this procedure is based on environmental green approach, allows time saving, and complete valorization of waste.

Acknowledgements: This research is funded by Nguyen Tat Thanh University, Ho Chi Minh City, Viet Nam

References

[1] Y. Zhao, H. Sun, L. Ma, A. Liu, *International Journal of Biological Macromolecules*, vol. 101, 2017, pp. 680-689
[2] M.S. Al-Aamri, N.M. Al-Abousi, S.S. Al-Jabri, T. Alam, S.A. Khan. *Journal of Taibah University Medical Sciences*, vol. 13, 2018, pp. 108-112
[3] R. Pathan, P.R. Gali, P. Pathan, T. Gowtham, S. Pasupuleti, *Asian Pacific Journal of Tropical Disease*, vol. 2, 2012, pp. s328-s331
[4] S. Mollayi, M. Farzaneh, F. Ghanati, H.Y. Aboul-Enein, A. Ghassempour, *Physiological and Molecular Plant Pathology,* vol. 93, 2016, pp. 93-98
[5] F. Donsì, G. Ferrari, *Journal of Biotechnology*, vol. 233, 2016, pp. 106-120
[6] S. Mandal, M. Mandal, *Asian Pacific Journal of Tropical Biomedicine*, vol. 5, 2015, pp. 421-428
[7] P. Tiwari, *Industrial Crops and Products*, vol. 82, 2016, pp. 141-148
[8] Thinh P V, Trinh N D, Thuong N T, Ho V T T, Quynh B T P and Bach L G 2018 *Solid State Phenomena* **279** 230–4
[9] Pham H Q, Huynh T T, Nguyen A V, Mai A T N, Phan V T T, Bach L G, Trinh N D and Ho V T T 2018 *Solid State Phenomena* **279** 181–6
[10] Pham T V, Nguyen T T, Nguyen D T, Thuan T V, Bui P Q T, Viet V N D and Bach L G 2019 *Journal of Nanoscience and Nanotechnology* **19** 1122–5
[11] Tran T, Nguyen H, Nguyen D, Nguyen T, Tan H, Nhan L, Nguyen D, Tran L, Do S and Nguyen T 2018 *Processes* **6** 206
[12] Z. Liu, B. Deng, S. Li, Z. Zou, *Industrial Crops and Products*, vol. 124, 15 November 2018, Pages 353-362
[13] H.S. Kusuma, A. Altway, M. Mahfud, *Journal of Industrial and Engineering Chemistry*, vol. 58, 2018, pp. 343-348
[14] Qun Chen, Zhuoyan Hu, Fiona Yan-Dong Yao, Hanhua Liang, LWT. *Food Science and Technology*, Volume 66, March 2016, Pages 538-545
[15] M. Turk, C. Mathe, A.S. Fabiano-Tixier, D. Carnaroglio, F., *Comptes Rendus Chimie*, vol. 21, 2018, pp. 622-627
[16] F. Chen, J. Jia, Q. Zhang, L. Yang, H. Gu, *Industrial Crops and Products*, vol. 112, 2018, pp. 327-341.
[17] T.T. Hien, N.P.T. Nhan, N.D. Trinh, V.T.T. Ho, Long Giang Bach, *Solid State Phenomena* vol. 279, 2018, pp. 217-221
[18] P. Putnik, D.B. Kovacevic, A.R. Jambrak, F.J. Barba, Giancarlo Cravotto, Arianna Binello, Jose Manuel Lorenzo and Avi Shpigelman, *Molecules* 2017, 22, 680; doi:10.3390/molecules22050680.
[19] M. Boukroufa, C. Boutekedjiret, F. Chemat, *Resource Efficient Technologies*, vol. 3, 2017, pp. 252-262

IOP Conf. Series: Materials Science and Engineering **542** (2019) 012039 doi:10.1088/1757-899X/542/1/012039

Finite Element Analysis of Fatigue Crack of Square Bird-Beak X-joints Under Brace Axial Loading

P Adhikari[1], F Huang[1] and B Cheng[1]

[1]Department of Civil Engineering, Shanghai Jiao Tong University, Shanghai, China
E-mail: adhikarip17@sjtu.edu.cn

Abstract. This paper presents a numerical study of the fatigue crack propagation of square bird-beak hollow section welded X-joints. A pair of axial forces were applied at the brace ends, while the chord ends were free. Finite element modelling was done in ABAQUS while FRANC3D was used for the computation of stress intensity factor and crack propagation. The initial crack inserted at the site of maximum stress concentration. Paris' law was used to compute the fatigue life. The results obtained from the finite element analysis compared with the experimental data, and the predicted crack growth was found to be in good agreement with the test observations. For such joints, the surface cracks length vs. fatigue crack growth curve proposed to ensure practicability in real applications.

1. Introduction

The bird-beak X-joints are an innovative type of weld square hollow section joints. Under the cyclic loading, such joints are most vulnerable to fatigue failure, affecting the safety of components and structures. These members are, most of the time, subject to variable load cases, resulting in a mixed mode fatigue case. Many researchers have devoted their career to study the fatigue crack growth behavior of the welded joints. Different methods are introduced to study this problem: hot spot stress method[1]; effective notch method[2]; continuum damage model[3]; and linear elastic fracture mechanics (LEFM)[4]. Mashiri et. al. [5] tested the static behavior of welded square hollow section (SHS) T-connections. Stress intensity factor (SIF) estimation of weld toe semi-elliptical crack in offshore tubular joints was proposed by Lee & Bowness [6]. Chen [7] and Tong [8] investigated the stress concentration factors (SCFs) and fatigue behavior of Square bird beak joints using the experimental method.

Moreover, many researchers are focusing on fatigue behavior and stress concentration field and are concentrating over the application of LEFM, making this the one of the most useful assessment technique, as the initial crack in the real structures is inevitable[4], [9]–[12]. Under the cyclic load, the initial crack tends to occur in the stress concentration filed, and the crack will be lengthened over time. Such process of lengthening is the crack propagation. There have been many methods to study this behavior, one of the most adopted methods is the LEFM. Crack propagation analysis using LEFM is usually conducted with Paris law[4], with the reliable SIF(K) along the crack front points. Previous researches have shown that it is difficult to calculate SIFs from the theoretical analysis. Therefore, the numerical simulation is usually utilized to obtain the SIFs and predict the crack growth.

Various researchers have proposed different initial crack shapes and its propagation paths. Generally, semi-elliptical crack with the fixed aspect ratio is assumed throughout the crack growth process. However, the crack shapes and aspect ratios change with the crack advancement and loading cycle[9].

IOP Conf. Series: Materials Science and Engineering **542** (2019) 012039 doi:10.1088/1757-899X/542/1/012039

As a result, the FEM and BEM are the most suitable methods to predict the crack growth propagation over fatigue load. Seifi and Omidvar[10] studied the modified compact test specimen (MCT) under mixed mode I + II loading effect experimentally and computationally. Experimental and simulation results of mixed mode I +II+ III loading in the crack were discussed [13]. Fatigue crack behavior of base metal and butt weld was investigated by Zong et. al.[14]. Corbani et. al.[9] proposed a line and quarter ellipse method to predict the crack shape under pure out of plane bending. Ju, X. and Tateishi, K.[12] studied the behavior of through-thickness crack under out of plane bending experimentally and numerically. Yang et. al.[15] researched the fatigue crack growth of SHS T-joints using the LEFM theory and boundary element method (BEM), which is in accordance with the experimental results. Extended finite element method (XFEM) is used to predict the crack growth in the 3D problem and reveal the sturdiness and versatility of this approach by Pathak et. al.[16]. Branco and Antunes[17] analyzed the evolution of crack shape in mode-I fatigue growth in Middle Cracked Tension specimens that included different parameters such as specimens' thickness, Poisson's ratio, the exponent of the Paris' law and cracks closure. Evaluation of SIF values and crack growth based on SIF-based flaw propagation procedure is proposed by Carpinteri et. al.[18]. Kikuchi et. al.[19] used s-version of FEM to model Fatigue Crack Growth(FCG) process under mixed mode loading. Moreover, extensive experimental and numerical analysis of static and fatigue performance has been conducted on different type of joints, such as T-[5], [8], K-[6], X-[20]–[22], TX-[23], DX-[24], and XX-[25] joints.

In this study, an un-cracked finite element model was built by ABAQUS, and then the initial crack is inserted in FE model. The fatigue crack growth of square bird beak joints (SBBJ-X) at the weld toe under brace axial loading is investigated. Paris' Law and maximum tensile stress criteria are used to predict the crack growth, and the results are compared with the experimental observations. Moreover, the variation of SIF on the initial crack front is also analyzed.

2. Fatigue test specimens and test setup

Two square bird-beak joints (SBBJ-X) are formed. These specimens are designed based on the three non-dimensional parameters, brace to chord width ratio ($\beta = b_1/b_0$), chord-wall slenderness ratio($2\gamma = b_0/t_0$), brace to chord wall thickness ratio($\tau = t_1/t_0$). b_1 and b_0 represent the sectional width of chord and brace respectively. Similarly, t_0 and t_1 correspond to the thickness of the chord and brace respectively. Chords of length L_0 and braces of the length L_1 are used. The geometric configuration of the specimen is shown in figure. 1 and the geometric dimensions of the joints are enlisted in Table 1.

Chord and brace members are connected by 80% partial joint penetration (PJP) groove-fillet welds. The cold-formed steel tubes, which are of Q420 grade, are used to fabricate the joints which strictly conforms to Chinese standard GB/T 6728-2002 [26]. Standard coupons were tested for the uniaxial test. Measured mechanical properties are 494 MPa, 597 MPa and 202 GPa for yield stress (*Sy*), ultimate tensile stress (*Su*) and Young's modulus (*E*) respectively.

Before the test, to conform to Chinese standard GB/T 6728-2002 [26] specimens were prepared for the test as (1) rigid steel platform was fixed onto the ground for fix support, (2) lower brace end and rigid steel platform was connected via a pin connection, (3) fatigue actuator and upper brace end was connected by a pin connection. (4) Then, a pair of axial tension force was applied at the brace ends in

Table 1. Dimensions and Loading Conditions

Specimen	Chord			Brace			Non-dimensional Parameters			Axial- Load (kN)(P)
	b_0 (mm)	t_0 (mm)	L_0 (mm)	b_1 (mm)	t_1 (mm)	L_1 (mm)	$\beta\ (=\frac{b_1}{b_0})$	$2\gamma\ (=\frac{b_0}{t_0})$	$\tau\ \left(=\frac{t_1}{t_0}\right)$	
SBBJ-X-1	200	12	1200	90	8	500	0.45	16.67	0.67	63
SBBJ-X-2	200	9.5	1200	120	4.5	500	0.6	21.05	0.47	63

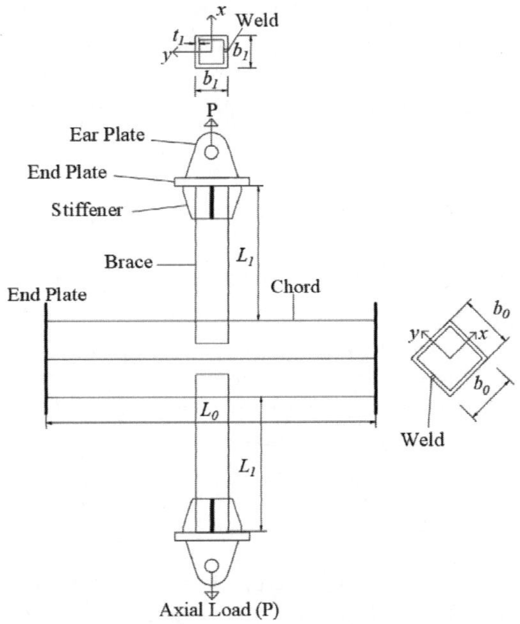

Figure 1. The geometry of SBBJ-X specimens

the direction of brace axis. Load within the stress range is implemented by using the actuator of maximum capacity 200 kN with the load ratio of 0.1.

Fatigue load cycle was recorded on loading machine itself. Crack initiation was determined by human eye aided with the magnifying glass. Thin layer of white paint was sprayed at the potential initial crack location (i.e., at the junction surface) to assist the crack detection. During the fatigue tests, human inspection had performed to record surface crack length in every 2h until the end of the test. Two 12 megapixels HD cameras were always focused on the potential crack initiation location to avoid human error during the test.

3. Fatigue Life Prediction

FRANC3D software is used for modeling and re-meshing of the specimen, and also for the prediction of crack growth. This software was developed in the Cornell University by Fracture Analysis Group (FAG) [27]. Initially, Boundary Element Method (BEM) was used to predict the fatigue crack growth. After reformation and redesign of the software replacing BEM with finite element method to work in conjunction with FE solver were developed by fracture analysis consultant [28]. This software can calculate the new crack front based on displacement provided by the commercially available FE solver. Then the crack is inserted in the un-cracked FE model, and the model is re-meshed with advancing front meshing algorithm to simulate the further crack growth. Kink angle, which can be determined by four criteria: maximum strain energy release rate criterion (SERR); generalized stress criterion (GEN); maximum shear stress criterion (MSS); maximum tensile stress criterion (MTS), is used to determine the direction of crack growth in this model. The maximum tensile stress criterion is used to determine the kink angle[28]. Similarly, constant amplitude model, one of the models in FRANC3D, is used for the crack growth analysis. Subsequently, the Paris' crack growth model, which can predict the crack propagation behavior accurately, is used to calculate the extension of crack front points[4]. And the SIF

and Paris's constant are of importance to the crack propagation behavior. Extended crack front location can be calculated based on the equivalent stress intensity factor (Keq), which is determined by equation (1) in mixed mode loading[29]:

$$K_{eff} = \sqrt{K_I^2 + (\alpha K_{II})^2 + (\beta K_{III})^2} \quad (1)$$

Where $\alpha = KIc / KIIc = 1.155$ and $\beta = KIc / KIIc = 1.0$, which are material independent parameters[10] [12]. Fracture toughness and threshold fracture toughness of the Q420 grade steel is taken as Kc = 1900 MPa $mm^{1/2}$ and K_{th} = 148MPa $mm^{1/2}$ respectively as the recommendation from BS7910[30]. The governing universal fatigue crack threshold for common steel is considered in the equation (2) which is as follows.

$$K_{th} = 170 - 214\, R\ (Unit: MPa^{\frac{1}{2}}) \tag{2}$$

An increment in each step of crack extension (δ_m) is manually provided at the location of maximum equivalent SIF. Among both available methods (manual and automatic), for the consideration of the crack growths, maximum crack growth in each step should not be more than 30% of the cumulative crack length. And the propagation at the discrete points along the crack front is approximated using the Paris' equation, in which the maximum crack extension was manually keyed by equation (3).

$$\delta = \delta_m \left(\frac{K_{eq}}{K_{eq,m}}\right)^m \tag{3}$$

The cubic spline curve fitting approach is implemented for fitting the crack front points using one of the curve fitting tools in FRANC3D. It should be noted that FRANC3D smoothens the entire crack front, making simulated crack front different from the experimental observations [9]. Meanwhile, the simulated crack front leaves some error from the theoretical statement which states crack growth direction is always perpendicular to the crack front [9].
The number of fatigue cycle is determined by the Paris' Law as:

$$da/dN = C\,(\Delta K_{eff})^m \tag{4}$$

Where a is the crack length, N is the number of cycle of propagation life. C= 2.18×10^{-13} and m =3 respectively for the Q420 steel are material constants. da/dN is expressed in unit mm/cycle and ΔK_{ef} in $Nmm^{-3/2}$.
Thus, Fatigue life in each step is calculated as follows [4]:

$$\Delta N = \frac{D_m}{C\ \Delta K_{eff}^m} \tag{5}$$

Since the crack front is tested under the brace axial loading, there exists Mode II, and Mode III loading of SIF at crack presents at the weld toe. These are considered as the effective SIF [31] in the simulation process as:

$$\Delta K_{eff} = \sqrt[4]{\Delta K_I^4 + 8\Delta K_{II}^4 + \frac{(8\Delta K_{III}^4)}{1-\nu}} \tag{6}$$

4. Model Generation

The loading, boundary conditions, construction and geometric dimension of these type bird-beak X-joints are all of the symmetry. Thus, one-eighth of SBBJ-X specimen is modeled in ABAQUS, as shown in figure 2. Sub-model is defined as shown in the enlarged view in figure 2. The brace axial force is applied in the FE model, and the stress distribution is analyzed in ABAQUS to locate the node with maximum stress from where the crack initiates being, as shown in figure 3. Through the static analysis, the location of the initial crack for FRANC3D can be manually determined, while the initial crack length for FRANC3D can be obtained from the experimental results. Moreover, the initial crack is aligned parallel to the welding direction and is perpendicular to the chord surface.

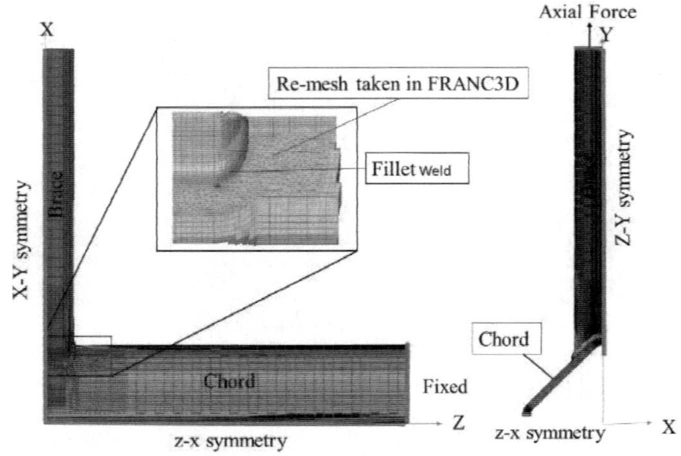

Figure 2. Meshing and Boundary Conditions

The flaws presented in the typical steel structures are of a semi-elliptical shape which is used for the simulation of the initial shape of the crack using a special module in FRANC3D[28]. Crack front template around the crack tip is defined with three rings and eight circumferential elements. Quarter-point Wedge elements are used around the crack front followed by twenty-node hexahedral elements, while the remaining portion is re-meshed with tetrahedron element by using front advancing re-meshing algorithm technique. The crack is then inserted into the un-cracked FE model as shown in figure 4.

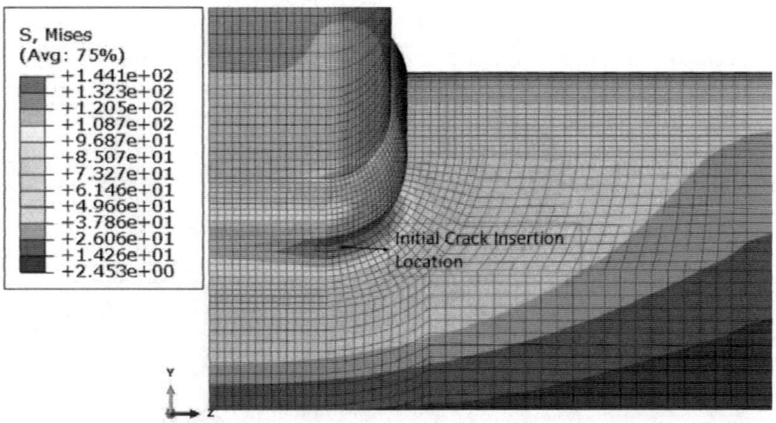

Figure 3. Stress Analysis of Un-Cracked Model

After the insertion of crack into FRANC3D, load ratio of 0.1 is taken into account in the simulation. The input file created by FRANC3D is used for stress analysis in ABAQUS. The obtained displacements are used to calculate the SIF at corresponding crack front locations. The value of the crack extension is manual where the maximum crack occurs at the location of maximum SIF, but is as explained in equation (3) in other location. The newly formed cracks obtained from the crack extension is subjected to un-cracked model and the process repeats until the crack propagates to reach the symmetrical boundary condition. This crack propagation of the specimen is shown in figure 5.

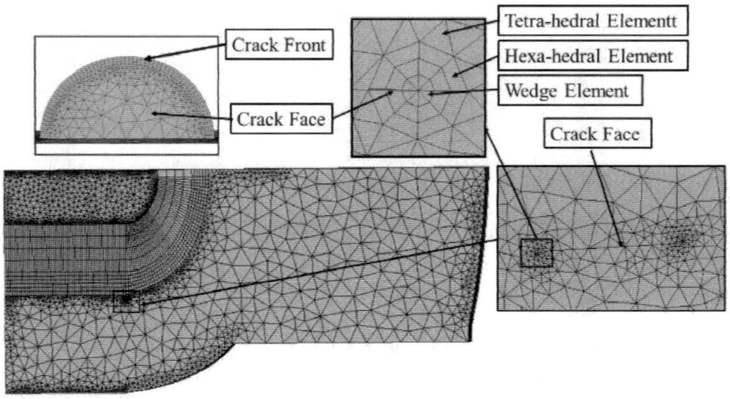

Figure 4. Initial Crack Location and Shape of the Initial Crack

5. Results and Discussion

The stress intensity factors are calculated in SBBJ-X joints with different initial crack lengths under brace axial loading, as shown in figure 6. It shows that with the initial crack length in SBBJ-X-1 specimen being placed parallel with the welding, average SIF in the Mode I is increased by 12% with the crack length increasing from 4 mm to 8mm, while it is increased to 22% with the crack length getting to 18 mm. Similarly, in case of SBBJ-X-2 specimen, the initial crack is placed parallel with the welding, average SIF in the Mode I is increased by 23% with the increase of crack length from 4mm to 12mm, while it is increased to 32% with the crack length getting to 19 mm. Based on the SIF curve, it is observed that the SIF near the crack front ends have the higher values and middle has lower values, which means there is comparatively more growth in the ends than that of through-thickness direction. L This growth continues until the symmetrical end of the specimen is reached. Once the crack reach the symmetrical end, small portion of crack undergoes compression resulting in the crack closure which, to date, cannot be further accounted by FRANC3D. In order to address such limitations, Corbani et. al.[9] proposed a new method of eliminating this shortcoming by combining the crack front as a straight line and quarter ellipse. However, this has not been performed in this paper and can be a matter of new research.

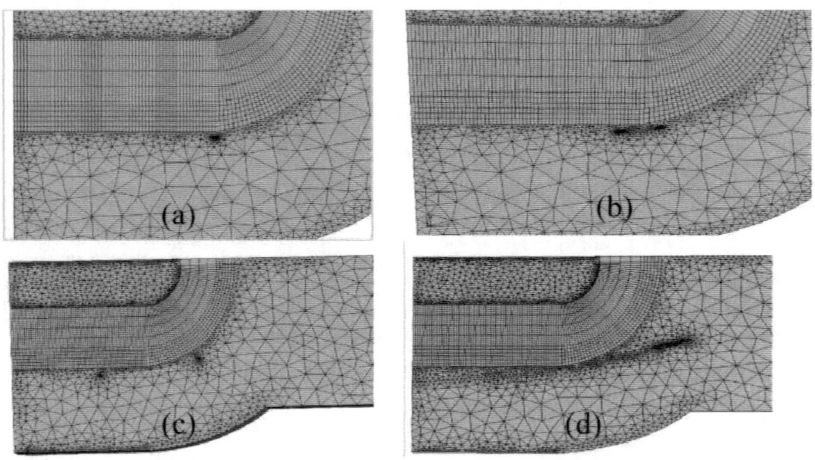

Figure 5. Crack State at (a) 1st Step (b) 50th Step (c) 100th Step (d) 141st Step (Last Step)

The result obtained from figure 6 are similar to the fact that the largest stress intensity factors are obtained on the longer length of the crack, which is also stated by Masoudi et al. [31]. figure 7 shows the fatigue life estimation against different initial crack length. There is an increase in SIF with the increase in initial crack length which signifies that the fatigue life decreases with an increase in crack length. As in the figure 7, for SBBJ-X-1 specimen, with the increase in the initial crack length from 4 mm to 8 mm, the fatigue life decreases by 19%, while it decreases by 44% with the initial crack length increasing from 4 mm to 18 mm. Similarly, in case of SBBJ-X-2, there is also a decrease in fatigue life

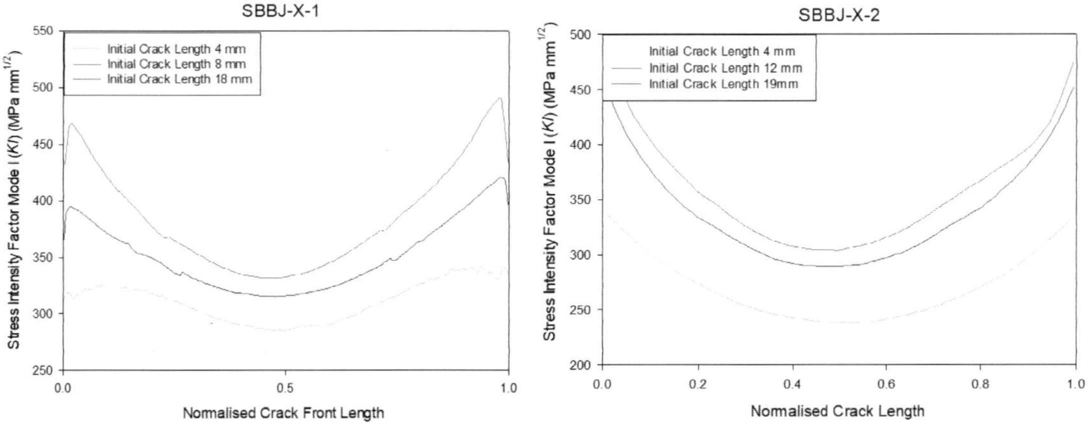

Figure 6. Normalized Crack Length vs SIF(KI) for a) SBBJ-X-1 b) SBBJ-X-2

by 27% and 39% for an increase in the initial crack length from 4 to 12 mm and 4 to 19 mm respectively, as shown in figure 7.

The effect of the initial crack length over two specimens are examined for the fatigue crack growth which illustrates the effect of an initial crack length over the fatigue life and cracks growth as shown in figure 7. As seen in figure 6, the stress intensity factor is higher in the longer length of the initial crack length, and it is lower in the smaller initial crack length. This behavior of SIF implies that shorter initial crack length will have a longer fatigue life than that of longer initial crack length. The numerical simulation of crack complies with the experimental results.

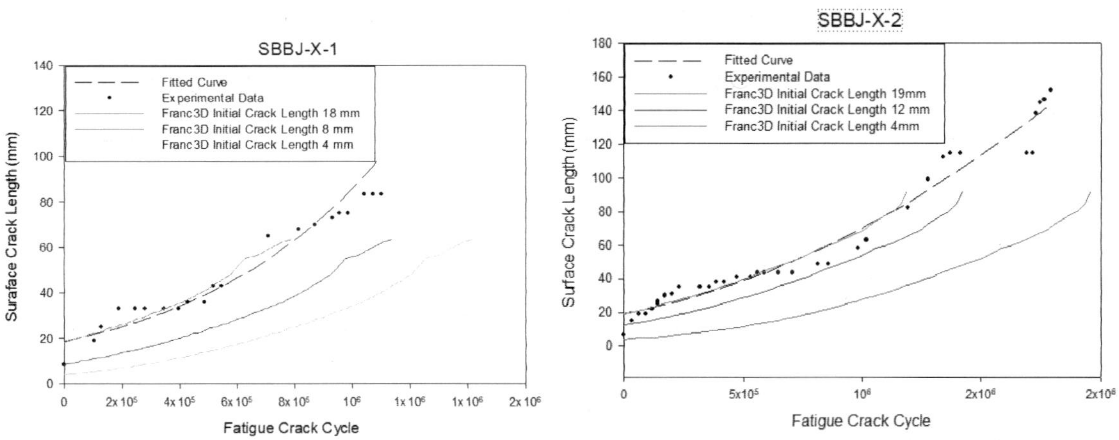

Figure 7. Fatigue life vs. Surface Crack Length for a) SBBJ-X-1 b) SBBJ-X-2

6. Conclusion

In this paper, the effects of brace axial loading on the crack propagation of SBBJ-X specimens are investigated using a FEM approach. It demonstrates that SIF has evident effects on the fatigue behavior of the joints. Fatigue life is obtained by using the linear elastic fracture mechanics. The initial crack length has an eminent effect on fatigue life. Having control over the initial flaw increases the fatigue life, thus, will enhance the life of the structure. These numerical results are in good agreement with the available test results. This analysis demonstrates the effectiveness of finite element analysis in such steel joints in evaluating the fatigue life for the future works.

References

[1] Radaj D, Sonsino C M and Fricke W 2006 *Woodhead publishing*
[2] Zhang G and Richter B 2000 *Fract. Engng. Mater. Struct.* **23** 499–508
[3] Maimí P, Camanho P P, Mayugo J A and Dávila C G 2007 *Mech. Mater.* **39** 897–908
[4] Paris P C and Erdogan F *J. Basic Engng.* **85** 528–34
[5] Mashiri F R, Zhao X L and Grundy P 2002 *J. Struct. Eng.* **128** 22–31
[6] Lee M 2002 *Int. J. Fatigue* **24** 861–75
[7] Cheng B, Qian Q and Zhao X L 2015 *Eng. Struct.* **99** 677–84
[8] Tong L, Xu G, Yan D and Zhao X L *J. Constr. Steel Res.* **118** 49–59
[9] Corbani S, Castro J T P, Miranda A C O, Martha L F, Carter B and Ingraffea A 2018 *Eng. Fract. Mech.* **188** 493–508
[10] Seifi R and Omidvar N 2013 *Mar. Struct.* **34** 1–15
[11] Lewandowski J and Rozumek D 2016 *Theor. Appl. Fract. Mech.* **86** 342–50
[12] Ju X and Tateishi K 2012 *Int. J. Steel Struct.* **12** 85–92
[13] Richard H A, Schramm B and Schirmeisen N H 2014 *Int. J. Fatigue* **62** 93–103
[14] Zong L, Shi G and Wang Y 2015 *Constr. Build. Mater.* **77** 419–29
[15] Yang Z M, Lie S T and Gho W M 2007 *Constr. Steel Res.* **63** 1184–93
[16] Pathak H, Singh A and Singh I V 2013 *Int. J. Mech. Sci.* **76** 112–31
[17] Branco R and Antunes F V. 2008 *Eng. Fract. Mech.* **75** 3020–37
[18] Carpinteri A, Brighenti R, Huth H J and Vantadori S 2005 *Int. J. Fatigue* **27** 59–69
[19] Kikuchi M, Wada Y and Suga K 2011 *Procedia Eng.* **10** 427–32
[20] Cheng B, Li C, Lou Y and Zhao X L 2018 *Thin-Walled Struct.* **123** 57–69
[21] Chen Y, Feng R and Wang J *JCSR* **106** 234–45
[22] Chen Y, Wu Y, Ruan X and Feng R 2015 *Thin Walled Struct.* **97** 171–85
[23] Vegte G J Van Der and Wardenier J 1998 *J. Construt. Steel* **47** 141–68
[24] Liu G, Zhong B, Tian X, Chen P and Mu W 2016 *Ocean Eng.* **127** 258–68
[25] Liu S L, Zhu Z Q and Dong X 2013 *Adv. Mater. Res.* **639–640** 774–7
[26] GB/T 6728-200 2002 *National Standards of the People's Republic of China, Beijing, China*
[27] Cornell Fracture Group [Online] 2018 Available: *http://cfg.cornell.edu/software/.* [Accessed: 03-May-2018].
[28] Inc F A C 2016 *FRANC3D, Reference manual for version 7.*
[29] Zhang X B, Ma S, Recho N and Li J 2006 *Eng. Fract. Mech.* **73** 1925–39
[30] BS-7910 2005 *BSI Stand. Publ.* **3** p. 306
[31] Masoudi N R, Shariati M and Farhangdoost K 2016 *Tribol. Int.* **94** 118–25

IOP Conf. Series: Materials Science and Engineering **542** (2019) 012040 doi:10.1088/1757-899X/542/1/012040

Effect of Zn on microstructure evolution of Mg-Gd-Y-Nd-Zn-Zr alloy during heat treatment

T Li[1], D J Chen[2], W Liu[1], J W Yuan[2], and J Qu[1]

1 China United Test &Certification Co., Ltd.,Guobiao(Beijing)Testing & Certification Co., Ltd., No.2, Xinjiekouwai Street, Xicheng district, Beijing, 100088, China
2 State Key Laboratory of Non-ferrous Metals and Processes, GRIMAT Engineering Institute Co., Ltd., No.11, Xinkedong Street, Huairou district, Beijing, 101407, China
E-mail: liting@gbtcgroup.com

Abstract. The effect of Zn on the microstructure evolution of Mg-7Gd-3Y-1Nd-xZn-0.5Zr (x=1.0, 2.0) (wt.%) alloys during homogenization and ageing heat treatment has been investigated systemically. The results indicate that as-cast 1 Zn alloy is composed mainly of α-Mg, Mg_5(RE, Zn) eutectic phase, $(Mg,Zn)_3$RE eutectic phase, stacking faults and block-like compounds rich in RE. Increasing of Zn content result in the disappearance of Mg_5(RE, Zn) eutectic phase, but the volume fraction of 14H-typ LPSO phase is increasing, which has made it even more difficult to dissolve into the matrix during homogenization. After homogenization at 520 °C for 48 h, the 14H LPSO phase remains in the alloys with 2 wt.% Zn, but only compounds rich in RE can be seen in the alloy with 1 wt.%Zn. After ageing at 240°C for 18 h, the coherent β′ phase can be found both in 1Zn and 2Zn alloys, resulting in the increase of mechanical properties. The β′、 $β_1$ and stacking faults can be found in 1Zn and 2Zn alloys after ageing at 240°C for 100 h. However, it can be concluded that Zn can suppress the formation of LPSO phases but impede the precipitation of β′ and $β_1$ during the aging process. The as-aged alloy with 1% Zn addition shows the optimal mechanical properties.

1. Introduction

Magnesium alloys are attractive structural materials for automobiles, aircrafts, 3C products and biological fields, due to their low density, high specific strength and good damping capacity. Among these alloys, Mg–RE (rare earth) alloys with remarkable ageing responses is one of the most promising high strength Mg alloy. [1-4] Recently, Zn additions into Mg-RE alloys have widely been confirmed that will introduce LPSO structures, which can further improve the mechanical properties especially the ductility of magnesium alloys. The superior mechanical performance of Mg-RE-Zn alloy has been reported extensively. Kawamura et al.[5] firstly developed $Mg_{97}Zn_1Y_2$ alloys with high strength above 600 MPa and 5% elongation at room temperature through rapidly solidified powder metallurgy. Yamasaki et al. [6] reported a high strength hot-extruded Mg–2.3Zn–14Gd (wt.%) alloy with 345 MPa tensile proof strength and 6.3%. elongation. L. Zheng[7] produced Mg–10Gd–6Y–2Zn–0.6Zr alloy with the ultimate tensile strengths of 432 MPa and the elongations are 18% and 5% respectively.
Although there are many studies about the effect of Zn on the microstructure evolution and properties, there are still some disputes about the strengthening mechanism of Zn additions to Mg-RE alloys. T. Honma [8] considered that the Zn addition will enhance precipitations of the β′ phases and LPSO structures. The formations of these two phases minimize the total energy in the systems, leading to active movement of dislocations on the basal plane, which gives rise to the unusual elongation of Mg–2.0Gd–1.2Y–0.2Zr alloy. Yun Li[9] suggested that formation of LPSO phase increased the efficiency

IOP Conf. Series: Materials Science and Engineering **542** (2019) 012040 doi:10.1088/1757-899X/542/1/012040

of power dissipation and impeded the dynamic recrystallization, which causes Zn-containing alloys had stronger softening effects. Jinghuai Zhang[10] suggested that the kinking of LPSO structure is beneficial for both work hardening and plasticity, and 14H LPSO structure contributes more to the improvement of ductility while SF is more effective in increasing strength.
High strength Mg-7Gd-5Y-1Nd-0.5Zr alloys was fabricated in our previous work[11,12]. To further improve the ductility, Mg-Gd-Y-Nd-Zn-Zr alloy was developed by Zn addition recently[13-15]. In this alloy system, Zn was supposed to be mainly composed β_1 phases and LPSO structures but does not partition into β' phases[8], whereas Nd is reported that can impede the LPSO formation but enhance the solid solubility effect and peak-aging hardness[16]. Thus, the influence of Zn content on the microstructure and mechanical properties of Mg-Gd-Y-Nd-Zn-Zr alloys during heat treatment as well as the strengthening mechanism are essential to be studied.
The aim of this paper is to understand the influence of the Zn content in the Mg-Gd-Y-Nd-Zn-Zr system during heat treatment based on the microstructural observations and mechanical properties test. In addition, the precipitates in peak-aged alloys have been investigated for better understandings of the age-hardening mechanism.

2. Experimental procedures

Two alloys with nominal composition of Mg-7Gd-3Y-1Nd-1Zn-0.5Zr and Mg-7Gd-3Y-1Nd-2Zn-0.5Zr were prepared in high frequency induction melting furnace. The as-cast ingots were homogenized at 520 °C for 48h and aged at 210 °Cfor 26 h. Microstructure observations were performed through scanning electron microscopy (SEM) equipped with EDS and transmission electron microscopy (TEM). Disc samples (3 mm diameter) for TEM analyses were ground to 60 μm thick and then twin-jet electro-polished in a solution of 15 mL perchloric acid and 285 mL ethanol at -30 °C and 0.01 A. The microstructures were observed by JEM-2010 HRTEM, HAADF-STEM images were taken using a FEI Tecnai F20 electron microscope. The mechanical properties of these two alloys were tested using the SANS machine at a speed of 2 mm/min at ambient temperature.

3. Results and discussion

3.1. As-cast microstructures

Fig. 1 shows the XRD patterns of the as-cast alloys. The as-cast 1Zn alloy is mainly composed of α-Mg, Mg_5(RE, Zn) (Mg, Zn)$_3$RE phase, and Mg_{12}(RE, Zn) phases, as seen from Fig. 1. Otherwise, in as-cast 2Zn alloy, disappearance of Mg_5(RE, Zn) phase can be recognized from the results. The intensity of the diffraction peaks for Mg_{12}(RE, Zn) is much stronger in as-cast alloys.

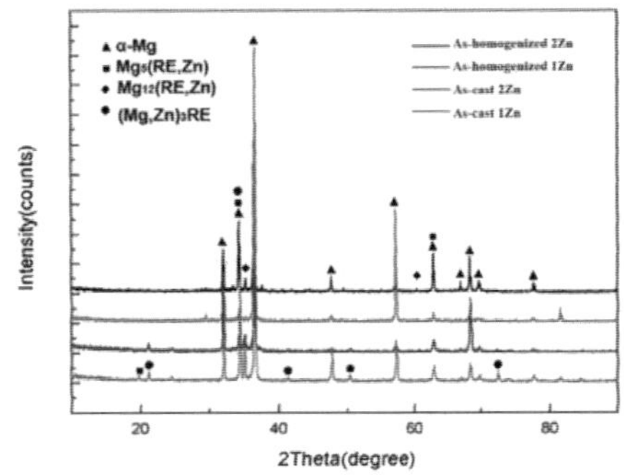

Figure 1. XRD patterns of as cast 1Zn and 2Zn alloy

Fig.2 shows as-cast microstructure of 1Zn and 2Zn alloys. As seen in Fig.2a, the eutectic phase (indicated by A and B), and square shaped compounds (indicated by C) rich in Gd and Y distribute along the grain boundaries in 1Zn alloy. Besides, a number of needle-like structures near grain boundaries which distribute along the same direction can also be found. According to XRD and EDS results, the eutectic phase A and B can be designated as Mg_5(RE, Zn) and (Mg, Zn)$_3$ RE, respectively. In 2Zn alloy, it is found that the characteristics of the eutectics such as (Mg, Zn)$_3$ RE and compounds rich in Gd/Y are same to that of the phases observed in the as-cast 1 Zn alloy, which is consistent with the XRD patterns. The difference is only that the disappearance of Mg_5(RE, Zn) phases. The addition of 2% Zn has made more needle-like structure precipitated throughout all the grains. Moreover, besides these similar phases, (Mg, Zn)$_{12}$RE phase can be observed in 2Zn alloy, which is not found in the microstructure of 1Zn alloy as shown in Fig.2a, although the diffraction peaks for Mg$_{12}$(RE, Zn) in 1Zn alloy can be found as shown in Fig.1.

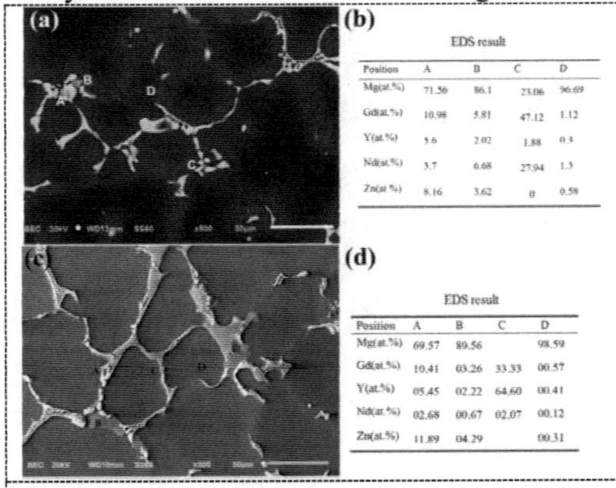

(b) EDS result

Position	A	B	C	D
Mg(at.%)	71.56	86.1	23.06	96.69
Gd(at.%)	10.98	5.81	47.12	1.12
Y(at.%)	5.6	2.02	1.88	0.3
Nd(at.%)	3.7	6.68	27.94	1.3
Zn(at.%)	8.16	3.62	0	0.58

(d) EDS result

Position	A	B	C	D
Mg(at.%)	69.57	89.56		98.59
Gd(at.%)	10.41	03.26	33.33	00.57
Y(at.%)	05.45	02.22	64.60	00.41
Nd(at.%)	02.68	00.67	02.07	00.12
Zn(at.%)	11.89	04.29		00.31

Figure 2. (a) SEM image of as-cast 1Zn alloy; (b) EDS results of the compounds marked by A, B C, D in Fig.2a; (c) SEM image of as-cast 2Zn alloy; (b) EDS results of the compounds marked by A, B C, D in Fig.2c

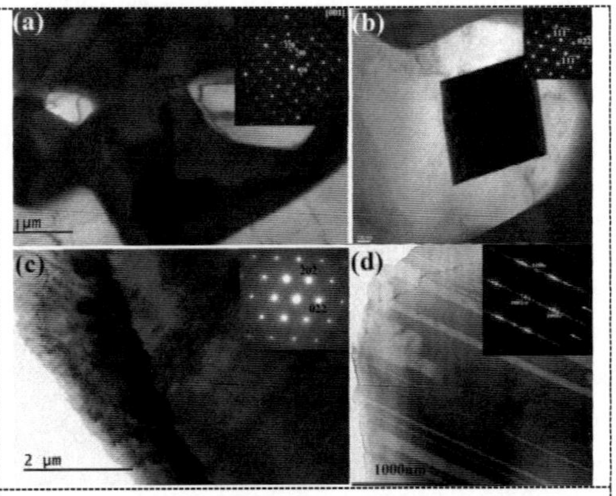

Figure 3. (a) bright field TEM image and corresponding SAED pattern of eutectic phase B in 1Zn alloy; (b) bright field image and corresponding SAED pattern of secondary phase C; c) bright field image and SAED pattern of A in 2Zn alloy; d) bright field image and SAED pattern of LPSO in 2Zn alloy

TEM and SAED analysis for eutectic phase B and compounds C was conducted, as shown in Fig.3. The SAED pattern shown in Fig.3a indicates the skeletal eutectic phase B formed along grain boundaries can be deduced as f.c.c. crystal structure with a=2.22 nm. The square-shaped compounds C shown in Fig. 1(a) is determined to be f.c.c. crystal structure with a lattice parameter of a=0.53 nm. It is obvious that the eutectic phase B and compounds C is similar to that formed in Mg-RE series alloy in our previous research [13]. Moreover, (Mg, Zn)$_3$RE phase observed in 1Zn alloy that is distinguish with 2Zn alloy was analyzed by TEM and HRTEM, as shown in Fig.3c. The (Mg, Zn)$_3$RE phase in 1Zn and 2Zn alloy have been determined to present a f.c.c. with a=0.73 nm. Furthermore, TEM image and SAED pattern of LPSO structure in 2Zn alloy are shown in Fig. 3d, The SAED pattern viewed along $[11\bar{2}0]_\alpha$ axis of LPSO phase indicate that the present LPSO phase formed during solidification is 14H-type with a lattice parameter of a=1.112nm， c=3.647nm. The orientation relationship between 14H-type LPSO structure and matrix can be described as $(0001)_{14H}//(0001)_\alpha$, $[01\bar{1}0]_{14H}//[\bar{1}\bar{1}20]_\alpha$. This 14H-type structure considered to be the same as the 14H-type structure reported by Zhu[18].

3.2. *As-homogenized microstructures*

Fig. 4 shows the differential scanning calorimetry (DSC) curves of the as-cast alloys. As seen from Fig.4, the melting points of eutectics are evaluated to be approximately 537.1°C, 529.3 °C for 1Zn and 2Zn alloy, respectively. With increase of Zn content, the melting point of eutectics is decreased and the endothermic peak area increase significantly Zn, which is consistent with XRD results shown in Fig.1. According to the DSC results, the subsequent homogenization heat treatment for both 1Zn and 2Zn alloys is performed at 520 °C for 32 h.

IOP Conf. Series: Materials Science and Engineering **542** (2019) 012040 doi:10.1088/1757-899X/542/1/012040

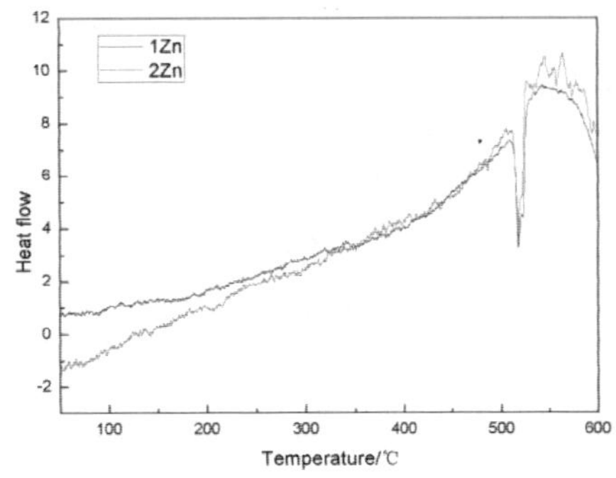

Figure 4. The differential scanning calorimetry curves of 1Zn and 2Zn alloy

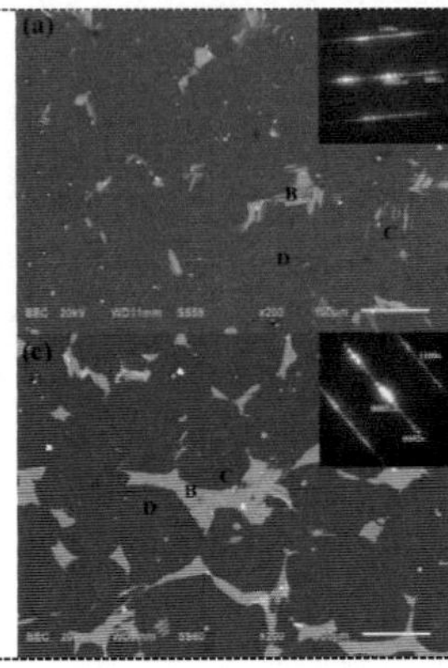

(b)

EDS Result

Position	B	C	D
Mg(at.%)	86.21	29.26	96.48
Gd(at.%)	04.09	23.79	01.60
Y(at.%)	03.21	44.15	00.94
Nd(at.%)	00.79	01.62	00.34
Zn(at.%)	05.69	01.18	00.64

(d)

EDS Result

Position	B	C	D
Mg(at.%)	87.03	23.17	96.66
Gd(at.%)	04.19	22.67	01.39
Y(at.%)	02.90	51.34	00.69
Nd(at.%)	00.87	01.28	00.36
Zn(at.%)	05.01	01.54	00.90

Figure 5. (a) SEM image of as-homogenized 1Zn alloy; (b) EDS results of the compounds marked by B C, D in Fig.1a; (c) SEM image of as-homogenized 2Zn alloy: (b) EDS results of the compounds marked by B C, D in Fig.1c

Fig.5 shows the microstructure of as-homogenized alloys. After homogenization heat treatment at 520 °C for 32 h, the eutectic phase has been completely dissolved into matrix. However, the LPSO along grain boundaries both in 1Zn and 2Zn alloy can be seen in Fig.5a and Fig.5c. With increasing of Zn content, the volume fraction of LPSO phase is increasing, which has made it even more difficult to dissolve into the matrix during homogenization. As shown in the illustrations in Fig.5a and Fig.5c, 14H LPSO structure can be deduced from the SAED pattern. Thus, the undissolved block shaped phase is LPSO with 14H-type structure. The as-homogenized microstructures of 1Zn and 2Zn alloys are mainly composed of α-Mg, secondary phase rich in RE and 14H-type LPSO phase. The research by Zhu et al. [18] revealed that the stable structure of $Mg_{12}YZn$ in the solid solution Mg–Y–Zn alloy was 14H-LPSO. For 1Zn alloy, needle-like structures near grain boundaries transformed into block LPSO phase, and LPSO phase in 2Zn alloy is much larger after homogenization. The similar phenomenon was reported by Yamasaki et al.[19], and there are many other standpoints about the formation mechanism of 14H-LPSO during homogenization heat treatment, such as precipitation from α-Mg matrix, or transformed from different type of LPSO, or decomposition of eutectic phases. However, in the present study, it is hard to distinguish the mechanism of 14-H LPSO phase coarsening, and further research has to be studied to understand the coarsening mechanisms of the 14H-LPSO during homogenization.

3.3. As-aged microstructures

To reveal the microstructure of as-aged alloys, TEM was conducted for 1Zn alloy and 2Zn alloy. Fig. 6(a) shows a low-magnification TEM image of 1Zn alloy after 240 °C/18 h ageing, taken with the incident beam parallel to $[0001]_\alpha$. The microstructure is mainly composed of fine precipitates formed on $(11\bar{2}0)_\alpha$ planes. The corresponding SAED pattern taken from both the matrix and precipitates shown in Fig. 6(b) can be indexed consistently with b.c.o. structure with a=0.64 nm, b=2.22 nm and c=0.52 nm. The bright field image and SEAD pattern both suggested that β′ phases have formed during aging process in 1Zn alloy. Besides β′ phase which has the same structure with that in EW75 alloy in our previous work, some clusters marked by arrows has found in matrix. These clusters has a hexagonal shape, Nie has reported and thought they are short-range order [20]. In future work, the crystal structure of this cluster will be further investigated.

The TEM image viewed along the $[11\bar{2}0]_\alpha$ zone axis for the 1Zn alloy after 240 °C/100h ageing are shown in Figs. 6(d). Viewing from the $[11\bar{2}0]_\alpha$ direction, all the β′ precipitates are parallel to $[0001]_\alpha$ direction. In addition, compared with 1Zn alloy, the microstructure of 2Zn alloy shown in Fig. 6c and Fig.6f indicate that with increasing of Zn content, the fewer β′ phases has formed in matrix. This result can be ascribed to the larger amount of residual LPSO phase in 2Zn alloy. As a second phase, the composition of 14H-type LPSO is approximate $Mg_{12}Zn_1RE_1$. The residual LPSO phase in homogenized alloy consumed Gd, Y and Nd atoms, which result in the fewer β′ phases precipitated from matrix during ageing.

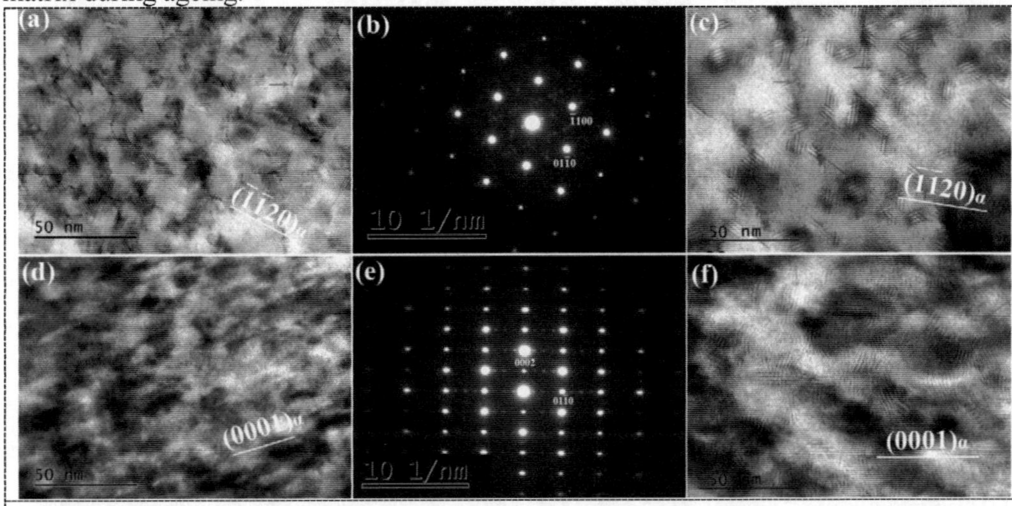

Figure 6. TEM images of samples aged at 240 ℃ for 18h (a) bright field image of 1 Zn alloy taken along $[0001]_\alpha$; (b) SAED pattern corresponding to a; (c) bright field image of 2 Zn alloy taken along $[0001]_\alpha$; (d) bright field image of 1 Zn alloy taken along $[11\bar{2}0]_\alpha$; (e) SAED pattern corresponding to d; (f) bright field image of 2 Zn alloy taken along $[11\bar{2}0]_\alpha$

To investigate precipitates formed after continued ageing at 240 °C for 100 h in 1Zn alloy, HAADF-STEM technique was adopted. The HAADF-STEM images for 1Zn alloy after 240 °C/100 h ageing are shown in Fig. 7(a) and 7(d), respectively. Viewing from $[0001]_\alpha$ zone axis, a large amount of coarse β′ precipitates can be observed in Fig. 7(a). The size of β′ precipitate along $[0001]_\alpha$ direction is about 100nm. The FFT pattern inserted in Fig. 7(c) shows that the structure of β′ precipitates not change during ageing. Besides β′ phase, another phase with a pair of facets parallel to $\{1\bar{1}00\}_\alpha$ planes and the other pair of facets attached to β′ can be found, which was marked in Fig.7(a). Closer examination revealed that the habit of this precipitation is $(1\bar{1}00)_\alpha$. FFT recorded from such precipitates, as shown in Fig. 7(c), could be indexed consist with β_1 phase(f.c.c, a=0.74nm) . An HAADF-STEM image for 1Zn alloy taken with the incident beam parallel to $[11\bar{2}0]_\alpha$ is shown in Fig. 7(d). Besides β′ and β_1 phase, some needle-like structures perpendicular to $[0001]_\alpha$ direction can be observed. FFT pattern of this structure is shown in Fig.7f, strong streaks indicate the existence of stacking fault. HAADF-STEM observation (Fig. 7d) reveals that solute elements such as Zn and Gd concentrate in stacking faults. The high resolution TEM image of SF (Fig. 7f) shows a stacking sequence of ABABACBCBC of the closely packed plane. The interaction of β′ and SF can also be found, and our further work will be done to investigate the interface between β′, β_1 and SF structures.

Based on the Z-contrast mechanism in HAADF technique, we can associate the fact that the RE and Zn concentrate in β′, β_1 phase and SF structure, and the content of RE in β_1 phase is higher than that in β′ phase. Precipitations in 2Zn alloy after ageing at 240 °C for 100 h are shown in Fig. 7(b) and 7(e), respectively. A HAADF-STEM image taken with the incident beam parallel to $[11\bar{2}0]_\alpha$ in Fig. 7(b) and 7(e) show that with increasing of Zn content, the fewer β′ and β_1 phases formed in matrix, but higher density of SF can also can be found. This result can be ascribed to the larger content of Zn in 2Zn alloy, thus the SF formed easily. Due to the SF contains Gd, Y and Nd atoms, which result in the fewer RE in matrix in 2Zn alloy and fewer β′ and β_1 phases formed in matrix.

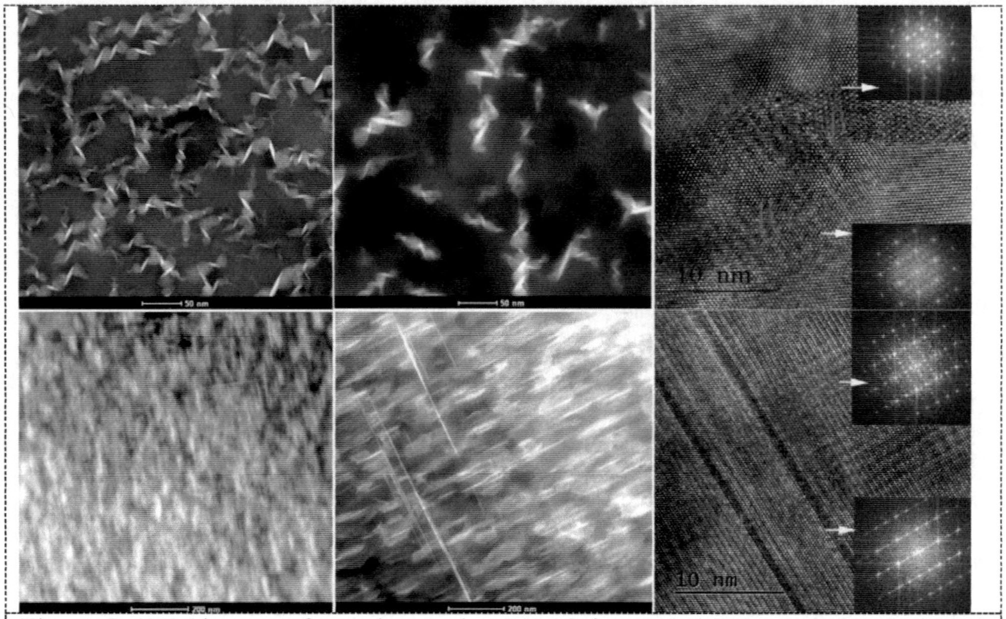

Figure 7. TEM images of samples aged at 240 °C for 100h (a) HAADF-STEM image of 1 Zn alloy taken along $[0001]_\alpha$; (b) HAADF-STEM image of 2 Zn alloy taken along $[0001]_\alpha$; (c) HRTEM image of 1 Zn alloy taken along $[0001]_\alpha$; (d) HAADF-STEM image of 1 Zn alloy taken along $[11\bar{2}0]_\alpha$; (e) HAADF-STEM image of 2 Zn alloy taken along $[11\bar{2}0]_\alpha$; (f) HRTEM image of 2 Zn alloy taken along $[11\bar{2}0]_\alpha$

3.4. Mechanical properties
The mechanical properties such as yield strength (YS), ultimate tensile strength (UTS) and elongation for 1Zn and 2Zn alloy before and after heat treatment has been tested. The as-cast 2Zn alloy exhibits lower mechanical properties and the UTS, YS and elongation are 172MPa 128MPa, and 2.8, respectively. After homogenization and 100 h ageing heat treatment, the UTS, YS and elongation for 1Zn alloy are 298 MPa, 233 MPa and 10.3, respectively. The UTS, YS and elongation for 2Zn alloy are 274 MPa, 198 MPa and 12.5, respectively. The lower strength for as-aged Mg-7Gd-5Y-1Nd-2Zn-0.5Zr alloy is attributed to the fewer β′ and β_1 phases formed during ageing. However, the higher density of SF result in the higher elongation for as-aged 2Zn alloy.

4. Conclusions

In this study, the effect of Zn on the microstructure evolution of Mg-7Gd-3Y-1Nd-xZn-0.5Zr (x=1.0, 2.0) (wt.%) alloys during homogenization and ageing heat treatment has been investigated systemically, the main conclusions can be summarized as follows:

1) As-cast 1 Zn alloy is composed mainly of α-Mg, Mg_5(RE, Zn) eutectic phase, $(Mg,Zn)_3RE$ eutectic phase, stacking faults and block-like compounds rich in RE. Increasing of Zn content result in the disappearance of Mg_5(RE, Zn) eutectic phase, but 14H-type LPSO phase can be found in as-cast 2Zn alloy.

2) After homogenization heat treatment of 520 °C for 32 h, the eutectic phase has been completely dissolved into matrix. However, the 14-H type LPSO phases distributed along grain boundaries both in 1Zn and 2Zn alloy. The variation of morphology and size of LPSO during homogenization can be found.

3) After 240 °C/18 h ageing, β′ phase has precipitated from matrix. Some hexagonal shaped clusters with a hexagonal structure have also been found in 1Zn alloy. After continued ageing at 210 °C for 100 h, a large volume fraction of coarse β′ precipitates can be observed. Besides, β_1 phase with higher RE content can also be found.
4) With increasing of Zn content, the fewer β′ and β_1 phases formed in 2Zn alloy, but higher density of SF can also be found. As a result, the strength of as-aged 2Zn alloy is lower but elongation is higher.

5. References

[1] Aderhold J, Davydov V Yu, Fedler F, Klausing H, Mistele D, Rotter T, Semchinova O, Stemmer J and Graul J 2001 *J. Cryst. Growth* **222** 701
[1] Lorimer G W, Apps P J, Karimzadeh H, and King J F, Mater. Sci. Forum 419–422 279
[2] Smola B, Stulikova I, Vonbuch F, Mordlike B L, Mater. Sci. Eng. A 324 113
[3] Peng Q M, Hou X L, Wang L D, Wu Y M, Cao Z Y, Wang L M, Mater. Des. 30 292
[4] Smola B, Joska L, Brezina V., Stulíková I., Hnilica F., Mater. Sci. Eng. C 32 659
[5] Kawamura Y, Hayashi K, Inoue A, Mater. Trans. 42 1171
[6] M. Yamasaki, T. Anan, S. Yoshimoto, Y. Kawamura, Scripta Mater. 53 799
[7] Zheng L, Liu C M, Wan Y C, Yang P W, Shu X. J. Alloys Compd. 509 8832
[8] Honma T, Ohkubo T, Kamado S, Hono K. Acta Mater., 55 4137
[9] Li Y, Xiao W L, Wang F, Hu T, Ma C L. J. Alloys Compd., 745 33
[10] Zhang J H, Leng Z, Liu S J, Li J Q, Zhang M L, Wu R Z. J. Alloys Compd. 509 7717
[11] Xia X S, Chen Q, Zhao Z D, Ma M L, Li X G, Zhang K. J. Alloys Compd., 623 62
[12] Li T, Du Z W, Zhang K, Li X G, Yuan J W, Li Y J, Ma M L, Shi G L, Fu X, Han X L, J. Alloys Compd., 574 174
[13] Li M, Zhang K, Du Z W, Li X G, Li Y J, Ma M L, Shi G L, Yuan J W, Li T, Liu J B. Mater. Charact. 109 (2015) 66–72.
[14] Li M, Zhang K, Du Z W, Li X G, Li Y J, Ma M L,. Trans. Nonferrous Met. Soc. China 26(2016) 1835−1842.
[15] Li M, Zhang K, Li X G, Li Y J, Ma M L, Shi G L, Yuan J W, Li T, Liu J B. Mater. Sci. Eng.A 626 415
[16] Zhang X G, Meng L G, Fang C F, Peng P, Ja F, Hao H, Mater. Sci. Eng. A 586 19
[17] Metallic materials-tensile testing-part 1: method of test at room temperature (ISO6892-1: 2009, MOD), GB/T 228.1-2010, p. 35.
[18] Zhu Y M, Morton A J, Nie J F: Acta Mater. 60 6562
[19] Yamasaki M, Sasaki M, Nishijima M, Hiraga K, and Kawamura Y: Acta Mater. 55 6798
[20] Nie J F, Wilson N C, Zhu Y M, Xu Z. Acta Mater. 106 260-271

Acknowledgments

This work was supported by the Natural Science Foundation of China (No. 51501015, No. 51871195), Jinqiao Seed Foundation of Beijing, and National Basic Research Program of China (No. 2013CB632202)

Fragment Velocity Test System Based on PVDF

Xuanquan Wang[1], Ping Song[1*], Yayu Zhai[1] and Xiaoxiao Chen[1]

1 Key Laboratory of Biomimetic Robots and Systems (Ministry of Education)
Beijing Institute of Technology, Beijing, 100081, China
E-mail: sping2002@bit.edu.cn

Abstract. In order to solve the problem of low measurement accuracy, large environmental interference and high cost of production in fragment velocity test, this paper proposed the fragment velocity test system by using PVDF piezoelectric film, NI-PXI platform and LabVIEW graphical programming language. The test system mainly includes PVDF film, adjusting circuit, acquisition module and terminal control software. The adjusting circuit amplifies the weak signal of the PVDF sensor. The acquisition module is responsible for reading the data collected by the data acquisition card for subsequent analysis and storage. The terminal control software executes data multi-channel acquisition, storage and display. The test system is stable and reliable with high accuracy through experimental verification.

Keyword. PVDF Film, Fragment Velocity, LabVIEW, PXI Platform

1. Introduction

Fragment velocity is one of the important indicators to measure the degree of damage. As the basic information to improve the capability of destruction, the technology of fragment velocity measurement plays a vital role in the development of destructive weapons [1]. However, the current fragment measurement method still has many problems such as cumbersome testing process, low test accuracy, and poor repeatability[2].

At present, many people at home and abroad research on the fragment velocity test system, for example, Xu[3] established fragment velocity test platform which combining laser vertical target and film switch array. Also, designed a signal processing circuit of thin film switching array and propose a CPLD pulse width coding algorithm for multi-channel signals. Tian[4] presented the construction method with a circular screen of light screen velocity measuring device, the circular ring light source and received array placed at the same center of circular are designed to form two cylindrical detection light screens with equal intervals to measure accurately the velocity of fragments which flying in any direction. Zhang[5] proposed band-pass filtering properties and multi-resolution analysis based on discrete wavelet transform, combined wavelet threshold de-noising method for wavelet filtering of fragmented over-target signals. Liu[6] proposed a fragment velocity method based on photogrammetry technology . By analyzing the sequence of projectile motion images captured by high-speed cameras, they restored the equation of motion of the projectile and obtained the flight velocity of the projectile. The effective measuring area of WTS03 light curtain target developed by Czech Prototypa Company[7]was 950mm×950mm, and the velocity of the projectile can be measured in the range of 50-3000m/s. The accuracy of the measuring velocity reaches 0.2%. The Skyscreen III light curtain target produced by USA Oehler Company [8] was mainly used for the velocity measurement of bullets. It

measured an effective area of 84 square inches and could measure speeds of 1000-4000 feet per second.
This study designed a fragment velocity test system based on PVDF film[9-10]. The signal collected by PVDF sensor firstly is processed by an adjusting circuit. Secondly the processed signal is collected by PXI platform. Lastly, we achieve multi-channel acquisition, storage and display of data finally through terminal control software.

2. Overall Design

Test equipment includes PVDF piezoelectric film, guide rail, support structure, signal adjusting circuit, PXI platform and skyscreens. It comprises a total of two piezoelectric films placed front and back, and the PVDF piezoelectric film is divided into several piezoelectric measuring units. The PVDF film is fixed on the guide by the support structure, and the distance between the two films can be adjusted by the guide. The system records penetration time and penetration point when the fragment penetrates the front and back PVDF films respectively. Using the time it takes to fly through the two PVDF piezoelectric films, the fragment velocity can be analyzed and calculated. The actual velocity is measured by skyscreens. The sub-region PVDF film can effectively solve the short circuit problem compared with other single films which using to velocity measurement. The main device is shown in Figure.1 below.

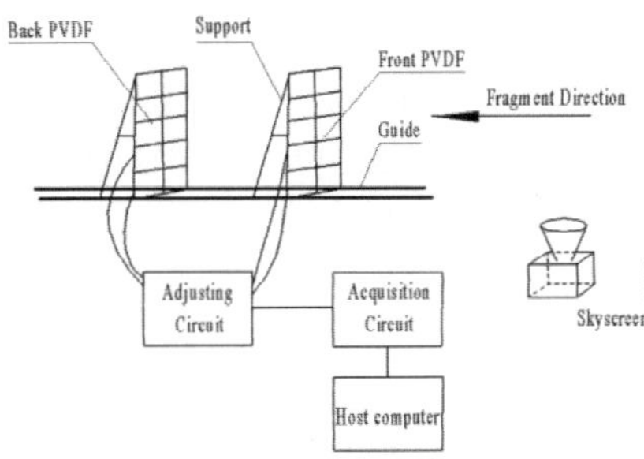

Figure 1. Fragment velocity measuring device.

3. Hardware System Design

The hardware mainly extracts and regulates the signal and transmits it to the host computer, mainly including signal adjusting circuit and PXI platform deploy.

3.1. Signal Adjusting Circuit

PVDF film output impedance can generally reach $10^{13}\Omega$, it can equivalent to a charge source under the influence of external force, and the output signal is charge. Therefore, the input impedance of the preamplifier needs to be high and the noise is low. After consulting, we choose TI's LF356 operational amplifier as the core device.
The first stage is a charge amplifier in the preamplifier circuit. Its important role is converting the charge signal into a voltage signal, amplifying the weak signal produced by the piezoelectric sensor, converting the high output impedance of the voltage signal to low output impedance, and improving the load capacity of the circuit. The second stage is a low pass filter, and it is used to reduce high frequency noise. Signal adjusting circuit design is shown in Figure.2.

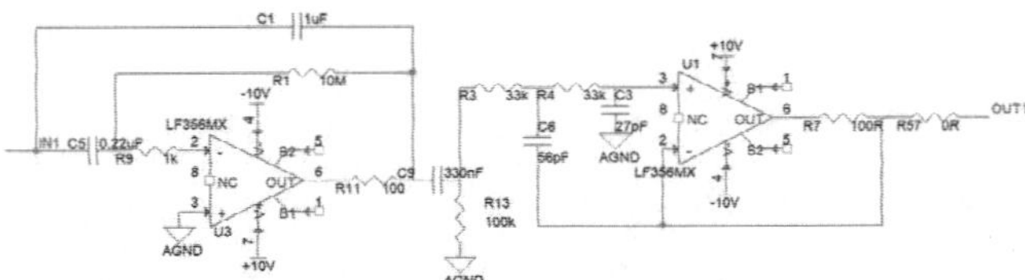

Figure 2. Signal adjusting circuit.

IOP Conf. Series: Materials Science and Engineering **542** (2019) 012041 doi:10.1088/1757-899X/542/1/012041

3.2. PXI Platform Deploy
Hardware selection is shown in Table 1:

Table 1. List of velocity test system PXI platform hardware.

Name	Quantity	Description
PMA 1115	1	Windows 7, 15-inch LCD touch screen display with 1024 * 768 resolution
PXIe 1082 Chassis	1	8 slots, maximum bandwidth8GB/s, VCVO on board clock
PXIe 6853 Capture Card	2	Multi-function DAQ, 16AI, 1.25MS/s, 4AO, 48DIO, 4CTR
PXIe 8880 Controller	1	Intel Xeon 8 Core, 2.3 GHz
SCB-68A Quick-Reference Label	4	68-pin shielded junction box M Series and X Series Devices

4. Software System Design

The software part is developed by LabVIEW graphical programming language. According to the requirement of the test system, we realize multi-channel data acquisition, storage and display in our test system.

4.1. GUI
The graphics user interface of the system is shown in Figure.4, which includes the acquisition board channel deployment, trigger setting and real-time waveforms display. The specific operation process is to set the related parameters of the signal acquisition card. After triggering, the system starts real-time multi-channel acquisition and stores the collected data in a binary file format to the computer hard disk.

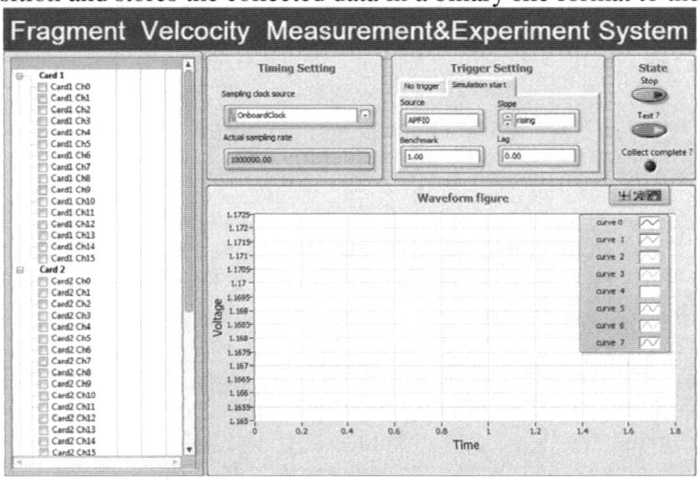

Figure 3. Operation interface diagram.

4.2. Program Design
When the fragment penetrates the PVDF film, a weak charge signal is generated. The signal is processed by the signal adjusting circuit, and transmitted to the acquisition card. Finally, the program realizes multi-channel data acquisition, storage and display.

- Acquisition module: In this process, we set the channel parameters, trigger settings, sampling clock settings, read function. This system selects multi-channel analog signal input and sets the input voltage range to ±10V. The trigger mode is triggered by the rising edge of the digital signal, the sampling rate is 1MB/S, and the read function category is analog multi-channel multi-sampling 2D DBL. As shown in Figure.4.

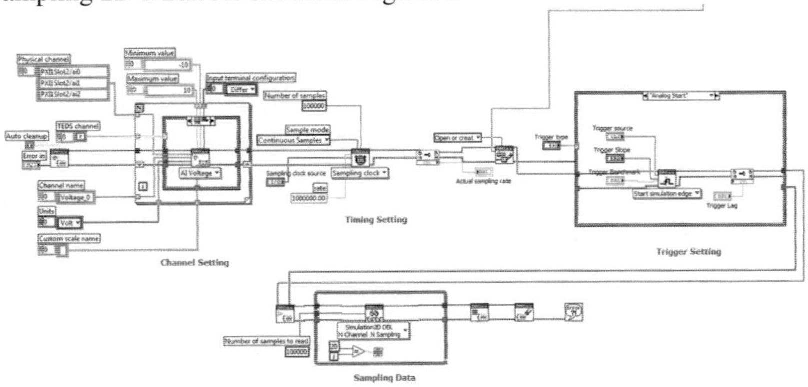

Figure 4. Acquisition module.

- Storage and display module: we store the data as a binary file (.tdms). In this process, we set the storage path, storage format and the read file storage path to achieve data playback and real-time waveform display. As shown in Figure.5.

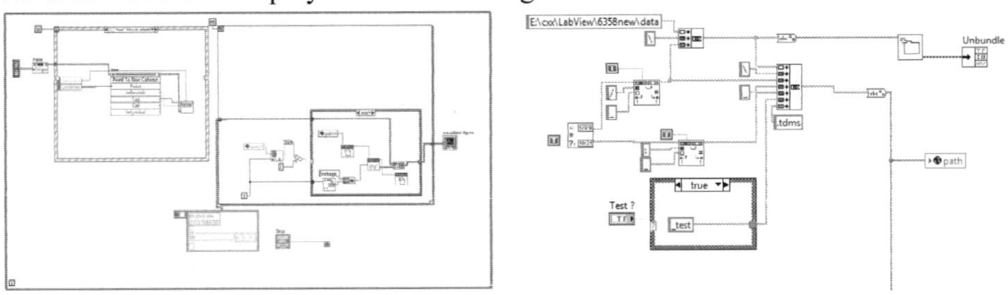

Figure 5. Storage and display module.

5. Analysis of Test Results

The system is tested in real experiments. Figure.6 is an amplified view of the signal front-end pulse collected in a single test. It can be clearly seen from the figure that the time when the fragments penetrate the PVDF film on the front and back sides respectively. The velocity of the fragment can be calculated according to the time difference between the front and back moments.

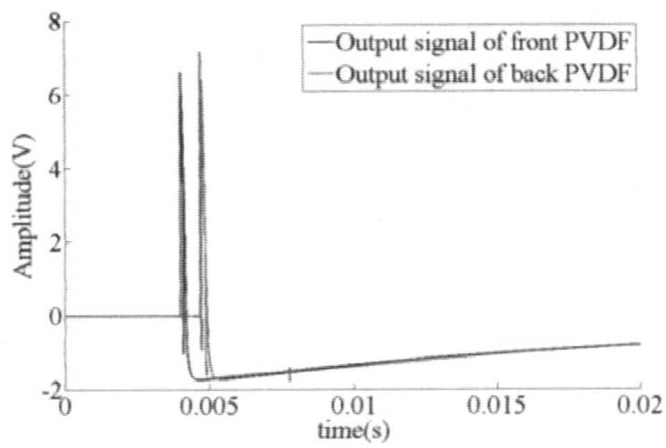

Figure 6. Single test result.

When testing the performance of the test system, ten routine tests are conducted. After each test, if the film's damaged area exceeds 60% of the whole film or the film is short circuit, we will replace it. The results are showed in Table 2. In the data processing, we use 20% of the maximum value of the filtered signal pulse as the time when the fragment penetrate the PVDF. Then we calculate the time difference between the fragments reaching the two PVDFs .The velocity is the ratio of the distance between the membranes and the time of flight. According to the final results, the average error is 1.28%, meeting the test requirement.

Table 2. Test data results.

Test Times	Test Value (m/s)	Actual Value (m/s)
1	722	737.3
2	724.2	734.4
3	720.5	731.9
4	729.1	73.4.4
5	733.7	735.5
6	732.7	734.6
7	726.7	734.2
8	723.7	737.3
9	725.1	735.1
10	715.1	732
Average Value	725.3	734.7
Standard Deviation	5.61	1.93

6. Conclusion

In this paper, a fragment velocity test system is proposed. This system uses PVDF film as sensor, and is designed on the PXI platform. This system mainly includes PVDF piezoelectric film, guide rail, support structure, signal conditioning circuit, PXI platform. It realizes the integration of signal acquisition, calculation processing, display and storage, and improves the system's anti-interference, compatibility and scalability. Through experimental tests, all functional modules run normally. The system can accurately determine the parameters and performance indicators of the fragments. Compared with the traditional fragment velocity test methods, our system has the advantages of simple operation, friendly man-machine interface and high measurement accuracy.

7. References

[1].Li L. 2017 Investigation on some Key Measuring Technologies of Fragment Warhead Power Field. *Nanjing University of Science and Technology.*
[2].Chen L. 2010 Researches on key technology of fragments velocity measuring and data acquisition system. *National University of Defense Technology.*
[3].Xu S. 2018 Research on Fragment Speed Testing Technolog. *Xi'an Technological University.*
[4].Tian H, Jin P, Tian Y, Yuan Y. 2018 *Journal of Test and Measurement Technology*, 32(04): 353-357.
[5].Zhang B, Li J, Zhao D, Liu J, Li Y, Shi X. 2016 *Acta Armamentarii*, 37(03): 489-495.
[6].Liu Z, Zhang Y, Zhao J, Du B. 2015 *Journal of Ballistics*, 27(04): 47-51.
[7].http://www.prototypa.cz/en
[8].http://www.oehler-reseach.com
[9].Jia Y, Chen X, Ni Q, 2013. *Sensors*, 13:8669-8678.
[10].Huang Z, Gao S, Li L, Chen H. 2000 *Journal of Nanjing University of Science and Technology*, (05): 466-468.

Acknowledgments
This work was partially supported by the National Defense Basic Scientific Program of China (JCKY2016208B008). We are also very grateful to the reviewers for their useful opinions and suggestions, which have improved representativeness.

Response Surface Methodology for Optimization Studies of Microwave-assisted hydrodistillation of essential oil from Vietnamese *Citrus aurantifolia* (Lemon fruit)

Thien Hien Tran[1], Tri Duc Lam[1], Van Tien Nguyen[1], Tan Phat Dao[1], Le Thi Hong Nhan[3], Tran Quoc Toan[4] , Dai-Viet N. Vo[5], Tran Anh Vy[6], Le Minh Bui[1*]

[1]NTT Institute of High Technology, Nguyen Tat Thanh University, Ho Chi Minh City, Vietnam
[2]Faculty of Chemical Engineering and Food Technology, Nguyen Tat Thanh University, Ho Chi Minh City, Vietnam
[3]Department of Chemical Engineering, HCMC University of Technology, VNU-HCM, Ho Chi Minh City, Vietnam
[4]Institute of Natural Products Chemistry, Vietnam Academy of Science and Technology, Ha Noi, Vietnam
[5]Faculty of Chemical and Natural Resources Engineering, Universiti Malaysia Pahang, Malaysia
[6]Department of Chemical and Biochemical Engineering, Gachon University, South Korea
*Corresponding author: blminh@ntt.edu.vn; labams2013@gmail.com

Abstract. In this study, microwave-assisted hydrodistillation extraction technique was applied to extract the essential oil from Vietnamese Citrus aurantifolia (Lemon fruit). The optimal conditions for microwave-assisted extraction of lemon fruit oil was determined by response surface methodology. A central composite design (CCD) was used to estimate the effects of three independent variables plant material to extraction time (B: 30, 60, 90 min), microwave power (C: 300, 450, 600 W), and solvent ratio (A: 2, 3, 4 mL/g)., and on the extraction yield of Lemon fruit oil. The optimal extraction conditions of Citrus aurantifolia oil was plant material to solvent ratio 2.80 mL/g, extraction time 63.29 min and microwave power 477.82W. The maximum Lemon fruit oil yield was 2.427%. Under the extraction condition, the experimental values matched with the anticipated results by analysis of variance. It showed high fitness of the model used and the benefit of response surface methodology for optimizing and indicate the expected extraction condition. .

1. Introduction

Nowadays, the application of the microwave-assisted method in order to extract constituents from plant material has revealed enormous research potential and interest. [1-4]. A previous study demonstrated that using the microwave extraction method is a choice that could be advanced than the conventional methods due to the high levels of product purity, reduced extraction time and more efficient heating. [5-7]. To achieve higher quality and higher yield of oil extraction from different plant materials, it is important to select suitable techniques and optimize extraction parameters. Given that a certain response, such as oil yields, could be significantly influenced by various individual experimental conditions and interactions of such variables, the response surface methodology may prevail as an appropriate method to identify the optimal conditions. RSM involves establishing a regression equation that describes the relationship between controllable experimental parameters, their interactions and desired response. Thanks to the widespread application of RSM, development of new and improved products has been rapid and increasingly efficient [8-15]. Citrus fruits play an important role in commercial crops grown in around the world. It has gained attention not only for their medicinal value but also for their nutritional characteristics. Citrus species have a broad spectrum of biological activities, including antioxidant, analgesic, antiviral, antibacterial and so on. Citrus fruits are one of the best sources of ascorbic acid and different bioactive compounds such as limonoids, coumarins, and flavonoid [16]. Thus, the objectives of this work was to use the microwave-assisted hydrodistillation techniques for the extraction of essential oil from Vietnam *Citrus aurantifolia* (Lemon fruit) commonly used in the food, pharmaceutical and cosmetic industriesThe microwave-assisted extraction parameters of essential oils from the Lemon fruit was examined and optimized using a three-level, three variable central composite design (CCD).

2. Materials and Method

2.1 Plant sample preparation

Citrus aurantifolia (Lemon fruit) was collected from Tien Giang province, Vietnam. It was then soaked with water to eliminate impurities and was dried. After that, the peels were separated from the lemon sarcocarp. All raw materials are stored in a non-hygroscopic bag, which is stored in a cooler (LC-1416B, Alaska, Vietnam) at temperatures under 10°C.

2.2 Extraction method

The lemon peels is extracted to obtain the essential oil by a system consisting of a microwave oven MW71E (manufactured by SAMSUNG, Vietnam), which acts as a source of heat for the extraction process and a hydro-distillation apparatus (Clevenger type apparatus, Germany) used to condense and separate the extracted oil and water phase. Finally, the lemon essential oil was anhydrated using sodium sulfate purchased from Sigma Aldrich (US).

2.3 Experimental design with RSM

The response surface methodology was chosen to optimize the effect factors of lemon oil on the effect of the ratio of water and lemon peels (A), time extraction (B), and microwave power (C) to maximize essential oil yield (Y). The central composite design (CCD) offers the experiment matrix designs comprising variation levels of experimental parameters, which were displayed in Table 1. Parameter A was set in range of 2:1–4:1 mL/g, B was set from 30 to 90 min, and C was fixed in range of 300 to 600 W. The raw materials in the RSM experiments were cut to a size of about 1-2 mm. For the statistical dissect, Design-Expert® software version 11, Minneapolis was adopted to assume the experimental design and to check complex multinomial to analogue the data.

Table 1. Encodes the values of the optimization factors

Levels	Independent factors		
	A (mL/g)	B (Min)	C (W)
Minimum point (-1)	2	30	300
Central point (0)	3	60	450
Maximum point (+1)	4	90	600

IOP Conf. Series: Materials Science and Engineering **542** (2019) 012042 doi:10.1088/1757-899X/542/1/012042

3. Results and Discussion

The results of 20 experimental runs with the RSM model were completed design and shown in Table 2. The table 2 results displayed was the yield oils extracted would be severely changed when the survey parameters changed (ratio, time, and microwave power). In Table 2, we see not only the actual experimental results shown, but also the predicted results made by DX11 software.

Table 2. Results of the actual and predicted values for the RSM model

	Independent variables			Y (%)			Independent variables			Y (%)	
	A	B	C	Actual	Predicted		A	B	C	Actual	Predicted
1	2	30	300	1.90	1.90	11	3	10	450	1.85	1.86
2	4	30	300	1.85	1.84	12	3	111	450	2.00	1.99
3	2	90	300	2.00	2.00	13	3	60	198	1.90	1.91
4	4	90	300	1.95	1.95	14	3	60	702	2.10	2.09
5	2	30	600	2.05	2.05	15	3	60	450	2.45	2.42
6	4	30	600	1.95	1.95	16	3	60	450	2.40	2.42
7	2	90	600	2.10	2.11	17	3	60	450	2.45	2.42
8	4	90	600	2.00	2.00	18	3	60	450	2.40	2.42
9	1.3	60	450	2.15	2.14	19	3	60	450	2.40	2.42
10	4.7	60	450	2.00	2.01	20	3	60	450	2.40	2.42

Table 3. ANOVA for Quadratic model

Source	Sum of Squares	Degree of freedom	Mean Square	F-value	Prob. > F	Comment
Model	0.8915	9	0.0991	249.91	< 0.0001	significant
A-A	0.0223	1	0.0223	56.34	< 0.0001	
B-B	0.0223	1	0.0223	56.34	< 0.0001	
C-C	0.0397	1	0.0397	100.17	< 0.0001	
AB	0.0000	1	0.0000	0.0000	1.0000	
AC	0.0012	1	0.0012	3.15	0.1061	
BC	0.0012	1	0.0012	3.15	0.1061	
A^2	0.2101	1	0.2101	530.02	< 0.0001	
B^2	0.4352	1	0.4352	1097.88	< 0.0001	
C^2	0.3125	1	0.3125	788.38	< 0.0001	
Residual	0.0040	10	0.0004			
Lack of Fit	0.0006	5	0.0001	0.1891	0.9542	not significant
Pure Error	0.0033	5	0.0007			
Std. Dev.= 0.0199		**Mean = 2.11**	**C.V. % = 0.9413**			**R^2 = 0.9956**
Adjusted R^2 = 0.9916		**Predicted R^2 = 0.9893**	**Adeq Precision = 40.9550**			

The Table 3 displayed results of ANOVA for Quadratic model for the process of extracting lemon oil. In this table, F-value, P-values and Lack of Fit F-value are parameters that exhibit the interaction of the conditions. The Model F-value of 249.91 implies the model by DX11 software is significant. There is only a 0.01% chance that an F-value this large could occur due to noise. In this case A, B, C, A^2, B^2, C^2 are significant model terms. The Lack of Fit F-value of 0.19 implies the Lack of Fit is not significant relative to the pure error. The Predicted R^2 of 0.9893 is in agreement with the Adjusted R^2 of 0.9916. The 40.955 value shows an signal and this model can be applied to navigate the design space. In addition, the experimental model was considered fit when calculated residuals follow a random pattern as shown in Figure 1A. Figure 1B presented that data points corresponding to anticipated and actual values are scattered across the 45 degree line with close proximity, suggesting that the actual results are accurately anticipated from the factor values. Thus, based on the data analysis of the oil yield from experiments, the RSM software results in optimized parameters as A = 2.80:1 (mL/g), B = 63.29 (min), and C = 477.82 (W) to obtain the yield of 2.427% with 96.2% reliability with the quadratic model described by the following equation: Yield of lemon oil = $2.42 - 0.0404A + 0.0404B + 0.0539C - 0.0125AC - 0.0125BC - 0.1207A^2 - 0.1738B^2 - 0.1473C^2$ (2)

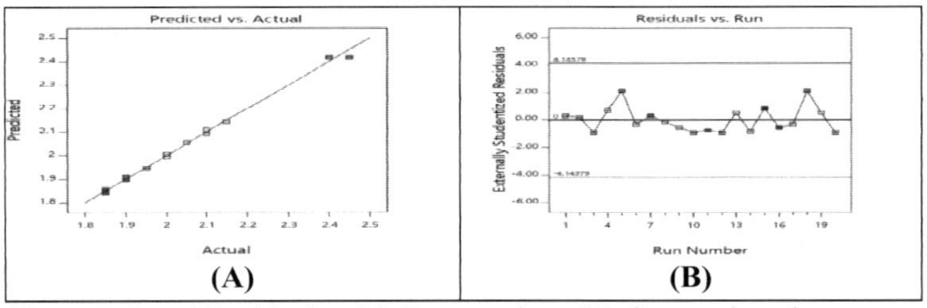

Figure 1 (A) Normal plot of Residuals with Run number **(B)** Comparison between actual values and predicted values

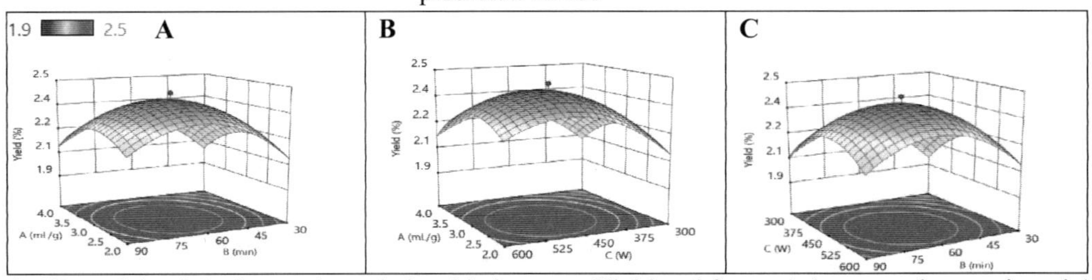

Figure 2. 3D models of the interaction relationship between Yield and factor A (water and turmeric ratio), B (extraction time), C (microwave power)

Based on the optimized parameters, Figure 2 shows the mutual interactions of the factors and the yield of the lemon oil obtained. In general, the efficiency of the attained oil increases proportionally with condition parameters. However, as these conditions exceed the optimal point (2.80:1 mL/g, 63.29 min, and 477.82 W); the obtained lemon oil content ceases to rise, and eventually, starts diminishing. The parameters that DX11 software predicted were made to extract lemon essential oil and the results are shown in Table 4. With that result, the DX11 software predicted accurate results in comparison with empirical values.

Table 4. Compare lemon essential oil yield results between DX11 software and actual experiments

	A (mL/g)	B (min)	C (W)	Yield (%)
DX11 software	2.80:1	63.29	477.82	2.472

Experiment	2.80:1	63.30	478.00	2.500

4. Conclusion

Response surface methodology has demonstrated to be useful in determining the result of three independent variables including extraction time, microwave power and plant material to solvent ratio on the lemon essential oils yield and for anticipating the optimal operational conditions. To conclude, the parameters that affect the extraction of the lemon extract using microwaves during extraction were optimized by the RSM method. The yield of lemon oil is at 2.472% with 96.2% confidence when the water-to-material ratio reached 2.80:1 mL/g, at the extraction time of 63.29 min and microwave power of 477.82 W. Through the agreement of the experimental and the analysis of variance and anticipated results, it can be assumed that the generated model was suitable for the simulation of microwave-assisted extraction of lemon with any combination of tested variables.

Acknowledgements: This research is funded by Nguyen Tat Thanh University, Ho Chi Minh City,

References

[1] Pham H Q, Huynh T T, Nguyen A V, Mai A T N, Phan V T T, Bach L G, Trinh N D and Ho V T T 2018 *Solid State Phenomena* **279** 181–6
[2] Thinh P V, Trinh N D, Thuong N T, Ho V T T, Quynh B T P and Bach L G 2018 *Solid State Phenomena* **279** 230–4
[3] Pham T V, Nguyen T T, Nguyen D T, Thuan T V, Bui P Q T, Viet V N D and Bach L G 2019 *Journal of Nanoscience and Nanotechnology* **19** 1122–5
[4] Tran T, Nguyen H, Nguyen D, Nguyen T, Tan H, Nhan L, Nguyen D, Tran L, Do S and Nguyen T 2018 *Processes* **6** 206
[5] Hien T T, Nhan N P T, Trinh N D, Ho V T T and Bach L G 2018 *Solid State Phenomena* **279** 217
[6] Liu Z, Deng B, Li S and Zou Z 2018 *Industrial Crops and Products* **124** 353–62
[7] Sparr Eskilsson C and Björklund E 2000 *Journal of Chromatography A* **902** 227–50
[8] Van Tran T, Bui Q T P, Nguyen T D, Le N T H and Bach L G 2017 *Adsorption Science & Technology* **35** 72–85
[9] Van Tran T, Bui Q T P, Nguyen T D, Ho V T T and Bach L G 2017 *Water Science and Technology* **75** 2047–55
[10] Bach L G 2018 *Vietnam Journal of Science and Technology* **54** 123
[11] Van Thuan T, Quynh B T P, Nguyen T D, Ho V T T and Bach L G 2017 *Surfaces and Interfaces* **6** 209–17
[12] Bach L G, Van Tran T, Nguyen T D, Van Pham T and Do S T 2018 *Research on Chemical Intermediates* **44** 1661–87
[13] Nhan N P T, Hien T T, Nhan L T H, Anh P N Q, Huy L T, Nguyen T C T, Nguyen D T and Bach L G 2018 *Solid State Phenomena* **279** 235–9
[14] Arvindekar A U and Laddha K S 2016 *Industrial Crops and Products* **83** 587–95
[15] Chen C, Shao Y, Tao Y and Wen H 2015 *LWT - Food Science and Technology* **64** 1263–9
[16] Caballero B, Finglas P M and Toldrá F 2016

Application of Box–Behnken design with Response Surface Methodology for Modeling and Optimizing Microwave-assisted Hydro-distillation of Essential Oil from *Citrus reticulata* Blanco Peel

Thien Hien Tran[1,2], Tan Phat Dao[1,2], Duy Chinh Nguyen[1], Tri Duc Lam[1], Sy Trung Do[3], Tran Quoc Toan[4], Nguyen Thi Thanh Huong[5], Dai-Viet N. Vo[6], Long Giang Bach[1,2], Trinh Duy Nguyen[1,*]

[1]NTT Hi-Tech Institute, Nguyen Tat Thanh University, Ho Chi Minh City, Vietnam

[2]Faculty of Chemical Engineering and Food Technology, Nguyen Tat Thanh University, Ho Chi Minh City, Vietnam

[3]Institute of Chemistry, Vietnam Academy of Science and Technology, Ha Noi, Vietnam

[4]Institute of Natural Products Chemistry, Vietnam Academy of Science and Technology, Ha Noi, Vietnam

[5]Institute for Tropical Technology, Vietnam Academy of Science and Technology, Ha Noi, Vietnam

[6]Faculty of Chemical and Natural Resources Engineering, Universiti Malaysia Pahang, Malaysia

*Corresponding author: ndtrinh@ntt.edu.vn; labams2013@gmail.com

Abstract.
In this research, microwave-assisted hydro-distillation and Response Surface Methodology were selected for optimizing factors affecting the yield of essential oil from *Citrus reticulate* Blanco peels. This combined process was successfully modeled and optimized using a Box–Behnken design with response surface methodology (RSM). The effects of the material and water ratios, extraction time and microwave power on the extraction essential oils, efficiency of *Citrus reticulata* Blanco Peel were investigated. The optimal conditions for the *Citrus reticulate* Blanco peels oils yield were found to be a material and water ratios were 3.1:1mL/g, extraction time was 91.24 min, microwave power at 478 (W). Furthermore, analysis of variance indicated that the proposed quadratic model successfully interpreted the experimental data with coefficients of determination of the predicted R^2 of 0.9909 concurred with the adjusted R^2 of 0.9826. A confirmation test of the optimal conditions verified the validity of the model, yielding an essential oil efficiency of 5.5

Keywords: *Citrus reticulate* Blanco peels, Essential Oil, Microwave-assisted hydro-distillation, Response Surface Methodology

1. Introduction

Citrus reticulata Blanco (Rutaceae) is a well-known crop with high economic and culinary value. The fruit is edible and the peel could be extracted for essential oils which play an important role in many industries, from food chemistry to pharmacology and pharmaceutics. Citrus essential oils are composed of different compounds with diverse biological activities including antioxidant, anti-inflammatory, anxiolytic, antimicrobial, and antifungal [1-3]. Such benefits have made *Citrus reticulate* Blanco peels oil increasingly popular and therefore development of new methods for oil extraction, such as supercritical CO_2 and microwave extraction, is necessary to replace traditional methods [4-6].

To obtain high quality compounds, the extraction technique of choice is crucial. The extraction process could be affected by various experimental conditions including temperature, time, and amount of solvent. To investigate the effect of such parameters and optimize the process, response surface methodology (RSM) with Box-Behnken design (BBD) was employed. RSM is a statistical technique that allows for evaluation of the effects of several process variables and their interactions on selected responses [7-14]. This approach is often used to generate second degree polynomial model, which is consequently used in optimizing an experiment process. The method is also advantageous since only a small number of experimental runs is required, minimizing time and costs of experiment. Considering the growing interest on the production of these essential oils, the goal of the present study was to extract essential oil compounds from *Citrus reticulate* Blanco peels using the optimized microwave-assisted hydro-distillation (MAHD). In the present study, parameters (material and water ratios, extraction time, microwave power) are optimized for maximum yield of *Citrus reticulate* Blanco peels oils.

2. Materials and Methods

2.1. Material and chemicals

Fresh ripe fruits were purchased from local market in Ben Tre Province, Viet Nam. All samples were collected from the same source and stored in a dry place. The fruits were cleaned, peeled and ground with a home grinder and distilled directly by steam. Anhydrous sodium sulfate (Na_2SO_4) was purchased from Sigma Aldrich (US). Deionized water was used as a solvent to extract *Citrus reticulate Blanco* peels oil by Milli-Q purification system (Millipore, USA).

2.2. Microwave-Assisted Hydro-distillation

A Clevenger type apparatus was connected to a domestic microwave oven MW71E (manufactured by SAMSUNG, Vietnam) for microwave assisted hydro-distillation operation. The power source has the maximum output power of 800W and voltage of 250v-50Hz. In this operation, parameters including materials and water ratio, microwave power and extraction time were determined by Design Expert 11 software.

2.3. Experimental Design

The central composite design was adopted to determine the optimal parameters of MAHD. ANOVA analysis, calculation of coefficients and plotting was carried out with Design Expert 11 software. Predicted yields and actual yields were also compared to evaluate the fitness of the model to the data.

3. Results and Discussion

The experiments were designed according to the design method of complex CCD center. Experimental results (20 experiments) and predictions by Design-Expert 11 are shown in Table 1.

Table 1. Experimental and predicted extraction efficiencies under different conditions for the RSM model

	Experimental Parameters			Y (%)			Experimental Parameters			Y (%)	
	A	B	C	Actual	Predicted		A	B	C	Actual	Predicted
1	2.0	60	300	2.8	2.89	11	3	39.60	450	2.8	2.87
2	4.0	60	300	4.8	4.63	12	3	140.5	450	3.4	3.27
3	2.0	120	300	4.6	4.52	13	3	90.00	198	4.4	4.41
4	4.0	120	300	2.2	2.37	14	3	90.00	702	4.8	4.73
5	2.0	60	600	2.2	2.07	15	3	90.00	450	5.6	5.54
6	4.0	60	600	4.6	4.72	16	3	90.00	450	5.6	5.54
7	2.0	120	600	4.6	4.81	17	3	90.00	450	5.4	5.54
8	4.0	120	600	3.6	3.56	18	3	90.00	450	5.4	5.54
9	1.3	90	450	3.6	3.56	19	3	90.00	450	5.6	5.54
10	4.7	90	450	4.0	3.97	20	3	90.00	450	5.6	5.54

The results were analyzed using Analysis of Variance (ANOVA). Table 1 presents the experimental parameters produced by CCD design and associated yields corresponding to four maximum points, four minimum points, six axial points, and six center points. RSM software showing no significant difference when comparing actual and predicted values. To determine the factors effects in the model are statistically significant, the p-values less than 5% indicate model terms are significant, whereas the lack of fit (LOF) was not significant at the 5% α-level only for the quadratic model described by the following as in equation: $Y= 5.54 + 0.1225A + 0.1178B + 0.0932C - 0.9750AB + 0.2250BC + 0.2750\,BC - 0.6245\,A^2 - 0.8720\,B^2 - 0.3416C^2$

Table 2. Analysis of variance results for ANOVA model

Source	Sum of Squares	dF	Mean Square	F-value	p-value		Comment
Model	24.92	9	2.77	120.40	< 0.0001	Significant	SD = 0.1517
A	0.2049	1	0.2049	8.91	0.0137		Mean =4.28
B	0.1896	1	0.1896	8.24	0.0166		CV (%) =3.54
C	0.1196	1	0.1186	5.16	0.0465		R^2= 0.9909
AB	0.1186	1	7.61	330.65	<0.0001		AP =32.2665
AC	7.61	1	0.4050	17.51	0.0018		Adj R^2 = 0.9826
BC	0.4050	1	0.6050	26.30	0.0004		Pred R^2= 0.9358
A^2	5.62	1	5.62	244.35	<0.0001		
B^2	10.96	1	10.96	476.40	< 0.0001		
C^2	1.68	1	1.68	73.13	< 0.0001		
Residual	0.2300	10	0.0230				
Lack of Fit	0.1767	5	0.0353	3.31	0.1073	not significant	
Pure Error	0.0533	5	0.0107				
Cor Total	25.15	19					

The results in Table 2 displayed the Analysis of Variance results for the quadratic model of essential oils yield. Three independent factors in the ANOVA table include: water and material ratio (A), extraction time (B), microwave energy level (C), interaction terms (AB, AC, BC) and second-order (A^2, B^2, C^2) terms. The model is significant as demonstrated by significant model F-value of 120.40. There is only a 0.01% chance that an F-value of this large could occur due to noise. Moreover, low p-value of all variables indicated that the response factor A, B, C, AB, AC, BC, A^2, B^2, C^2 are recognized as significant model terms at 95% confidence level. The Lack of Fit F-value of 3.31 implies that the LOF was not significant relative to the pure error. There is 10.73% chance that a LOF this large could occur due to noise. The predicted R^2 of 0.9909 concurred with the adjusted R^2 of 0.9826. A ratio greater than 4 is desirable. Ratio of 32.2665 indicated an adequate signal. Therefore, this model could be used to navigate the design space.

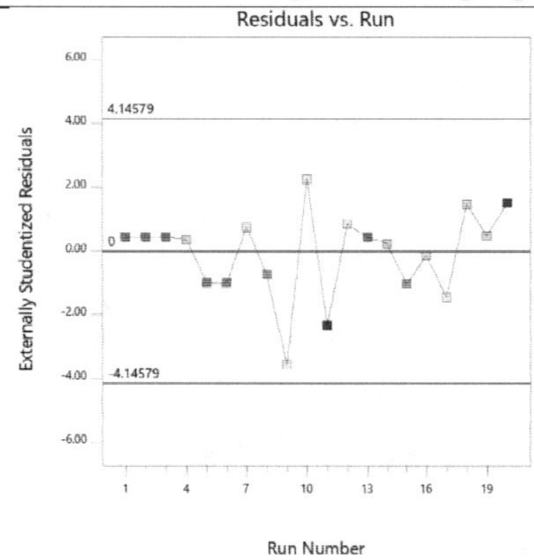

Figure 1. Normal plot of Residual vs Predicted

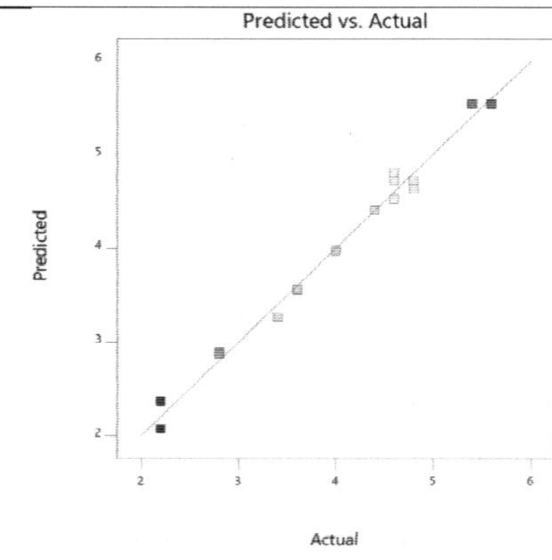

Figure 2. Comparison of predicted and experimental

IOP Conf. Series: Materials Science and Engineering **542** (2019) 012043 doi:10.1088/1757-899X/542/1/012043

reponse for yield of essential oil	extraction efficiencies

To validate the resulting model, residuals of 20 runs and yields of oil were plotted in the figure 1(a) and 1(b), respectively. Figure 1 indicated that residuals of experimental yields clearly follow a random pattern. This suggested that there was no violation of assumptions regarding independence of variables and constant variance. Figure 2, which plotted predicted versus against actual values, also indicated a close proximity of scattered data points to the 45 degree line suggesting the reasonable predicting accuracy of the model. The interaction between the essential oil yield and three process factors are shown in Figure 3. Based on the graphs, it could be observed that general trends of the three factors are similar, when the quantities of microwave power, extraction time, material and water ratio gradually increased, the yield of the *Citrus reticulate Blanco* peels oil obtained is also increased. However, this only holds to a certain point where the yield stops rising and eventually, starts diminishing. Optimization of the estimated statistical model yielded following optimal conditions: material and water ratios 3.1:1, extraction time 91.24 min, microwave power 478W.

The data from Table 3 shows the optimum conditions resulted from optimization. Accordingly, material and water ratios of 3.1, the time of 91.24 minutes and 482.06 W operating power yielded highest efficiency of 5.55%. This number approximates to the actual yield, conucted with almost identical conditions, of 5.5%

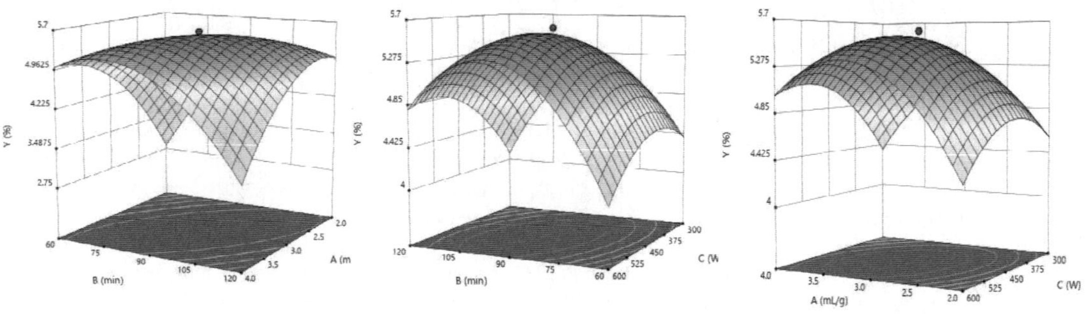

(a) (b) (c)

Figure 3. 3D response surface plots of interaction of Y with **(a)** B-extraction time and A-material and water ratio, **(b)** B-extraction time and C-microwave power, **(c)** A-material and water ratio and C-microwave power

Table 3. The experimental results using optimum condition comparison with predicted results

	Material and water Ratios (g/mL)	Extraction time (min)	Microwave power (W)	Yield of essential oil (%)
Predicted	3.1:1	91.2365	478.01	5.55228
Actual	3.1:1	91.24	478	5.5%

4. Conclusions

In this study, A Box–Behnken design with Response Surface Methodology was successfully applied to the extraction essential oil from *Citrus reticulata* Blanco Peel using the microwave-assisted hydro-distillation technology. Under the optimization performed by RSM, we found that the optimal level of three influencing conditions for extraction of the *Citrus reticulate* Blanco peels oil using microwave oven heat are as follows: the ratio of water to raw materials of 3.1:1 mL/g, the extraction time at 91.24 min, and microwave power of 478W with yield of 5.5523 %. The model fitted the experimental data well, with a coefficient of determination, R^2, of

0.9909 and an Adj-R^2 of 0.9826. In addition, the p-value of this model was less than 0.0001, which indicates that the model is very significant.

Acknowledgments: This research was funded by the Nguyen Tat Thanh University, Ho Chi Minh City, Vietnam

References

[1] S.S. Ferreira, A.M. Silva, F.M. Nunes, Citrus reticulata Blanco peels as a source of antioxidant and anti-proliferative phenolic compounds, Industrial Crops and Products, vol. 111, 2018, pp. 141-148
[2] H. Zhang, Y. Yang, Z. Zhou, Phenolic and flavonoid contents of mandarin (Citrus reticulata Blanco) fruit tissues and their antioxidant capacity as evaluated by DPPH and ABTS methods, Journal of Integrative Agriculture, vol. 17, 2018, pp. 256-263.
[3] J. Yang, L. Wen, Y. Zhao, Y. Jiang, M. Tian, H. Liu, J. Liu, B. Yang, Structure identification of an arabinogalacturonan in Citrus reticulata Blanco 'Chachiensis' peel, Food Hydrocolloids, vol. 84, 2018, pp. 481-488
[4] M. Vinatoru, T. J. Mason, I. Calinescu, Ultrasonically assisted extraction (UAE) and microwave assisted extraction (MAE) of functional compounds from plant materials, TrAC Trends in Analytical Chemistry, vol. 97, 2017, pp. 159-178
[5] A.F. Vega, N.R. Corona, E. Palou, A.L. Malo, Estimation of mass transfer coefficients of the extraction process of essential oil from orange peel using microwave assisted extraction, Journal of Food Engineering, vol. 170, 2016, pp. 136-143
[6] C.H. Chan, R. Yusoff, G.C. Ngoh, F.W.L. Kung, Microwave-assisted extractions of active ingredients from plants, Journal of Chromatography A, vol. 1218, 2011, pp. 6213-6225
[7] T.V. Tran, Q.T.P. Bui, T.D. Nguyen, N.T.H. Le, L.G. Bach, A comparative study on the removal efficiency of metal ions (Cu^{+2}, Ni^{+2}, and Pb^{2+}) using sugarcane bagasse derived $ZnCl_2$-activated carbon by the response surface methodology, Adsorption Science & Technology 2017, vol. 35(1–2), pp. 72–85
[8] V.T. Tran, D.T. Nguyen, V.T.T.Ho, P.Q.H. Hoang, P.Q Bui, L.G. Bach, Efficient removal of Ni^{2+} ions from aqueous solution using activated carbons fabricated from rice straw and tea waste, Journal of materials and Environmental Sciences, vol. 8, 2017, pp. 426-437
[9] T.V. Tran, Q.T.P. Bui, T.D. Nguyen, V.T.T. Ho, L.G. Bach, Application of response surface methodology to optimize the fabrication of $ZnCl_2$- activated carbon from sugarcane bagasse for the removal of Cu^{2+}, Water Science & Technology, vol. 75, no.9, 2017, pp. 2047-2055
[10] L.G. Bach, T.V. Tran, T.D. Nguyen, T.V. Pham, S.T. Do, Enhanced adsorption of methylene blue onto graphene oxide-doped XFe_2O_4 (X = Co, Mn, Ni) nanocomposites: kinetic, isothermal, thermodynamic and recyclability studies, Research on Chemical Intermediates, vol. 44, no. 3, 2018, pp 1661–1687
[11] T.V. Thuan, B.T.P. Quynh, T.D. Nguyen, L.G. Bach, Response surface methodology approach for optimization of Cu^{2+}, Ni^{2+} and Pb^{2+} adsorption using KOH-activated carbon from banana peel, Surfaces and Interfaces, vol. 6, 2017, pp. 209-217
[12] T.T. Hien, N.P.T. Nhan, D.T. Nguyen, V.T.T. Ho, L.G. Bach, Optimizing the Pomelo Oils Extraction Process by Microwave-Assisted Hydro-Distillation Using Soft Computing Approaches, Solid State Phenomena, 2018, Vol. 279, pp. 217-221.
[13] N.P.T. Nhan, T.T. Hien, L.T.H. Nhan, P.N.Q. Anh, L.T. Huy, N.T.C. Trinh, D.T. Nguyen, L.G. Bach, Application of Response Surface Methodology to Optimize The Process of Saponification Reaction from Coconut Oil in Ben Tre – Vietnam, Solid State Phenomena, vol. 279 , 2018, pp 235-239
[14] T.H. Tran, H.H.H. Nguyen, D.C. Nguyen, T.Q. Nguyen, H. Tan, L.T.H. Nhan, D.H. Nguyen, L.D. Tran, S.T. Do, and T.D. Nguyen, Microwave-Assisted Extraction: Application to the Optimization of Essential Oil Extraction from Vietnamese Basil (Ocimum Basilicum) Using Response Surface Methodology, Processes, vol. 6, 2018, 206; doi:10.3390/pr6110206.

IOP Conf. Series: Materials Science and Engineering **542** (2019) 012044 doi:10.1088/1757-899X/542/1/012044

Research on the influence mechanism of friction factor of gear based on mathematical statistics principle

H Ge[1] and C Wang[2]

1 School of Mechanical Engineering, University of Jinan, Jinan 250022, China
Corresponding author and e-mail: H Ge, gehao03118@126.com

Abstract. This paper first introduces the development status of gear friction factor research.Secondly, the advantage of the experimental method based on mathematical statistics is discussed.In addition, the mathematical statistics principle is used to design the orthogonal experiment to verify the existing formula of gear friction factor calculation.The correlation of each parameter was studied according to the experimental data.The deficiency of calculation formula is analysed and the prospect of future research is put forward.

1. Research status of gear friction factor

Gear mechanism is the most widely used transmission mechanism. Scientists have never stopped studying the gear mechanism. Because of its complicated mechanism, tooth surface friction is always one of the hot and difficult points in the study. The main effects of gear surface friction on gear transmission include: reduction of gear transmission efficiency, aggravation of gear tooth failure (tooth surface wear, tooth surface pitting, tooth surface cementation, etc.), system vibration and noise[1][2].Therefore, the study of tooth surface friction has great economic benefits and is of great significance to improve the performance of gear transmission system.

In theory, involute gear transmission does not generate friction, because the two gears are pure rolling at the nodes, and the gears are rigid.However, in actual generation, the gear is not rigid, and there is inevitable elastic deformation.The elastic deformation makes the gear meshing no longer pure rolling in theory, and the actual machining accuracy and installation precision cannot achieve the theoretical accuracy, which also makes the actual meshing different from the theory, so there will be gear friction.Since Martin first proposed the use of Reynolds equation to analyse gear friction in 1916.At present, the formula of gear friction is based on elastohydrodynamic lubrication theory.For example, the formula for calculating the friction factor of the numerical solution of linear contact isothermal total membrane elastomeric presented by Dowson and Higginson is as follows[3]:

$$f_s = \frac{u_2 - u_1}{F_n h_{\min}} \int_{-\partial}^{+\partial} \eta dx \quad (1)$$

Calculation formula of friction factor of gear for thermal elastohydrodynamic lubrication of non-newtonian fluid proposed by Xu[4]:

$$\mu = e^f P_h^{b_2} \left|SR\right|^{b_3} V_e^{b_6} v_0 R^{b_8} \quad (2)$$

It should be pointed out that the current calculation formula is mainly established for steady state, and the parameters do not change with time, which is different from the actual motion process of gear.

2. Orthogonal experimental design based on mathematical statistics theory

Orthogonal experiment is one of the experimental design methods.In brief, the experimental design is to study how to arrange experiments scientifically and obtain more comprehensive information with less consumption.The experiment is well designed, with half the effort.On the contrary, it will not only

IOP Conf. Series: Materials Science and Engineering **542** (2019) 012044 doi:10.1088/1757-899X/542/1/012044

fail to achieve the expected effect, but also consume a lot of manpower, material resources and time.Therefore, how to carry out test design is an important part of successful test.

Orthogonal experimental design is a scientific and reasonable arrangement of experiments based on the principle of mathematical statistics.From the results, the optimal conditions of the process can be found quickly, and it has the advantages of judging which factors are the main factors and the mutual influence between the influencing factors.Orthogonal experimental design is a common technique for experimental optimization.The so-called experimental optimization refers to an optimization method for optimal design under the guidance of the optimization thought.For example, if there are 3 influencing factors in a test and each influencing factor has 3 levels, then there are 3^3=27 permutations in the comprehensive test on 3 factors and 3 levels.Orthogonal experimental design is to select some representative test points from the comprehensive test points to carry out experiments.It can achieve the experimental effect and reduce the number of experiments.3 factors 3 level test using orthogonal test design only requires 9 tests.

China began to study this subject in 1950s and gradually applied it to industrial and agricultural production.It is proved that orthogonal experimental design is an important method of experimental design technology and brings great economic benefits.Orthogonal experimental design has become a necessary technology for engineers and researchers.

3. The formula of gear friction factor is verified by orthogonal experiment

As mentioned above, XU proposed the calculation formula of friction factor of non-newtonian thermal elastohydrodynamic lubrication gear:

$$\begin{cases} \mu = e^f P_h^{b_2} |SR|^{b_3} V_e^{b_6} v_0 R^{b_8} \\ f = b_1 + b_4 |SR| P_h \log_{10}(v_0) + b_5 e^{-|SR|P_h \log_{10}(v_0)} + b_9 e^s \end{cases} \quad (3)$$

Where μ is friction coefficient ; SR is the sliding velocities(V_S)and entraining velocity(V_e) of tooth surface meshing point, calledSlide-to-roll ratio ,SR = V_S/V_e;R is radius of curvature ; $P_h$ismax. Hertzian pressure;$V_0$is the dynamic viscosity of the lubricant; S is surface roughness;bi is a constant related to gear oil, using a common gear oil (75W90),bi(i = 1， 2， ···， 9) = (-8.92， 1.03， 1.04， −0.35， 2.81， −0.10， 0.75， −0.39， 0.62).

According to the formula analysis, it is considered that SR, radius of curvature, max. Hertzian pressure, surface roughness and entraining velocity have great influence on the friction factor of gear. So take 5 levels for each of these five factors, listed in table 1.

Table 1. Factors and levels

	Radius of curvature (R), m	Entraining velocity (Ve), m/s	Slide-to-roll ratio (SR)	Surface roughness (S), m	Max. Hertzian pressure (Ph), GPa
1	0.01	1.00	0.01	0.05	0.50
2	0.02	5.00	0.05	0.10	1.00
3	0.03	10.00	0.10	0.20	1.50
4	0.04	15.00	0.15	0.30	2.00
5	0.08	20.00	0.20	0.40	2.50

This experiment is a 5 factor and 5 level test, each factor occupies one column, withoutconsidering the interaction, so there are a total of 5 columns.Therefore, orthogonal table $L_{25}(5^5)$ was selected. According to the orthogonal table, the parameters with multiple levels were filled in successively, and the formula was used for calculation. The experimental results and the analysiswere included in table 2 together.Complete orthogonal table in appendix .Details of the testcan be found in [5].

Table 2. Test results and intuitive analysis

	R	*Ve*	SR	*S*	Ph
K1	0.054606	0.040223	0.004612	0.032484	0.016061
K2	0.039714	0.043447	0.021807	0.029857	0.024697
K3	0.032482	0.036307	0.047862	0.028842	0.037468
K4	0.030071	0.034220	0.054392	0.043491	0.052797
K5	0.027115	0.029792	0.055315	0.049313	0.052966
Range	0.027491	0.013655	0.050703	0.020471	0.036905
Principal→Subordinate SR>P_h>R>S> *Ve*					

The intuitive analysis of the test results, that is, range analysis, generally speaking, the range of each column is not equal.This indicates that the change of the level of each factor has different effects on the test results. The greater the difference, the greater the change of the value of this factor within the test range, which will lead to greater changes of the test index in the value.Therefore, the column with the maximum range value is the factor whose level has the greatest influence on the test results, which is the most important factor.For this experiment, it can be seen that the range is from large to small: SR> P_h>R>S> V_e.In other words, SR is the biggest influence on the friction factor of gear, and V_e is the smallest.The method of range analysis is simple in operation, small in calculation and relatively intuitive, but with poor accuracy, it cannot accurately estimate the influence of the test results of various factors.In order to make up for the deficiency of range analysis, variance analysis was further carried out, and the specific results were shown in table 3.

Table 3. Analysis of variance table

	sums of squared deviations(SSj)	degree of freedom（df）	SSj/df	F	significance test
R	9.67E-05	4	2.416E-05	4.319	un-significance
Ve	2.24E-05	4	5.595E-06		
SR	4.07E-04	4	1.018E-04	18.190	significance
S	6.63E-05	4	1.658E-05	2.962	un-significance
Ph	2.19E-04	4	5.471E-05	9.778	significance

Analysis of variance is a very practical and effective statistical test method, which can be used to test the significance of the influence of related factors on the test results.For example, a typical analysis of variance is how to find out whether different catalysts have significant effects on the rate of a product when reflecting time, temperature, pressure and other conditions.

As shown in table 3, $F_{0.05}(4,4)=6.39$ was obtained from the F-distribution table.$F_{SR}>F\alpha$, $F_{Ph}>F\alpha$,so the slide-to-roll ratio SR and Max. Hertzian pressures Ph have significant influence on the friction factor of gear, and the slide-to-roll ratio SR has the biggest influence on the friction factor.Similarly, $F_R<F\alpha$, $F_{Ve}<F\alpha$, $F_S<F\alpha$, so the radius of curvature R, entraining velocity Ve and surface roughness S have no significant influence on the friction factor of gear, and entraining velocity Ve has the least influence.

4. Conclusions

Through the test and analysis of the formula, it can be seen that the sslide-to-roll ratio SR and the Max. Hertzian pressures Ph have a great influence on the friction factor of gear, and the radius of curvature R, entraining velocity Ve and the surface roughness S have little influence.According to the practical experience, Max. Hertzian pressures Ph has a great influence on the friction factors.However, surface roughness S has little effect on the friction factor, but it does not accord with it.In addition, the entraining velocity Ve has the least influence on the friction factor, and the slide-to-roll ratio SR has

the greatest influence on the friction factor. But the slide-to-roll ratio SR= V_S/Ve ratio shows that the sliding speed V_S has a great influence on the friction factor.
To sum up, this formula is basically in line with the law of gear friction factor, but there are some deficiencies, which need further correction and supplement.In this paper, the principle of mathematical statistics is used to design the orthogonal experiment, and the formula is verified.This process is a qualitative analysis of each parameter.It is helpful to understand the influencing factors and grasp the research direction.It provides new ideas for future experimental research.

[1] Masatoshi Yoshizaki,ChotaroNaruse,RyozoNemoto,et al 1991 *Study on frictional loss of spur gears(concerning the influence oftoothform,load,tooth surface roughness,and lubricating oil)* Tribology Transactions Vol.34 (1) pp 138-146
[2] Y. Michlin, V. Myunster 2002 *Determination of power losses in gear transmissions with rolling and sliding friction incorporated* Mechanism and Machine Theory Vol.37 (2) pp 167-174
[3] Hamrock B J,Dowson D 1979 *Elastohydrodynamic lubrication of elliptical contacts for materials of low elastic modulus*,2-starved conjunction Journal of Lubrication Technology
[4] Xu H,Kahraman A,Anderson N E ,2007. *Prediction of mechanical efficiency of parallel-axis gear pairs*, Transaction of the Asme Journal of Mechanical Design, 129(1), 58-68.
[5] YYLi et al 2008 *Experiment Design And Data Processing* (Beijing :Chemical Industry Press) pp124-150

Acknowledgments
The authors wish to acknowledge the financial support of National Natural Science Foundation of China (Grant No. 51475210), A Project of Shandong Province Higher Educational Science and Technology Program (Grant No. J17KA027) and major research project of Shandong province (Grant No. 2018GGX103035) during the course of this investigation.

Appendix

Complete orthogonal table

Text NO	R	Ve	SR	s	Ph	μ
1	0.01	1.00	0.01	0.05	0.50	0.0006
2	0.01	5.00	0.05	0.10	1.00	0.0052
3	0.01	10.00	0.10	0.20	1.50	0.0121
4	0.01	15.00	0.15	0.30	2.00	0.0172
5	0.01	20.00	0.20	0.40	2.50	0.0195
6	0.02	1.00	0.05	0.20	2.00	0.0088
7	0.02	5.00	0.10	0.30	2.50	0.0137
8	0.02	10.00	0.15	0.40	0.50	0.0067
9	0.02	15.00	0.20	0.05	1.00	0.0094
10	0.02	20.00	0.01	0.10	1.50	0.0011
11	0.03	1.00	0.10	0.40	1.00	0.0089
12	0.03	5.00	0.15	0.05	1.50	0.0094
13	0.03	10.00	0.20	0.10	2.00	0.0106
14	0.03	15.00	0.01	0.20	2.50	0.0017
15	0.03	20.00	0.05	0.30	0.50	0.0018
16	0.04	1.00	0.15	0.10	2.50	0.0118

IOP Conf. Series: Materials Science and Engineering **542** (2019) 012044 doi:10.1088/1757-899X/542/1/012044

17	0.04	5.00	0.20	0.20	0.50	0.0058
18	0.04	10.00	0.01	0.30	1.00	0.0007
19	0.04	15.00	0.05	0.40	1.50	0.0049
20	0.04	20.00	0.10	0.05	2.00	0.0069
21	0.08	1.00	0.20	0.30	1.50	0.0100
22	0.08	5.00	0.15	0.40	2.00	0.0093
23	0.08	10.00	0.10	0.05	2.50	0.0062
24	0.08	15.00	0.05	0.10	0.50	0.0011
25	0.08	20.00	0.01	0.20	1.00	0.0005
K1	0.054606	0.040223	0.004612	0.032484	0.016061	
K2	0.039714	0.043447	0.021807	0.029857	0.024697	
K3	0.032482	0.036307	0.047862	0.028842	0.037468	
K4	0.030071	0.034220	0.054392	0.043491	0.052797	
K5	0.027115	0.029792	0.055315	0.049313	0.052966	
Range	0.027491	0.013655	0.050703	0.020471	0.036905	
Principal→Subordinate SR> Ph>R>S> Ve						

IOP Conf. Series: Materials Science and Engineering **542** (2019) 012045 doi:10.1088/1757-899X/542/1/012045

Research on gear box with adjustable center distance

H Sun[1] C Wang[2] and Y M Ma[3]

1 School of Mechanical Engineering, University of Jinan, Jinan 250022, China
2 School of Mechanical Engineering, University of Jinan, Jinan 250022, China
3 School of Mechanical Engineering, University of Jinan, Jinan 250022, China
E-mail: C Wang , me_wangc@ujn.edu.cn

Abstract. The invention discloses a test gear with adjustable center distance gear. The invention comprises gearbox housing, a sliding bearing seat and a fixing device. The sliding bearing seat comprises a set of T guide rail and two groups of moving sliders. A bearing seat is arranged on the movable slide block to complete the changing process of the gear center distance. The movable slider structure has a sliding groove which opens two fastening bolt holes on the right. When the bearing seat moves to the desired position, the slider installed with the bearing seat is fastened to the gear box shell by fastening bolts, and the change of the center distance of the gear box is completed. The invention has the advantages of simple structure, convenient operation, and structural design does not affect the oil mixing and lubrication of the gear when it is working, and meets the needs of different gear center distance experimental devices.

Keywords. Sliding bearing block; Gear box; Center distance variable

1. Introduction

The gear has the advantages of high transmission ratio and high stability, so gear mechanism is the most widely used transmission device. However, in the actual production and daily life, the effective life of gear mechanism and its transmission efficiency have great influence on our work. Therefore, we usually analyze the service life and transmission efficiency of gears. So it is necessary to design the test-bed for the gears under test, which requires the gears that mesh with each other to be installed in the gearbox [1-3].

At present, DCS-150 gear test-bed, CL-100 gear test-bed, JG-150 gear test-bed and ZHD82-A closed gear test-bed can't change the center distance at will. But in practice, the test stand we built has been reused. Therefore, it is impossible to choose different gear boxes for the tested gear in order to meet the installation requirements of different center distance gears in the design of test bed. In order to satisfy the requirement of the test bench for different center distance gears, we designed a test box with adjustable center distance gears. The invention has the advantages of simple structure and

IOP Conf. Series: Materials Science and Engineering **542** (2019) 012045 doi:10.1088/1757-899X/542/1/012045

convenient operation, and does not affect the oil stirring lubrication of the gear box in the working process.

2. Summary of the invention

Cylindrical gears include spur gears, helical cylindrical gears, arc tooth cylindrical gears, etc. In actual production, gear is the main component of mechanical transmission, its failure will directly affect the normal operation of the machine. In order to make the gear mechanism work reliably, it is necessary to measure the bending strength, contact strength and check bearing capacity of the gear teeth, so as to analyze the mechanism of gear failure and provide a basis for the design of gear transmission. Such a complex process can only be carried out on the gear test-bed. Based on this, we design and invent a new method "A gear box with adjustable center distance".

The purpose of the invention is to provide an experimental gear box with distance. It meets the installation requirements of different center distance gears in the design of the test bed. Its structure is simple, easy to operate, reduces capital investment and cost investment, easy to manage, and saves human resources. It hasn't affected the gear oil lubrication when gearbox is working [4].

As shown in table 1, the invention mainly comprises a gear box shell (1), a T-type guide rail (2), a moving slider (3), a bearing seat (4), ~~a~~ mounting bolts (5, 7), a fastening bolt (6) and a positioning block (8). (Where are numbers of position in the Table?)

Table 1. Center distance adjustable gearbox parts

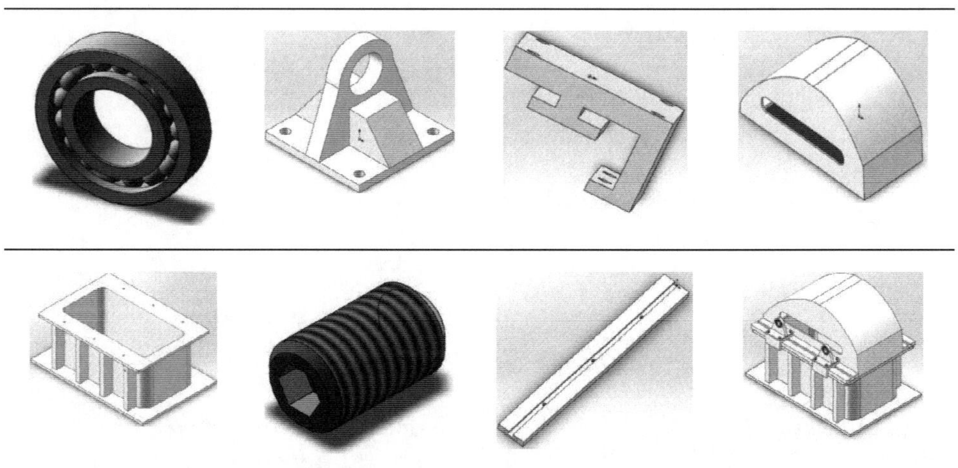

The T-type guide rail is mounted on the gearbox housing through mounting bolts. Moving slider is combined with the T-type guide rail. The bearing block is installed on the movable slide block through mounting bolts. Sliding chute is designed with chutes [5]. The chute can hook the gearbox housing. Fastening bolts secure the moving slider to the gearbox housing when moving the slider to a proper position. When moving the slider, it needs to ensure its parallelism. In order not to interfere with the lubrication of the gears in the gearbox case, a positioning block is used to ensure the parallelism of the moving slider.

3. Specific operation method

In the design of the gearbox cover. we used two elliptical grooves on the side wall of the gearbox cover. The specific size is based on the size of the required gearbox. The elliptical groove can facilitate the shifting of gear shaft while changing different center distance gears. There is no need to change the corresponding gear box cover. The operation process is simpler and more convenient.

As shown in Figure 1, in the replacement of gears with different center distance, the specific operation is as follows: First of all, we need to loose the fixed bolts of the fixed slide rail to change the position of movable bearing block. At the same time changing the movable bearing seat, we use external location blocks to locate the moving bearing block. The objective is to ensure the parallelism of the movable bearing seat. We need to position the block together with two movable bearing bases corresponding to the same gear shaft, and push the positioning block to carry out the position transformation of the movable bearing seat. After the position transformation is completed, the fixed bolts are tightened to complete the replacement of the gears with different center distance.

In the movable gear box, because the lower box and the upper box are fixed, the lower box plays a supporting role, the upper box plays a role in preventing oil spill and dust fall. However, the main component of the movable gear box is the movable bearing seat. Its strength and hardness are directly related to the service life of the movable gear box and the accuracy of the experimental data.

In order to test its strength, we have carried out a static analysis of the bearing seat, and its family. F fixed force, deformation during observation period. First, as shown in Table 2, we need to select our material and test its properties, where we choose annealed stainless steel. Its material properties include elastic model, Poisson's ratio, mass density, tensile strength, yield strength, coefficient of thermal expansion, thermal conductivity, specific heat, etc.

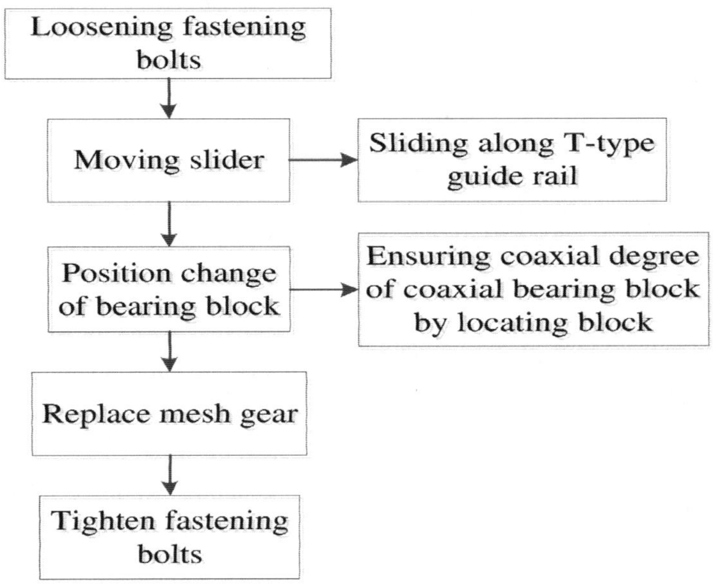

Figure1. Operation process of mesh gear replacement

Table 2. Material properties

IOP Conf. Series: Materials Science and Engineering **542** (2019) 012045 doi:10.1088/1757-899X/542/1/012045

Attribute	**Numerical value**	**Dimension**
Modulus of elasticity	2.07e+011	N/m^2
Poisson ratio	0.27	
Mass density	7860	kg/m^3
Tension strength	685000000	N/m^2
Yield strength	292000000	N/m^2
Coefficient of thermal expansion	1.7e-005	1/K
Thermal conductivity	16.3	W/(m • K)
Specific heat	502	J/(kg • K)

By fixing the bottom end of the bearing seat and exerting an axial force, the strength and deformation of the bearing seat are analyzed. As shown in Figure 2, the displacement and strain of the bearing pedestal are observed after applying an axial force of 100N.m. The biggest change was found at the top of the bearing pedestal and its displacement was 1.014e-005mm. The largest strain is the middle part of the bearing block, with a maximum of 8.463e-008.

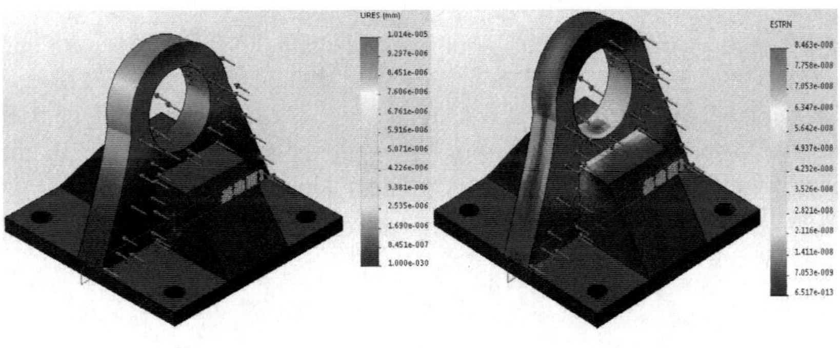

Figure2. Force analysis of bearing block

Because the slider plays a very important role in the adjustable center distance gearbox, it is mainly used to fix the sliding bearing seat on the gearbox after changing the position, so as to ensure that the sliding of the bearing seat will not occur in the operation of the gearbox, which will lead to gear meshing errors, and then lead to the failure of gear testing. Because of its importance, the material of slider is alloy steel. Its hardness, modulus of elasticity, coefficient of thermal expansion and yield strength all meet our needs. Through the model analysis of the alloy steel material in slide block in SolidWorks, the minimum safety factor is obtained.

4.Conclusion

Gear box is the most common device in gear experiment. Different gears are tested according to different center distance. But most of the gearboxes are one-to-one, that is, a set of gears correspond to a set of gearboxes. This not only increases the test cost, but also reduces the test efficiency and increases the actual size.

The error is tested. Through the improvement of the gearbox, the variable center distance gear box we invented mainly includes four parts: the lower box, the upper box, the movable bearing seat and the fixing device of the bearing seat. By moving the bearing seat mounted on the lower box, the center distance can be changed, the test cost can be reduced and the test efficiency can be improved [6-7].

References

[1] Shijie Zhou. A gearbox with adjustable gear center distance [P]. Chongqing:CN207278784U, 2018-04-27.

[2] Haoliang LvPenghui Wu, etc. High-power cylindrical gear test rig with stepless adjustable center distance[P]. Zhejiang: CN10652541A, 2017-03-22.

[3] Chunxiao Ma, Jun He. Gearbox and agricultural harvesting machinery [P]. Zhejiang: CN205047790U, 2016-02-24.

[4] Wenyu Zhao, Xiaochao Zhang. Transmission device with adjustable center distance between input and output shafts [P]. Jiangsu: CN103398162A, 2013-11-20.

[5] Xiaofeng Liu. Gear reducer processing method [P]. Jilin: CN101972920A, 2011-02-16.

[6] Yongjie Han. Gear lubrication test bed [D]. Harbin Engineering University, 2009.

[7] Ruian Zhu. Adjustable center gear box [J]. Mechanical manufacturing, 1990 (12): 11.

Acknowledgement

The authors wish to acknowledge the financial support of National Natural Science Foundation of Chi na (Grant No. 51475210), A Project of Shandong Province Higher Educational Science and Technolog y Program (Grant No. J17KA027) and major research project of Shandong province (Grant No. 2018G GX103035) during the course of this investigation. The authors would also like to thank the editor and anonymous reviewers for their suggestions for improving the paper.

IOP Conf. Series: Materials Science and Engineering **542** (2019) 012046 doi:10.1088/1757-899X/542/1/012046

Literature review of RV reducer test bed

Pan Shilin[1], Wang Cheng[2]
1 School of Mechanical Engineering, University of Jinan, Jinan 250022, China
2 School of Mechanical Engineering, University of Jinan, Jinan 250022, China
e-mail: m17569610369@163.com

Abstract. The RV reducer test bench belongs to the mechanical transmission test bench. It is a test equipment for testing the comprehensive performance of the reducer. Various tests are carried out on the test bench to achieve the purpose of rationality of design and manufacture and performance evaluation of the tested RV reducer. With the further development of national industrialization and the continuous improvement of automation, the performance of the RV reducer has been put forward higher requirements, but also the accuracy of the test bench has been put forward higher requirements, the research of the RV reducer test bench has more practical significance.

1 Abroad Research Status

The research on mechanical transmission test-bed began earlier abroad. Loading system, driving device, control system and layout of the test-bed have been studied extensively by many companies and research institutes,The more famous research units are the Lewis Research Center of NASA, the Central Institute of Machinery Manufacturing and Design of the former Soviet Union, the German SCHENCK Company, the Japanese Electric Power Company, the American General Dynamics Company, the French Skoda Company and so on[2-4].Gleason Corporation in the United States in the 1950s, the former Soviet Union in the 1960s on the design of a loading system through the wheel train as a test bed. The control system of the test-bed designed by Toyota in the 1970s in Japan adopts a microcomputer, which improves the reliability and universality of the whole test device. Subsequently, in the 1980s, Toyota improved the original test-bed system, using four microcomputers to control the test-bed, the entire control system includes a correction system and monitoring devices. The control system of the test-bed can not only be a terminal of the main computer, but also be an independent system. It has the advantages of decentralized functions and convenient adjustment. With the development of computer technology, the company began to use the main computer, minicomputer, test-bed control microcomputer three-level computer control system, used to engine laboratory, transmission system laboratory for centralized control and data processing. The Department of Mechanical Engineering of the University of Illinois in the United States designed and manufactured a comprehensive gear test rig with compound planetary transmission. The test-bed has many functions and can be used to test the dynamics of gear transmission, the efficiency of planetary gear train and gear lubrication[1].

In the field of researching the transmission precision and performance of RV reducer, the most representative research group abroad is J.G. Blanche of Hughes Aircraft Company of America and the sunshine of Japanese academia. Their in-depth research and exploration in related fields have obviously surpassed other experts of the same trade.

In the field of torsional stiffness research, some famous research groups are as follows: Park.J.S.Stiffness and others focus[5] on the topological structure of the reducer, but also establish the relevant performance model; Kim Kyoung-Hong, Lee Chun-Se and others[6] in South Korea use the finite element method to simulate the contact of cycloidal wheels with springs, and make the torsional stiffness and actual.

2 Domestic research status

RV reducer prototype has been successfully developed, and its main technical performance indicators have been tested static and dynamic with the help of enterprise testing equipment. In recent years, experts and scholars from Dalian Jiaotong University, Harbin University of Technology, Chongqing University and other well-known colleges and universities have carried out more in-depth theoretical research on the transmission accuracy and dynamic characteristics of RV reducer, and conducted some appropriate experimental exploration, but they are not perfect and systematic. In 2013, Shi Zhaoyao of Beijing University of Technology developed a multi-parameter comprehensive test and analysis system for transmission characteristics of RV reducer, and put it into practice[7].

At present, there are several well-known research groups: Zhang Dawei[8] and Zhang Yinghui[9] from different perspectives, the former reasonable use of relevant basic knowledge to create a torsional characteristic analysis model, the other group fully consider the curvature and other factors, they simplified the torsional stiffness of the previous scholars on the basis of their work. It also avoids some abnormal situations and makes the overall analysis model more precise. But they are all studied and analyzed in a linear way, and some references at home and abroad show that the stiffness of the gear will change nonlinearly with the load[10].

Because the RV reducer in the use of different environments, different forces in the transmission will change the torsional stiffness of the gear, so it is necessary to further study the specific factors on the torsional stiffness of the product how to affect[11].

In 1965, Cui Xilie's factory built an asynchronous generator loading test bench based on the principle of motor feedback braking. The winding motor was used as the asynchronous generator to load the test bench, which can realize automatic load change. In March 1976, a joint test group consisting of Lanzhou Institute of Petroleum Machinery and Lanzhou General Machinery Factory was used to test the double circular arc gear reducers on two pumping units on a closed test bench. In 1984, Wang Naixin discussed the working principle of the electric closed torque loading system, and divided the loading methods of the test bench into three categories: mechanical loading method, hydraulic closed loading method and electric loading method[12].In 1995, Liu Ruifeng and Shi Xiangdong designed a new type of mechanical transmission test bench to test the efficiency of the reducer. The test-bed adopts DC closed-loop loading technology, two DC motors of the same type are connected in parallel with the same power supply, and the two motor shafts are connected by the mechanical transmission (such as gear transmission) under test. The load motor is in the state of power generation by using the speed difference between the two motors, and the resistance torque is produced[13].In 1998, Yan Qingdong of Beijing University of Science and Technology designed a gearbox gear life test bench, which was loaded by a self-developed electrical mechanical loader[[14].The loader consists of a loading motor, a harmonic reducer and a planetary mechanism. When loading, the shaft of the harmonic reducer drives the sun wheel of the planetary device in the transmission box to rotate and load.In the same year, Xu Yinqin analyzed the layout of the reducer test bench under different conditions. After comparing the different layout methods, it was found that the test cost of the same type of reducer was very different with different layout methods[15].For example, on the basis of the basic layout of the test bench, adding a reducer which can be operated in reverse as an accelerator can reduce the output torque by increasing the output speed, which can solve the problem of excessive loading device in the high-power test bench.In 2000, aiming at the shortcoming of single function of the existing mechanical transmission test-bed, the building block type test-bed structure should be adopted by Fuqiang et al. Through different assemblies of test devices, various tests for different types of transmission devices were carried out[16].In recent years, with the rapid development of computer technology and communication technology, the control system of the test-bed is also developing towards intelligent direction. In 2003, in order to improve the performance of the test-bed, Zhang Ling designed a gear reducer efficiency test-bed based on computer control, using computer control technology to replace the traditional secondary instrument test technology.Practice has proved that this kind of computer control method is feasible[17].In 2010, Han Min et al. carried out the efficiency test of harmonic gear reducer by using Labview software and advanced virtual instrument technology on the basis of the original reducer test-bed technology. The high-precision and high-efficiency test was realized at low cost, which provided a new method for the efficiency test of the reducer[18].

In recent years, many universities in China have carried out research on mechanical transmission test rig.For example, Chongqing University, Hefei University of Technology[19], Xi'an University of Technology[20], Zhejiang University[21], Harbin University of Engineering[22] and so on, these schools have

designed various types of test-bed, and have done a lot of research on the mechanical structure and loading mode of the test-bed, accumulated rich practical experience and made great progress.
According to the existing literature, the test and research of RV reducer are mostly carried out for a single type of product, and no test system has been developed for the test and research of different types of RV products in the same series; there is also a lack of uniform test standards in the test operation and implementation; there is also a lack of precision transmission of RV reducer. Research on important characteristics such as degree and torsion rigidity still lacks theoretical support.

3 Main test direction of RV reducer test bench

3.1. Torsion stiffness test

The research focuses on Meshing Stiffness Analysis and modeling, considering not only the elasticity of the main components, but also the meshing stiffness of different contact positions between cycloid wheel and pin teeth, and the deformation coordination under multi-tooth contact[23-24].It is worth mentioning that the RV reducer is also a kind of two-stage closed planetary transmission system including cycloidal transmission. Its low-speed stage is cycloidal transmission, but the transmission structure is quite different from the single-stage cycloid, and the torsional stiffness characteristics are also different. Therefore, scholars at home and abroad have studied the stiffness characteristics of this kind of reducer.Based on Hertz theory, the meshing stiffness model of cycloid pin wheel is established in the literature[8]. The meshing stiffness calculation of involute gear is simplified by Ishikawa formula.Zhang Yinghui[9] further considered the influence of the curvature of the cycloid gear profile, avoiding the case that the meshing stiffness of single pair of teeth is negative, so that the meshing stiffness model is more refined.Park[25] set up a semi analytical model for the torsion stiffness of the reducer based on the topological characteristics of the reducer.Kim Kyoung-hong[26] considers the elasticity of the bearing at the joint of the cycloid wheel and the crankshaft in the reducer, and the bearing is equivalent to a flexible ring with a certain stiffness. The results show that the stiffness of the bearing is the main factor affecting the torsional stiffness of the whole machine[27]. The study of the above literature makes the torsional stiffness model of the whole machine gradually refined, but the bearing stiffness is treated in a linear manner. The related literature points out that the bearing stiffness will change nonlinearly with the load. In RV reducer, the change of bearing's stiffness will be caused by the change of bearing's force at different meshing positions. The influence of this change on the torsional stiffness of such reducer needs further study.

3.2 Rotary drive accuracy test

At present, the error modeling methods of cycloidal planetary transmission mechanism are more mature. The common methods are geometry method[28-29], gear tooth contact analysis method[30] and differential method[31].These methods are rigorous in theory and accurate in analysis results, which provide an effective reference for the precision research of RV reducer. However, these methods only consider the influence of the tooth profile error of cycloidal gear, do not take into account the error of other components except cycloidal gear, and the calculation is complicated. Therefore, it is more complicated to calculate the rotary transmission error of RV reducer with multi-crankshaft, double cycloid wheel and complex input-output mechanism, as well as the machining and assembly errors and clearances of high-speed stage and other components.Domestic scholars also discussed the influence of cycloid profile error on the output error of cycloid planetary mechanism. For example, in reference[32], starting from the manufacturing and assembly process of reducer, the analytical calculation method of the rotary transmission error of cycloid pin-wheel transmission mechanism was studied by applying the basic principle of gear meshing.Reference[33] assumes that the meshing point and the normal direction of the cycloid wheel and the pin tooth do not change when the error exists. On this premise, the influence of the cycloid tooth profile error on the output error of the mechanism is deduced by geometric method. Japanese scholars, such as Rigao Shining, used the lumped mass method in dynamic modeling for reference, set up the mass-spring equivalent model of RV reducer by setting the components of RV reducer as rigid body and replacing the clearance between components with springs, and then set up the processing, assembly error, clearance and relative position of each component to its theoretical position. The small displacement caused by the deviation is equivalent to the equivalent error in the direction of each spring. Finally, the error analysis model of the whole rotary transmission of the reducer is established based on the principle of static balance. The model considers the structural characteristics of three cranks and double cycloidal wheels in the RV reducer, and the machining and assembling errors of the high-speed stage and other components. It is an effective method for precision analysis of this kind of reducer.Reference [34-35] combined with virtual prototyping technology, the influence of original error and dynamic angle error of RV reducer was discussed.

3.3 Mechanical transmission comprehensive performance test stand

With the development of mechanics and the continuous improvement of scientific research level, higher requirements have been put forward for the testing and testing methods of mechanical transmission products[37].A great deal of research has been done on the scheme design and performance improvement of the mechanical transmission performance test bench and related test system. In the aspect of test bench, literature[38] takes cycloidal pinwheel reducer as test object, develops a set of hardware system for performance test of reducer, tests the torque and speed of input and output of reducer, and improves the follow-up instrument circuit of previous sensors by adopting signal synchronization circuit, thus improving the precision of system test. Reference[39-40] designed an electric enclosed test bench for reducer. The input and output of the test bench are loaded by electric motors. The power loss is reduced by using AC frequency conversion loading electric enclosed method. The variation of input and output torque of reducer with motor speed is measured under loading conditions. Reference[41] also adopts the design of the electric closed test-bed. Taking the worm and worm gear reducer as the object, a number of performance indexes (efficiency, wear, temperature rise) are measured.

References

[1] Lu Xingzhong, General situation and development of gear transmission experimental platform[J].Chongqing University,1986.
[2] Huangfu Yuliang,Research on measurement and control system of reducer test bed[D].Taiyuan University of Technology, 2007.
[3] Qin Xiaofeng. Design of measurement and control system for mechanical transmission test-bed [D]. Chongqing University, 2007.
[4] Cheng Heming. Test rig for reducer of mixed gear train with large speed ratio [D], Wuhan University of Technology, 2011.
[5] Park J S. Stiffness and Error Analysis of the Cycloid Drive [D].Korea: College of Engineering, Seoul National University,2006.
[6] Kim Kyoung-Hong,Lee Chun-Se,Ahn Hyeong-Joon. Torsional rigidity of a cycloid drive considering finite bearing and hertz contact stiffness[C]// Proceedings of the ASME 2009 International Design Engineering Technical Conferences and Computers and Information in Engineering Conference.San Diego, California, USA,2009, 1-6.
[7] Wang Xiaoling, Shi Zhaoyao.Development of the measuring instrument for transmission error of RV reducer.Proceedings of the National Symposium on Small Modulus Gear Technology [C] Beijing University of Technology, 2013, 26-27.
[8] Zhang Dawei, Wang Gang, Huang Tian,.RV reducer dynamics modeling and analysis of structural parameters of [[J]. mechanical engineering, 2001, 37 (1): 69-74.
[9] Zhang Yinghui, Xiao Junjun, He Weidong. Calculation of meshing stiffness of needle-pendulum drive with RV reducer for robot [[J].Journal of Dalian Jiaotong University, 2010, 31 (2): 20-23.
[10] Liu Jingya,Matsumura Shigeki,Chen Bingkui,et al. Torsional stiffness calculation of double-enveloping cycloid drive[J]. Journal of Advanced Mechanical Design,2012, 6(1): 2-14.
[11] Qin Xiaofeng. Design of measurement and control system for mechanical transmission test-bed [D]. Chongqing University, 2007.
[12] Wang Naixin.Principle and Test of Electric Closed Drive Test Bench[J].Journal of Northwest Institute of Light Industry,1984,1:22-39.
[13] Liu Ruifeng, Shi Xiangdong, Li Jinshan.An attempt to combine mechanical and electrical design of DF-430 mechanical transmission test bench[J].Journal of Jilin Institute of Technology, 1995,16(1): 69-73.
[14] Yan Qingdong, Xiang Changle, Feng Yongcun.Power closed mechanical transmission gear life test bench[J].Journal of Beijing University of Technology, 1998,18(3): 302-306.

[15] Xu Yin Qin. Structural analysis and design of retarder test-bed [[J]. Journal of Hebei Institute of electrical and mechanical engineering, 1998,15 (2): 18-22.
[16] Zhu Sener, Ying Fu Qiang, Lu Congda. Development of Building Block Mechanical Drive Comprehensive Test Bench [J].Journal of Zhejiang University of Technology, 2000, 28:258-261.
[17] Zhang Ling.A gear reducer efficiency test-bed based on computer control[J].Journal of Tianjin Vocational University of Higher Vocational Education,2003,12(2): 26-28.
[18] Han Min, Liu Kun, Wu Kaichun, Xu Lina. Research on Efficiency Testing System of Harmonic Gear Reducer [J].Machine Tool and Hydraulic, 2010,38(12): 90-92.
[19] Lin Juguang, Yang Hongbin, Lei Zhi, Huang Gongwei. Development of the main reducer assembly test bench [J]. Modular machine tool and automatic processing technology, 2007, (8): 88-90.
[20] Ma Zhaofeng. Study on simulation loading technology of electric closed mechanical transmission test stand [D]. Xi'an University of technology, 2003.
[21] Jin Weifeng. Fatigue test technology and test system software development of drive axle [D]. Zhejiang University, 2002.
[22] Wang Zhengbo.Study on the Efficiency Measurement Method of Closed Power Transmission Test Device and Reducer[D].Harbin Engineering University, 2010
[23] Liu Jingya, MATSUMURA Shigeki, CHEN Bingkui, et al. Torsional stiffness calculation of double-enveloping cycloid drive. Journal of Advanced Mechanical Design, 2012, 6(1): 2-14.
[24] Zhang Xiuyan, Dai Xiaojun. Meshing stiffness analysis of four ring-plate-type pin-cycloid gear planetary drive. Applied Mechanics and Materials. 2012, 229-231:499-502.
[1]Park, J. S. Stiffness and error analysis of the cycloid drive: [Ph.D. thesis]. Korea: Seoul National University, 2006.
[25] Kim Kyoung-hong, Lee Chun-se, Ahn Hyeong-Joon. Torsional rigidity of a cycloid drive considering mite bearing and hertz contact stiffness. Proceeding of the ASME 2009 International Design Engineering Technical Conferences&Computers and Information in Engineering Conference, 2009: 1-6.
[26]Guo Yi, Parker Robert G Stiffness matrix calculation of rolling element bearings using a finite element/contact mechanics model. Mechanism and Machine Theory, 2012(51): 32-45.
[27]J.G Blanche, D.C.H. Yang.cloid drives with machining tolerances. ASME J. Mech. Des.1989, 111:337-344.
[28]D.C.H. Yang, J.G Blanche. Design and application guidelines for cycloid drives with machining tolerances. Mech. Mach. Theory, 1990, 25(5): 487-501
[29]Alberto Demenego, Daniele Vecchiato, et al. Design and simulation of meshing of a cycloidal pump. Mech. Mach. Theory, 2002, 37: 311-332.
[30] Lozica Ivanovic, Goran Devedzic, et al. Modeling of the meshing of trochoidal profiles with clearances. ASME J. Mech. Des., 2012, 134: 1-9.
[31]Yao Wenxi. Transmission error analysis of cycloidal pinwheel planetary reducer. Mechanical transmission, 1998, 22 (4): 14-18.
[32]Li Lixing, He Weidong. Research on High Precision RV Transmission for Robots. Journal of Dalian Railway Institute, 1999, 20 (2): 1-11.
[33]Zhu Bin, Qin Wei, Sun Wei, et al. Dynamic rotary transmission error analysis of 2 K-V cycloid pin wheel reducer. Mechanical transmission, 2011, 35 (4): 12-15.
[34]Zhu Bin.Dynamic Transmission Accuracy Simulation Analysis of 2K-V Cycloidal Pinwheel Reducer Based on Multibody System Dynamics: "Master's Degree Dissertation". Chongqing: Chongqing University, 2011.
[35]Cui Yahui, Xue Longquan, both Xiang Ru. Loading characteristic analysis of open electric power flow type mechanical transmission test bench. Mechanical transmission, 1995, 19 (2): 33-35.
[36]Wen Hu, Xiang Zhongfan, Yao Hongmei a new type of multi-functional mechanical transmission test bench design. Mechanical transmission, 2004, 28 (2): 58-60.
[37]Hu Qingyu.Study on the Performance Testing System of Cycloidal Pin Wheel Reducer: [Master's Dissertation].Changchun: Jilin University, 2007.
[38]Lu Guiyou, Zhang Huijuan, Ding Bo. Design of Electric Loading Mechanical Transmission Experimental Bench. Journal of Natural Science, Jilin University of Technology, 1999, 29 (3): loo-1021.
[39]Dai Junkang.Research on Comprehensive Performance Test-bed of Automobile Main Reducer Assembly: [Master's Degree Dissertation].Hefei: Hefei University of Technology, 2008.
[40]Xu Hongping, Ying Fu Qiang, Song Lingling. Design and Research of Multifunctional Test Bench for Mechanical Transmission System. Mechanical and Electrical Engineering, 2002,19(3): 8-10.

Structural design and analysis of a new axial magnetic suction planetary gear train

Wentao Yu[1], Cheng Wang[2] and Zhiwen Lin[3]

1 School of Mechanical Engineering, University of Jinan, Jinan 250022, China
2 School of Mechanical Engineering, University of Jinan, Jinan 250022, China
3 School of Mechanical Engineering, University of Jinan, Jinan 250022, China
E-mail: me_wangc@ujn.edu.cn

Abstract. By analyzing the working principle of magnetic gear in the axial magnetic field, the design method of the active wheel and the driven wheel structure of the magnetic gear reducer in axial magnetic field is proposed. The influence of the magnetic tooth thickness on the transmission performance for the axial magnetic field coupled magnetic gear is theoretically explained. The gear reducer permanent magnet adopts a block structure, and the tooth increase coefficient is obtained by establishing a mathematical model. The finite element simulation analysis method is used to verify the rationality and feasibility of the design method, and the data is analyzed by the experimental method to verify the correctness of the simulation method

1. Introduction

Magnetic gears of axial permanent magnet magnetic gears realize a pair of variable speed transmission with multiple transmission ratios by axial coupling of magnetic field. It has the advantages of automatic centering, no backlash, high torque density, definite peak torque, fast dynamic response, and overload protection when axial coupling occurs during overload. It was the earliest successful application in magnetic pump to realize the non-leakage transportation of toxic or corrosive liquid or gas. With the emergence of rare earth permanent magnet materials and the development of magnetic drive technology, various types of permanent magnet coupler drive mechanisms such as permanent magnet clutch, magnetic coupling, and permanent magnet motor have been widely used in petroleum, chemical industry, medicine and textiles, which have special transmission requirements.

Japanese scholar Ikuta put forward the earliest new magnetic gear transmission technology, and initiated the research direction of using rare earth permanent magnetic materials in the field of mechanical transmission. It has some disadvantages, such as small magnetic pole coupling area, few magnetic poles and large variation of air gap curvature, and can't transmit large torque and the movement is transferred unstably, so the overload protection is difficult to realize [1-3]. Professor D. Howe of England proposed a new radial topology based on magnetic field modulation principle, which mainly includes high-speed rotor, adjusting and adjusting. The magnetic ring and the three parts of the low speed rotor increase the maximum torque and torque density. At the end of the 20th century, Hefei University of Technology proposed several torque algorithms for magnetic coupling devices and magnetic gears. At the beginning of this century, the Institute of Electrical Engineering, Chinese Academy of Sciences, proposed parallel-axis magnetic gear mechanism and applied to artificial heart valve [4,5].

How to reduce the eddy current loss of magnetic field in magnetic tooth coupling, reduce magnetic leakage and axial friction, and achieve greater transmission torque is one of the important factors affecting the industrialization. In this paper based on the working principle of the axial coupling magnetic field permanent magnetic gear, the influence factors of the structure design of the magnetic suction gear on the transmission torque are analyzed in different aspects, the design technology of the magnetic suction gear are expounded.

2. Principle analyses

2.1. The radial magnetic gear design

As the Figure 1 shows the outer circumference of the disc set with magnets or magnetic substances that are on the surface of disc are magnetized to N and S magnetism. The pole face of the magnetic gear is small, furthermore, the gear can drive normally with the common principle that opposite magnetic poles attract each other, so, in this regard, the force of transmission is small. Figure 2 shows a process that installs the elastic rubber magnet or magnet piece on the gear along the involute tooth surface, the gear teeth are the same polarity, and it works with the principle that the same polarity resists each other [6, 7].

There are more leakage fluxes in the single approach of gear engagement. meanwhile, accompanied with a train of shortcomings, such as asymmetrical magnetic field intensity generated by gear tooth; non uniform bonding force of gears; etc. Therefore these design flaws lead to a couple of disadvantages such as lack of impetuses and instability during the gear engagement. Owing to the inferior impact resistance of the magnets, the gear is easily damaged in such a condition that gears contact with each other in the case of overloading by face of teeth.

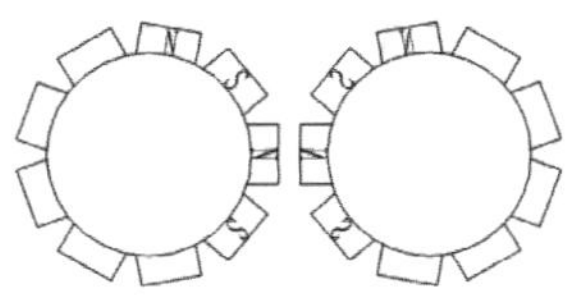

Figure 1. Magnetic gear

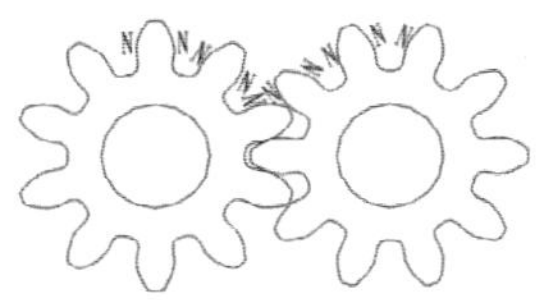

Figure 2. Magnetic gear

2.2. Axial magnetic suction gear design

2.2.1. Working principle. As shown in the Figure.3, the planetary gear and the center wheel are coupled by axial magnetic field, the tooth profile of the magnetic tooth block is outer the circle, and the magnetic pole of the magnetic tooth block with the axial coupling surface of the magnetic center wheel is N, S, N, S, N.... the magnetic center wheel and the magnetic planetary gear produce axial attraction force through the opposing magnetic tooth block with the different magnetic poles of the two gears, the deflection opposing magnetic tooth block with the same magnetic poles of the two gears produce axial mutual repulsion force. And the magnetic center wheel and the magnetic planetary gear self-aligning through the interaction force between the magnetic tooth blocks to generate an incomplete axial coupling region

Figure 3. Magnetic suction planetary gear train

The tooth profile of the magnetic tooth block of solar round is inside the circle, and the magnetic poles S, N, S, N, S.... of the magnetic tooth block on the axial coupling surface of the magnetic planetary gear and the magnetic solar round are on the outer side of the circle. The magnetic solar round and the magnetic planetary gear produce axial attraction force through the opposing magnetic tooth block with the different magnetic poles of the two gears, the deflection opposing magnetic block with the same magnetic poles of the two gears produce axial mutual repulsion force. And the solar round and the planetary gear self-aligning through the interaction force between the magnetic tooth blocks to generate an incomplete axial coupling region. According to the different magnetic poles of opposing magnetic tooth block in axial direction, the interaction force of the magnetic presents a change period of generate-increase-reduce-disappear.

The deflection opposing magnetic tooth block with the same magnetic poles of the two gears produce axial mutual repulsion force, and presents a change period of disappear-reduce-increase-generate. The interaction force of magnetic teeth block through the magnitude of attraction and mutual exclusion compensation each other, so as to achieve stable transmission torque.

2.2.2. Structural design. The magnetic center wheel and the magnetic planetary gear are connected in parallel, the magnetic solar round is connected in series with the magnetic center wheel and the magnetic planetary gear respectively.The magnetic center wheel and the magnetic planetary gear have the same size of the magnetic tooth blocks, the distance between the axis of the magnetic planetary gear and the axis of the magnetic center wheel same as magnetic planetary gear axis distance from solar round axis distance is L. Magnetic solar round radius is R, magnetic center gear radius is r_1, magnetic planetary gear radius is r_2, the height of magnetic tooth block is h, the position relationship as follows :

$$2h+r_1+r_2=R \quad (1)$$

$$L= h+r_1 \quad (2)$$

2.3 Axial suction change period of gear

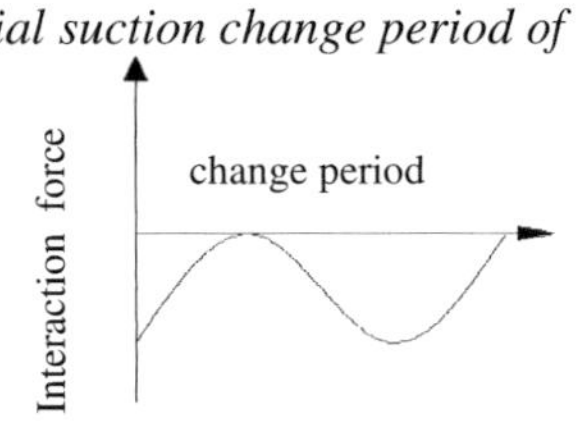

Figure 4. Suction change period of deflection opposing magnetic tooth block

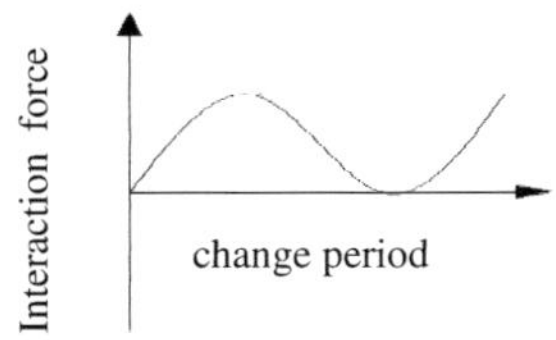

Figure 5. Suction change period of opposing magnetic tooth block

Figure.4 and Figure.5 show the function images relative to the interaction force and change period of opposing magnetic tooth block or deflection opposing magnetic tooth block. the derivative of the torque of axial magnetic suction gear drive is a even function with continuity, periodicity and fixed peak value, and The compensation of the mutual interaction force of magnetic teeth block by the attraction and repulsive force to compensate each other to achieve stable transmission torque, which satisfies the stability and continuity of the drive system.

Table 1. The rotation angle change regulation of gears

Rotation angle of magnetic solar round	Revolution angleof magnetic planetary gear	Autorotation angle of magnetic center wheel
0	0	0
-7	-2	7
-14	-4	14
-21	-6	21
-28	-8	28
-35	-10	35
-42	-12	42

The magnetic center wheel as the driving wheel, the magnetic planetary gear as the driving wheel or the driven wheel connects the magnetic solar round to rotate, and the rotation angles between the three of them in the incomplete coupling region as shown in the Table1. By the data analysis, the difference of the relative rotation angles between the magnetic center wheel and the magnetic planetary wheel with the magnetic solar round and the planetary wheel is equal, and the magnitude of the three magnetic gear blocks is equal. Thus, the area of incomplete coupling area is the same, which is identical to the stability of transmission.

3. Simulation

3.1. Relationship of transmission torque and tooth thickness

IOP Conf. Series: Materials Science and Engineering **542** (2019) 012047 doi:10.1088/1757-899X/542/1/012047

Axial magnetic field coupling drive is based on the principle of magnetism, so the internal magnetic field structure of permanent magnet is one of the important factors of magnetic suction gear drive. As shown in the Figure.6, the volume of the permanent magnet will change, and the transmission performance of the magnetic suction planetary gear train will change too.

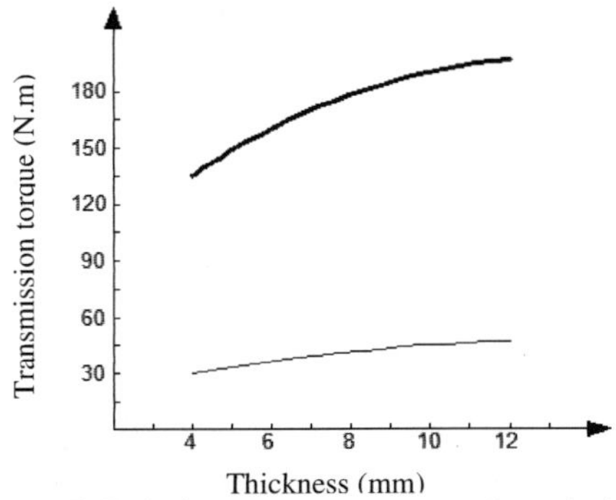

Figure 6. Relationship of torque and tooth thickness

When the magnetic tooth thickness changes from 4mm to 8mm, the transmission torque of the permanent magnet planetary gear increases approximately linearly with the magnetic tooth thickness, and when the magnetic tooth thickness changes from 8mm to 12mm, the increase rate of the transmission torque decreases.

3.2 Relationship of transmission torque and gear ratio in various pole pairs

As shown in the Figure.7-10, the pole pairs is the number ratio of magnetic planetary gear and magnetic central wheel of magnetic suction planetary gear train, the transmission ratio is magnetic tooth blocks number ratio of the magnetic center wheel and the magnetic solar round of the magnetic suction planetary gear train.

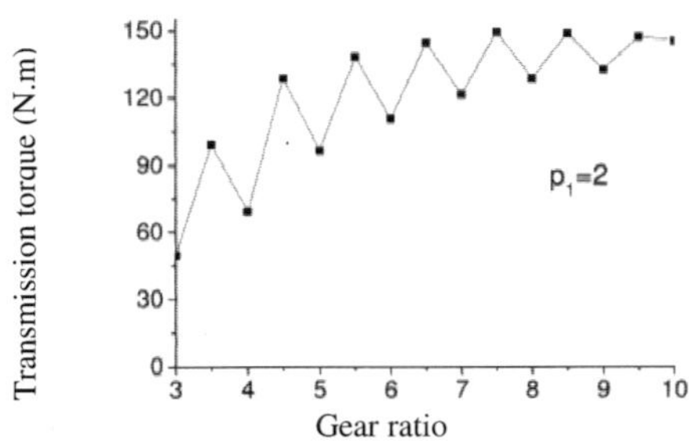

Figure 7. Relationship of transmission torque and gear ratio

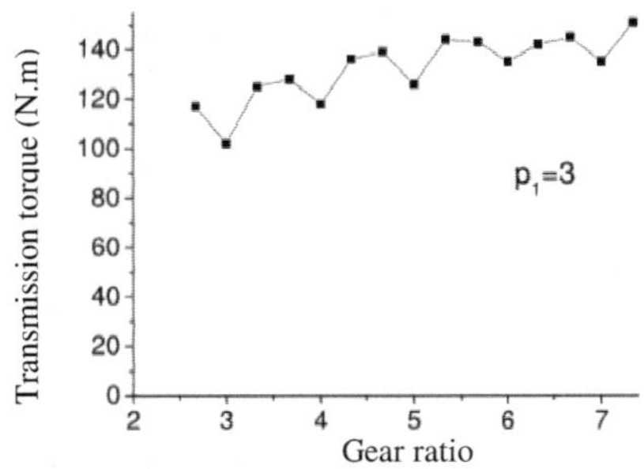

Figure 8. Relationship of transmission torque and gear ratio

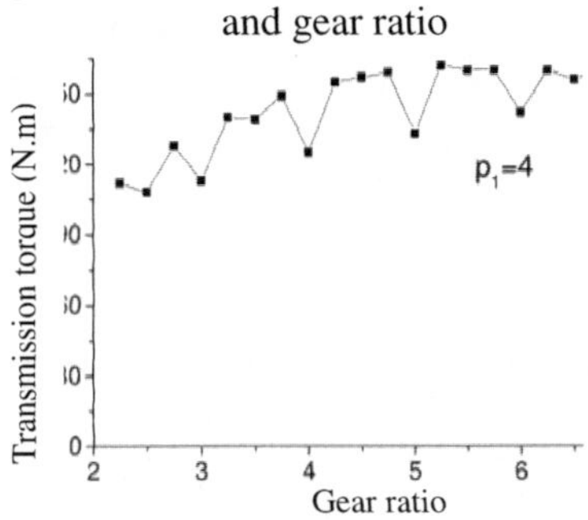

Figure 9. Relationship of transmission torque and gear ratio

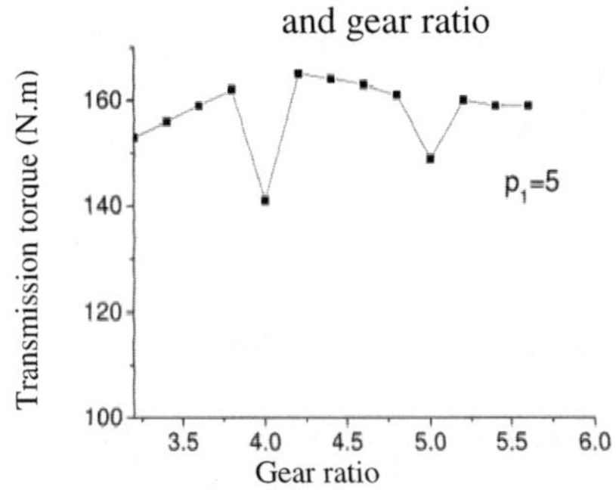

Figure 10. Relationship of transmission torque and gear ratio

When the transmission ratio of magnetic suction gear is integer, the transmission torque ripple is greater than fraction; and the gear ratio is integer, the maximum torque is less than the fraction.

For transmission torque curves in various pole pairs and on the occasion of loading, the more poles in magnetic planetary gear and the magnetic center wheel, the more periodic distributing the incomplete axial

IOP Conf. Series: Materials Science and Engineering **542** (2019) 012047 doi:10.1088/1757-899X/542/1/012047

magnetic coupling field is, while the less the ripple of transmission torque is. The effective match of pole pair number of axial magnetic suction gear and elaborate choice of circumferential width of magnetic tooth blocks can restrain ripple of transmission torque effectively.

3.3 Relationship of transmission torque and span

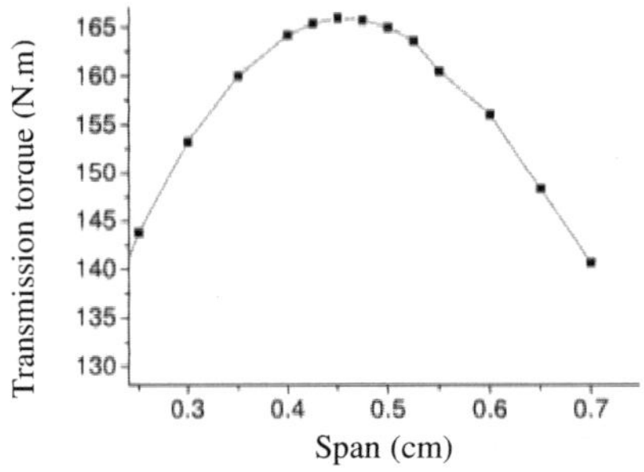

Figure 11. Relationship of transmission torque and span

As shown in the Figure 11, The span is acycle of radial distancce of two magnetic teeth blocks of magnetic suction gear , when the span change from infinitesimal to 0.45 cm, the torque density and transmission torque with the increase of span to increase, and the span change from 0.45cm to 0.7 cm the torque density and transmission torque with the increase of span to decrease. Therefore, under certain conditions of other factors, the span of two magnetic tooth blocks is 0.45cm, and the transmission torque reaches the optimal value.

4. Conclusions

Magnetic suction planetary gear train gear is based on the face to face of magnetic tooth blocks axial coupling to transfer torque. According to the simulation analyses, magnetic tooth thickness is 8mm of planetary gear and solar round, the span of two magnetic tooth blocks is 0.45cm, the torque density is the largest and transmission torque reaches the optimal value. When the transmission ratio of the magnetic center wheel and solar round is integer and pole pairs is 5,the the ripple curves of transmission torque reaches to the most stable, and the lagest value.

References

[1] Mei Shunqi Zhang Zhiming Xu Qiao 2011 *Research progresses of design methods for axial magnetic coulping drive mechanisms J. China Mechanical Engineering* .**22** 19 (2011).

[2] E Gouda and S Mezani 2011 *Comparative Study between Mechanical and Magnetic Planetary Gears J. IEEE Transactions on Magnetics* pp 439-450.

[3] Han Xiangji 2012 *Calculation of magnetic transmitting torque based on magnetic charge integration method J. Journal of Jilin Institute of Technology* **15** 4(2012).

[4] S Kikuchi and K Tsurumoto 1987 *A new magnetic gear using permanent magnet J. IEEE Trans. Magnetics,* pp 3622-3524.

[5] S Kikuchi and K Tsurumoto. 1994 *Trial construction of a new magnetic skew gear using permanent magnet J. IEEE Trans. Magnetics*, pp 4767-4769.

[6] K. Atallah, D Howe 2001 *A novel high-performance magnetic gear J. IEEE Transactions Magnetics* pp 2844-2846.

[7] P O Rasmussen and O Nielsen 2005 *Development of a high performance magnetic gear J. IEEE Transactions on Industry Applications* pp764-770.

Acknowledgement

The authors wish to acknowledge the financial support of National Natural Science Foundation of China (Gra nt No. 51475210), A Project of Shandong Province Higher Educational Science and Technology Program (Gr ant No. J17KA027) and major research project of Shandong province (Grant No. 2018GGX103035) during th e course of this investigation.

Research on Rolling Bearing Fault Identification Method Based on LSTM Neural Network

Pan Luo[2],Yumei Hu[1]

1. State Key Laboratory of Mechanical Transmission, Chongqing University, Chongqing, 400024;
2. School of Automotive Engineering, Chongqing University, Chongqing, 400024
E-mail: 1120593594@qq.com

Abstract: In order to simplify the fault detection process, improve the efficiency of fault detection and recognition accuracy, a rolling bearing fault recognition based on LSTM neural network is proposed. In this model, there is no need to perform any preprocessing on the original data. As long as the neural network model training is completed, the original signal can be detected and identified automatically by the model. In order to verify the performance of the model, the test results of the same fault data set are compared with the fault recognition model based on traditional machine learning. The results show that the fault recognition model based on LSTM neural network has obvious superior performance and higher recognition reliability. Its recognition accuracy rate reaches 98.00%, and the recognition accuracy of the fault recognition model based on traditional machine learning is only 94.20%.

Key words: fault recognition; LSTM recurrent neural network; feature learning; machine learning; vibration analysis

1.Introduction

Rolling bearings are the most widely used components in gearboxes, and their operating conditions directly affect the overall performance of the gearbox. Therefore, accurate and efficient detection of bearing faults has important practical significance. In recent decades, researchers have proposed many machine-based fault detection models. Among them, Fengqi Wu proposed a support vector machine trained by a small amount of raw data, and then directly diagnosed the fault [1], which verified the feasibility of the machine learning model in fault recognition. Since the recognition accuracy of machine learning fault recognition model depends heavily on input features, the recognition accuracy of this method is not high. In order to improve the accuracy of machine learning fault recognition model, many feature extraction methods have been proposed by researchers, such as Zaeri R extracted mathematical and statistical features of wavelet coefficients from raw data as input data of neural network model [2], Kateris D directly extracted the mathematical statistics of the original data as the input data of the neural network model [3]. Jianguo Wang proposed the Empirical Mode Decomposition (EMD) of the original data, and the energy of each EMD component as the input data of support vector machine classifier[4]. The above feature extraction methods can improve the accuracy of the fault recognition model through machine learning. However, it is necessary to carefully design the feature engineering, which requires a lot of time, labor costs, and requires the technician to have signal processing and engineering practical experience.

In recent years, deep learning technology has been able to implement end-to-end learning for raw data due to its powerful feature extraction ability, which has attracted widespread attention[5]. Because of the characteristics of in-depth learning, some researchers have applied it to mechanical fault recognition, such as L Jiang and Z Ge proposed a fault recognition model of self-coding machine. The model can automatically extract data features and improve the efficiency of fault recognition, but the accuracy of the model still has much room to further improve [6] . Xiaoxi Ding and Qingbo He proposed a fault detection model based on convolution neural network. Although the model has good recognition accuracy, it needs to construct a complex data preprocessing process of wavelet scale energy map first.

In order to overcome the defects of the model in fault recognition, improve the efficiency and accuracy of fault recognition, a rolling bearing fault recognition model based on LSTM neural network is presented in this paper. The model first constructs a training input sequence by intercepting the sample data of length 1024 through a one-dimensional time window which window length and step size are both of 32, and then training the model. After training, the input data sequence is constructed by intercepting the same length of signal to be detected in the same window, and finally the model performs fault identification and classification on the input sequence. Finally, four commonly used machine learning models for fault recognition are selected, including SVC, KNC, DTC and ANN. The differences of recognition accuracy and speed between the four models and the LSTM neural network model adopted in this paper are compared.

2 Machine Learning Model and LSTM Model

2.1Feature Engineering

Feature engineering is an engineering activity that extracts data features from raw data by certain methods for model or algorithm use. Vibration signals depend on the operation state of machinery and contain abundant information about the operation state of machinery, so they are often used for fault detection. Currently what are used for fault recognition are time-based mathematical statistical features such as: Kurtosis, Crestor Factor, RMS, etc. The feature calculation formulas used in this paper are as follows [3]:

$$RMS = \sqrt{\frac{1}{N}\sum_{i=1}^{N} x_i^{\,2}} \tag{1}$$

$$Kurtosis\ Indicator = \frac{\sum_{i=1}^{N}(x_i - \mu_x)^4}{N\sigma_x^{\,4}} \tag{2}$$

$$Crest\ Indicator = \frac{\max|x_i|}{\sqrt{\frac{1}{N}\sum_{i=1}^{N} x_i^{\,2}}} \tag{3}$$

$$Clearance\ Indicator = \frac{\max|x_i|}{\left(\frac{1}{N}\sum_{i=1}^{N}\sqrt{|x_i|}\right)^2} \tag{4}$$

$$Shape\ Indiactor = \frac{\sqrt{\frac{1}{N}\sum_{i=1}^{N} x_i^2}}{\frac{1}{N}\sum_{i=1}^{N} |x_i|} \tag{5}$$

$$Impulse\ Indicator = \frac{\max|x_i|}{\sqrt{\frac{1}{N}\sum_{i=1}^{N} |x_i|}} \tag{6}$$

In order to improve the performance of machine learning fault recognition model, it is usually necessary to design feature engineering. Due to the difficulty in interpreting the data features extracted from feature engineering, the design difficulty of feature engineering is increased to some extent.

2.2Machine learning

At present, machine learning technology is a common method in mechanical fault recognition. Usually we think that the occurrence of abnormal data indicates that the machinery has failed during the operation. In the process of fault detection, we need to extract data features from the original data obtained in the process of mechanical operation by feature engineering, and then use machine learning models such as support vector machine [1], clustering analysis [8], and neural network [2] to identify faults.

The main difference between machine learning fault recognition model and traditional methods is that the input data can be quickly detected and identified after the model training is completed by the feature extracted from the feature engineering mentioned in the previous section. Compared with traditional fault recognition methods, machine learning fault recognition model has greatly improved the recognition efficiency. However, the data preprocessing process of Feature Engineering limits the performance of machine learning model in fault recognition.

2.3Feature learning

Feature learning refers to the collection of related technologies such as feature conversion from raw data. It is significantly different from feature engineering. Feature engineering is to extract data features through a series of formulas, and then select features through the principles of divergence and correlation, and finally select efficient features for recognition and classification. Feature learning is to automatically learn the relevant features through a series of transformations of the original data by the model itself, and finally classify the data. In feature learning, the model will iterate over and over again to correct empirical errors to extract effective data features. Therefore, the features extracted from feature learning model can better express data. That means we can get the optimal features for recognition tasks.

2.4 LSTM model

In the traditional neural network model, since the nodes between the layers are independent of each other, the connection between the input data cannot be effectively established. However, sequence data is generally considered to be related to each other, which makes traditional neural networks unable to process sequence data efficiently. In the Recurrent Neural Network (RNN), the output of the current time of the neuron is affected by the output of the neurons at different times before, that is, the model stores and uses the input of the neuron at different times. Thus it can establish the relationship between the input data at different times. Its network neural unit structure is shown in Figure 1.

The RNN model contains input units, the input data is marked as $X = \{x_1, x_2, \ldots, x_t, \ldots, x_{32}\}$ (x_t representing the input of the time t neuron), and the output unit is marked as $Y = \{y_1, y_2, \ldots, y_t, \ldots, y_{32}\}$ (y_t representing the output of the time t neuron), and the update-output unit is marked as $C = \{c_1, c_2, \ldots, c_t, \ldots, c_{32}\}$ (c_t representing the update status of the time t neuron)). There are two information flows in RNN. One is from the input unit to the hidden unit, the other is from the hidden unit

IOP Conf. Series: Materials Science and Engineering **542** (2019) 012048 doi:10.1088/1757-899X/542/1/012048

to the output unit. There is only one simple activation function in the RNN neuron unit. If the information flow is too long, the RNN will have the disadvantageous effect of gradient disappearance or gradient explosion when it is trained by Back propagation Through Time (BPTT)[9]. To overcome the shortcoming of RNN neural network, LSTM neural network is proposed by researchers. LSTM neural network solves this problem by using four threshold structures instead of simple state update units widely used in RNN[10]. Due to the existence of these four threshold structures, the LSTM neural network's memory storage capacity is greatly improved. The LSTM neuron structure is shown in Figure 2, the $H = \{h_1, h_2, \ldots, h_t, \ldots h_T\}$ (h_t representing the hidden output of T neurons).

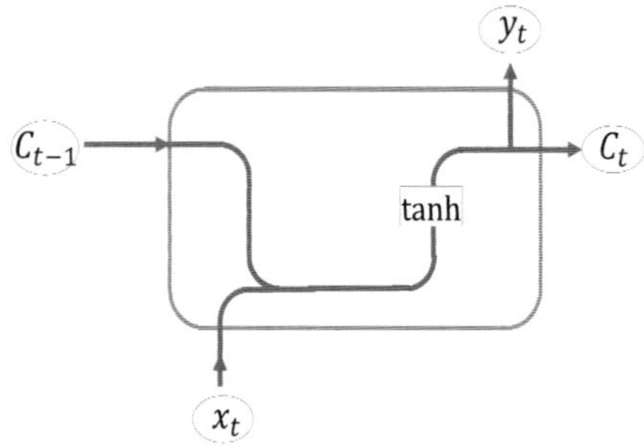

Figure 1 RNN unit structure

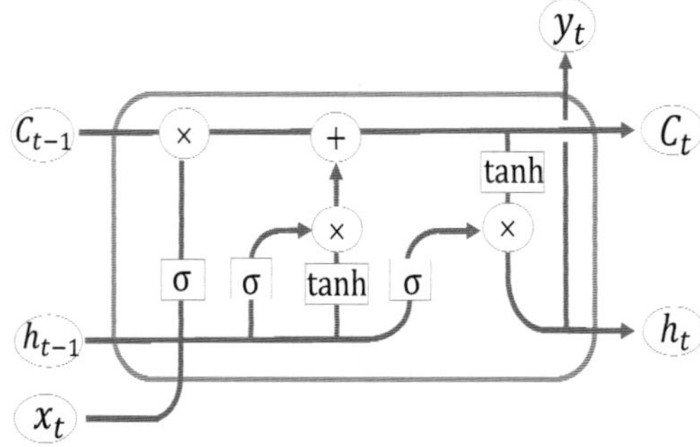

Figure.2 LSTM Cell Structure

The LSTM formula is as follows::

$$f_t = \sigma(W_f \cdot [h_{t-1}, x_t] + b_f) \tag{7}$$

$$i_t = \sigma(W_i \cdot [h_{t-1}, x_t] + b_i) \tag{8}$$

$$o_t = \sigma(W_o \cdot [h_{t-1}, x_t] + b_o) \tag{9}$$

$$\tilde{C}_t = \tanh(W_c \cdot [h_{t-1}, x_t] + b_c) \tag{10}$$

$$C_t = f_t \cdot C_{t-1} + i_t \cdot \tilde{C}_t \tag{11}$$

$$h_t = o_t \cdot \tanh(C_t) \tag{12}$$

f, i, o and *C* respectively represent forgotten gate, input door, output door and update door.

σ_x and $\tanh_x$ are activation functions, which are defined as follows:

$$\sigma_x = \frac{1}{1+e^{-x}} \tag{13}$$

$$\tanh_x = \frac{e^x - e^{-x}}{e^x + e^{-x}} \tag{14}$$

To classify the extracted features, the final output of the LSTM network is used as input to the Softmax layer. The Softmax layer output formula is as follows

$$y_t = W_y \cdot h_t + b_y \tag{15}$$

$$pred_i = \frac{e^{w_i y_i + b_i}}{1 + \sum_{i=1}^{N} e^{w_i y_i + b_i}} \tag{16}$$

In the formula: w_i is the i-th weight matrix.

3.Experiment and result comparison

3.1 Data set description

The data in this paper are from the gearbox fault data set opened by the Case Western Reserve University Bearing Data Center. The test data were sampled at 12 kHz at three different locations under four different load conditions (0 hp, 1 hp, 2 hp, 3 hp). This paper selects the drive side measurement data with load of 3hp, which contains 10 sets of data (9 sets of fault data and 1 set of normal data). The types of faults are divided into three categories according to bearing diameters: 0.07 inches, 0.14 inches, and 0.21 inches. Each category is further classified into three types of sub-faults: ball failure, inner ring failure, and outer ring failure. Each set of data contains 100 samples (divided into one sample every 1024 points, each sample length is 1024), in which training data and test data each account for 50%. In order to facilitate training, one-hot coding of fault types is carried out, as shown in Table 1.

For feature engineering, six features are extracted from each sample by formulas (1)-(6) to construct a one-dimensional input feature vector; for feature learning, a one-dimensional time window with a window length and a step size of 32 is used to perform 32 intercepts for each sample to construct LSTM neural network fault identification model input data.

In this paper, four different machine learning models are used to identify faults: Artificial Neural Network (ANN), Decision Tree Classifier (DTC), Support Vector Machine Classifier (SVC), and K Nearest Neighbor Classifier (KNC). Since the performance of the above model is affected by its hyperplane parameters, in order to obtain reliable test results, the hyperplane parameters of each model are optimized by the grid search method [11], and the model with the best performance is selected as the final test model.

Table 1 raw data table

fault size	fault type	markup symbol	fault number
0	normal	normal	0
0.07	ball failure	b-0.07	1
0.07	inner ring failure	ir-0.07	2
0.07	outer ring failure	or-0.07	3
0.14	ball failure	b-0.14	4
0.14	inner ring failure	ir-0.14	5
0.14	outer ring failure	or-0.14	6
0.21	ball failure	b-0.21	7
0.21	inner ring failure	ir-0.21	8
0.21	outer ring failure	or-0.21	9

The LSTM neural network model used in the feature learning method has four layers, which are divided into input layer, hidden layer (two-layer LSTM structure, each layer length is 32), and output layer (Softmax layer). The model structure is shown in Figure 3. The number of each LSTM neural network unit is set to 128, the training set (Batch size) is set to 50, and the Epochs is set to 150. In order to avoid over-fitting of the neural network, Dropout operation is performed in each hidden layer, and the Dropout rate is set to 0.5[12]. Before the training data is input into the network, the random shuffling operation is first performed, but the test data is not operated.

The LSTM neural network fault detection model proposed in this paper is based on the TensorFlow_gpu-1.7 version. The running platform is a desktop computer (Win7 64-bits system, 3.4GHZ CPU, GTX-950 GPU, 12G memory).

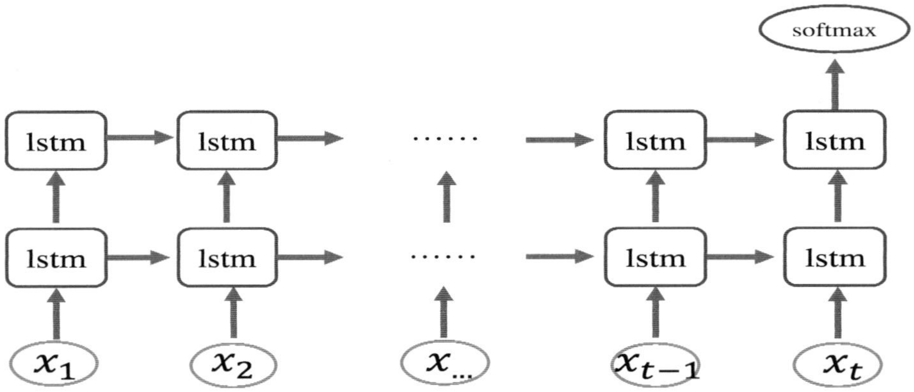

Fig.3 Model Structure

3.2 Comparison of results

3.2.1 Evaluation indicators

In order to measure the performance difference between the fault recognition model proposed in this paper and the machine learning fault recognition model, four performance evaluation indicators are selected in this paper: Accuracy, Precision, Recall and F1_score. The calculation formula are as follows:

$$Accuracy = \frac{|TP|+|TN|}{|TP|+|FP|+|TN|+|FN|} \tag{17}$$

$$Precision = \frac{|TP|}{|TP|+|FP|} \tag{18}$$

$$Recall = \frac{|TP|}{|TP|+|FN|} \tag{19}$$

$$F_1\_score = \frac{2*Precision*Recall}{Precision+Recall} \tag{20}$$

In the formula: | TP | denotes the total number of correctly judged faults; | TN | denotes the total number of correctly judged normal states; | FP | denotes the total number of incorrectly judged faults; | FN | denotes the total number of incorrectly judged normal states

When the model is used to detect fault, it is difficult to avoid incorrect judgments in the recognition results. Therefore, we require a reliable fault recognition model with good recognition accuracy. When the fault recognition model need identify a large number of data, the number of misjudgments in the recognition results may increase. To avoid the occurrence of this phenomenon, "Precision" as a evaluation index is introduced. In addition, if only real faults occur and some real faults are neglected in the recognition results, the "Precision" will be improved, but the recall rate will be reduced, and F1_score will comprehensively measure the precision and recall rate. Therefore, the above four indicators should be optimized simultaneously for the fault identification model to ensure that the model can identify faults while reducing the probability of false diagnosis and real fault omission.

3.2.2Comparing the results of different models

The initial performance table and optimization performance table of the four machine learning fault detection models selected in this paper are shown in Table 2 and Table 3. A comparison of Table 2 and

Table 3 shows that the performance of the optimized machine learning model has been improved to varying degrees. The SVC model has the highest performance improvement, and the four performance indicators after optimization are higher than those of other models, and the recognition accuracy is 94.20% from the initial 78.20%. Figure 5 shows that the optimized SVC model greatly improves the accuracy of identification of health conditions and fault conditions, and can better distinguish between health status and fault status. However, comparing Fig. 4 and Fig. 5, it can be found that the three faults at 0.14 inches are still difficult to identify, and the optimized SVC model has no obvious improvement effect. This may be because the features extracted based on feature engineering at this size cannot contain enough effective information. Therefore, the different faults under this size are judged by the SVC fault identification model, and the accuracy is low, which affects the overall performance index of the model. .

For the LSTM fault identification model, in order to ensure the performance of the LSTM neural network fault identification model during the test, a checkpoint strategy is adopted in the training process. That is, when the accuracy of the LSTM neural network fault recognition model reaches the current highest value in the test data set, the model is saved once, and finally the result with high test accuracy is selected as the final test result. Its performance comparison table and confusion matrix are as follows. Table 4 shows that the LSTM neural network fault identification model has better performance than the optimized SVC fault identification model. Its performance evaluation indicators are higher than the SVC fault identification model, in which: the recognition accuracy rate is up to 98.00%. Figure 5 and Figure 6 show that compared with the SVC fault identification model, the LSTM neural network can distinguish fault signals and normal signals more accurately, and there is basically no misjudgment. Although the accuracy of individual fault type recognition has not reached the average value of model identification accuracy, its lowest accuracy still reaches 90%, which is still much higher than the minimum accuracy of the SVC model. At the same time, the recognition accuracy of the three types of faults that are prone to confusion in the SVC model is significantly improved in the LSTM neural network fault identification model. In summary, the LSTM neural network fault identification model proposed in this paper has better identification reliability. We can visualize the working state of LSTM model in the process of feature extraction. Figure 7 is a scatter plot of feature extraction of LSTM model at different times.

Table 2 Initial Performance Table

score	SVC	KNC	DTC	ANN
Acuracy	78.60%	87.60%	93.40%	89.00%
Precision	79..80%	88.13%	93.82%	89.20%
Recall	78.60%	87.60%	93.40%	89.00%
F1_score	76.02%	87.46%	93.54%	88.80%

Table 3 Optimization Performance Table

score	SVC	KNC	DTC	ANN
Accuracy	94.20%	90.00%	93.40%	92.60%
Precision	94.56%	90.23%	93.82%	92.70%
Recall	94.20%	90.00%	93.40%	92.60%
F1_score	94.31%	90.01%	93.54%	92.56%

Table 4 Final Performance Table

score	Accuracy	Precision	Recall	F1_score
SVC	94.20%	94.56%	94.20%	94.31%
LSTM	98.00%	98.02%	98.00%	97.99%

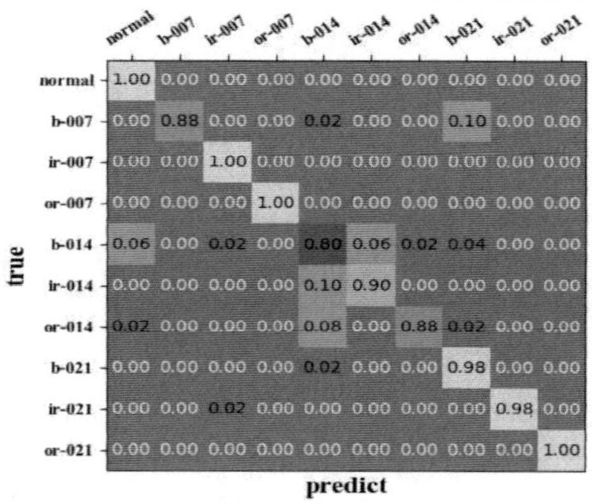

Fig.4 SVC Initial Confusion Matrix

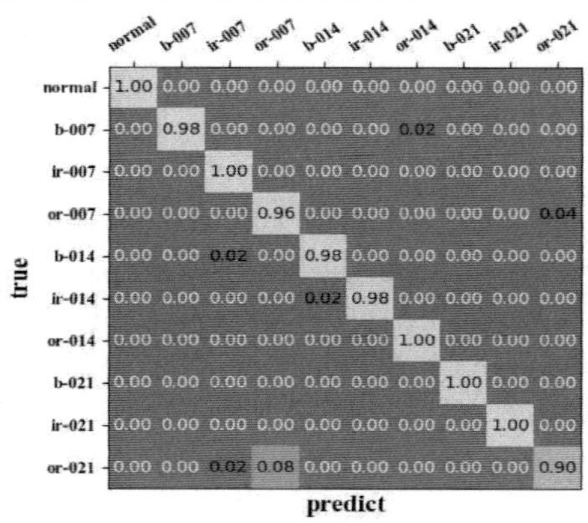

Fig.5 SVC Optimization Confusion Matrix

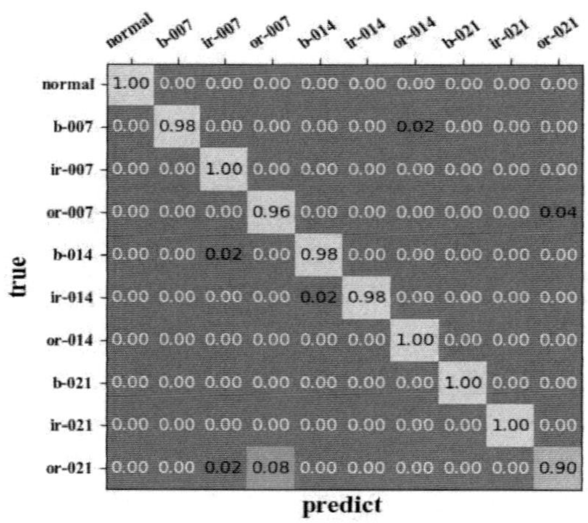

Fig.6 LSTM Confusion Matrix

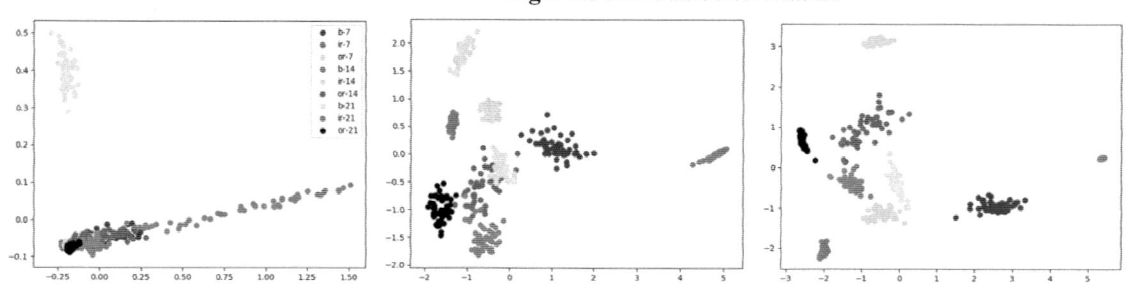

(a) (b) (c)

Fig.7 Scatter plots at different times of the extracted features for data set E by LSTM((a) is the initial time; (b) is the middle time; (c) is the end time)

From figure 7, we can clearly see the working process of LSTM model in extracting features. On the right side of Figure 7, we can see that all kinds of fault types have clear boundaries and no overlapping categories. This intuitively shows that LSTM model has a very strong feature extraction capability. So LSTM has excellent fault recognition ability.

4.Conclusion

1) In this paper, a rolling bearing fault recognition model based on LSTM neural network is proposed. When the fault detection and recognition is carried out, the original data need not be pre-processed, and it has high fault recognition efficiency and speed.

2) The fault recognition model based on LSTM neural network can fully exploit the potential information of the original data, better express the original data, and has excellent recognition reliability.

3) Compared with the experimental results of machine learning model in gear box fault data set, the results show that the proposed LSTM-based neural network fault recognition model has better performance and provide a new method and idea for gear box fault recognition.

References

[1] WU F Q, MENG G. Fault Classification of Rotor Vibration Signals Based on Support Vector Machines[J]. *Journal of Vibration Engineering,* 2006, 19(2): 238-241.

[2] Zaeri R, Ghanbarzadeh A, Attaran B, et al. Artificial neural network based fault diagnostics of rolling element bearings using continuous wavelet transform[C]// *International Conference on Control, Instrumentation and Automation. IEEE*, 2012:753-758.

[3] Kateris D, Moshou D, Pantazi X E, et al. A machine learning approach for the condition monitoring of rotating machinery[J]. Journal of Mechanical Science & Technology, 2014, 28(1):61-71.

[4] WANG J G, LIU Y L, QIN B, et al. Fault Diagnosis Based on EMD and Multi-Features Support Vector Machines[J]. *Mechanical Design and Manufacture*, 2015(10):64-67.

[5] Lecun Y, Bengio Y, Hinton G. Deep learning[J]. *Nature*, 2015, 521(7553):436.

[6] Jiang L, Ge Z, Song Z. Semi-supervised fault classification based on dynamic Sparse Stacked auto-encoders model[J]. *Chemometrics & Intelligent Laboratory Systems*, 2017, 168.

[7] Ding X, He Q. Energy-Fluctuated Multiscale Feature Learning With Deep ConvNet for Intelligent Spindle Bearing Fault Diagnosis[J]. *IEEE Transactions on Instrumentation & Measurement*, 2017, 66(8):1926-1935.

[8] Peng L, Jiao B, Liu H, et al. The Fault Diagnosis of Wind Turbine Gearbox Based on Improved KNN[J]. *Advances in Energy Engineering*, 2015, 3.

[9] Zhao H, Sun S, Jin B. Sequential Fault Diagnosis based on LSTM Neural Network[J]. *IEEE Access*, 2018, PP(99):1-1.

[10] Hochreiter S, Schmidhuber J. Long Short-Term Memory[J]. *Neural Computation*, 1997, 9(8):1735-1780.

[11] Bergstra J, Bengio Y. Random search for hyper-parameter optimization[J]. *Journal of Machine Learning Research*, 2012, 13(1):281-305.

[12] Cheng G, Peddinti V, Povey D, et al. An Exploration of Dropout with LSTMs[C]// *INTERSPEECH*. 2017:1586-1590.

IOP Conf. Series: Materials Science and Engineering **542** (2019) 012049 doi:10.1088/1757-899X/542/1/012049

Environmental effects on Chestnut wood (Castanea sativa Mill.) treated with different surface coatings

Marina Chavenetidou[1], Dimitris Birbilis[2] and Konstantinos V Kakavas[2*]

[1]Laboratory of Forest Utilization, School of Forestry and Natural Environment, Aristotle University of Thessaloniki, 54124, Thessaloniki, Greece
[2*]Laboratory of Interfaces, Chemical Technology and Biotechnology, Department of Wood and Furniture Design and Technology, School of Technological Applications, Technological Educational Institute of Thessaly, 11 V. Griva, Karditsa, Greece, author to whom any correspondence should be addressed, email: kakavas@teithessaly.gr

Abstract. This study investigated the influence of climatic factors on the deterioration of chestnut wood after various manipulations with surface coatings after exposure to ambient conditions for two years. The exposure of chestnut wood surfaces to atmospheric precipitation namely snow, rain and humidity, whether coated or uncoated with hydrophobic substances of low effectiveness led to discolouration due to leaching of water-soluble extracts possessed by the wood. The handling of chestnut wood with surface coatings of hydrophobic protectives (WR) such as teak oil in an organic solvent, epoxy resin based stain, transparent polyurethane varnish with UV protection factor, surface alkyd-based white varnish and exposure to ambient atmospheric conditions showed that discolouration was created due to the leaching of the water-soluble extracts, contained in the chestnut wood for surfaces coated or uncoated with hydrophobic substances of low effectiveness. It was therefore concluded that environmental factors cause discolouration of chestnut wood whether treated or untreated with hydrophobic substances of low effectiveness.

1. Introduction

Chestnut wood is rich in water-soluble extracts (Kakavas *et al.*, 2018) which were found to affect its properties, such as hydroscopicity and density (Galinanes *et al.,* 2015) and also its utilization in the cases where chestnut wood is used for constructions and exposed to ambient conditions. Sivrikaya *et al.* (2009) and Moheby and Saeiam (2015) described discoloration of the wood as being caused by the effect of the chemical reaction of the metal connectors (mainly iron and iron compounds) with tannins or polyphenols contained in the water-soluble extracts of the wood. Another expected effect is the discolouration which comes from the washing of the extracts contained in the wood (Sudeshna *et al.*, 2013). These discolourations are detected in the wood or even cause discolouration to materials which are in contact with the wood. Discolouration of the wood can also be observed in species-rich in soluble extracts, when the wood is exposed to high relative humidity or in contact with water, because of the washing of the extracts (Pandey, 2005; Nzokou*et al.*, 2006). Thaler et al. (2014) in a research on wood specimens that had been in outdoor use for 35 years, after extraction with methanol or water observed that the specimens lost between 11% and 15% of total mass by leaching of extracts. However, extraction only slightly decreased the durability of the leached wood. This indicates that other components in addition to water and methanol soluble extracts could be responsible for durability. In another research conducted on 17 samples of *Castanea sativa*, the performance of wood-based materials after a certain period of times was tested (Žlahtič and Humar 2017). The results showed that, after several applications to improve moisture performance, the water exclusion efficacy of wood appeared to be decreasing over time. Less durable wood in comparison to durable or wood treated with preservatives was mostly affected by all ageing factors. The purpose of this research was to study the effect of environmental atmospheric conditions on the chestnut wood treated with different surface coatings.

2. Methods

The material studied was chestnut wood planks intended for coating and originated from Arnaia Chalkidiki. The dimension of the wood samples of 1.7 x 9 x 19.5 cm (thickness x width x length) were created with a random distribution of the radial and tangential surfaces, as well as the proportion of heartwood and sapwood. The applied handling was in the form of surface coatings[§] and includes:

I. Uncoated chestnut wood (Control)
II. Chestnut wood coated with hydrophobic protective (W.R.) (teak oil in an organic solvent)
III. Chestnut wood coated with epoxy resin based stain (transparent varnish)
IV. Chestnut wood coated with transparent polyurethane varnish with organic UV-absorber
V. Chestnut wood coated with surface alkyd-based white varnish

Surface coatings II, III, IV and V were applied by brush in two layers. The wood samples were exposed to the environment in an appropriate context, inclined 45^0 to the horizon and southwest exposure, at an altitude approximately 16 m, on the roof of the building of Department of Forestry and Natural Environment, University of Thessaloniki, Greece. The durations of the exposure were 3, 6, 12 and 24 months. The wood sample numbers and codes used for each treatment and the duration of the exposure are shown in Table 1.

Table 1. Samples number, code and duration of exposure for each operation (A front side, B buck side).

Duration of exposure (months)	Handling/Code				
	I	II	III	IV	V
0	$0A/I_{0A}$	$0A/II_{0A}$	$0A/III_{0A}$	$0A/IV_{0A}$	$0A/V_{0A}$
3	$3A/I_{3A}$,$3B/I_{3B}$	$3A/II_{3A}$,$3B/II_{3B}$	$3A/III_{3A}$,$3B/III_{3B}$	$3A/IV_{3A}$,$3B/IV_{3B}$	$3A/V_{3A}$,$3B/V_{3B}$
6	$6A/I_{6A}$,$6B/I_{6B}$	$6A/II_{6A}$,$6B/II_{6B}$	$6A/III_{6A}$,$6B/III_{6B}$	$6A/IV_{6A}$,$6B/IV_{6B}$	$6A/V_{6A}$,$6B/V_{6B}$
12	$12A/I_{12A}$,$12B/I_{12B}$	$12A/II_{12A}$,$12B/II_{12B}$	$12A/III_{12A}$,$12B/III_{12B}$	$12A/IV_{12A}$,$12B/IV_{12B}$	$12A/V_{12A}$,$12B/V_{12B}$
24	$24A/I_{24A}$,$24B/I_{24B}$	$24A/II_{24A}$,$24B/II_{24B}$	$24A/III_{24A}$,$24B/III_{24B}$	$24A/IV_{24A}$,$24B/IV_{24B}$	$24A/V_{24A}$,$24B/V_{24B}$

*2.1The CIEL*a*b* colour system.*

The CIEL*α*b* colour system was used in the estimation of wood colour variation. According to the

IOP Conf. Series: Materials Science and Engineering **542** (2019) 012049 doi:10.1088/1757-899X/542/1/012049

three-dimensional CIEL*α*b* colour space, every colour can be indicated as a point in the Euclidean space defined by three coordinates which are used to correlate each colour with subjective colour perception. Coordinate L* for lightness shows the position on the black to white axis meaning L=0 for black and L=100 for white, whereas the chroma value α* represents the position on the red to green axis implies positive values for red shades and negative values for green shades, and coordinate b* on the yellow to blue axis implies positive values for yellow shades and negative values for blue shades. Loss of lightness is observed as a smaller value of L*, α large value of α*indicates more red chroma, and large values of b* mean a strong presence of yellow chroma. On the negative parts of the chroma scales α* and b*, a great negative value of α* expresses an intense green colour, while for a bright blue, b* becomes large with a negative sign. Zero values for α* or b* imply lack of redness, that is, greenness, yellowness or blueness. A difference in a colour called △E* shows the distance between two points in the colour coordinate system and it is estimated from the slit up values △L*, △α* and △b*.According to the CIEL*α*b* colour system [ISO 7724:3-1984(E)], the equation for the total colour difference ΔE* is: $\Delta E^* = \sqrt{\Delta L^2 + \Delta a^2 + \Delta b^2}$. The most frequently used formulae are CIE 1976, CMC formulae (tolerances 2:1 or 1:1) and CIE 1994, which express the colour difference with an optimum accuracy to visual observation (Clarke et al. 1984). The smallest colour difference which can be seen by the human eye in agricultural or horticultural applications is △E*= 1.0 to 3.0 (Voss & Hale, 1998; Terziev & Boutelje, 1998). An accurate estimation for wood colour variation would depend strongly on the wood specimen, wood lightness, colour and surface pattern, as well as on the direction of colour change.

3. Results and discussion

After the completion of the exposure time, the exposed samples were collected, transported to the laboratory, air dried and then photographed by the front and rear surfaces (Figures 1-10).

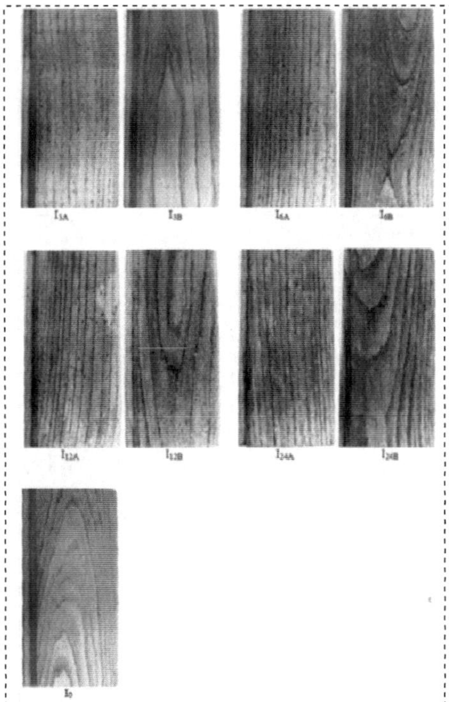

Figure 1. Front surface of exposed samples (Handling I)

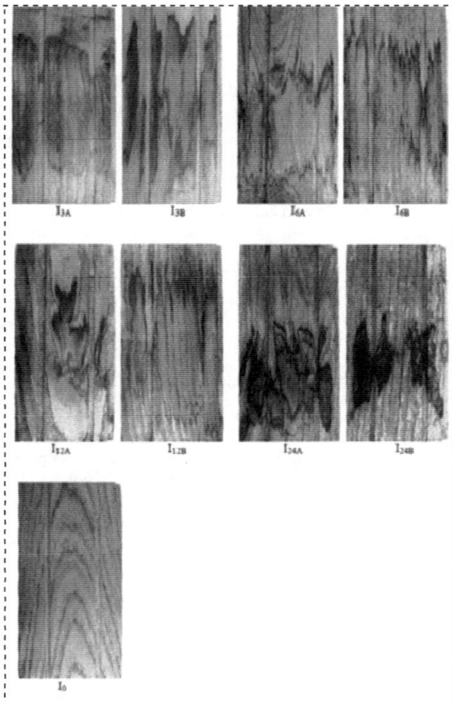

Figure 2. Back surface of exposed samples (Handling I)

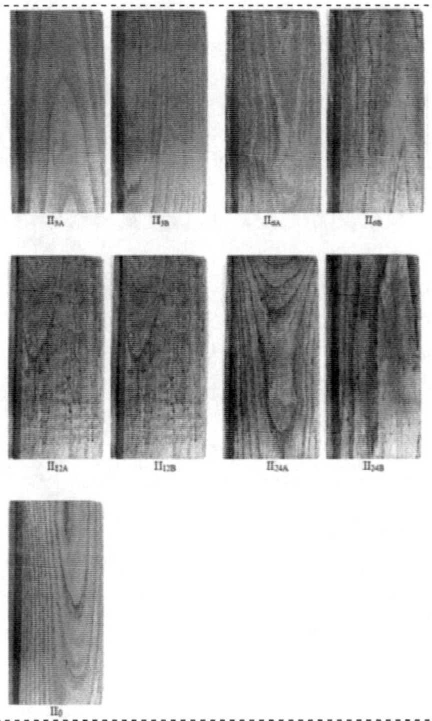

Figure 3. Front surface of exposed samples (Handling II)

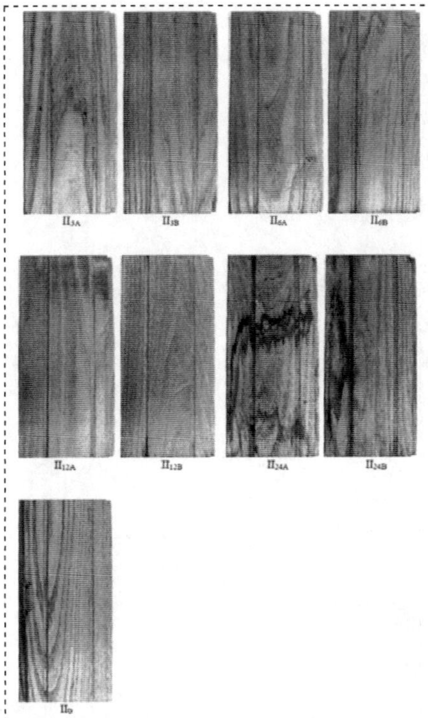

Figure 4. Buck surface of exposed samples (Handling II)

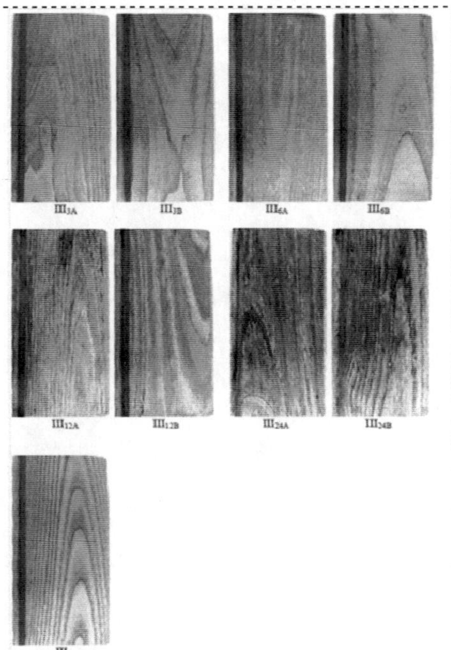

Figure 5. Front surface of exposed samples (Handling III)

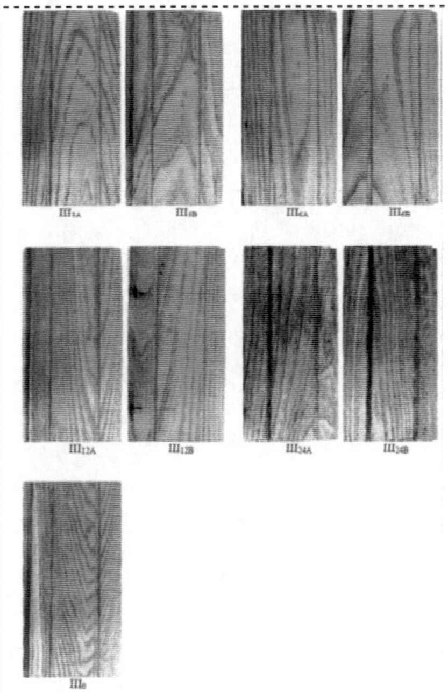

Figure 6. Back surface of exposed samples (Handling III)

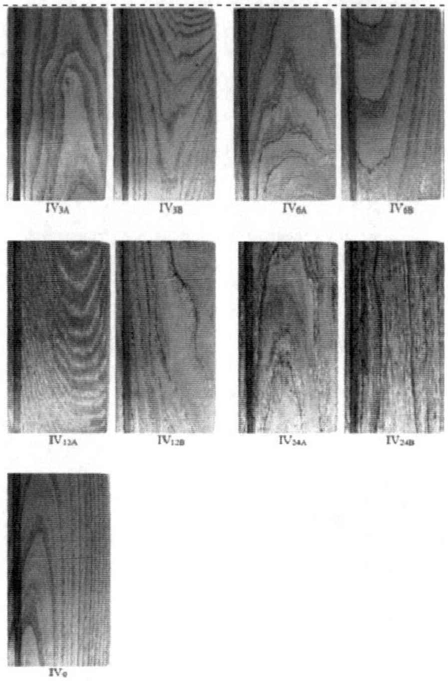

Figure 7. Front surface of exposed samples (Handling IV)

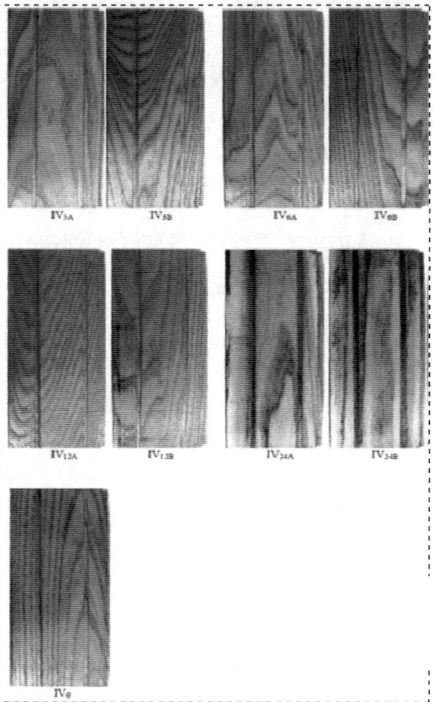

Figure 8. Back surface of exposed samples (Handling IV)

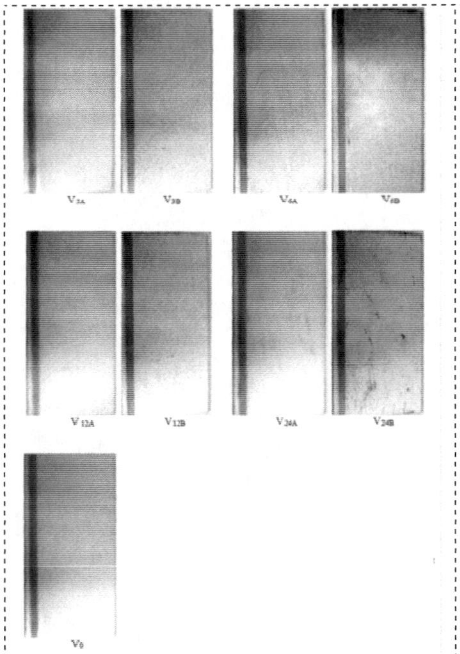

Figure 9. Front surface of exposed samples (Handling V)

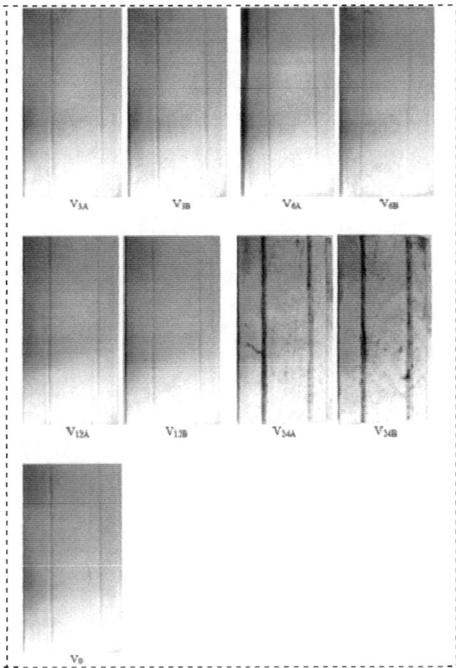

Figure 10. Back surface of exposed samples (Handling V)

In the handlings, I, II, III and IV, the deterioration of chestnut wood samples which were exposed to atmospheric conditions was increased with duration of exposure. The discolouration observed on the exposed front surfaces was evenly distributed over the surface of the sample. The elution of the extracts resulted in creating discolouration at positions on the back surface of the test samples mainly in the handlings I and II. In handling I (uncoated chestnut wood), the alteration was observed in the first months of the research (Figure 1) while in handling II lesion was detected in samples exposed to a period longer than 12 months (Figure 3 and Table 2). Similar changes in the rear surfaces of the samples, to a much lesser extent were identified in the handlings III and IV only in the samples that were exposed to a period longer than 12 months (Figures 5 & 7 and Table 2) while in handling V, no surface degradation was observed even in samples exposed for a period of two years (Figure 9). The accretion of air pollutants on some wood samples (samples V_{3A}, V_{6A}, V_{12A} and V_{24A}) could be removed by gentle cleansing with water without altering the paint.

Table 2.a. Handling II and ΔE correlation

Handling II (teak oil in an organic solvent)		
Code	ΔE (front side, A)	ΔE (back side, B)
II_{0A}	16,358	17,481
II_{3A}	18,318	19,762
II_{3B}	19,808	20,960
II_{6A}	20,001	19,934
II_{6B}	20,271	19,461
II_{12A}	26,280	27,577
II_{12B}	28,634	29,025
II_{24A}	45,322	48,888
II_{24B}	46,533	49,517

Table 2.b. Handling III and ΔE correlation

Handling III [epoxic resin based stain (transparent varnish)]		
Code	ΔE (front side, A)	ΔE (back side, B)
III_{0A}	12,249	13,238
III_{3A}	17,706	17,853
III_{3B}	17,467	17,953
III_{6A}	19,608	18,711
III_{6B}	18,530	19,728
III_{12A}	25,433	24,104
III_{12B}	26,877	25,215
III_{24A}	40,798	41,858
III_{24B}	41,257	42,733

Table 2.c. Handling IV and ΔE correlation

Handling IV (transparent polyurethane varnish with organic UV-absorber)		
Code	ΔE (front side, A)	ΔE (back side, B)
IV_{0A}	15,737	14,261
IV_{3A}	16,461	17,586
IV_{3B}	17,720	18,780
IV_{6A}	19,987	18,731
IV_{6B}	18,347	19,553
IV_{12A}	21,513	22,856
IV_{12B}	22,570	22,439
IV_{24A}	35,906	36,644
IV_{24B}	34,009	35,669

Table 2.d. Handling V and ΔE correlation

Handling V (Surface alkyd-based white varnish)		
Code	ΔE (front side, A)	ΔE (back side, B)
V_{0A}	0,0	0,0
V_{3A}	1,877	1,070

V_{3B}	1,970	1,105
V_{6A}	1,916	1,205
V_{6B}	1,738	1,217
V_{12A}	2,539	1,373
V_{12B}	2,915	1,848
V_{24A}	12,722	10,890
V_{24B}	14,611	11,567

4. Conclusions

The chestnut wood surfaces exposed to air pollutants whether coated or uncoated with hydrophobic substances of low effectiveness led to discolouration due to the leaching of water-soluble extracts contained in the chestnut wood.

Acknowledgments

The authors would like to acknowledge all the researchers who worked in this project.

References

1. Clarke FJJ, McDonald R., Rigg B. (1984). Modification to the JPC79 colour-difference formula. *JSDC*. 100: 128-132.

2. Galinanes C., Freime MS, Gonzalez–AlvarezJ. (2015). UV protection effects of phenolic extracts from chestnut fruit and forest industries residues. *J. Wood Prod*.73:731–739.

3. Kakavas K. *et al*. (2018). Chemical Properties of Greek Stump Chestnut (Castanea sativa Mill.). *Natural Products Chemistry & Research*. 6:4. DOI: 10.4172/2329-6836.1000331.

4. Moheby B., Saeiam. (2015). Effects of geographical directions and climatological parameters on natural weathering of fir wood. *Construction and Building Materials*.http://dx.doi.org/10.1016/j.conbuildmat.2015.07.0490950-0618/2015.94:684–690.

5. Nzokou P., Kamdem DP. (2006). Influence of Wood Extractives on the Photo-discoloration of Wood Surfaces Exposed to Artificial Weathering. *Industrial applications*. 31(5): 425-434.

6. Pandey KK. (2005). Study of the effect of photo-irradiation on the surface chemistry of wood. *Polymer Degradation and Stability*.doi:10.1016/j.polymdegradstab.2005.02.009.90: 9-20.

7. Sivrikaya H., Hafizoglu H., Yasav A., Aydemir D. (2009). Natural Weathering of Oak (Quercus petrae) and Chestnut (*Castanea sativa*) Coated with Various Finishes. *Color research and application*. 36(1):72-78.

8. Sudeshna S., Duygu K., Yaman B., Vakhtangm., Legault.(2013). Boreal forest conifer extracts: potential natural additives for acrylic polyurethane coatings for the protection of heat-treated jack pine. *J. Coat. Technol. Res*.,10 (1): 109–122. DOI 10.1007/s11998-012-9435-5.2013.

9. Terziev & Boutelje. (1998). Effect of felling time and Kiln-drying on color and susceptibility of wood to mold and fungal stain during an above-ground field test. *Wood and fiber*.30(4): 360-367.

10. Thaler, N., Žlahtič, M. & Humar, M. (2014). Performance of recent and old sweet chestnut (*Castanea sativa*) wood. *International Biodeterioration & Biodegradation*, (94): 141-145.

11. Voss, D.H. & Hale, W.N. (1998). A comparison of the three editions of the Royal Horticultural Society colour chart. *HortScience*. 33(1): 13-17.

12. Xiaoyur IPZ. (2017). Measurement of three wood materials against weathering during long natural sunlight exposure. *Measurement.* 102: 179–185. http://dx.doi.org/10.1016/j.measurement.2017.01.0340263-2241.

13. Žlahtič, M. & Humar, M. (2017). Influence of artificial and natural weathering on the moisture dynamic of wood. "Wood, weathering, moisture". *BioResources*, 12 (1): 117-142

IOP Conf. Series: Materials Science and Engineering **542** (2019) 012050 doi:10.1088/1757-899X/542/1/012050

Meso-scale finite element analysis on creep behaviour of asphalt mixture considering interface effect

X Qiu[1] , S L Xiao[1], J X Xu[1], Q Yang[1] and C L Li[1]
1 Road and Traffic Engineering Research Center, Zhejiang Normal University, Jinhua, 321004, China
E-mail: xqiu@zjnu.cn

Abstract. This paper investigated the creep behaviour of asphalt mixtures using a randomly generated meso-scale finite element model that takes into account the effect of asphalt-aggregate interface. Simulation analysis on the three-point bending creep tests of asphalt mixture were performed and validated. Furthermore, parameters studies were conducted to evaluate the effect of gradation and angularity of coarse aggregate, air void content, and asphalt-aggregate interface strength. The results indicated that mesostructural parameters and interface properties have an important effect on the creep behaviour of asphalt mixture. The creep performance of medium-sized asphalt mixture is better than that of fine-sized asphalt mixtures. The larger the angularity index of coarse aggregate is, the better ability asphalt mixture has to resist deformation. As to the suspended dense structure asphalt mixtures, with the increase of void ratio, the creep deformation increase gradually, and the excessive void ratio would lead to the fracture of the interface. Asphalt-aggregate interface with higher interface strength would result in better resistance to of the mixture to permanent deformation.

1. Introduction

Asphalt mixture is one of the most widely used materials in pavement construction. At meso-scale, the asphalt mixture can be regarded as a multi-phase material composed of aggregate, asphalt, air void and asphalt-aggregate interface. Among them, the asphalt-aggregate interface has a significant influence on the failure mechanisms and mechanical properties of the materials [1]. Failures often start from interface, then gradually propagate, and finally form macro cracks leading to the destruction of the asphalt mixture.

It is well known that the properties of the materials are closely related to compositions and structures. Grasping the intrinsic relationship between compositions, structures and performances is the key to understand the fundamental properties of the materials. Based on this point of view, the complex mechanical behavior of asphalt mixture at macro-scale should be the reflection of its meso-scale compositions and structures. The traditional theoretical analysis and numerical simulation often regard the asphalt mixture as a macroscopic homogeneous material, ignoring the influence of the mesoscopic components of the materials, and cannot reflect the failure process of the asphalt-aggregate interface. Therefore, it is important to create a meso-scale model to study the macro-scale performance of asphalt mixtures. Due to the anisotropy of the material, the mechanical behaviors of the asphalt mixture exhibits highly nonlinear under the action of loadings. It is not only time-consuming, laborious and expensive to study the creep behaviors of asphalt mixture only by means of laboratory experiments, but also difficult to reveal the influence law of the asphalt mixture from a perspective meso-scale analysis. With the development of numerical simulation technology and the improvement of calculation ability, more and more scholars are devoting themselves to the numerical study of the relationship between meso-structure and macro-performance of asphalt mixture[2-4]. However, due to the complexity of structure and properties of asphalt mixture, further investigations still needed to clarify the mechanism of asphalt-aggregate interface interaction and its effect on the overall performance of the mixture. Therefore, it is an important issue to establish a more accurate and reasonable meso-mechanical model to explore the mechanical properties of asphalt mixture.

The objective of this paper is to develop a meso-scale finite element model that considers the interface effect to describe the creep behavior of asphalt mixture and to investigate the influence of factors such as gradation and angularity of coarse aggregate, air void content, and asphalt-aggregate interface parameters on the overall performance of the materials. The study would give an insight into the complex mechanical analysis of asphalt mixtures by considering the asphalt-aggregate interface effect.

2. Modeling Methodology

The AC-20 asphalt mixtures with a nominal maximum aggregate size of 19mm was prepared for materials parameters acquisition and numerical verification. The numerical simulations were conducted using the commercial finite element software ABAQUS 6.14. The model that considers each component of asphalt mixtures was established by using the random aggregate generation technique and cohesive element embedding technique, in which aggregates larger than 2.36 are classified as coarse aggregate, fine aggregates (passing 2.36 mm sieving) combined with asphalt and mineral filler are defined as asphalt mortar, and the asphalt-aggregate interface is represented by the embedded cohesive element between the coarse aggregate and asphalt mortar.

2.1 Model generation

The random aggregate generation technique was adopted to simulate the meso-structure of asphalt mixture [5]. Due to the limitation of computational resources and huge time consumption, this study used a two-dimensional model instead of a three-dimensional model. In order to simulate damages that occur in the asphalt-aggregate interface, it is necessary to embed a zero-thickness cohesive elements between the coarse aggregate and the asphalt mortar. However, it is impossible to embed the zero-thickness cohesive elements into the interface by manual operation. Alternatively, by using Python programming process on the derived ABAQUS input file that records all information of nodes, elements and sets of the model, the boundary between the coarse aggregate and asphalt mortar can be identified, and the cohesive element can be automatically embedded to simulate the interface by a series of operations. The key to embed the zero-thickness cohesive element is to split the initial mesh nodes, and then make the coordinates of the newly generated nodes equal to the original nodes, so that the thickness of the embedded cohesive element equal to zero. Another important issue is that the node numbers of newly generated cohesive elements should be arranged in counter-clockwise. Otherwise, errors will be reported during the simulation processing. The interface can be highlighted by visualizing the operation in the software as shown in figure 1.

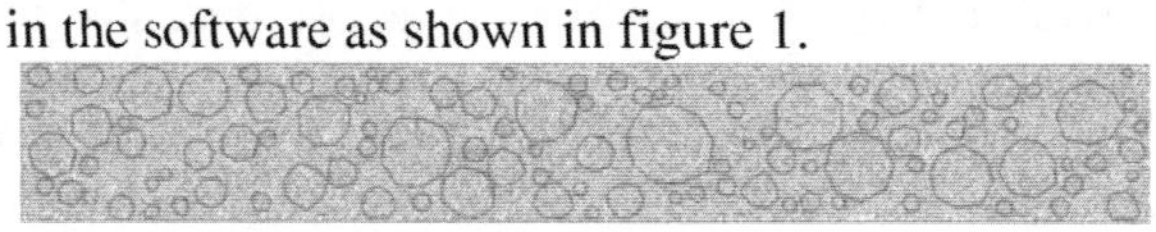

Figure 1. Cohesive elements on the asphalt-aggreagte interface

By combining the random aggregate model with cohesive elements, the meso-scale numerical model that considers the characteristics of the coarse aggregate, asphalt mortar and asphalt-aggregate interface was obtained. The three-point bending creep test can be equivalent to a simple supported beam for numerical simulation as shown in figure 2. The size of the beam is 250mm × 35mm, and the supported beam span is 200mm. A vertical downward concentrated load is applied at the midpoint of its top surface. The global mesh size of the beam is 0.5mm. Both the coarse aggregate and the asphalt mortar are meshed using reduced integration plane strain (CPE4R) element, and the asphalt-aggregate interface is meshed with zero-thickness cohesive elements (COH2D4).

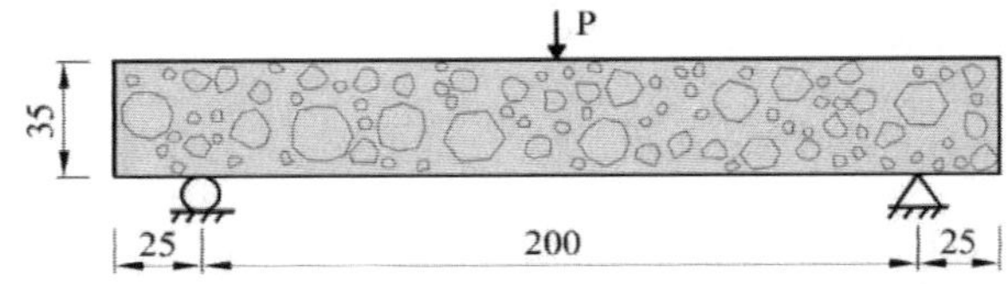

Figure 2. Model geometry and boundary conditions (dimensions in mm)

2.2 Model parameters

The coarse aggregate is considered as a linear elastic material with elastic modulus of 100 MPa and Poisson's ratio of 0.3. The asphalt mortar is defined as a viscoelastic material, and its model parameters were determined from the laboratory three-point bending beam creep test of asphalt mortar. The Burgers model was used to characterize the viscoelastic properties. And the parameters of Burgers model were transformed into the prony parameters (in table 1) for numerical simulation according to Boltzmann linear superposition principle and Laplace mathematical transformation.

Table 1. Prony parameters of asphalt mortar

Burgers model				Prony series			
E_1 (Mpa)	η_1 (MPa·s)	E_2 (MPa)	η_2 (MPa)	g_1	τ_1 (s)	g_2	τ_2 (s)
26.591	2738.247	2.978	209.793	0.9108	17.364	0.0892	2824.2

A two-parameter bilinear cohesive law was chosen to describe the behaviour of asphalt-aggregate interface. This study selected the interface properties from previous researches to realistically build the computational model. The material parameters used in this study are listed in table 2.

Table 2. Parameters of the asphalt-aggregate interface

Interface strength (MPa)	Critical fracture energy (mJ/mm^2)
0.5	1.558

2.3 Model validation

The model was established to simulate the creep behaviour of the AC-20 asphalt mixture. The simulation result was compared with the experimental result, and the correlation coefficients of the two curves was calculated. As can be seen from figure 3, the correlation coefficient between the experimental and simulation results is 0.881, indicating that the numerical model has provided a reliable solution for the simulation of creep behaviour of asphalt mixture. The two curves are basically the same at the initial creep stage, and gradually begin to differ after 100s. Finally, the two curves tend to be consistent. The differences may due to simplification of the aggregate distribution by the random aggregate model. In addition, the numerical simulation is a two dimension model, while the real state is a complex three-dimension structure, which may cause some difference in the simulation results.

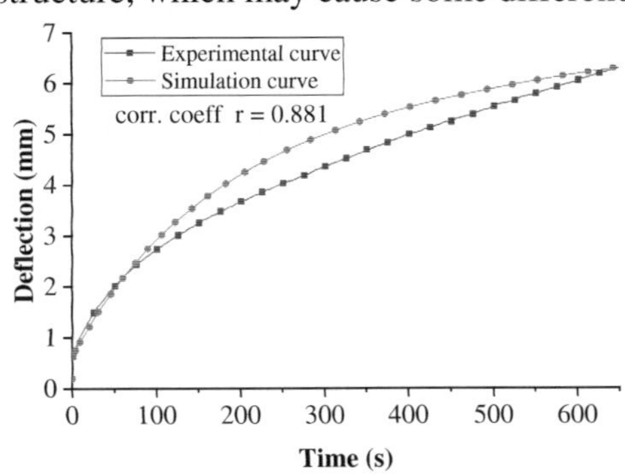

Figure 3. Comparison between the simulation curve and experimental curve

The SDEG (scalar stiffness degradation) is defined in Abaqus filed output, which represents damage status of interface elements adopted in cohesive element. When the interface stress reach to its

IOP Conf. Series: Materials Science and Engineering **542** (2019) 012050 doi:10.1088/1757-899X/542/1/012050

maximum strength, the SDEG value begins to be greater than 0. And when the interface is completely damage, the SDEG value reaches 1. In order to better characterize the cracking phenomenon of the interface, the coarse aggregate is hidden in the visualization operation. The distribution SDEG contours is shown in figure 4, where the red part indicates that the asphalt-aggregate interface has completely cracked. It shows that the location of the complete failures of the interface are mainly near the coarse aggregate with larger particle size near the middle of the specimen.

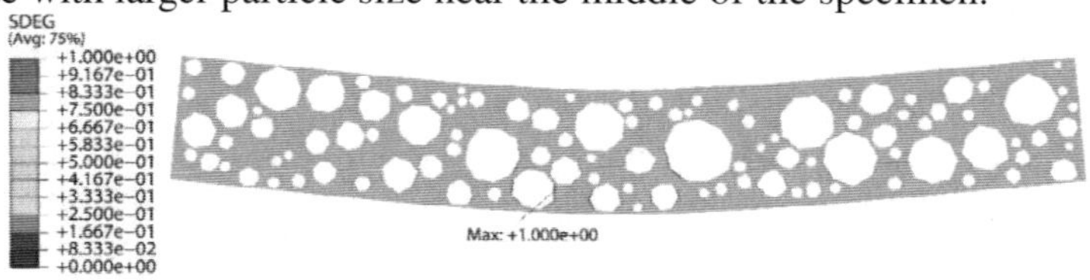

Figure 4. SDEG contours of the AC-20 asphalt mixture

According to the creep deformation curve, the finite element simulation can reflect the three-point bending creep behavior of asphalt mixture, reflecting a good agreement with the experimental results. At the same time, the cohesive zone model adopted in this paper can better reflect the failure characteristics of interface damage and provide a basis for further meso-mechanical analysis of asphalt mixture.

3. Results and analysis

3.1 Effect of coarse aggregate gradation

In order to study the effect of aggregate gradation on the creep behaviours of asphalt mixture, three types of gradations including AC-13, AC-16 and AC-20 were generated, as shown in figure 5. Median recommended aggregate gradations were chosen for each type of asphalt mixture in accordance with the Chinese specification JTG F40-2004.

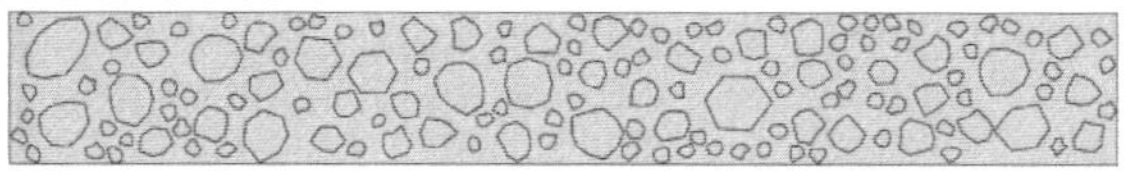

(a) AC-13

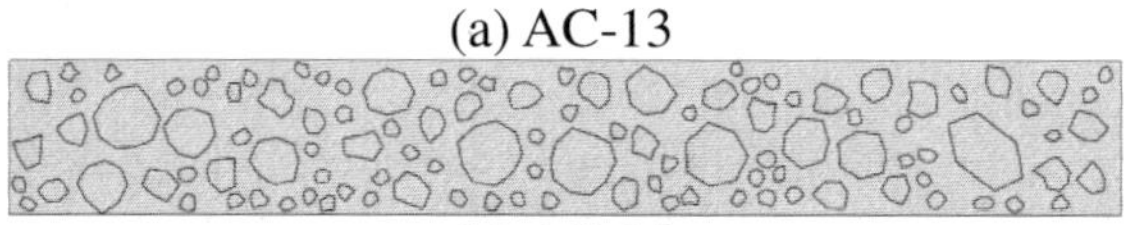

(b) AC-16

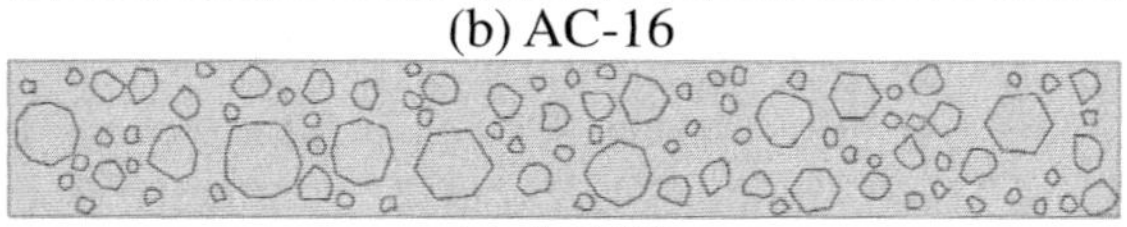

(c) AC-20

Figure 5. Numerical specimens with different gradations

Figure 6 shows the deflection versus time curves for different aggregate gradations. Deflection reflects the creep deformation of specimens over time, and it can be divided into elastic deformation and viscoplastic deformation. The elastic deformation can be restored, while some viscoplastic deformation cannot be restored, thus becoming the permanent deformation. The smaller the deflection is, the greater the resistance the asphalt mixture has to permanent deformation. As can be seen, AC-13 asphalt mixture has the largest deflection, followed by AC-20, and the smallest is AC-16. Therefore, the creep performance of medium-sized graded asphalt mixture (AC-16 and AC-20) is better than that of fine-sized graded asphalt mixture (AC-13), and the creep performance of AC-16 is better than that of AC-20. This indicates that under the condition of guaranteeing aggregate skeleton, appropriately reducing the content of coarse aggregate can avoid the problems of compaction difficult and void fraction caused by the large size of coarse aggregate, thus improving anti-deformation ability of asphalt mixture.

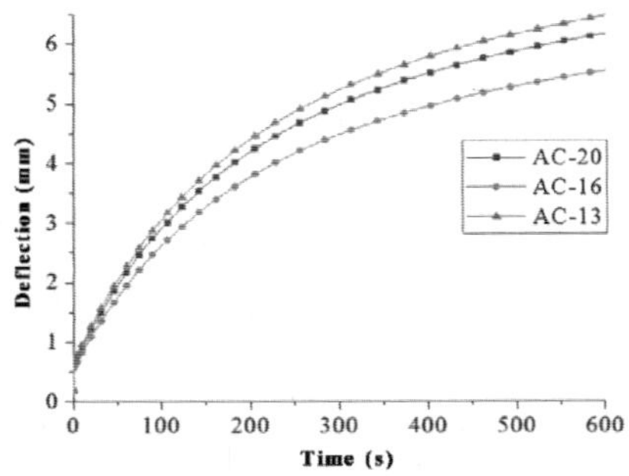

Figure 6. Creep curves under different aggregate gradations

3.2 Effect of coarse aggregate angularity

In order to investigate the influence of aggregate irregularity on creep behavior of asphalt mixture, the angularity index (*AI*) was selected as the control factor, which can be calculated as in equation (1). Different shapes of the aggregate with various angularity indexes were considered to generate numerical specimens. When the aggregate is irregular, the angular index is greater than 1. Angularity indexes for different shapes of the aggregate are shown in table 3.

$$AI = \frac{C^2}{4\pi S} \tag{1}$$

Where,

C —— Perimeter of coarse aggregate (mm),

S —— Area of coarse aggregate (mm^2).

Table 3. Angularity index of coarse aggregates with different shapes

	Circle	Square	Hexagon	Octagon	Decagon
Angularity index	1	1.273	1.103	1.055	1.033

Figure 7 shows the creep results of the asphalt mixtures under different aggregate angular indexes. It can be seen from the figure that the deflection of the mixture decreases with the increase of angularity indexes. The creep deformation at the angularity index of 1.273 is 15.8 % lower than that of angularity index of 1.000. A more irregular aggregate can result in more per unit contact areas (length) between the asphalt and aggregate, hence a better mechanical interlocking, physical adhesion and chemical bonding will be obtained, thus enhancing the overall strength of the material. This means that improving the angularity of coarse aggregate can greatly enhance the deformation resistance of asphalt mixture. Therefore, choosing crushed stone has a better creep performance than that of pebble stone.

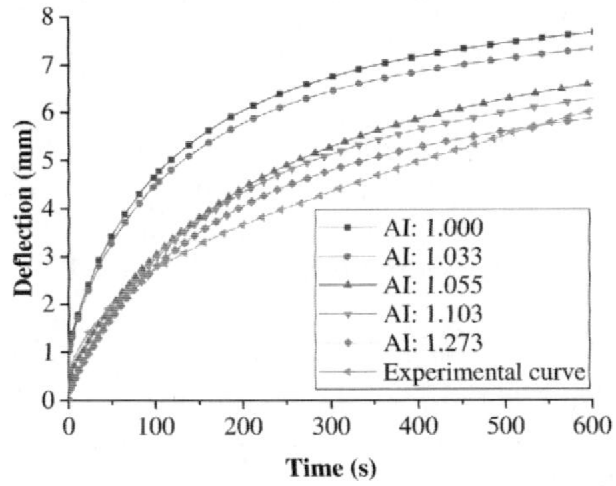

Figure 7. Creep curves under different angularity indices of aggregate

3.3 Effect of air void content

In order to investigate the effect of air voids on the creep behaviour of asphalt mixture, air void content varies from 1% to 4% were discussed. It should be noted that too fine a mesh will lead to high computational cost and convergence problems. The air void is reasonably assumed as random polygon shape with an equivalent diameter of 1mm in the numerical model.

Figure 8 shows the results of creep tests under different air void contents. As the air void content increases, the deflection increased, which means that the ability to resit deformation is decreased. This may be due to the fact that the air void in the model can be regarded as a material like asphalt mortar but without bonding strength and cannot withstand any stress, so the overall strength of the asphalt mortar is reduced, and the creep deformation is increased. That is, at a larger void content, asphalt mixture is more prone to yeild compression deformation, resulting in rutting problem, especially under high temperature conditions.

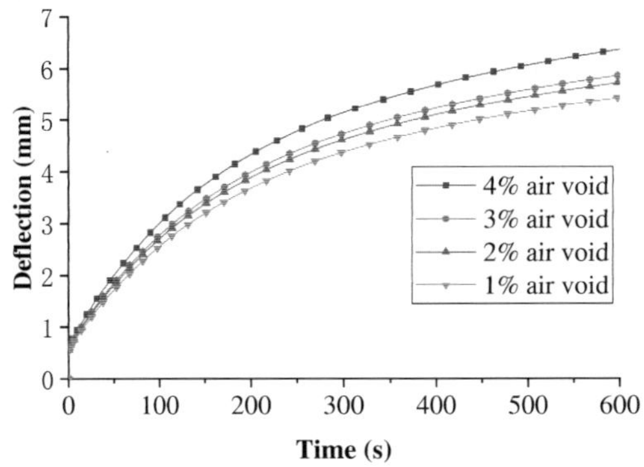

Figure 8. Creep curves under different air void contents

At the same time, part of the air voids that near aggregate correspond to some initial crack at the interface, which may easily lead to the development of the cracks, as shown in the lower left corner of figure 9. In case of large air void content, voids may be linked together with the increase of deformation, as shown in right lower corner of figure 9, which may cause the expansion of the micro-creaks of the asphalt-aggreagte interface and eventually lead to the macro-creaks of asphalt mixture. Therefore, it is necessary to control the air void content of asphalt mixture to ensure its durability and stability.

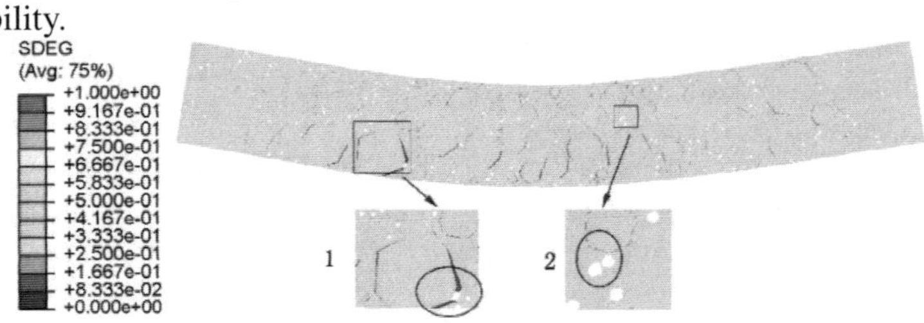

Figure 9. Adverse effects of voids in the specimen

3.4 Effect of asphalt-aggregate interface strength

The asphalt-aggregate interface is the weakest link in the overall strength of asphalt mixture. Keeping the fracture energy of the interface unchanged, various of interface bonding strength including 0.1 MPa, 0.3 MPa, 3 MPa, 10 MPa and 1000 MPa were chosen to further investigate the creep properties of the asphalt mixtures.

Figure 10 shows the creep curves of asphalt mixture under different interface strength. As can be seen, with the increase of interface strength, the deflection decreases gradually, which indicates that if the bonding strength between aggregate and asphalt is improved, the creep deformation of asphalt mixture under will be reduced and the ability to resist permanent deformation can be enhanced. When the interface strength range from 0.3MPa to 0.8 MPa, the numerical results are closed to the experimental results by the end of the creep tests. While the interface strength is 1000 MPa which is far greater than the actual interface strength, the deflection shows the lowest value at 3.5mm at the end of the creep test. Interface with a strength of 1000 MPa can be considered as completely bonded interface, or perfect interface. However, the aggregate is more commonly surrounded by a weak interface due to the surface texture of the aggregate and the difficulty in fully mixing and compaction of asphalt mixture [6]. Therefore, interface effect cannot be neglected when establish a numerical simulation to study the macro-mechanical behaviour of asphalt mixture.

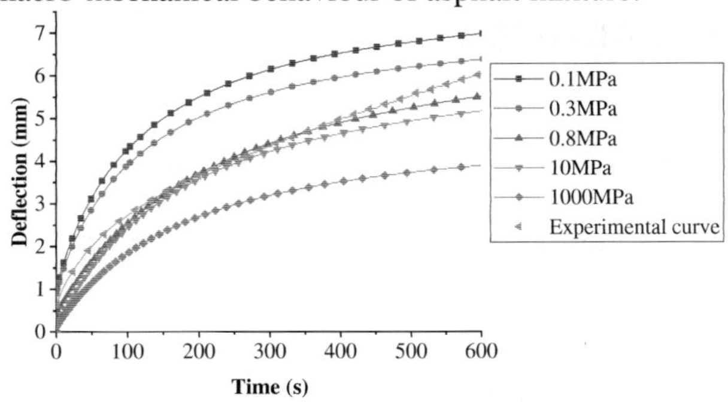

Figure 10. Creep curves under different interface strength

4. Conclusions

The following major findings were achieved from this study.

(1) The creep property of medium-sized asphalt mixture is better than that of fine-sized asphalt mixture. Reducing the content of coarse aggregate to a certain degree can avoid the problem of compaction difficult, thus improving the ability of an asphalt mixture to resist permanent deformation.

(2) The greater angularity index of coarse aggregate, the longer the perimeter per unit area, and the larger the area (length) of asphalt in contact with aggregate, which implies more microtextural features and stronger mechanical interlocking between the asphalt and the aggregate. Therefore, a better ability to resist deformation is obtained.

(3) For the suspend-dense mixtures, with the increase of air void content, a larger energy dissipation and creep deformation was obtained, which indicates that the fracture performance of the asphalt-aggregate interface and the resistance to deformation is reduced.

(4) The asphalt-aggregate interface has a significant effect on the overall behavior of asphalt mixture. The greater bonding strength the interface has, the stronger the resistance the mixture has to permanent deformation.

Acknowledgement

The authors are grateful for the financial supports from the Natural Science Foundation of Zhejiang Province (LY18E080020) and the National Science Foundation of China (51408550).

Reference

[1] Gao Y M, Dong M S, Li L L, Wang L and Sun Z B 2015 *Constr. Build. Mater.* **96** 591-598.

[2] Dong Z J, Gong X B, Zhao L D and Zhang L 2014 *Constr. Build. Mater.* **53** 665-673.

[3] Wang H, Wang J and Chen J 2014 *Eng. Fract. Mech.* **132** 104-119.

[4] López C M, Carol I and Aguado A 2008 *J. Mech. Mater. Struct.* **41** 583-599.

[5] Guo J 2015 *Numerical simulation of asphalt concrete specimen cracking process based on the microscopic material interface features* Nanjing: Southeast University p 115.

[6] Zhu X Y 2013 *Constr. Build. Mater.* **49** 797-806.

IOP Conf. Series: Materials Science and Engineering **542** (2019) 012051 doi:10.1088/1757-899X/542/1/012051

Assessment of fatigue damage in asphalt mixture using an acoustic emission approach

X Qiu[1], Y J Wang[1], Q Yang[1], J X Xu[1] and W H Cheng[1]

1 Road and Traffic Engineering Research Center, Zhejiang Normal University, Jinhua, 321004, China
E-mail: xqiu@zjnu.cn

Abstract. Fatigue failure is the main damage forms of asphalt mixture. Acoustic emission (AE) is an effective technique for continuously monitoring the development of fatigue damage in asphalt mixture. Firstly, the AE characteristic and mechanical behaviour of asphalt mixture and their relationship were investigated by performing four-point bending fatigue test and AE test. Furthermore, the parameters of continuum damage mechanics model that describes the deterioration process of asphalt mixture were calibrated in according with the AE technique. Finally, simulation analysis on the four-point bending fatigue test of asphalt mixture is performed. The results indicated that the AE parameters could reflect the changes of mechanical characteristic. The variation of AE energy and dissipated energy had a similar tendency. There is a strong correlation between the result of continuum damage mechanics model and mechanical test. The numerical results show that the definition of damage variable based on AE energy is reliable and reasonable. This research would provide a new idea for the application of AE technique on diagnosing the fatigue damage evolution process of asphalt mixture.

1. Introduction

Asphalt mixture is one of the most widely used materials in pavement construction. The repeated traffic loads combined with the effects caused by environmental factors could lead to the pavement degradation, which, in the long-term, may cause the fatigue failure of the pavement during their service life. Therefore, exploring the material fatigue performance is an important issue. In depth study of the characteristics and evolution rule of material fatigue damage is the key to improve the pavement service life. The core of this problem is to find a damage factor that can represent the evolution process of damage.

Nowadays, there are plenty of research works reported in the literature about the definition of damage factor. Such as: stiffness modulus, the equivalent strain rate, the number of cycles, dissipative energy, pseudo strain energy density function and so on[1,2]. However, asphalt mixture is a composite material, its failure process is complicated. The above indicators just reflect the change in some macroscopic physical properties. A reasonable method should reflect the local micro-damage and structure degradation process of asphalt materials in real time and dynamic. Acoustic emission testing is a nondestructive detecting technology, and it is able to perceive the failure process of material by transient elastic wave. The waveform of AE signal is collected by the detection device and can be processed and transformed into the various basic acoustic emission parameters. These parameters reflect the activity and development trend of acoustic emission sources during the acoustic emission process. The AE technique has been widely used for the non-destructive evaluation of engineering materials, such as coal, rock and cement concrete. At the same time, the parameters, such as AE energy, ringing counts and AE events, has been utilized to evaluate the damage process of these materials[3,4]. AE technique also can be used to study the performance of asphalt mixture. Now, most of the research about asphalt mixture focused on the crack initiation and propagation of asphalt mixture in low temperature, and these results revealed that AE technique holds great potential for the low temperature performance of asphalt mixture[5,6]. Thus, using AE technique to describe the asphalt mixture fatigue damage evolution process is an important research work.

The main objective of this paper is to establish the damage factor by AE technique, and explore the fatigue failure characteristics of asphalt mixture. All of these are based on the research of the failure process and corresponding acoustic emission characteristics. The study would give an insight into the asphalt mixture fatigue damage evolution process by using AE technique.

2. Materials and methods

2.1 Bitumen and aggregate

The 70 penetration grade bitumen was used in the investigation. The type of aggregate is limestone. The AC-20 asphalt mixture with a nominal maximum aggregate size of 19mm were prepared for material performance test. Table 1 summarizes the detailed property parameters of the bitumen and aggregates.

Table 1. Properties of bitumen and aggregate

Bitumen			Aggregate		
Density *(kg/m^3)*	*Viscosity at 135°C* *(Pa·s)*	*Viscosity at 175°C* *(Pa·s)*	*Particle size* *(mm)*	*Apparent relative density* *(g/cm^3)*	*Water absorption* *(%)*
1.034	0.405	0.082	9.5-20	2.715	0.25
			4.75-9.5	2.711	0.30
			2.36-4.75	2.710	0.43
			0-2.36	2.716	1.38

2.2 Sample preparation

The asphalt mixture design in the study was designed using the Marshall mixture design method. The void fraction is 4.5%, and other relevant parameters are presented in table 2.

Table 2. Marshall test results at optimal binder content

Properties	*Bulk density* *(kg/m^3)*	*Theoretical maximum gravity* *(kg/m^3)*	*Volume of air void* *(%)*	*Voids in mineral aggregate* *(%)*	*Voids filled with asphalt* *(%)*	*Marshall stability* *(kN)*	*Flow value* *(0.1mm)*
Test results	2.406	2.520	4.5	13.6	66.8	13.42	37.1

The standard testing sample like a beam with the dimension of 380 mm × 63 mm × 50 mm were cut from the rutting plate with the dimension of 400 mm × 300 mm × 70 mm. In order to guarantee the reliability of test results, four specimens were prepared, according to the requirements of parallel tests in the test specifications.

2.3 Four-point bending fatigue test

In this research, the UTM-30 mechanics testing system was used for the four-point bending (4PB) fatigue test. The 4PB fatigue test, as described in the AASHTO T321-17, were performed under the stress-controlled mode of loading with the stress amplitudes of 5000 kPa at a temperature of 10 ℃. A repeated sinusoidal load was applied at a frequency of 10 Hz without rest periods. According to the specifications, the stiffness at the 50th cycle of loading is considered the initial stiffness (E_0), and the number of load repetitions required to reduce initial stiffness of a specimen to 50% was traditionally used as the fatigue failure point.

2.4 Acoustic emission testing

The Micro-II Express digital AE system was used for acquiring sound emissions in asphalt mixture specimens. The AE sensors used were resonator sensors with an operating frequency range of 25-750 kHz and the signals were pre-amplified using a 40dB preamplifier. To further amplify the signals and help AE sensors stick to the specimen, grease was used for gaining good acoustic coupling on the surface of the sensor. Depending on the results of the environmental noise tests, the threshold of 30 dB was adopted, and the mechanical and electrical noises were screened. The sampling rate for the system was set to 1MHz to ensure all sensor output data was being recorded. The time parameter, includes peak definition time (PDT), hit definition time (HDT) and hit lock-out time (HLT), is intimately linked to the specimen size and wave propagation. In consideration of these factors, these time parameters were set to 30μs, 130μs and 300μs, respectively. Once the input parameters were set and the load was applied to the specimen, the waveform of AE signal is collected by the AE system, and several AE parameters were calculated from the waveform. These AE parameters were used for the material performance evaluation.

3. Results and analysis

3.1 Mechanical properties of asphalt mixture

During the test, some mechanical parameters, such as initial flexural stiffness(S_0), cumulative dissipated energy (U_f), stress (σ) and corresponding strain (ε), were recorded and used for the asphalt mixture performance evaluation. Figure 1(a) showed the time history of stress and corresponding strain during the 1st replicate of the 4PB fatigue test. It reflects the process of strain accumulation under the constant stress cyclic loading of 5000kPa. Figure 1 (b) illustrated that the test is terminated after the stiffness modulus of asphalt mixture decreases to 50% of the initial value. It reflects the process of degeneration of the material property.

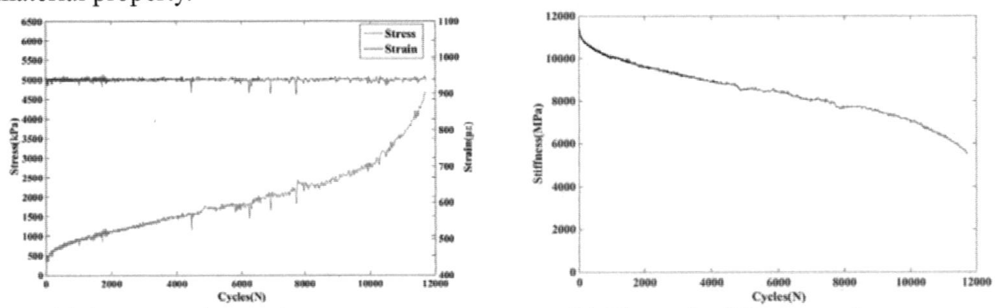

(a) Stress-cycles-strain curve (b) Flexural stiffness–cycles curve

Figure 1. Evolution curves of mechanical parameter

However, the fatigue damage process of asphalt mixture cannot be well reflected by the attenuation of stiffness modulus alone. It is well known that the energy is represented by the area under the stress-strain curve. If the loading and unloading curves coincide, the energy put into the material is recovered after the load is removed. As shown in figure 2, the strain is not recovered when the load is removed from the material, which means that there is energy lost in the material, energy that was dissipated through mechanical work, heat generation, or damage to the material to its original shape. Hence, the area of an ellipse is the dissipated energy of the material caused by one load cycle[7]. The total dissipated energy is the sum of each loop dissipated energy during the test.

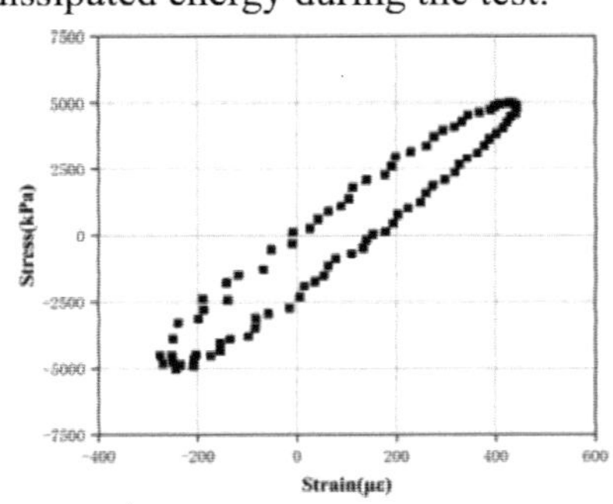

Figure 2. Dissipative energy calculation model

Dissipation energy which based on thermodynamics theory is an important parameter to describe fatigue damage. It gives full expression to the influence of cyclic stress and strain. Hence, many researchers suggest that the damage factor of asphalt mixture can be defined on the basis of dissipation energy, and it can accurately reflect the change of mechanical properties of asphalt mixture. Liu[8] proposed to the definition of damage factor by using the ratio of dissipation energy at a certain time to total dissipation energy, which can be calculated as in equation (1).

$$D_i = \frac{U_i}{U_f} \quad (1)$$

Where,

D_i—— the cumulative damage value of asphalt mixture after the i th cycle,

U_i—— the cumulative dissipation energy value of asphalt mixture after the i th cycle,

U_f—— the cumulative dissipation energy value of asphalt mixture at the end of test.

3.2 Acoustic emission characteristics of asphalt mixture

The release rapid of local energy in the form of transient elastic wave, generally known as acoustic emission. The process of deformation, crack initiation and expansion of asphalt mixture until the ultimate failure all generates the AE signal. The waveform of AE signal is collected by the detection device and can be processed and transformed into the various basic acoustic emission parameters. Among them, the AE energy and amplitude are always used in damage analysis because it reflects the strength of the signal.

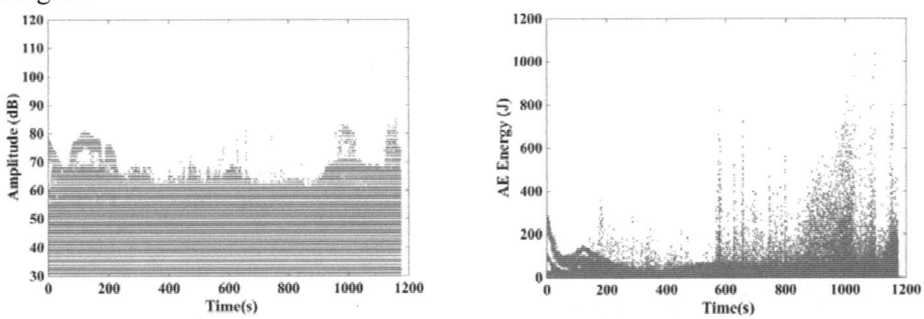

(a) Amplitude-time curve (b) AE Energy-time curve

Figure 3. Evolution curves of AE parameters over time

The initial cracks or voids within the asphalt mixture will be compacted and closed during the initiation of loading. As shown in figure 3, it is found that the value of AE Energy and Amplitude is high in the initial stage. This corresponds to where the initial deformation has been formed. It indicated that the material caused damage at this stage. In addition to, the state parameter revealed the paroxysmal characteristics at about 600 seconds. That might be because the internal structure of materials is adjusting in this moment. Prior to test termination, peaks in the AE energy parameter predominantly appeared. In the area of 1000 to 1200 seconds, the strain has increased drastically. At this time, the acoustic emission energy can reach the peak value of 1036 J, and the corresponding signal amplitude exceeds 85 dB. It means that the material has been badly damaged. Based on the foregoing, there is no denying that the variation of AE parameter is linked closely with material mechanical behavior.

3.3 Relationship Between Dissipative Energy and Acoustic Emission Energy

It is also known that the release of strain energy produced the AE signal. The AE energy reflects the strength of the signal. At the same time, dissipated energy is also a form of strain energy, and is linked closely with material damages. Thus, it is not difficult to find that there is a relationship between AE energy and dissipative energy, and the AE energy is closely related to damage. In fact, the variation of cumulative AE energy and cumulative dissipated energy had a similar tendency, as shown in figure 4.

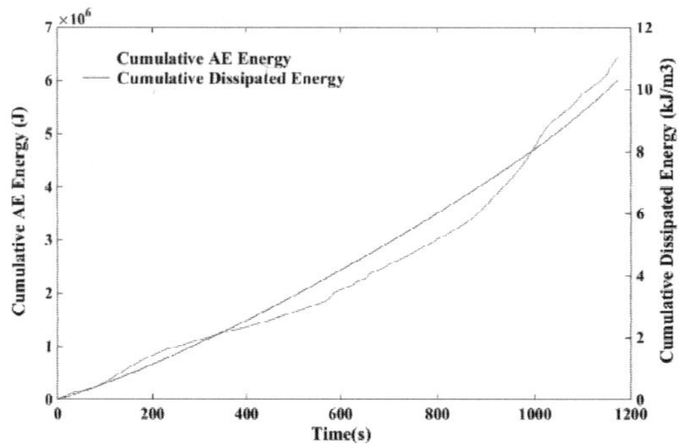

Figure 4. Evolution curve of AE energy and dissipated energy over time

The results showed that the energy of an acoustic emission signal is proportional to the dissipative energy associated with the damage. Therefore, the equation (2) which reflected the linear relationship between cumulative AE energy and cumulative dissipated energy was proposed.

$$W = 1744 \times U \tag{2}$$

Where,

W—— the cumulative AE energy,

U—— the cumulative dissipation energy.

Using equation (1) and (2), the damage factor D of asphalt mixture can be calculated from AE energy through equation (3).

$$D = \frac{W(\varepsilon)}{W_m} \tag{3}$$

Where,

$W(\varepsilon)$ —— the cumulative AE energy in a certain amount of strain,

W_m—— the cumulative AE energy for a specimen that is failure.

3.4 Damage evolution properties

As for cyclic loading tests, damages happen all the time. The damage constitutive laws which is based on the definition of the effective stress concept is established as in equation (4).

$$\sigma = (1 - D)E\varepsilon \tag{4}$$

Where,

σ—— the nominal stress,

D—— the damage factor,

E—— the elastic stiffness of the undamaged material,

ε—— the total strain.

There are many methods of the definition of damage factor. Among them, the continuum damage mechanics (CDM) is a combination of the internal state variable theory and the thermodynamics of irreversible processes. This approach is appealing because it has the intrinsic simplicity and versatility, as well as its consistency. The Weibull distribution is an important method in the CDM. The rate of evolution of a scalar damage factor D is given by the Weibull distribution function as in equation (5).

$$\frac{dD}{d\varepsilon} = f(\varepsilon) \tag{5}$$

Where f (ε) is the Weibull distribution function given by the function as in equation (6).

$$f(\varepsilon) = \frac{\alpha}{\beta} \frac{(\varepsilon - \varepsilon_0)^\alpha}{\beta} \exp\left[-\frac{(\varepsilon - \varepsilon_0)^\alpha}{\beta} \right] \tag{6}$$

Where,

α and β —— the parameters related to material size and deformation characteristics,

ε_0—— the initial strain, which is obtained through test.

By integrating equation (5), the expression of damage factor can be obtained as in equation (7).

$$D = 1 - \exp\left[-\frac{(\varepsilon - \varepsilon_0)^{\alpha}}{\beta}\right] \tag{7}$$

The parameters (α and β) of the Weibull distribution function were usually estimated by the constitutive laws. However, the damage factor with AE energy provides another alternative. It can be used to calibrate the parameters in the Weibull distribution function. By fitting test data, the fitted value for α and β are 1.45 and 5630, respectively. The correlative degree between test result and Weibull distribution function is 0.94, as shown in figure 5. It presents that Weibull distribution function is only consistent with the front part of the 1st replicate test. This is due to the definition of fatigue failure criterion.

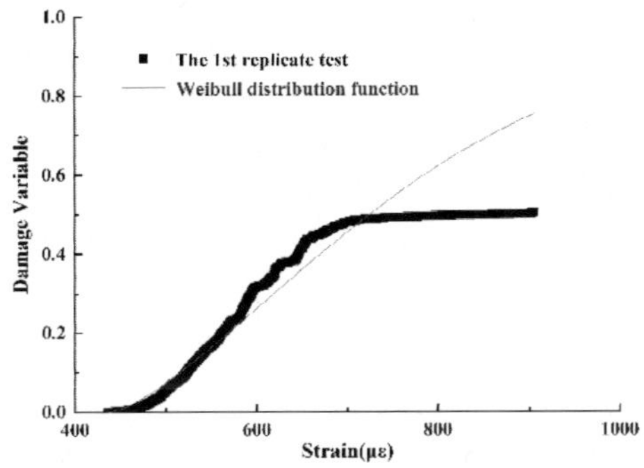

Figure 5. Fitting curve of damage factor with AE parameters

There is a strong correlation between the result of continuum damage mechanics model and mechanical test. It means that the damage factor based on AE energy is reliable and reasonable. Thus, a damage factor equation was applied to describe the asphalt mixture damage with AE energy, which can be calculated as in equation (8). Because of the definition of fatigue failure criterion, damage factor has an upper limit of 0.5, and so it has been corrected by a constant.

$$D = \frac{W(\varepsilon)}{2 \times W_m} = 1 - \exp\left[-\frac{(\varepsilon - \varepsilon_0)^{1.45}}{5630}\right] \tag{8}$$

3.5 Finite element analysis

The FE simulation reproduces the boundary conditions that the specimen is subjected to in the 4PB device. As mentioned previously, the beam was loaded with a sine waveform, and other conditions were set in accordance with test. In addition, 12000 MPa and 0.4 are the material properties of Young's Modulus and Poisson's ratio respectively. The mesh used is based on a regular three-dimensional square grid with elements of 5 mm.

A user defined material subroutine (UMAT) was developed to calculate the bending responses with a commercial finite element software ABAQUS 6.14. In particular, the modulus of material for every cycle had to be redefined. At the end of each step, the maximum strain would be used to calculate the damage factor based on the Weibull's function model, and the modulus would have been corrected. Figure 6 showed the curves from experiments and FEM computation. It was found that the experimental results are consistent with the finite element results. The mechanical properties of the asphalt mixture decreased with the rise of the strain levels. The numerical computational method provides a reference for the research of fatigue damage in asphalt mixture.

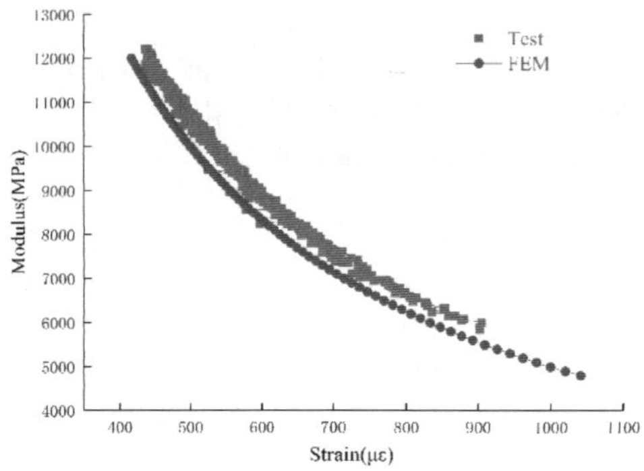

Figure 6. Comparison of test curve with FEM simulation curve

4. Conclusions

The findings and conclusions are as follows:

(1) The variation of AE parameter is linked closely with material mechanical behavior. AE energy as an important parameter can be utilized to evaluate the fatigue failure characteristics of asphalt mixture.

(2) Based on the AE technique, a novel calibration method to determine the CDM model parameters is proposed.

(3) An effective numerical computational method is developed to explore the evolutional law of damage in asphalt mixture.

Acknowledgement

The authors are grateful for the financial supports from the Natural Science Foundation of Zhejiang Province (LY18E080020) and the National Science Foundation of China (51408550).

Reference

[1] Virgili A, Partl M N, Grilli A and Santagata F A 2008 *Fatigue Fract. Eng. Mater. Struct.* **31** 967-979.

[2] Zhang J, Fan Z, Fang K, Pei J and Xu L 2016 *Constr. Build. Mater.* **102** 384-392.

[3] Kong B, Wang E, Li Z, Wang X, Liu J and Li N 2016 *Rock Mech. Rock Eng.* **49** 4911-4918.

[4] Geng J, Sun Q, Zhang Y, Cao L and Zhang W 2017 *Constr. Build. Mater.* **149** 9-16.

[5] Behnia B, Dave E V, Buttlar W G and Reis H 2016 *Constr. Build. Mater.* **111** 147-152.

[6] Li X and Marasteanu M 2010 *Eng. Fract. Mech.* **77** 1175-1190.

[7] Ghuzlan K A and Carpenter S H 2006 *Can. J. Civ. Eng.* **33** 890-901.

[8] Liu X S, Ning J G, Tan Y L and Gu Q H 2016 *Int. J. Rock Mech. Min. Sci.* **85** 27-32.

IOP Conf. Series: Materials Science and Engineering **542** (2019) 012052 doi:10.1088/1757-899X/542/1/012052

Preparation and High Temperature Tensile Mechanical Properties of Multilayer Metal Hot Forging Die Specimens

Huajun Wang[1], Ye Gao[2], Liang Yao, Zhenzhen Qi, Chundong Zhu and Qingyang Liu

School of Materials Science and Engineering, Wuhan University of Technology, Wuhan 430070 China

Email: [1] wanghuajunhb@163.com; [2] 15105423272@163.com

Abstract. The bonding strength of cladding is the key to the normal use of multilayer metal hot forging die. This paper presents a method of multilayer metal specimens prepared by PTAW. The specimens are selected H13 steel as the substrate, M2 steel as the transition layer, Ni60A alloy as the high temperature cladding, and the high temperature tensile test is performed on the Gleeble 3500 machine. The research results show that the substrate and cladding have good metallurgical bonding and the cladding hardness is higher than the substrate. Compared with H13 specimens, tensile strength of cladding metal specimens increase, elongation and yield strength decrease. All tensile fractures appeared on the H13 substrate during the experiment. It indicates that the cladding strength and bonding strength are higher than the substrate and the high temperature strength of cladding metal hot forging die is overall superior to H13 hot forging die.

Keywords: Multilayer metal hot forging die; PTAW; Gleeble high temperature tensile; Mechanical property; Fracture

1. Introduction

Hot forging die bears cyclic mechanical load during billet deformation process. It will be easily damaged due to high temperature oxidation, wear and thermo-mechanical fatigue [1][2]. Based on the temperature distribution and service requirements of hot forging die, one or more layers of high performance material claddings are formed on the surface of hot forging die, then a special multilayer structure obtained, namely multilayer metal hot forging die (MMHFD) [3][4]. This method can improve high temperature properties of hot forging die, such as wear resistance and impact fatigue resistance, and uses in the field of life extension and repair of hot work dies. Currently gradient structure of hot work die has become a hot spot. Xu [5] used Stellite6+WC powder on SM400B to prepare the welding cladding and FGM cladding respectively by laser cladding technology. They found that the functional gradient cladding has less crack sensitivity. Cong [6] used laser remelted specimens to conduct tensile tests, and found that laser repair technology of Biomimetic could improve structure property and tensile strength. Soodi [7] tested the mechanical properties of FGM and sandwich structure under different formulations by room temperature tensile test. BERTI [8] conducted thermal fatigue test for H11 steel by hot forging die service environment simulator, and studied the effects of die surface temperature gradient and hot forging process parameters on the thermal fatigue properties of H11.

In order to study the tensile properties of MMHFD specimens and the bonding strength between cladding and substrate, we propose a method of transverse stretch specimen preparation process and conduct transverse stretching experiments of H13 homogeneous, bimetal (H13+Ni60A) and multilayer metal (H13+W6Mo5Cr4V2+Ni60A) specimens by Gleeble 3500 thermal simulation testing machine. Comparing the high temperature mechanical properties of the above three specimens.

2. Materials and experiments

2.1 Material Design of Multilayer Metal Forging Die Specimen

The temperature field of homogeneous hot forging die can be divided into Temperature Fluctuation Zone (600-700°C), Temperature Equilibrium Zone (400-600°C) and Cooling Zone (200-400°C). According to the temperature characteristics, we chose Ni60A alloy for Fluctuation Zone, W6Mo5Cr4V2 (M2) high-speed steel for Equilibrium Zone, and H13 steel for Cooling Zone. We used M2 steel as a transition layer [9]. Table 1 shows the thermophysical properties of three materials [10].

Table 1. Thermophysical properties of three materials

Material	Coefficient of Linear Expansion ($10^{-6}K^{-1}$)	Thermal Conductivity ($W \cdot m^{-1} \cdot K^{-1}$)	Specific Heat Capacity ($J \cdot Kg^{-1} \cdot K^{-1}$)	Elasticity Modulus (GPa)
H13	10.3-12.2	31.2-28.6	506-587	185-190
M2	10.8-11.27	27.1-26.5	—	230-250
Ni60A	13.9-14.4	26.34-25.56	—	180-195

It is very difficult to prepare specimens conforming to national standards by surfacing. This study proposed a new method of preparing tensile specimens of bimetal and multilayer metal (Figure 1). We machined the groove on the H13 substrate before welding, then selected the "S" path on the groove for PTAW. We chose M2 steel as transition layer and Ni60A alloy as surface cladding. Wire cutting specimens referred to the national standard GBT228-2002.

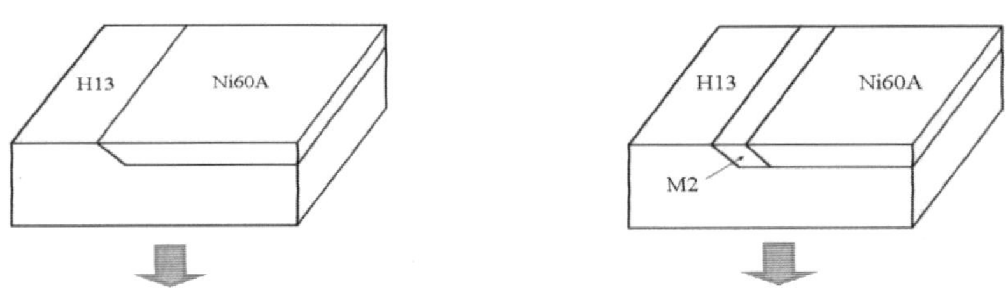

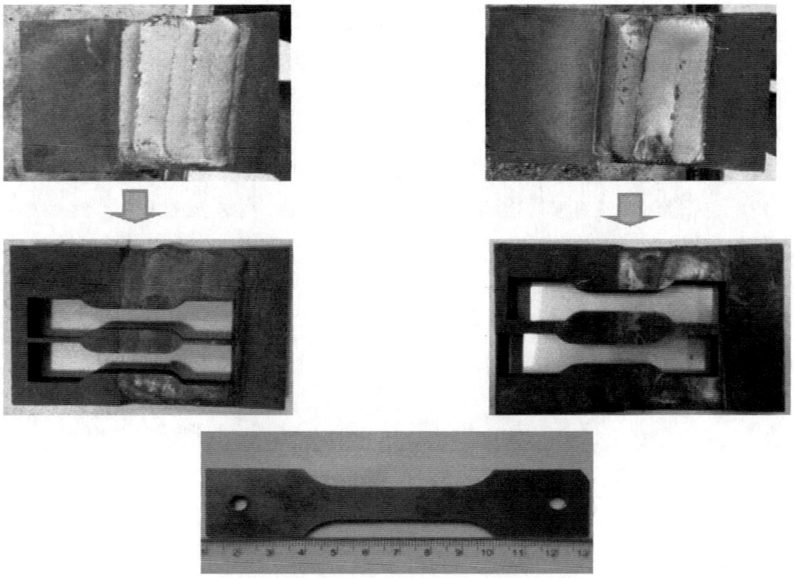

Figure 1. Preparation process of bimetal and multilayer metal tensile specimens

2.2 Experiment in mechanics

Schematic drawings of the high temperature tensile specimens show in Figure 2. J, SH and DH represent homogeneous, bimetal and multilayer metal specimen respectively. Parameters of high temperature tensile test were as follows, the heating temperature were 700 °C, the heating rate was 100 $°C·s^{-1}$, the holding time was 30 s, the deformation rate was 0.02 $mm·s^{-1}$.

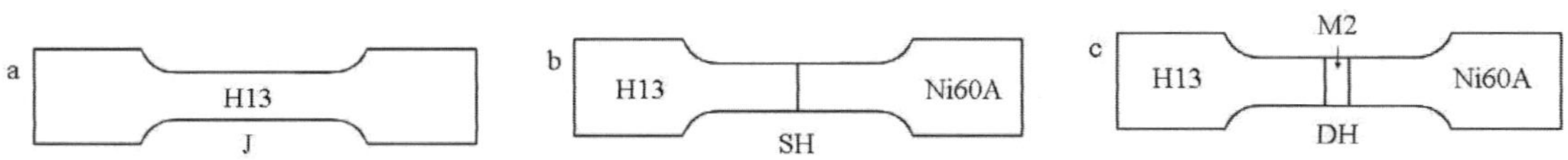

Figure 2. Schematic drawings of the high temperature tensile specimens (a) homogeneous specimen; (b) bimetal specimen; (c) multilayer metal specimen

3. Result and discussions

3.1 Analysis of microstructure and hardness

Figure 3 shows microstructure from H13 substrate to Ni60A cladding (400×). There are globular or blocky perlite and reticulation secondary carbide in substrate (Figure 3a). The heat source of welding makes temperature raise in the zone, leading to the precipitation of secondary cementite and formation of reticulation cementite (Figure 3b). The metallurgical bonding between the fusion interface of the substrate and the transition layer has no obvious cracks or pores (Figure 3c). The transition layer is grown in planar crystal form on the surface of the substrate and then grown in a dendritic form, shows as cast dendrite microstructure (Figure 3d). There is no apparent interface between Ni60A cladding and transition layer, which means it has excellent transition performance (Figure 3e). Ni60A cladding microstructure contains γ-Ni solid solutions, carbides and borides (Figure 3f).

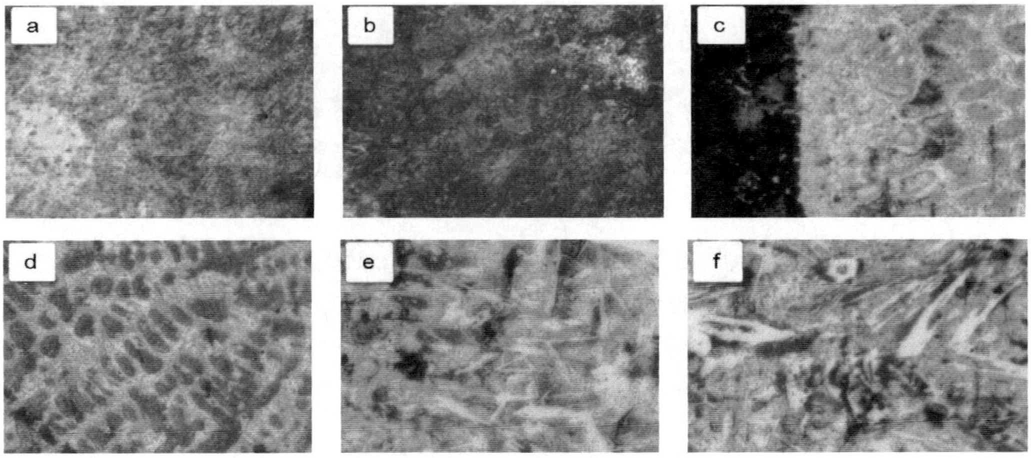

Figure 3. Microstructure from H13 substrate to Ni60A cladding (400×)
(a) substrate; (b) substrate HAZ; (c) fusion interface between substrate and transition layer; (d) transition layer; (e) fusion interface between transition layer and nickel cladding; (f) nickel cladding

Figure 4 shows the microhardness curves of bimetal and multilayer metal. The microhardness of cladding is higher than substrate. The hardness of bimetal and multilayer metal vary the same on the substrate side. Affected by the quenching of heat source, the hardness of heat affected zone increases to 750HV.

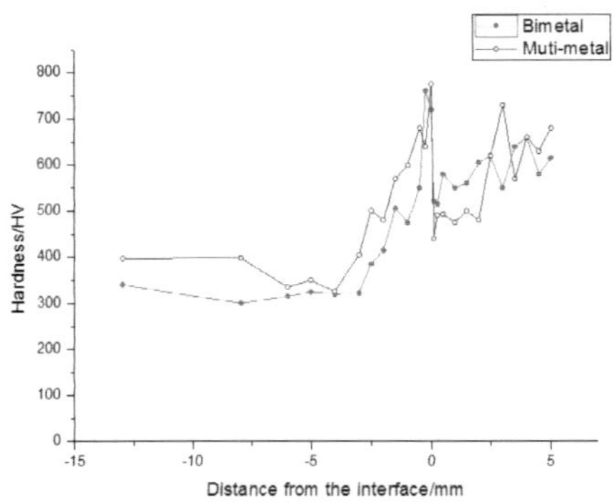

Figure 4. Microhardness curves of bimetal and multilayer metal

3.2 High temperature tensile property

Table 2 shows mechanical performance of high temperature (700 °C) tensile specimens. The plasticity of homogeneous specimen is the best, whose elongation reaches 96.5%. The multilayer specimen reaches 56% and bimetal specimen reaches 44.3%. Elongation of cladding metal specimen is significantly reduced compared with homogeneous specimen; elongation of multilayer metal specimen increases compared with bimetal specimen. The result shows that middle cladding can reduce stress effectively.

Figure 5 shows the high temperature tensile stress-strain curves of specimens. There is no yield point in bimetal tensile curve, no significant plastic deformation before breaking, and no obvious necking phenomenon (Figure 6b). It indicates that bimetal tensile fracture is brittle. There are two yield points of multilayer metal in tensile process. The reason may be that heat affected zone is reannealed and yielded.

Table 2. Mechanical performance of high temperature (700 °C) tensile specimens

Specimen	Elongation (%)	Yield point(MPa)	tensile strength(MPa)
Homogeneous	96.5	128.3	336.4
Bimetal	44.3	—	409.3
Multilayer metal	56	30.6 / 123.3	348.8

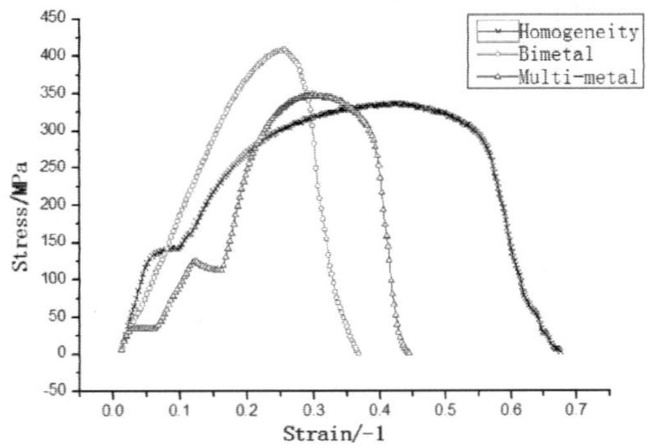

Figure 5. High temperature tensile stress-strain curves of specimens

3.2.1 High temperature stretching fracture analysis

Figure 6 shows the fractures of high temperature tensile specimens. The fracture of homogeneous specimen appears in the center area of specimen, with obvious necking phenomenon and serious oxidation in fracture area (Figure 6a). The fracture appears on the H13 substrate 5-7mm from fusion line, instead of the bimetal specimen center (Figure 6b). The fracture of multilayer metal specimen appears 5mm from the fusion line (Figure 6c).

The fractures appear on H13 substrate instead of cladding. That means the high temperature strength of cladding is better than H13 substrate. The fractures don't appear on fusion area, which means the bonding strength between substrate and cladding and the tensile strength of fusion area is better than H13 substrate.

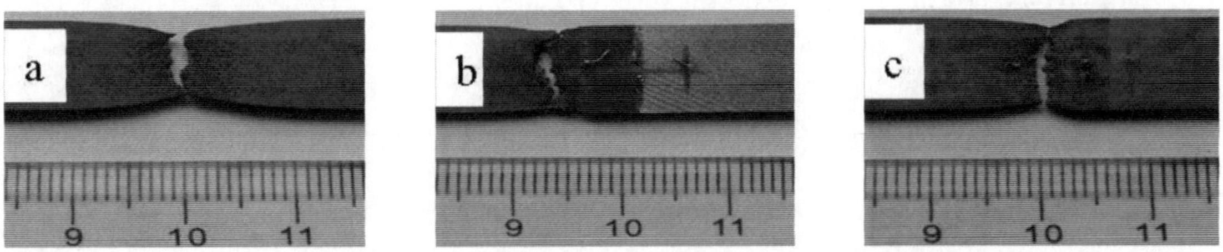

Figure 6. Fractures of high temperature tensile specimens (a) homogeneous specimen; (b) bimetal specimen; (c) multilayer metal specimen

4. Conclusion

(1) Propose a solution of preparing multilayer metal specimens by PTAW.

(2) Multilayer metal hot forging specimens prepared by PTAW have good metallurgical bonding between cladding and substrate, and the microhardness of the cladding is higher than substrate.

(3) Tensile strength of the cladding metal are higher than homogeneous metal, elongation and yield strength decrease during the high temperature tensile test.

(4) The cladding strength and bonding strength are higher than H13 substrate. High temperature strength of cladding metal forging die is better than H13 hot forging die.

Acknowledgements

This work was financially supported by the Natural Science Foundation of China (51475346), Natural Science Foundation of China (51875427) and Wuhan University of Technology Graduate Outstanding Thesis Cultivation Project (2017-YS-001). The authors gratefully acknowledge Ms. Bing Xie for her assistance in experiment.

References

[1] Summerville E, Venkatesan K and Subramanian C 1995 Wear Processes in Hot Forging Press Tools. *Materials & Design* vol 16 pp 289-94

[2] Kang J H, Park I W, Jae J S and Kang S S 1998 A study on die wear model of warm and hot forgings. *Metals & Materials International* vol 4 pp 477-83

[3] Wang H J, Wu Y Z, Wang H C, Sun Y Z and Wang G 2011 Design method and verification for long life hot forging die. *Material Research Innovations* vol 15 pp 377-80

[4] Wang H C, Long M L 2004 FEM analysis of the integrated stress exerted on forging dies. *Die & Mould Industry* vol 3 pp 43-7

[5] Xu G J, Kutsuna M, Liu Z J and Sun L Q 2006 Characteristic behaviors of clad layer by a multi-layer laser cladding with powder mixture of Stellite-6 and tungsten carbide. *Surface and Claddings Technology* vol 201 pp 3385-92

[6] Cong D L, Zhou H, Ren Z N, Zhang Z H, Zhang H F, Meng C and Wang C W 2014 The thermal fatigue resistance of H13 steel repaired by a biomimetic laser remelting process. *Materials & Design* vol 55 pp 597-604

[7] Soodi M, Masood S H and Brandt M 2014 Tensile strength of functionally graded and wafer layered structures produced by direct metal deposition. *Rapid Prototyping Journal* vol 20 pp 360-68

[8] Berti G A, Monti M. 2005 Thermo-mechanical fatigue life assessment of hot forging die steel. *Fatigue & Fracture of Engineering Materials & Structures* vol 28 pp 1025-34

[9] Wang H J, Yao L, Lu J F, and Peng J B 2015 Cyclic load analysis on multilayer metal hot forging die during hot forging. *Forging & Stamping Technology* vol 40 pp 1-4

[10] Feng D, Lin G Y, Liu J, Zou Y M and Peng D S 2010 Mathematical model of thermal deformation behaviors for h13 steel. *Iron & Steel* vol 45 pp 52-6

IOP Conf. Series: Materials Science and Engineering **542** (2019) 012053 doi:10.1088/1757-899X/542/1/012053

Thermal and hydraulic performance of micro pin fin heat sinks with different pin fin shapes

Peng Wang[1]**, Liang Chen** [2,3*]

1 Tianjin Key Laboratory for Civil Aircraft Airworthiness and Maintenance Airworthiness and Maintenance, Civil Aviation University of China, Tianjin 300300, China

2 Department of Mechanical & Electrical Engineering, Xiamen University, Xiamen, 361005, China

3 Shenzhen Research Institute of Xiamen University, Shenzhen 518000, China

E-mail: brightc66@163.com

Abstract. Efficient cooling of high heat flux devices has promoted the development of micro pin fin heat sinks. In this study, four micro pin fin heat sinks (MPFHSs) with different pin fin shapes, i.e., circular, square and diamond and streamline ones are fabricated by laser micromilling methods. Their single-phase heat transfer and fluid flow performance are explored to assess the pin fin shape effect. Convective flow tests are conducted at Reynolds number of 150-750 at two heat fluxes using deionized water as the coolants. It was found that the streamline MPFHS presented the best heat transfer performance. On the contrary, the diamond fin pin heat sink presented the worst heat transfer performance, whereas it featured the smallest pressure drop among the four MPFHSs. The diamond micro pin fins showed the smallest thermal resistance at a certain pressure drop, and should be selected as the optimum one for the MPFHS applications when the overall thermal and hydraulic performance is considered.

1. Introduction

The rapid development in the microelectronic technologies has promoted the miniaturization and high speed operation of electronic devices, e.g. projector, LED graphic chip, high power laser, etc. It induces a tremendous increase in power density, and results in serious thermal crisis [1]. Efficient cooling technologies are thus critical for the safe and reliable operation of electronic devices. Among them liquid cooling with micro pin-fin heat sinks (MPFHSs) is a promising solution [2]. The micro pin-fin heat sink features large surface area per given volume and low flow resistance, compact dimensions and small working fluid inventory requirements. Besides, the micro pin fins are helpful for the flow mixing and thermal and hydraulic boundary layer re-development, which introduces excellent heat transfer performance for solving high heat flux problems in electronic components [3]. Therefore, they have attracted increasing attentions in both research and industrial areas in recent years [4–6]. Many studies have been conducted to explore thermal and hydraulic performance of micro pin-fin heat sinks, and the applications in microelectronic chips[7], high-power LED lighting [8], turbine blades[9] have been also implemented.

Most of the works are focused on the heat transfer and pressure drop characteristics of a certain type of micro pin fins. The research groups of Peles et al. [10] comprehensively explored the convective flow performance of circular micro pin fins. The end-wall effect was found to contribute to a delay in flow separation for MPFHS. Besides, very low thermal resistances can be reached using a pin-fin heat sink. Lee et al. [6] experimentally studied the fluid flow and heat transfer of oblique pin fin heat sinks. A parametric study involving the oblique angle, fin pitch is also carried out. Notable heat transfer augmentation was reached for these oblique pin fin heat sinks with maximum heat transfer performance enhancement at 47% compared to conventional straight microchannels. Liu et al. [11] investigated the convective liquid flow and heat transfer performance of micro square pin fin heat sink using water. High heat transfer coefficient was obtained with water single-phase flow in micropin-fin arrays, suggesting

IOP Conf. Series: Materials Science and Engineering **542** (2019) 012053 doi:10.1088/1757-899X/542/1/012053

that micropin-fin heat sinks can be very effective meeting the needs of many high-heat-flux electronic cooling applications.

Besides, the effects of arrangement patterns and aspect ratios on the thermal performance of MPFHSs are also explored. Jeng et al. [12] experimentally studied the pressure drop and heat transfer of a square pin-fin array in a rectangular channel with different relative longitudinal pitch, relative transverse pitch and pin fin arrangements (in-line or staggered). It was found that the staggered pin-fin array presented larger pressure drop. The optimal inter-fin pitch and transverse pitch are also given. Roth et al. [13] found that the staggered pin fin arrangement enhanced heat transfer for 56% compared to corresponding in-line arrangement, which should be selected for efficient cooling applications. Moreover, Rasouli et al. [14] explored pitch and aspect ratio effects on single-phase liquid flow and heat transfer of micro pin fin heat sinks together with flow visualization. Unsteady vortex shedding in the micro pin fins were also found, which played a notable role on the heat transfer performance.

From the above literature review, most of the previous studies of micro pin fins were focused on the thermal and hydraulic performance of a certain type of micro pin fins, or MPFHSs with different arrangement patterns and aspect ratios. The micro pin fin shapes may play a notable role on the thermal and hydraulic characteristics of MPFHSs, which has been pointed out in the studies of silicon MPFHS with circular, rectangular, hydrofoil and cone-shaped micro pin-fins by Kosar and Peles [3] . Nevertheless, the information for the MPFHSs with different pin fin shapes are far from sufficient, especially for the MPFHSs with metal micro pin fins. To this aim, we fabricated four kinds of micro pin fins with different shapes, i.e., square, diamond, circular and streamline ones in copper to build copper micro pin fin heat sinks. Fluid flow and heat transfer performance of these four MPFHSs were explored and compared for the design optimizations.

2. Experiments

2.1 Fabrication of micro pin fins

The micro pin fins are fabricated by a laser micromilling method on pure red cooper plates (99.9% Cu) with the dimensions of $45 \times 20 \times 2$mm (length $\times$ width $\times$ height). Laser micromilling, as a non-contact and highly precise microfabrication technique, can fabricate a part via laser ablation by controlling the materials removal. A prototype pulsed fiber laser (IPG, No: YLP-1-100-30-30- HC-RG, Russia) system is utilized, and the cross machining route and loop multiple-pass reciprocating scanning strategy are adopted to process the micro pin fins. The laser is set to produce 100 ns pulses with a 1064nm fundamental wavelength (λ) at a repetition rate of 20kHz. All the micro pin fins are fabricated at the same processing parameters, i.e., the laser output power of 27W, scanning speed of 250mm/s and scanning times of 20.

Fig.1 shows the fabricated micro pin fin heat sinks with different pin fin shapes, i.e., square, diamond, circular and streamline ones. They are in staggered arrangements. All the micro pin fins are fabricated in the center of copper plate with the dimensions of 17×45mm ($W_{ht} \times L$, W_{ht} and L are the width of heat transfer area and length of heat sink, respectively). The geometric dimensions of these micro pin fins are shown in Table.1.

Table 1 Geometric parameters of four micro pin fins samples

Pin shape	Circular	Square	Diamond	Streamline
Length of a micro pin fin, L_{fin}, (mm)	0.5	0.5	1.5	0.5
Width of a micro pin fin, W_{fin}, (mm)	0.5	0.5	0.4	0.5
Longitudinal pitch, S_L, (mm)	1.5	1.5	3.43	1.5
Transverse pitch, S_T, (mm)	1.5	1.5	0.92	1.5
Cross-sectional area of a single micro pin fin, A_{fin}, (mm^2)	0.196	0.25	0.3	0.297
Number of rows, N_{row}	13	30	30	30
Number of pin fins in a single row, N_s	35	22	22	22

Total number of micro pin fins, N	660	660	455	660

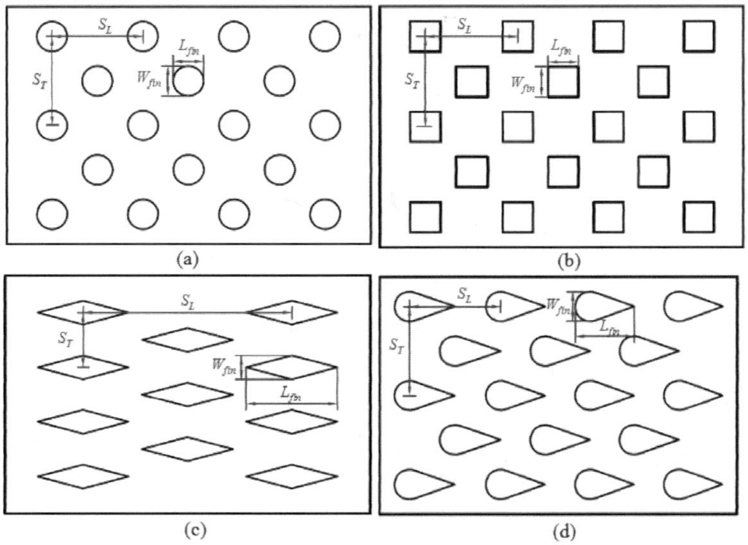

Figure 1 Illustrations of the four micro pin fins: (a) circular; (b) square; (c) diamond (d) streamline

2.2 Description of the test system and procedure

The convective flow test system is similar to that in the previous study [5], which is described here briefly. The experimental test loop is composed of a water tank, a magnetic gear pump, a filter and valves, a micro turbine type flowmeter, a constant temperature water bath and heat exchanger, MPFHS test section and heating modules, a microscope together with high-speed camera for flow visualization and data acquisition system. The MPFHS test section consists of the micro pin fin samples, copper block heating module, inlet and outlet plenums, a Pyrex glass cover plate, a polyetheretherketone (PEEK) flow housing, an assembling top plate and insulating components, as shown in Fig.2. The heating copper block with cartridge heaters inside consists of an upper rectangular section of 20x45mm, which is identical to footprint dimensions of micro pin fin samples. The micro pin fin sample is soldered on top of the copper block using a 0.5mm thin layer of solder (Pb-Sn-Ag-Sb, k_s=50W/m·K) to minimize the contact thermal resistance. Any possible leakage of coolants around the periphery of heating copper block is prevented by utilizing RTV silicone rubber in the interface of copper block and flow housing. Good insulation of heating copper block is maintained as it is wrapped by the PEEK insulating block with low thermal conductivities. The assemblies of cover plate, glass cover, microchannel sample, flow housing and insulating block are clamped together by screws with an O-ring to ensure the good contact between the glass cover and the top surface of micro pin fins.

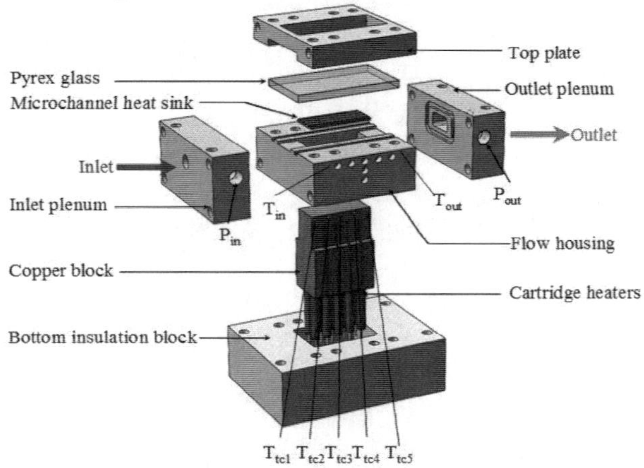

Figure 2. Schematic of the cross-section of test section.

The test section is heated by ten cartridge heaters with the maximum total power of 1000W. Five K-type shielded thermocouples of the diameter of 1mm are embedded in the copper block at a distance of 2mm below the top surface of copper block to measure the stream-wise wall temperature distribution of micro pin fins. The locations were 2.5, 12.5, 22.5, 32.5, 42.5mm, respectively, from the channel inlet position. The inlet, outlet temperatures are monitored by type-K thermocouples, and the inlet and outlet pressure during the test are measured by two pressure transducers with a response time of 3 ms. All the temperatures and pressure drop data are collected by an Agilent 34970A data acquisition system.

In this study, deionized water is utilized as the test coolant. The water is fully degassed via vigorous boiling for about half an hour prior to each experimental run. Tests are conducted for all the MPFHS samples at inlet temperature of 33°C and two heat fluxes levels (133 and 267kW/m^2) in atmospheric conditions. The supplied power and inlet temperature are maintained to be constant, and differentf coolant flow rates are supplied by adjusting the flowmeter. When the wall temperatures are fluctuated within 0.5°C, the steady state is thought to be reached. Then all temperature and pressure data are recorded at 1 s interval for 2 min. The heat transfer and pressure drop data can thus be calculated by the averaged values of measurement results for all the period of 2 min.

2.3 Theoretical background and data reduction

The sensible heat gained by the coolant can be determined from the energy balance,

$$q_{eff} = \rho c_p \dot{V}(T_{out} - T_{in}) \tag{1}$$

where $\dot{V}$ is the volumetric flow rate, T_{in} and T_{out} is the inlet and outlet coolant temperatures, respectively. c_p and ρ are the specific heat and density of coolant, respectively.

The average convective heat transfer coefficient can be obtained by Newton's law of cooling,

$$h = \frac{q_{eff}}{(\overline{T_w} - \overline{T_f})A_{ch}} \tag{2}$$

where $\overline{T_f}$ is the average of the inlet and outlet coolant temperatures,

$\overline{T_w}$ is the average wall temperature of the five thermocouples locations along the microchannel,

$$\overline{T_w} = \frac{T_{w,tc1} + T_{w,tc2} + T_{w,tc3} + T_{w,tc4} + T_{w,tc5}}{5} \tag{3}$$

The wall temperatures of microchannels ($T_{\text{w,tci}}$) at the thermocouple locations are derived from one dimensional Fourier's heat conduction equation. They are given in detail in a thermal resistance network in the previous study [15], and are given as follows

$$T_{w,tci} = T_{tci} - q_{eff}\left(\frac{l_{Cu}}{k_{Cu}A_t} + \frac{l_{hs}}{k_m A_t} + \frac{l_s}{k_s A_t}\right) \tag{4}$$

where T_{tci} is the local thermocouple reading(i=1-5), k_{Cu} , k_s , k_m were the thermal conductivities o f copper, solder and the MPFHS base, respectively. l_{Cu} and l_s are the distances from the thermocoupl e location to the top heating surface of copper block and thickness of solder, respectively. l_{hs} denotes t he distance between the backside surface of MPFHS sample and bottom surface of micro pin fins. A_{ch} is the total heat transfer area of MPFHS, which can be calculated using the approaches detailed in Ref. [5].

After the cacaulation of convective heat transfer coefficient, the Nusselt number cna be deterimined by,

$$Nu = \frac{hD_h}{k_f} \tag{5}$$

in which D_{h} is the hydraulic diameter, which is cacaulated as follows,

$$D_h = \frac{4A_c L}{A_t} \tag{6}$$

where A_t is the total surface area in contact with coolants, A_c is the minimum cross-section free flow area, L is the length of MPFHS sample.

The Reynolds number is expressed as

$$\mathrm{Re} = \frac{\rho u D_h}{\mu} \tag{7}$$

where ρ , μ , u are the density, dynamic viscosity of the fluid, average flow velocity, respectively.

The total thermal resistance of MPFHS is defined as

$$R_t = \frac{T_{\max} - T_{in}}{q_{eff}} \tag{8}$$

where $T_{\max}$ is the maximum measured temperature of the five thermocouples in the stream-wise direction, T_{in} is the inlet coolant temperature.

Uncertainties in individual temperature measurements are ±0.3℃ for the type-K thermocouples. The maximum errors for the flow meter and pressure transducer are 2% and 0.1% of full scale, respectively. The supply power measured by the digital power meter yield an uncertainty of 1%. Using the standard error analysis method [16], the maximum uncertainties in the Nusselt number and pressure drop can be estimated to be 7.5%, 5.4%, respectively.

3. Results and discussion

3.1 Heat transfer performance

Fig.3 shows the average Nusselt number as a function of the varation of Reynolds number in both heat fluxes cases. Generally, all the MPFHS samples presented a monotonic increase trend in Nu with increasing Re, as the thermal boundary layer tended to be thinner when the fluid velocity increased. For the three MPFHSs with circular, square and diamond micro pin fins, when the flow velocity was low with small Reynolds number, there are no notable differences among circular, diamond and square pin fin heat sinks. This can be related to that these MPFHSs featured thick boundary layers in small flow rates, and the cross-sectional shapes did not played s significant role on the heat exchange process. Nevertheless, when the Reynolds number increased to larger than about 250, the Nu~Re curves of the circular, diamond and square pin fin heat sinks tend to diverge from each other. The cross-sectional pin fin shapes showed a remarkable role on the heat transfer performance. The square pin fin heat sinks showed a fast rise in the Nu, and outperformed other two MPFHSs. The square pin fins seemed to facilitate the heat transfer process from pin fin walls to coolants. The following is the circular ones, and the diamond MPFHS presented the worst heat transfer performance.

For the entire test range of flow rates, the streamline MPFHS presented the best heat transfer performance among all the four MPFHSs. For example, at the test conditions with Re of about 400 in heat flux of 133kW/m^2, the streamline MPFHS present a Nu of 9.6, which is 32% larger than the square one, and 68% larger than the circular one. The streamline MPFHS enhanced the heat transfer for even 142% compared to the diamond one. As can be seen in Fig.1, the streamline micro pin fins featured circular portions in the head and triangular tails. The circular upstream head may induce intense flow disruptions and separations, which contributed to more flow mixing and enhancement in the convection heat exchange. Moreover, the triangular tails reduced the occurrence of vortices , as the separated flow can flow quickly along the long pin fin surface to the downstream. Therefore, they present the good heat transfer performance.

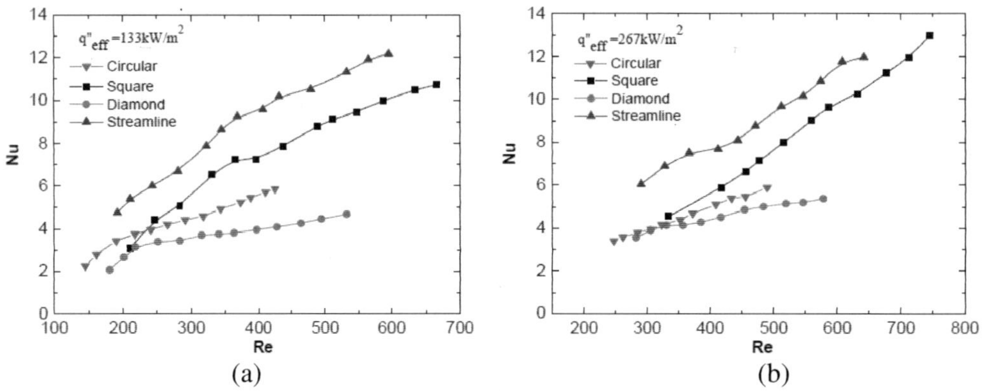

Figure 3 Average Nusselt number for MPFHS samples with different pin fin shapes: (a) q''eff =133kW/m^2; (b) q''eff =267kW/m^2.

3.2 Pressure drop

Fig.4 shows the pressure drop characteristics of four micro pin fin heat sinks at two levels of heat fluxes. All the samples presented a monotonic increase trend in pressure drop with increasing Reynolds number, as larger flow rate cases resulted in larger friction loss during the fluid flow in micro pin fins. This is consistent to the results of many other reports [6, 11]. For all the four MPFHSs with different cross-sectional shapes, it can be noted from Fig.4 that the circular pin fins showed the largest pressure drop penalty. This may be attributed that the circular pin fins without no protrusions or corners induced faster local flow velocities inside the micro pin fin flow passages, which resulted in the larger frictional drag and large pressure drop. Following is the streamline and square ones. The diamond fin pin heat sinks presented the smallest pressure drop. For example, at the test conditions with Re of about 400 in heat flux of 133kW/m^2, the $\triangle P$ of diamond MPFHS was just 1.25kPa, which is 42% smaller than the square one, 58% smaller than streamline one, and even 80% smaller than the circular one. The notable reduction in $\triangle P$ for diamond MPFHS can be related to the fact that the small protruded head facilitated to separate the main flow into two portions of flow, which reduced the collision between coolants and pin fins. This reduced the shear forces between coolants and pin fin walls, and induced the smallest form drag. Therefore, the smallest pressure drop can be noted for diamond micro fin pin heat sinks.

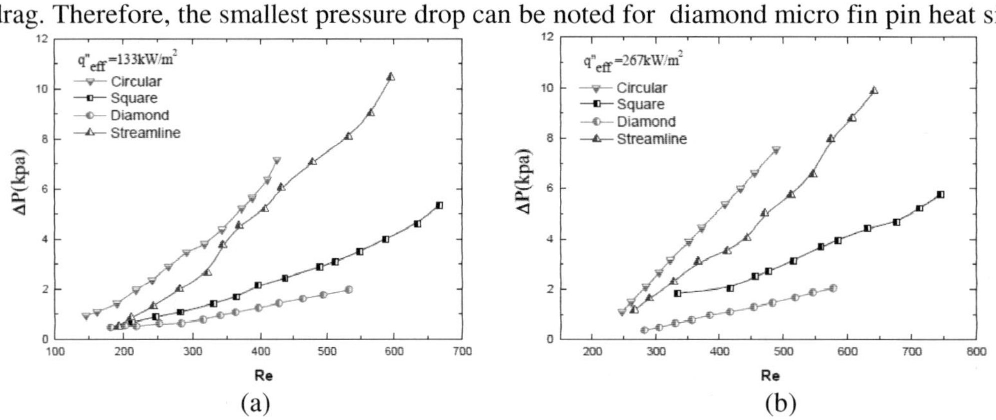

Figure 4 Measured pressure drop for MPFHS samples with different pin fin shapes: (a) q''eff =133kW/m^2; (b) q''eff =267kW/m^2.

3.3 Thermal resistance and overall performance comparison

To combine the evaluation of heat transfer and pressure drop characteristics of the heat sinks, the total thermal resistance was often assessed and compared with each other under the constant of pressure drop [10, 17]. Fig.5 shows the thermal resistance of all the MPFHS samples versus the pumping power in both heat fluxes tests. When the pressure drop increased together with the flow rates and Reynolds numbers, the total thermal resistance of all the MPFHS samples firstly decreased quickly, and then flattened out at large pressure drop which is linked to large flow rates conditions. As the heat transfer performance of the MPFHS samples increased considerably at large flow rates cases as mentioned before, the thermal resistance can be lowered down with the rise in flow rates.

For the four MPFHS samples with different pin fin shapes, it can be noted from Fig.5 the circular one generally presented the worst thermal performance with the largest thermal resistance, as it accompanied with the largest pressure drop. Followed is square and streamline ones. Despite that the streamline MPFHS presented the best heat transfer performance, its large pressure drop may play a negative role on the overall performance of MPFHS. The diamond micro pin fin heat sinks, however, present the smallest thermal resistance at the same pressure drop. Such behavior is in line with the pressure drop trends, and can be associated to the dominance of pressure drop in the $R_t \sim \triangle P$ curves [10]. Even though the diamond MPFHS showed the smallest heat transfer coefficients, it present much smaller pressure drop than other three samples. The significant decrease of pressure drop for the diamond MPFHS contributed to the lowest thermal resistance in the $R_t \sim \triangle P$ curves. Therefore, the diamond micro pin fins should be selected as the optimum one for the MPFHS applications when the overall thermal and hydraulic performance is considered. Nevertheless, if the heat transfer performance is of the prior considerations, the streamline one should be selected as the optimum one.

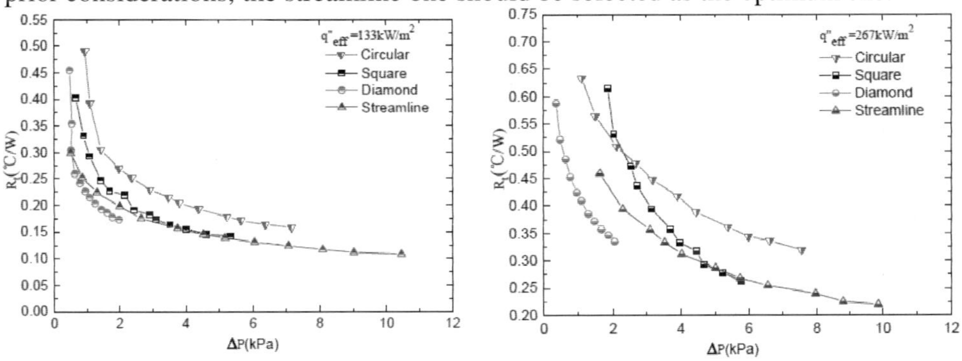

(a) (b)

Fig. 5. Thermal resistance against pressure drop for MPFHS samples with different pin fin shapes: (a) q''eff =133kW/m^2; (b) q''eff =267kW/m^2.

4. Conclusions

In this study, micro pin fin heat sinks with four different micro pin fin shapes were developed. Their thermal and hydraulic characteristics were assessed using water at two heat fluxes. It was found that the streamline MPFHS presented much lager Nusselt number, and presented the best heat transfer performance. The micro pin fin shapes firstly showed no notable effects on the heat transfer performance of circular, square and diamond MPFHS at small flow rates, but they began to play a significant role on the Nu when the Re increased to be larger than about 250. The circular pin fins showed the largest pressure drop penalty, followed by streamline and square ones. The diamond fin pin heat sinks presented the smallest pressure drop. The overall thermal and hydraulic performance in terms of total thermal resistance as a function of pressure drop generally followed the order of diamond, streamline, square and circular MPFHSs. The diamond micro pin fins should be selected as the optimum one for the MPFHS applications when considering the overall thermal and hydraulic performance.

Acknowledgments

The research was financially supported by and the Science and Technology Planning Project of Guangdong Province, China (No. 2017A010104002), and the open fund of Tianjin Key Laboratory for Civil Aircraft Airworthiness and Maintenance Airworthiness and Maintenance.

References

[1] Tang H, Tang Y, Wan Z, Li J, Yuan W, Lu L, Li Y, Tang K. Review of applications and developments of ultra-thin micro heat pipes for electronic cooling. 2018 *Appl. Energy* **223**,383–400
[2] Mohammadi A, Koşar A. Review on Heat and Fluid Flow in Micro Pin Fin Heat Sinks under Single phase and Two phase Flow Conditions. 2018 *Nanoscale Microscale Thermophys. Eng.* **22**,153-197
[3] Kosar A, Peles Y, Micro Scale pin fin Heat Sinks—Parametric Performance Evaluation Study. 2007 *IEEE Trans. Compon. Packag. Tech.* **30,** 1-11.
[4] Qu W, Siu-Ho A, Experimental study of saturated flow boiling heat transfer in an array of staggered micro-pin-fins. 2009 *Int. J. Heat Mass Transfer* **52**, 1853-1863.
[5] Wan W, Deng D, Huang Q, Zeng T, Huang Y. Experimental study and optimization of pin fin shapes in flow boiling of micro pin fin heat sinks. 2017 *Appl. Thermal Eng.* **114**, 436–449
[6] Lee YJ, Singh PK, Lee PS. Fluid flow and heat transfer investigations on enhanced microchannel heat sink using oblique fins with parametric study. 2015 *Int. J. Heat Mass Transfer* **81**, 325–336
[7] Kandlikar SG, Review and Projections of Integrated Cooling Systems for Three-Dimensional Integrated Circuits. 2014 *ASME J. Electronic Packag.* **136**, 024001
[8] Jang D, Yook S, Lee K, Optimum design of a radial heat sink with a fin-height profile for high-power LED lighting applications. 2014 *Appl. Energy* **116**, 260–268
[9] Marques C, Kelly KW. Fabrication and Performance of a Pin Fin Micro Heat Exchanger. 2004 *ASME J. Heat Transfer* **126**, 434-444
[10] Kosar A, Peles Y. Thermal-Hydraulic Performance of MEMS-based Pin Fin Heat Sink. 2006 *ASME J. Heat Transfer* **128**, 121-131
[11] Siu-Ho A, Qu W, Pfefferkorn F. Experimental Study of Pressure Drop and Heat Transfer in a Single-Phase Micro-Pin-Fin Heat Sink. 2009 ASME J. Electronic Packag. **129,** 479 -487.
[12] Jeng T, Tzeng S.Pressure drop and heat transfer of square pin-fin arrays in in-line and staggered arrangements. 2007 *Int. J. Heat Mass Transfer* **50,** 2364–2375.
[13] Roth R, Lenk G, Cobry K, Woias P. Heat transfer in freestanding microchannels with in-line and staggered pin fin structures with clearance. 2013 *Int. J. Heat Mass Transfer* **67**, 1–15
[14] Rasouli E, Naderi C, Narayanan V. Pitch and aspect ratio effects on single-phase heat transfer through microscale pin fin heat sinks. 2018 *Int. J. Heat Mass Transfer* **118**, 416–428
[15] Deng D, Tang Y, Liang D, He H, Yang S. Flow boiling characteristics in porous heat sink with reentrant microchannels. 2014 *Int. J. Heat Mass Transfer* **70**, 463–477
[16] Taylor JR. An Introduction to Error Analysis, second ed., University Science Books, 1997.
[17] Shen S, Xu JL, Zhou JJ, Chen Y. Flow and heat transfer in microchannels with srough wall surface. 2006 *Energy Conver. Manage.* **47**, 1311–1325

IOP Conf. Series: Materials Science and Engineering **542** (2019) 012054 doi:10.1088/1757-899X/542/1/012054

A new type of rotary paddle switch

Xu Tian Ming[1] and Wang Cheng[2] and Yu Wen Tao[3]

1 School of Mechanical Engineering, University of Jinan, Jinan 250022, China
2 School of Mechanical Engineering, University of Jinan, Jinan 250022, China
3 School of Mechanical Engineering, University of Jinan, Jinan 250022, China
E-mail: me_wangc@ujn.edu.cn

Abstract. The purpose of the invention is to provide a new type of rotary paddle switch which can be used to measure the material level in large area and to monitor the material level in the cylinder. It includes driving and measuring devices and racks;Driving device tansmits the motor's input torque to different space position;The measuring device is equipped with position sensors to observe the times of rotating inside and outside the bearing and output the signal to the computer;The longer measuring device and the shorter measuring device each measure a plane;The frame is made up of two parallel and concentric metal discs, which are used to install driving device and measuring device.The beneficial effect of this invention:the measurement of material level in large range is realized, and the monitoring of the material level in the cylinder is satisfied;It has advantages of high degree of automation .

1. Background information

At present, rotary paddle switch is commonly made up of motors and baffles and position switches which have several drawbacks inclouding short life span,higher cost and even poor working efficiency.Therefore,traditional rotary paddle switch is neither suitable for large-scale measurement nor meet the need of working efficiently [1,2] .In regard of this, those devices can not intuitively observe the accumulation of materials at every very moment.consequently,a new type of rotary paddle switch is designed by myself to overcome shortcoming and to promote woring efficiency.

2. Technical solution

The driving device is driven by the motor 1 which drives the central bull gear 11, the center bull gear 11 drives eight pinion gears 12 which are evenly circling the center bull gear 11 ,Each pinion 12's axle is fixed to a pinion pulley 4, each pinion pulley 4 radial outer side is fitted with another outer pulley 13,between two pulleys ,there is installed with a belt 5,As a result,A complete driving device is completely formed to transfer the motor's input torque to different space positions [3,4].The measuring device is composed by bearing 14, bearing 17 , outer shaft sleeve 9, central sleeve 10, BM50-R angle sensor , outer short block 6 and center gear 7.The frame consists of upside circular rack16 and lower circular rack 8. those are two concentric metal discs which are parallel to each other.Further, the measuring device consists of outer short block 6, central block 7 and bearing 14, bearing 17 and outer sleeve 9 , central sleeve 10, more than BM50-R angle sensor.meanwhile, Bearing 14 and outer ring 17 are fixed with outer sleeve 9 and central shaft block 10 separately.Additionly,inner ring of bearing 14 and bearing 17 are fixed with outer pulley13 which is away from pinion 12 .A position sensor is set on the outer ring of bearing to monitor whether there is rotation between the outer ring and the inner ring [5].That is measuring device.Further, the racks are made up of two parallel and concentric metal discs , the upper disc frame 16 and lower disc rack 8, the driving device is centered on the common axis of

IOP Conf. Series: Materials Science and Engineering **542** (2019) 012054 doi:10.1088/1757-899X/542/1/012054

the upper circular rack 16 and lower circular rack 8, and the measuring device is mounted on lower rack8 and eight outer pulley13 which are away from pinion12.Further, the central longer measuring device and the shorter peripheral measuring device are equipped with bearings, using the static friction force of the inner and outer ring of the bearing to rotate the baffle.It will not rotate when there is resistance;The central longer measuring device and the shorter peripheral measuring device each measure a plane separately.additionally,The measuring device is equipped with position sensors to observe the times of rotating between inside and outside ring of bearing and output the signal to the computer.The effect provided in the invention is merely the effect of one embodiment, not the total effect, and one of the above technical programmes has the following advantages or beneficial effects:1. material level can be measured more widely with higher reliability.2. cut the cost and figure of measuring device that have to be used.3.The sensor distribution at different locations makes it easier to see where the material is piling up.

3. Description of figures

Figure 1 is the front view of the whole structure ;Figure 2 is the vertical view of whole structure ;Figure 3 is a side view of the overall structure;Figure 4 is the three-dimensional structure diagram;Figure 5 is the position sensor, to be precise, a BM50-R angle sensor. as is depicted in these Photoes:1—motor, 2—bull gear shaft, 3—pinion shaft, 4—pinion pulley, 5—belt, 6—outer short block, 7—central long block, 8—lower circular rack, 9—outer sleeve(installed with position sensor), 10—central sleeve(installed with position sensor), 11—central bull gear, 12— pinion, 13—outer pulley, 14—bearing, 15—bearing, 16—upside circular rack, 17—bearing, 18— position sensor (BM50-R).

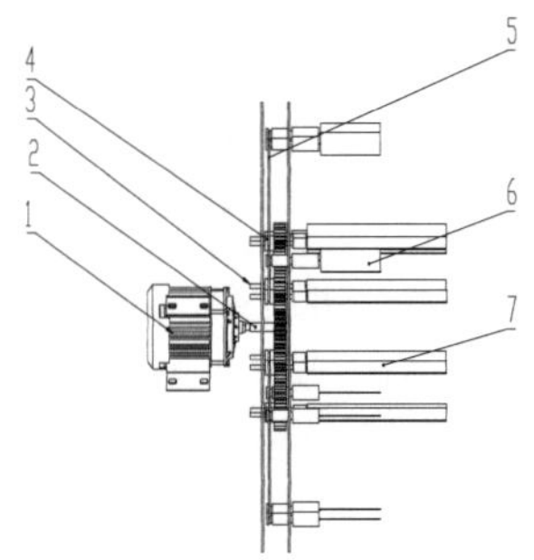

Figure 1. Front view of the whole structure

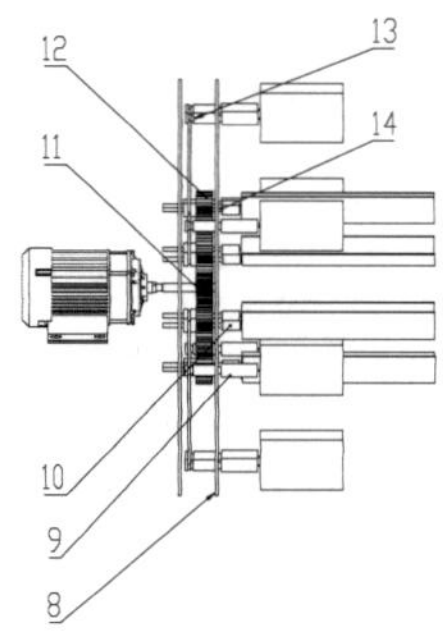

Figure 2. Vertical view of whole structure

IOP Conf. Series: Materials Science and Engineering **542** (2019) 012054 doi:10.1088/1757-899X/542/1/012054

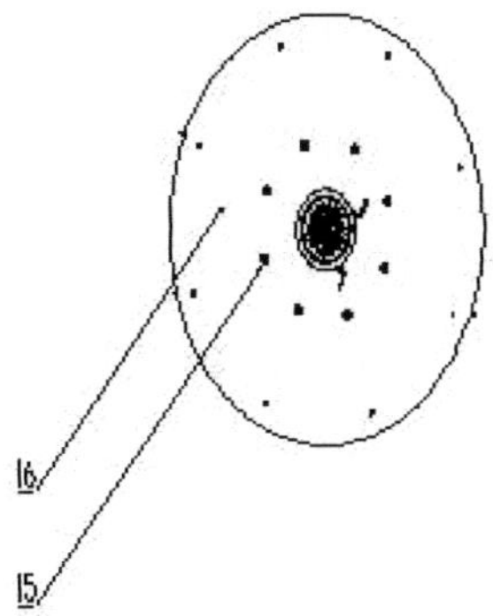

Figure 3. Side view of the whole structure

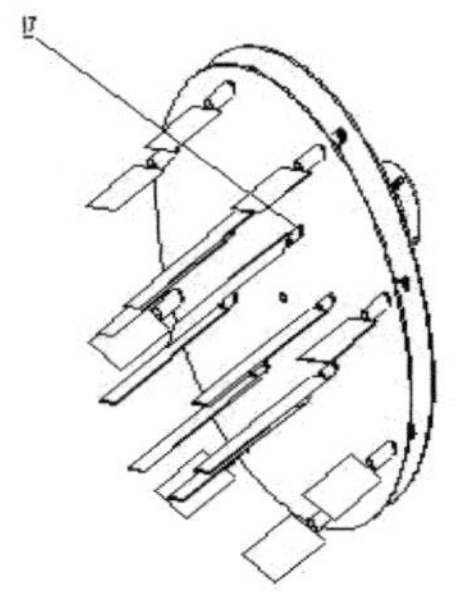

Figure 4. Three-dimensional structure diagra

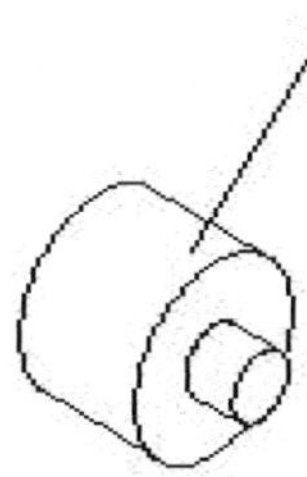

Figure 5. Position sensor

4. Concrete implementation

A new type of rotary paddle switch is shown in Figure 1, motor of the driving device which is a part of this mechine is connected by a shaft coupling to the bull gear shaft 2, the bull gear shaft 2 is coupled with the center gear 11 by the key and the underside surface of central bull gear 11 is parallel to the lower circular rack 8 , the distance between these two parts is 10mm.The central sleeve 10(installed with position sensor)and bearing 14 and center longer block 7 are all components that are used to make up of longer measuring device,compared with shorter peripheral measuring device it is more easily contact the material.Combining with Figure 1 and Figure 3, it is clear that the pinion axle 3 is distributed evenly around bull gear shaft 2 and installed on the lower circular rack 8 and upside circular rack 16. and the lower circular rack 8 and the upside circular rack 16 are equipped with bearings 15 and 17. it's inner circle is in trasition fit with the pinion shaft 3, and the pinion shaft 3 is fixed with pinion 12, each pinion 12 has a fixed pinion pulley 4.As shown in Figure 2, the slot of the outer belt pulley 13 is located at the same height as the slot height of the pinion pulley 4 and there is installed with belt 5 between the two pulleys, thus it eables the motor 1 to drive the central bull gear 11 .The center bull gear 11 meshes with the pinion 12 that surround bull gear evenly. The pinion pulley 4 drives belt 5 while belt 5 drives the outer pulley 13 and these components also rotates with the motor 1 [6-8].The outer pulley is composed of two parts, the upside part is rotating and the underside part is cylindrical rack.Among them, the rotating part can rotate freely in the center hole of the cylindrical rack.The cylindrical rack is connected with the lower circular tack 8, while the upper part of the rotating part is longer and it can pass through the cylindrical rack 8 and protrude from the lower part of the rack, The outer surface of protruding part is interference fitted with the bearing inner ring and the bearing outer ring matches the outer sleeve9 of the outer block 6.Because of the long length of the sleeve, there is plenty of space to fix sensors inside the sleeve to see if the inner and outer rings are rotating.When no material impede block, the inner and outer rings without load do not rotate,On the contrary,When too much material impedes the blocks , inner and outer rings start to rotate,The sensor sends a signal to the computer and can alarm or slow down transporting speed according to the program that designed bu user.As such the principle of a longer central measuring device is similar to the short one . Additional, central block 7 stationary fixes with the central sleeve 10,The inner surface of central hole of central sleeve 10 matches the outer ring of bearing 14 in transition fit， The inner ring of bearing 14 is connected with the pinion shaft 3 in trasition fit， Because the central sleeve 10 very long, Therefore, there is enough space to install position sensors to detect inner and outer circles.This is a new type of rotary paddle switch, the signals measured by measuring devices at different positions are sent to the computer by sensors, because the longer center measuring device and the shorter peripheral measuring device each measure a plane;Therefore,You can intuitively observe the way that how materials are piling in the cylinder resorting on computer.

5. Conclusions

This newly developed rotary paddle switch can not only easily address the problem of limited measuring range ,but also it grant a great benefit to measuring efficiency.Additionally,it enables users to intuitively observe the accumulated state of powder by watching screen of computer which receive signal from sensors.finally,it's outstanding measuring function greatly reduce the necessity of installing many traditional rotary paddle switches and help to shrink the cost during the industrial manufacturing process.

6.Reference

[1] Bing Wang,Zhanqiang Liu,Qinghua Song,Yi Wan,Zhenyu Shi. Proper selection of cutting parameters and cutting tool angle to lower the specific cutting energy during high speed machining of 7050-T7451 aluminum alloy[J]. Journal of Cleaner Production,2016,129.

[2] Congbo Li,Xingzheng Chen,Ying Tang,Li Li. Selection of optimum parameters in multi-pass face milling for maximum energy efficiency and minimum production cost[J]. Journal of Cleaner Production,2017,140.

[3] Wang Chih-Wen, Chen Ching-Jui, Lee Cheng-Tao, Wu Ting-Kuo, Cheng Chao-Kai. Rotary paddle level switch[P]. : US9111704,2015-08-18.

[4] DAVIS JOHN STEVEN (GB). A rotary paddle switch for monitoring the level of a particulate solid within a storage container[P]. : GB2411004,2005-8-17.

[5] SAVICKI JR GERALD R (US), COLEY BENJAMIN (CA), DUVAL FRANCOIS (CA), SHAANAN GAD (US). Paddle switch assembly[P]. : US7026564,2006-4-11.

[6] KIDMAN BRENT L (US). Wide-strap paddle swipUN,LEE, SUNG BEOM. PADDLE TYPE LEVEL SWITCH[P]. : KR101401898,2014-05-29.

[7] Srbinovski Igorce,Buttner Thomas E.,James Stephen C.PADDLE SWITCH SYSTEM[P]. : US2014262705,2014-09-18.

[8] LIM, SUNG WON, OH, JONG WON. Gear shift paddle switch[P]. KR20180038218,2018-04-16.

7.Acknowledgement

The authors wish to acknowledge the financial support of National Natural Science Foundation of China (Grant No. 51475210), A Project of Shandong Province Higher Educational Science and Technology Program (Grant No. J17KA027) and major research project of Shandong province (Grant No. 2018GGX103035) during the course of this investigation.

IOP Conf. Series: Materials Science and Engineering **542** (2019) 012055 doi:10.1088/1757-899X/542/1/012055

Experimental study on process of carbon fiber composites based on high temperature epoxy system curing

Mingfan Liu, Zhipei Tong, Lei Gao, Peng Zhang, Yongqi Cheng and Yousong Sun

Material and energy institute, Guangdong University of Technology, Guangzhou, Guangdong 510006, China

Corresponding author e-mail: zp1990@sina.com

Abstract. The carbon fiber reinforced composite materials used in sliding friction pairs was studied in the paper. A type resin matrix of epoxy CYD-128 with 4-4' Diamond dipheny methane(DDM) was prepared, and the curing dynamics parameters under different heating rates were studied by DSC thermal analysis, the curing step as 100℃/2h+160℃/2h was assured according to the β extrapolation method; The exsistance and the content change of the characteristic functional groups in the cured epoxy resin were analysed by the Infrared spectrum of the cured epoxy resin; The carbon fibers were processed with vacuum impregnation technology after glued, and the tensile fracture of the sample was observed as well as the micro morphology by SEM. The tests showed that the tensile strength of carbon fiber composite had grown 18.9% and the compression strength increased 25.9% by vacuum impregnation, and the carbon fibers were closely combined with the epoxy resin.

1. Introduction

Sliding friction pairs are important parts of mechanical equipments, Conventional friction pairs materials have the disadvantages of high manufacturing cost, high friction coefficient,insufficient bearing capacity and low efficiency[1]. Polymeric composites are frequently used in applications where traditional fluid lubrication cannot be used. These composites usually consist of a reinforcement,resin matrix, one or more solid lubricant added for binding components together, for environmental stabilization, etc. . Carbon fiber (CF) has the advantages of high strength, excellent chemical stability and low price. It was the good choice for reinforcement materials. Epoxy resin (EP) materials are the most commonly used resin matrix. Solid lubricants often use MoS_2 and graphite, etc.. [2]

Excellent properties of CF composite materials are closely related to resin matrix. The high-temperature curing agents for EP commonly used in the preparation of CF composites are: acid anhydrides and aromatic amines [3]. Compared with the acid anhydride curing agent, the aromatic amine curing agent has the characteristics of low viscosity, short curing time, easy control of the reaction process, and excellent mechanical properties of the cured product [4]. However, the EP of the room temperature curing system has a short storage period, a high viscosity, and a poor wettability of the glue, thereby weakening the interface bonding strength between the fiber and the substrate. Moreover, its low glass transition temperature is not suitable for sliding friction transmission applications. Therefore, it is very urgent to study a suitable high temperature curing preparation process..In the published paper[5], special research on DDM (4-4 'diaminodiphenylmethane) is less, and it is often necessary to blend with m-phenylenediamine to lower the melting point. In this paper, preparation process of CF composites based on high temperature epoxy system curing was experimented primarily.

2. Fabrication of samples

2.1. Experimental materials

The resin-based fiber reinforced composite materials used in the experiment are mainly composed of CF and EP, and others are added to improve the properties of the composite. CF reinforcement material： CF plain weave cloth, T300-1K, produced by Toray Co. Japan(Table 1). EP (Table 2), industrial grade, produced by Yueyang Baling Petrochemical Co., Ltd., Curing materials： 4-4' diamino Diphenylmethane（DDM）(Table 3), analytical grade, Light yellow crystalline flake at room temperature; produced by Aladdin Reagent Co..Other materials： Epoxy reactive diluent 501 ; dibutyl phthalat;Amine . The detailed Composite material formula are shown in Table 4.

Table 1. The product specification of Japan Toray 1K CF.

Warp yarn	Weft yarn	Weaving type	Thickness	Width(cm)	Densit(g/m^2)
T300-1K	T300-1K	Plain	0.15	100	119

Table 2. The essential parameter of EP CYD-128.

Type	epoxy value	Epoxy equivalent	Viscosity(25℃) mPa·s
CYD-128	0.53mol/g	184~194	11000~14000

Table 3. The essential parameter of DDM.

Type	molecular formula	Active hydrogen equivalent	Melting point
DDM	C13H14N2	49.57	89~91℃

Table 4. Composite material formula.

Epoxy	DDM	Active diluent	Plasticizer	Graphite	MoS_2	French chalk	Chopped Fiber	Plain weave carbon fabric
50	12.5	2.5	5	3.25	5	8.25	13.5	Several layers based on the sample

2.2. Fabrication of samples

2.2.1 Preparation of resin samples. DDM has poor compatibility with EP. Before being used,

Weighed the EP, curing agent, modified components, etc. according to the specific formula ratio. The EP and the modified component are mixed and stirred uniformly at room temperature, and the DDM curing agent is heated to 120 °C to melt, in order to prevent the liquid DDM from cold crystallization again, the other components are pre-baked at low temperature for 20 min, and then DDM is evenly mixed. Finally, the vacuum defoaming treatment is carried out at room temperature for 30 min.

2.2.2. Preparation of resin matrix samples. Slowly pouring the prepared colloid into the self-made mold, and heating and keeping warm according to a certain curing step; after cooling, it is naturally cooled to room temperature, and after demolding, it is processed to stretch and compress the sample.

2.2.3. Preparation of resin-based fiber reinforced composite materials samples. The CF cloth is cut into a certain size, and pre-treated according to the sequence of: acetone cleaning,nitric acid heating oxidative,coupling treatment. Using hand lay-up method, uniformly coating the glue on the surface of the CF cloth, stacking and placing it in the mold, pressing to the required thickness, the mold is placed in an oven and heated in accordance with the same curing step of the resin body. After the curing, it is naturally cooled. then demolding, processing and polishing to the stretching and compressive samples.

2.3. Experimental equipment

Q2000 DSC, American TA Co.; Nicolet 6700 Fourier Infrared Spectrometer, American Thermo Fisher Co.; CMT5504 Microcomputer Control Electronic Universal Tensile Machine, Shenzhen Xinsansi Material Co.; NDJ-1E Rotating Viscometer, Shanghai Changji Co.; S-3400N SEM, Japan Hitachi Co.; vacuum drying oven, Shenzhen Kuaike Co..

2.4 Performance test methods

2.4.1. DSC test. About 10 mg sample was taken, nitrogen flow rate was 50.0 ml/min, temperature rise interval: 20°C to 300°C, and heating rates were 5 K/min, 10 K/min, 20 K/min, respectively.

2.4.2. Fourier infrared spectroscopy test. Take a little epoxy cured product, grind it into powder, sample according to KBr tableting method, wave frequency interval 500-4000cm^{-1};

2.4.3. Colloid viscosity test. Select 4# rotor, 12r/min, Take appropriate amount of colloid and immerse the rotor in the glue;

2.4.4. Mechanical properties. Tensile strength and compressive strength of the resin matrix samples are tested on the test machine according to GB/T2568-1995 and GB/T2569-1995 test requirements. The Tensile and compressive strength of CFRP were tested on the test machine according to the test requirements of GB/T1447-2005 and GB/T1448-2005, respectively, wherein the compressive sample pressure was perpendicular to the direction in which the CF cloth was laid.

3. Experiment results and analysis

3.1. Epoxy curing process by DSC test

The curing process of the resin and the degree of curing directly determine the properties of the composite material. For a certain resin and curing agent system, it is especially important to select suitable curing process. In this study, based on a set of heat flow curves at different heating rates, the extrapolation T-β method was used to determine the gelation temperature, peak temperature, cure termination temperature and reaction heat ΔH at the heating rate β=0, thus determining the best curing process [6]. The DSC test results of the colloid system are as follows:

Figure 1 is the heat flow curve when the heating rate is β = 5, 10, 20K/min, The result can be seen : the reaction process is an exothermic reaction, with the increase of the heating rate, The curing initiation temperature Ti, The peak temperature Tp and the termination temperature Te also increase, the heat flow curve continuously moves toward the high temperature direction, and the curing exothermic peak becomes high and wide, and the reaction exotherm is also generally increased. This is because, with the increase of the heating rate, the thermal inertia of the system becomes larger, and the resulting measurement temperature difference is larger, resulting in the movement of the curve [7].

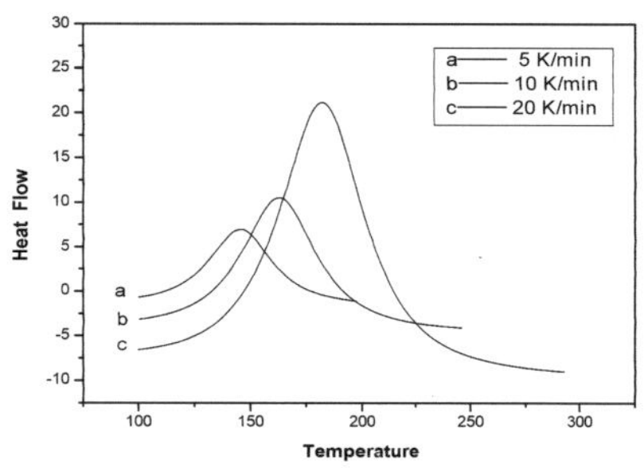

Figure 1. The DSC curves at different heating rate.

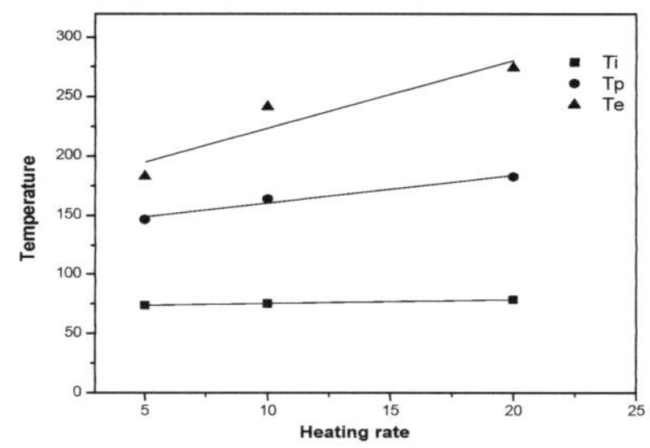

Figure 2. The linear regression relationship of temperature and heating rate

Table 5. The thermo dynamic parameters at different heating rate.

β/(K/min)	Ti (℃)	Tp (℃)	Te (℃)	ΔH (J/g)
5	73.8	146.4	182.9	385.6
10	75.3	163.7	241.3	402.2
20	78.4	182.5	274.2	448.0

In Table 5, the thermodynamic parameters of the curing system at different heating rates are listed according to the DSC heat flow curve.Using the data in Table 5, Ti, Tp, and Te curve (Figure 2) can be obtained by linear fitting.It can be seen from Fig.2 that when β=0, the initial solidification temperature Ti_0=72.2℃, the peak temperature Tp_0=137℃, and the termination temperature Te_0=166.5℃. It is concluded that the resin system begins to crosslink between 72.2℃and 137℃. Due to the explosion of EP, it can not be raised to high temperature in a short time. It should be heated slowly and kept for a period of time. Related literature [8], considering the cycle of the curing reaction and the regularity of the data, the curing step is determined to be: 100°C/2h+160°C/2h. According to the calculation, the degree of solidification of the product is over 95% on this curing system, which satisfies for the use of the resin matrix for the CF composite material.

3.2. Glass transition temperature of Resin cured product

Due to the poor thermal conductivity of polymer composites, it is important to increase the glass transition temperature Tg of composites as much as possible for composites used in sliding friction pairs. In order to reflect the effect of curing conditions on Tg of the cured EP, the DDM/CYD-128 epoxy system was cured according to three different heating steps, and the vitrification temperature of each system was tested. Among them, system1 is cured at 100°C/2h+150°C/2h, system2 is cured at 100°C/2h+160°C/2h, system3 is 100°C/2h+160°C/2h+180°C/1h. Tg of the cured product were measured to be 107.8 ° C, 147.82 ° C, and 148.45 ° C, respectively.

The effect of the curing process on Tg of the cured product is:The increase of the curing step temperature can further promote the reaction of the EP with the curing agent molecule, so that the degree of curing is higher, and the cross-linking density of the EP is increased.For tighter cross-linked networks, more energy is needed to trigger large-scale coordinated segmental motion at the glass transition [7],Therefore, in the system2, Tg of the resin is improved as compared with the system1. After the system3 was subjected to 100°C/2h+160°C/2h, the degree of reaction was consistent with that of the system 2, and the curing was complete. Therefore, after the 180°C/1h curing process, the Tg of the cured product did not change significantly.

3.3. Infrared spectroscopy

It was known that the molecular structure determines the characteristic wavelength of its infrared spectrum. In this paper, infrared spectroscopy is used to analyze and identify the existence of characteristic functional groups such as epoxy, amine and hydroxyl, and quantity based on the position and shape of the characteristic peak.

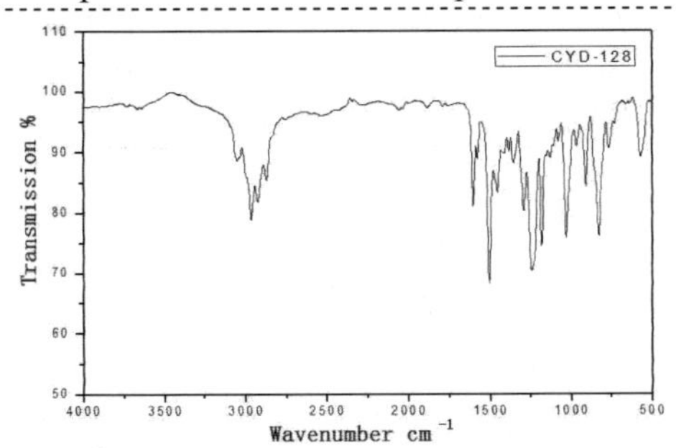

Figure 3. The Infrared spectrum of EP.

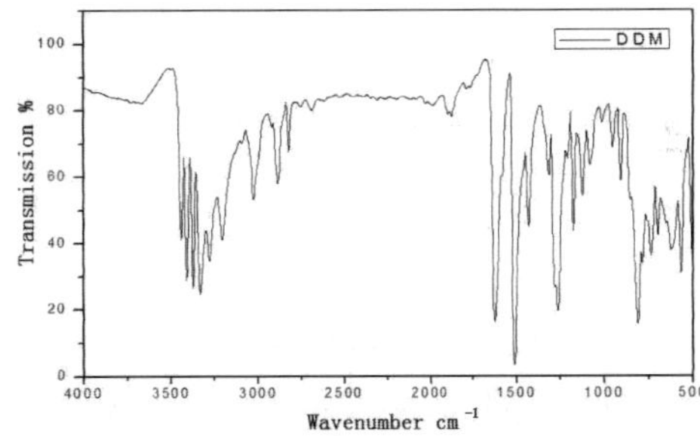

Figure 4. The Infrared spectrum of DDM.

It can be seen from Fig. 3 that in the region around 913 cm^{-1}, the appearance of the characteristic peak of the epoxy group corresponds to the CYD-128 spectrum. On the infrared spectrum of the curing agent DDM in Figure 4 Symmetrical and antisymmetric stretching vibration peaks (double peaks) of nN-H appear at 3407 cm^{-1} and 3446 cm^{-1}, and there are spikes on the low side of the nN-H absorption band, This phenomenon is consistent with the characteristics of aromatic diamines[11].In Fig.5, there are infrared spectra of cured products of three different degrees of cure, respectively. It can be seen that as the curing reaction proceeds, the epoxy group is gradually opened, the amine group is continuously consumed, and the corresponding band intensity is correspondingly weakened.

Compared with Figures 4 and 5, Figure 5 shows a hydroxyl peak at around 3500 cm^{-1}. This indicates that the epoxy group undergoes a ring opening reaction with an amine group, and a hydroxyl group is continuously formed. In terms of quantity, System 1> System 3, Comparing the ratio of the ratio of the epoxy group to the hydroxyl group, it can be inferred that the degree of cure $\alpha 1>\alpha 3$.

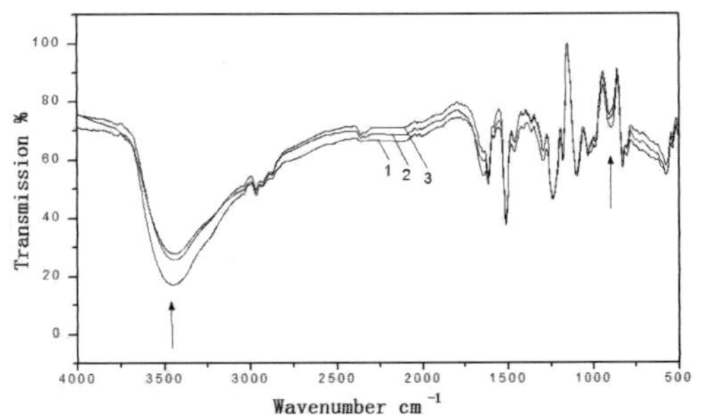

Figure 5. The Infrared spectrum of EP systems with different degree of cure.

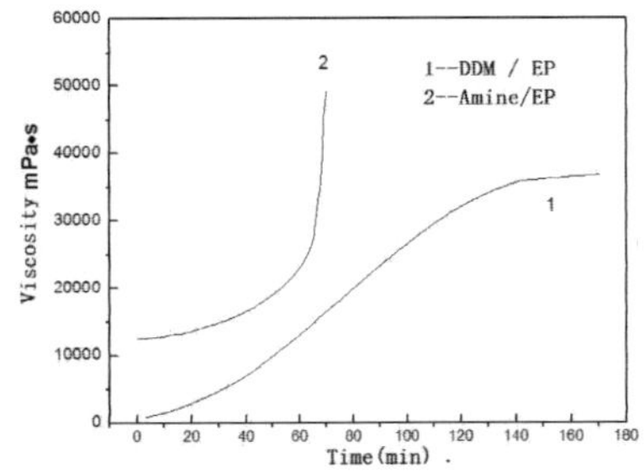

Figure 6. The Viscosity-Time curve of two expoy resin systems.

3.4. Viscosity curve

Figure 6 is a plot of viscosity versus time for DDM/epoxy system (high temperature cure) and amine/epoxy system (room temperature cure).It can be seen from the Figure 3 that the viscosity of the room temperature curing system is 12000mPa. s slowly rises to 20000mPa. s within 60min, but then its viscosity increased sharply, This indicates that the EP begins to carry out the crosslinking reaction. It can also be seen the the viscosity of the high-temperature curing system tends to increase linearly with time from 0 to 140 min. After this time , the viscosity change rate becomes slower and is basically stable at about 35000 mPa.s, indicating that the gelation process of the system is basically completed at this time, and because the aromatic amine curing agent has low reactivity at room temperature, its viscosity does not change much. This shows that the DDM/epoxy system not only has low viscosity, but also has a relatively long storage period at room temperature.

3.5. Mechanical properties of resin-based fiber reinforced composite materials

*3.5.1. Effect of vacuum dipping on mechanical properties.*It can be seen from the data in Table 6 that the CF reinforced epoxy composite material is superior to the resin matrix itself by 4 to 5 times. After resin coating , the tensile strength and compressive strength of the composite were increased by 18.9% and 25.9%, respectively, under the condition of pressure of -101 KPa for 1 h.

Table 6. Effect of CFRP mechanical properties.

Sample	Tensile Strength / MPa	Compressive strength / MPa
Epoxy Body	88.7	106.3
CFRP	474.3	407.2
CFRP(vacuum dipping)	563.6	512.5

After vacuum treatment, the bubbles generated by the glue and the small molecule gas generated by the reaction are easily discharged, which facilitates further penetration of the colloid into the CF tow, strengthens the bonding of the EP, and thereby improves the mechanical properties of the composite.

*3.5.2. Macroscopic fracture morphology observation.*In Fig. 7, the left and right sides are respectively the fracture portions of the tensile specimen of the composite which has not been vacuum treated and subjected to vacuum treatment.Careful observation of the fracture of the left side of the sample, the length of the fiber break is different, the CF tow is exposed to the air, and the layer peeling phenomenon occurs at the fracture; while the fracture of the right side of the sample is relatively flat, and the fiber cloth layer is firmly bonded. The left side of Fig.8 is a compressed sample of the composite after vacuum treatment. The fracture surface is at an angle to the direction of compression. It is known that the fracture occurs between the fibers; the right side is a composite sample compressed without vacuum treatment, and the surface is flat. The fracture occurs between the CF layers during the failure.

Figure 7. The fractures of CFRP tensile samples.

Figure 8. The fractures of CFRP compression samples.

(a)Vacuum treated 1h. (b) Not vacuum treated

Figure 9. The SEM photographs of CFRP tensile sample fracture.

*3.5.3. SEM analysis of tensile fracture.*As can be seen from Figure 9, after the vacuum treatment, the CF tow is filled with EP, the fibers are very tightly bonded, and the fracture is neat; and the CF without vacuum treatment is scattered and detached from the resin matrix. It is inferred that the cohesive force formed between the vacuum-treated composite fibers is large, the fibers are broken at one place during the break, and the fractures are flat; while the untreated CFs are insufficiently combined with the resin body, and the fibers are pulled out.

IOP Conf. Series: Materials Science and Engineering **542** (2019) 012055 doi:10.1088/1757-899X/542/1/012055

4. Conclusion

The optimal curing process of DDM/epoxy system is 100°C/2h+160°C/2h by DSC extrapolation method , and the ratio of CYD-128:DDM=4:1 is suitable by test. At the same time, the glass transition temperature of the resin matrix is greatly affected by the curing process, and the Tg reaches 147.82°C.

Vacuum treatment can improve the wetting effect of fiber and glue, and further improve the bonding strength between CF and epoxy. When the pressure is -101KPa, the tensile strength of CF composite increases by 18.9% , its compression strength increased by 25.9%.

Compared with the room temperature curing system, the high temperature curing system has the characteristics of low viscosity and long storage period, and its effective use period is more than 2 hours, which improves the practical operability of the hand lay-up method and is beneficial to the fiber reinforced composite material.

Acknowledgment

This research was got the financial support by Guangdong Science and Technology Plan Project (2014A010105042, 2017A010102021). The authors express their sincere thanks for these.

References

[1] Sun YS, Zhang Q, Gao L , Zhou XH, Cheng YQ and Zhang P 2016 Experimental study on tribological properties of carbon/polytetrafluoroethylene hybrid fabric reinforced composite under heavy loads and oil lubrication,Tribol. Int. 94 p 82-86.

[2] Zhang Q Y, Zhang P, Cheng YQ and Sun YS 2012 Study on mechanical properties of CF/glass fiber synergistically reinforced epoxy resin composites Glass Steel/Composites 5 p 38-41.

[3] Wang DZ 2006 Epoxy resin and its application(Beijing: Chemical Industry Press) p 48-50.

[4] Zhang CZ 2001 High performance aromatic amine curing epoxy resin system Fiber composite materials 3 p 5-7.

[5] Li HY, Liu N, Fei M, Liu YZ and Yang N 2012 Study on the properties of bismaleimide modified epoxy/diaminodiphenylmethane curing system Insulation Materials 6 p 46-50;

[6] Ding JP, Liao D and Fan XY 2011 Study on the Curing Reaction of Epoxy Resin for Composites by Non-Isothermal DSC Method Materials and Reports Research 7 p 139-143.

[7] Rosu D，Mustata F and Cascaval C N 2001 Investigation of the curing reactions of some multifunctional expoxy resins using differential scanning calorimetry Thermochim Acta 370 p 105.

[8] Liu W 2012 Effect of Curing Conversion Rate on Structure-Performance Relationship of Epoxy Materials (Shanghai: Fudan University) p 33-34.

IOP Conf. Series: Materials Science and Engineering **542** (2019) 012056 doi:10.1088/1757-899X/542/1/012056

Deodorization of Calophyllum Inophyllum oil using Steam distillation method

Nguyen Dinh Phuc[1*], Luong Minh Thanh[1], Tri Duc Lam[2]

[1]Faculty of Chemical Engineering and Food Technology, Nguyen Tat Thanh University, Ho Chi Minh City, Vietnam
[2]NTT Hi-Tech Institute, Nguyen Tat Thanh University, Ho Chi Minh City, Vietnam
*Corresponding author: ndphuc@ntt.edu.vn; labams2013@gmail.com

Abstract. Deodorization is a crucial refining stage with an important effect on the quality of oil refined. We used steam distillation method to remove the smell of calophyllum inophyllum oil. We investigated the effect of various factors such as vaporization temperature, oil volume to the deodorization process. In laboratory conditions, at vaporization temperature of 140^{0}C and oil volume of 200mL, the optimal time of deodorization process is 21 hours with 95% oil's smell removed. In the other hands, deodorized oil still remains the benefit compents. The aim of this study was to eliminate unpleasant pungent smell of calophyllum inophyllum oil. In the other hands, deodorized oil still retain the beneficial component and properties eminece. Thereby, creating a substrate with great potential in pharmaceuticals and cosmetics. Also, deodorized calophyllum inophyllum oil will help increase the economic value of traditional calophyllum inophyllum oil.

1. Introduction

Natural products represent a source of biologically active molecules that have an important role in drug discovery and food products [1-5]. The Calophyllum Inophyllum is widely distributed in the southern coastal belt of India, East Africa, Australia and Southeast Asia. In Vietnam, the calophyllum inophyllum tree was found most in the south of Vietnam. Calophyllum Inophyllum oil is non-edible fat oil and contains 50-73% of lipids [6-8]. It was extracted from calophyllum inophyllum seeds. The total lipids included 92.0% of neutral lipids, 6.4% glycolipids and 1.6% phospholipids. Calophyllum Inophyllum oil contains fatty acids such as neutral lipid, glycolipids and phospholipids which have a ability to penetrate into skin and regenerate cell of the skin.

The previous studies show that Oil also has the capacity to cure physical and chemical burns, radio-dermatitis and anal fissures. Recently some scientists have discovered anti-HIV compounds from calophyllum inophyllum seed report by Patil et al (1993) and Taylor et al (1994), Spino et al (1998) [9,10]. In addition, it also plays an important medicine to treat skin diseases and different purposes ranging from treatment of burns, cuts, scars to sciatica, sores to neuralgia, eczema and so on. Moreover, many researches have discovered application in the cosmetics field in order to prevent skin aging and regenerate damaged skin and [11,12]. Besides, the lastest studies described Calophyllum Inophyllum oil biological activity such as antiseptic, anti-inflammatory, antioxidant and so on [13].

Of various steps in the refining process, deodorization is critical for its capabilities to remove free fatty acids (FFAs), oxidation products and volatiles, which in turn improves the quality of the refined oil [14]. The aim of this process is the elimination of undesired volatile odoriferous elements in oils including free fatty acids, aldehydes, ketones, and carbohydrates [15]. Since traditional methods of deodorization such as steam distillation (SD) and molecular distillation (MD) requires high temperatures (190 - 275◦C) and low pressure (0.2-1 kPa), it is suggested that heating may induce a serious of chemical changes such as cyclization, cis-trans isomerization, and double bond migration. At present, three types of deodorizers including batch, semi-continuous and continuous are applied in the practice. Deodorizated calophyllum inophyllum oil will be a potential substrate in the cosmetic and pharmaceutical. In the other hands, deodorization will increase the economic value calophyllum inophyllum oil. However, there are very little researchs on calophyllum inophyllum oils deodorization as well as deodorized calophyllum inophyllum oils effective have not been reported and compared systematic. Thus, the optimization of the parameters in deodorization process is a large challenge for removing maximally undesirable components but still remaining maximally desirable valuable minor components. The objective of this study is optimizing deodorization process by steam distillation method. This process was conducted based on physicochemical parameters such as vaporization temperature, oil volume.

2. Material and methods

2.1. Sample preparation

Calophyllum Inophyllum oil was provided by AoTa Joint Stock Company. Chemicals used for analysing and evaluating of calophyllum inophyllum oil indicators: Sodium thiosulfate pentahydrate (Na2S2O3.5H2O) 99.5% (Xilong Scientific Co., Ltd), Potassium hydroxide (KOH) 85% (Xilong Scientific Co., Ltd) , Hydrochloric acid (HCl) 36-38% (Xilong Scientific Co., Ltd), Phenolphthalien (C20H14O4), Starch soluble (C6H10O5), Xylong Scientific Co., Ltd., 99% Potassium Iodide (KI), anhydrous Na2SO4 99% (Xilong Scientific Co., Ltd), Acetic acid glacial (CH3COOH) 99.5% (Quangdong Guanghua Sci-Tech Co., Ltd) made in China. Chloroform LR (CHCl3) 99% purchased from Alpha Chemika made in India. Diethyl ether (C4H10O) 99.5% purchased from VN - Chemsol Co.

2.2. Deodorization process of Calophyllum inophyllum oil by steam distillation

The steam is generated by heating the flask 1 (distilled water flask) then for steam to escape through the flask 2 (oil flask) which was heated at a temperature of 50^0C. Variable parameters such as evaporation temperature 130^0C – 150^0C, volume of deodorant 100mL – 300mL. This process is carried out for every 2 hours will take out the sample once to check out until 90% - 95% oil's odor was removed, recording time of deodorization process. The deodorizated mixture will be filtered water. We continute anhydrous by Na_2SO_4. Finally, we obtain the deodorized calophyllum inophyllum oil

2.3. Chemical analysis of calophyllum inophyllum oils by GC-MS analysis

Chemical composition of the oil sample was analyzed with following procedure. First, 25 μL essential oil was introduced into 1.0 mL n-hexane, followed by dehydratation with Na_2SO_4. The GC machine was Agilent 6890 N with MS 5973 inert, HP5-MS column, and head column pressure of 9.3 psi. Operating conditions are as follows: injection volume 1.0 μL, split 1:100, carrier gas He, flow rate 1.0 mL/min, injection temperature 250 °C. The temperature progressed from 50 °C for to 80 °C at 2 °C/min, from 80 °C to 150 °C at 5 °C/min, from 150 °C to 200 °C at 10 °C/min, from 200 °C to 300 °C at 20 °C/min and was kept at 300 °C for 5 min.

3. Result and discussion

3.1. Effect of vaporization temperature and Effect of ratio of vaporation/oil volume

The influence of temperature to the deodorization process is show in figure 1. The figure 1 shows that higher temperatures gave higher effective deodorization, due to increased speed of motion of a molecule and volatile components. For calophyllum inophyllum oil decrease in the deodorization time from 25 hours to 22 hours was observed when increasing the temperature from 130^0C to 140^0C. But a further temperature increase to 145^0 C resulted in an increase in the deodorization time to 23 hours. This may be because the vaporization of water is too fast should not enticing the smell component of the calophyllum inophyllum oil. Therefore, the proper temperatures for deodorization process by steam distillation is 140^0C

The influence of ratio of vaporation/oil volume to the deodorization process is show in figure 2. The figure 2 shows when increase oil volume then lower effective deodorization. When increase oil volume then time deodorization process also increase. When increasing the volume of calophyllum inophyllum oil then the time deodorization also increasing turns sequence 1.7/100; 1.7/150; 1.7/200; 1.7/250; 1.7/300 (mLH2O/minute)/mL oil and 20; 21; 21; 21.5; 22 hours, respectively. That shows

200mL calophyllum inophyllum oil with 21 hours time deodorization is optimal. Thus the opitimal volume of calophyllum inophyllum oil is 200mL

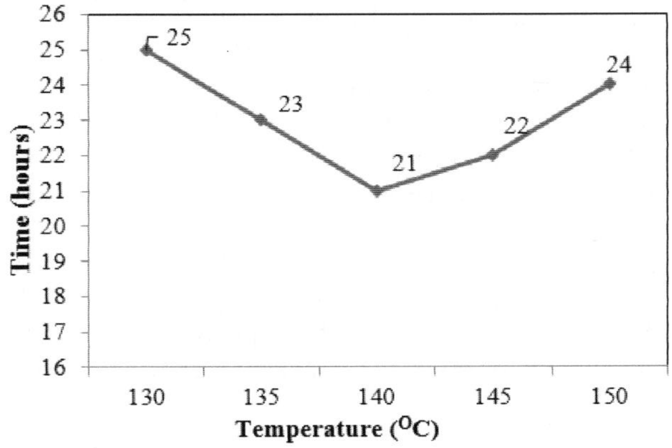

Figure 1. Effect of temperature

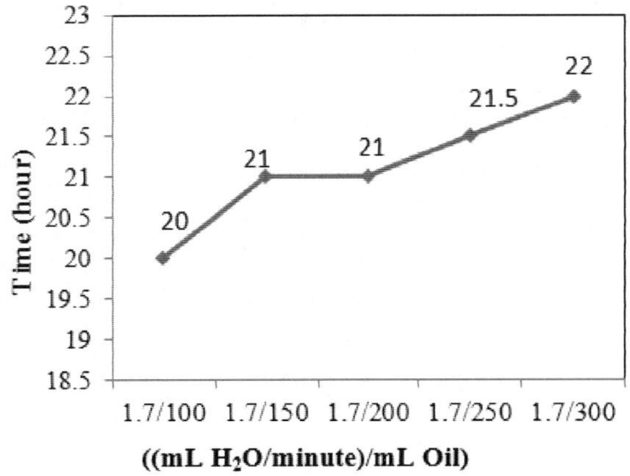

Figure 2. Effect of ratio of vaporation/oil volume

3.2. Effect of distillation method

The influence of distillation method to the deodorization process is show in figure 3. The figure 3 show deodorization by steam distillation method (21 hours) more effective hydro distillation method (26 hours) at the same con. Because the hydro distillation method the process of evaporation of the water hampered by calophyllum inophyllum oil and this method needs more heat. In the other hand, high temperature makes the calophyllum inophyllum oil denaturation lead to affects the quality of the calophyllum inophyllum oil

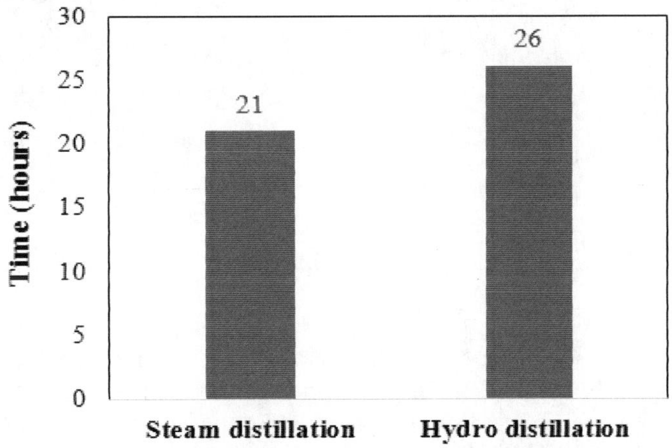

Figure 3. Effect of distillation method

3.3. Physicochemical properties of calophyllum inophyllum oil before and affter deodorized

Table 1. Physicochemical properties of Calophyllum Inophyllum Oil

STT	Property	Original Oil	Deodorizated Oil
1	Density at 25^{0}C (kg/m^3)	1.013	0.946
2	Kinematic Viscosity (cst)	92.2	94.8
3	Free Fatty Acid (mgKOH/g)	21.385	24.346
4	Saponification Value (mgKOH/g)	63.580	61.710
5	Peroxide Value (m Eq/kg)	0.385	0.201
6	Este Value	59.784	57.456
7	AV Value (mgKOH/g)	3.796	4.254

8	Glycerol content (%)	3.4162	3.2832
9	Water content (%)	0.4	0.4

The index such as AV, PV, EV, and SV of crude and deodorized cqlophyllum inophyllum oils were presented in table 1. Physicochemical index of calophyllum inophyllum oil before and affter deodorized with no significant changes. Crexi et al (2010)[16] suggested that AV is usually associated with free fatty acids (FFA) and undesirable flavor present in fats and oils. However, The AV index of crude calophyllum inophyllum oil was 3.796 mgKOH/g after deodorization it was increase to 4.254 mgKOH/g. That shows some of the AV index as well as FFAs of calophyllum inophyllum oil do not affect the smell of oil. The PV of crude calophyllum inophyllum oil was 0.385 mEq/kg and it was reduced to 0.201 mEq/kg after deodorization with steam distillation method. In addition, the PV index reduction also show of the freshness of the deodorized calphyllum inophyllum oil higher original oil. The SV index was reduced from 63.580 mgKOH/g rest was 61.710 mgKOH/g. This reduction in SV was explained for effectiveness removing phosphatides and soap during the deodorization process, describe by Menegazzo et al (2014)[16]

3.4. Calophyllum Inophyllum oil chemical analysis by GC-MS method

After determining the optimal parameters (temperature, volume of calophyllum inophyllum oil). Acid composition of product was determined by GC-MS. Results of GC-MS analysis is show in figure 4. The figure 4 show content acid of calophyllum inophyllum oil change is not significant after deodorization process. This show deodorized calophyllum inophyllum oil retains the acid composition and properties eminence.

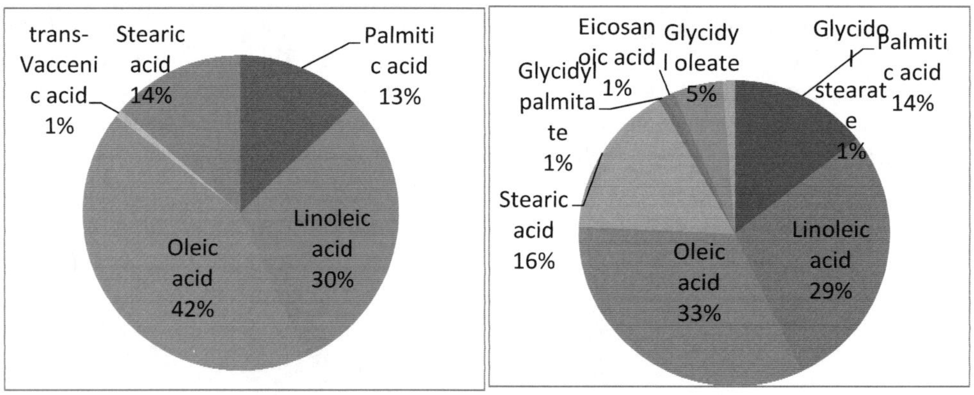

Figure 4. Acid component of calophyllum inophyllum oil before and after deodorized

4. Conclusion

The results of this study were achieved: optimizing parameters (vaporization temperature, oil volume) of the deodorization process by steam distillation at laboratory conditions. The physicochemical index and acid content of calophyllum inophyllum oil were determined by two titration methods according to the standard TCVN and GC-MS. Deodorized calophyllum inophyllum oil will be a potential substrate in the cosmetic and pharmaceutical. In the other hands, deodorized oil still retain the beneficial component and properties eminece. This study, we not yet able to of the parameters optimization of the deorization process of calophyllum inophyllum oil by steam distillation method and could not detect odor-causing components of oil. If I have time I will continue to investigate and further research to find out the optimal parameters as also odor-causing components of oil.

Acknowledgments: This work was supported by grants from Nguyen Tat Thanh University, Ho Chi Minh City, Viet Nam.

References

[1] Vo T-S, Ngo D-H, Bach L G, Ngo D-N and Kim S-K 2017 *Process Biochemistry* **54** 188–94
[2] Mai H C, Le T T T, Diep T T, Le T H N, Nguyen D T and Bach L G 2018 *Asian Journal of Chemistry* **30** 293–7

IOP Conf. Series: Materials Science and Engineering **542** (2019) 012056 doi:10.1088/1757-899X/542/1/012056

[3] Hien T T, Nhan N P T, Trinh N D, Ho V T T and Bach L G 2018 *Solid State Phenomena* **279** 217–21

[4] Nhan N P T, Hien T T, Nhan L T H, Anh P N Q, Huy L T, Nguyen T C T, Nguyen D T and Bach L G 2018 *Solid State Phenomena* **279** 235–9

[5] Tran T, Nguyen H, Nguyen D, Nguyen T, Tan H, Nhan L, Nguyen D, Tran L, Do S and Nguyen T 2018 *Processes* **6** 206

[6] Ong H C, Mahlia T M I, Masjuki H H and Norhasyima R S 2011 *Renewable and Sustainable Energy Reviews* **15** 3501–15

[7] Iinuma M, Tosa H, Tanaka T and Yonemori S 1994 *Phytochemistry* **35** 527–32

[8] Crane S, Aurore G, Joseph H, Mouloungui Z and Bourgeois P 2005 *Phytochemistry* **66** 1825–31

[9] Singh A, Yadav M, Srivastava R, Singh N, Kaur R, Gupta S K and Singh R K 2016 *Medicinal Chemistry Research* **25** 2842–59

[10] Patil A D, Freyer A J, Eggleston D S, Haltiwanger R C, Bean M F, Taylor P B, Caranfa M J, Breen A L and Bartus H R 1993 *Journal of Medicinal Chemistry* **36** 4131–8

[11] Spino C, Dodier M and Sotheeswaran S 1998 *Bioorganic & Medicinal Chemistry Letters* **8** 3475–8

[12] Dweck A C and Meadows T 2002 *International Journal of Cosmetic Science* **24** 341–8

[13] Mohibbeazam M, Waris A and Nahar N 2005 *Biomass and Bioenergy* **29** 293–302

[14] Song G, Zhang M, Peng X, Yu X, Dai Z and Shen Q 2018 *LWT* **96** 560–7

[15] Akterian S 2009 *Journal of Food Engineering* **91** 29–33

IOP Conf. Series: Materials Science and Engineering **542** (2019) 012057 doi:10.1088/1757-899X/542/1/012057

Extraction of Jasmine Essential Oil By Hydrodistillation method and Applications On Formulation of Natural Facial Cleansers

Nguyen Dinh Phuc[1], Le Hoang Phuong Thy[1], Tri Duc Lam[2], Vo Hoang Yen[1], Nguyen Thi Ngoc Lan[1]

[1]Faculty of Chemical Engineering and Food Technology, Nguyen Tat Thanh University, Ho Chi Minh City, Vietnam
[2]NTT Hi-Tech Institute, Nguyen Tat Thanh University, Ho Chi Minh City, Vietnam
*Corresponding author: ndphuc@ntt.edu.vn; labams2013@gmail.com

Abstract. Jasmine essential oil has been studied and widely used for many centuries in many countries throughout the world. However, relatively little is explored about jasmine essential oil application. Here, we produced essential oil of jasmine by hydrodistillation method. The effect of distillation time, distillation temperature, feed size, water-raw material ratio to the yield of the distillation process was analyzed and recorded. With 6 hours, 120^{0}C, grind fresh and ratio of 2:1 the yield was 0.092%. Antibacterial compounds of Jasmine essential oil were determined by measuring Gas Chromatography-Mass Spectrometry equipment. Jasmine essential oils are applied to develop the facial cleanser product that is safe for human. Jasmine facial cleanser will be a new product in Viet Nam's cosmetic with many potentials.

1. Introduction

As a traditional medicinal herb and valuable natural spice, essential oil has many significant pharmacological effects [1-4]. Facial cleanser plays the functional role in eliminating dirt and different unwanted substances from the surface of human face skin. More recently, facial cleanser has progressed immensely in terms of popularity and production. The product does not only remove dust from the skin but also assists in acne treatment, and help relax on a daily basis [5-7]. However, cleanser surfactants can create immediately after wash tightness (AWT), as well as irritation, erythema, and barrier damage [8]. Jasmine (Jasminum Sambac) has been designated in folk medicines in many countries due to its multipurpose activities. . Hongratanaworakit T. (2010) described the effect of jasmine oil and presented evidence for its application in aromatherapy for the mood improvement in humans and relief of depression [9]. In addition, Winai Sayowan et al (2013) has examined that jasmine oil has stimulatory influences on the function of the nervous system [10]. The material of jasmine oil could aid their utilization in perfume or cosmetics industry [11].
Conventionally, the essential oil is taken from plant raw material by different extraction methods [12,13]. including solvent extraction and steam distillation. However, these methods have drawbacks such as loss of volatile compounds, low yield, and accruement of toxic solvent residues [14,15]. In the industrial system, this process is still applied for different reasons such as simplicity of installations, ease of method performing. Despite the promising application of jasmine plant in cosmetics, facial cleanser extracts of jasmine has not been attempted in formulation and extraction. Therefore, this

study aim to utilize jasmine essential oils, in conjunction with natural ingredients, to formulate a facial cleanser for treatment of acne, humectants and for washing. To be specific, we first determine the conditions during hydrodistillation to optimize the yielded extract of jasmine flowers. Following that, optimum parameters of jasmine essential oil in a facial cleanser formula is determined.

2. Material and methods

2.1. Materials

Jasmine flowers were taken in a garden in Tien Giang Province, Vietnam. The maximum preservation period is 5 days at a temperature from 10 – 15^0C. All materials for facial skin cleanser were provided by 3C cosmetics company including water (aqua), emulsifiers wax, lauryl glucoside, Cocamidopropyl betaine, jasmine essential oil and almond oil, vitamin E, B3, preservative PE 1090, cetyl alcohol, SCI foaming, and stearic acid.

2.2. Description of hydrodistillation process

Hydrodistillation is an advanced method for extracting essential oils from plants because of its ability to maintain the original quality of plants. Flowers was removed from leaves and wilt, weighed to 250 gr, ground and put into a 1000 ml globe. The hydrodistillation process is carried out at the flower-water ratio of 1:1 to 1:5, distillation temperatures from 110^0C to 150^0C and distillation time of 4 hours to 8 hours to find the optimal conditions. After the distillation process had completed, the essential oil is collected, dehydrated by Na_2SO_4 and stored in the jar.

Table 1. Processing parameters affecting the essential oil yield

Description	Condition
Distillation time (hour)	4h, 5h, 6h, 7h, 8h
Temperature (^{o}c)	110^0C, 120^0C, 130^0C, 140^0C, 150^0C
Raw material size	grind fresh, fresh, dry grind
Material/water radio (g/ml)	1:1, 1:2, 1:3, 1:4, 1:5

2.3. The process for the preparation facial cleanser

The extract components are weighted accurately. Becher 1 contains water phase consist of the water-soluble components (water, Lauryl Glucoside, SCI, vitamin B3) are heated at 70^0C at a speed of 20 rpm for 15 minutes. Becher 2 contains oil phase consisting of the oil-soluble components (almond oil, emulsifiers wax, cetyl alcohol, acid stearic) are heated at temperatures of 45 - 50^0C at a speed of 20-24 rpm for 15 minutes. Two mixture is combined with glycerin, vitamin E, PE 1090, Cocamidopropyl betaine, Jasmine essential oil. The facial cleanser will be examined for pH, viscosity and organoleptic about the degree of cleansing.

2.4. Analysis the component of jasmine essential oil by GC-MS

GC/MS (Gas Chromatography-Mass Spectrometry) **[16]** is one of the most modern chromatography methods available today with high sensitivity and specificity and are used in the study and analysis of combined. GC-MS is an analytical method that combines the features of gas-chromatography and mass spectrometry to identify different substances within a test sample (SCION SQ 456-GC).

2.5. Analysis product achievement

The facial cleanser will be measuring for pH (Potential of Hydrogen) by LAB 850 device, viscosity with a Brookfield model and degree of cleansing is assessed organoleptic by 30 people.

3. Result and discussion

3.1. Jasmine essential oil distillation process

In a laboratory scale, a systematic study of the extraction temperature and time, parameters size, solvent composition, and solvent-to-material ratio was taken in order to discover the optimal level of hydrophilic antioxidants, lipophilic antioxidants, and fragrance compounds. As shown in Figure 1, extracts from dried ground samples was non-existent and fresh ground sample yielded 0.092%, indicating that solvent-solute diffusion out of the flowers and diffusion of the water solvent into the material can be restricted by the extraction method. In terms of the extraction yield, the conditions considered in this work material size of grind fresh was accepted as optimal **[17]**. Figure 2 shows that when the time extraction is longer, the yield of essential oil from 4h to 6h increases. However, when extending the time from 6h to 8h, the amount of oil obtained was almost unchanged because it has almost reached the limit oil under the conditions **[18]**. Moreover, the extended time extraction and maintained exposure to high-temperature limits throughput at a large scale and potentially enhances the loss of solvent by evaporation. It is therefore recommended that an extraction time of 6 hours is employed.

Figure 3 shows the water-material ratio for 1:1 – 1:5 in the hydrodistillation process. At 1:1 ratio, the yield of distillation is lest. At 2:1 ratio, the yield has reached a max of the distillation process. All other extraction conditions were kept constant. Because water vapor is absorbed in the epidermis contains essential oils which break down the oil sacs and attracts the oil by steam. Less water cannot melt adhesives, essential oils do not escape. Using as enough water for distillation as greater the diffusion capacity of the oil into the water. The optimal solvent-to-material ratio is chosen as 2:1. Figure 4 shows the extract yield of the extraction temperature. Because of diffusion coefficients and increased solubilities. When raising the temperature from 110 to 120°C, there is an increase in recovery from 0.37% to 0.92%. However, the temperature rises to 150°C following in a reduction in the recovery back to 0.36%. An extraction temperature of 120°C was preferred as optimal for overall extraction. The results show that the optimum harvesting conditions satisfy the distillation conditions. The oil was found to grounds, distilled at 120^{0}C in 6 hours, the ratio of the solvent to 2:1. The extraction efficiency is 0.1% at laboratory scale. The content of compounds found in jasmine essential oils will be expressed by GC-MS gas chromatography.

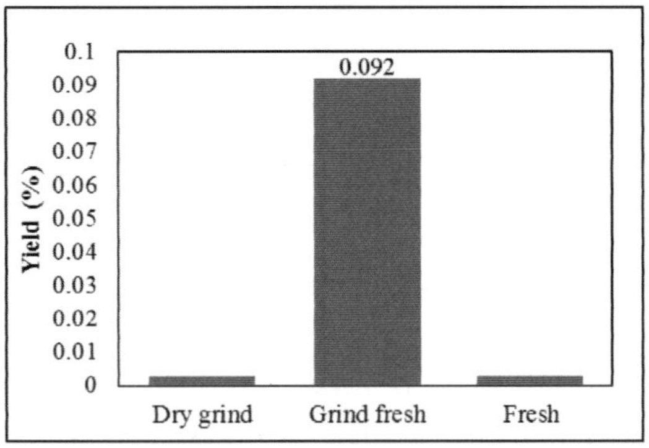

Figure 1. The amount of essential oil obtained on-size material.

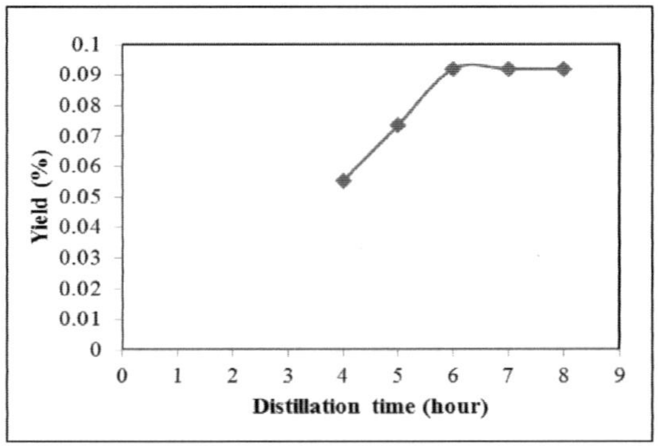

Figure 2. Jasmine extract yield as a function of extraction time

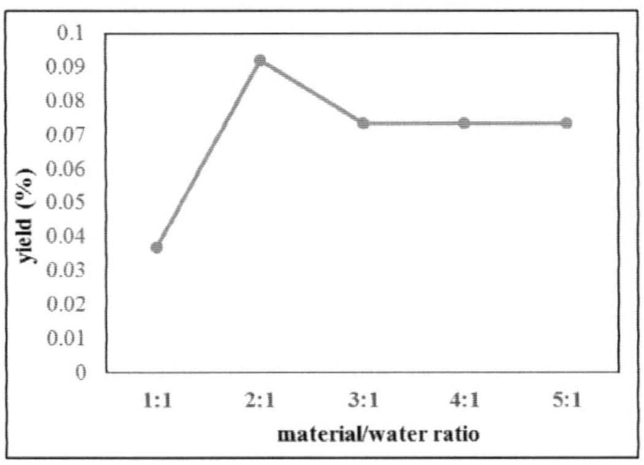

Figure 3. Jasmine extract yield as a function of extraction time

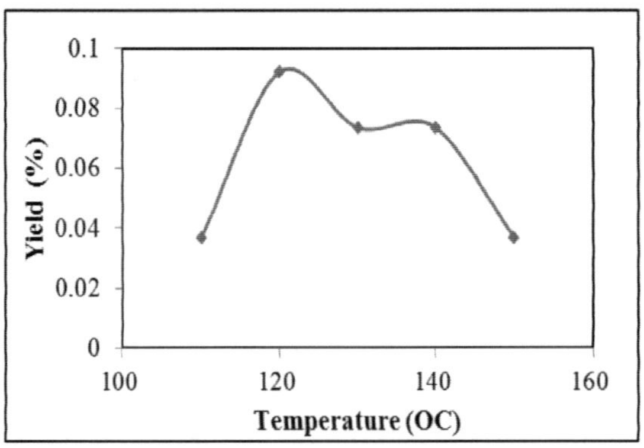

Figure 4. Jasmine extract yield as a function of extraction time

3.2. The relative contents of these compounds in Jasmine essential oil by GC-MS.

The table 2 shows that the components of Jasmine essential oil are found by hydrodistillation process. Including 16 in research of Naves và Grampoloff (1942) has determined the chemical composition of essential oil of Jasmine. GC/MS results show some components have high concentrations such as linalool (8.522 benzyl acetate (1.443%); the cis-3-Hexenyl benzoate (12.160%); α-epi-Muurolol (10.697%). These are the substances to creates the special scent of jasmine and the applications are in the production technology of cosmetics and flowers as linalool and benzyl acetate.

Table 2. Compositions of jasmine essential oil.

	Name	**Content (%)**
1	Cis-Linalool oxide	1.898
2	Linalool	8.552
3	Bezyl acetate	1.443
4	Levomenthol	5.321
5	Cis-3-hexenyl benzoate	12.160
6	α-epi-Muurolol	10.697
7	α-Cadinol	19.990

3.3. Formulation of Natural Facial Cleansers

This research has found appropriate parameters is showed in table 3. Thefacial cleanser has a pH of 6,78. The sample wasn't layered and oxidated during for 6 months. The sample for measuring viscosity by NDJ-9s recent number of the device is 16100 mPa.s. Based on our acquaintance with the Jasmine essential oil, the sample ability is to remove dirt and moisturize to the skin. It is reduced irritation and dryness, through organoleptic estimation.

Table 3. The ingredients of facial cleanser.

Name	**Mass (gram)**	**Content (%)**
Aqua	24,5	49%
Lauryl Glucoside	7	14%
Emulsifiers Wax	2,5	5%
Cetyl alcohol	1,5	3%
SCI	1	2%
Cocamidopropyl betaine	5,5	11%
Jasmine essential oil	1	2%
PE1090	0,5	1%
Vitamin E	1	2%
Axit Stearic	0,5	1%
Almond oil	1	2%
Vitamin B3	1	2%
Glycerin	3	6%

4. Conclusion

In this study, the optimal distillation conditions of Jasmine flowers are determined, distillation temperature is 120°C in 6 hours, the water-material ratio of 2:1. The result that the yield is 0.092% in laboratory-scale. The content of the compounds is found in the essential oil of jasmine will be performed by GC-MS analysis. Our Jasmine facial cleanser has pH is 6.78 suitably for using on face skin and organoleptic estimation has the ability to cleanse the skin from 30 people. Jasmine facial cleanser will provide a new product in Viet Nam's cosmetic. However, the product should be evaluated about antibacterial and antioxidant properties, so that it can be developed into a commercial product.

Reference

[1] Hien T T, Nhan N P T, Trinh N D, Ho V T T and Bach L G 2018 *Solid State Phenomena* **279** 217–21

[2] Nhan N P T, Hien T T, Nhan L T H, Anh P N Q, Huy L T, Nguyen T C T, Nguyen D T and Bach L G 2018 *Solid State Phenomena* **279** 235–9

[3] Tran T, Nguyen H, Nguyen D, Nguyen T, Tan H, Nhan L, Nguyen D, Tran L, Do S and Nguyen T 2018 *Processes* **6** 206
[4] Mai H C, Le T T T, Diep T T, Le T H N, Nguyen D T and Bach L G 2018 *Asian Journal of Chemistry* **30** 293–7.
[5] YS Choi, HS Suh, MY Yoon, S U Min, JS Kim, J Y.Jun, *Journal of Dermatological Treatment* **21** 201-205 2010/05/01 2010.
[6] Jananya Sampuntharat, P Boonkoom *International Journal of Clinical Dermatology & Research,* **3** 60-64 2015.
[7] K P Wilhelm, A B Cua, H H Wolff, and H I Maibach, *Journal of Investigative Dermatology,* **101** 310-315 1993.
[8] THongratanaworakit **5** 157-162 2010.
[9] WSayowan, V Siripornpanich, T Hongratanaworakit, N Kotchabhakdi, and NRuangrungsi, *J Health Res* **27** 2013.
[10] S Rathod And S GDeshpande *India Journal Pharmaceutical Sciences,* 2008.
[11] L Wang and CLWeller *Trends in Food Science & Technology* **17** 300-312, 2006
[12] D A J Starmans and H H Nijhuis *Trends in Food Science & Technology* **7** 191-197 1996.
[13] M Khajeh, Y Yamini, N Bahramifar, F Sefidkon, and M Reza Pirmoradei *Food Chemistry,* **91** 639-644 2005.
[14] PMasango *Journal of Cleaner Production* **13** 833-839 2005.
[15] N S Radulovic and M R Đorđevic *Chemistry & Biodiversity* **11** 2014.
[16] N E Durling, O J Catchpole, J B Grey, R F Webby, K A Mitchell, LY Foo *Food Chemistry,* **101** 1417–1424, 2007.
[17] E Akhihiero, B Ayodele, and G Akpojoto**r** **66** 2013.

IOP Conf. Series: Materials Science and Engineering **542** (2019) 012058 doi:10.1088/1757-899X/542/1/012058

Synthesized $BiVO_4$ was by the co-precipitation method for Rhodamine B degradation under visible light

Trinh Duy Nguyen[1,2,*], Vu Dai Cao[1,3], Vinh Huu Nguyen[1], Linh Xuan Nong[1,3], Tuan Duy Luu[1,4], Dai-Viet N. Vo[5], Sy Trung Do[5], Tri Duc Lam[1]

[1]NTT Hi-Tech Institute, Nguyen Tat Thanh University, Ho Chi Minh City, Vietnam
[2]Department of Chemical Engineering, Pukyong National University, Busan, South Korea
[3]Faculty of Chemical Engineering and Food Technology, Nguyen Tat Thanh University, Ho Chi Minh City, Vietnam
[4]Department of Chemical Engineering, HCMC University of Technology, VNU-HCM, Ho Chi Minh City, Vietnam
[5]Faculty of Chemical and Natural Resources Engineering, Universiti Malaysia Pahang, Malaysia
[6]Institute of Chemistry, Vietnam Academy of Science and Technology, Ha Noi, Vietnam
* Corresponding author: ndtrinh@ntt.edu.vn, labams2013@gmail.com

Abstract. Recently, $BiVO_4$ photocatalysts has been received much attention in field of catalysts. Because it can be used to degrade harmful organic catalysts in visible light, irradiation produces CO_2, H_2O and less harmful organic matter. In this study, we have successfully synthesized a $BiVO_4$ photocatalysts via co-precipitation method in the presence of urea and different calcined temperatures. The survey calcined temperatures as 300ºC; 350ºC; 400ºC and 450ºC. The obtained materials were characterized by Scanning electron microscope (SEM) and X-ray diffraction (XRD). The photocatalytic activity was evaluated by the photocatalytic degradation of rhodamine B (RhB) degradation under visible compact Philip lamp (40W) light irradiation. The result indicates that all samples calcined are monoclinic scheelite structure of $BiVO_4$. The $BiVO_4$-350ºC sample performed the best in the photodegradation of RhB.

1. Introduction

The global energy crisis and environmental pollution problem resulting from rapid population growth and industrial development [1-6]. It have increased the demand for renewable energy source and environmentally friendly technologies. Especially solar energy is renewable energy source. It is widely used as Solar panel, water heating, Water treatment and Other potential. For this research, photocatalytic techniques using semiconductor materials provide an ideal solution in terms of solar conversion and removal of pollutants. There are a lot of visible-light responsive photocatalysts such as CdTe, TiO_2, ZnO, Cu_2O, Fe_2O_3, WO_3, $BiVO_4$, have been researchers reported [7-14].

Among them, bismuth vanadate ($BiVO_4$) has attracted significant interest of researchers due to its outstanding features, as resistance to corrosion, low band gap, non-toxicity and is a photocatalyst for pollutant decomposing under visible-light irradiation [15-18]. It has three crystalline phases: monoclinic scheelite (s–m) and tetragonal zircon (s–t) and tetragonal scheelite structure (z–t). However, the monocrystalline scheelite crystal has the best photocatalytic activity in the three main crystalline structures of $BiVO_4$ [19-21]. At present, The $BiVO_4$ can be synthesized by hydrothermal, co-precipitation and metalorganic decomposition to make for fabrication of monoclinic $BiVO_4$ crystals [22-24]. In this study, we synthesized BiVO4 by co-precipitation method in the presence of urea. Ure pile plays an important role in determining morphology, crystalline and crystalline dimensions. In this study, $BiVO_4$ were synthesized via a simple combination of urea assisted co-precipitation and post-calcination processes to further enhance its photocatalytic activity.

2. Materials and Methods

2.1. Materals

Bismuth (III) nitrate pentahydrate ($Bi(NO_3)_3$. $5H_2O \geq 98\%$); Amomonium metavanadate ($NH_4VO_3 \geq 99\%$) and Rhodamin B $\geq 95\%$) were purchased from Sigma-Aldrich; Urea ($H_2NCONH_2 \geq 99\%$) and nitric acid (HNO_3, 65-68%) were obtained from Xilong Chemical Co., Ltd. (China).

2.2. Synthesis of $BiVO_4$

$BiVO_4$ was synthesized by the co-precipitation method in presence of urea, followed by annealing process. Typically, 20.0 mmol of $Bi(NO_3)3.5H_2O$ was dissolved in 200 mL of HNO_3 (2M) for about 30 min under stirred to give solution A. 20.0 mmol of NH_4VO_3 was dissolved in 200 mL of water to give solution B. The solution B was stirred for about 230 min at 70°C until a uniform transparent yellow solution. The solution B was drop by drop into solution A, a dark yellow solution was obtained. A certain amount of urea were added to the mixture to obtain the molar ratio of Bi:V: urea = 1:1:3.125. The obtained mixture was vigorously stirred for 30 min then heated at 85°C (RT) to evaporate the water for overnight. Finally, the obtained powder was calcined at 300, 350 ,400 and 450°C for 3h with a heating rate of 5 °C/min from room temperature to the annealing temperature. The as-calcined sample was denoted as RT, 300°C, 350°C ,400°C and 450°C.

2.3. Characterization

The phase structures of $BiVO_4$ were characterized by the powder X-ray diffraction (XRD) patterns on a D8 Advance Bruker powder diffractometer with a Cu Kα excitation source at a scan rate of 0.030 °/s in the 2-theta range of 5-80°. The morphology of the as-prepared $BiVO_4$ were visualized by scanning electron microscope (SEM, JEOL JSM 7401F) with an accelerating voltage of 3 kV.

2.4. Photocatalyts Test

Photocatalytic activity of products were prepared experimentally during the decomposition of toxic organic pollutants as Rhodamine B (RhB). In each sample experimental, a 0.1 g amount of photocatalyst was added into 100 mL of RhB solution with a concentrations of 1×10^{-5} M and useing a compact Philip lamp (40W, Vietnam) as the visible-light resource. The mixture is stirred with a magnetic stirrer at 550 rpm. Before lighting, the mixture was continuously stirred in the dark for 1 hour to ensure adsorption balance. After, the suspension was exposed to visible light irradiation while being vigorously stirred. The suspension (5 ml) was taken out at 45g min intervals and centrifuged to remove the photocatalyst particles for the analysis of the RhB. The concentration (C) of RhB in the solution after time was monitored using a Thermo Evolution 60S UV-Visible Spectrophotometer. The wavelength were recorded is 554 nm. The C/C_0 ratio was calculated to evaluate the photocatalytic efficiency of the as-prepared m-$BiVO_4$ samples.

3. Results and Discussion

3.1. XRD analysis

We are investigating the effect of calcined temperatures on the formation of the $BiVO_4$ crystalline phase and all samples are calcine at different temperatures. Fig. 1 shows the XRD patterns of the $BiVO_4$-T sample calcined at 300°C; 350°C, 400°C; 450°C for 5h and RT. No significant characteristic peaks for $BiVO_4$ because of the sample stirred and heated at 85°C as show fig 1a. All sample calcined temperatures with characteristic peaks with 2θ at 18.5, 29, 30.6, 35 and 46° and the diffraction peaks could be indexed to the pure monoclinic scheelite $BiVO_4$ phase (JCPDS card no. 14-00688). The similar XRD patterns of $BiVO_4$ were also conform according to researchers [25.26].

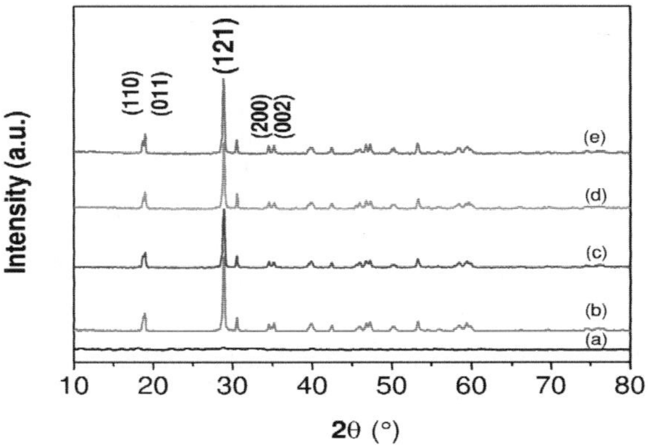

Figure 1. XRD patterns of the different $BiVO_4$-T samples (a) RT; (B) 30; (C) 350; (D) 400 and 450

3.2. SEM analysis

Fig. 2 shows the SEM images of all sample by the co-precipitation method at calcined temperatures and RT. When the sample was calcined as shown in fig 2(b-e), it shows the structure of the nanoparticles as well as grape-like morphology and the grape of create cluster. However, fig 2a shows the RT sample Form the adhesive sheet together. Because, it has a urea concentration that does not completely burn at 85°C. According to the above SEM observations, we can thus assume that calcined temperature is a crucial role in determining the crystal structure. Moreovers, it can be deduced from the above results that the morphology as well as size of the grape could be easily controlled by varying urea concentration and calcined temperatures time.

Figure 2. SEM images of (a) RT; (B) 300; (C) 350; (D) 400 and 450

3.3. Photocatalytic activity

We examined photocatalytic degradation of RhB dye solution under visible-light . Fig 3a shows more than 83.4% concentrations degradation of RhB is observed over sample 350 within 270 min under visible-light. The photocatalytic activity on the $BiVO_4$-T (T is the calcined temperatures) decreases in the following order: $BiVO_4$-350, $BiVO_4$-300 , $BiVO_4$-400 and $BiVO_4$-450, the rate constant (k) as show at table 1. The results of the photocatalytic activity showed that $BiVO_4$-350 was the best sample at concentrations 1×10^{-5} mol L^{-1} in RhB.

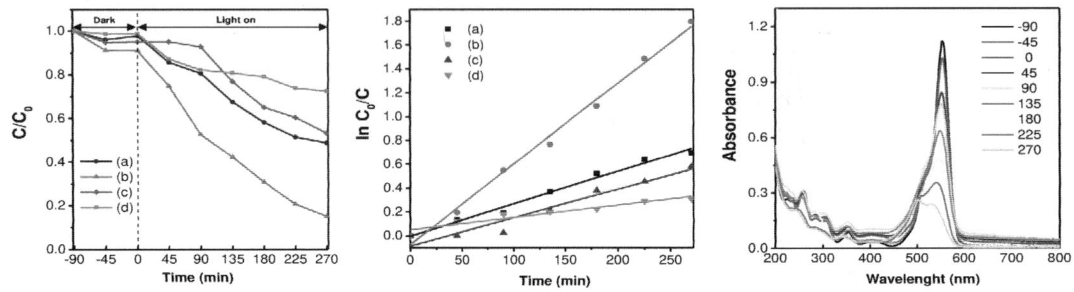

Figure 3. (A) Photocatalytic activity of $BiVO_4$-T crystallites with different T values: (a) 300; (b) 350; (c) 400; (d) 450, (B) $\ln(C_0/C)$ versus time and (C) Absorption changes of RhB aqueous solution in the presence of $BiVO_4$-350

Table 1. Lattice parameters (Å); Volume (Å^3); Crystallite size(nm); the rate constant (K) and Correlation coefficients (R^2) of $BiVO_4$-T

Sample	Lattice parameters (Å)[a]			Volume[a]	Crystallite size[a]	Kapp[b] ($\times 10^{-3}$)	$R^{2\,b}$
	a	b	c	(Å^3)	(nm)	(min)	
300	5.1823	11.6972	5.0973	308.9924	29.47	1.4637	0.9857
350	5.1798	11.6967	5.1037	309.2230	28.13	2.5729	0.9928
400	5.1849	11.69167	5.0923	308.6988	30.33	2.7137	0.9389
450	5.1864	11.6963	5.0882	308.6665	29.64	1.4094	0.9147

[a] Data obtained by XRD data; [b] Data obtained by the relationship between $\ln(C_0/C)$ and irradiation time t (min)

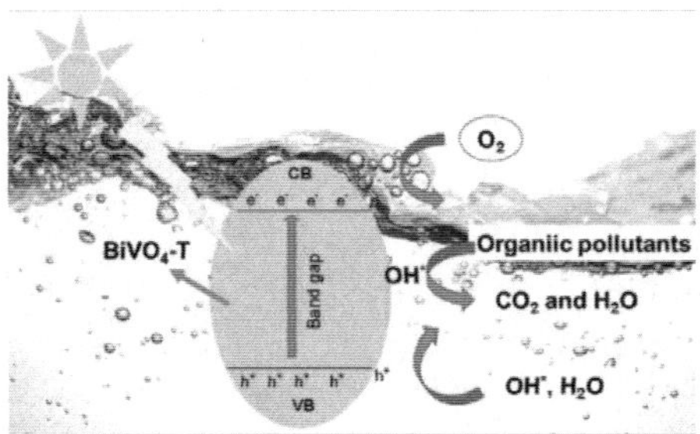

Figure 4. Schematic illustration for Photocatalytic activity of $BiVO_4$

Fig. 4. Show as The schematic illustration for Photocatalytic activity of $BiVO_4$. This reaction would be by the following equations:

$$BiVO_4 \xrightarrow{hv \geq 2.4\ eV} e^-_{cb} + h^+_{vb} \quad (1)$$

$$h^+_{vb} + H_2O \rightarrow OH^{\cdot} + H^+ \quad (2)$$

$$h^+_{vb} + OH^{\cdot} \rightarrow OH^{\cdot} \quad (3)$$

$$e^-_{cb} + h^+_{vb} \rightarrow t + L \quad (4)$$

The scheelite $BiVO_4$ with a band gap of 2.4 eV and Wavelength respectively is 554 nm (Ep. (1)). Ion OH- can react with h^+_{vb} on valency to produce additional $OH^{\cdot}$ (Ep. (3)). The other hand whereas, e-cb tend to re-associate with h + vb attached to release heat (t) or light (L) (Ep. (4)). The $OH^{\cdot}$ with high oxidation that can decompose compounds Organic with the final product is CO_2, H_2O and Inorganic ions are less toxic.

4. Conclusions

We have successfully synthesized a $BiVO_4$ photocatalysts via co-precipitation method. The calcined temperature is a crucial role in determining the crystal structure. The result indicates that all samples calcined are monoclinic scheelite structure of $BiVO_4$. The photocatalytic activity results indicated the $BiVO_4$-350 sample to have good photocatalytic activity. The degradation efficiencies of RhB achieved

83.4 % after 270 min irradiation. Moreover, the present study motivates us to explore Urea concentration increase and calcined temperature to better results.

Acknowledgments: This research is funded by Nguyen Tat Thanh University, Ho Chi Minh City, Vietnam

References

[1] Kim D W, Bach L G, Hong S-S, Park C and Lim K T 2014 *Molecular Crystals and Liquid Crystals* **599** 43–50
[2] Thieu C A, Bach L G and Cao X T 2016 *Molecular Crystals and Liquid Crystals* **635** 18–24
[3] Nhung N T H, Quynh B T P, Thao P T T, Bich H N and Giang B L 2018 *IOP Conference Series: Earth and Environmental Science* **159** 012015
[4] Vinh N H, Hieu N P, Van Thinh P, Diep N T M, Thuan V N, Trinh N D, Thuy N H, Long Giang B and Quynh B T P 2018 *Journal of Nanoscience and Nanotechnology* **18** 6859–66
[5] Huynh T T, Van Nguyen A, Pham H Q, Vinh N H, Bach L G and Thanh Ho V T 2018 *Journal of Nanoscience and Nanotechnology* **18** 6928–33
[6] Thanh Ho V T, Tuan P D, Bach L G, Tuong V T T and Hwang B-J 2018 *Journal of Nanoscience and Nanotechnology* **18** 6934–41
[7] Bach L G, Cao X T, Ho V T T, Islam M R and Lim K T 2015 *Molecular Crystals and Liquid Crystals* **618** 120–8
[8] Cao X T, Bach L G, Islam M R and Lim K T 2015 *Molecular Crystals and Liquid Crystals* **618** 111–9
[9] Bach L G, Cao X T, Quynh B T P, Ho V T T and Lim K T 2017 *Molecular Crystals and Liquid Crystals* **644** 183–9
[10] Bach L G, Nguyen N G and Ho V T T 2016 *International Journal of Photoenergy* **2016** 1–7
[11] Nguyen V, Nguyen T, Bach L, Hoang T, Bui Q, Tran L, Nguyen C, Vo D-V and Do S 2018 *Catalysts* **8** 487
[12] Sun S and Wang W 2014 *RSC Adv.* **4** 47136–52
[13] Li G-S, Zhang D-Q and Yu J C 2009 *Environmental Science & Technology* **43** 7079–85
[14] Li L, Chu Y, Liu Y and Dong L 2007 *The Journal of Physical Chemistry C* **111** 2123–7
[15] A. M, J. M, Ashokkumar M and Arunachalam P 2018 *Applied Catalysis A: General* **555** 47–74
[16] Hu Y, Fan J, Pu C, Li H, Liu E and Hu X 2017 *Journal of Photochemistry and Photobiology A: Chemistry* **337** 172–83
[17] Shang M, Wang W, Zhou L, Sun S and Yin W 2009 *Journal of Hazardous Materials* **172** 338–44
[18] A. M, J. M, Ashokkumar M and Arunachalam P 2018 *Applied Catalysis A: General* **555** 47–74
[19] Nguyen H V, Thuan T V, Do S T, Nguyen D T, Vo D V N and Bach L G 2018 *Applied Mechanics and Materials* **876** 52–6
[20] Nguyen D T and Hong S-S 2017 *Journal of Nanoscience and Nanotechnology* **17** 2690–4
[21] Jiang H, Endo H, Natori H, Nagai M and Kobayashi K 2008 *Journal of the European Ceramic Society* **28** 2955–62
[22] Madhusudan P, Ran J, Zhang J, Yu J and Liu G 2011 *Applied Catalysis B: Environmental* **110** 286–95
[23] Martínez-de la Cruz A and Pérez U M G 2010 *Materials Research Bulletin* **45** 135–41
[24] Zhao Z, Dai H, Deng J, Liu Y and Au C T 2013 *Chinese Journal of Catalysis* **34** 1617–26
[25] Zhao Z, Dai H, Deng J, Liu Y and Au C T 2013 *Solid State Sciences* **18** 98–104
[26] Tian G, Pan K, Fu H, Jing L and Zhou W 2009 *Journal of Hazardous Materials* **166** 939–44

IOP Conf. Series: Materials Science and Engineering **542** (2019) 012059 doi:10.1088/1757-899X/542/1/012059

A Facile Synthesis and Properties of Bismuth Vanadate ($BiVO_4$) Photocatalyst by Hydrothermal Method

Nguyen Duy Trinh[1,2*], Ho Huy Hoang[1,3], Nong Xuan Linh[1,3], Nguyen Huu Vinh[1], Hien Thi Vu[4], Hai Thanh Nguyen[5], Dai-Viet N. Vo[6], Sy Trung Do[7], Lam Tri Duc[1]

[1]NTT Hi-Tech Institute, Nguyen Tat Thanh University, Ho Chi Minh City, Vietnam
[2]Department of Chemical Engineering, Pukyong National University, Busan, South Korea
[3]Faculty of Chemical Engineering and Food Technology, Nguyen Tat Thanh University, Ho Chi Minh City, Vietnam
[4]Faculty of Hydro-Meteology, HCMC University of Natural Resources and Environment, Ho Chi Minh City, Viet Nam
[5]Son La Hydro Power Company, Son La Province, Vietnam
[6]Faculty of Chemical and Natural Resources Engineering, Universiti Malaysia Pahang, Malaysia
[7]Institute of Chemistry, Vietnam Academy of Science and Technology, Ha Noi, Vietnam
*Corresponding author: ndtrinh@ntt.edu.vn; labams2013@gmail.com

Abstract. In this study, $BiVO_4$ photocatalysts were synthesized by hydrothermal method using $Bi(NO_3)_3.5H_2O$ and NH_4VO_3 as raw materials followed by calcination at different temperatures in the range from 350 °C to 600 °C. The as-synthesized $BiVO_4$ samples were characterized by a number of physicochemical techniques including X-ray diffraction (XRD), Raman analysis, Scanning Electron Microscopy (SEM), and UV-Visible (UV-Vis) light diffuse reflectance spectrophotometry. The effect of temperatures calcination on structure, surface morphology, visible-light photocatalytic activity and light absorption performance of $BiVO_4$ was discussed in details.

1. Introduction

In chemistry, the traditional photocatalysts comprise mainly metal oxides such as TiO_2, ZnO, SnO_2 and CeO_2. These photocatalysts are semiconductors with relatively low band-gap and are therefore suitable for absorbing natural or artificial light and exhibiting photochemical activity [1-10]. Optical catalysts play an important role in many areas such as environmental improvement, photovoltaic applications, bio-applications, clean energy production, sensors, nanoscale electronic components and solar battery. One of such photocatalysts is the $BiVO_4$, which has been attracting a lot of attention due to its remarkable features, including low band gap energy, good dispersion, non-toxicity and high photochemical activity under visible light irradiation. Several synthesis methods have been utilized in order to prepare $BiVO_4$ powder of high quality with predominantly solid-state including aqueous-based and hydrothermal methods [11-15].

In this study, a simple solution-based preparation method under acidic conditions has been selected and adopted, involving the calcination of obtained $BiVO_4$ precipitates at different temperatures. Subsequently, an in-depth analysis of important physical parameters of as-obtained products allows for a coherent correlation with the different photocatalytic O_2 evolution. The insights gained from this analysis contribute to the better understanding of the visible-light water oxidation mechanisms and more efficient optimization of photocatalytic materials. The objective of this study was to synthesize

IOP Conf. Series: Materials Science and Engineering **542** (2019) 012059 doi:10.1088/1757-899X/542/1/012059

$BiVO_4$ materials by hydrothermal method. Following that, the as-synthesized samples were calcined at different temperatures ranging from 350 °C to 600 °C. Structural properties of material samples calcined at different calcination temperatures from 350 °C to 600 °C were characterized using analytical methods including XRD, SEM and UV-Vis-DRS.

2. Materials and Method

2.1. Materials

Chemicals used in this study include Ammonium metavanadate (NH_4VO_3, ≥ 98%, Sigma-Aldrich), Bismuth (III) nitrate pentahydrate ($Bi(NO_3)_3.5H_2O$, ≥ 98.0%, Sigma-Aldrich), Nitric acid (HNO_3, 65-68%), and ammonia solution (NH_4OH 25-28%, Xilong). Chemicals for analysis: Rhodamine B (RhB, Jihuada), Distilled water (from Lasany's two-storey water machine).

2.2. Synthesis of $BiVO_4$ photocatalyst by hydrothermal method

The typical preparation process comprises six-step as follows: (1) 4 mmol of $Bi(NO3)_3.5H_2O$ was dissolved in 20 mL of 4M HNO_3 solution; (2) another solution was prepared by dissolving a stoichiometric of NH_4VO_3 in 40 mL of hot water; (3) The two solution was mixed together and stirred for about 1 hour to achieve a stable and homogeneous state; (4) The pH of the mixture is adjusted with NH_4OH or HNO_3 solution; (5) The mixture was sealed in an autoclave, and subjected to heat at 160 °C for 24 hour under pressure. At the end of the process, the precipitate was filtered, washed with distilled water and dried overnight at 110 °C; (6) The obtained solid was calcined in air at different temperature (350 °C, 400 °C, 450 °C, 500 °C and 600 °C) for 2 hour. To investigate the effects of calcination temperature on the formation of crystalline and phase structure of $BiVO_4$, $BiVO_4$ samples synthesized by hydrothermal method were calcined at different temperature (350 – 600 °C) to analyze the structural crystals of $BiVO_4$.

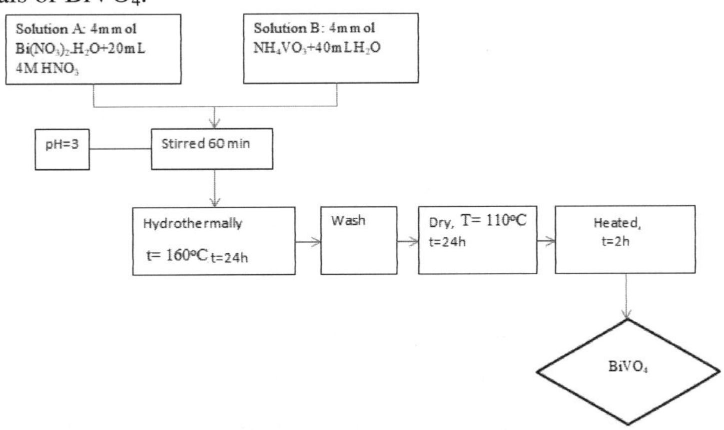

Scheme 1. Process of synthesis of materials by hydrothermal method.

In this study, $BiVO_4$ was synthesized by hydrothermal followed by calcination process at different temperatures (from 350 °C to 600 °C) as shown in Table 1.

Table 1: *Statistical samples and aggregate conditions*

Calcinated Sample	Heat (°C)	Time (h)	Sample characteristics	Efficiency (g)
BVO-350 °C	350 °C	2	Yellow powder	83%
BVO-400 °C	400 °C	2	Yellow powder	90%
BVO-450 °C	450 °C	2	Yellow powder	81%
BVO-500 °C	500 °C	2	Yellow powder	83%
BVO-600 °C	600 °C	2	Yellow powder	78%

2.3. Characterization

Samples were analyzed on Jeol JSM 7401F Scanning Electron Microscope instrument, at the Institute of Chemistry, Vietnamese Academy of Science and Technology. X-ray diffraction (XRD) powder analysis was carried out with a Rigaku D/max-RB12KW X-ray diffractometer using Cu K_α radiation (λ = 1.5418 Å) in the 2θ range of 4–80° at room temperature. The XRD was developed by William L.Bragg in 1913, in which the Bragg equation (1) is considered as a condition for diffraction: $d =$

IOP Conf. Series: Materials Science and Engineering **542** (2019) 012059 doi:10.1088/1757-899X/542/1/012059

$\frac{n\lambda}{2sin\theta}$, where λ is the X-ray wavelength (1.5418Å). The diffraction angle θ corresponds to the diffraction peaks on the diffraction pattern defined θ = arcsin (nλ / 2dhkl), n, or diffraction patterns, takes the values of 1, 2 or 3 .

3. Results and Discussion

The crystalline structure of $BiVO_4$ was confirmed by XRD patterns as shown in the Figure 1. The results showed that all samples had strong diffraction peaks at 28° and 30.6°, and split diffraction peaks at 18.5°, 34°, and 46°. This result corresponds to the standard of the $BiVO_4$ crystalline monoclinic scheelite crystal (m-$BiVO_4$, JCPDS card no. 14-00688) and is consistent with previous studies [15]. In addition, all samples show the characteristic peak splitting diffractions at 2θ = 18.5, 34 and 46 ° [16]. Even though monoclinic phase was observed at every calcination temperature, splitting of the these characteristic peaks becomes more pronounced as the calcination temperature increases, which likely signifies the increasing crystallinity degree of the (s-m) phase in $BiVO_4$ samples [16,17].

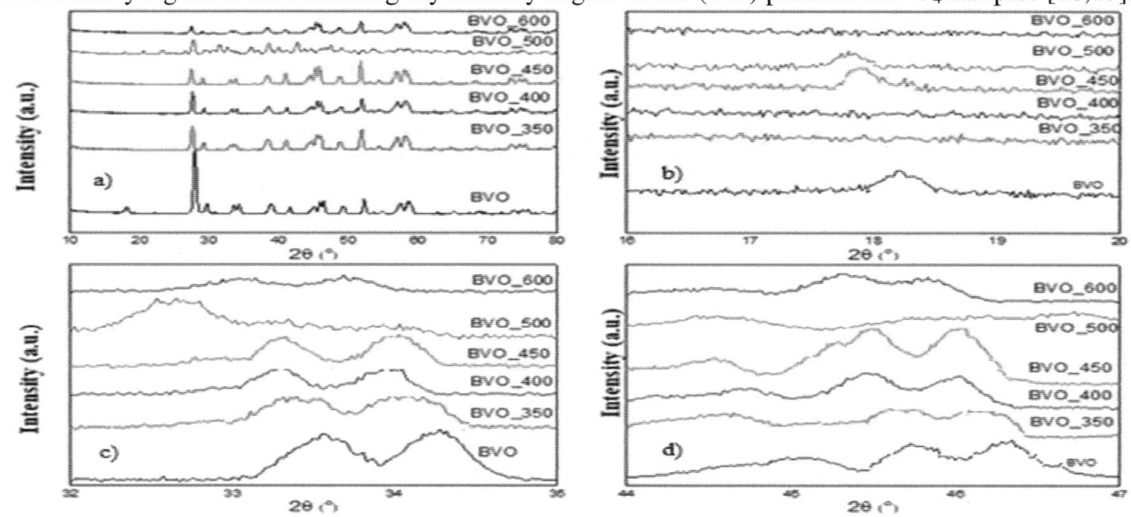

Figure 1. a) XRD diagram of $BiVO_4$ samples, b) XRD is at 2° from 16° to 20°, c) The XRD detail diagram at 2θ from 33° to 34.5°, d) The XRD detail diagram 2θ from 44° to 47°.

Crystal shape, particle size and particle distribution of the material were observed through SEM images which are given in the Figure 2. Visually, materials and structures have flattened shape which is 1 μm in size and is 1 μm in flange size. At calcination temperatures of 450 °C, , the form of the materials is spherical and rod-shaped.

Figure 2. SEM images of the samples $BiVO_4$ in: BVO-450°C.

The vibrational properties of the $BiVO_4$ crystals were studied through Raman scattering spectrum at room temperature as shown in the Figure 3. The results show that the samples exhibited specific vibration levels for the m-s crystalline $BiVO_4$ and the thermal increases through the hydrothermal method, the intensity of these oscillations decreases, affecting the crystalline structure of the material. In the Figure 3, Raman spectra of $BiVO_4$ powders were excited by a red (648 nm) laser and corresponding spectra. Raman spectroscopy can provide structural information and is also a sensitive method for the investigation of the crystallization, local structure, and electronic properties of materials. Raman bands at around 210, 324, 366, 640, 710, and 826 cm^{-1} were observed for all samples. These are the typical vibrational bands of $BiVO_4$ [15,18]. The structural information of

$BiVO_4$ is given by the band centered at 210 cm^{-1} . The asymmetric and symmetric formations of VO_4 tetrahedron are given by the bands centered at 324 and 366 cm^{-1}, respectively. The Raman band at 640 cm^{-1} can be assigned to the asymmetric stretching vibration of the shorter V-O bond. The stretching modes of two vibrational modes of V–O bonds are determined by the bands centered at 710 and 826 cm^{-1}. These two bands provide valuable information about the structural variations among the powder samples calcined at different temperatures. It has been observed a positive shift in the vibrational mode of V-O, which varies from 825.30 to 827.52 cm^{-1}. The Raman stretching frequencies and the respective metal-oxygen bond lengths share an inverse relationship which describes that a higher stretching frequency will correspond to a lower metal-oxygen bond length. By utilizing the expression for bond length [18,19]: $V(\ cm^{-1}) = 21349 \exp(-1.9176\ R(^{o}A))$ (2), where V is the stretching Raman frequency for V–O, it can be seen that the bond length varies from 1.6964 ^{o}A to 1.6950 ^{o}A for 350 ^{o}C and 600 ^{o}C sample, respectively [20].

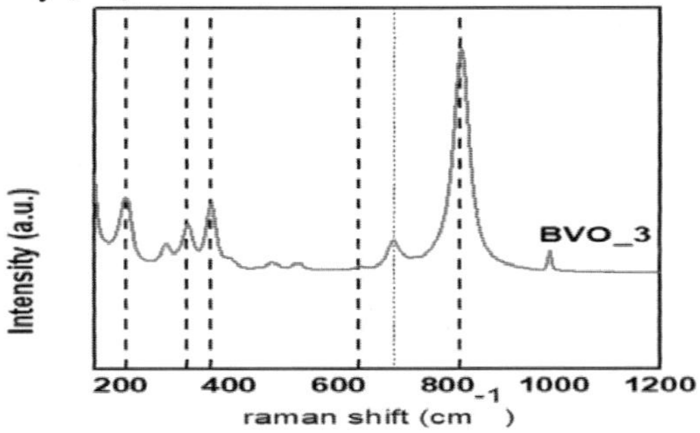

Figure 3. Location of the sample Raman scattering spectrum

The photocatalytic performance of samples was strongly associated with the optical absorption property. Therefore, UV-vis diffuse reflectance spectra (DRS) was utilized to further investigate the absorption band the band gap energy values of the crystals. It can be seen from Figure 4 that all of the five samples showed absorption bands in the visible light region, and the absorption edges of the samples exhibit a slight red shift when the Bi is present. The optical band gap energies of the five samples were estimated from the absorption spectra by using the following relationship: $\alpha h\nu = A(h\nu - Eg)^{n/2}$ where α, $h\nu$, A and Eg are the absorption coefficient, the photon energy, constant, and the optical band gap energy, respectively. The value of n depends on whether the transition is direct (n = 1) or indirect (n = 4), and it was determined to be unity on the absorption spectra. The band gap energy was determined by extrapolating the linear part of $(\alpha h\nu)^2$ vs. $h\nu$ plot of the energy axis as showed in Figure 4. The 2.42 eV value for pure m-$BiVO_4$ is in agreement with the literature [21].

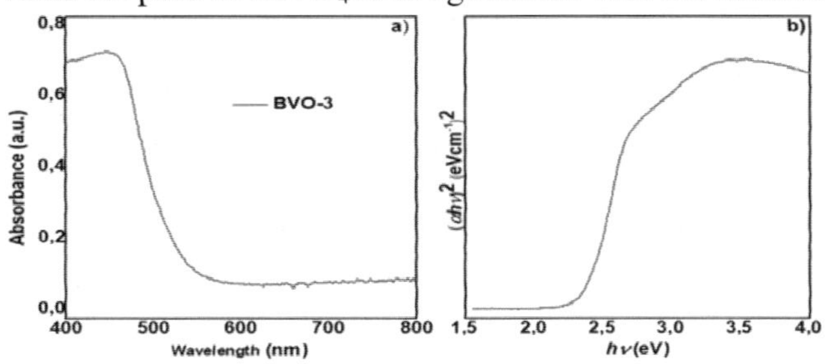

Figure 4. UV-vis diffuse reflectance spectra (DRS) of $BiVO_4$ and Plot of the transformed Kubelka–Munk functions vs. the energy of light.

4. Conclusion

In this study, $BiVO_4$ photocatalyst was synthesized by a simple hydrothermal followed by calcination process at different temperatures (350 °C, 400 °C, 450 °C, 500 °C, 600 °C). The structure of materials were evaluated by chemical and physical methods such as XRD and SEM modern. The results indicate that the calcining at different temperature affects the formation of the crystal form of $BiVO_4$. UV-vis diffuse reflectance spectra (DRS) was utilized to further investigate the absorption band the band gap energy values of the $BiVO_4$ crystals

IOP Conf. Series: Materials Science and Engineering **542** (2019) 012059 doi:10.1088/1757-899X/542/1/012059

Acknowledgments: This research is funded by Nguyen Tat Thanh University, Ho Chi Minh City, Vietnam

References

[1] Zhang M, Zhen Y, Sun F and Xu C 2016 *Materials Science and Engineering: B* **209** 37–44
[2] Yang S G, Pakhomov A B, Hung S T and Wong C Y 2002 *IEEE Transactions on Magnetics* **38** 2877–9
[3] Bach L G, Cao X T, Ho V T T, Islam M R and Lim K T 2015 *Molecular Crystals and Liquid Crystals* **618** 120–8
[4] Cao X T, Bach L G, Islam M R and Lim K T 2015 *Molecular Crystals and Liquid Crystals* **618** 111–9
[5] Bach L G, Nguyen N G and Ho V T T 2016 *International Journal of Photoenergy* **2016** 1–7
[6] Bach L G, Quynh B T P and Lim K T 2016 *Molecular Crystals and Liquid Crystals* **635** 12–7
[7] Bach L G, Nguyen T D, Thuong N T, Van H T T and Lim K T 2019 *Journal of Nanoscience and Nanotechnology* **19** 1192–5
[8] Bach L G, Cao X T, Quynh B T P, Ho V T T and Lim K T 2017 *Molecular Crystals and Liquid Crystals* **644** 183–9
[9] Chu M, Nguyen L, Mai X, Thuan D, Bach L, Nguyen D and Nguyen D 2018 *Metals* **8** 851
[10] Nguyen V, Nguyen T, Bach L, Hoang T, Bui Q, Tran L, Nguyen C, Vo D-V and Do S 2018 *Catalysts* **8** 487
[11] Nguyen H V, Thuan T V, Do S T, Nguyen D T, Vo D V N and Bach L G 2018 *Applied Mechanics and Materials* **876** 52–6
[12] Nguyen D T and Hong S-S 2017 *Journal of Nanoscience and Nanotechnology* **17** 2690–4
[13] Huynh T T, Van Nguyen A, Pham H Q, Vinh N H, Bach L G and Thanh Ho V T 2018 *Journal of Nanoscience and Nanotechnology* **18** 6928–33
[14] Luo J, Fu P, Qu Y, Lin Z and Zeng W 2018 *Physica E: Low-dimensional Systems and Nanostructures* **103** 71–5
[15] Zhang X, Huang Y, Ma F, Zhang Z and Wei X 2018 *Journal of Physics and Chemistry of Solids* **121** 85–92
[16] Yu J and Kudo A 2006 *Advanced Functional Materials* **16** 2163–9
[17] Walsh A, Yan Y, Huda M N, Al-Jassim M M and Wei S-H 2009 *Chemistry of Materials* **21** 547–51
[18] Thalluri S M, Martinez Suarez C, Hernández S, Bensaid S, Saracco G and Russo N 2014 *Chemical Engineering Journal* **245** 124–32
[19] Saison T, Chemin N, Chanéac C, Durupthy O, Ruaux V, Mariey L, Maugé F, Beaunier P and Jolivet J-P 2011 *The Journal of Physical Chemistry C* **115** 5657–66
[20] Dong L, Zhang X, Dong X, Zhang X, Ma C, Ma H, Xue M and Shi F 2013 *Journal of Colloid and Interface Science* **393** 126–9
[21] Thalluri S R M, Martinez-Suarez C, Virga A, Russo N and Saracco G 2013 *International Journal of Chemical Engineering and Applications* 305–9
[22] Kiwi J and Nadtochenko V 2005 *Langmuir* **21** 4631–41

IOP Conf. Series: Materials Science and Engineering **542** (2019) 012060 doi:10.1088/1757-899X/542/1/012060

Visible Light Induced Enhanced Photocatalytic Degradation of Industrial Effluents (Rhodamine B) Using $BiVO_4$ Nanoparticles

Nguyen Duy Trinh[1,2*], Ho Huy Hoang[1,3], Nong Xuan Linh[1,3], Nguyen Huu Vinh[1], Hien Thi Vu[4], Hai Thanh Nguyen[5], Sy Trung Do[6], Dai Viet N-Vo[7]

[1]NTT Hi-Tech Institute, Nguyen Tat Thanh University, Ho Chi Minh City, Vietnam
[2]Department of Chemical Engineering, Pukyong National University, Busan, South Korea
[3]Faculty of Chemical Engineering and Food Technology, Nguyen Tat Thanh University, Ho Chi Minh City, Vietnam
[4]Faculty of Hydro-Meteorology, HCMC University of Natural Resources and Environment, Ho Chi Minh City, Viet Nam
[5]Son La Hydro Power Company, Son La Province, Vietnam
[6]Institute of Chemistry, Vietnam Academy of Science and Technology, Ha Noi, Vietnam
[7]Faculty of Chemical and Natural Resources Engineering, Universiti Malaysia Pahang, Malaysia
*Corresponding author: ndtrinh@ntt.edu.vn; labams2013@gmail.com

Abstract. This study investigates the photodegradation of the organic dye Rhodamine B (RhB) under visible light irradiation by $BiVO_4$ photocatalysts synthesized hydrothermal method. A RhB solution (100 mL, 15 ppm) was degraded in 20 min using the $BiVO_4$ photocatalysts with visible light irradiation. The effect of temperatures calcination on visible-light photocatalytic activity and light absorption performance of $BiVO_4$ was discussed in details. The results showed that $BiVO_4$ sample calcined at 450 °C exhibited the highest photocatalytic performance on the degradation of a Rhodamine B solution. The RhB degradation by the BiVO4 catalyst is 62.38% after 200 min of simulated solar irradiation.

1. Introduction

The prevalence of wastewater generated from industries and household activities poses a serious problems to the modern society. Among common contaminants in water, organic dyes, especially Rhodamine B, have caused increasing concern [1-5]. Characteristically, the dye is easily dissolvable in water and is commonly utilized in textile, food, and pharmaceutical industry [6-8]. In addition, Rhodamine B is water tracer fluorescent and has been known to cause allergies to the skin, eyes, and respiratory system. Such adverse effects have underlain studies exploring the degradation mechanism of RhB [9,10] and raised dye control as an important issue to be tackled with globally. In the literature, previous studies have mostly dealt with degradation mechanism under visible illumination, suggesting photocatalytic oxidation an efficient technology to remove organic pollutants.

Heterogeneous photocatalysis using semiconductors is one of the most promising techniques for environmental remediation. Photocatalytic degradation has various advantages including simplicity, efficiency, low cost and low secondary pollution [11-14]. Under proper irradiation, semiconductors

IOP Conf. Series: Materials Science and Engineering **542** (2019) 012060 doi:10.1088/1757-899X/542/1/012060

could act as a photocatalyst for the degradation of pollutants. Among semiconductors, bismuth vanadate ($BiVO_4$) prevails as a potent photocatalyst owing to its capability to be activated under solar radiation [15-17]. The catalyst is also nontoxic, inexpensive, chemically stable and highly resistant to photocorrosion. The most photoactive structure is the monoclinic scheelite-like crystalline structure with a band gap of 2.4 eV. As a result, BiVO4 has already been studied for the photo-oxidative degradation of many organic compounds such as phenol, organic dyes and metal organic compounds. The aim of this study is to examine the photocatalytic activity of the the prepared $BiVO_4$ materials against the degradation of Rhodamine B dye under visible light irradiation. Different samples of $BiVO_4$ materials were synthesized by hydrothermal method at different temperature ranging from 350 ^oC to 600 ^oC.

2. Materials and Method

2.1. Materials

NH_4VO_3 ($\geq$ 98%, Sigma-Aldrich), $Bi(NO_3)_3.5H_2O$ ($\geq$ 98.0%, Sigma-Aldrich), HNO_3 (65-68%), NH_4OH (25-28%, Xilong), Rhodamine B (RhB, Jihuada), and distilled water (from Lasany's two-storey water machine). The $BiVO_4$ was synthesized by hydrothermal method: First, a mixture containing 4 mmol of $Bi(NO3)_3.5H_2O$ and 20 mL of 4M HNO_3 solution was prepared. Second, the mixture was introduced into a stoichiometric of NH_4VO_3 in 40 mL of hot water, followed by one hour of stirring to obtain a stable and homogeneous solution. The pH of the mixture is adjusted with NH_4OH or HNO_3 solution. The mixture was autoclaved and heated at 160 ^oC for 24 hour under pressure. The produced precipitate then was filtered, washed with distilled water and dried overnight at 110 °C and then calcined in air at different temperature (350 ^oC, 400 ^oC, 450 ^oC, 500 ^oC and 600 ^oC) for 2 hours.

2.2. Photocatalytic test

The $BiVO_4$ photocatalyst was tested for photocatalytic performance against RhB (Sigma-Aldrich) under visible light irradiation. Rhodamine B (RhB) is commonly used in colored papers, temporary hair dyes, fabric dyes and many other applications. However, RhB is toxic and is responsible for diseases such as vomiting, cyanosis, affecting the nervous system of humans. In this study, the light source used is compact Philip 60W (LED light, cree L6, 60W).

The evaluation process of $BiVO_4$ photocatalytic through the following steps : First, 0.1 g of synthesized $BiVO_4$ was introduced into 100 mL of RhB solution. The suspension was then magnetically stirred in the dark for 1h to reach the adsorption-desorption equilibrium. Following that, the suspension was exposed to irradiation in 2 hours. Periodically, 3 mL of aliquot was taken and centrifuged at 7000 rpm/min to remove the particles. The concentration were analyzed by UV-Visible Spectrophotometer. The diagram illustrating the experiments is shown in Figure 1.

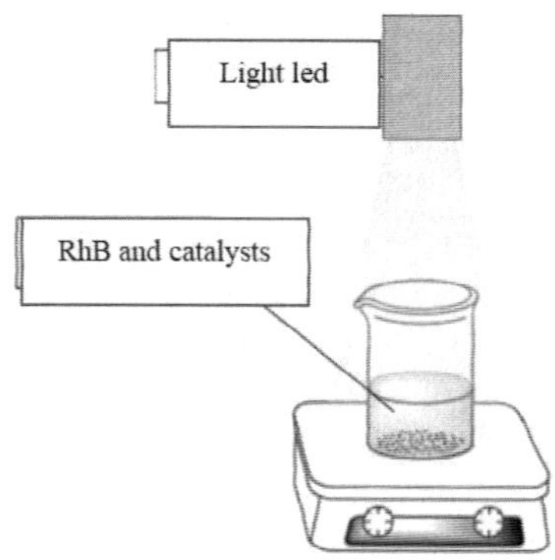

Figure 1. Process diagram of visible light induced enhanced photocatalytic degradation of Rhodamine B using $BiVO_4$ nanoparticles

3. Results and Discussion

According to the UV–vis diffuse reflectance spectrum produced by UV–vis DRS, the band gap of the BiVO4 catalyst was estimated to be approximately 2.42 eV. The activity of photocatalytic materials are evaluated based on decomposition reaction of organic pigments RhB using visible light. Figure 2 shows the photocatalytic decomposition of RhB on $BiVO_4$ samples synthesized by hydrothermal

IOP Conf. Series: Materials Science and Engineering **542** (2019) 012060 doi:10.1088/1757-899X/542/1/012060

method. BVO_450 sample catalyst exhibited the highest activity in the photocatalytic decomposition of RhB under visible light. Figure 3 shows the result of constant optical decomposition rate (k) RhB on the catalyst sample determined from the first-order kinetic model: -ln(C/Co) = kt . As shown in Figure 2, when the RhB solution was illuminated for 120 minutes without a catalyst, about 8% RhB was decomposed. Also according to Figure 3, the peak absorption intensity of 664 nm of RhB solution under photocatalytic treatment of synthetic $BiVO_4$ materials at different calcination temperatures showed a significant decrease. Notably, the as- synthesized samples calcined at 450 °C (Figure 2) is the highest processing efficiency, 62.38% of RhB in solution was decomposed (Table 1).

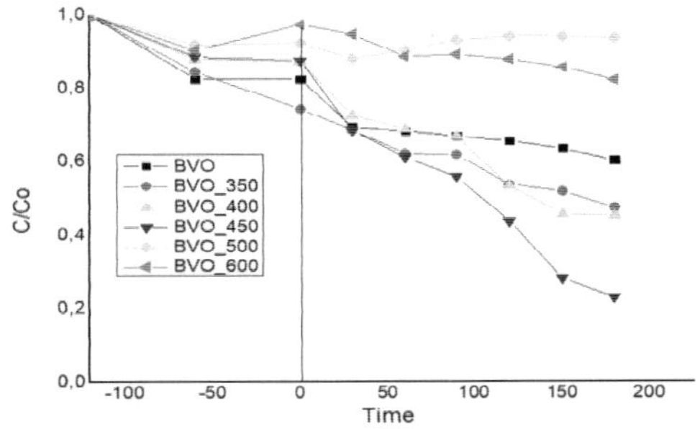

Figure 2. Photolysis of RhB on $BiVO_4$ samples at different calcination temperatures.

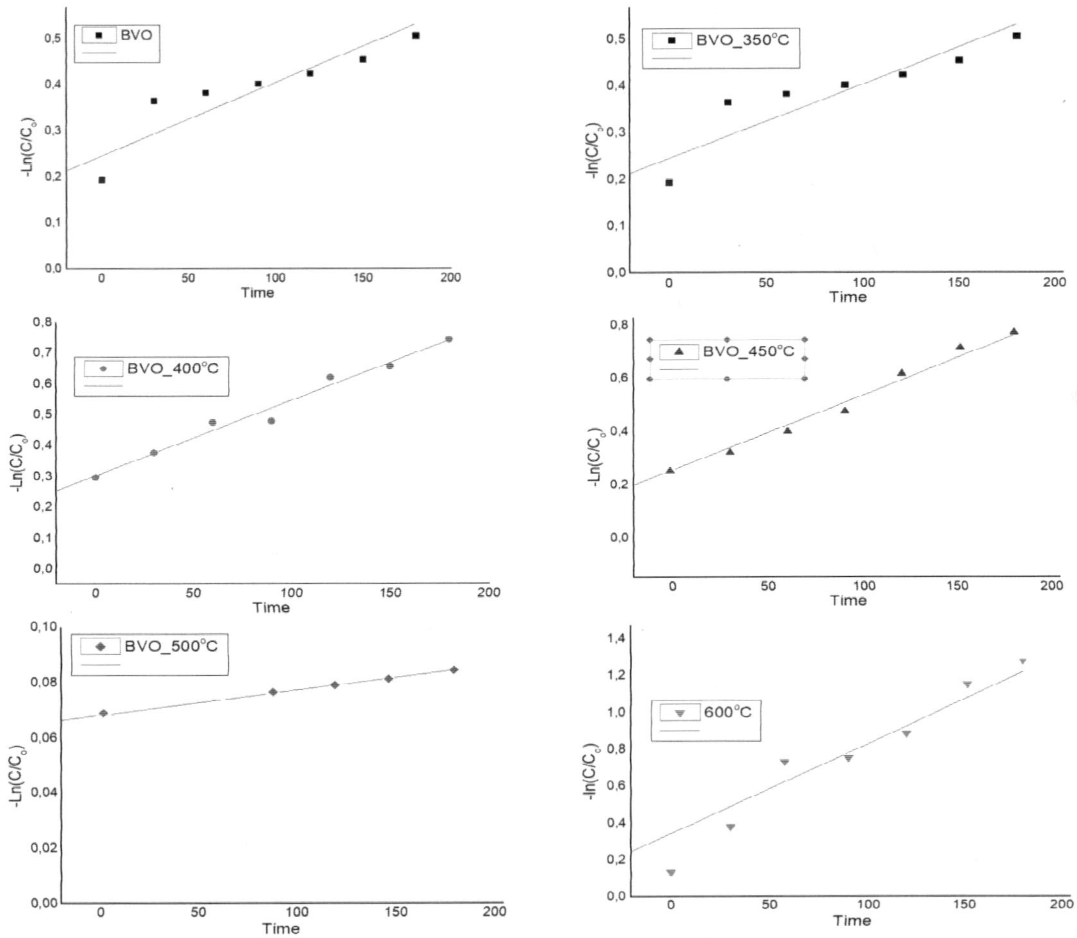

Figure 3. -ln (C/C_o) over time of photo-catalytic decomposition of RhB on $BiVO_4$ samples synthesized by hydrothermal method.

Table 1. Optical and catalytic performance results of samples kinetics $BiVO_4$

Name sample	Calcination temperature (°C)	Degradation rate constants K (min)	R^2	Efficiency H%
BVO	350	0.00244	0.9173	52,52
	400	0.00281	0.9242	54,7
	450	0.00487	0.9497	62,38
	500	0.0023	0.9317	45,31
	600	0.0142	0.922	43,22

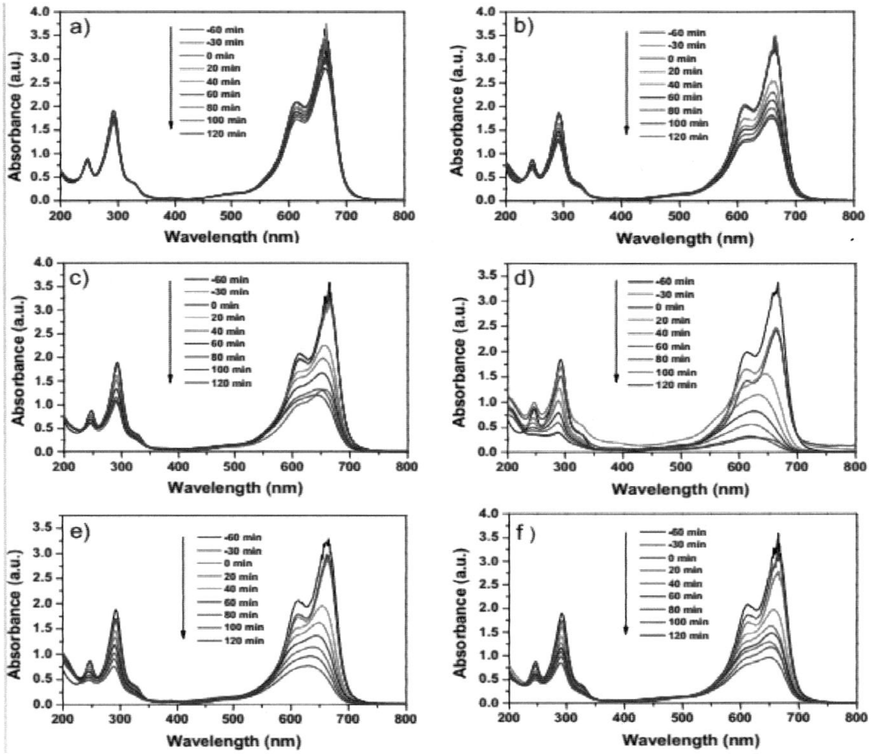

Figure 4. UV-vis change of RhB on BiVO4 samples at different calcination temperatures: a) BVO, b) BVO_350 °C, c) BVO_400 °C, d) BVO_450 °C, e) BVO_500 °C, f) BVO_600 °C

Figure 4 presents the change of absorption peak intensity on the absorption spectrum of the RhB solution over time with $BiVO_4$ calcined at different temperatures. Photocatalytic activity in visible light region indicates absorbance of visible light. Visually, absorption curve presents in the range of 550-800 nm for most catalyst samples and the intensity of characteristic absorption peak at 652 nm diminished rapidly with time and had gradually hypsochromic shifts of absorption bands from 652 nm to 735 nm, which is commensurate with other reported results [18]. Except for Figure 4a, where weak absorption in the region of 550–800 nm of pure $BiVO_4$ is presented, all remaining $BiVO_4$ samples showed enhanced absorption in this region upon calcination at different temperatures in range of 350 °C to 600 °C. In particular, Figure 4d showed strongest absorption in the visible-light regions with the intensity of characteristic absorption peak at 664 nm. On the contrary Figure 4a and b showed weakest absorptions in the visible-light regions with the intensity of absorption peak at 652 nm and 653 nm. In addition, the peak was slightly shifted to longer wavelength, possibly due to an increase of $BiVO_4$ particle size. This result shows that the sample of $BiVO_4$ prepared at calcination temperatures of 450 °C has a larger surface area than all other, thus enhancing the absorption of pigments on the surface of the material and thereby increasing the activity photochemistry, which is consistent with those reported in literature.

4. Conclusion

The photocatalytic activity of BiVO4 was tested for the photodegradation of rhodamine B under visible light irradiation. The as- synthesized samples calcined at 450 °C gave the highest processing efficiency with 62.38% of RhB in solution was decomposed. This result opens the possibility of wide application of $BiVO_4$ material specially applied as catalyst for environmental treatment, especially when it comes to removing RhB dye from wastewater by degradation.

Acknowledgments: This research is funded by Vietnam National Foundation for Science and Technology Development (NAFOSTED) under grant number 104.05-2017.315.

References

[1] Kim D W, Bach L G, Hong S-S, Park C and Lim K T 2014 *Molecular Crystals and Liquid Crystals* **599** 43–50
[2] Nhung N T H, Quynh B T P, Thao P T T, Bich H N and Giang B L 2018 *IOP Conference Series: Earth and Environmental Science* **159** 012015
[3] Bach L G, Van Tran T, Nguyen T D, Van Pham T and Do S T 2018 *Research on Chemical Intermediates* **44** 1661–87
[4] Nie K, Yang H, Gao Z and Wu J 2018 *Materials Science in Semiconductor Processing* **83** 12–7
[5] Van Tran T, Bui Q T P, Nguyen T D, Le N T H and Bach L G 2017 *Adsorption Science & Technology* **35** 72–85
[6] Vinh N H, Hieu N P, Van Thinh P, Diep N T M, Thuan V N, Trinh N D, Thuy N H, Long Giang B and Quynh B T P 2018 *Journal of Nanoscience and Nanotechnology* **18** 6859–66
[7] Traven V F, Cheptsov D A, Dolotov S M and Ivanov I V 2018 *Dyes and Pigments* **158** 104–13
[8] Singh S, Parveen N and Gupta H 2018 *Environmental Technology & Innovation* **12** 189–95
[9] da Silva Filho S H, Vinaches P and Pergher S B C 2018 *Materials Letters* **227** 258–60
[10] Fayaz F, Bach L G, Bahari M B, Nguyen T D, Vu K B, Kanthasamy R, Samart C, Nguyen-Huy C and Vo D N 2019 *International Journal of Energy Research* **43** 405–16
[11] Nguyen V, Nguyen T, Bach L, Hoang T, Bui Q, Tran L, Nguyen C, Vo D-V and Do S 2018 *Catalysts* **8** 487
[12] Nguyen H V, Thuan T V, Do S T, Nguyen D T, Vo D V N and Bach L G 2018 *Applied Mechanics and Materials* **876** 52–6
[13] Nguyen D T and Hong S-S 2017 *Journal of Nanoscience and Nanotechnology* **17** 2690–4
[14] Bach L G, Cao X T, Ho V T T, Islam M R and Lim K T 2015 *Molecular Crystals and Liquid Crystals* **618** 120–8
[15] Lin H, Zhang H, Pang D, Zhou Y and Yi Z 2019 *Journal of Alloys and Compounds* **774** 651–5
[16] Ullah H, Tahir A A and Mallick T K 2018 *Applied Catalysis B: Environmental* **224** 895–903
[17] Zhang X, Huang Y, Ma F, Zhang Z and Wei X 2018 *Journal of Physics and Chemistry of Solids* **121** 85–92
[18] Iwasaki M, Hara M, Kawada H, Tada H and Ito S 2000 *Journal of Colloid and Interface Science* **224** 202–4

IOP Conf. Series: Materials Science and Engineering **542** (2019) 012061 doi:10.1088/1757-899X/542/1/012061

Research Progress of Blade Coatings in Wwind Turbines

Meiling Tang[1]*, Lizhe Huang[2], Jianyu Wang[1] and Duojiao Guan[1]

1 Shenyang Institute of Engineering, Shenyang, P. R. China
2 Huaneng New Energy Co., Ltd. Liaoning Branch, Shenyang, P. R. China
E-mail: 398280075@QQ.com

Abstract. With the expansion of wind turbine capacity, blade technology is evolving. Correspondingly, the preparation of the surface coatings of the blade should develop rapidly. Since most of the wind farms in China are built in deserts and oceans, the protection of the blades is particularly important. Blade coatings protect the blade's normal operation and extend its service life. The work discussed the preparation materials for wind turbine blade coatings and the difficulties faced by the development of coatings in China.
Keywords: coating; resin; wind turbine blade; protective material

1. Introduction: current status and development trends

With the global consumption of non-renewable energy, there is an urgent need for a renewable energy source to replace primary energy such as oil. As a low-cost technology conducive to greenhouse gas emission reduction, wind energy is mainly used. Wind power as an important alternative energy source has received attention and rapid development in recent years, becoming the third largest source of electricity after thermal power and hydropower [1]. At present, China has developed into the world's largest wind power market. It is estimated that by 2020, China's installed capacity of wind power will reach 100 million kilowatts. On the contrary, the coatings of domestic generator set have lagged, and the coatings used in wind turbines in operation are from foreign branded products. Besides high price, it is only suitable for the European maritime climate and cannot adapt to the environment of China's wind power plants. Therefore, the development of wind turbine coatings is a promising project.

2. Performance of the coatings

Wind power blades are the core part of the wind turbine equipment. In order to improve power generation and reduce costs, it should be guaranteed to operate well [2]. Large wind turbine blades are expensive to be installed and take a lot of time. The general maintenance time is about ten years after they are put into operation. Therefore, there is a high standard for coatings that protect the blades. At present, China's wind farm is land-based (about 90%). These units are built in deserts and dust-prone areas, with strong light and large temperature difference between day and night. Other wind power plants are built in the ocean due to the huge wind resources of the ocean. It can be predicted that the offshore wind power technology will develop rapidly. The preparation of the coating should also consider the environment at sea [3]. In summary, the coating should have the following properties:
(1) The substrate and the coating should have excellent adhesion.
(2) The blade coating should be able to withstand high and low temperature changes.
(3) Good weather resistance. It can resist the erosion of blades by sand and rain, organic solvents, hydraulic oils and lubricating oils.

(4) It should have good mechanical strength, flexibility and elastic deformation to prevent cracking or even falling off of the protective film.
(5) It can withstand the irradiation of ultraviolet light, without the change of gloss, powdering, peeling and mildew in a decade.
(6) With excellent practical application, the high-thickness film can be obtained by one spray. Suitable for large-area spraying, it should have fast drying speed, short construction period and high production efficiency.
(7) It should have good anti-icing [4].
(8) When the main engine of the wind power system is in a fire, the flame-retardant coating applied to the special part of the blade surface can isolate the heat of the fire source to prevent the blade from burning and prolong its service life [5].
(9) The coating should be smooth. During the working process of the wind turbine blade, energy loss occurs due to friction with the air, thereby reducing the conversion rate of wind power generation. In order to improve the conversion of wind power generation, it is necessary to ensure the smoothness of the wind turbine blades.

3. Selection of raw materials for coatings

3.1. Selection of pigments and fillers
Pigments: They are used for coloring and covering, with certain anti-corrosion and UV resistance [6]. After comparison of tinting strength, hiding power, weather resistance and corrosion resistance, rutile titanium dioxide is the best pigment and filler.
Fillers: Ultra-fine powder filler can improve the weather resistance of the coating, delay the pulverization and discoloration, and enhance the mechanical properties of the coatings.

3.2. Selection of additives
Adding appropriate additives can improve the coating performance, aging resistance and storage of the coatings.
Matting agent: Make the layer have a suitable gloss and partially improve the abrasion resistance of the film.
Dispersant: Promote the wetting and dispersion of the pigments and fillers, as well as the storage stability and water resistance of the coatings.
Defoamer: Residual bubbles or dry bubbles can affect the appearance of the film and become the center of corrosion. The active ingredient in the defoaming agent interferes with or destroys the stabilizing effect of the bubble to achieve defoaming, thus improving the service life of the coatings [7].
Flatting agent: External pollutants, chemical changes and physical changes during the drying process of the coatings affect the flow and leveling of the coatings to reduce the protective properties of the coatings.
Anti-aging agent: Polymer materials are degraded by the external environment such as light and oxidation, which affects their service life.

3.3. Selection of curing agents
The curing agents should be selected to have excellent weather resistance and gloss retention, no yellowing and good elasticity, such as aliphatic isocyanates.

3.4. Selection of resin
3.4.1 Polyurethane. Polyurethane resin is widely used in the preparation of wind power blade coatings because of its excellent oil and abrasion resistance, chemical resistance and strong adhesion. When polyurethane is used to configure the coating, it is preferred to use aliphatic and alicyclic polyisocyanates instead of aromatics that are easily yellowed [8]. At present, the most used coatings in China are two-component solvent-based polyurethane coating and two-component water-based polyurethane coating [9].

3.4.2 Silicone resin. Silicone resin is a high polymer having a silicon-oxygen bond (Si-O-Si) as a main chain between organic substance and inorganic substance. Therefore, this material has the dual characteristics of organic materials and inorganic materials, with excellent heat resistance, chemical resistance and weather resistance. [10] Silicone coatings use organic silicone resins or modified silicone resins as effective film-forming materials. Therefore, the denatured silicone resin has good electrical insulation and weather resistance of the silicone resin. Besides, it compensates for the insufficiency of the silicone resin, such as shortening of curing time, adhesion, and improvement of mechanical properties.

3.4.3 Fluororesin. Fluororesin coatings have excellent properties, with good weather resistance and low wear. In the design of wind farms, the age requirement of wind power blade coatings is more than 20 years, which is satisfied by fluororesin. Other coatings, such as aliphatic coatings and acrylic silicone coatings, do not last for more than 10 years. Although the price of fluororesin is relatively high, it has advantages in terms of price and durability in the long run.

3.4.4 Polyacrylate material. Acrylic resin has the advantages of weather resistance, light resistance, corrosion resistance, good adhesion, and strong adhesion to the substrate; however, the disadvantages are water resistance, poor solvent resistance and non-wear resistance, so they are used as primers. By adding an appropriate amount of acrylic polymer thereto, the weather resistance and stain resistance can be improved [11].

Now, the coatings of water-based wind power blade have been developed. Using water as solvent, the vaporization temperature of water is relatively higher than that of general solvents, which may cause water to evaporate after applying the coating. Therefore, the construction environments affect its performance, and the ambient humidity should be adjusted to achieve the highest performance. [12]

At present, the protective coatings commonly used in large-scale wind turbine blades are epoxy zinc-rich coatings, inorganic zinc-rich coatings, etc. as primers to provide electrochemical protection. The epoxy-based thick anticorrosive coating and the chlorinated rubber serve as an intermediate layer to provide shielding. Acrylic modified polyurethane coatings and polysilane oxide coatings are used as finishes for corrosion protection and decoration [13, 14]. These are excellent anti-corrosive materials with stable chemical properties.

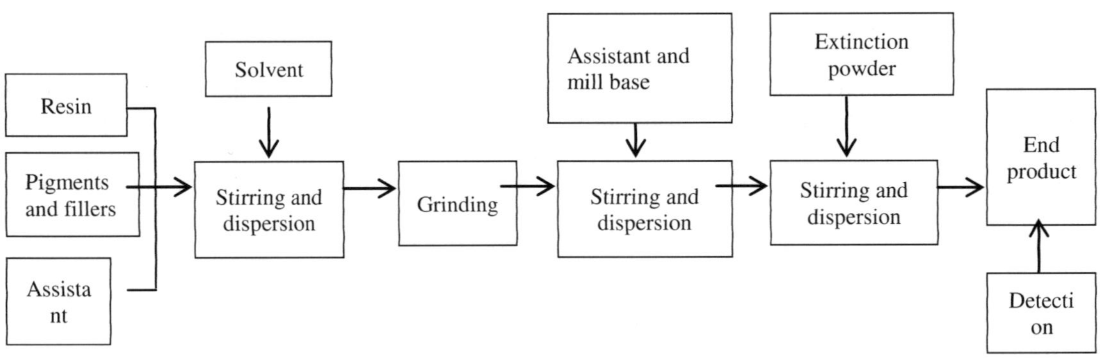

Fig. 1 Preparation of coating on wind turbine blade

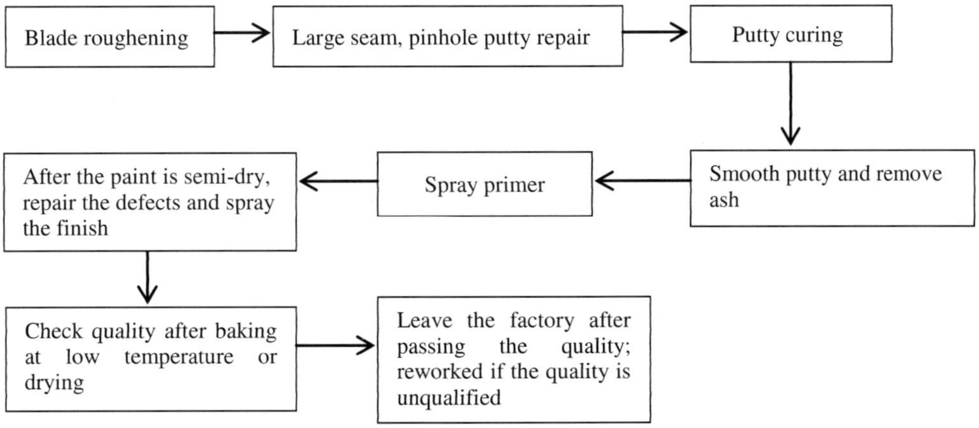

Fig. 2 Construction of coating on wind turbine blade

4. Prospects

The significance and importance of studying wind power blade coatings is to prolong the service life and safety of wind turbine blades. Besides, it is to reduce the production cost of blades, the maintenance cost, and human resources waste caused by blade damage during operation. The research on blade coating should combine the properties of adhesion, wear resistance, weather resistance, heat and humidity resistance, mechanical strength, etc. Thus, it can be used on wind turbine blades with long-term effective protection.

5. Dilemma

China's coatings on wind turbine blade are almost imported from abroad. It is imperative to develop the domestic coatings manufacturing industry. Due to the special nature of China's environmental conditions, foreign coatings cannot fully meet China's requirements. From supply and competition, assurance, selectivity and economy need to be improved [15].

(1) With weak foundation and few implementation schemes, it is almost monopolized by several coating systems.

(2) The high price of imported coatings has led to a large investment in wind power, with great economic pressure.

(3) It has more waste and more pollution, contrary to the concept of saving resources and protecting the environment advocated by the wind power industry.

Acknowledgments

The study is supported by the Guidance Plan Project of 2018 Liaoning Province Key Research and Development Project (2018220015) and Huaneng New Energy Co., Ltd. Liaoning Branch.

References

[1] Jianhua Q, Zhuozhen L, Runpei Z, Xiaoda B, Xing Y and Jianhui F 2012 *Fiber Reinforced Plastics/Composites* **3** 70

[2] WANG Xiao, WANG Hua-jin, ZHAO Wei, LI Zhi-shi, JIANG Qing-huai and WANG Ming-qiang 2016 *Surface Technology* **45(6)** 28

[3] Kaldellis J K and Zafirakis D 2011 *Renewable Energy* **36(7)** 887

[4] Zhao Qingjun, Li Li, Yao Hailong, Lv Zhanqiang, Zhang Peng, Zhang Yongle, Luo Shengkai and Bai Jiangwei 2011 *MODERN PAINT AND FINISHING* **14(4)** 38

[5] Huifan Y, Peng Z, Zhongjie L, Ruifang C and Qingjun Z 2013 *Wind Energy* **2** 88

IOP Conf. Series: Materials Science and Engineering **542** (2019) 012061 doi:10.1088/1757-899X/542/1/012061

[6] Liang Haiyan and Zhang Chaoqun 2018 *Paint & Coatings Industry* **48(11)** 29
[7] JI Xing-hong 2018 *China Coatings* **33(9)** 70
[8] Jin Z, Qinghua Z, Zhenhuan L, Weirong Z, Xiaoli Z and Fengqiu C 2012 *Polymer Materials Science & Engineering* **11** 182
[9] Zhihan Z, Jianping S and Deyong Z 2011 *Advanced Materials Industry* **12** 74
[10] Haibo W and Feiyun H 2018 *China Chemical Trade* **10(24)** 206
[11] Ruipeng W 2013 *GLOBAL MARKET INFORMATION GULDE* **8** 121
[12] Jiahong S 2013 *World Plastics* **31(1)** 62
[13] Hui W 2015 *Shanghai Coatings* **53(8)** 26
[14] Wenyu Z 2014 *Shanghai Coatings* **10** 38
[15] Tian Zhaohui and Li Huaming 2009 *CHINA COATINGS* **24(11)** 6

Photochemical Synthesis of Composite Noble Metal Nanoparticles for Fuel Cell Catalysts

Yingnan Dong[1]*, Shanshan Li[1], Mei Zhu[2], Jian Tang[1], Wei Niu[1] and Xiaoguo Bi[1]

1 Shenyang Institute of Engineering, Shenyang, P. R. China
2 Sino-Platinum Metals Co., Ltd., Kunming, P. R. China
E-mail: dyn4870@163.com

Abstract. Direct methanol fuel cell is an ideal way for energy use because of its low noise, simple structure, rich fuel source, low price, easy storage and transportation. However, in the commercial application, it is necessary to overcome high catalyst cost and short life. The work used photochemical reduction and seed crystal growth to reduce Pt^{II} and Pd^{II} on the surface of Ag nanoparticles to form Ag@Pt-Pd core-shell composite nanoparticles with particle size of about 5~10 nm. The results showed that the synthesized composite nanoparticles have good electrocatalytic activity, with good research and application in the electrode catalysts of fuel cells.
Keywords: Ag@Pt-Pd, core-shell composite nanoparticles, electrocatalytic activity, fuel cell

1. Introduction

In recent years, precious metal composite nanoparticles have been used in energy, environment, biomedicine and other fields due to their unique structure, and optical, electrical and catalytic properties[1]. It has become the focus of scientific research. Our research group, engaged in the synthesis and application of precious metal composite nanoparticles, has synthesized Au@Ag, Au@pd, Au@Pt, Pd@Pt, Ag@Pt, etc. Wherein, Au@Pt, Pd@Pt and Ag@Pt show good electrocatalytic activity with further research value[2-4]. However, there are still few reports on the synthesis of three or more components of precious metal nanoparticles at home and abroad, which is still a challenge.

The work used photochemical reduction and seed crystal growth to reduce Pt^{II} and Pd^{II} on the surface of Ag nanoparticles, thus forming Ag@Pt-Pd colloid particles. The morphology and size of Uv-vis and TEM were characterized. The results showed that the synthesized colloid was a nanocrystal with core-shell composite structure. The electrochemical properties of the synthesized nanoparticles were studied by using methanol oxidation reaction (MOR) as the evaluation reaction. It was found that the electrocatalytic activity of the synthesized Ag@Pt-Pd composite nanoparticles is due to the traditional Pt nanoparticles, which has good application in fuel cell catalysts.

2. Experiment

2.1. Photochemical preparation of Ag nanoparticles

Add a certain amount of $AgNO_3$, trisodium citrate, absolute ethanol and acetone in a certain proportion in a volumetric flask to constant volume with water. Then, mix by shaking. The solution was transferred to a quartz reactor and placed on an ultraviolet transilluminator for irradiation. The UV-Vis spectral curve of the colloidal solution was recorded at intervals of time until two adjacent spectral curves substantially coincided, indicating the complete photochemical reaction. The same method was used to synthesize Pt nanoparticles for comparative experiments.

2.2. Photochemical preparation of Ag@Pt-Pd composite nanoparticles

A certain amount of the Ag colloid was synthesized by Method 1, and the prepared Pt^{II}-Pd^{II} solution was added to the quartz reactor in a certain ratio. The total molar volume of trisodium citrate, ethanol, acetone and precious metals was kept constant, and the molar ratio of Ag to Pt Pd was 1:1. The Pt and Pd molar ratios were 1:4, 1:1 and 4:1, respectively. After uniform oscillation, it was placed on an ultraviolet transilluminators for irradiation. After a period, the UV-Vis spectral curve of the colloidal solution was measured until the spectral curves substantially coincided. That is to say, the photochemical reactions were completely carried out, labeled as Ag@Pt-$Pd_{1:4}$, Ag@Pt-$Pd_{1:1}$ and Ag@Pt-$Pd_{4:1}$, respectively.

2.3. Ag@Pt-Pd/C composite nanoelectrochemical analysis

The obtained Ag@Pt-Pd and Pt nanoparticles were loaded on XC-72 carbon black for electrochemical testing. Three-electrode system was used in the electrochemical test, with Pt as the counter electrode, SCE as the reference electrode, the electrolyte of 0.5 $mol\cdot L^{-1}$ KOH+2.0 $mol\cdot L^{-1}$ CH3OH, and the temperature of 30-60 °C. The dissolved oxygen in the electrolyte was removed by high purity N_2 gas before the test, and the cyclic voltammetry curve was recorded under N_2 atmosphere. With the test potential range of -0.6 to 0.8 V and the potential scan rate of 50 $mV\cdot s^{-1}$, the time-current curve (I-t) was tested at the potential of -0.3 V.

In the test, a TEM image of nanoparticles was taken using H-800 transmission electron microscope. Shanghai Jinghua 759MC UV-Vis spectrophotometer was used to record the UV-Vis spectra of nanocolloids. Electrochemical analysis of nanoparticles was performed by Wuhan Corrtest Instrument CS350 Electrochemical Workstation. The precious metal precursor was made of 4N grade. The analytical reagents were used with secondary reverse osmosis demineralized water.

3. Results and discussion

3.1. Characterization of Ag@Pt-Pd composite nanoparticles

Fig. 1a shows the UV-Vis absorption spectrum of Ag colloidal particles. The UV-Vis absorption spectrum of Ag colloidal particles has a distinct characteristic absorption peak around 400 nm, and the colloidal solution turns bright yellow. The Ag@Pt-Pd colloid is brownish black, and the spectrum is a downwardly inclined line from left to right with no characteristic absorption peak. This is similar to the Pt-Pd colloid reported in the literature (See Fig. 1b)[5]. The ultraviolet-visible spectrum of precious metal nanoparticles is generated by the plasmon resonance of the nanoparticle surface, mainly reflecting the surface characteristics of the particles. The plasmon resonance absorption peak of Ag surface has disappeared in the Ag@Pt-Pd colloidal UV-Vis absorption spectrum. There is no Ag on the surface of the colloidal particles. The Ag colloid is encapsulated by Pt and Pd atoms deposited on the surface. The synthesized Ag@Pt-Pd colloid should be a core-shell composite nanostructure.

Fig. 2 shows the TEM image of Ag@Pt-Pd composite nanoparticles. Ag@Pt-Pd colloids are monodisperse near-spherical nanocrystals with the particle size of 5~10nm. Ag@Pt-$Pd_{1:4}$, Ag@Pt-$Pd_{1:1}$ and Ag@Pt-$Pd_{4:1}$ have average particle diameters of 5.6, 6.8 and 9.4 nm, respectively, larger than the average particle size (4.8 nm) of Ag seed crystals synthesized in this experiment. The average particle size of Ag@Pt-Pd increases with the increased proportion of Pd. There is a clear contrast

IOP Conf. Series: Materials Science and Engineering **542** (2019) 012062 doi:10.1088/1757-899X/542/1/012062

between the center of the particle and the surrounding, which may be caused by the difference in imaging between Ag and Pt with Pd in TEM characterization.

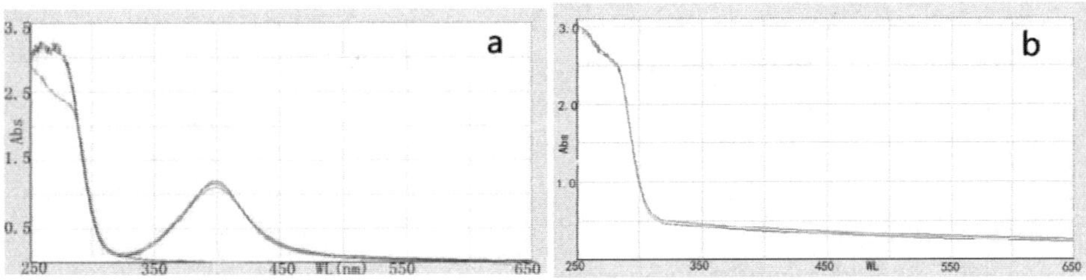

Fig. 1 UV-Vis absorption spectrum of Ag and Ag@Pd-Pt colloidal particles

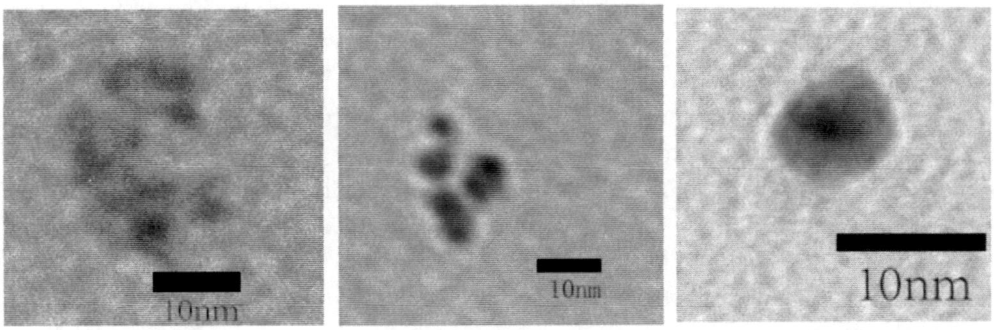

a Ag@Pt-Pd=4:1 b Ag@Pt-Pd=1:1 c Ag@Pt-Pd=1:4

Fig. 2 TEM image of Ag@Pt-Pd composite nanoparticles

3.2. Electrochemical analysis of Ag@Pt-Pd composite nanoparticles

Fig. 3 shows the cyclic voltammogram of Ag@Pt-Pd composite nanoparticles in methanol dissolution. Table 1 shows the current data of oxidation peak of Ag@Pt-$Pd_{1:4}$/C, Ag@Pt-$Pd_{1:1}$/C, Ag@Pt-$Pd_{4:1}$/C and Pt/C at different temperatures.

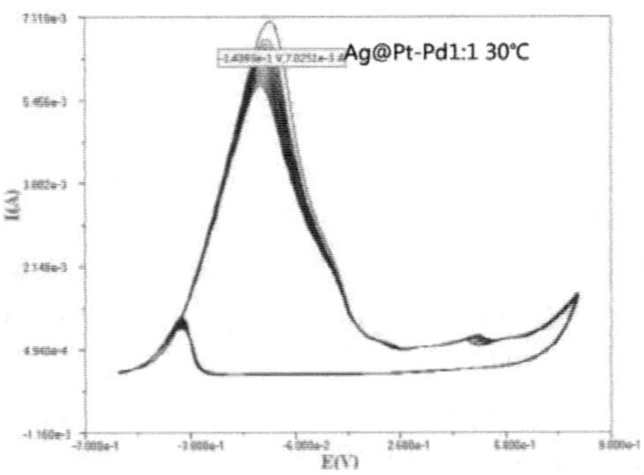

Fig. 3 Cyclic voltammogram of Ag@Pt-Pd composite nanoparticles

Table 1 Current peak mA/$mg_{Preciousmeta}$ of linear cyclic voltammetry of Ag@Pt-Pd/C catalyst.

	30°C	45°C	60°C
Pt/C	144.6	284.2	322.6
Ag@Pt-$Pd_{1:4}$	262.2	491.3	692.8
Ag@Pt-$Pd_{1:1}$	522.6	582.0	719.4
Ag@Pt-$Pd_{4:1}$	359.4	508.8	652.9

The results showed that for MOR, the electrocatalytic activity Ag@Pt-Pd was higher than that of Pt nanoparticles. Wherein, Ag@Pt-$Pd_{1:1}$ had the optimal activity, about 3.5 times Pt nanoparticles at 30 °C. The peak currents of Ag@Pt-$Pd_{1:4}$ and Ag@Pt-$Pd_{4:1}$ also reached 1.8 times and 2.5 times that of monometal Pt, respectively. We speculated that, on the one hand, the heterogeneous catalytic reaction mainly occurred on the surface of the solid catalyst. The Ag@Pt-Pd composite nanoparticles Pt and Pd atoms in this experiment were deposited on Ag nanoparticles to form a thin shell. This improved the utilization of Pt and Pd, beneficial to the high catalytic activity of the nanoparticles.

On the other hand, we found that the electrocatalytic activity of Ag@Pt-Pd composite nanoparticles was closely related to the molar ratios of Pt and Pd, which is because the valence electron arrangement of Pt was $5d^9 6s^1$. Influenced by the relativistic effect, 6s electron contraction led to the increased electron cloud density. The barrier of d-electron hybridization of Pt and Pd atoms was reduced, which facilitated the interaction of Pt atoms with Pd atoms to form a large conjugated system. At the same time, the d% (46) of Pd was slightly larger than that (44) of Pt, so the 4d electrons of Pd could partially flow into the 5d orbit of Pt. The electron donating of the Pd core to the Pt shell improved the amount of d electron holes of Pt and Pd in the shell layer, which was the main reason for the good electrocatalytic activity of Ag@Pt-Pd composite nanoparticles[6-8]. With the increase of temperature, the peak current of each catalyst increased significantly, and the peak current difference of each catalyst gradually decreased. This was due to the fact that MOR was more likely to occur with increasing temperature, and the difference in activity of each catalyst was somewhat reduced.

In addition, the results showed that the cyclic voltammetry curve of Ag@Pt-Pd composite nanoparticles was coincident with a continuous cyclic voltammetry scan of 30 min (See Fig. 3), and the peak current slightly decreased. The electrocatalytic activity of Ag@Pt-Pd composite nanoparticles was relatively stable, and the peak current drop should be the result of continuous oxidation of methanol. Therefore, Ag@Pt-Pd composite nanoparticles have good prospects in the electrode catalysts of fuel cell.

4. Conclusions

1) In the work, Ag nanoparticles were used as seed crystals, with Pt^{II} and Pd^{II} added. Ag@Pt-Pd ternary precious metal nanoparticles with different metal molar ratios were synthesized by photochemical seed crystal growth.

2) TEM characterization showed that the synthesized Ag@Pt-Pd composite nanoparticles were monodisperse near-spherical nanoparticles of 5~10 nm. Combined with UV-Vis spectroscopy, we speculated that the synthesized Ag@Pt-Pd nanoparticles should be a core-shell composite structure.

3) Ag@Pt-Pd ternary precious metal nanoparticles were supported on carbon black Vulcan XC-72 carrier, and their electrocatalytic activity for methanol was analyzed by cyclic voltammetry. The results showed that Ag@Pt-Pd nanoparticles had excellent catalytic activity. Especially at the experimental temperature of 30°C, the peak current of Ag@Pt-$Pd_{1:1}$ reached 522.6 mA / $mg_{Preciousmeta}$, which was about 3.5 times that of single metal Pt. The effective Pt and Pd utilization and the electron donating of Pd to Pt by Ag core were the main reasons for the high electrocatalytic activity of Ag@Pt-Pd composite nanoparticles. Ag@Pt-Pd core-shell composite nanoparticles have good electrocatalytic activity with prospects in the electrode catalysts of fuel cell.

Acknowledgments

The study is supported by the Guidance Plan Project of 2018 Liaoning Province Key Research and Development Project (2018220016).

References

[1] YU Peikai and LI Yuechan 2018 *Journal of Shenyang Normal University(Natural Science Edition)* **36(4)** 369
[2] Yingnan Dong, Xikun Yang and Shanshan Li 2014 *Rare Metal Materails and Engineering* **43(12)** 2985
[3] Yingnan Dong, Shou'an Dong and Chun Tang 2013 *Rare Metal Materails and Engineering* **42(4)** 789
[4] Yingnan Dong, Shou'an Dong and Chun Tang 2009 *Chem J Chin Univ* **30(7)** 1397
[5] Toshima N, Harada M and Yamazaki Y 1992 *J. Phys. Chem.* **96** 9927
[6] Dong S A, Tang C and Zhou H 2004 *Gold Bull* **37** 187
[7] NING Xiao-Hua, XU Shu-Ping and DONG Feng-Xia 2009 *Chem. J. Chinese. Univ.* **30(1)** 159
[8] Haiqiang Zhao, Weihong Qi, Xinfeng Zhou, Haofei Wu and Yejun Li 2018 *Chinese Journal of Catalysis* **39(2)** 342

IOP Conf. Series: Materials Science and Engineering **542** (2019) 012063 doi:10.1088/1757-899X/542/1/012063

Efficient analytical method to obtain the responses of a gear model with stochastic load and stochastic friction

Yining Fang[1,2], Ming J Zuo[2] and Yue Li[1]

1 Laboratory of Science and Technology on Integrated Logistics Support, National University of Defense Technology, Changsha, China
2 Department of Mechanical Engineering, University of Alberta, Edmonton, Canada
E-mail: liyue@nudt.edu.cn

Abstract. The friction, which is widely existed in practical, is seldom considered when modelling the gear system with stochastic load. Due to the variation of the temperature and lubrication condition, friction is a stochastic factor to a gear model. In this paper, a gear model with stochastic load, stochastic friction, and some other deterministic factors is considered. Due to the effects of stochastic factors (i.e., load and friction), the gear system faces more vibration and noise than the case with all deterministic factors. Thus, to analyse the variation of responses in the gear dynamical model, the corresponding dynamic equation needs to be solved. However, the statistic characteristics of dynamic responses are hard to obtain by numerical methods. Thus, an efficient analytical method is proposed, and then, an approximate analytical solution of the dynamic equation can be obtained in this paper. By the obtained solution, the vibration and noise of gear systems can be well investigated. Simulation results are provided to demonstrate the superior performance of the proposed method.

1. Introduction

Gear system which is one of the most widely adopted mechanical parts shows significant important to the development of mechanical theory [1]. Therefore, the research about dynamics and vibrations of gear systems attracts lots of research attention. The gear models under deterministic domain have been investigated in decades. However, stochastic load is one of the main sources of gear vibration and noise in practical [2]. The gear system dynamic modelling with stochastic load attracts more and more attention. In [3], a gear model under stochastic load with transmission error, backlash, and periodic gear mesh stiffness was introduced. Wen et al. [4] investigated a gear dynamic model under stochastic load with backlash, time-varying mesh stiffness (TVMS), and constant damping coefficient. However, besides the load, the internal factors (e.g., friction, backlash, etc.) of a gear system may also have random variation and affect the system's dynamic behaviour greatly [5]. For example, friction, as a main cause of vibration on gear system's transient state, is observed as stochastic in lots of scenarios [6]. Therefore, the stochastic internal factor should be considered when modelling gear systems.

Under a gear system model, there are two kinds of forms, analytical expression and numerical solutions, to demonstrate the responses of the gear system. Therefore, some methods have been proposed to obtain the responses of a gear system [7]. For example, Runge Kutta-Monte Carlo (MC) method and statistical Newmark method are proposed to obtain the numerical solutions. Stochastic averaging method and path integration (PI) method are proposed to obtain the analytical expression [7]. Obtaining numerical solution is costly since a large number of samples are required for reasonable accuracy. By contrast, it would save much time if we could obtain the analytical solution [8]. Therefore, we focus on obtaining the analytical solution in this paper.

Several researchers have explored the analytical solution to a gear dynamic model under stochastic load. Naess et al. [9] derived the analytical solution to a gear system considering constant stiffness, constant damping coefficient, and backlash under the excitation of white noise. In [4], the analytical solution to a gear system considering constant damping coefficient, TVMS, and backlash under stochastic load was derived. However, only deterministic internal factors are considered in these works. A gear model with a stochastic internal factor under stochastic load cannot be solved by existing analytical methods.

In this paper, we consider the gear dynamic model with stochastic load, stochastic friction (i.e., the stochastic internal factor), and other deterministic internal factors (e.g., damping, TVMS, backlash, etc.). Then, a method is proposed to derive an approximate analytical solution of the corresponding dynamic equations. By the proposed method, the dynamic characteristics (e.g., vibration, noise, etc.) of gear systems can be well investigated. Compared with the MC method, our proposed method can achieve similar accuracy responses with much less time cost.

The remaining pars of this paper are organized as follows. The gear system with stochastic load and stochastic friction is modelled in Section 2. The method to obtain the responses of the gear model is proposed in Section 3. The simulation results are given in Section 4. Section 5 draws conclusions.

2. Gear dynamic model

In this work, a gear model considering TVMS, backlash, and friction is introduced and is shown in figure 1. In this model, both the load and the friction (i.e., the single stochastic internal factor) are stochastic. Therefore, the considered model is formulated as:

$$J_1\ddot{\theta}_1 = T_1 - FR_{b1} - F_f X_1 \tag{1}$$

$$J_2\ddot{\theta}_2 = T_2 + FR_{b2} + F_f X_2 \tag{2}$$

where J_i, T_i, X_i, and R_{bi} are the moment of inertia, external torque, arm length of friction, and base circle of gear $i \in \{1,2\}$, respectively, F represents the total force between the contact teeth, F_f is the sliding friction [2]. Let x, $\dot{x}$, and $\ddot{x}$ denote the relative angular displacement, the relative angular velocity, and the relative angular acceleration, respectively. Thus, we have

$$x = \theta_1 R_{b1} - \theta_2 R_{b2} \tag{3}$$

where θ_i is the angular displacement of gear $i \in \{1,2\}$.

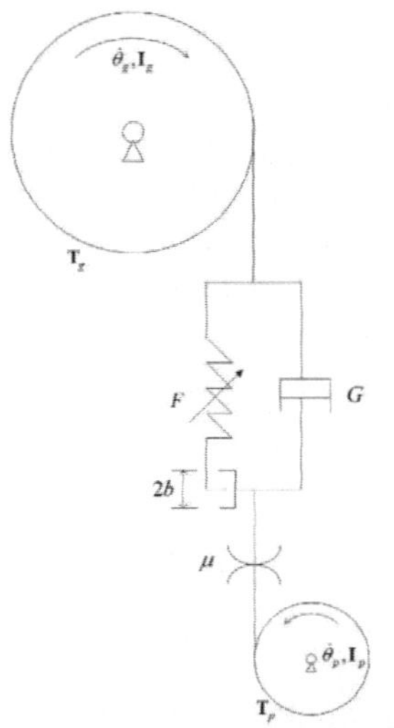

Figure 1. Model of the gear system

Simplify equation (1) and equation (2), and then, a normalized equation is obtained and shown as:

$$\ddot{x}+cL(\theta_1,\mu,t)\dot{x}+kL(\theta_1,\mu,t)g(x)=f(t) \tag{4}$$

where c denotes the damping coefficient which is considered as a constant, k denotes the TVMS, μ represents the friction coefficient, $g(x)=\begin{cases} x-b, & x>b \\ 0, & -b\le x\le b \\ x+b, & x<-b \end{cases}$ is the function of backlash, $f(t)$ is the external load, and $L(\theta_1,\mu,t)$ is a function which is caused by friction. The expression of $L(\theta_1,\mu,t)$ can be obtained as [2]:

$$L(\theta_1,\mu,t)=\chi_1+\mu\left(\chi_2\theta_1+\chi_3\right) \tag{5}$$

where $\chi_j(j=1,2,3)$ relates to gear design parameters. Note that $\chi_j(j=1,2,3)$ can be considered as constant. μ is a summation of a deterministic part μ_0 and a random part $\xi_1(t)$. The external load $f(t)$ is modeled as a combination of a constant deterministic part f_0, a periodical deterministic part $f_1\cos(\varphi t)$, and a random part $\xi(t)$ [4]. About the random part, it is generally set to a Gaussian white noise. Therefore, the expressions of the external load are given as follows.

$$f(t)=f_0+f_1\cos(\varphi t)+\xi(t) \tag{6}$$

$$E(\xi(t))=0 \tag{7}$$

$$E[\xi(t)\xi(t+\tau)]=r\delta(\tau) \tag{8}$$

where φ is a constant frequency, t denotes time, r is the variance of the random part $\xi(t)$, and $\delta(\tau)$ is the Dirac Delta function.

From equation (5), we can see that $L(\theta_1,\mu,t)$ depends on θ_1. Considering the deterministic nonlinear and time-varying characteristics, θ_1 is also stochastic because of $f(t)$ as given in equation (6). To obtain the responses of the gear system, we need to derive the analytical solution of the stochastic differential equation (SDE), i.e., equation (4).

3. The proposed method to derive the analytical solution

The existing methods can obtain the analytical solution of the differential equation which $L(\theta_1,\mu,t)$ is independent with θ_1. Note that $L(\theta_1,\mu,t)$ depends on θ_1 in our work. Therefore, equation (4) cannot be solved by existing methods directly. To obtain the analytical solution of equation (4), a method is proposed in this paper. The basic idea of the proposed method contains the following two steps:

1) Obtain the tentative analytical solution. We first introduce a method to transform the SDE equation (4) into a form that can be solved by PI method. Then, we derive a tentative analytical solution using PI method.
2) Adjust results by adding a modification function. Due to the previous transformation, errors may be brought into the tentative analytical solution, and thus, a modification function is applied to adjust the tentative analytical solution. Supervised learning is used to obtain the modification function.

3.1. Obtaining the tentative analytical solution

By considering the deterministic part of the load, equation (9) is obtained.

$$\ddot{x}+cL(\theta_1,\mu,t)\dot{x}+kL(\theta_1,\mu,t)g(x)=\hat{f}(t) \tag{9}$$

where $\hat{f}(t)=f_0+f_1\cos(\varphi t)$ is the deterministic part of the load. The solution of equation (9), which are x (considered as μ_1) and $\dot{x}$ (considered as μ_2), can be obtained by solving equation (9). Note

IOP Conf. Series: Materials Science and Engineering **542** (2019) 012063 doi:10.1088/1757-899X/542/1/012063

that, according to equation (3), $\hat{\theta}_1$ and $\hat{\theta}_2$ can be obtained according to the obtained x in equation (9). After that, we get $\lambda_L(t)$ as expressed in equation (10) by substituting $\hat{\theta}_1$ to equation (3).

$$\lambda_L(t) = \chi_1 + \mu\left(\chi_2\hat{\theta}_1 + \chi_3\right) \tag{10}$$

By replacing $L(\theta_1, \mu, t)$ by $\lambda_L(t)$ in equation (4), we can obtain

$$\ddot{x} + c\lambda_L(t)\dot{x} + k\lambda_L(t)g(x) = f(t) \tag{11}$$

Based on this transformation, all parameters in equation (11) become deterministic except for load, and thus, it can be solved by PI method now.

For the case that $g(x) \neq 0$, the system equation is shown in equation (11). Under the excitation of $\delta(t)$, an impulse function $h_x(t)$ satisfies the following equation.

$$\frac{d^2h_x(t)}{dt^2} + c\lambda_L(t)\frac{dh_x(t)}{dt} + k\lambda_L(t)h_x(t) = 0 \tag{12}$$

The general solution to the homogeneous equation (12) can be obtained as

$$h_x(t) = \frac{v}{\beta}e^{\alpha t}\sin(\beta t) \tag{13}$$

where $\alpha = -\frac{c\lambda_L(t)}{2}$ and $\beta = \frac{\sqrt{4k\lambda_L(t) - (c\lambda_L(t))^2}}{2}$. Let σ_1^2 and σ_2^2 denote the variances of x and $\dot{x}$, respectively. And σ_{12} denotes the covariance of x and $\dot{x}$. Then, we can obtain σ_1^2, σ_{12}, and σ_2^2 as follows, where $G_0 = 2\pi r$.

$$\sigma_1^2 = \frac{G_0 v^2}{4\beta^2\left(\alpha^2 + \beta^2\right)}\left\{-\frac{\beta^2}{\alpha} - e^{2\alpha\Delta t}\left(-\frac{\alpha^2 + \beta^2}{\alpha} + \alpha\cos(2\beta\Delta t) + \beta\sin(2\beta\Delta t)\right)\right\} \tag{14}$$

$$\sigma_{12} = \frac{G_0 v^2}{4\beta^2}e^{2\alpha\Delta t}\left[1 - \cos(2\beta\Delta t)\right] \tag{15}$$

$$\sigma_2^2 = -\frac{G_0 v^2}{4\alpha\beta^2}\left\{\beta^2 - e^{2\alpha\Delta t}\left[\alpha^2 + \beta^2 + \alpha\beta\sin(2\beta\Delta t) - \alpha^2\cos(2\beta\Delta t)\right]\right\} \tag{16}$$

For the case that $g(x) = 0$, the system equation is shown as

$$\ddot{x} + cL(\theta_1, \mu, t)\dot{x} = f(t) \tag{17}$$

Similarly, we have equation (18) by eliminating the random term of load $f(t)$ in equation (17).

$$\ddot{x} + cL(\theta_1, \mu, t)\dot{x} = \hat{f}(t) \tag{18}$$

Then, we can obtain the results by solving equation (18). By substituting the obtained results to equation (10), we can obtain $\tilde{\lambda}_L(t)$. The following equation is obtained by replacing $L(\theta_1, \mu, t)$ to $\tilde{\lambda}_L(t)$ in equation (17).

$$\ddot{x} + c\tilde{\lambda}_L(t)\dot{x} = f(t) \tag{19}$$

The general solution to the homogeneous equation (19) is given as:

$$h_x^{(2)}(t) = \frac{v}{a_2}\left(1 - e^{a_2 t}\right) \tag{20}$$

where $a_2 = c\tilde{\lambda}_L(t)$ and v represents the mean of initial velocity. Then, σ_1^2, σ_{12}, and σ_2^2 can be obtained as:

$$\sigma_1^2 = \frac{G_0 v^2}{a_2^2}\left(\Delta t - \frac{1}{a_2}\left(\frac{3}{2} - 2e^{-a_2\Delta t} + \frac{1}{2}e^{-a_2\Delta t})\right)\right) \tag{21}$$

$$\sigma_{12} = \frac{G_0 v^2}{a_2^2}\left[\frac{1}{2} - e^{-a_2\Delta t} + \frac{1}{2}e^{-2a_2\Delta t}\right] \tag{22}$$

$$\sigma_2^2 = \frac{G_0 v^2}{2a_2}\left(1 - e^{-2a_2\Delta t}\right) \tag{23}$$

3.2. Updating the tentative analytical solution

Due to the transformation in Section 3.1, the obtained tentative analytical solution is not the accurate solution of equation (4). Therefore, we try to derive the modification function and then apply it to the tentative analytical solution to obtain the accurate solution of equation (4).

Generally, the exact analytical solution of equation (4) is hard to derive. The numerical responses obtained by MC method are regarded as accurate results. Therefore, a modification function is required to demonstrate the rule of errors according to the samples obtained by MC method and the tentative analytical solution. A supervised learning algorithm is applied to obtain the modification function. Finally, we adjust the tentative analytical solution by the modification function and then obtain the final probability distributed function (PDF). The obtained final PDF can be regarded as the analytical solution (i.e., PDF) of the SDE equation (4).

Therefore, a modification to μ_1, μ_2, σ_1, σ_2, and σ_{12} should be taken. The modification function, denoted $\eta(t)$, is based on the results of MC method and the tentative analytical solution. For different cases, $\eta(t)$ may have different forms, such as polynomial function, exponential function, and so on. Supervised learning is used to find $\eta(t)$. Supervised learning is used to find a mapping between a set of input samples and the corresponding output and this mapping is then applied to predict the outputs under other input data [10]. The samples obtained by the tentative analytical solution would be the input data and the samples obtained by MC method would be the corresponding output. We define the input samples and the corresponding output as $\chi(t)$ and $\zeta(t)$, respectively. Then, we try to seek a function $\eta(t): \chi(t) \to \zeta(t)$. To evaluate $\eta(t)$, the risk function ε is defined as [11]:

$$\varepsilon = \int_{t_1}^{t_2} \left|\frac{\zeta(t) - \eta(t)\chi(t)}{\zeta(t)}\right| dt \tag{24}$$

To get a proper $\eta(t)$, the value of ε should be minimized. In this paper, the structural risk minimization method is adopted to get $\eta(t)$. After obtaining the modification function $\eta(t)$, the modified results can be obtained by

$$\tilde{\sigma}_1(t) = \sigma_1(t)\eta_1(t) \tag{25}$$

$$\tilde{\sigma}_2(t) = \sigma_2(t)\eta_2(t) \tag{26}$$

$$\tilde{\sigma}_{12}(t) = \sigma_{12}(t)\eta_{12}(t) \tag{27}$$

IOP Conf. Series: Materials Science and Engineering **542** (2019) 012063 doi:10.1088/1757-899X/542/1/012063

where $\tilde{\sigma}_1(t)$, $\tilde{\sigma}_2(t)$, and $\tilde{\sigma}_{12}(t)$ are the modified values of $\sigma_1(t)$, $\sigma_2(t)$, and $\sigma_{12}(t)$, respectively, and $\eta_1(t)$, $\eta_2(t)$, and $\eta_{12}(t)$ are the corresponding modification functions. Then, the improved PDF expressions of x and $\dot{x}$ with a high accuracy are obtained.

4. Simulation results

The parameters of the gear system used in this validation are from [12] and friction coefficient parameters are set as $\mu_0 = 0.04$ and $r_1 = 0.08$. For the stochastic load in equation (6), the parameters are set as $f_0 = 2\times10^4$, $f_1 = 2\times10^3$, $\varphi = 100$, and $r = 2\times10^3$.

4.1. Accuracy evaluation

The instantaneous marginal PDF of x and $\dot{x}$ at 0.3s (chosen by random) are shown in figure 2. According to the simulation results, the PDFs of the two methods are very close. This agreement validates the analytical solution which is obtained by the proposed method is accurate.

The joint PDF of x and $\dot{x}$ are shown in figure 2. It illustrates the marginal PDFs of x and $\dot{x}$ at $t = 0.76s$ (chosen by random). It is observed that the joint PDF from the proposed method agrees well with that from MC method.

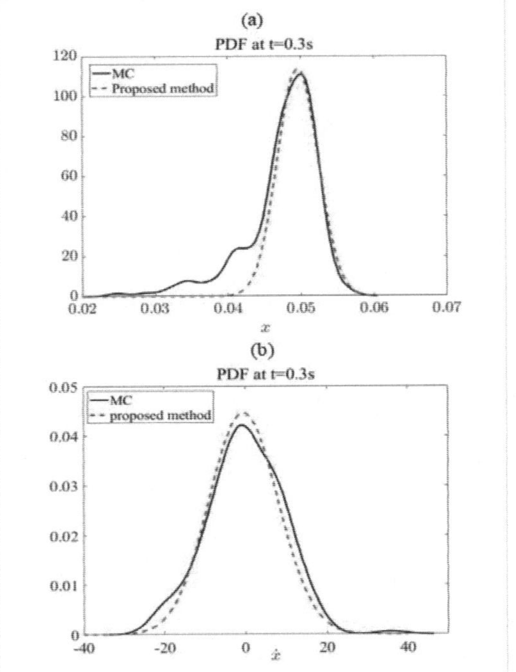

Figure 2. Marginal distribution of (a) x (b) $\dot{x}$ at $t = 0.3$ s.

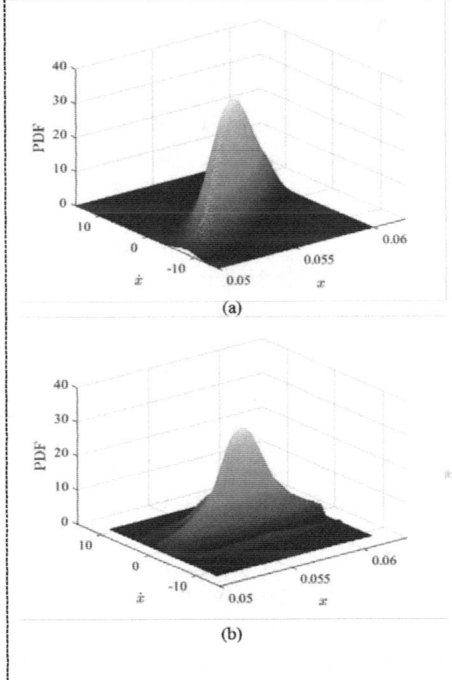

Figure 3. Joint distribution at 0.76s (a) proposed method (b) MC method.

4.2. Accuracy evaluation

Suppose H is the total computational cost for single simulation under stochastic load, H_1 is the time cost by solving the equation under deterministic load (obtaining μ_1 and μ_2), H_2 is the time cost of obtaining a learning sample by MC method, and N_m denotes the number of samples. Note that H_2 is much smaller than H. Table 1 gives the total computational cost comparison between MC method and the proposed method. The time cost of obtaining analytical solution from [0, 1.6] s by the proposed method and MC method is also given in table 1.

Table 1. The total computation cost comparison.

MC method	$H \times N_m$	127836.4 s
Proposed method	$H_1 + H_2 \times N_m$	22337.1 s

From Table 1, the proposed method saves 83.5% computational cost. In addition, once the modification function is obtained, it can be applied to the longer responses. Therefore, with the increasing of simulation length of gear system, our proposed method could save more time in calculation.

5. Conclusions

This paper proposes an analytical solution to a spur gear dynamic model with stochastic load and stochastic friction. In the proposed method, a tentative analytical solution is derived using PI method and then updated using a supervised learning algorithm. The simulation results show that our proposed method can achieve a similar accuracy results under a much smaller time cost compared with the MC method. This analytical solution can be a useful tool in investigating gear systems' vibration and noise.

Acknowledgments

This research is supported by the National Natural Science Foundation of China (Grant No. 51675523), Natural Sciences and Engineering Research Council of Canada (NSERC), and Future Energy Systems under Canada First Research Excellent Fund (CFREF).

References

[1] Shao Y and Chen Z 2013 Dynamic features of planetary gear set with tooth plastic inclination deformation due to tooth root crack, *Nonlinear dynamics* **74(4)** 1253–1266.
[2] Fang Y, Liang X and Zuo MJ 2018 Effects of friction and stochastic load on transient characteristics of a spur gear pair, *Nonlinear dynamics* **93(2)** 599–609.
[3] Theodossiades S and Natsiavas S 2000 Non-linear dynamics of gear-pair systems with periodic stiffness and backlash, *Journal of Sound and Vibration* **229(2)** 287–310.
[4] Wen Y, Yang J and Wang S 2014 Random dynamics of a nonlinear spur gear pair in probabilistic domain, *Journal of Sound and Vibration* **333(20)** 5030–5041.
[5] Shen Y, Yang S and Liu X 2006 Nonlinear dynamics of a spur gear pair with time-varying stiffness and backlash based on incremental harmonic balance method, *International Journal of Mechanics sciences* **48 (11)** 1256–1263.
[6] Fang Y, Liang X and Zuo MJ 2017 Effect of sliding friction on transient characteristics of a gear transmission under random loading, *Proceedings of 2017 IEEE International Conference on Systems, Man, and Cybernetics (SMC)* 2551–2555.
[7] Wei S, Han Q and Chu F 2016 Review on research of uncertain dynamics for gear system, *Journal of Mechanical Engineering* **52(1)** 1–12.
[8] Liang X, Zuo MJ and Feng Z 2018 Dynamic modeling of gearbox faults: A review, *Mechanical Systems and Signal Processing* **98** 852–867.
[9] Naess A, Kolnes F and Mo E 2007 Stochastic spur gear dynamics by numerical path integration, *Journal of sound and vibration* **302(4-5)** 936–950.
[10] Cord M and Cunningham 2008 Machine learning techniques for multimedia: case studies on organization and retrieval, *Springer Science & Business Media*.
[11] Alpaydin E 2014 Introduction to machine learning, *MIT press*.
[12] Yang D and Sun Z 1985 A rotary model for spur gear dynamics, *Journal of mechanisms, transmissions, and automation in design* **107(4)** 529–535.

IOP Conf. Series: Materials Science and Engineering **542** (2019) 012064 doi:10.1088/1757-899X/542/1/012064

Study on Base Resin of Fire-proof and Anticorrosive Steel Structure Coatings

Jun Shi [1] and Yuan quan Yang[2]

1 Postgraduate, Shenyang Jian Zhu University, Shenyang, China
2 Doctoral student, Dalian University of Technology, Dalian, China
E-mail: a 1571802091@qq.com, baquarius0109@163.com

Abstract. Study on the effect of waterborne epoxy resin on coating properties. The experimental results show that the basic and mechanical properties of the paint film will be enhanced with the appropriate increase of curing agent. When the amount of waterborne epoxy resin: curing agent = 1:1.25, the film performance is the best. When the curing agent content continues to increase, it will affect the performance of the coating; The amount of distilled water in waterborne epoxy resin also affects the performance of the paint film. When the amount of waterborne epoxy resin: distilled water consumption = 1:1.25, the film performance is the best, When the amount of distilled water continues to increase, the viscosity, impact strength and hardness of the paint film begin to decrease.

1.Introduction

As the main film-forming material, Base resin is the basis of fire-proof and anticorrosive coatings, and plays a decisive role in the performance of coatings. Therefore, the selection of Base resin is particularly important for the preparation of coatings.[1] Commonly used steel structure coating resins include the following: waterborne alkyd resin, acrylate emulsion, chlorinated emulsion and waterborne epoxy resin. Waterborne alkyd resin is one of the important development directions of alkyd resin.[2] However, it has the disadvantages of easy hydrolysis of lipid bonds in the main chain, poor storage stability and poor water resistance. [3]Acrylate emulsion has been widely used in steel structure fireproof coatings, with excellent decorative properties and film forming properties, but has the disadvantages of hot sticky, cold and brittle, and solvent resistant. Compared with solvent acrylic resin, the performance of acrylate emulsion in various aspects is reduced.[4]

Among various waterborne resins, waterborne epoxy resins have many unique advantages: First of all, the waterborne epoxy resin has strong adaptability and high adhesion to many substrates, and its coating film has small shrinkage, high hardness, good wear resistance and excellent electrical, [5]waterborne epoxy resin has good environmental protection performance, does not contain organic solvents or low volatile organic compounds, and does not pollute the air. It is very suitable for the current requirements of environmental protection in China. [6] moreover , it is water-based, low-cost,

IOP Conf. Series: Materials Science and Engineering **542** (2019) 012064 doi:10.1088/1757-899X/542/1/012064

tasteless, non-flammable, storage, transportation and use of the process of safety greatly improved. At the same time, it has strong operability and good construction performance. It can be washed directly with water and solidified at room temperature and humidity. It has a reasonable curing time while ensuring high crosslink density.[7] Finally, it has good thermal stability, convenient source, low price and wide application, which is incomparable with conventional waterborne acrylic and waterborne polyurethane coatings.

2.Test materials and methods

2.1 Test materials

The base resin of the test is primarily composed of a water-dispersed epoxy resin R O1 , a water-soluble epoxy curing agent ZO1 and distilled water at a room temperature in a certain proportion by a curing agent emulsification method. wherein R01 is a kind of pure epoxy Water-dispersible epoxy resin with resin properties. The product is stable at -5-40 ° C and is a new generation of high performance waterborne epoxy resin. Free of free surfactant, good stability, good compatibility with epoxy resin, excellent curing performance. The main performance standards for raw material water-dispersed epoxy resin RO1 and water-soluble epoxy curing agent ZO1 are shown in Tables1 and.2.

Table1. Water-dispersed epoxy resin R O1 main performance standard

project	Test Methods	index
Appearance	Visual inspection	Milky white uniform fluid
Solid content%	GB/1725-1989	50±2
Epoxy equivalent (solids)	Acetone hydrochloride method	220-280
Viscosity mpa•s/25°C	NDJ-79 Rotary Viscometer	100-450
pH value	Acidity meter	2-7
proportion	GB1756-79	1.01-1.08

Table 2. Main Performance Standards of Water-Soluble Epoxy Curing Agent ZO1

project	Test Methods	index
Appearance	Visual inspection	Light yellow uniform fluid
Solid content%	GB/1725-1989	40±2
Epoxy equivalent (solids)	Alcohol ether method	210±60
Viscosity mpa•s/25°C	NDJ-79 Rotary Viscometer	Less than 10000
pH value	Acidity meter	7-10
proportion	GB1756-79	1.00-1.09

2.2 Test Methods

The aqueous epoxy resin system usually uses a curing agent emulsification method to prepare an aqueous epoxy resin emulsion. The epoxy resins in these systems generally do not require pre-emulsification, and the aqueous epoxy curing agent is mixed and emulsified prior to use, so the curing agent must be a crosslinking agent and an emulsifier. Waterborne epoxy curing agent is based on polyamine, polyamine curing agent addition, grafting, chain extension and end-capping. Surface active non-ionic surfactant chain is introduced into its molecule, which is suitable for the good emulsification of low molecular weight liquid epoxy resin. The advantage of waterborne epoxy resin system prepared by curing agent emulsification is that the epoxy resin is directly emulsified by curing agent before use. Moreover, the storage stability and freeze-thaw stability of epoxy resin emulsion need not be considered. The disadvantage is that the formulated emulsion has a short pot life, so it should be used now.

3 .Results and discussion

3.1 Effect of the ratio of epoxy resin to curing agent on the properties of paint film

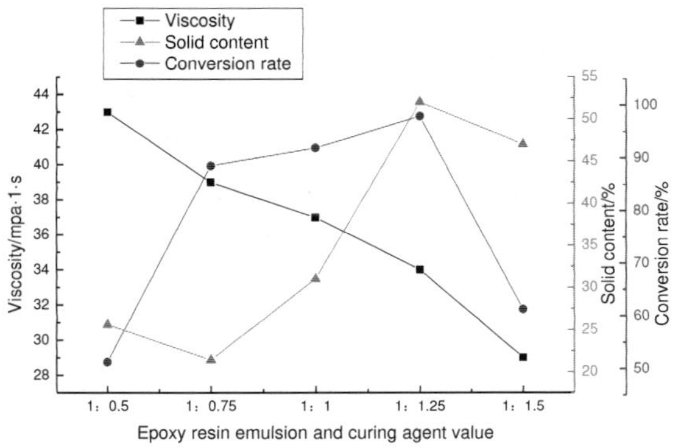

(a) Effect of the ratio of epoxy resin emulsion and curing agent on the basic properties of paint film

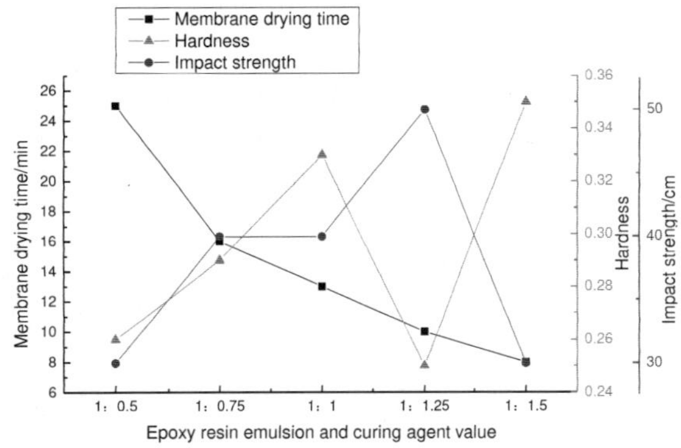

(b) Effect of the ratio of epoxy resin emulsion and curing agent on the mechanical properties of paint film

Figure1. Effect of the Ratio of Epoxy Resin to Curing Agent on the Properties of Paint Film

Figure1 reflect the effect of the epoxy film and curing agent on the properties of the film. As can be seen from Figure 2.1, shows that the viscosity of paint film first increases and then decreases with the decrease of the ratio of epoxy resin to curing agent. When the ratio of the epoxy resin emulsion to the curing agent is 1:1.25, the solid content of the curing agent is as high as 52%, and the amount of the curing agent is continuously increased to find that the solid content of the emulsion starts to drop sharply. Studies have shown that increasing the amount of curing agent can increase the solid content of the coating. When the curing agent is excessive, it will have the opposite effect. Studies have shown that increasing the amount of curing agent can increase the solid content of the coating. When the ratio of epoxy emulsion to curing agent was 1:1.25, the conversion rate was 98%, and then began to decrease. The results show that the conversion of coatings can be improved by increasing the amount of curing agent properly, and the conversion will begin to decrease when the curing agent is used excessively. The mechanical properties of the paint film, as the ratio of the epoxy resin emulsion to the curing agent decreases, the film drying time is gradually reduced, indicating that the increase of the curing agent can reduce the film drying time; The hardness of the coatings increased first and then decreased. When the ratio of epoxy emulsion to curing agent was 1:1.25, the hardness of the coatings was 0.36. The results show that the hardness of the base resin can be enhanced by increasing the amount of curing agent properly. When the curing agent is overused, the hardness will begin to decrease. The impact strength of the coating will increase first and then increase slowly, and then increase. When the ratio of epoxy emulsion to curing agent is 1:1.25, the impact strength of the coating will be 50cm. It is indicated that an appropriate increase in the amount of the curing agent can enhance the impact strength of the base resin, and when the curing agent is used in an excessive amount, the impact strength thereof starts to decrease The test results show that the best combination of basic performance and mechanical properties of the epoxy paint film is the ratio of the epoxy resin emulsion to the curing agent is 1:1.25.

3.2 Effect of the ratio of epoxy resin to distilled water on the properties of paint film

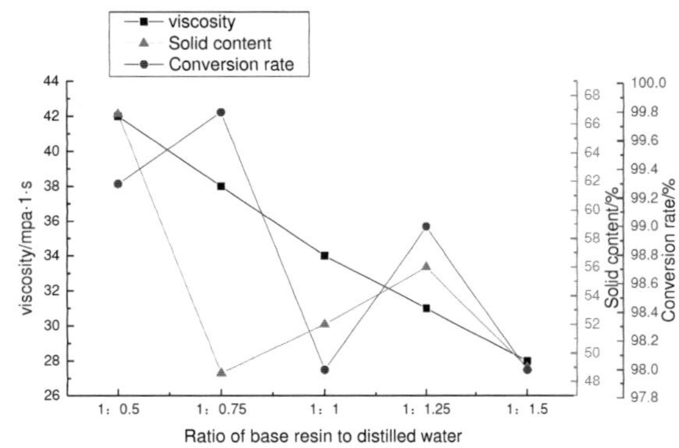

(a) Effect of the ratio of binder resin to distilled water on the basic properties of paint film

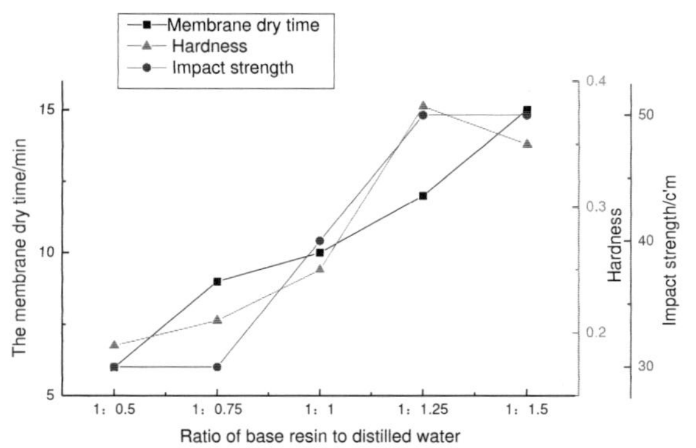

(b) Effect of dosage ratio of binder resin to distilled water on mechanical properties of paint film.

Figure2. Effect of the Ratio of Epoxy Resin to Distilled Water on the Properties of Paint Film

Figure 3.1 reflect the effect of the epoxy film and water on the properties of the film When the ratio of base resin to distilled water is 1:0.5, its viscosity is the best, When the amount of distilled water continues to increase, the viscosity of the paint film begins to decrease, the solid content shows a sharp decrease and then increases. When the ratio of base resin to distilled water is 1:0.5, the solid content is up to 66.7. When the amount of distilled water is increased to 1.25 times the base resin, the solid content reaches 56%, and then begins to decline.. It indicates that the appropriate increase of the amount of distilled water can increase the solid content of the coating, but the excess of distilled water will reduce the solid content; When the ratio of resin to distilled water was 1:1.25, the conversion rate reached 99%, and then began to Decrease. It indicates that the appropriate increase of the amount of distilled water can increase the conversion rate of the coating, but when the distilled water is used in excess, the conversion rate is lowered. With the decrease of the ratio of base resin to distilled water, the dry time of the film increases gradually. It indicates that the increase of the amount of distilled water will prolong the film drying time of the coating; the hardness will increase first and then decrease. When the ratio of base resin to distilled water is 1:1.25, the hardness is up to 0.38. The results show that increasing the amount of distilled water properly can improve the hardness of coatings, but excessive use of distilled water will reduce the solid content of coatings hardness. The impact strength first increases slowly, then increases substantially, and finally stabilizes gradually. It indicates that the appropriate increase of the amount of distilled water can improve the impact strength of the coating, but the excessive use of distilled water will reduce the impact strength of the coating. The test results show that the basic performance and mechanical properties of the paint film are the ratio of the base resin to the distilled water of 1:1.25.

4. Conclusions

(1) The results show that when the ratio of binder resin to curing agent is 1:1.25, the film has the highest solid content and conversion resistance, and has better viscosity, hardness and lower film drying time.

(2) The results show that when the ratio of base resin to distilled water is 1:1.25, the solid content and conversion rate of the paint film are the highest, and the viscosity and hardness are good, but the film drying time increases with the increase of distilled water.

Acknowledgement

The financial support from the University Innovation Team Project of Liaoning Province (No. LT2015024) and the Key Technology Research and Development Project of Shenyang (No.17-48-2-00) are also gratefully acknowledged.

References

[1] SINGH A P,GUNASEKARAN G, SURYANARAYANAC, et al. Fatty acid based waterborne air drying epoxy ester resin for coating applications [J] .Progress in Organic Coatings,2015,87:95-105.

[2] RAHMAN O U, KASHIF M, AHMADS Nanoferrite dispersed waterborne epoxy acrylate; Anticorrosive nanocomposite coatings [J].Progress in Organic Coatings,2015,80:77-86 .

[3] Zou Ming. Preparation and properties of ultra-thin expanded steel structure fireproof and anticorrosive coatings [D]. [Master's thesis] Beijing: Beijing University of Chemical Technology, 2011.

[4] Liu Wanpeng, preparation of epoxy modified waterborne acrylic resin ultra-thin fire retardant coating for steel structure [D]. Master's degree thesis Shenyang: Shenyang University of Technology, 2010.

[5] LI Y, YANG Z, QIU H, et al. Self-aligned graphene as anticorrosive barrier in waterborne polyurethane composite coatings [J] .Journal of Materials Chemistry A, 2014, 2(34):14139 – 14145 .

[6] Guo W, Qiao X, Huang Y, et al. Study on energy saving effect of heat of heat-reflective insulation coating on envelopes in the hot summer and cold winter zone [J].Energy Build,2012.
[7] Ji Wang,Zhong-Jun Shu,Zhen Chen. The protective effect of a fire-retardant coating on the insulation failure of PVC cable [J]. Engineering Failure Analysis,2013,34.

IOP Conf. Series: Materials Science and Engineering **542** (2019) 012065 doi:10.1088/1757-899X/542/1/012065

Analysis of the Correlation Degree between Safety Condition and Influencing Factors of Heavy Machinery

Hongqi Luo[1] and Xia Liu[2*]

1 School of Material and Mechanical Engineering, Beijing Technology and Business University, Beijing, China
2 Quality Management Branch of China National Institute of standardization, Beijing, China
E-mail: lhqfrq@126.com
*Corresponding author
Abstract. It is important to research the relationship of safety condition and its influencing factors in heavy machinery industry. The influence level of factors on the death toll in ten thousand special equipment is analysed, which includes the main business income of heavy machinery, standard quantity, operators of special equipment, number of safety supervision organization and accident frequency. The correlation degree between the accident frequency and the death toll in ten thousand special equipment is 0.699, while the correlation degree for standard number, special equipment operators and safety supervision organization is 0.540, 0.514 and 0.512, respectively. The accident frequency has the most significant impact on the death toll in ten thousand special equipment, followed by the standard number, special equipment operators and safety supervision organization, while the main business income has the least impact. It shows that promoting local legislation and standards are conducive to improving the safety condition of heavy machinery in China.

1. Introduction
Heavy machinery industry is a combination of metallurgical machinery, heavy forging machinery, mining machinery, material handling machinery manufacturing and large-scale casting and forging manufacturing. According to the standard of GB/T 4754-2011 "Classification of National Economic Industries", the industrial sub-categories of heavy machinery industry have been changed from the original metallurgical equipment, mining equipment, lifting and transportation equipment to metallurgical equipment, mining machinery, light and small lifting equipment, cranes and production vehicles, continuous handling equipment, Elevator escalators and elevators and other material handling equipment. The death toll in ten thousand special equipment is used to represent the index of safety condition, while the influencing factors mainly include the main business income, the accident frequency, the number of safety supervision institutions and the number of relevant standards.

2. Statistics Selection and Data Collection
Choosing different statistical indicators for the safety condition of heavy machinery and its influencing factors will have some influence on the calculation results, but there is generally the same proportion relationship between different calibre data[1].

The main business income is used to represent the main economic indicators of heavy machinery industry, which refers to the business income of enterprises engaged in the production and operation activities of the industry. The relevant data of special equipment was selected to represent the number of safety accidents and casualty data indicators through referring the "China Heavy Machinery Industry

IOP Conf. Series: Materials Science and Engineering **542** (2019) 012065 doi:10.1088/1757-899X/542/1/012065

Yearbook" from 2009 to 2017[2]. Special equipment refers to boilers, pressure vessels (containing gas cylinders, the same below), pressure pipes, elevators, cranes, passenger ropeways, large recreational facilities and special motor vehicles in the field (factory). Boilers, pressure vessels (containing gas cylinders) and pressure pipes are special equipment of pressure bearing type. Elevators, lifts, passenger ropeways and large-scale recreational facilities are special mechanical and electrical equipment. The number of special equipment operators, the number of ten thousand of equipment deaths, the accident frequency and the number of special equipment safety supervision organization were selected as the analysis elements. The work of standardization is to unify repetitive things and concepts in social practice such as economy, technology, science and management by formulating, promulgating and implementing standards so as to achieve the best order and social benefits. In other words, standards are the summary and refinement of scientific, technological and empirical results, rules and regulations that need to be followed within the corresponding scope are determined after consultation and discussion among experts from all concerned. Heavy machinery industry standards include the standards of heavy machinery (metallurgical machinery), mining machinery, and material handling machinery (crane transport machinery) and large casting and forging.

By referring "China Heavy Machinery Industry Yearbook" and "China Price Statistics Yearbook", it can collect the statistical data of the safety condition of heavy machinery in each year and the affecting factors, and the chain index of price. The collected price index are deduced from each other and the fixed base price index for each year are obtained. Because each value in the statistical data is calculated according to the current price, the inflation situation in recent years is different and the present price and moisture of each year are also different. The period of higher inflation is much higher than that of lower inflation. In order to accurately reflect the growth rate of the main business income of heavy machinery, price index should be used to refer to each other [3]. Since the collected price index is chain index of price, it is the ratio of price level based on the previous year. It is expressed by formula as $\frac{a_1}{a_0}, \frac{a_2}{a_1}, \frac{a_3}{a_2}, \ldots \frac{a_n}{a_{n-1}}$. In order to make the data being compared, the chain index of price cannot meet the requirements. The base price index is the ratio of the level of the reporting period to that of a fixed base period, reflecting the development of prices over a longer period of time. It can be seen from the formula that the fixed-base price index is equal to the continuous product of the corresponding chain index of price, which is expressed by the formula (1) as follows[4-5].

$$\frac{a_n}{a_0} = \frac{a_1}{a_0} \times \frac{a_2}{a_1} \times \frac{a_3}{a_2} \times \cdots \times \frac{a_{n-1}}{a_{n-2}} \times \frac{a_n}{a_{n-1}} \tag{1}$$

According to the above conversion relations, the collected price index can be calculated mutually, and the fixed base price index for each year can be obtained. By dividing the current price index by the fixed base, the comparable price of main business income can be obtained. According to the national chain index of price index from 2009 to 2017, a fixed base price index of 100% in 2008 can be calculated. The data after processing are shown in Table 1.

Table 1. Data Processing Results of Safety Conditions and Indicators of Influencing Factors

Year	Death toll in ten thousand special equipment	Main business income (Billion Yuan)	Operators of special equipment (ten thousand people)	Standard quantity	Safety supervision organization	Accident frequency
2016	0.33	1324.42	1099.52	1175	4132	233
2015	0.36	1296.66	1047.64	1111	3799	257
2014	0.39	1290.81	963.79	1047	3573	283
2013	0.46	1172.12	829.79	1035	3236	227
2012	0.517	1037.62	706.09	955	3189	228
2011	0.595	889.91	657.6	951	3144	275
2010	0.67	710.38	562.23	935	3092	296
2009	0.76	577.78	520.02	768	3049	380
2008	0.82	493.51	380.92	719	3011	307

IOP Conf. Series: Materials Science and Engineering **542** (2019) 012065 doi:10.1088/1757-899X/542/1/012065

3. Analysis of the Correlation Degree

3.1. Dimensionless Processing of Raw Data

Dimensionless data processing is a unified transformation of the content of each index, so that the order of magnitude of each index is basically the same, in order to eliminate the impact of excessive disparity between different index dimensions and numerical standards, and avoid unequal "weight" situation [5-8]. In this paper, interval relativity is used to deal with the problem which is shown in formula (2).

$$X_i(k) = \frac{X_i(k) - \min_k X(k)}{\max_k X(k) - \min_k X(k)} \tag{2}$$

Among them, k=1, 2... N; I = 0, 1, 2 ... M. It should satisfy $0 \le |X_i(k)| \le 1$.

The results of dimensionless processing are shown in Table 2. X_0 denotes the death toll in ten thousand special equipment, X_1 denotes the main business income, X_2 denotes the operators of special equipment, X_3 denotes the standard quantity, X_4 denotes the number of safety supervision organization, and X_5 denotes the accident frequency.

Table 2. Dimensionless data of safety conditions and influencing factors

X_0	X_1	X_2	X_3	X_4	X_5
0	1	1	1	1	0.039216
0.061224	0.966589	0.9278	0.859649	0.702944	0.196078
0.122449	0.959553	0.81112	0.719298	0.501338	0.366013
0.265306	0.816712	0.62465	0.692982	0.200714	0
0.381633	0.654833	0.4525	0.517544	0.158787	0.006536
0.540816	0.47707	0.38503	0.508772	0.118644	0.313725
0.693878	0.261007	0.25231	0.473684	0.072257	0.45098
0.877551	0.101423	0.19357	0.107456	0.033898	1
1	0	0	0	0	0.522876

3.2. Calculation of correlation degree

According to the grey system theory, the safety condition of heavy machinery is regarded as a grey system, and each influencing factor is an element in the system. The death toll in ten thousand special equipment was selected as the reference series, which is $X_0 = \{X_0(k) | k = 1,2,3,..., n\}$, and the number of influencing factors was listed as the comparative series, which is $X_i = \{X_i(k) | k = 1,2,3,..., n\} (i = 1,2,3,...m)$. Among them, n is the influencing factor. The correlation coefficient can be calculate between the influencing factors and the parent series according to formula (3) [9-11].

$$\zeta_i(k) = \frac{\min_i \min_k |X_0(k) - X_i(k)| + \sigma \max_i \max_k |X_0(k) - X_i(k)|}{|X_0(k) - X_i(k)| + \sigma \max_i \max_k |X_0(k) - X_i(k)|} \tag{3}$$

The formula $\min_i \min_k |X_0(k) - X_i(k)|$ is the minimum difference of two levels, $\max_i \max_k |X_0(k) - X_i(k)|$ is the maximum difference of two levels. The symbol of σ is the resolution coefficient, and its range of values is between 0 and 1, usually being 0.5. The calculation results are shown in Table 3.

Equal weight correlation degree is obtained according to formula (4):

$$r_i = \frac{1}{n} \sum_{k=1}^{n} \zeta_i(k) \tag{4}$$

The calculation results are as follows: r_1=0.481, r_2=0.514, r_3=0.540, r_4=0.512, r_5=0.699. The order of sub-series is main business income, operators of special equipment, standard quantity, number of safety supervision organization and accident frequency. The correlation order of x_0 is $r_5 > r_3 > r_2 > r_4 > r_1$, which shows that the sub-sequence $\{x_5\}$ has a better correlation with the parent sequence, $\{x_3\}$, $\{x_2\}$, $\{x_4\}$ takes the second place, and $\{x_1\}$ is the worst. That is to say, the accident frequency has the most significant impact on the death toll in ten thousand special equipment, followed by the standard number, special equipment operators and safety supervision agencies, and the main business income is the minimum.

Table 3. Relevant coefficient of various factors and safety conditions of heavy machinery

$\zeta_1(k)$	$\zeta_2(k)$	$\zeta_3(k)$	$\zeta_4(k)$	$\zeta_5(k)$
0.333	0.333	0.333	0.333	0.927
0.356	0.366	0.385	0.438	0.788
0.374	0.421	0.456	0.569	0.672
0.476	0.582	0.539	0.886	0.653
0.647	0.876	0.786	0.692	0.571
0.887	0.762	0.940	0.542	0.688
0.536	0.531	0.694	0.446	0.673
0.392	0.422	0.394	0.372	0.803
0.333	0.333	0.333	0.333	0.512

It should further straighten out the relationship between safety technical specifications and regulations and standards, and establish a benign interaction mechanism between them. It should continue to promote the integration and optimization of safety technical specifications and focus on solving the problems of excessive, detailed and complex safety technical specifications. Efforts should be made to promote the formulation of local legislation and standards, and to enhance the pertinence and operability of laws and regulations.

4. Conclusion

Through analysing the correlation degree, it is concluded that besides the factors of accident number, the number of relevant standards and the increase of the number of safety supervision institutions play an important positive role in the death toll in heavy machinery accidents. It should improve the dynamic supervision system, further clarify the responsibilities of safety supervision agencies at all levels, and should study and define the relationship between safety supervision and comprehensive administrative law enforcement, strengthen the training of safety supervision personnel, which is useful to improve the treatment of grass-roots safety supervision personnel, and reduce the responsibility risk of safety supervision personnel.

Because of the randomness of the collected statistical data and the unsatisfactory treatment of some statistical indicators, there will be some deviations in the calculation results. It is necessary to further improve the construction of information network, realize the interconnection, interoperability and sharing of data, and improve the reliability and accuracy of data.

5. References

[1] Zhang Qinwen, Wang Xuemen. Grey System Analysis of Rural Economy - Model, Method and Application [M]. Beijing: *Academic Journal Press*. 1989

[2] Executive Editorial Committee of China Heavy Machinery Industry Yearbook. China Heavy Machinery Industry Yearbook [J]. Beijing: *China Machine Press*, 2009-2017

[3] Department of Urban Social and Economic Survey, National Bureau of Statistics. China Price Statistics Yearbook [J]. Beijing: *China Statistics Press*, 2018

[4] Ouyang Tao, Xie Xiaoyun. Statistics [M]. Changsha: *National Defense University of Science and Technology Press*. 2002

[5] Zhang Li, Liu Shuqing. Grey correlation analysis in the study of relationship between pig blood biochemical parameters and weight gain [J]. *Journal of Domestic Animal Ecology*, 2000, 21 (2): 10-12

[6] Lin Shaoxia, Lin Changhu, He Tengbing, et al. Application of grey relational degree method in analysis of factors affecting agricultural non-point source contaminated soil [J]. *Bulletin of Soil and Water Conservation*, 2010,30(4):171-174

[7] Li Yuefen, Tang Jie, Li Yanmei. Evaluation of grassland soil quality by principal component analysis and grey correlation analysis [J]. *Global Geology*, 2004, 23 (2): 169-174

[8] Wu Xinmin, Pan Genxing. Relevance analysis of factors affecting heavy metal pollution in urban soils [J]. *Acta Pedologica Sinica*, 2003.40 (6): 921-928

[9] Zhao Qian, Liu Zhaoye, Liu Chunlei. Grey relational analysis of new wheat varieties [J]. *Chinese Agricultural Science Bulletin*, 2007, 23 (9): 259-262

[10] Zhang Aihua, Ren Gongchang, Yan Xiangyang. Fault diagnosis method of CNC machine tool spindle system based on grey relational degree analysis [J]. *Machinery Design & Manufacture*, 2010 (11): 160-161

[11] Kou Zhiqiang, Li Qianqian, Wu Shaohui, et al. Selection of Sesame Varieties by grey correlation method [J]. *Guangdong Agricultural Sciences*, 2008 (7): 23-25

Acknowledgments
Project is supported by Research and Application of Common Technologies of National Quality Foundation (2017YFF0207200).

Development of solar-light-driven photocatalyst for energy and environmental application

Na Liu[1], Qi Zhu[1], Nan Zhang[1], Jiayun Wang[2] and Yingnan Yang[1,*]

1 Department of Bioindustrial Sciences, Graduate School of Life and Environmental Sciences, University of Tsukuba, Tsukuba, Ibaraki 305-8572, Japan
2 Shanghai In-Ou AgroScience Co., Ltd., Shanghai, China
*Corresponding author's e-mail: yo.innan.fu@u.tsukuba.ac.jp

Abstract. Efficient utilization of solar energy could relieve many energy and environmental problems. Overall photocatalytic water-splitting using solar energy has received enormous attention for production of renewable hydrogen from water. Recently, a novel Ag_3PO_4 photocatalyst with band gap of 2.43 eV, possessing remarkable photocatalytic ability for organic dye degradation has been reported. In this work a highly efficient photocatalyst $P/Ag/Ag_2O/Ag_3PO_4/TiO_2$ has been synthesized and characterized by SEM, TEM, XRD, EDS, XPS, UV-vis and PL respectively. The results suggest that a highly efficient photocatalyst $P/Ag/Ag_2O/Ag_3PO_4/TiO_2$ has been synthesized successfully and performed strong absorption band in visible-light region, high separation efficiency of photo excited electron-hole pairs and small band gap (approximately 2.2 eV). On the other hand, Rodamine B (RhB) and methanol are representative organic compounds that can be used as sacrificial reagent for increasing water-splitting efficiency. For the purpose of identifying the possibility of generating oxygen and hydrogen by the as-prepared $P/Ag/Ag_2O/Ag_3PO_4/TiO_2$ under simulated solar irradiation, the accumulated amount of oxygen and hydrogen was evaluated. The results showed that the optimal conditions of water-splitting process by using the mixture of RhB and methanol under solar light is 20% of methanol, 1.0 ppm of RhB and 0.100 g/L of catalyst dosage. In summary, as-prepared $P/Ag/Ag_2O/Ag_3PO_4/TiO_2$ composite has the potential for environmental and energy application under solar light.

1. Introduction

Photocatalytic water splitting has received enormous attention for production of renewable hydrogen from water since the discovery of Honda-Fujishima effect in 1970s [1]. In the past decades, one of the most prominent photocatalysts that have been used is TiO_2. Although TiO_2 has excellent characteristics as a photocatalyst, the wide band gap of pure TiO_2 (3.2 eV) limits its utilization in the visible light region [2]. As such, regarding the development of photocatalysts that work not only under UV light but also under visible-light illumination to more efficiently utilize solar energy, a highly efficient photocatalyst $P/Ag/Ag_2O/Ag_3PO_4/TiO_2$ has been synthesized successfully in our previous reports and it performed strong absorption band in visible-light region, high separation efficiency of photo excited electron-hole pairs and small band gap (approximately 2.2 eV) [3]. It has been demonstrated with remarkable photocatalytic efficiency in decomposing the organic matters and sterilization of bacteria [4]. However, regarding its photocatalytic water-splitting ability, there is no research being conducted. It's predicted that $P/Ag/Ag_2O/Ag_3PO_4/TiO_2$ should also have high ability for photocatalytic water-splitting process under solar light irradiation.

Usually, the direct splitting of water into H_2 and O_2 has a very low efficiency due to its uphill reaction along with rapidly recombination of photo excited electrons and holes as well as rapid reverse reaction [5]. To provoke water-splitting process for hydrogen production by modified TiO_2 photocatalyst from distilled water, adding sacrificial reagent (electron donors), such as dyes or alcohols, to react with the photo-excited holes in the valence band can inhibit the photo-excited electron-hole pair recombination. Rodamine B (RhB), as a representative dye that can be an indicator of organic pollution, has been reported efficient for increasing gas generation as sacrificial agent in photocatalytic water-splitting process [6]. Considering the solar energy utilization and environment decontamination, production of hydrogen by photocatalytic water splitting from dye wastewater appears to be an exciting idea. On the other hand, using methanol as a sacrificial agent was reported efficient in provoking overall water-splitting process [7]. In addition, it is reported that initial concentration of sacrificial agent, mass of photocatalyst and pH of the reaction solution can significantly affect the efficiency of photocatalytic water-splitting process [2].

Hence, the main purpose of this study is to identify the possibility of generating gas by $P/Ag/Ag_2O/Ag_3PO_4/TiO_2$ under simulated solar light irradiation within the sacrificial reagent existed, such as RhB and Methanol, and then optimize the H_2 and O_2 generation by identifying the optimal reaction conditions.

2. Materials and methods

2.1. Preparation of $P/Ag/Ag_2O/Ag_3PO_4/TiO_2$ photocatalyst

At first, the transparent $P/Ag/Ag_2O/Ag_3PO_4/TiO_2$ sol was prepared according to our previous study [3]. Then, the solution was transferred into a Teflon-lined autoclave and heated in an oven at a certain temperature. After heat treatment, the resulting precipitates were separated from the rest of solution by 3 times centrifuge (9800 rpm) and dried at 60°C for 12 h. Finally, the $P/Ag/Ag_2O/Ag_3PO_4/TiO_2$ photocatalyst powder was successfully synthesized and stored in desiccator. All the agents used during this study were purchased from FUJIFILM Wako Pure Chemical Corporation, Japan.

2.2. Preparation of photocatalytic water-splitting reactant

Firstly, a certain amount of $P/Ag/Ag_2O/Ag_3PO_4/TiO_2$ powder and sacrificial reagent (RhB (40 mg/L) or methanol) were added into the water-splitting reactor and mixed with distilled water to achieve 150 mL reaction solution. After that, the reactor containing 150 mL reactant was agitated with a magnetic stirrer in dark for 30 min and then purged with N_2 gas for 3 times. After the purge step was completed and the photo reaction will be started under a simulated solar lamp (XC-100, SERIC., Ltd.) irradiation (550 W/m^2) for 3.0 h. During the RhB-H_2O reaction system, the effects of initial RhB concentration (0.5, 1.0 and 2.0 ppm), catalyst dosage (0.050, 0.100, 0.200 g/L) and pH (3.0, 7.0 and 11.0) on the photocatalytic activity were systematically investigated. As for methanol-H_2O reaction system, the effects of methanol percentage (10%, 20% and 30%) and catalyst dosage (0.025, 0.050 and 0.100 g/L) were evaluated.

2.3. Water-splitting activity investigation and evaluation

During the water-splitting test, gas and liquid samples were taken out every 30 min. The gas sample was directly examined by Gas Chromatography (GC-8A, SHIMAZU, Japan) to measure the amount of accumulated oxygen and hydrogen, while, for the liquid sample, firstly the photocatalyst powder was separated from the mixed reactant by centrifugation (8000 rpm), and then the optical absorption of the clarified solution was analyzed to calculate the pseudo-first-order kinetic apparent rate constant of RhB degradation by using a spectrophotometer (UV-1600, Shimadzu) at a wavelength of 554 nm which corresponds to the maximum absorption wavelength of RhB molecules.

3. Results and Discussion

3.1. Optimization of O_2 generation in RhB solution

IOP Conf. Series: Materials Science and Engineering **542** (2019) 012066 doi:10.1088/1757-899X/542/1/012066

In order to identify the possibility of generating O_2 by $P/Ag/Ag_2O/Ag_3PO_4/TiO_2$ photocatalyst, water splitting experiment was conducted under simulated solar light irradiation. As results shown in Figure 1 (a), O_2 evolution could be detected both in distilled water and RhB aqueous solution. And when adding sacrificial reagent RhB into distilled water, the ability of generating O_2 had been significantly enhanced which implied that RhB is an efficient sacrificial reagent.

To investigate the optimal experiment conditions for photocatalytic O_2 generation by $P/Ag/Ag_2O/Ag_3PO_4/TiO_2$ in RhB solution, the influence of initial RhB concentration, catalyst dosage and reaction pH were taken into consideration (Figure 1 (b-d)). From Figure 1 (b) we can conclude the initial RhB concentration had obviously effects on generating O_2 and the highest O_2 evolution was obtained when initial RhB concentration was 1.0 ppm. Ultimately, the accumulated O_2 achieved 164.00 μmol after 3.0 h reaction which was higher than that of 0.5 ppm and 2.0 ppm RhB concentration. Figure 1 (c) shows the dependence of photocatalytic O_2 generation on the dosage of photocatalyst. It can be found that at first with increasing the catalyst dosage, the amount of evolved O_2 also increased. Then the accumulated O_2 reached maximum at 0.100 g/L of dosage, while it largely decreased with further increasing the dosage to 0.200 g/L, indicating 0.100 g/L was the optimal dosage for the RhB-H_2O reaction system. Besides, the photocatalytic O_2 production was also affected by reactant pH (Figure 1 (d)). When the experiment was conducted at pH 7.0, the highest amount of accumulated O_2 (259.66 μmol) was achieved in comparison with pH at 3.0 or 11.0, suggesting the generation of O_2 by the $P/Ag/Ag_2O/Ag_3PO_4/TiO_2$ photocatalyst was more favourable under natural condition.

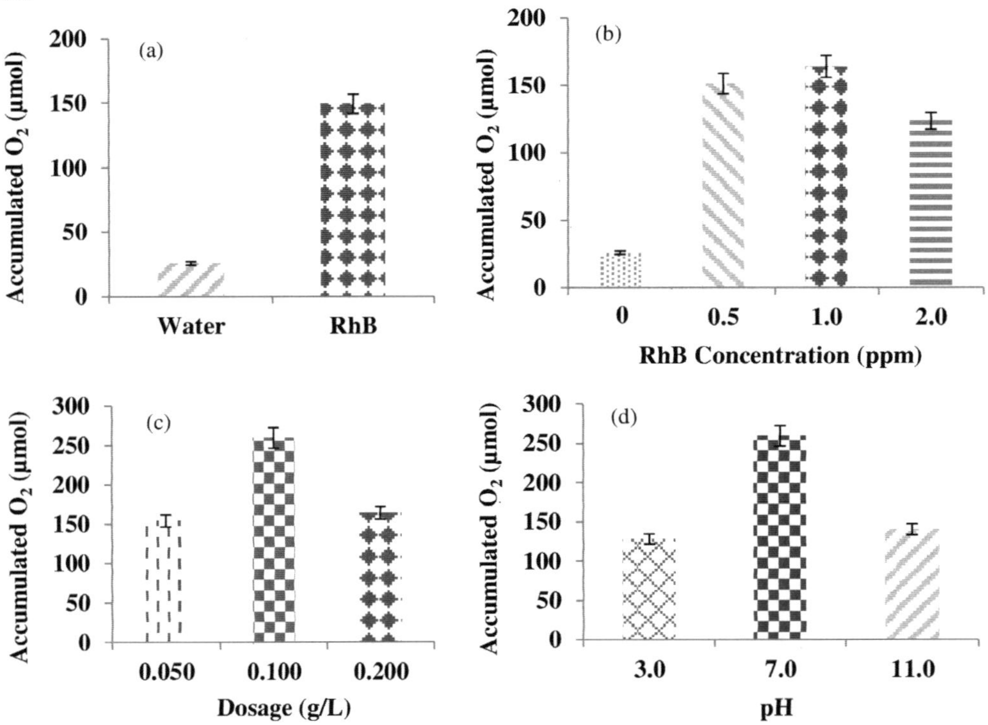

Figure 1. Accumulated O_2 by $P/Ag/Ag_2O/Ag_3PO_4/TiO_2$ (a) in pure water and Rh B solution under different (b) Rh B concentration, (c) catalyst dosage and (d) pH after 3.0 h solar light irradiation.

In addition, through all the experiments, the RhB degradation was also achieved along with the water-splitting and the pseudo-first-order kinetic parameters was calculated according to the RhB degradation efficiency, which is summarized in Table 1. The variation of RhB degradation rate under different conditions showed a similar tendency with O_2 evolution and the faster decomposition of RhB is, the more O_2 produced. The *k* value reached maximum (0.1410) under the same optimal conditions with the accumulated O_2 results. Thus, the optimal conditions for both O_2 production and RhB degradation by $P/Ag/Ag_2O/Ag_3PO_4/TiO_2$ photocatalyst during RhB-H_2O reaction system was the initial RhB concentration at 1.0 ppm, catalyst dosage at 0.100 g/L and pH at 7.0.

Table 1. First order kinetic results of RhB degradation under different experiment conditions.

Factors	Values	Other conditions	k (min^{-1})
Initial RhB concentration (ppm)	0.5	Catalyst dosage: 0.200 g/L	0.0788
	1.0	Initial pH: 7.0	0.0808
	2.0		0.0424
Catalyst dosage (g/L)	0.050	Initial RhB concentration: 1.0 ppm	0.0839
	0.100	Initial pH: 7.0	0.1410
	0.200		0.0808
Initial pH	3.0	Initial RhB concentration: 1.0 ppm	0.0504
	7.0	Catalyst dosage: 0.100 g/L	0.1410
	11.0		0.0739

3.2. Optimization of O_2 and H_2 generation in methanol solution

In previous study, the optimal conditions of generation O_2 has been confirmed. However, the overall water-splitting process was not achieved. For the purpose of identifying the possibility of overall water-splitting process and increasing the production of O_2 and H_2 generation by $P/Ag/Ag_2O/Ag_3PO_4/TiO_2$ under simulated solar irradiation, the accumulated amount of O_2 and H_2 evolution was investigated in methanol solution. From Figure 2, it was confirmed that the overall water-splitting was realized in a methanol-H_2O system by using the as-prepared composite. Both O_2 and H_2 were successfully detected and the amount of accumulated O_2 and H_2 gradually increased with time profile under solar light irradiation.

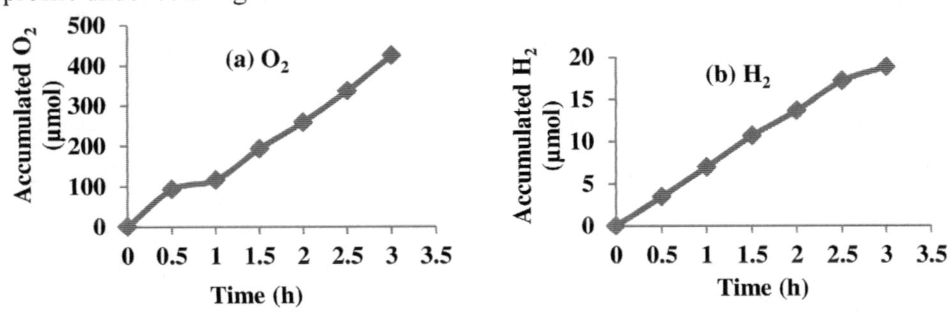

Figure 2. O_2 (a) and H_2 (b) evolution by $P/Ag/Ag_2O/Ag_3PO_4/TiO_2$ under solar light irradiation in methanol solution (Dosage: 0.050 g/L, pH=7.0, 20% Methanol).

To further enhance the overall water-splitting efficiency, the percentage of methanol in reaction solution was optimized (Figure 3 (a)). Compared to 10% and 30% of methanol, using methanol percentage of 20%, ultimately the highest amount of accumulated O_2 and H_2 achieved after 3.0 h reaction which were 424.55 μmol and 18.77 μmol, respectively. This result was also better than the optimal O_2 generation using RhB as sacrificial reagent that achieved 259.66 μmol of O_2. Figure 3 (b) exhibited the influence of $P/Ag/Ag_2O/Ag_3PO_4/TiO_2$ dosage on O_2 and H_2 evolution in methanol solution. It indicated that the dosage of catalyst had obviously effects of generating O_2 and H_2. The best performance occurred when catalyst loading was 0.050 g/L. This result is not same with the optimal production of O_2 generation in RhB-H_2O system which the optimal dosage of catalyst was 0.100 g/L. Thus, both of the optimal dosage of $P/Ag/Ag_2O/Ag_3PO_4/TiO_2$ (0.050 g/L and 0.100 g/L) should be investigated when use the mixture of RhB and methanol as sacrificial reagents. As shown in Figure 3 (c), when the RhB and methanol were involved in the reaction system simultaneously, more O_2 and H_2 was observed at the 0.100 g/L of catalyst dosage. Under this condition, the amount of accumulated O_2 and H_2 reached to 524.99 and 26.49 μmol after 3.0 h, respectively, which achieved the highest production through this study. Therefore, the optimal conditions of overall water-splitting process by using $P/Ag/Ag_2O/Ag_3PO_4/TiO_2$ under solar light irradiation was 0.100 g/L of catalyst dosage with the mixture of 1.0 ppm of RhB and 20% of methanol.

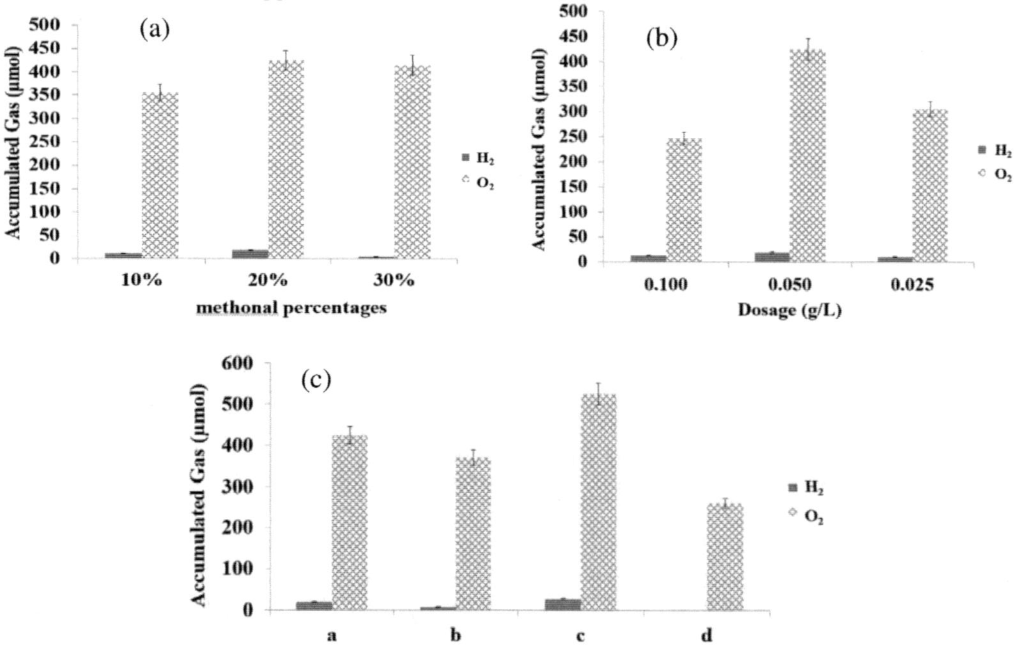

Figure 3. Influence of (a) methanol percentages, (b) catalyst dosage and (c) sacrificial reagent with the best dosage of $P/Ag/Ag_2O/Ag_3PO_4/TiO_2$ on O_2 and H_2 generation after 3.0 h (a: 20% methanol + 0.050 g/L, b: 20% methanol + 1.0 ppm Rh B + 0.050 g/L, c: 20% methanol + 1.0 ppm Rh B + 0.100 g/L and d: 1.0 ppm Rh B + 0.100 g/L)

3.3. Overall water-splitting process in long-time reaction under optimal conditions

The long-time photocatalytic water-splitting experiment was finally carried out to verify the stability and durability of the as-prepared composite under the optimal conditions. As displayed in Figure 4, there is no obvious loss of photocatalytic activity even after 36 h, implying the outstanding stability of the composite in the long-time photocatalytic reaction. Only a slight decrease of generation rate was observed after 12 h which was mainly caused by the consumption up of the RhB sacrificial reagent. To further compare with other photocatalysts as shown in Table 2, the O_2 and H_2 yield rates of the as-prepared photocatalyst was much higher than reported values. Therefore, the above results substantiate that the $P/Ag/Ag_2O/Ag_3PO_4/TiO_2$ composite is an effective photocatalytic water-splitting material with reasonable stability.

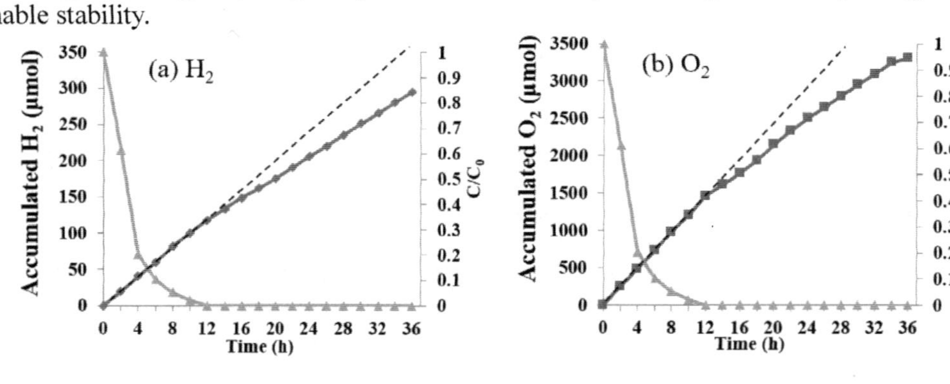

Figure 4. (a) H_2 and (b) O_2 evolution with RhB degradation by $P/Ag/Ag_2O/Ag_3PO_4/TiO_2$ under solar light irradiation in mixed solution after 36 h (Dosage: 0.100 g/L, 20% Methanol and 1.0 ppm RhB).

Table 2. Photocatalysts comparison for H_2 and O_2 evolution from aqueous solutions.

Photocatalyst	Light source	Activity (μmol/h/g)		Reference
		H_2	O_2	
$TiO_2 \cdot NiOx$	Hg-P	6	2	[8]
$CaTiO_3 \cdot NiO_x$	Hg-Q	30	17	[9]
$SrTiO_3 \cdot Rh$	Xe-P	27	14	[10]
$P/Ag/Ag_2O/Ag_3PO_4/TiO_2$	Simulated solar light	589	11667	This research

4. Conclusion

In summary, the $P/Ag/Ag_2O/Ag_3PO_4/TiO_2$ photocatalyst exhibited remarkable photocatalytic water-splitting activity under solar light irradiation. The generation of O_2 from RhB solution showed better performance than that from pure water and further the overall water-splitting process was achieved by using methanol as sacrificial reagent. The optimal conditions of overall water-splitting process by using the mixture of RhB and methanol under solar light irradiation is 20% of methanol, 1.0 ppm of RhB and 0.100 g/L of catalyst dosage. By taking the advantages of dye waste decontamination and H_2 generation, the as-prepared solar-light-driven $P/Ag/Ag_2O/Ag_3PO_4/TiO_2$ composite has great potential for environmental and energy application.

Acknowledgement

This research was funded by Japan Society for the Promotion of Science KAKENHI JP18J20620.

References

[1] Wang S, Yun J H, Luo B, Butburee T, Peerakiatkhajohn P, Thaweesak S, Xiao M and Wang L 2017 *J. Mater. Sci. Technol.* **33** 1-22.
[2] Ahmad H, Kamarudin S K, Minggu L J and Kassim M 2015 *Renew. Sust. Energ. Rev.* **43** 599-610.
[3] Hu X, Zhu Q, Wang X, Kawazoe N and Yang Y 2015 *J. Mater. Chem.* A **3** 17858-65.
[4] Zhu Q, Hu X, Stanislaus M S, Zhang N, Xiao R, Liu N and Yang Y 2017 *Sci Total Environ.* **577** 236-44.
[5] Henderson M A 2011 *Surf Sci Rep.* **66** 185–297.
[6] Zhang L, Wang W, Sun S, Sun Y, Gao E and Xu J 2013 *Appl. Catal. B-Environ.* **132** 315-20.
[7] Choi H J and Kang M 2007 *Int. J. Hydrog. Energy* **32** 3841-48.
[8] Kudo A, Domen K, Maruya K and Onishi T 2011 *Chem. Phys. Lett.* **133** 517.
[9] Mizoguchi H, Ueda K, Orita M, Moon S C, Kajihara K, Hirano M and Hosono H 2009 *Mater. Res. Bull.* **37** 2401.
[10] Domen K, Naito S, Soma S, Onishi M and Tamaru K 2010 *J. Chem. Soc. Chem. Commun* **12** 543-4.

Synthesis and characterization of the microcellular injection molded PA6/flax and the PA6/graphene nanocomposites

Ming-Wei Shao[1], Jung-Jie Huang[1], Yu-Xuan Chen[2] and Shyh-Shin Hwang[2,*]

1 Da-Yeh University, Chang-hwa, Taiwan.

2 Chien-Hsin University of Science and Technology, Jhongli, Taiwan.

E-mail: stanhwang@uch.edu.tw

Abstract. Polymer composites combine two materials together to produce a unique property which is not inherent to the individual materials. The most important reason to use composites is the enhancement of the mechanical properties. However, there are several applications that can be used by composites, such as the heavy metal adsorption of waste water by flax. This study has investigated the effects of flax (1, 3, 5 wt%) and graphene loading (0.5, 1, 3 wt%) on the tensile strength, thermal, and heavy metal adsorption properties of the microcellular injection molded PA6 composites. The fillers used, flax and graphene, were micro and nano materials in size, respectively. The results showed that the dispersion with 0.5 wt% of graphene loading, as seen in the TEM pictures, had the best dispersion for the PA6/graphene nanocomposites. This amount of graphene also had the best tensile strength in three different loadings (0.5, 1.0, 3.0 wt%). Tensile strength was related to the filler dispersion in the matrix. Good dispersion resulted in good tensile strength, and the elongation decreased by increasing the flax loading. The 5 wt% flax loading and 3 wt% graphene loading of the composites showed the largest storage modulus for the PA6composites. The 1 wt% of the flax loading showed the highest degradation temperature for the PA6/flax micro-composites, and the 3 wt% for the PA6/graphene nanocomposites. The cell size decreased and the cell density increased with the addition of graphene to the PA6 composites. Heavy metal of the Cr(IV)adsorption test showed that the PA6/GP nanocomposites had a better adsorption than that of the PA6/flax composites.

1. Introduction

The filler has several functions, e. g., reinforcements [1], flame resistance [2], cost reduction [3], thermal stability [4], and a toughness modifier [5].Since the introduction of polymer which replaced the metal and non-metal in many household applications however, polymer strength is not good enough to sustain the external load as compared to those of the metal and non-metal materials. Hence,fiber reinforced polymers were used to improve the matrix strength. There are several types of reinforced fillers, e.g. glass fiber, carbon fiber[6], and aramids fiber. They have wide applications, especially in the aerospace and automobile industry. However, those fibers are not environmentally-friendly. So the natural fibers as the reinforcement of the polymer are more popular. People had used fibers extracted from animals or plants before the synthetic fiber was discovered in 1885. The most common natural fibers are from cotton, hemp, silk, and fur. Hemp fiber like flax, cannabis, and jute, possess the antibacterial characteristics and are green fibers. Flax fibers are mostly composed of cellulose and hemicellulose. Flax fiber is extracted from the bast beneath the surface of the stem of the flax plant. Flax fiber is soft, lustrous, and not easy to corrode. Zampaloni et al., [7]

used marijuana, flax, sisal, mestha, and coir fiber as the reinforcement for the PP matrix and compared the mechanical property between them. Maleic anhydride (MA) was used as the grafted agent. The results showed that marijuana, flax, and sisal had better properties that those of mestha and coir fiber. A coupling agent of silane [8, 9] was used to modify the flax fiber since it is hydrophilic.
Graphene (GP),two-dimensional, materials have attracted tremendous research interest due to their remarkable physio-chemical properties including electrical, mechanical, and thermal. Thus, a combination of different beneficial properties makes GP an ideal candidate for polymer nanocomposites and nanotechnology.GP dramatically improves the properties of polymer based composites at a very low loading point, and the most fascinating attribute is the very high surface conductivity leading to the formation of numerous electrically conductive polymer composites [10-9].Polymer/GP nanocomposites are formed when the polymer and GP nanofiller (graphene oxides [13], reduced graphene oxides [14], chemically reduced graphene oxide, and thermal reduced graphene oxides etc.) are mixed via various methods, such as in situ, polymerization, solution blending, and melt blending. Among these different processes, most of the polymer/GP nanocomposites were produced by melt blending because it is a commercially feasible, lower cost, environmentally friendly, and straightforward approach that does not involve any solvents or monomers [15]. Melt blending is accepted as being the most researched and developed method within the industry.
Water pollution is one of the most serious problems worldwide. In particular, the toxic metals in drinking water pose a huge challenge to the globalmunicipalities. Heavy metals are toxic to plants, animals, human beings, and aquatic life, hence the reason why they have attracted so much attention.Sources of heavy metal pollutants include industrial wastewater from mining, metal processing, and tannery, pharmaceutical and chemical engineering industries, and if the metal ions are not removed fromthe wastewater they may bring long-term risks to the environment. Water treatment such as precipitation, ionic exchange, and adsorption [16-18] has been previously used to remove heavy metals from the aqueous solution. Among them, adsorption receives considerable interest in heavy metal removal because of its high adsorption efficiency, easy handling, cost effectiveness, and the availability of different adsorptive materials [19-21].
In this research, we have investigated the microcellular injection molding process as a commercially useful and valuable technique for improving the shrinkage/warpage of the plastics. The study has focused on the effects of the fillers loading on the structure, thermal, tensile, and the heavy metal adsorption properties of the microcellular injection molded PA6/flax and the PA6/GF composites. The primary purpose of the research was to investigate the processing benefits and property variances by micro and nano composites.

2. Materials and methods

2.1. Materials

Micro-sizedflax fiber with an L/D ratio of 2000 and a diameter of 10μm was supplied by the Shan-yuan Co. of Taiwan. Graphene, HGN-8, with a diameter of 8 μm and a thickness of 100 nm, and micromechanical exfoliation from Graphite was used. PA6, type 2100 with a melt flow index of 45 g/10minwas purchased from the Nan-Ya Plastics Corporation, Taiwan.

2.2. Compounding and foaming process

The PA6/flax and the PA6/graphene composites were preparedthrough the blending of the flax and graphene with the PA6resin using a twin-screw extruder. The loadings of the flaxwere 1, 3, and 5 wt%,and 0.5, 1.0, and 3.0wt% for the graphene. Then, the micro compositeswere molded separately usingthe conventional and Mucell®injection molding processes. TheMucell®injection molding process was performed on a 100 ton ARBURG-420C injection molding machine, equipped with a microcellular foamingsystem. Nitrogen was used as a physical blowing agent in the foaming process. The injection molding setup is as shown in Figure 1. The processing parameters for both the conventional and Mucell® moldingare presented in Table 1.

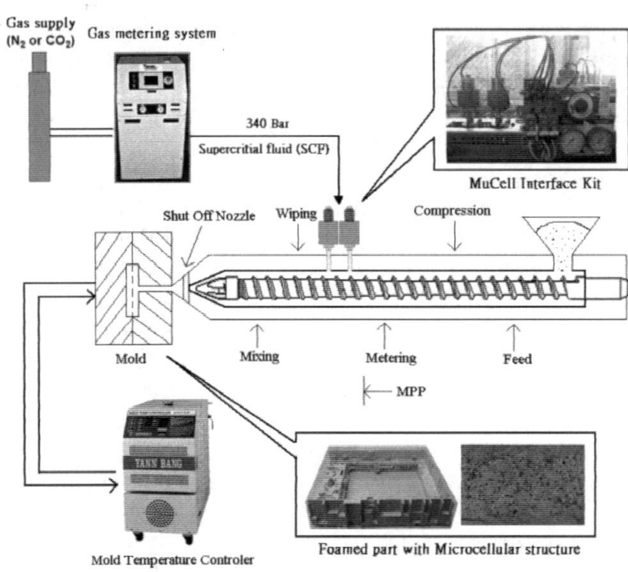

Figure 1. Microcellular injection molding setup.

Table 1. Process parameters both for conventional and Mucellinjectionmolding processes

Process parameters	PA6/flax and GP (Unfoamed)	PA6/flax and GP (Foamed)
Shot size (cm^3)	19	17
Melt temperature (°C)	290	290
Mold temperature (°C)	90	90
Injection speed (cm^3/s)	140	150
Injection pressure (bar)	1000	1000
Packing pressure (bar)	700	50
Back pressure (bar)	50	110
Screw speed (m/min)	20	20
Cooling time (s)	25	25
SCF content (wt %)	–	0.4
SCF flow rate (kg/h)	–	0.17

2.3 Instrumentations

The cellular structures of the foamed compositeswere examined usinga JEOL JSM6360scanning electron microscope (SEM) system. Specimens were cut into small pieces and sputter coated with gold before examining with the SEM. The transmission electron microscopy (TEM), was obtained on a JEOL JSM 6500F, and Model JEOL JEM2010, 200 kV, respectively. The sample thickness of 80 nm was prepared by using a Leica Ultracut-UCT to verify the physicochemical properties of the layer materials.Differential scanning calorimetry (DSC) analysis was used to study the thermal properties of the composites. The thermal analysis was performed using DSC technique (DSC821e, Mettler Toledo). Temperature range was 30–300 ℃ at the heating rate of 4°C/min. The thermal decomposition temperature was determined by thermosgravimetric analysis (TGA) (SII TG/DTA6200), and the experiment was performed by samples (10 mg) of air gas flow from 40°C to 900°C at a heating rate of 10°C min^{-1}.Dynamic mechanical analysis (DMA) was performed using Perkin Elmer DMA 8000.

The molded samples were subjected to the tensile (ASTM Standard D638-02) examination. The reported mechanical properties data were the average of five successful tests.The tensile tests of the ASTM standard samples were carried out using a tensile test analyzer andan HT-9102M system(Hong-Da Company).

3. Results and discussion

This study aims to investigate the relationships and comparison between the processing, morphology, and the properties of the foamed PA6/flax and the PA6/GPnanocomposites produced bythe microcellular injection molding process.

3.1. Morphology of PA6/graphene nanocomposites and PA6/flax composites

The morphology of the PA6/GP was characterized by the XRD, TEM, and SEM analysis. The morphology of the PA6/flax was characterized by the SEM only. The XRD patterns of the nanocomposites are as shown in figures2(A) and (B). The XRD determines the change of the interlayer distance of the layered material. The characteristic basal peak of grapheneshowed at 26.47 degreecorresponding to 0.34 nm layer spacing at (0 0 2) direction. A higher content of graphene increases the intensity peak of the polymer which shows the graphene is more oriented [23]. From the XRD results, it is evident that the graphene is intercalated in the polymer matrix [24].

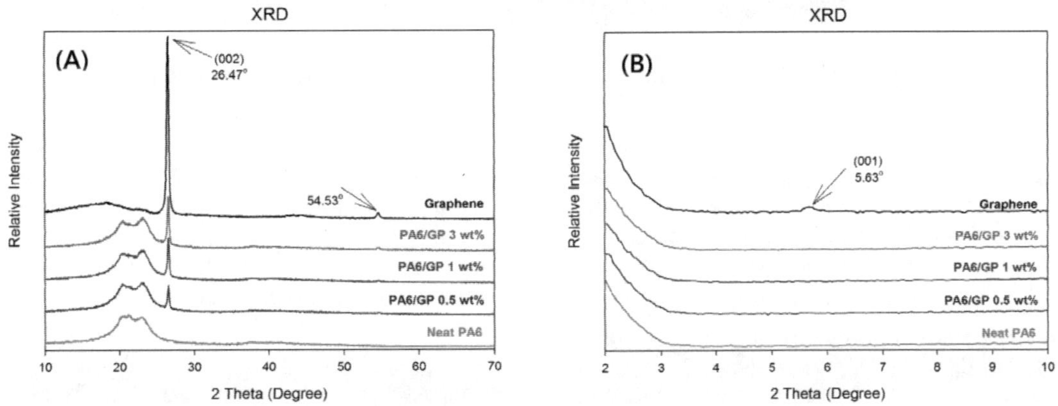

Figure 2: XRD pattern of PA6/graphene nanocomposites.

The XRD analysis alone is not sufficient to prove either intercalated or exfoliation morphology,thus the nanocomposites morphology was confirmed by the TEM analysis. The TEM pictures of the PA6/GP0.5 and 3.0 wt% nanocomposites (Figure 3 (A), (B), (C) and (D)) showall possible platelet morphologies, dramatically intercalated and stacked structures in a polymer matrix. The graphene thickness is around 100 nm.

Figures 4a and 4b show the typical SEM images of the fracture surfaces perpendicular to the melt flow direction obtained from the foamed PA6/flax and the PA6/GP composites. From the images, it is confirmed that the effects of the flax and GP on both the cell density and size of the foamed samples. The cell size of the neat PA6 is large and the quantity is in a small amount. The cell size is 47.89 μm for the neat PA6 and is 64.67, 32.09, and 48.92 μm for the flax loading of 1, 3, and 5 wt%, respectively (Figure 3a).The cell size is 47.89 μm for the neat PA6, and is 28.64, 16.72, and 14.36 μm for the GP loading of 0.5, 1.0, and 3.0 wt%, respectively, (Figure 3b). The cell size and density are in Tables 2 and 3 for the PA6/flax and PA6/GP nanocomposites, respectively. The GP nanocomposites havea smaller cell size, in that it has a more specific surface than that of the flax composites. When the GP content increased, the cell size decreased, and the cell density increased. Further, the decrease of the cell size also affected the tensile strength. This can be confirmed with the tensile strength from Figure 5. These results indicate that the flax rod and the GP layer served as the heterogeneous nucleating sites during the foaming process. In this heterogeneous nucleating system, filler particles and cells nucleate at the boundary between the polymer matrixes. The remarkable increase in the cell density of the composites is attributed to in the effectiveness of the nucleating sites either on the flax or GP layer surfaces. However,

IOP Conf. Series: Materials Science and Engineering **542** (2019) 012067 doi:10.1088/1757-899X/542/1/012067

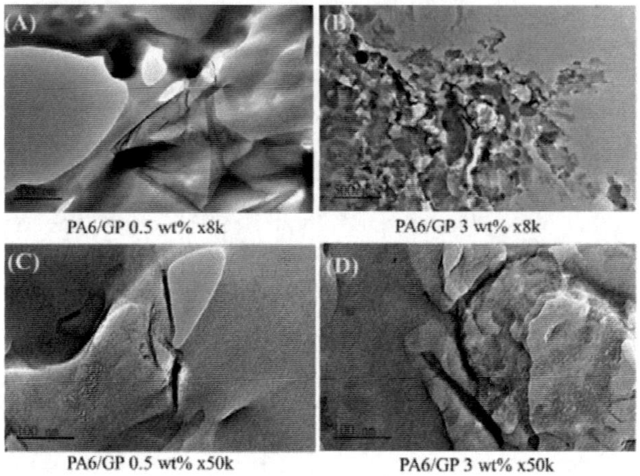

Figure 3.TEM images of PA6/graphene nanocomposites with different graphene loadings and magnifications.

the cell size leveled off when the MMT content was greater than 1 wt%. Upon the foam injection molding, bubble nucleation occurred at around 2 or 3 sec after the pressure started descending. In order to evaluate the effect of clay on the bubble nucleation and growth, the pressure profile has to be carefully controlled for every experiment to realize the same pressure conditions. The variability of the pressure profile at around 2 to 4 sec was less than 0.2 MPa among all the experiments.

Figure 4.SEM images of the fracture surfaces perpendicular to the melt flow direction. (a) PA6/flax composites (b) PA6/GP nanocomposites.

Table 2. Cell size and cell density for PA6/flax composites.

Loading	Cell Size (μm)	Cell Density (Quantity/cm^3)
0 wt%	47.89	1.33 x10^{12}

IOP Conf. Series: Materials Science and Engineering **542** (2019) 012067 doi:10.1088/1757-899X/542/1/012067

1wt%	64.67	4.70 x10^{11}
3 wt%	32.09	2.69 x10^{12}
5 wt%	48.92	8.36 x10^{11}

Table 3. Cell size and cell density for PA6/GP nanocomposites.

Loading	Cell Size (μm)	Cell Density (Quantity/cm^3)
0 wt%	47.89	1.40 x10^{12}
0.5wt%	28.64	3.86 x10^{12}
1 wt%	16.72	2.22 x10^{13}
3 wt%	14.36	4.66 x10^{13}

3.2. Tensile Property

The purpose of adding the filler into the polymer is the tensile strength enhancement of the polymer matrix for most of the composites. The shape (rod, plate, and sphere) and the L/D ratio of the filler playan important role on the strength enhancement. The interface (untreated or treated filler) between the filler and matrix is also an important issue. Figures5(a) and (b) shows the load-displacement of the foamed PA6/flax composites and the PA6/GP nanocomposites, respectively. Figure 5(a) showsthe decreasedtensile strength and elongation with the addition of the flax filler. The tensile strength was1637.9, 1641.6, 1542.0, and 1414.0 Nfor the neat PA6, 1, 3,and 5wt% flax, respectively. Elongation was 66.4, 42.0, 26.9, and 16.0% for the neat PA6, 1, 3, and 5wt% flax, respectively. However, it has a different trend for the PA6/GP nanocomposites. The tensile strength increased then leveled off at 3 wt of the GP loading.The tensile strength was 1453.35,1644.1,1493.47, and 1266.17 Nfor the neat PA6, 0.5, 1,and3 wt% of the GP loading, respectively.

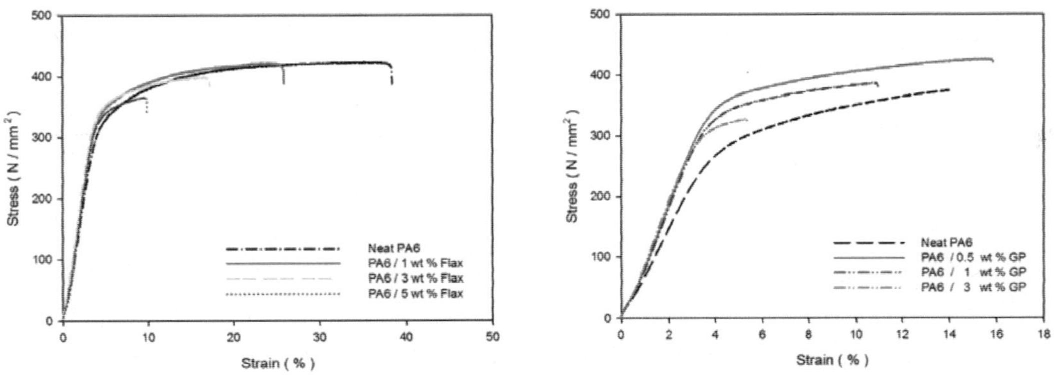

Figure 5. Load displacement curves of PA6/flax and PA6/GP nanocomposites.

3.3 Thermal Property

3.3.1 TGA

The TGA results of the nanocomposites are as shown in Figures 6 (a) and (b). The thermal stability ofthe PA6/flax composites andthe thermal decomposition temperature (Td) were increased as compared to that of the pure polymer. The peak degradation temperature of the PA6/flax compositeswas found to increase to 4.7°C, from 470.3 to 475.5°C. Whereasthe flaxfiller acts as a thermal retardant, also the enhanced thermal stability were attributed to the homogeneous dispersion of the flaxfiller in the polymer matrix. While the thermal decomposition temperature decreased for the PA6/GP nanocomposites.This is caused by the poor interface bonding between the GP and the PA6 matrix.

IOP Conf. Series: Materials Science and Engineering **542** (2019) 012067 doi:10.1088/1757-899X/542/1/012067

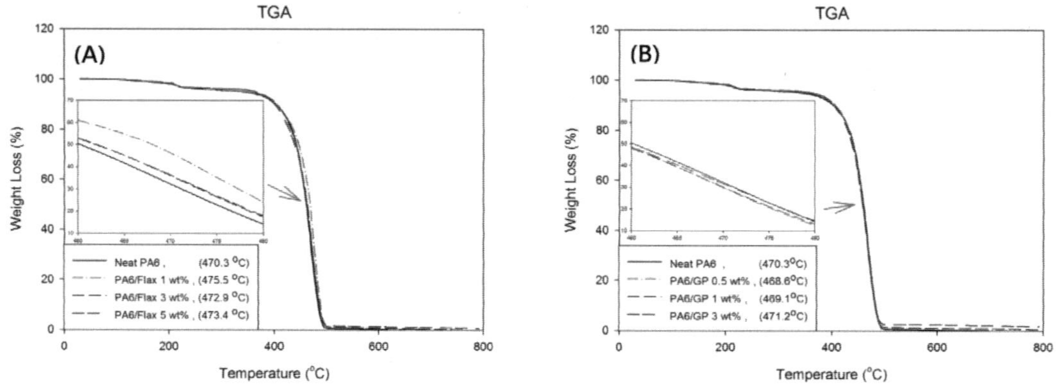

Figure 6.TGA results of (a) PA6/flax and (b) PA6/GP nanocomposites.

3.3.2 DSC

The DSC thermograms of the nanocomposites are shown in figures 7, and 8,and the melting temperature (*Tm*) and the cooling crystal temperature (Tcc) of the polymer nanocomposites. It was observed that the polymer nanocomposites Tm were almost similar after the addition of either the flax or GP in both the polymer composites. But the Tcc were higher for the GP filler than that of the flax filler indicating the GP helped the crystallization. However, the increased concentration of the micro or nano-fillers leads to the decrease of the crystallinity of the nanocomposites. This is caused by the foaming process [25].

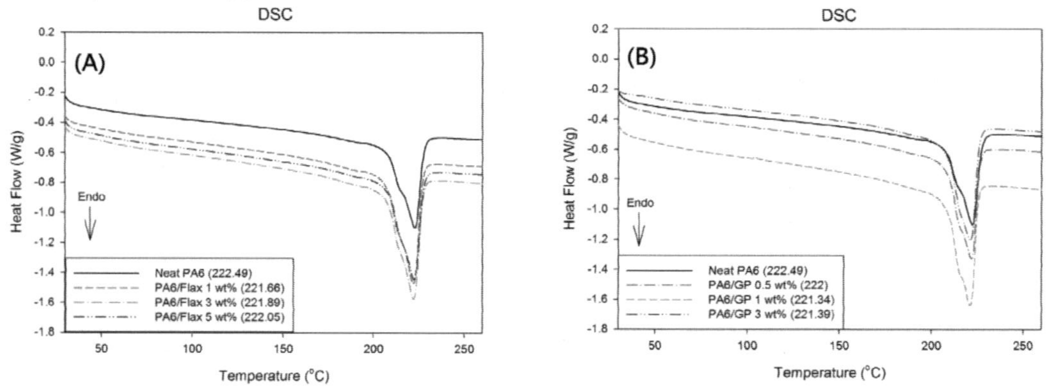

Figure 7. Heating curves of DSC for (a) PA6/flax and (b) PA6/GP nanocomposites.

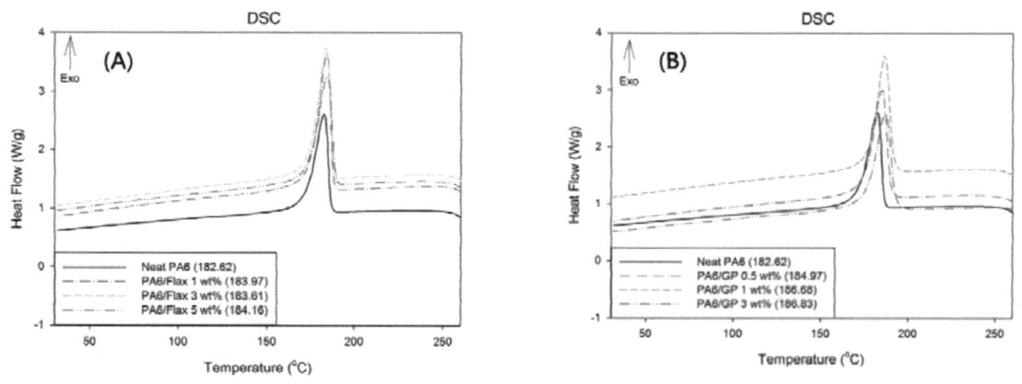

Figure 8. Cooling curves of DSC for (a) PA6/flax and (b) PA6/GP nanocomposites.

Table 4 .Thermal analysis of PA6/flax and PA6/GP nanocomposites

Filler (wt%)	Tm (℃)	ΔH (J/g)	Tcc (℃)	ΔHc (J/g)	Xc (%)
Neat PA6	222.5	36.7	182.6	47.6	5.7
Flax 1	221.6	53.5	184.0	62.9	5.0
Flax 3	221.9	50.2	183.6	61.3	5.9
Flax 5	222.	48.1	184.2	55.8	4.2
GP 0.5	222	48.6	185.0	52.5	2.0
GP 1.0	221.34	49.94	186.68	53.62	1.96
GP 3.0	221.39	48.8	186.83	51.27	1.34

3.4 DMA

Figures 9 and 10 show the storage modulus and tan δ for the PA6/flax and the PA6/GP nanocomposites, respectively. The storage modulus can exhibit the stiffness of the material and tan δ indicates the glass transition temperature of the composites. The storage modulus is 2431, 2780, 2835, and 3065MPa at 60℃for the neat PA6, 1, 3, 5 wt% flax, respectively. The storage modulus increases as the flax loading increases. While it is 2399, 2622, 2775, 3221 MPaat 60℃for the neat PA6, 0.5, 1, 3 wt% GP, respectively. When comparing these two composites, the PA6/GP nanocomposites have a larger storage modulus than that of the PA6/flax composites. The glass transition temperature is not apparent for the PA6/flax composites as shown in Figure 9B. This can be verified in the TMA analysis section. The glass transition temperature is 70.4, 62.9, 60.6, and 58.6℃for the neat PA6, 0.5, 1, 3 wt% GP, respectively, as shown in Figure 10b.

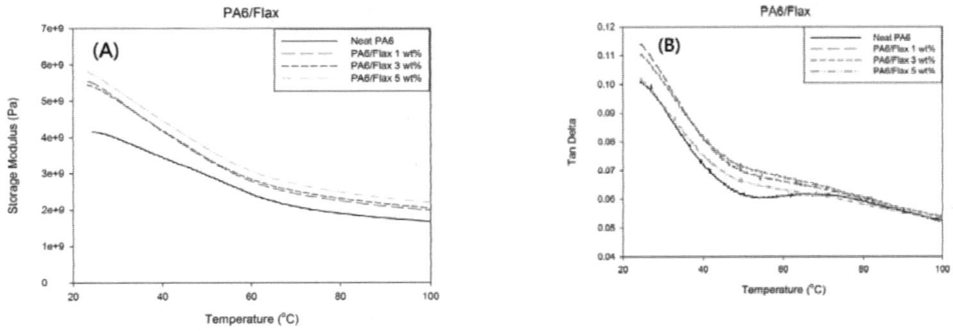

Figure 9. Storage modulus and Tanδof PA6/flax composites.

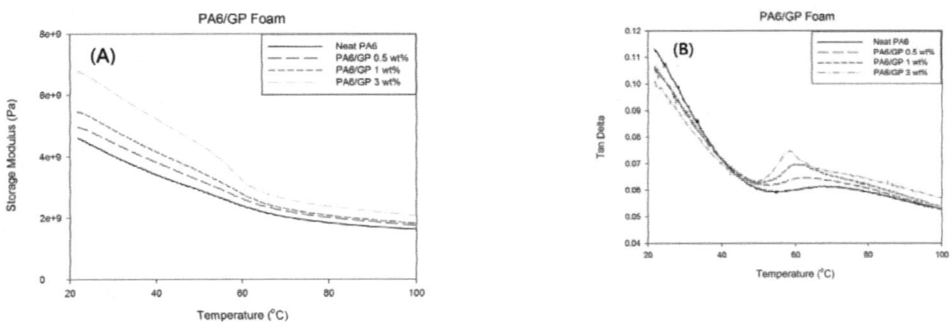

Figure 10. Storage modulus and Tanδ of A6/GP nanocomposites

IOP Conf. Series: Materials Science and Engineering **542** (2019) 012067 doi:10.1088/1757-899X/542/1/012067

3.5 Heavy metal adsorption

This adsorption study is based on the Langmuir isotherm. The Langmuir model has been established on the theory that a max adsorption arises when a saturated monolayer of solute molecules is existent on the adsorbent surface[26], also that the energy of the adsorption is continuous, and that there is no movement of the adsorbate particles on the surface plane. The effect of the filler content and the initial concentration on the percentage removal of the Cr(IV) by the PA6/flax composites and the PA6/GP nanocomposites areas shown in Figure 11, respectively. Figure 11a shows the effect of the filler content on the heavy metal removal. The results show that the GP has a better Cr(IV) adsorption than that of the flax. There were five different initial concentrations of the Cr(IV) chosen for the study, i.e. 10, 20, 30, 40 and 50 mg/l with a constant dose of 0.05 g of the adsorbent composites and nanocomposites. The ability of the PA6/flax and the PA6/GP composites to adsorb the Cr(IV) is attributed to the presence of the exchangeable ions, such as Si-O and Al-O, and also the swelling ability. The PA6/flax and the PA6/GP composites were able to adsorb almost 42% and 49%, respectively, of the Cr(IV) at 20 mg/l. The adsorption levels off when the concentration is higher than 20 mg/l, both for the PA6/flax and the PA6/GP nanocomposites.

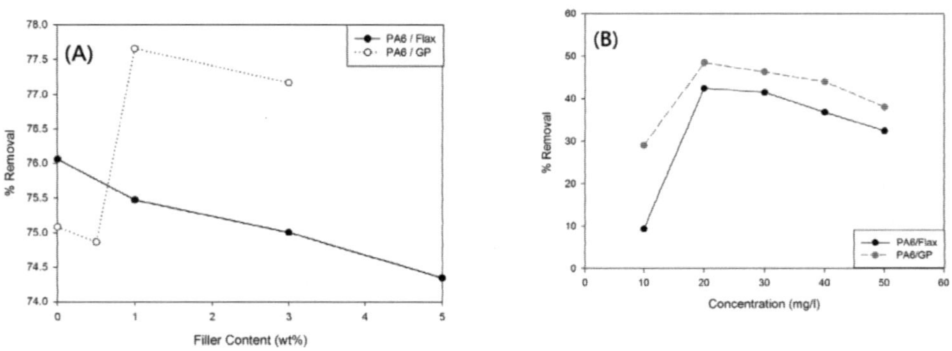

Figure 11. Heavy metal Cr (IV) adsorption test of PA6/flax and PA6/GP nanocomposites.

Acknowledgments
This study was supported by Ministry of Science and Technology, Taiwan under grant MOST 105-2923-E-231-001-MY3.

[1] TsaiT-Y, LinM-J, ChuangY-C, ChouP-C, 2013 *MCP,* **138**, *1, 230-237.*
[2] Chang, M-K, Hwang, S-S, Liu, S-P, 2014 *JIEC*,**20**, 4,25,1596.
[3] Katz H S, Mileski J V, 1988, *Handbook of Fillers for Plastics*, (New York: Springer) 5, pp 55
[4] Blanco I, Abate L, Antonelli M L, Bottino F A, Bottino P, 2012 *Express Polym Lett* **6**(12), 997-1006.
[5] Petrucci R and Torre L, 2017, Filled Polymer Composites, *Modification of Polymer Properties*,edCarlos Federico Jasso-Gastinel, José M. Kenny (New York: William Andrew), chapter 2, pp 28-2.
[6] LiuY, ZwingmannB,Schlaich M, 2015 *Polym*, **7**(10), 2078-2099.
[7] ZampaloniM, PourboghratF,YankovichSA, RodgersBN, MooreJ, DrzalLT, Mohanty AK, Misr M, 2007 *Compos Part Appls*, **38**(6), 1569-1580.

[8] YanjunX, CallumAS, Zefang X, Holger M, CarstenM, 2010*Compos Part A-Appls*, **41**(7), 806-819.
[9] Valadez-GonzalezA, CervantesR,OlayoPJ, HerreraF,1999 *Compos Part B-Eng*, **30**(3), 321–331.
[10] Reina A, Jia X T, Ho J, NezichD, Son H B, Bulovic V, Dresselhaus M S, Kong J, 2009 *Nano Lett*, **9**, 30.
[11]NovoselovKS, GeimK, MorozovSV, Jiang D, ZhangY, Dubonos S V, Grigorieva IV, Firsov AA, ,2004 *Science*, **306,** 666.
[12] WuMQ, XuY, YaoZ, LiuA, Shih G, 2010 *ACS Nano*, **4**, 1963.
[13] HummersWS,OffemanRE, 1958 *JACS*,**80** (6): 1339.
[14] WuZ-S, RenW, GaoL, LiuB, JiangC, ChengH-M, 2009 *Carbon*, **47**, 493.
[15] BunekarN, TsaiT-Y, YuY-Z, 2016 *Mater Today-Proc*, **3**, 1415.
[16] HuaM, ZhangS, PanB, Zhang W,LvL, Zhang Q,2012 *J. Hazard. Mater,* **211–212**, *317–331*
[17] TaghipourKolaeiZ,TanziM,YouseA, and EisazadehH, 2012*J. VinylAddit. Technol*, **18**, *52–56.*
[18] MittalB,and MishraS,2014 *CarbohydPolym*, **101**, 1255.
[19] VeliS,AlyüzB, 2007*J Hazard Mater*, **149**, 1, 226-233.
[20] ŞölenerM,TunaliS,ÖzcanA,GedikbeyT, 2008 *Desalination*, **223**, 1-3, 308-322.
[21] RožićM,Cerjan-StefanovićS, KurajicaS,VančinaV, 2000 *Water Res*, **34**, 14, 3675–3681.
[22] HastutiB, Masykur A, HadiS, 2016 *IOP Conf. Ser.: Mater. Sci. Eng.*,**107**, 012010.
[23] LiX-Y, WangZ-Z, WuL-X, TsaiT-Y, 2016 *RSC Adv.6*, 74903-74912.
[24] TsaiT-Y, HsuC-C, Chao W-K, HuangS-H, ChanC, WuT-C, WangP-H, 2015 *JCCS*,**62**, 6, 562-568.
[25] Mi H-Y, JingX, TurngL-S, 2015 *J. Cell. Plast.*,**51**, n2, 165-196.
[26] Langmuir, I, Langmuir Isotherm, 1916 *JACS*, **38**, 2221–2295.

IOP Conf. Series: Materials Science and Engineering **542** (2019) 012068 doi:10.1088/1757-899X/542/1/012068

Influence of the Pressure of Open-Close Water-Cooled Moulds on Casting/Mould Interface Heat Transfer and Casting Solidification Process

Q H Yao ,Y D Zeng* , L T He,J Zhang

College of Materials Science and Engineering, Inner Mongolia University of Technology, Huhehaote,010051, China
*Corresponding author e-mail: zyidan@sina.com

Abstract. In this paper, the experimental model for a self-designed open-close water-cooled mould was utilized to measure the temperature curves of various points inside the casting and mould under different mould pressure. The Interface heat transfer coefficient(IHTC) of the casting/water-cooled mould interface was calculated by using the Beck nonlinear estimation back-calculation method, and on this basis, the rule of how pressure affects heat transfer coefficient was obtained. Then the coefficient can be defined as the boundary condition(BC) to simulate the solidification process of the experimental casting. The results showed that the simulation outcome considering the influence of pressure was completely different from those of general water-cooled moulds, which can be proved by the about 35% solidification time decrease, visibly reduced shrinkage porosity area in the casting and more accurate simulation results. The anatomical results of the experimental casting also confirmed this.

1. Introduction

Before putting the casting into practical production or designing new technologies, engineers can use computers to carry out pouring tests through solidification numerical simulation (referred to as solidification simulation) in order to find out potential defects or evaluate the rationality of new technologies. This can greatly save the time and cost of experiments and trial production and also improve the production yield. Therefore, before practical pouring, producers will generally require solidification simulation to verify the technology. This is why its accuracy becomes so important.

As a new casting technology, open-close water-cooled mould casting has received much attention these years. Its main advantage lies in the outstanding cooling performance and it can realize light press during solidification. Thus, the resulting casting enjoys fine structure, few defects and excellent performance. Recent years also saw its successful application examples in large steel ingots production. As such, manufacturers hope solidification simulation will provide a guidance for process design and flawless products.

Different from common ones, the open-close water-cooled mould can perform light press on castings, changing the contact status of the casting and mould, and casting a great influence on the heat exchange between them. From the perspective of solidification simulation, that means the boundary condition, which is a significant prerequisite for accurate simulation results, changes[1]. Inappropriate boundary conditions may lead to results completely deviating from the real one, misleading subsequent analysis and researches. This effect is quite obvious in the numerical simulation of water-cooled moulds with boundary heat transfer coefficients as boundary conditions. So, using original water-cooled mould heat transfer coefficients will cause significant errors to the numerical simulation result. Realizing how important the coefficient is, the researchers have done a lot in this respect, including improving the calculation method and model of the coefficient to improve calculation efficiency and accuracy[2,3], and exploring the rules of the influence of casting and mould materials[4], casting process[5], interface state[6,7] and multiple factors on the heat transfer coefficient of the casting/mould interface. For a new technology as open-close water-cooled mould casting, it is necessary to study how boundary conditions will change under pressure, so as to improve the accuracy of numerical simulation.

In this paper, the experimental model of a self-designed open-close water-cooled mould was utilized to measure the temperature curves of various points inside the casting and mould under different mould pressure. The heat transfer coefficient of the casting/water-cooled mould interface was calculated by using the heat transfer back-calculation method, and on this basis, the rule of how pressure affects heat transfer coefficient was obtained, which can help rasing the accuracy of the coefficient. This is important for the solidification simulation of open-close water-cooled moulds applied in practical production.

2. Experiment

Based on the similarity theory, an experimental model imitating actual production was designed (Fig. 1). The casting used was a 150*80*80 rectangular parallelepiped, and the position of thermocouples placed on the casting and the water-cooled mould was shown in Fig. 1. Four temperature measuring points were put in the mold and casting: point 1, 2 and 3 were inside the casting, 5, 10 and 40 mm from the casting/water-cooled mould interface respectively; point 4, which was designed to measure the wall temperature of the water-cooled mould, was located on the inner wall of the mould and was 1 mm from the interface.

The pressure was set to 0, 70, 150, 200 and 300kg, in order to study how interface heat transfer was affected by pressure if applied a light press. The cooling water flows at a rate of 0.39kg/s. Walls of the water-cooled mould can move to the left and right of other mould walls, and the push rod can control the pressure exerted to the casting/mould interface.

The study used the heat transfer back-calculation method to obtain the heat transfer coefficient of the casting/water-cooled mould interface. Based on different mould pressures, the solidification temperature curve of the experimental casting was measured. Then by the Beck nonlinear estimation back-calculation method[8], the forementioned curve can be used to gain the target coefficient.

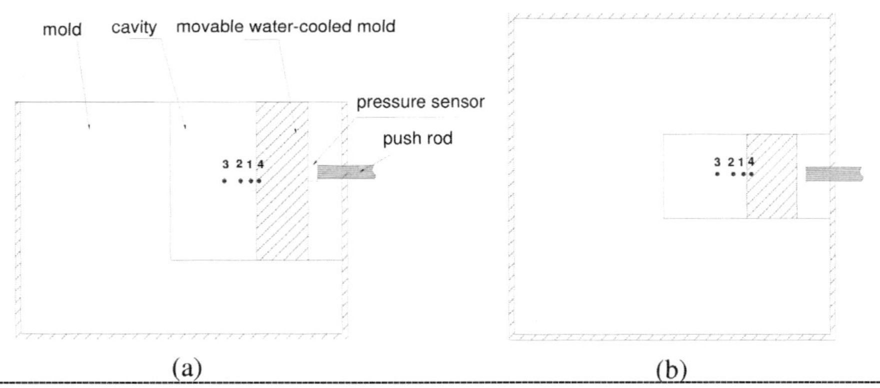

Figure 1. Experimental apparatus: (a) Front sectional view. (b) Horizontal sectional view.

3. Experimental results and analysis

3.1. Curves of measured temperature

Fig. 2 shows the temperature curve measured from Point 1~3 under the pressure of 0, 70, 150, 200 and 300kg. Without pressure (Fig. 2a), the cooling of the casting is slow, but after applied pressure (Fig. 2b~e), the process accelerates, and the time shortens as the pressure increases. Taking temperatures of Point 1~3 inside the casting when cooled down to 400 s as an instance. Without any pressure, the

IOP Conf. Series: Materials Science and Engineering **542** (2019) 012068 doi:10.1088/1757-899X/542/1/012068

temperature are 564.5, 568.1, 577.4 °C respectively and become 324.7, 330.3, 341.7 °C after applied a 300kg pressure. It can be seen that the interface pressure can greatly affect the solidification process of the casting

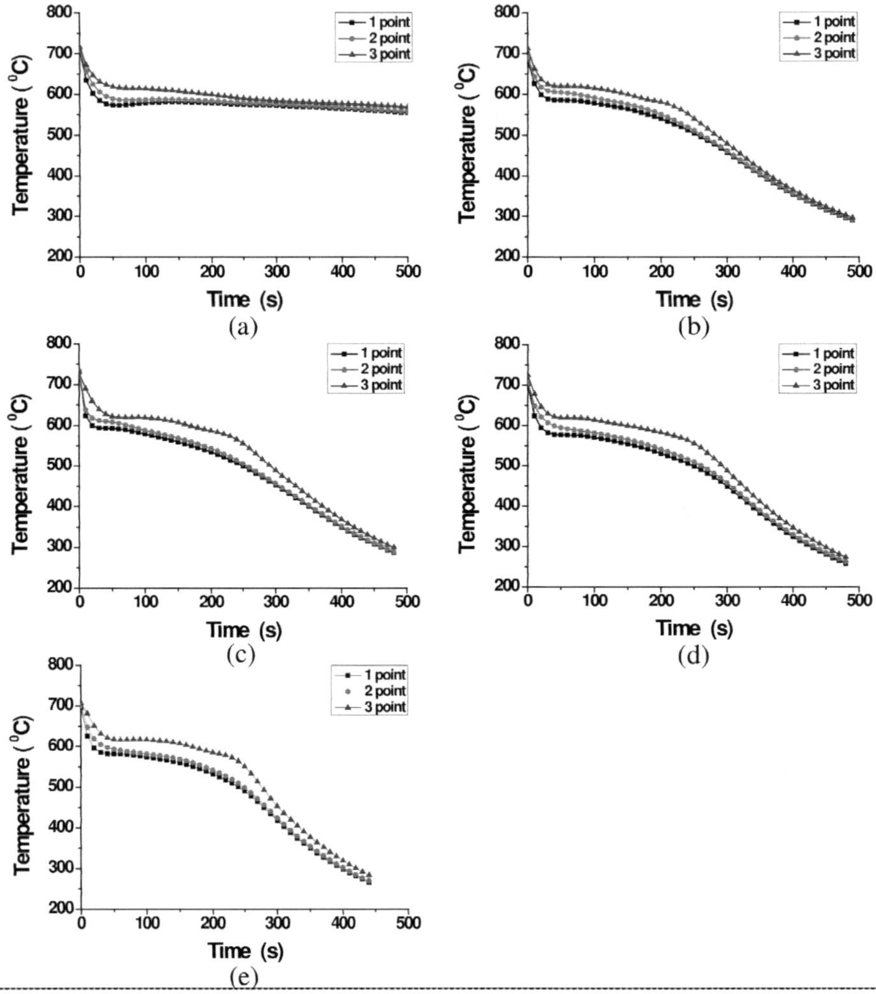

Figure 2. Measured temperature at different pressures

(a) pressure 0kg (b) pressure 70kg (c) pressure 150kg (d) pressure 200kg (e) pressure 300kg

3.2. Influence of Pressure on Interface Heat Transfer Coefficient and Heat Flow

Fig. 3 shows the interface heat transfer coefficient and heat flow under abovementioned conditions. It can be seen that without pressure (Fig. 3a), the interface heat transfer coefficient first decreases rapidly, and then slowly after the casting is cooled by 150 s. After applying pressure (Fig. 3b~e), the coefficient also drops rapidly first, but then rises and slowly decreases again. As the pressure increases, the interface heat transfer coefficient increases and rebounds to higher peaks, from 781.6 to 1002. 4 W/(m^2·K) Consistent with the heat transfer coefficient, the interface heat flow also increases significantly after the pressure is applied. Compared with the interface heat transfer coefficient when cooled to 400 s, the heat transfer coefficient, with no pressure, decreases to 291.8 W/(m^2·K), while those with pressure are 412.8, 303, 621.3 and 824.2 W/(m^2·K), respectively, all higher than 291.8 W/(m^2·K). That is because increased interface pressure will push closer the casting and the mold. In respect of the three heat transfer modes existing at the casting/mould interface, the proportion of heat conduction increases, resulting in a larger heat transfer coefficient.

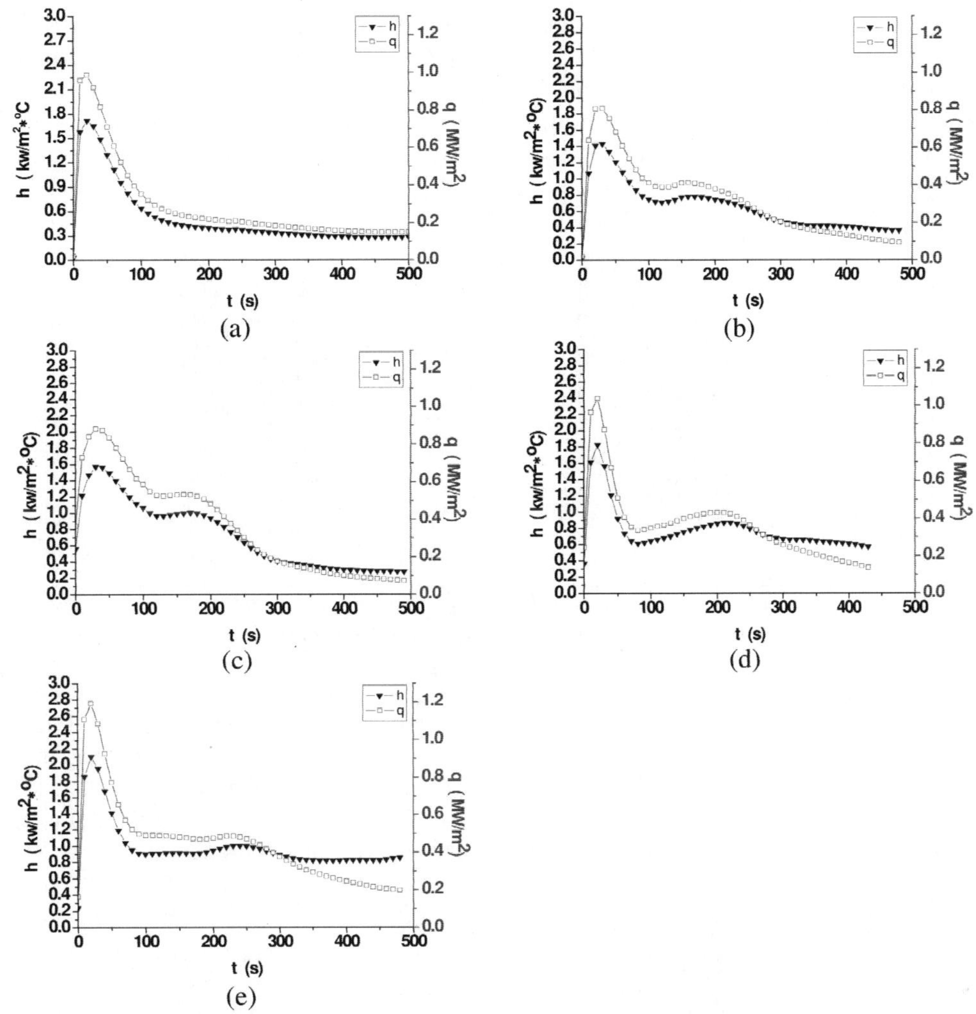

Figure 3. IHTC at different pressures
(a) pressure 0kg (b) pressure 70kg (c) pressure 150kg (d) pressure 200kg (e) pressure 300kg

4. The Result of Solidification Simulation

In order to study the change of the solidification simulation results after the heat transfer coefficient is

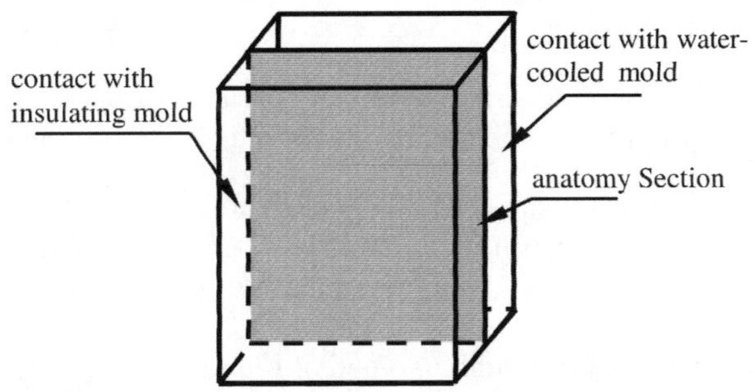

Figure 4. Anatomy section of the casting

affected by the pressure, the simulation process and the anatomy are carried out on the experimental casting. Fig. 4 depicts the anatomy section.

Fig. 5 is the solidification time numerical simulation results of different points on the anatomy section, with the heat transfer coefficient of no pressure applied and pressure applied as boundary conditions, respectively. The color bar on the left side of Fig. 5 is the scale of the solidification time. Different colors are used to indicate the solidification time at each point on the casting. In the respect of solidification time, the casting with pressure applied takes less time than the one without, and has rather smaller final solidification region (the red part).

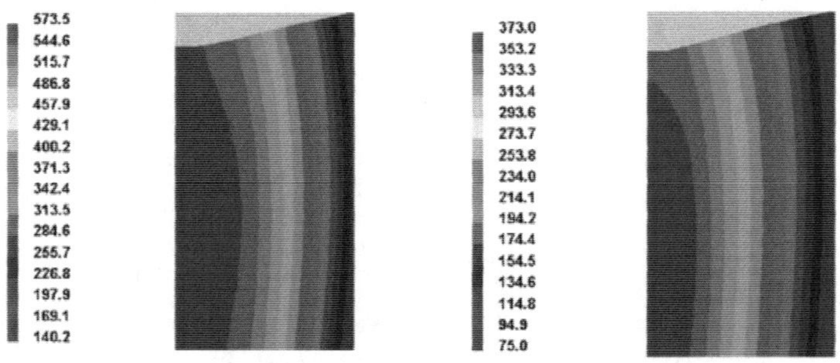

(a) (b)

Figure 5. Solidification time of the casting
(a) no pressure applied as BC (b) pressure applied as BC

Figure 6 are numerical simulation results of the shrinkage porosity range of the casting under the abovementioned boundary conditions. Colors on the left-side bars represent how large are the shrinkage porosity areas. Obviously, the area of the casting without pressure is larger than the one with, and has larger severely defected area (the blue part). The reason is that the pressure increases the heat transfer coefficient, thereby pacing up the cooling process. Accordingly, the temperature gradient is increased, and the mushy zone, reduced, so the shrinkage porosity defect area shrinks.

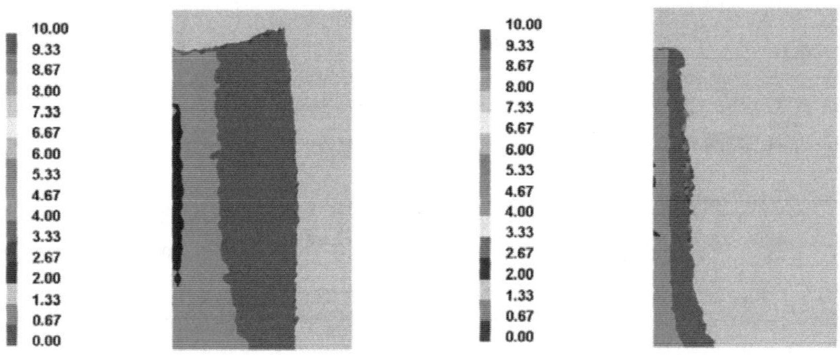

(a) (b)

Figure 6. Shrinkage porosity area of the casting
(a) no pressure applied as BC (b) pressure applied as BC

Fig. 7 shows the actual casting sections(Fig. 4) with and without pressure .The shrinkage at the top of the casting in Figure 7b is significantly larger than that in Figure 7a, indicating that the internal dispersed shrinkages are less than the latter, which is consistent with the numerical simulation results of Figure 6.

It can be concluded from these experimental results that pressure can greatly affect the heat transfer coefficient. For open-close water-cooled mould casting, which can realize light press, if directly setting the heat transfer coefficient ignoring the influence of pressure (namely with a pressure of 0 kg) as the boundary condition for solidification simulation, the results will be drastically deviate from reality, not to mention guiding actual production.

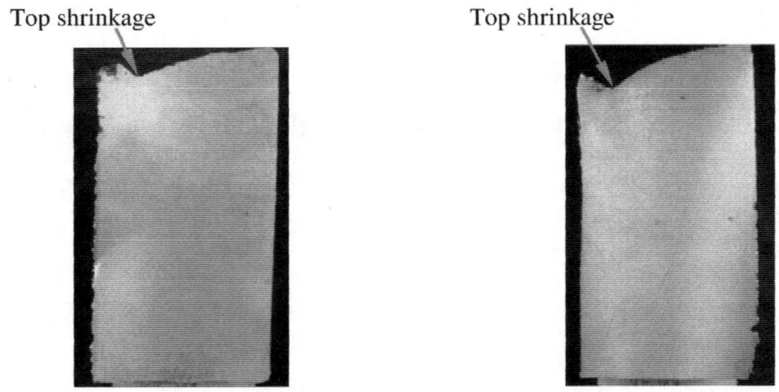

(a) (b)

Figure 7. Top Shrinkage on the Section of the Experimental Casting
(a) no pressure applied (b) pressure applied

5. Conclusions

(1) In the process of applying pressure, the casting/mould interface heat transfer is different from the interface heat transfer in common casting process. Under pressure, the heat transfer coefficient of the casting/mould interface will rapidly increase to the maximum, and then drop as the solidification process goes on. As the interface pressure rises, the heat transfer coefficient between the casting and mould will enlarge.

(2) The interface pressure has a critical influence on the heat transfer coefficient of the casting/mould interface. Under experimental conditions, the coefficient augments by nearly 41% after the pressure applied. Leaving out the influence of pressure will cause a big error to numerical simulation results.

(3) The interface pressure affects the solidification process of the casting. Larger pressure leads to shorter solidification time and significantly reduced shrinkage porosity area.

References

[1] Griffiths W D 2000 *Metall. Mater. Trans B* **31** 285-295.

[2] Blackwell B and Beck J V 2010 *International Journal of Heat and Mass Transfer* **53** 753-759.

[3] Duda P 2015 *International Journal of Heat and Mass Transfer* **84** 1101-1112.

[4] Ferreira I L, Spinelli J E and Nestler B, et al 2008 *Materials Chemistry and Physics* **111** 444-454.

[5] Prabhu K N, Kumar S T and Venkataraman N 2002 *Journal of Materials Engineering and Performance* **11(3)** 265-273

[6] Coates B and Argyropolulos S A 2007 *Metallurgical and Materials Transactions* **38(B)** 243-255.

[7] Zeng Y D,Yao Q H and Wang F 2016 *Materials Science Forum* **893** 174-180

[8] Beck J V and Blackwell B F 1985 *Inverse heat conduction Ill-posed problems*, Wiley, New York.

IOP Conf. Series: Materials Science and Engineering **542** (2019) 012069 doi:10.1088/1757-899X/542/1/012069

Comparison of Calibration Methods Used for Rietveld Quantitative Phase Analysis of Cement and Fly Ash

Y Q Wei[1,2], W Yao[1] and C Y Wang[2]

1 Key Laboratory of Advanced Civil Engineering Materials of Ministry of Education, Tongji University, Shanghai, China
2 School of Civil Engineering, East China University of Technology, Nanchang, China
E-mail: yq.wei@tongji.edu.cn

Abstract. The overestimated contents of crystalline phases in polycrystalline powder samples containing amorphous phases must be calibrated when they are done by using Rietveld phase analysis (RPA) for quantitative X-ray diffraction patterns of samples. It has been indicated that there are two methods for this calibration are verified with nice accuracy involving internal standard method (ISM) and G-factor method (GFM) that have been applied to the RPA of both cement and fly ash that contain amorphous mine phases. However, the consistency between these two methods in quantitative determination of phase contents in silicate materials has not yet been discussed. Therefore, the accuracy and consistency between them in quantifying standard mixtures, cement and fly ash were evaluated experimentally. Results indicate that (1) both ISM and GFM gained the very nice accuracy and consistency for the ternary standard mixture, and also the good consistency for cement clinker and fly ash; (2) relatively, GFM is suitable for systems with any content of amorphous phases while ISM is preferably suitable for the system with much more amorphous phase contents, e.g. more than 50% in weight (denoted as wt.% henceforth), by adding the suggested 10wt.% of Al2O3 as internal standard;(3) ISM is not suggested to be used for quantifying the system containing less than 10w.t % of amorphous phases, and the wisely higher addition of internal standard, such as 20%, should be employed when it applied to quantifying the system with amorphous phases from 10wt.% to 50 wt.% in content.

1. Introduction

Conventional X-Ray diffraction (XRD) is commonly used for the phase determination of polycrystalline powder samples. The specific determination methods include traditional internal standard method, absorption diffraction method, peak method and reference intensity method. In most cases, these methods can only achieve semi-quantitative analysis of phases [1], because of the overlapped peaks of different phases, the preferred orientation and the information that is not fully utilized in the diffraction pattern. The Rietveld Quantitative Phase Analysis (RQPA or RPA) method [2] effectively solves the above problems by means of crystal structure refinement, physical correction and full-spectrum fitting, and it makes XRD convenient and reliable in quantitative analysis of polycrystalline systems [3]. The RPA method was first introduced by Taylor [4] in 1990 to the quantitative analysis of phases of Portland cement and, and then widely recognized in the quantitative analysis of cement mineral phases with the gradually improvement of both the Alite (C_3S) superstructure model [5, 6] and other crystal structure models, and also the refinement procedures. However, The premise for applying the RPA method to the quantitative phase analysis of polycrystalline systems is the fact that the sample does not contain amorphous phases, otherwise the crystal phase content will be overestimated. Previous studies [1, 7-9] have shown that cement contains amorphous phases, so it is necessary to correct the quantitative results of cement when using RPA. Aranda [10] introduced and used for the first time the Internal Standard Method (ISM) to quantify the mineral phase of clinkers containing amorphous phases when using RPA, verifying that the absolute accuracy of the amorphous phase content quantified by ISM can reach 1%. After that, this method was applied to the quantitative phase determination of cement and fly ash, and also the measurement of their hydration degree by Korpa [8] and Palomo [11]. Another method for correcting the quantitative results of RPA is the G-factor method (GFM), also known as the external standard method, which was first proposed by O'Connor et al. [12] in 1988 but used delayedly recent years by Jansen et al. [13-16] for quantitative phase analysis of cement, fly ash and early hydration products of cement due to the need for real-time analysis. When Jansen [16] used the GFM to calibrate the quantitative phase results of cements, it was found that the amorphous phase was not detect obviously. This is in contradiction with the results of Yao Wu et al. [9] who found that cements used were detected to contain distinct amorphous phase with higher content by using ISM to quantitatively study the phase composition of cement clinkers. Therefore, a comparison between ISM and GFM was conducted to test the consistency of the two calibration methods in application to calibrating phase content of cement clinker and the fly ash. This paper firstly quantitatively analyzed binary and ternary standard mixture systems (without amorphous phase) to test both accuracy and accuracy of the two methods. And then the consistency of the two methods in calibrating quantitative results for cement clinker and fly ash systems. The results will help us fully understand the respective characteristics of the two methods, and select the appropriate method to quantitatively characterize cement, fly ash and its hydration process according to the actual situation.

2. Experimental

2.1. Materials

Cement clinker was provided by Hailuo Cement Co., Ltd. , and Fly ash (Class Ⅰ) provided by Huaneng Thermal Power Plant. The chemical compositions of them were determined by quantitative X-ray fluorescence spectrometry (XRF), and the results were shown in Table 1. Zinc oxide (ZnO) as working chemical was provided by Tianjin Chemical Reagent Research Institute, which is in white powder, and its purity is higher than 99.95%. Alumina (Al_2O_3) as spectrum pure in white powder, and dihydrate gypsum (CH_2$) as analytical reagent with purity higher than 99.99%. were provided by Sinopharm Chemical Reagent Co., Ltd. The XRD spectra of the three chemicals were plotted in Figure 2.

Table 1. Chemical composition and mass absorption coefficient of the clinker and fly ash used here

Sample	MAC [17] (*cm*^{2}*/g*)	Fly ash (*wt.%*)	Clinker (*wt.%*)
Na_2O	25.00	0.45	0.61
MgO	28.69	1.23	0.79
Al_2O_3	31.66	28.98	5.48
SiO_2	35.90	54.70	23.1
P_2O_5	39.43	0.18	0.07
SO_3	44.26	0.58	0.62
K_2O	122.32	1.65	0.76
CaO	124.78	4.48	64.8
MnO	211.70	0.06	0.08
Fe_2O_3	214.68	5.24	3.15
LOI/C	4.51	2.45	0.55

2.2. Sample preparation

It takes the block cement clinker and weighs 20 g of cement clinker intermediate particles and milled it by hand for 20 min; then, the milled clinker, fly ash, zinc oxide, aluminium oxide and dihydrate gypsum are respectively sieved through a 45 μ m sieve. In order to ensure the authenticity of the sampling information, the fly ash and clinker sieve residue need to be dry-milled, and the milling and sieving steps should be repeated until all the samples are sieved out.Then mixing 2 g of Al2O3 and 2 g of ZnO and marking it as 5A+5Z ; mixing 2g of Al2O3 , 2g of ZnO and 1g of CSH2 , as 4A+4C+2S ; weighing 4g of cement clinker , as C; mixing 3.2g of cement clinker and 0.8g of Al2O3, as C+20A ; weighing 4g of fly ash, as FA ; mixing 3.6g of fly ash and 0.4g of Al2O3 , as FA+10A; then all sample are placed into agate mortars respectively with added acetone, and wet-milled for 10 min. Then placed them into a constant temperature and humidity drying oven at a temperature of 25 oC and relative humidity of 0% for 3 h.Finally, the sample placed into a standard sample tank with backpressure loading method. The mixing, preparation and loading steps of each sample need to be repeated three times.

2.3. Data collection and analysis

The instrument for collecting data is a Japanese D/Max-RB type X-ray diffractometer with CuK α radiation, Ni plate filtering, operating voltage of 40 KV, tube current of 100 mA, and optical system DS = SS = 1 ° , RS = 0. 3 mm, graphite monochromator, scintillation detector, etc. The instrument uses a step-and-scan method with a step size of 0.024 °, a count time of 0.6s per step, and a diffraction angle ranging from 7 ° to 41 °. The acquired map was imported into the finishing fitting software Topas academic V 4.1 for refinement. The mineral crystal structure used was described in the literature [18]. During the finishing process, the sample displacement, preferential orientation and other corrections were performed as needed. Repeat three times for each sample and average as the final result.

3. Results and discussion

The standard mixture system was quantified by different calibration methods. The results are shown in Table 2.

Table 2. Quantitative results of standard mixtures by a variety of methods（Unit: wt.%）

		Phase content in binary system (MAC=40.22 cm^2/g, G=6.62E-11 cm^5)				Phase content in ternary system (MAC=44.85 cm^2/g,G=7.19E-11 cm^5)			
Phase	MAC(cm^2/g)	Known	GFM	RPA	ISM	Known	GFM	RPA	ISM
A	31.66	50.0	51.3	50.9	50.0	40.0	39.3	39.9	40.0
Z	48.78	50.0	49.5	49.1	46.5	40.0	38.9	39.5	39.6
CSH_2	63.38	0.0	0.0	0.0	0.0	20.0	20.3	20.6	20.6
Am/ Error			(0.8)	0.0	3.5	0.0	1.47	0.00	(0.2)

3.1. Analysis of calibration calculation error of internal standard method

Due to the dilution of the internal standard, the intensity of the diffraction peak of the sample decreases, and it was very likely that the diffraction pattern of the internal standard was too strong to mask some important information of the diffraction pattern of the sample; therefore, the content of the internal standard must be control, Bensted [1] gives a recommended dosage of about 10%-15% (by weight, the same below). However, the low internal standard content makes the traceability of the internal standard too sensitive, it will cause a large error when the quantitative accuracy of the internal standard cannot be guaranteed. The following will focus on the sensitivity of internal standard tracer, that is the calibration error due to the low amount of standard addition.

In this paper, we use the literature [18] to generate the relationship between the internal standard post-refinement increment ($W_{Std\text{-}Rietveld}$ - W_{Std}) and the amorphous content (shown in Figure 1) , as shown in Figure 1-a, when the internal standard addition amount is 10%, if the absolute error of the RPA quantitative result ($W_{Std\text{-}Rietveld}$) is 1%, the maximum quantitative error of the amorphous phase content will be as high as 10%, which means that the internal standard method does not have a resolution for polycrystalline powders with an amorphous content of less than 10%, more clearly, the internal standard method cannot detect a sample having an amorphous content of less than 10%. Even with a 20% addition of the internal standard, the maximum quantitative error of the amorphous phase content is 6.0%. However, as the content of the amorphous phase in the sample increases, this error gradually decreases.

For samples with an amorphous phase content higher than 50% (shown in Figure 1-b), when the RPA quantitatively produces a 1% quantitative error, and the calibration error of the 20% internal standard dosage is reduced to 1.95%, if the tolerance is ±2%, then the accuracy of the measurement can be met; for amorphous content greater than 60%, the calibration error of 10% internal standard dosage is 1.9%, which also meets the accuracy requirements.

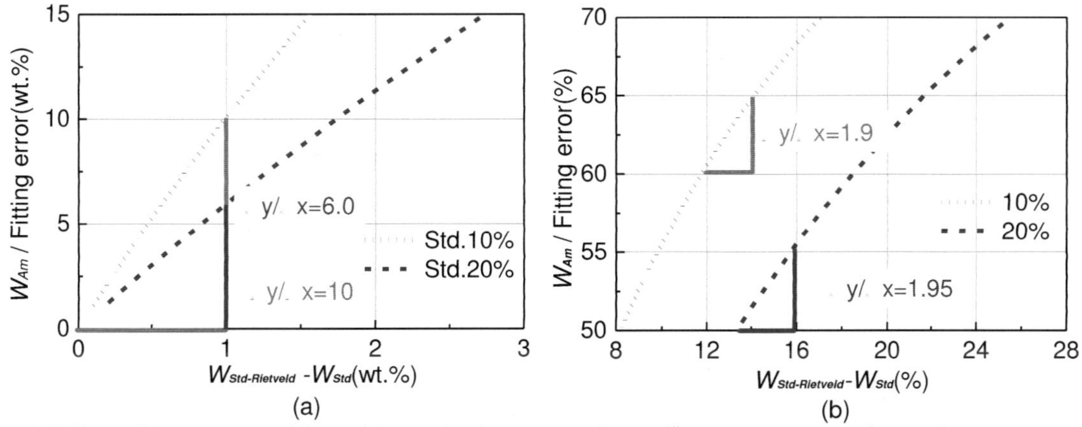

Fig. 1 Effect of Increment of internal standard content after refinement on amorphous phase content and fitting error

Therefore, it is necessary to estimate the amorphous content of the sample in advance to ensure the calibration accuracy. When the amorphous content is less than 50%, the internal standard method is not recommended; when the amorphous content is higher than 50%, it is necessary to add 20% of the internal standard for calibration; when the amorphous content is higher than 60%, it is necessary to add 10% of the internal standard.

3.2. Internal standard method and G-Value method for quantitative consistency of cement clinker

Cement clinker is a typical polycrystalline system. It has been studied [8, 16] that its amorphous phase content is between 0% and 20%, so if the internal standard method may cause a large error in the quantitative content of amorphous content. However, in order to compare these two methods, the internal standard method was still used and 20% of the standard alumina was added to the clinker for tracing or calibration, for comparing with the quantitative results without the standard and the results are shown in Table 3. In these two systems, the results of the G-value method (C system) and the internal standard method (C+20% A system) are in good agreement, and the amorphous content of the internal standard method does not appear large error, showing the accuracy and precision of RPA quantification. The absolute deviation between the quantitative and amorphous phases of each crystal phase is within 1%; from the quantitative results of the G-value method, it is known that the maximum deviation (1.7%) between each other occurs on the C3S phase, but it is considered that the absolute content of the crystal phase is high in the system (about 50%), so this result is acceptable. As expected, in the C system, due to the existence of the amorphous phase, compared with the G-value method, the quantitative results of RPA clearly overestimate the content of each crystal phase, and the overestimation is 18%-20.5%. In the C+20A system, the absolute deviation of the amorphous phase content quantified by the internal standard method and the G-value method is the largest, which is 1.7%. The main reason of the absolute deviation is that the G-value method determines the amorphous content is the 100% content minus the sum of the contents of the crystal phases calibrated by the G-value method, which was a cumulative value, and there is an accumulation of errors, but after being evenly distributed to each crystal phase, the absolute deviation of the crystal contents from each other is within 1%. It should be noted that the dilution effect on the sample is not shown in the quantitative amount, because the amount of the internal standard was controlled. This can also be seen intuitively in Figure 3. The characteristic peak of the internal standard is relatively low and is not the characteristic peaks of the main crystal phases in the clinker overlap.

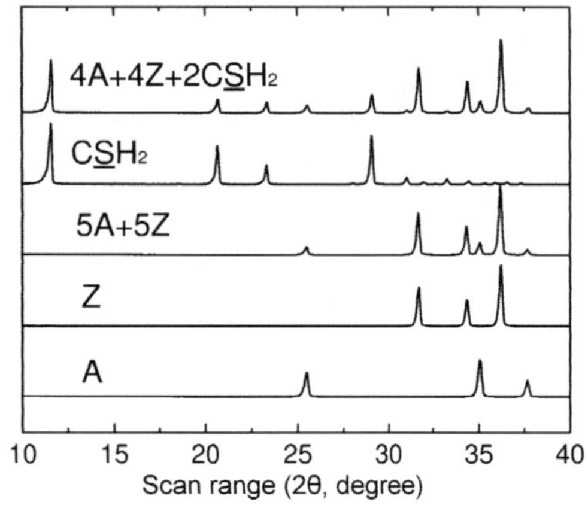

Fig. 2 XRD pattern of standards used and their mixtures

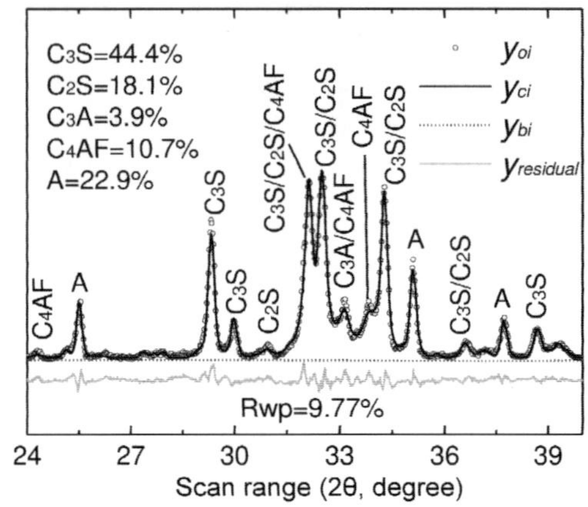

Fig. 3 Fitting example of the XRD pattern of C+20A

3.3. Internal standard method and G-value method for consistency of fly ash quantitative

Table 3. Quantitative results of fly ash by a variety of methods (Unit: wt.%)

	Content in FA MAC=48.52 cm2/g G=6.62E-11 cm5	Content in FA +10A MAC=46.83 cm2/g G=7.19E-11cm5	
Phases	GFM	GFM	ISM
Mullite	19.1	17.5	19.1
Quartz	7.9	8.2	8.9
Anhydrite	0.3	0.5	0.5
Hematite	1.7	1.4	1.6
Calcite	1.1	1.1	1.2
A	0.0	9.2[a]	26.2#
Am/Error	69.9	71.3	68.7

[a] stands for the content of A before corrected. 26.2# stands for the content of A by RPA

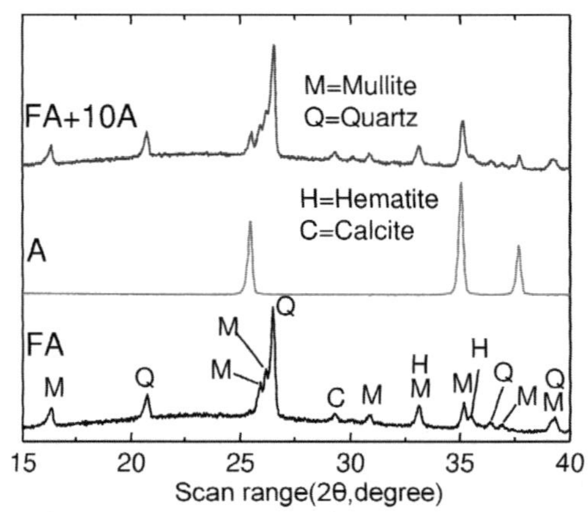

Fig. 4 XRD patterns of FA and FA+10A

Class I fly ash is low calcium fly ash, which is a vitreous ash-based material with a small amount of glass and it has been shown to have a glassy content of 66.6%-86% [8, 19, 20]; according to the analysis of 1.2.2, it is determined that the phase content of 10% alumina is calibrated and compared with the quantitative results without the standard system, the results are shown in Table 4. It can be seen that in both systems, the results of the G-value method (FA system) and the internal standard method (FA+10A system) are in good agreement, and the quantitative results of the crystal phase and the amorphous phase are absolutely deviated from each other within 1.2%; within the FA+10A system, due to the accumulation of errors, the maximum deviation of 2.6% of the phase content determined by the internal standard method and the G-value method still occurs in the amorphous content, which is different from the mullite in the fly ash. The apportioned deviation is 1.6%, the most likely reason is that the characteristic peak of the internal standard alumina at about 35.1 ° coincides with the characteristic peak of mullite (seen Figure 4), so a certain fitting error is caused, which can also be seen by the G-value method quantitative alumina content in the system, and the absolute deviation from the actual content is 0.8%.

Another interesting phenomenon is that regardless of the standard mixture system, cement clinker or fly ash, when a certain standard is added, the content of each crystal phase calibrated by G-value method is generally slightly smaller than the content calibrated by the internal standard method, this is a systematic reduction and the proportion of reduction is also close, and the reduction is gradually increasing from the standard system, clinker system to fly ash system; according to formula (14), it can be seen that there are two reasons for the reduction of this system, one is the decrease of the absorption coefficient of the sample mass, and the other is the increase of the G value; the possibility of the first cause should be smaller, because the larger value of the deviation (2.6%, FA+10A) does not appear in the system with a large change in mass absorption coefficient (C+20A, Total MAC values as shown in Table 2 and Table 3), and within the same system, the mass absorption coefficient is basically stable; the change in G value is the main reason. It can be seen that the G value of the two tests (shown in Table 2 and Table 3, the time interval between the two tests is 10 days) is still very different. In addition, in this paper, the order of sampling is, firstly, the standard mixed sample, next, the cement clinker, and then, the fly ash, which coincides with the above-mentioned decreasing order, indicating that the true G value gradually decreases with time. And the quantification is still using the first calibration of the G value, resulting in a systematic reduction in the quantitative results. In fact, Jansen [15] did find that the G value decreases over time when studying the G-value method. In view of this, it is necessary to shorten the G-value measurement time and the sample test time as much as possible when using the G-value method.

4. Conclusion

1) It is necessary to estimate the approximate content of the amorphous phase in the sample, which is helpful to select the appropriate calibration method and internal standard dosage to minimize the experimental error; For samples with an amorphous phase content of less than 10%, a G-value is recommended; When the amorphous phase content is 10%-50%, if the internal standard method is used, it is recommended that the internal standard content is 20%;The amorphous phase content is greater than 50%, and the recommended internal standard content is 10%.

2) For the standard sample mixed sample, the G-value method yields a fairly accurate quantitative result; in the binary standard mixed sample, the internal standard method will cause a large error of the overall system due to its small quantitative error and all passed to other phases; When the number of standard phases in the mixture increases, the systematic error caused by the internal standard will decrease due to the apportionment; In the ternary standard mixture system, both methods obtained consistent and accurate quantitative results.

3) For polycrystalline systems such as cement clinker and fly ash, both the internal standard method and the G-value method can obtain satisfactory quantitative results, and the quantitative results of the two are consistent, because more crystal phases can share the cumulative error.

4) It should be noted that the sample test interval needs to be shortened as much as possible, when using the G-value method, to avoid the actual G-value reduction, which would result in the quantify of the crystal phase content to be small and cause an increase in cumulative error or the amorphous phase content.

Acknowledgments

The authors would like to acknowledge the financial support for this study from the National Natural Science Foundation of China (51468002) and the National Key Research and Development Program of China (No. 2016YFC0700807).

References

[1] Bensted J and Barnes P 2008 *Structure and Performance of Cement (*London: Spon Press)
[2] Young R A. 1993 *The Rietveld Method* (New York: Oxford University Press)

IOP Conf. Series: Materials Science and Engineering **542** (2019) 012069 doi:10.1088/1757-899X/542/1/012069

[3] Ma L D 2004 *Morden X-Ray Diffraction of Polycrystal - Experimental Techniques and Data Analysis* (Beijing: Chemical Industry Press) p 399-435, 482-486.
[4] Taylor J C 1990 *Powder Diffr* **62** 9
[5] De la Torre Á G, De Vera R N and Cuberos A J M 2008 *Cem. Concr. Res.* **38** 1261
[6] De La Torre A G, Bruque S and Campob J 2002 *Cem. Concr. Res.* **32** 1347
[7] Le Saoût G, Kocaba V and Scrivener K 2010 *Cem. Concr. Res.*
[8] Korpa A, Kowald T and Trettin R 2009 *Cem. Concr. Res.* **39** 69
[9] Yao W, Wei Y Q and Wang W 2012 J. *Building Materials* **05** 58
[10] De La Torre A G, Bruque S and Aranda M A G 2001 *Journal of Applied Crystallography* **34** 19
[11] Fernandez-Jimenez A, de la Torre A G and Palomo A 2006 *Fuel* **85** 1960
[12] Connor B and Raven M 1988 *Powder Diffr.* **3** 2
[13] Jansen D, Goetz-Neunhoeffer F and Stabler C 2011 Cem. Concr. Res. **41** 602
[14] Jansen D, Goetz-Neunhoeffer F, Lothenbach B 2012 Cem. Concr. Res. **42** 134
[15] Jansen D, Bergold S T, Goetz-Neunhoeffer F 2011 J. Appl. Cryst. **44** 895
[16] Jansen D, Stabler C, Goetz-Neunhoeffer F 2011 Powder Diffraction **26** 31
[17] Prince E 2004 *International Tables for Crystallography: Vol. C: Mathematical, Physical and Chemical Tables* (Spinger)
[18] Wei Y Q 2016 *Quantitative Characterization and Evaluation of Hydration Progress of Composite Cementitious Systems* (Shanghai: Tongji University Press)
[19] Chancey R T, Stutzman P, Juenger M C G 2010 *Cem. Concr. Res.* **40** 146
[20] Taylor H F W 1997 *Cement Chemistry* (London: Thomas Telford Publishing)

IOP Conf. Series: Materials Science and Engineering **542** (2019) 012070 doi:10.1088/1757-899X/542/1/012070

Preparation and characterization of high-strength low-thermal-conductivity cement-based insulation materials

X X Zhang, Z P Wang*, A M She, Y Q Wei and Z Y Zhu

Key Laboratory of Advanced Civil Engineering Materials of Ministry of Education, Tongji University, Shanghai, China
E-mail: zxxlucky@126.com

Abstract. Thermal insulation materials with high strength and ultra-low thermal conductivity are ideal construction ones for residential infrastructure. However, high strength of materials regularly means high thermal conductivity. To solve this problem, a novel thermal insulation material was prepared by using cement as the binder, silica fume as the mineral admixture, aerogel as the coarse filler and hollow SiO_2 microsphere as the fine filler. The optimal proportion and preparation process were investigated by means of the orthogonal experiment and the single factor experiment method, respectively. On the other hand, their microstructures, in terms of aerogel particle size distribution and multi-scale composite structure were analysed by optical microscope, binarization image, scanning electron microscope. The results indicated that compared to conventional insulation materials, the produced materials in this study behaved much better performance in strength. The dry density, compressive strength and thermal conductivity reached 360 kg/m^3, 4.55 MPa and 0.055 W/m·K, respectively. The incorporation of hollow SiO_2 microsphere and silica fume led to the much thicker pore walls of the sample, and simultaneously the prolonging of stirring time resulted in the much smaller size of aerogel, which were considered as the two primary causes of its excellent mechanical performance.

1. Introduction

Lightweight cement-based insulation materials are manufactured by introducing lightweight fillers or foam into cement. Because of their advantages of superior durability, fire-resistant and safety, they are extensively used in thermal insulation projects[1, 2]. In lightweight cement-based insulation materials, low thermal conductivity depends on high porosity and low density, but high strength depends on low porosity and high density[3-5]. Therefore, it is difficult to prepare a cement-based insulation material with low dry density, low thermal conductivity, and high strength[6, 7]. The pore and pore wall structure are the main factors affecting compressive strength and thermal conductivity[8, 9]. Studies have shown that under the same porosity, small and uniform pore is beneficial to increase compressive strength[5]. The pore structure can be controlled by increasing the viscosity of the slurry[10], prolonging the stirring time, and accelerating the stirring rate. In addition, thick and compact pore wall can improve the mechanical properties of the insulation materials. Thick pore walls can be obtained by filling smaller high strength hollow particles between larger ones[11]. Compact pore walls can be developed by using high strength cement, low w/c ratio and large dosage of mineral admixture[12]. As a mineral admixture, silica fume is usually used in cement-based insulation materials in an amount of 5% to 20%, which has the function of assisting gelation and filling fine pores[13-18]. Aerogel has a significant reduction effect on thermal conductivity of thermal insulation mortar[19, 20]. Hollow SiO_2 microsphere has lightweight and high strength characteristic properties in thermal insulation mortar[21-25]. In this study, cement-based insulation materials with low dry density, low thermal conductivity and high-strength were prepared. Silica fume (content >60%) was used as mineral admixture, aerogel and hollow SiO_2 microspheres were used as porous lightweight fillers. An ultra-low w/c ratio was chosen to improve compressive strength. The effects of raw material composition and stirring time on the properties of thermal insulation materials were analysed, and the relationship between microscope morphology and macroscopic properties was studied.

2. Materials and methods

2.1. Materials and preparation

PII 52.5 cement: average particle size was 18 μm. Silica fume: average particle size was 0.3 μm. Hollow SiO_2 microspheres: average particle size was 25 μm, bulk density was 250 kg/m^3, compressive strength was 80 MPa, thermal conductivity was 0.05 W/m·K. Aerogel: average particle size was 2090 μm, bulk density was 75 kg/m^3, thermal conductivity was 0.02 W/m·K. Superplasticizer: water reduction rate was 28%, dosage was 5% (by cement mass). The raw materials particle size distribution and microscope morphology were shown in figure 1 and figure 2.

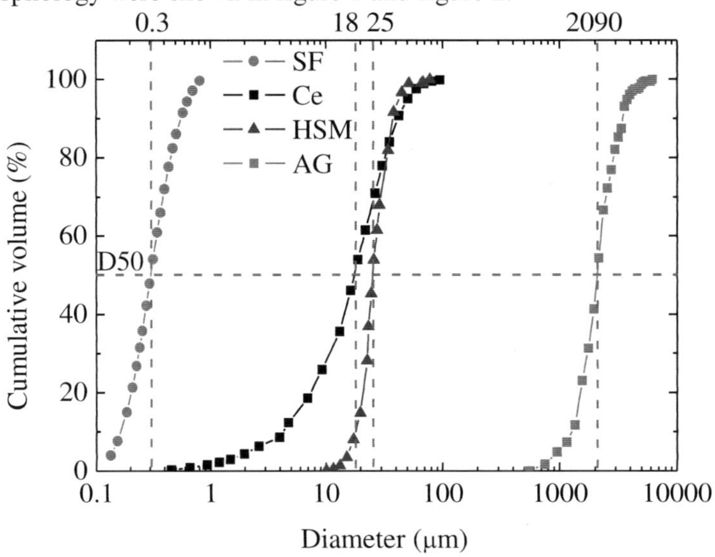

Figure 1. Particle size distributions of raw materials with Ce=cement, SF=silica fume, HSM=hollow SiO_2 microsphere, AG=aerogel.

Figure 2. Microscope morphology of raw materials with (a) Ce, (b) SF, (c) HSM, (d) AG.

2.2. Test methods

2.2.1. Sample preparation. Cement, silica fume and pre-wet hollow SiO_2 microspheres were added to the stirring pot for dry mixing for 30 s. Water (containing superplasticizer) was added to the dry mixture for stirring slowly for 60 s, and stirring quickly for 60 s. The aerogel was added to the slurry and stirred quickly for the corresponding time. The prepared slurry was poured in a mold and covered with a plastic film. After 48 h, specimens were released from the molds and cured at 20±2 °C and 75% RH for 28d.

The optimal recipe was explored by orthogonal experiment. The factors and levels of the orthogonal experiment were shown in table 1. The stirring time after the addition of the aerogel in the orthogonal experiment was 90 s.

Table 1. Orthogonal experimental factor and level.

Level	A SF/Ce /(wt. [a])	B HSM/Ce /(wt. [a])	C AG/Ce /(wt. [a])	D W [b]/Ce /(wt. [a])
1	0.6	1.25	0.45	2.18
2	1	1.75	0.6	2.26
3	1.4	2.25	0.75	2.34

[a] Weight.
[b] Water.

2.2.2. Performance Testing. Dry density, compressive strength and thermal conductivity of samples were tested according to Chinese national standard GB/T5486-2008, GB/T5486-2008 and GB/T10294-2008, respectively. The aerogel in the thermal insulation materials was analysed by MF-UA1010D optical microscope. The microscope morphology of the thermal insulation materials was analysed by Quanta 200F scanning electron microscope (SEM).

3. Results and discussion

3.1. Effect of raw material composition on macroscopic performance

The orthogonal test results and analysis of the cement-based insulation materials were shown in table 2 and table 3. It can be seen from table 3 that the optimal recipe under the dry density index was $A_1B_1C_3D_3$, and the main factors affecting the dry density were successively: aerogel> silica fume> water> hollow SiO_2 microspheres. The optimal recipe under the compressive strength index was $A_1B_3C_1D_1$, and the main factors affecting compressive strength were successively: silica fume> aerogel> water> hollow SiO_2 microspheres. The optimal recipe under the thermal conductivity index was $A_1B_1C_3D_3$, and the main factors affecting thermal conductivity were successively: aerogel> silica fume> water> hollow SiO_2 microspheres.

Table 2. Orthogonal experimental results of cement-based insulations.

Sample No.	ρ^{a} /(kg/m^3)	$f_{28d}{}^{b}$ /MPa	λ^{c} /(W/m·K)	f_{28d}/λ^{d} /(MPa/(W/m·K))
1	408	4.44	0.105	42.29
2	359	3.72	0.064	58.13
3	324	3.09	0.051	60.59
4	343	1.03	0.057	18.07
5	346	1.69	0.058	29.14
6	412	2.75	0.112	24.55
7	355	1.47	0.061	24.10
8	414	2.16	0.113	19.12
9	396	2.22	0.088	25.23

[a] Dry density.
[b] 28d compressive strength.
[c] Thermal conductivity.
[d] The ratio of 28d compressive strength to thermal conductivity.

Table 3. Orthogonal experimental analysis of cement-based insulations.

Effect	$k_1{}^{a}$	$k_2{}^{a}$	$k_3{}^{a}$	R^{b}	BR^{c}
ρ	364	367	388	25	A_1

IOP Conf. Series: Materials Science and Engineering **542** (2019) 012070 doi:10.1088/1757-899X/542/1/012070

/(kg/m^3)	369	373	377	9	B_1
	411	366	341	70	C_3
	383	375	360	23	D_3
	3.75	1.82	1.95	1.93	A_1
f_{28d}	2.31	2.52	2.69	0.38	B_3
/MPa	3.11	2.32	2.08	1.03	C_1
	2.78	2.65	2.09	0.69	D_1
	0.073	0.076	0.087	0.014	A_1
λ	0.074	0.078	0.084	0.009	B_1
/(W/m·K)	0.110	0.070	0.057	0.053	C_3
	0.084	0.079	0.074	0.010	D_3

[a] k_i=the average value of i level in three experiments.
[b] Range.
[c] The best result.

The effect of the composition of the raw materials on the properties of the insulation materials was shown in figure 3. It can be seen from figure 3 that SF/Ce, HSM/Ce, AG/Ce and W/Ce had an important influence on the properties of the insulation materials. With the SF/Ce value increasing, the dry density and thermal conductivity of the thermal insulation materials increased, and the compressive strength firstly decreased and then increased. As the HSM/Ce value increased, the dry density, compressive strength and thermal conductivity of the insulation materials all increased slightly. Compared to other three factors, the factor of HSM/Ce had the minimum ranges: 9 kg/m^3, 0.38 MPa and 0.009 W/m·K. The possible reason was that the compressive strength and thermal conductivity of the hollow SiO_2 microspheres were similar to those of the prepared thermal insulation materials. The dry density, compressive strength and thermal conductivity all decreased greatly with the AG/Ce value increasing, which was due to the ultra-high porosity of aerogel. As the W/Ce value increased, the dry density, compressive strength, and thermal conductivity of the insulation materials all decreased.

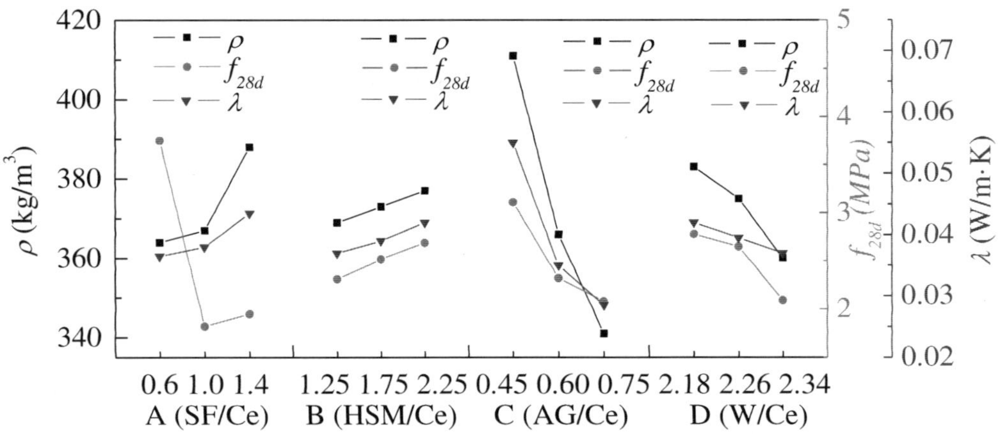

Figure 3. Effect of SF/Ce, HSM/Ce, AG/Ce, W/Ce on properties of thermal insulation materials.

The recipe and results of the optimal ratio under single index condition were shown in table 4. It can be seen from table 4 that the optimal recipe under different index had different properties. In order to obtain high-strength and low-thermal-conductivity insulation materials, the ratio of compressive strength to thermal conductivity (f_{28d}/λ) was introduced and used as an evaluation index. The f_{28d}/λ value reflects the compressive strength per unit of thermal insulation capacity. The f_{28d}/λ values in table 3 and table 4 ranged from 18.07 to 60.59 MPa/(W/m·K), among which the f_{28d}/λ value of sample No.3 was highest, being 60.59 MPa/(W/m·K). By Considering various factors, sample No.3 with the highest f_{28d}/λ value was the optimal ratio, being $A_1B_3C_3D_3$. The dry density was 320 kg/m^3, the compressive strength was 3.09 MPa and the thermal conductivity was 0.051 W/m·K.

Table 4. Optimal recipe and its results under single index condition.

Sample No.	A SF/Ce	B HSM/Ce	C AG/Ce	D W/Ce	Performance ρ /(kg/m^3)	f_{28d} /MPa	λ /(W/m·K)	f_{28d}/λ /(MPa/(W/m·K))
$A_1B_1C_3D_3$	0.6	1.25	0.75	1.21	297	1.52	0.048	31.67
$A_1B_3C_1D_1$	0.6	2.25	0.45	1.08	418	4.52	0.125	36.16
$A_1B_1C_3D_3$	0.6	1.25	0.75	1.21	297	1.52	0.048	31.67
3	0.6	2.25	0.75	1.21	324	3.09	0.051	60.59

3.2. Effect of stirring time on macroscopic performance

Using sample No.3, the effect of stirring time after the addition of aerogel on the performance of thermal insulation materials was studied, and the test results were shown in figure 4. As can be seen from figure 4, the dry density, compressive strength, thermal conductivity, and f_{28d}/λ value of the thermal insulation materials all increased with stirring time increasing. When the stirring time extended from 30 s to 120 s, the dry density varied from 277 kg/m^3 to 360 kg/m^3, increased by 30%; the compressive strength varied from 1.12 MPa to 4.55 MPa, increased by 300%; the thermal conductivity varied from 0.042 W/m·K to 0.055 W/m·K, increased by 31%; and the f_{28d}/λ value varied from 26.67 MPa/(W/m·K) to 82.73 MPa/(W/m·K), increased by 210%. It can be seen that as the stirring time increased, the dry density and thermal conductivity increased slightly, while the compressive strength and the f_{28d}/λ value increased significantly. In other words, prolonging stirring time was an effective method to improve compressive strength under the premise of ensuring low thermal conductivity. When the stirring time was 120 s, the thermal insulation materials had the highest f_{28d}/λ value. The dry density was 360 kg/m^3, the thermal conductivity was 0.055 W/m·K, and the compressive strength was 4.55 MPa.

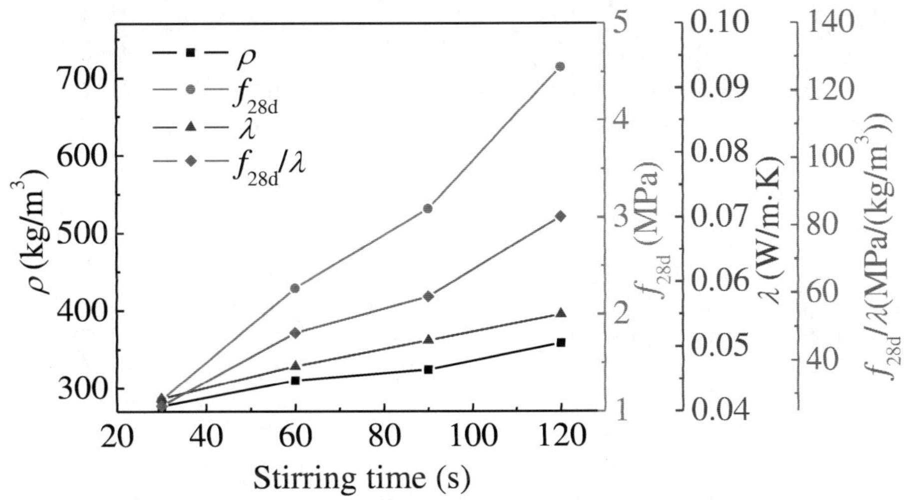

Figure 4. Effect of stirring time after adding AG on properties of thermal insulation materials.

3.3. Microstructure

The average particle size of the aerogel and hollow SiO_2 microspheres used in the work were 2090 μm and 25 μm. In order to investigate the distribution of porous lightweight fillers under different scales in the thermal insulation materials, optical microscope and SEM analyses of thermal insulation materials cross-section were carried out.

3.3.1. Aerogel morphology and particle size distribution in thermal insulation materials. The optical micrograph of the thermal insulation materials cross-section was binarized by image software, the pores left by the aerogel and the surrounding matrix were indicated in black and white respectively, as shown in figure 5. The stirring time had an important impact on the size, shape and position distribution of the aerogel particles in the thermal insulation materials. When the stirring time extended from 30s to 120s, the size of the aerogel became smaller, the edge of the aerogel varied from smooth to rough, and the position distribution of aerogel varied from large particles evenly distributing to small particles filling large particles' gap. These three changes were respectively beneficial to reduce stress concentration phenomenon, enhance interfacial adhesion, and improve particle gradation, thereby increasing the compressive strength.

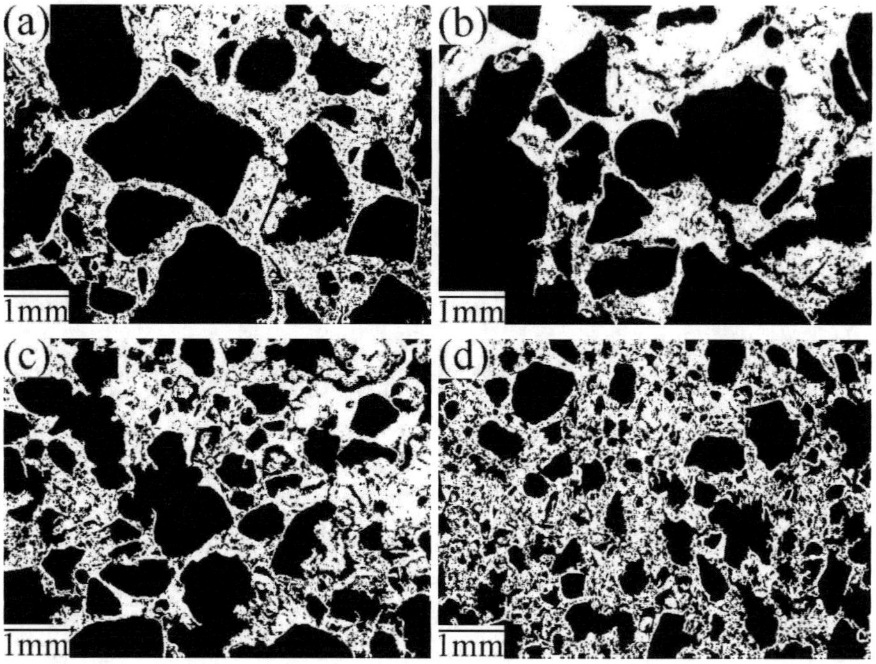

Figure 5. Binary graph of insulation materials cross-section with (a) stirring 30s, (b) stirring 60s, (c) stirring 90s, (d) stirring 120s.

The particle size analysis software was used to obtain the aerogel particle size, and its distribution was plotted as shown in figure 6. The stirring time had an important influence on the particle size distribution of the aerogel. When the stirring time extended from 30 s to 120 s, the average particle size of aerogel was reduced from 0.67 mm to 0.21 mm with a decrease of 69%, and the particle size distribution was changed from 45% of 0-0.6 mm, 38% of 0.6-1.2 mm, 17% of 1.2-1.8 mm, to 93% of 0-0.6 mm, 7% of 0.6-1.2 mm, 0% of 1.2-1.8mm. Among them, the particle numbers at size of 0~0.6 mm was increased by 48%, those at size of 0.6-1.2 mm and 1.2-1.8 mm were decreased by 31% and 17% respectively. The particle size distribution of aerogel concentrated to 0-0.6 mm. When the stirring time extended from 30 s to 120 s, the f_{28d}/λ value of the insulation materials increased from 26.67 MPa/(W/m·K) to 82.73 MPa/(W/m·K). It can be seen that prolonging stirring time made the aerogel particle size smaller, and the distribution gathers towards 0-0.6 mm, which eventually led to an increase of f_{28d}/λ value.

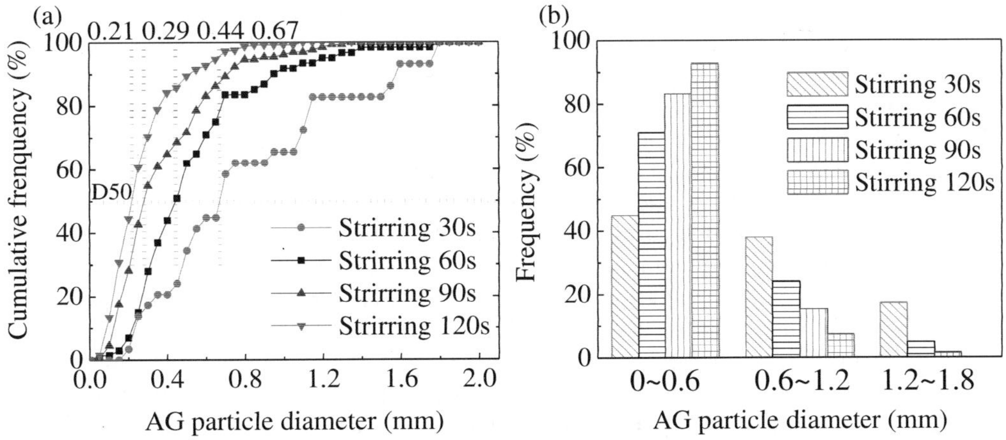

Figure 6. Aerogel particle size distribution in thermal insulation materials with (a) cumulative frequency distributions and (b) frequency distributions.

3.3.2. Cross-section microstructure. The SEM images of the cross-section of the thermal insulation materials were shown in figure 7(a). Figure 7(b), (c), and (d) were enlarged images of the square boxes in figure 7(a), (b), and (c), respectively.

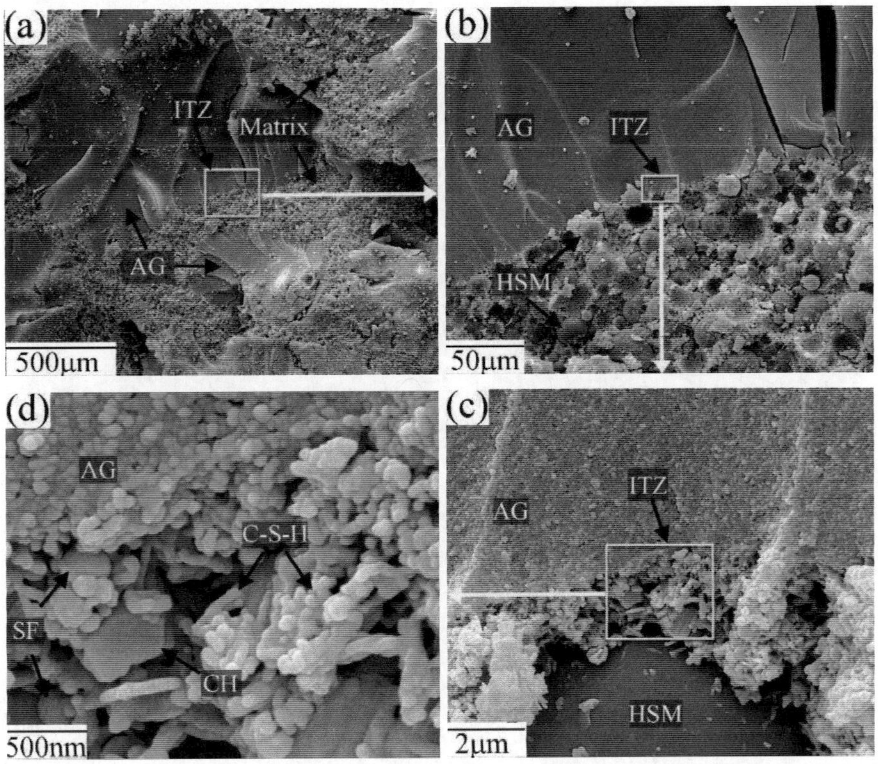

Figure 7. Multi-scale microstructure of thermal insulation materials.

As can be seen from figure 7(a), the aerogel (AG, smooth surface) distributed evenly in the matrix (Matrix, rough surface), and the interfacial transition zone (ITZ) of aerogel and matrix had no macroscopic crack and was linked tightly and closely. This may be one of the reasons for the high mechanical properties of thermal insulation materials. Combining figure 7(b) and (c), it can be seen that the matrix consisted of hollow SiO_2 microspheres (HSM) and cement hydration products, and the hollow SiO_2 microspheres were embedded in the hydration products tightly. Aerogels can be regarded as pores because of their high porosity, and matrix composed of hollow SiO_2 microspheres and hydration product can be viewed as compact pore wall (i.e. bubble spacing). Studies have shown that filling large pore gaps with small fillers can widen the bubble spacing and thicken the pore walls, thus increasing the compressive strength[11]. It can be reasonably assumed that the presence of the hollow SiO_2 microspheres herein increased the distance between the aerogels, thickened the pore walls, thereby improving the mechanical properties of the thermal insulation materials. It can be seen from figure 7(d) that the cement hydration products were mainly cigar-like C-S-H gels and micro-nano scale $Ca(OH)_2$ crystals; unhydrated silica fume particles presented in the interval of hydrated products. C-S-H gels were the main source of cement stone strength, smaller size $Ca(OH)_2$ crystals were beneficial to reduce the loss of strength, and unhydrated silica fume particles filling micro-nano pores was beneficial to improve the compactness of the pore walls. These three aspects reasons on the micro-nano scale led to better mechanical properties of the thermal insulation materials. The C-S-H gels bonded different sizes of aerogel and hollow SiO_2 microspheres to form a multi-scale microstructure: the aerogel and hollow SiO_2 microspheres particle size mainly distributed around 210m and 25m, despite the particle size distribution of overall was wide, that of each scale was concentrated. A good particle gradation was obtained by controlling particle size distribution, then a lightweight and high-strength insulation material could be achieved.

4. Conclusions

In this study, the effect of raw materials composition and stirring processing on the properties of cement-based thermal insulation materials was investigated. The microstructure of the thermal insulation materials was analysed. The results can be summarized as follows:

(1) The optimal material ratio obtained by orthogonal experiment was cement: silica fume: hollow SiO_2 microsphere: aerogel: superplasticizer: water = 1:0.6:2.25:0.75:0.05:2.34. The main factors affecting dry density, compressive strength and thermal conductivity were successively aerogel, silica fume and water content. The optimal stirring time after adding the aerogel was 120 s. The prepared thermal insulation material had a dry density of 360 kg/m^3, a compressive strength of 4.55 MPa, and a thermal conductivity of 0.055 W/m·K.

(2) The stirring time had the positive influence on compressive strength. When the stirring time extended from 30 s to 120 s, the aerogel average size was decreased by 69%, which led to a 300% increase in compressive strength.

(3) The produced thermal insulation material consisted of aerogel, hollow SiO_2 microspheres, cement hydration products, and unhydrated silica fume. The hydration product bonded porous fillers with different sizes to form a multi-scale microstructure. And by controlling the sizes of different particles, particle gradation could be properly improved.

Acknowledgements

The authors would like to acknowledge the financial support for this study from the National Natural Science Foundation of China (No.51468002) and the National Key Research and Development Program of China (No.2016YFC0700807).

References

[1] Samson G, Phelipot-Mardelé A and Lanos C, 2017 *Mag Concrete Res* **69** 1-16
[2] Aditya L, Mahlia T M I, Rismanchi B, Ng H M, Hasan M H, Metselaar H S C, Muraza O and Aditiya H B, 2017 *Renew Sust Energ Rev* **73** 1352-65
[3] Ramamurthy K, Nambiar E K K and Ranjani G I S, 2009 *Cement Concrete Comp* **31** 388-396
[4] Wei S, Chen Y, Zhang Y and Jones M R, 2013 *Constr Build Mater* **47** 1278-91
[5] Zeng Q, Mao T, Li H and Peng Y, 2018 *Energ Buildings* **174** 97-110
[6] Chen B and Liu N, 2013 *Constr Build Mater* **44** 691-698
[7] Jones M R, Ozlutas K and Li Z, 2017 *Mag Concrete Res* 1-11
[8] Nambiar E K K and Ramamurthy K, 2007 *Cement Concrete Res* **37** 221-230
[9] Kearsley E P and Wainwright P J, 2002 *Cement Concrete Res* **32** 233-239
[10] Jiang J, Lu Z, Niu Y, Li J and Zhang Y, 2016 *Constr Build Mater* **107** 181-190
[11] Shao N N, Liu Z, Xu Y Y, Kong F L and Wang D M, 2015 *Mater Lett* **161** 451-454

IOP Conf. Series: Materials Science and Engineering **542** (2019) 012070 doi:10.1088/1757-899X/542/1/012070

[12] Shi C, Wu Z, Xiao J, Wang D, Huang Z and Fang Z, 2015 *Constr Build Mater* **101** 741-751
[13] Nambiar E K K and Ramamurthy K, 2006 *Cement Concrete Comp* **28** 475-480
[14] Nambiar E K K and Ramamurthy K, 2006 *Cement Concrete Comp* **28** 752-760
[15] Reisi M, Dadvar S A and Sharif A, 2017 *Mag Concrete Res* **69** 1-13
[16] Koksal F, Gencel O and Kaya M, 2015 *Constr Build Mater* **88** 175-187
[17] Babu K G and Babu D S, 2003 *Cement Concrete Res* **33** 755-762
[18] Demirboğa R and Gül R, 2003 *Cement Concrete Res* **33** 723-727
[19] Jia G, Li Z, Liu P and Jing Q, 2018 *Journal of Non Crystalline Solids* **482** 192-202
[20] Jia G, Li Z, Liu P and Jing Q, 2017 *Mag Concrete Res* 1-54
[21] Zhang Q and Li V C, 2015 *Cement Concrete Comp* **60** 10-16
[22] Yun T S, Jeong Y J, Han T S and Youm K S, 2013 *Energ Buildings* **61** 125-132
[23] Oreshkin D, Semenov V and Rozovskaya T, 2016 *Procedia Engineering* **153** 638-643
[24] Gao T, Jelle B P, Sandberg L I and Gustavsen A, 2013 *Acs Appl Mater Interfaces* **5** 761
[25] Ruckdeschel P, Kemnitzer T W, Nutz F A, Senker J and Retsch M, 2015 *Nanoscale* **7** 10059-70

IOP Conf. Series: Materials Science and Engineering **542** (2019) 012071 doi:10.1088/1757-899X/542/1/012071

Study of zinc coating thickness gauge by Monte Carlo simulation

W Xu, J Y Kong, H G Liu and X D Wang

Key Laboratory of Metallurgical Equipment & Control Technology, Wuhan University of Science and Technology, Wuhan, China
E-mail: 18793815@qq.com

Abstract. This paper introduces how to utilize the Monte Carlo N-Particle Transport Code (MCNP) in calculating the thickness of zinc coating which distributed homogeneously in an analyzed specimen. Experiments were performed on the specimens that made of 2 mm thickness steel substrates which covered by intermediate thickness (40-175 gm^{-2}) zinc layer, with the purpose of analyzing the x-ray fluorescence spectra and comparing the changes of zinc fluorescence detection efficiency under different energy monochrome photons or different detector distance. The results show that when the energy of monochrome photons is equal to 10 keV as well as the distance of detector is equal to 4 cm, the fluorescence of zinc Kα has a very good linear relationship with the thickness of zinc coating. It indicates that the MCNP code can be applied for simulating the galvanizing thickness gauge. In addition, under the same other conditions, the experimental results were not affected by different incident spot sizes (0.01-1cm).

1. Introduction

In recent years, studies on the control of zinc coating thickness have been primarily focused on two areas: process control, and modeling and prediction. Studies on process control have investigated many aspects, including improvements to the air knife system[1], optimization of air knife control parameters[2], composition of the zinc bath[3], dipping duration[4], and withdrawal speed[5], while studies on modeling and prediction have investigated methods for the prediction and control of zinc coating thickness through mathematical models developed based on production line parameters[6][7]. However, from the perspective of automatic control of zinc coating thickness, the measurement of zinc coating thickness is a prerequisite and key factor in process control. Automatic thickness control can only be realized with the acquisition of real-time zinc coating thickness data through zinc coating thickness gauges and joint involvement in feedback control with the air knife system.

X-ray fluorescence (XRF) is a method of measurement in online zinc coating thickness gauges used in large-scale integrated steel corporations in China[8]. In thickness gauges that operate based on this method, the x-ray source, ionization chamber, high-pressure system, and cooling system are contained within a measuring head[9]. However, as x-rays have high penetrating power, when the conventional XRF measurement method is used, a simultaneous entry of fluorescence emitted by the galvanized steel sheet and scattered x-rays with a much higher energy than the emitted fluorescence into the measuring head[10–11]. This results in an elevated baseline in the measurement system and reduced measurement precision and accuracy[12]. Therefore, the development of novel x-ray fluorescence analytical techniques is necessary to meet the testing requirements for galvanized steel sheets.

2. Experimental

IOP Conf. Series: Materials Science and Engineering **542** (2019) 012071 doi:10.1088/1757-899X/542/1/012071

The Monte Carlo Neutron and Photon Transport (MCNP) Code[13] is a large-scale multipurpose Monte Carlo code developed by Los Alamos National Laboratory in the United States for the resolution of particle transport problems. By using the MCNP code, measurement model for the application of XRF in measurements of zinc coating thickness can be established.

Figure 1 is a schematic diagram of the measurement model. The base material was a stainless steel cylinder with a height of 0.2 cm and a radius of 15 cm, and its upper surface was located on the XZ plane. The zinc layer was homogeneously distributed over the base material, which consists of pure zinc. An annular ionization chamber filled with xenon gas was arranged above the galvanized steel plate h cm as a detector. The inside and outside diameters of ionization chamber were 2 cm and 40 cm respectively, and the height was 14 cm. The bottom of the ionization chamber near the galvanized steel plate end was 0.01 cm thick beryllium window. The x-ray source was monochromatic energy photons, and irradiates the galvanized layer with a conical shape. The radius of the beam spot was L/2.

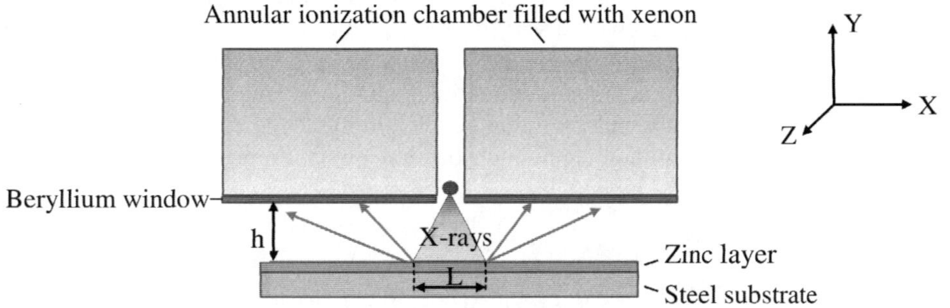

Figure 1. Schematic drawing of zinc thickness gauge.

The annular ionization chamber can effectively increase the window area of the detector, improve the detection efficiency as well as reduce the statistical error. It measures the zinc coating by collecting the fluorescence stimulated by x-ray beam from the source. After these x-rays are absorbed by the zinc coating atom, photons with certain energy are released from the atom. Due to the different thickness of zinc coating, the number of photons reflected back varies.

The simulation space was filled with air, which is composed of the elements C, N, O and Ar. The substrate is composed of the elements Si, Cr, Mn, Fe, Ni and Mo. To achieve a good approximation of actual conditions in the simulation, the concepts of ZAIDs (obtained by combining the atomic number and atomic weight) and atomic fractions (respective proportions of elements, with minus sign prefixes as required by the program)is introduced in the model construction process to indicate the actual material composition and proportions of the respective elements[14]. Table 1 shows the ZAIDs, atomic fractions and characteristic K x-ray energy of main elements. The Zn is the main target element of zinc coating measurement.

Table 1. ZAIDs, atomic fractions and characteristic K x-ray energy of main elements in analyzed specimen.

Material	Element	ZAID	Atom fraction	Characteristic Kα x-ray energy (keV)
Steel substrate	Si	14028	-0.01	1.740
	Cr	24052	-0.17	5.411
	Mn	25055	-0.02	5.894
	Fe	26056	-0.655	6.399
	Ni	28059	-0.12	7.471
	Mo	42096	-0.025	17.443
Coating layer	Zn	30066	-1.0	8.630
Air	C	6012	-0.000124	0.282
	N	7014	-0.755268	0.392

IOP Conf. Series: Materials Science and Engineering **542** (2019) 012071 doi:10.1088/1757-899X/542/1/012071

	O	8016	-0.231781	0.523
	Ar	18040	-0.012827	2.956

3. Results and discussion

3.1 Incident beam energy

Firstly, 30 keV monochrome photons were selected as the incident beam, and the half taper angle was 30°. The thickness of zinc coating was 10 μm. The tally type F1:P[15] was divided into 145 intervals according to the interval of 0.2 keV in the range of 1-30 keV. The calculated value represents the photons number of x-ray in each interval. The calculated results of MCNP are normalized to each source particle. The number of photons is counted by energy binning, which is similar to multichannel energy spectrum statistics.

Figure 2 shows the x-ray energy spectrum of zinc fluorescence which are excited by the incident beam of 30 keV and then collected by the beryllium window. The fluorescence energy of each element is consistent with the characteristic x-ray energy calculated in the table 1. Element Zn has the highest fluorescence intensity, and Zn Kα peak is about 7 times stronger than Zn Kβ peak. Iron and molybdenum in the substrate were also excited, indicating that the incident beam has penetrated the zinc coating. In addition to the photoelectric absorption effect, some photons generate Rayleigh and Compton scattering.

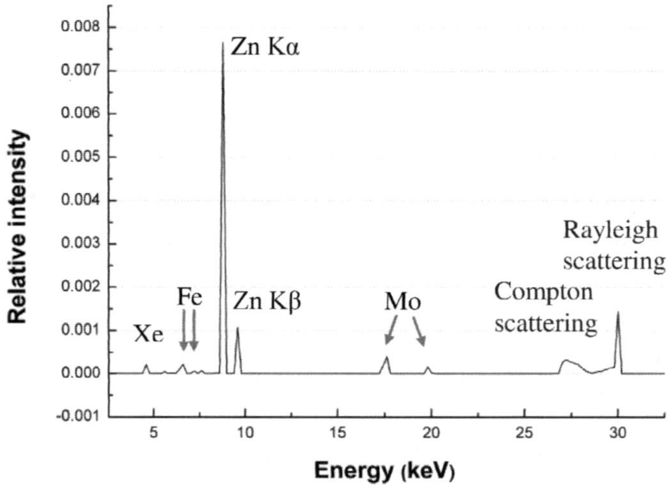

Figure 2. 30 keV energy spectrogram.

After confirming that the measurement model can effectively stimulate the fluorescence of zinc element and penetrate the coating, the most suitable source energy should be found. The characteristic x-ray energy of Zn Kα is 8.63 keV. Therefore, the energy of incident monochromatic photons must be greater than this value. For the convenience of calculation, the energy of the smallest monochrome photons is set to 10 keV. Figure 3 is a comparison of energy spectra of 10 keV, 20 keV, 30 keV and 40 keV. Figure 3 shows that the fluorescence intensity of Zn Kα produced by 10 keV monochrome photons is the highest.

With the increase of incident photons energy, the fluorescence of Zn Kα decreases gradually. This is due to more Rayleigh and Compton scattering of high energy incident photons. Therefore, the most suitable source energy in the measurement model should be 10 keV.

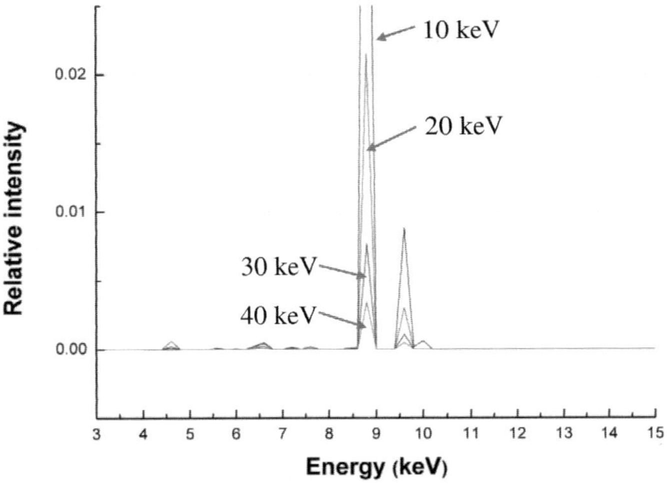

Figure 3. The comparison charts of Zn characteristic Kα fluorescence excited by 10-40 keV energy incident x-rays.

3.2 Detector distance

The distance (h) of detector is another important factor in the model construction process. Figure 4 is the contrast data of Zn Kα intensity collected by beryllium window at the distance of 1 cm to 10 cm respectively. The fluorescence intensity reaches the maximum when the detector distance is between 3 and 4 cm. When the detector is close to zinc coating, part of the fluorescence enter the blind area of the inner diameter of the detector and therefore is not collected. When the detector distance increases gradually, part of the fluorescence escape from the outside of the detector. Considering the actual production, the larger detector distance can effectively reduce the errors caused by the vibration of steel plate and the risk of instrument damage. Therefore, the most suitable detector distance in the measurement model should be 4 cm.

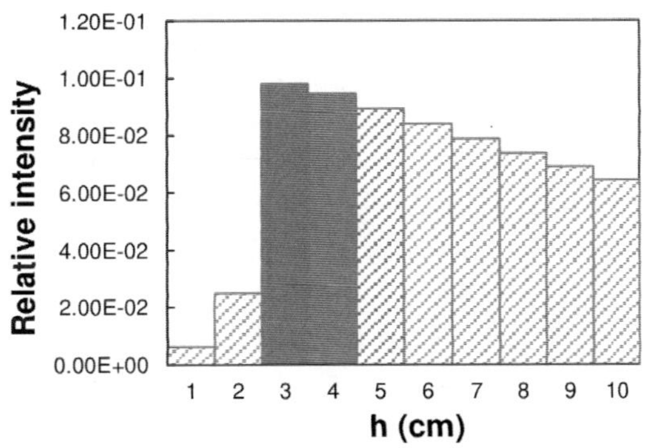

Figure 4. Comparison of fluorescence intensity at different detector distances (h).

3.3 Measurement curve

After the source energy and detector distance were determined, the zinc coating with intermediate thickness of 40-175 gm^{-2} was measured. Figure 5 (a) is the correlation between the thickness of zinc coating and the photon counting of Zn Kα. The linear correlation coefficient is 0.9997. The results show that under the above conditions, the thickness of zinc coating has a good linear relationship with the photon counting of Zn Kα. Figure 5 (b) is the comparative data of 20 keV monochrome photons. This is because the fluorescence of Zn Kα produced by high energy x-rays is less, and after passing through the thicker zinc coating, the fluorescence of Zn Kα is further reduced, so that the weak change trend could not be shown.

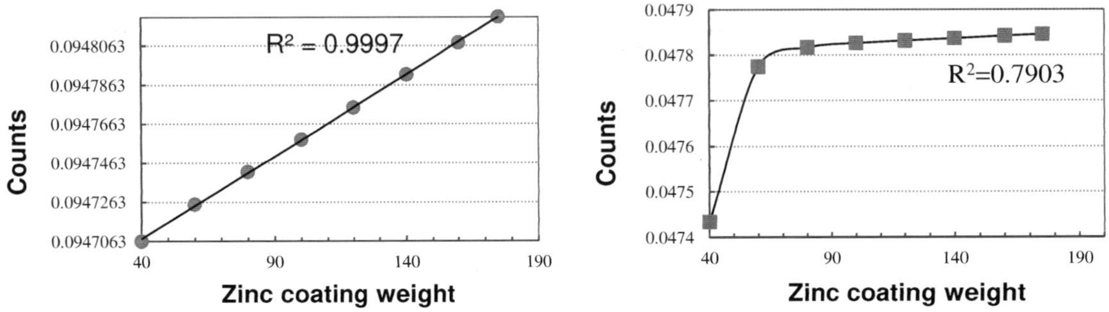

(a) 10 keV (b) 20 keV

Figure 5. Corresponding relationship between thickness of galvanized layers and photon counting of zinc Kα at h=4.

3.4 Incident beam spot size

Finally, in order to verify the influence of conical shape x-ray beam spot size on the results, the half taper angle was set to 26.6 ° and 88.855 ° (respectively corresponding spot radius L/2 is 0.01 cm and 1 cm) and carried out two groups of repetitive simulation experiments. The results are shown in table 2. As can be seen from table 2, when the spot size was between 0.01cm and 1cm, there was a good linear correlation between the thickness of zinc coating and Zn Kα fluorescence.

Table 2. Comparisons of linear correlation coefficients for different half taper angle. The zinc coating weight (d) was 40 gm^{-2}, 80 gm^{-2}, 120 gm^{-2} and 160 gm^{-2}, respectively.

Half taper angle (degree)	Number of zinc fluorescence photons				Linear correlation coefficient
	d=40	d=80	d=120	d=160	
26.6	0.0895045	0.0895292	0.0895513	0.0895761	0.9995
88.855	0.0001531	0.0001533	0.0001535	0.0001537	1

4. Conclusions

1) The MCNP model constructed in the present study passed 10 tally convergence checks, and all calculation errors were within the confidence interval. This shows that the tally process and results were true and effective.

2) Proposed physical model effectively simulates the XRF-based measurement process. It provides a new method for the design and simulation of zinc coating thickness gauge.

3) Under the condition that the source energy is 10 keV monochrome photons and the detector distance h=4, the measuring curves of zinc coatings on conventional steel sheets in the range of 40-175 gm^{-2} are obtained.

Acknowledgments

The authors would like to thank the anonymous reviewers for their valuable comments; Dr. Wenming Xia of the Department of Nuclear Science and Engineering, PLA Naval University of Engineering, for his assistance in the use of the MCNP program; and Dr. Yi Zhou of the College of Information Science and Engineering, Wuhan University of Science and Technology, for the care and support provided during the preparation of this paper.

This paper has been supported by the National Science Foundation of China [Grant No. 51174151], Hubei Province Major Science and Technology Innovation Project [2013AAA011], and the Key Technologies R & D Program of Hubei Province [Grant No. 2014BEC056].

References

[1] Dong Z D 2010 *Steel Roll.* **27** 56-8
[2] Zhao X and Zhao W H 2016 *Shanxi Metall.* **43** 83-5
[3] Li G, Guo J T, Liu Y S and Fu X T 2016 *Xinjiang Iron Steel* **38** 38-9
[4] Zhang S, Wang Y X and Zhang K 2016 *Plat. Finish.* **38** 40-3

[5] Jalel B N, Ali S, Chedly B and Foued H 2008 *Mater. Lett.* **62** 2150–2
[6] Chen C M, Lin J H, Hsu T W and Lin R R 2010 *Corros. Sci. Technol.* **9** 105–8
[7] Zhang Y, Shao F Q, Wu K K and Wang J S 2013 *J. Syst. Simul.* **25** 1060-4
[8] Wei W and Cheng G Y 2007 *Baosteel Technol.* **25** 54–7
[9] Ji X H, Wang D and Feng G 2012 *Sci. Technol. Baotou Steel* **38** 51–2
[10] An F L 1995 *Nucl. Electron. Detect. Technol.* **15** 224–7
[11] An F L 1997 *Nucl. Electron. Detect. Technol.* **17** 288–91
[12] Zheng Y C, Cen Y D and Tian RB 2012 *Electroplat. Pollut. Control* **32** 18–20
[13] Shultis J K and Faw E E 2011 *An MCNP Primer* [Online] pp 1-42
[14] Williams R G, Gesh C J and Pagh R T 2006 Compendium of material composition data for radiation transport modeling *Office of Scientific & Technical Information Technical Reports* (Washington: Pacific Northwest National Laboratory) pp 1-99
[15] Briesmeister J F 2000 *MCNPTM - A General Monte Carlo N-Particle Transport Code Version 4C* (New Mexico: Los Alamos National Laboratory) pp 1-38

IOP Conf. Series: Materials Science and Engineering **542** (2019) 012072 doi:10.1088/1757-899X/542/1/012072

Analysis of the causes of tube blocking in Jingbian gas field

Congdan Ye[1], Shunchao Gou[2], Yuhan Feng[2], Buneng Niu[2], Yun Ma[3,4]
1. College of Chemistry and Chemical Engineering, Xi'an Shiyou University, Xi'an, China
2. Production Logging Center, China Petroleum Logging Company Limited, Xi'an, China
3 College of Petroleum Engineering, Shaanxi Province Key Laboratory of Advanced Stimulation Technology for Oil & Gas Reservoirs, Xi'an Shiyou University, Xi'an, China
4. Engineering Research Center of Development and Management for Low to Ultra-Low Permeability Oil & Gas Reservoirs in West China, Ministry of Education, Xi'an, China
E-mail: mayun9401@xsyu.edu.cn

Abstract. A large number of blockage appeared in gas wells and pipeline appeared, which caused plugging, corrosion and the increasing of wellbore pressure difference, and seriously affect the normal production of gas well of problems. In this paper, the water quality over the years and natural gas production of the JB1# well in the Jingbian gas field were analyzed, also include the composition of blockage with the chemical volumetric method (CVM), X-Ray Diffraction (XRD) and x-ray fluorescence (XFS) method. Meanwhile, the blockage powder was leached in simulated acid work solution, and then the leaching solution was analyzed with CVM. The experimental results show that the produced water in the JB1# well has unstable water quality, high salinity, high Cl- content, and low pH value, high H_2S and CO_2 content in natural gas, so easy to produce corrosion and carbonate scaling. The main component of the blockage is a mixture of corrosion product Fe_2O_3 and a small amount of $CaCO_3$ scale. The corrosion product Fe_2O_3 is mainly generated by the mixture of dissolved oxygen, H_2S, CO_2 and other corrosion effects. The main cause of blockage in the JB1# well is the deposition of Fe_2O_3 and $CaCO_3$ scales caused by the corrosive effects of dissolved oxygen, H_2S and CO_2.

1. Introduction

With gas development entered the new stage, a mixing layer development had been more observed. The nature of the field produced water became more and more complicated, but due to a mount of scaling ion exists in produced water, when the changes in the conditions such as temperature, pressure and contact with the other horizon produced water, the scale ion would be easily deposited on the inner wall of the wellbore or the pipeline, formed some crystallization or non-crystallization precipitation. After a long time of accumulation, the phenomenon of plugging throttling, wellbore and pipeline would be seriously[1-2].At present, the common types of scaling blockages are mainly composite blockages caused by solid phase and corrosion inhibitor, sulfur deposition blockages, salt deposition blockages and hydrate blockages[3-4].At present, there are more than 100 Wells in Jingbian gas field, which seriously affected the normal production of gas wells due to the presence of a large number of blockage in trunk lines and pipelines, pipeline blockage and corrosion, pipeline perforation and equipment damage and failure, etc. All would result in the increase of pressure difference in the wellbore, the decrease of output, and even the shutdown of the well.

At present, scholars at home and abroad are mainly concerned with the study of scaling mechanism, influencing factors of scaling, prediction model of scaling trend and analysis of scaling rate in the process of gas field exploitation [5-6]. The severe cases of single well was selected as the research object, single well water quality of produced water was analyzed, also include the composition of blockage with the chemical volumetric method, x-Ray Diffraction (XRD) and x-ray fluorescence (XFS) method. Meanwhile, the blockage powder was leached in simulated acid work solution, and then the leaching solution was analyzed with the chemical volumetric method. On basis of these results, the causes of gas well blockage in Jingbian gas field was analyzed, which provide the basis for field implementation solution blocking and prevention, extend the service life of gas well and increase production capacity.

2 Materials and method

2.1 Material and medium

Reagents: petroleum ether, hydrochloric acid, sodium hydroxide, silver nitrate, barium chloride, sodium chloride are all analytically pure.

Instrument: ICP (HK-8100 inductively coupled plasma emission spectrometer, Beijing Huaco Yitong analytical instrument co., LTD.), XFS (XRF-1800, Shimadzu, Japan), XRD (X-ray diffraction D/MAX-2400, Science Corporation, Japan).

2.2 Analysis of produced water quality

According to the petroleum industry standard "Methods for Analysis of Oil and Gas Field Water" (SY/T 5523-2016)[7], using ICP (HK-8100 inductively coupled plasma emission spectrometer, Huaco Yitong, Beijing) to analysis of the quality of the produced water in the single well of JB1# in Jingbian gas field.

2.3 CVD and XFS analysis of blockage

The blockage was dried at 105 °C for 2 hours, and the water content was measured. The chemical composition of the blockage after drying was analyzed by XFS (XRF-1800, Shimadzu, Japan). After drying, the sample is finely ground in a mortar, and a certain amount is weighed and extracted with petroleum ether, and the solid phase is dried to determine the oil content.Then take a certain amount into a muffle furnace to calcine for 2 h at 550 °C and 950 °C respectively, and determine the calcination reduction of each temperature section. The blockage after calcination was dissolved in 1+1 HCl and filtered. The filtrate was made up to 500 mL, and the cation content was measured. The filter residue was used to determine the acid insoluble matter.

2.4 XRD analysis of blockage

The composition of the blockges was analyzed by XRD (X-ray diffraction D/MAX-2400, Science Corporation, Japan).

3 Results and discussion

3.1 Historical water quality and basic situation of detection well

The water quality composition of JB1# well and the basic situation of gas well production in 2018 are shown in Table 1 and Table 2.

Table 1 Compositions of the liquid phase Unit: ($mg \cdot L^{-1}$,except pH)

Date	pH	Cl^-	HCO_3^-	Mg^{2+}	∑Fe	Ca^{2+}	$\sum Na^+ + K^+$	salinity
Jul-15	5.64	38534.15	576.64	7106.54	306.65	10188.34	3754.41	56712.32
Mar-16	5.00	16941.56	80.11	0.00	16.55	6484.34	4011.62	23522.56
Aug-16	5.33	81031.61	298.39	2649.79	91.31	25120.14	25561.87	109191.24
Jan-17	5.60	29743.44	198.39	1674.96	29.60	9559.08	7399.75	41205.47
Aug-17	6.32	100646.10	410.05	4436.33	1619.08	32391.45	23352.89	139503.01

Table 2 Basic situation of JB1# wellbore production

Oil pressure (Mpa)	Casing pressure (Mpa)	Gas roduction ($10^4 m^3/d$)	Water production (m^3/d)	H_2S (mg/m^3)	CO_2 (%)	Total salinity (g/L)
3.20	8.15	3.50	1.49	1400.89	5.52	113.38

From the Table 1 and Table 2, the JB1# well has unstable water quality, low pH, high content of Cl^-, H_2S and CO_2 (5.52%), which is prone to corrosion and carbonate scaling.

3.2 CVD and XFS analysis of blockage

The original blockage of the JB1# well was dissolved in 1+1 HCl as it was, with less bubbles and a pungent smell of rotten eggs, indicating to a small amount of carbonate and a certain amount of H_2S. The appearance of the plug after calcination at 550 ° C and 950 ° C is as shown in Fig. 1 , and the appearance of the solution and the filtered acid insoluble matter after the plugged material is dissolved in 1+1 HCl after calcination is shown in Fig. 2 .

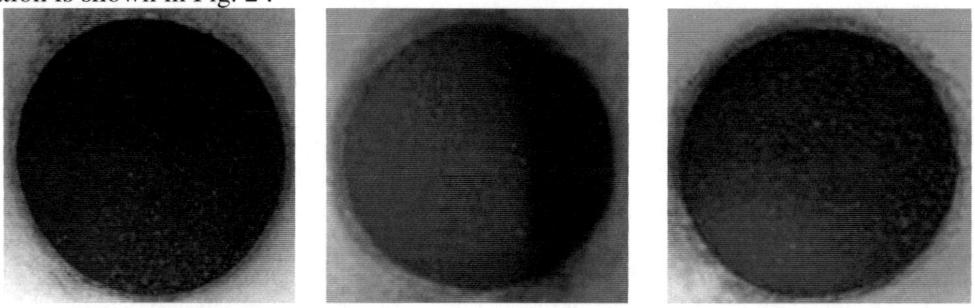

(a) Original blockage (b) 550°C (c)950°C

Fig.1 Comparison of appearance before and after calcination of blockages

(a) The solution after HCl dissolves (b) Appearance of HCl insoluble matter

Fig.2 Appearance of the solution after HCl dissolves and HCl insoluble matter

As shown in Fig. 1 and Fig. 2, both calcination and HCl dissolution have a red appearance, indicating that the plug may contain iron compounds. The original blockage contains a certain amount of H_2S, which may cause H_2S corrosion.

Chemical composition analysis of the blockage was conducted by CVM and XFS, the results are shown in Table 3 and Table 4.

Table 3 CVM analysis results of blockage Unit: w/w, %

indicators	water	organics	HCl insoluble matter	HCl soluble matter				
				Ca^{2+}	Mg^{2+}	Ba^{2+}	Sr^{2+}	∑Fe
Content(w/w, %)	0.87	18.95	12.60	0.57	0.25	0.30	0.16	38.01

Table 4 XFS analysis results of bockage Unit: w/w, %, based on oxide mass fraction

matter	SO_3	Fe_2O_3	BaO	SrO	CaO	SiO_2	MgO	MoO_3	Al_2O_3	P_2O_5	CuO	NiO	MnO	Cr_2O_3	ZnO
Content (w(M)/10^{-2}, %)	34.44	61.93	0.53	0.05	0.66	0.73	0.08	0.05	0.18	0.03	0.02	0.01	0.37	0.51	0.41

From the Table 3 and Table 4, the content of HCl insoluble matter in the blockage of JB1# well is low, while the content of Fe in HCl soluble matter is high and contains a small amount of $CaCO_3$ scale, so it is mainly caused by the clogging of inorganic substances caused by corrosion products generated by corrosion.The XFS analysis showed that the main elements were sulfur and iron oxide. Combining with Fig. 1 and Fig. 2 , it was found that the blockage after calcination was reddish, and the main components may be iron oxides (Fe_2O_3 and Fe_3O_4) depending on the color and the magnetic testing (A liittle of blockage has the magnetism, which is Fe_3O_4). In addition, it also contains a small amount of zinc, manganese, aluminum, chromate, silicate, trace amounts of magnesium, barium, molybdenum, phosphate and other substances.

3.3 XRD of blockage

The chemical composition of the plug was analyzed by XRD, and the results are shown in Fig.3.

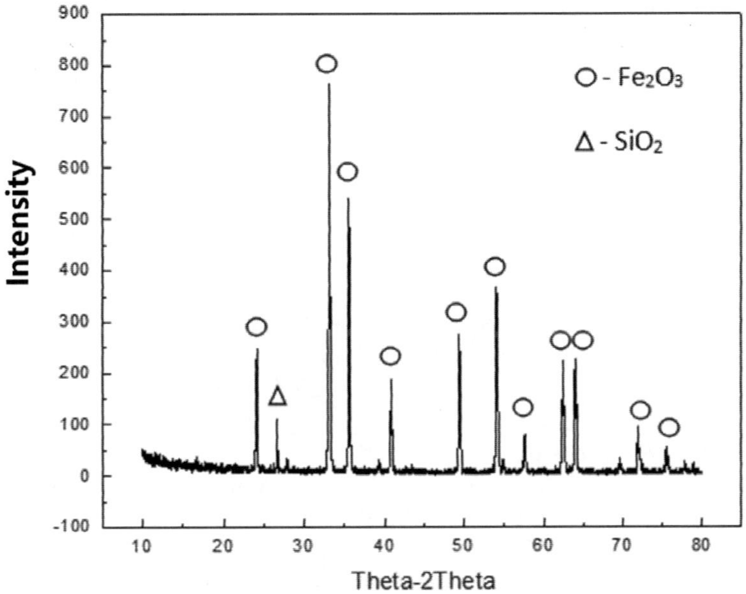

Fig.3 XRD analysis results of bockage

From the Fig. 3, the main component of the blockage is Fe_2O_3, and a certain amount of SiO_2. Where Fe_2O_3 comes from CO_2 and H_2S produce corrosion.

3.4 Plugging mechanism

The results of blockage analysis in JB1# well indicate that the main cause of blockage was the eposition of Fe_2O_3 generated by the corrosion of CO_2 and H_2S.When CO_2 is dissolved in water, it is extremely corrosive to steel[8], a galvanic cell is formed on the surface of steel, and oxygen corrosion occurs under dissolved oxygen conditions. The reaction process is as follows[9-11]:

$$CO_2+H_2O\rightarrow H_2CO_3\rightarrow H^++HCO_3^- \quad (1)$$

$$\text{Anodic reaction: } 2Fe\rightarrow 2Fe^{2+}+4e^- \quad (2)$$

$$\text{Cathodic reaction: } O_2+H_2O+4e^-\rightarrow 4OH^- \quad (3)$$

$$\text{Total reaction: } 2Fe+O_2+H_2O\rightarrow 2Fe(OH)_2\downarrow \quad (4)$$

The generated $Fe(OH)_2$ is very unstable and oxidized into magnetite (Fe_3O_4) and $Fe(OH)_3$ under the action of dissolved oxygen:

$$4Fe(OH)_2+2H_2O+O_2\rightarrow 4Fe(OH)_3\downarrow \quad (5)$$

$$Fe^{2+}+2FeOOH\rightarrow Fe_3O_4 \quad (6)$$

The generated $Fe(OH)_3$ was dehydrated to form Fe_2O_3:

$$2Fe(OH)_3+ nH_2O\rightarrow Fe_2O_3\cdot nH_2O+3H_2O \quad (7)$$

The corrosion products of CO_2 on steel are Fe_2O_3 and Fe_3O_4, which form a layer of a protective film on the metal surface, however, due to uneven or broken film formation, local corrosion such as etch pits, rim corrosion, and mesa corrosion often occur, the presence of Cl^- will interferes with the formation of this protective film, causing corrosion to continue and accelerating the scrapping of the pipe.When dissolved oxygen, H_2S and CO_2 or other corrosive impurities are present in the system, the corrosive effect on the metal will be more serious [12-13].

4 Conclusions

The JB1# well in the Jingbian gas field has unstable water quality, high salinity, high Cl^- content, and low pH value, high H_2S and CO_2 content in natural gas, so easy to produce corrosion and carbonate scaling.The main component of the blockage is a mixture of corrosion product Fe_2O_3 and a small amount of $CaCO_3$ scale.The corrosion product Fe_2O_3 is mainly generated by the mixture of dissolved oxygen, H_2S, CO_2 and other corrosion effects. The main cause of blockage in the JB1# well is the deposition of Fe_2O_3 and $CaCO_3$ scales caused by the corrosive effects of dissolved oxygen, H_2S and CO_2.

Acknowledgements

The authors are grateful for financial support from National Natural Science Foundation of China (51504193) and Key laboratory Research Project of Shaanxi Education Department (15JS090).

References

[1] Muhammad Shahzad Kamal, Ibnelwaleed Hussein, Mohamed Mahmoud, Abdullah S. Sultan, and Mohammed A.S. Saad 2018, *Journal of petroleum science and engineering,* **171** 127–9.

[2] A.S. Yaround, K. R. A. Khalik, and A. A. Khadom 2015 Journal of Loss Prevention in the Process Industries, 38 24-38

[3] Shenghou He 2008 *Natural gas industry* **28** 82-5.

[4] Gaurav R. Joshi, Karyn Cooper, Xiangli Zhong, Anthony B.Cook, Ehsan A.Ahmad, and Nicholas M.Harrison 2018 *Corrosion Science* 142 110-8.

[5] Yucheng Liu, Yinlong Zhang, and Jianmei Yuan 2014 *Engineering Failure Analysis* **45** 225-33.

[6] Tang Zhijian, Hong Seungkwan, Xiao Weizhong, and Taylor James 2006 *Corrosion Science*, 48 322-42.

[7] National Petroleum and Chemical Industry Bureau 2016 SY/T 5523-2016. Beijing: China Standard Press

[8] Xueyuan Zhang, Fengping Wang, and Yuanlong Du 2009 *Journal of Chinese Society for Corrosion and Protection*, **19** 72-8.

[9] Jingping Lin, Ellaway Mark, and Adrien Robert 2001 *Corrosion Science* **43** 2065-81.

[10] Nešic´, Srdjan 2007 *Corrosion Science* **49** 4308-38.

[11] Xiangzhen Zhang, Chaozhen Chen and Junqi Li 2015 *Surface Technology* **44** 122-5(137).

[12] M. A minullsalm and Z. N. Farhat 2013 *Tribology International* **68** 26-34.

[13] Hongchun Yang, Xiaoping Yang, Mingzhong Yang and Ge Lu 2007 *Drilling and Production Technology* **02** 117-9(159).

IOP Conf. Series: Materials Science and Engineering **542** (2019) 012073 doi:10.1088/1757-899X/542/1/012073

Ultrasonic Evaluation of Young's Modulus for Reduction of Brake Squeal Noise

N Kim[1]

1 School of Mechatronics Engineering, Korea University of Technology and Education, Cheonan, 31253, Korea
E-mail: nykim@koreatech.ac.kr

Abstract. Elastic properties of friction materials in brake linings are important design parameters because modulus variations may lead to noise generation of braking systems. This paper is to describe about ultrasonic estimation of Young's modulus of brake linings concerned with squeal noise problem for commercial trucks. Phase velocity of ultrasound propagating through brake lining was measured to calculate Young's modulus based on elastic relations of acoustic waves with material properties. Two kinds of brake lining samples were produced with the same composition by different manufacturing process, one of which is a traditional product and the other is an advanced one developed for brake noise reduction. Five rectangular samples were cut out from each type of brake linings and machined for ultrasonic test. Acoustic pulse of 500KHz was applied to each sample to calculate the time-of-flight(TOF) for the measurement of wave velocity and Young's modulus. The results of ultrasonic test indicate that the low noise brake samples have a less deviation of Young's modulus than the traditional brake samples. This agrees with previous researches and FEM analysis that Young's modulus variation of brake linings affects the squeal propensity and the instability of brake system.

1. Introduction

The braking system of a vehicle is one of the most fundamental safety-critical components used to decelerate a moving body. Due to the braking operation, the brake system generates an unwanted high frequency sound known as brake squeal that makes the driver uncomfortable. Among the different types of brake noise, squeal noise, because of its higher frequency contents between 1 kHz and 16 kHz, is the most troublesome and irritant one to drivers and the environment, and is expensive to the brakes and car manufacturers in terms of warranty costs [1-3]. There are six mechanisms of squeal formation namely: stick–slip, sprag–slip, negative friction velocity slope, hammering excitation, splitting the doublet modes and mode coupling of structures [4]. It is commonly accepted by researchers that squeal noise in a brake system may be generated by either one of the above mechanisms, or the combined depending on the brake design and operational conditions. From numerous papers on both linear and non-linear stability analyses of dynamical systems with friction-induced vibrations, it is obvious that the effect of the friction material stiffness on squeal noise is indispensable for the design of a quiet brake system. Brake linings or pads are a composite material bonded with various synthetic substances comprised of more than 10 different substances including metals, rubber and organic materials to improve friction property, increase strength, prolong life, and reduce noise. Thus, despite of the same composition, it can have a variation in density and Young's modulus depending on the manufacturing method. This local heterogeneity in material property, especially Young's modulus, may increase the propensity of squeal noise caused by friction-induced instability of vibration associated with mode coupling.

IOP Conf. Series: Materials Science and Engineering **542** (2019) 012073 doi:10.1088/1757-899X/542/1/012073

This paper investigates the variation of Young's modulus in brake linings responsible for squeal noise using ultrasonic method. Two kinds of brake lining samples for commercial trucks were prepared to measure and compare its Young's modulus. It is concluded from acoustic experiment that the variation of Young's modulus of low noise brake lining is half of the traditional linings, and is estimated effectively by ultrasonic method in non-destructive and quantitative way for the quality control of brake linings.

2. Fabrication of brake lining samples

The Brake system of trucks has a set of linings which is pressed against a rotating drum when it brakes and heated due to friction at the drum-lining interface in a short period of time. Due to these severe working conditions during operation, macro-cracks and friction-induced vibration with noise might develop on the disc surface in the radial direction. Brake lining is generally a combination of a steel back plate and a friction material fixed on the steel back plate as shown in Fig. 1. Heavy duty brake linings are currently manufactured by powder metallurgy(PM) method to compact and sinter the metal powder mixture containing about 20–30% non-metallic constituents [5,6]. In this process, the friction materials should be formed as uniform as possible even if the flowability of the powdered raw friction material is insufficient. However, conventional brakes have a problem of making brake noises caused by friction-induced vibration partly because of non-uniform distribution of friction materials. Therefore, it is necessary to evaluate local Young's modulus of brake pads/linings for the quality control of squeal noise. In this work, the variation of Young's modulus was investigated for an actual drum brake linings used for heavy-duty trucks as shown in Fig. 1(a) and Fig. 1(b) using ultrasonic nondestructive method.

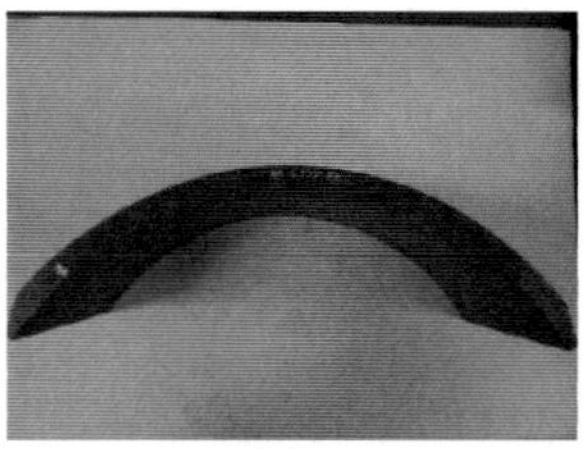
(a)

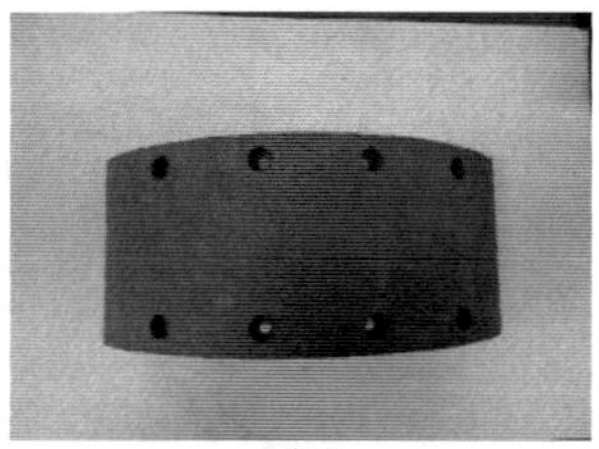
(b)

Figure 1. Drum brake lining of heavy-duty trucks, (a) side view, (b) top view.

Two kinds of brake linings were prepared with the same composition for this study, one of which is a product manufactured by conventional PM method, and the other made by an advanced PM method developed for noise reduction. From each brake lining, five rectangular samples (5mm in thickness and 30mm x 30mm in width) were cut out from different locations respectively as shown in Fig. 2(a) and machined for ultrasonic test. Total 10 specimens fabricated from a conventional brake lining and a newly developed brake lining are displayed all together in Fig 2(b) and numbered consecutively to identify where it comes from in each lining. In Fig. 2(b), five samples on the top (#1 ~ #5) were from a conventional brake lining, and the next five ones (#6 ~ #10) from a low-noise brake lining.

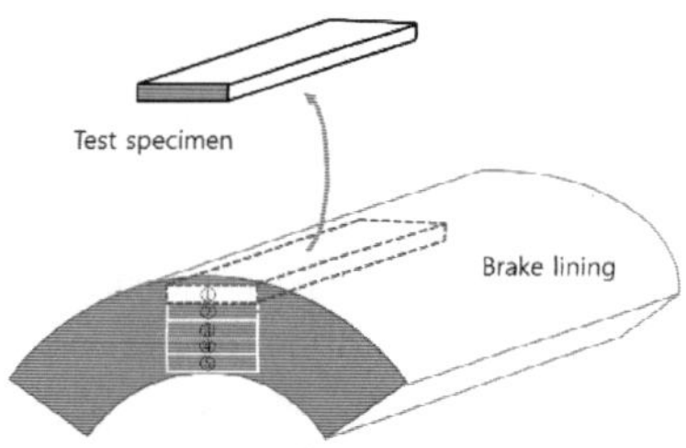

(a)

(b)

Figure 2. Fabrication of test specimens, (a) sampling method, (b) test specimens

IOP Conf. Series: Materials Science and Engineering **542** (2019) 012073 doi:10.1088/1757-899X/542/1/012073

3. Measurement of Young's modulus using ultrasonic method

Measurement of ultrasonic wave velocity was conducted to estimate the elastic modulus of brake lining specimens using the pulse-echo method and the time-of-flight (TOF) measurement. In the TOF method, acoustic transducer generates a pulse signal of ultrasound that propagates and reflects back from the back surface of samples, being detected after a delay. This duration of wave propagation depends on the thickness of the brake specimen as well as the elastic properties of the traveling medium (brake specimen). The configuration of experiment is illustrated in Fig. 3 to measure the TOF of ultrasonic pulse. At the start of a measurement, pulse generator (JSR Ultrasonics, DPR300) excites an electric pulse, which is converted in a ultrasonic transducer (Technisonics, C0008GP) to ultrasonic pulse of 500kHz propagating through the specimen at a velocity c. When it travels in the specimen, the pulse reflects multiple times at the boundaries of brake specimen and returns to the transducer repeating this round-trip travel as shown in Fig. 3(a). After these echoes were recorded in digital oscilloscope (Lecroy wave runner 530), the time interval between two maximum peaks of consecutive echoes was measured. Fig. 3(b) shows the pulse-echo measurement equipment of the TOF for brake lining samples, where an echo signal from the back surface is captured and used to calculate the TOF of the ultrasonic wave. The acoustic phase velocity c is expressed through the TOF and the thickness of the plate d by the simple relation of equation (1).

$$c = 2d/TOF \tag{1}$$

The TOF method is very easy and simple to use and works well unless it suffers from low signal-to-noise ratio (SNR) or signal shape distortion due to high attenuation and nonlinearity. For an isotropic material, the elastic modulus is related to the ultrasonic wave velocity (compression mode) c as the following [5,6].

$$E = \rho c^2 \tag{2}$$

where E is the Young's modulus, ρ is the density of material. The assumption made in equation (2) is not correct in a strict sense, but is reasonable if the friction materials of highly irregular shape and size are mixed well like poly-crystalline structure and relatively small comparing with the wave length of ultrasound. On each sample, five different points were selected for ultrasonic measurements including the center of the specimen and other four points around the center at 0, 90, 180, and 270 degrees. One hundred acoustic signals were collected at each point and averaged over the five measurement locations of the sample to obtain more reliable and accurate TOFs.

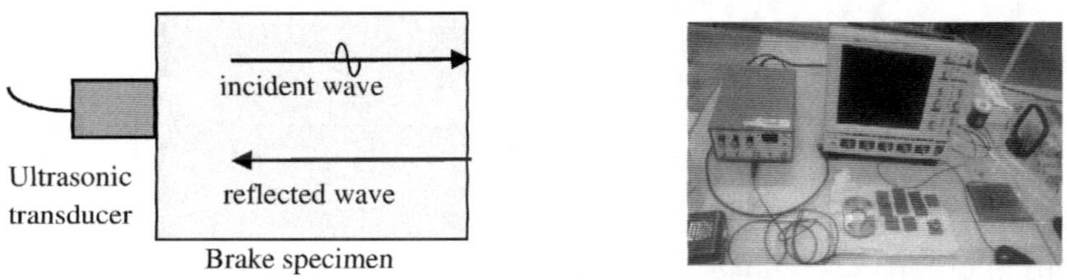

(a) (b)

Figure 3. Measurement of ultrasonic wave speed, (a) pulse-echo test in sample, (b) acoustic equipment - ultrasonic transducer, pulser/receiver and oscilloscope.

IOP Conf. Series: Materials Science and Engineering **542** (2019) 012073 doi:10.1088/1757-899X/542/1/012073

4. Experimental results

Once the TOF from reflection signals for brake lining specimen was obtained in ultrasonic test, the Young's modulus was calculated from the acousto-elastic relation of equation (2). Acoustic velocity and Young's modulus of five specimens of the conventional brake lining are listed in Table 1 along with basic geometric and physical properties. Maximum Young's modulus was 60.8Gpa and minimum was 57.1Gpa. Percentage standard deviation of five measurement values of Young's modulus was 2.5%. As for the low-noise brake lining samples, the maximum Young's modulus was found in Table 2 to be 60.8Gpa and the minimum 59.1Gpa. Percentage standard deviation of five measurement values of Young's modulus was 1.1%. It was found from experimental results that the low-noise specimens revealed a less deviation in Young's modulus than the traditional ones as it was reported in previous experimental and theoretical researches [2-4, 7].

Table 1. Young's modulus and wave velocity of conventional brake linings

Specimens (conventional lining)	Density [kg/m^3] ρ	Thickness [mm] d	Acoustic wave velocity[mm/sec] c	Elastic modulus[GPa] E	Standard Deviation [%] σ
#1	1,790	4.98	5,651	57.1	
#2	1,800	4.98	5,816	60.8	
#3	1,800	4.97	5,754	59.5	2.5 %
#4	1,810	4.91	5,697	58.7	
#5	1,810	4.98	5,644	57.6	

Table 2. Young's modulus and wave velocity of low-noise brake linings

Specimens (conventional lining)	Density [kg/m^3] ρ	Thickness [mm] d	Acoustic wave velocity[mm/sec] c	Elastic modulus[GPa] E	Standard Deviation [%] σ
#6	1,820	4.97	5,701	59.1	
#7	1,790	4.97	5,775	59.6	
#8	1,800	4.95	5,810	60.8	1.1 %
#9	1,830	4.97	5,757	60.6	
#10	1,830	4.97	5,724	59.9	

5. Conclusions

Young's modulus of drum brake linings for heavy-duty truck were estimated by ultrasonic method for the investigation of the variation of elastic constant responsible for squeal noise. Rectangular specimens extracted from two different types of brake linings were fabricated and subject to acoustic pulse excitation of 500 kHz for the measurement of ultrasonic wave velocity of lining samples. The specimens from the traditional brake lining had a variation in Young's modulus ranging from the maximum 60.8GPa to the minimum 57.1GPa with 2.5% standard variation. On the other hand, the specimens from low noise lining had Young's modulus from the maximum 60.8GPa to the minimum 59.1GPa with 1.1%

standard variation. From the experimental results, the samples from low noise lining were found to have more even distribution of Young's modulus than the traditional one, even though they had the same composition. This agrees well with previous researches that the variation of Young's modulus in brake lining has an effect on squeal noise, i.e., the low deviation of Young's modulus of brake lining reduces squeal noise. It is also concluded that ultrasonic measurement of Young's modulus is very effective, fast and economical to evaluate the quality assurance of brake linings for squeal noise reduction.

Acknowledgments
This work was supported by Research Project funded by Sangsin Brake Company and Korea University of Technology and Education.

6. References

[1] Rashid A 2014 *International J. of Vehicle Noise and Vibration* **10**(2014) 257
[2] Belhocine A and Ghazaly N M 2015 *Latin American Journal of Solids Structure* **12** 1432
[3] Nam J and Kang J 2013 *Trans. of the Korean Society of Noise and Vibration Eng.* **23** 717
[4] Ghazaly N M, El-Sharkawy M and Ahmed I 2014 *J. of Mechanical Design and Vibration* **1** 5
[5] SAE international Surface Vehicle Standard 2009 *SAE standard J2725_200901*
[6] Kim N and Yang S 2016 *J. of the Korean Soc. for Nondestructive Testing*, **36** 9
[7] Negesh S N, Siddaraju C, Prakash S V and Ramesh M R 2014 *Procedia Mat. Sci.*, **5** 295

IOP Conf. Series: Materials Science and Engineering **542** (2019) 012074 doi:10.1088/1757-899X/542/1/012074

Innovative Application of Digitized Design in Traditional Manufacturing: An Example in the Powder Metallurgy Industry

Lo Jih-Sheng [1,3] , Lee Kim C K [2] and Huang Shyh-Chour [1]

1 Dept. of Mechanical Engineering, National Kaohsiung University of Science and Technology, Taiwan
2 Dept. & Grad School of Industrial Design, Chaoyang University of Technology, Taiwan
3 Dept. of Digital Content and Technology, National Taichung University of Education, Taiwan
E-mail: ponylsen@mail.ntcu.edu.tw

Abstract. This research began with a series of in-depth interview with experts in the field of powder metallurgy, combined with literature review, a design process with a checklist of design quality for powder metallurgy cultural and creative products was established. We first verified this design process with the checklist by applying it to improve the design of an existing powder metallurgy cultural and creative product. Then we designed three cultural and creative products based on the established design process with the checklist of design quality and chose an icon of Hakka Lion among them to prepare a toolset and finished the test production of a sample, which all of our interviewed experts deemed as having excellent quality. Therefore, this research verified that the traditional technology of powder metallurgy could not only be utilized to produce mechanical parts, but it can also be applied in the cultural and creative industry to produce high value-added artifacts that are typically full of very complex curvature surfaces. In addition, the design process with the checklist of design quality for powder metallurgy cultural and creative products was also verified to be applicable. Besides, this research also introduced the technology of digitized design to the design and manufacturing process of an old powder metallurgy manufacturer successfully. Through which we enhanced the integrity and efficiency of this production process significantly. Hope that the result of this research will be helpful for the powder metallurgy industry in Miao-li County of Taiwan to find a way out.

1. Introduction

Powder metallurgy (the common abbreviation as PM will be utilized from now on) industry has developed in The Miao-li County for almost fifty years, there currently are more than thirty companies with a massive share of more than 70% among the total output of the PM industry in Taiwan[1][2]. In response to this massive trend, the Executive Yuan of Taiwan announced "Challenging 2008—A six-year plan of national key development" in 2002 [3]. In this planning, they vowed to take the development of cultural and creative industry as a key development policy of Taiwan government. Then a consecutive plan "Creative Taiwan—An action plan for the development of cultural and creative industry" was further announced in 2009 to deepen the result of this policy [4], which had created a boom of creative design in Taiwanese industries effectively. According to Newman's argument, digital transformation is the dominant trend in industries all over the world currently [5]. This research intended to combine the two leading trends in industries, i.e., the boom of cultural and creative artifacts as well as the proliferation of digitization on almost everything. We first interviewed some PM experts to acquire their specialized experience and knowledge; then we developed a design process with a checklist of design quality for cultural and creative PM products based on these interview results. Based on the design process and the checklist of design quality, we developed three cultural and creative PM product designs. Then we chose a Hakka Icon of Lion [6] among them to produce a test sample and our PM experts all rated this test sample as having good quality, which further verified the viability of our design process and checklist for design quality. The previous literature on applying PM technology in the field of cultural and creative artifacts can only barely found. As Chen [7] pointed out, the primary application of PM technology was in the field of mechanical parts, while the major developments of PM technology were in its material contents and micro phenomenon in the production process. Gummeson [8] also raised the issue of high-pressure water atomization in PM products. In the Taiwanese academia, there were only a few previous literatures on applying PM technology in the field of cultural and creative artifacts such as Cheng's article [9].

2. Aim of Research

This research was set out to figure out a comprehensive checklist for design review, in order to establish a firm basis of introducing both cultural and creative products as well as the digitization technology to the PM industry. In pursuing this objective, we began with interviewing PM experts to acquire the knowledge on constraints and rules of traditional design for PM products, then applied digitization technology to integrate this knowledge into the design process and the checklist for design review of PM products. We especially envisaged the enhancement of efficiency in communications between design and manufacturing engineers in arranging this design process and the checklist for design review. In summary, the scope of this research are as follows: (1). Exploring the design constraints of PM products to maximize design quality, (2). Developing the design guideline of PM cultural and creative products following the design constraints that we identified earlier. Since cultural and creative products are usually small in size, our design guideline can only fit with the powder compression model machine of 3 tons and the maximum product range under 30mm x 30mm, (3). Focusing on the merit of enhancing communication efficiency by applying digitization technology.

3. Research Method and Results

The traditional applications of MP were concentrated in the production of precision and simple geometrical form of mechanical parts mostly. The whole industry was generally in lack of the experience around any consumer commodities or any complicated curvature product forms. Under such an endangered situation of the PM industry in Taiwan, especially in Miao-li County, we saw it would be worthwhile if we could offer them with some help with our research. The logical structure of the methods utilized in this research is as depicted in Figure 1:

IOP Conf. Series: Materials Science and Engineering **542** (2019) 012074 doi:10.1088/1757-899X/542/1/012074

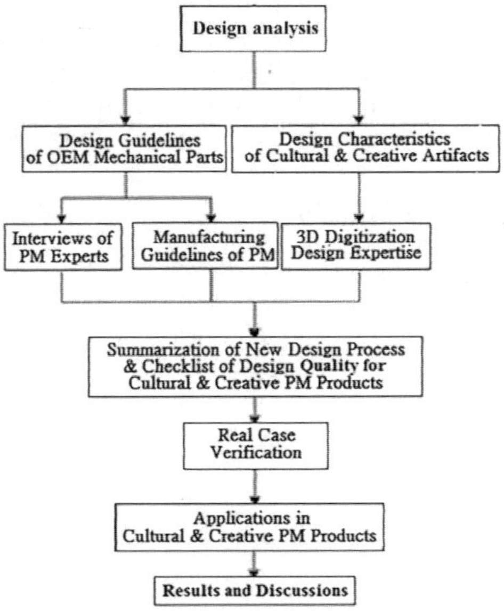

Figure 1. Logical Structure of Research Methods

(1) Design and manufacturing guidelines of OEM mechanical parts: In this stage, we collected engineer drawings of the OEM parts that were fabricated previously in the PM manufacturer which was collaborated in our research. Five types of PM parts that typically had some difficulties in forming or other fabrication processes were selected in this survey. We invited three PM experts from the PM manufacturer to develop a questionnaire to investigate the characteristics of these parts collaboratively. For instance, the questionnaire and the result of the first part of question b was shown in Figure 2:

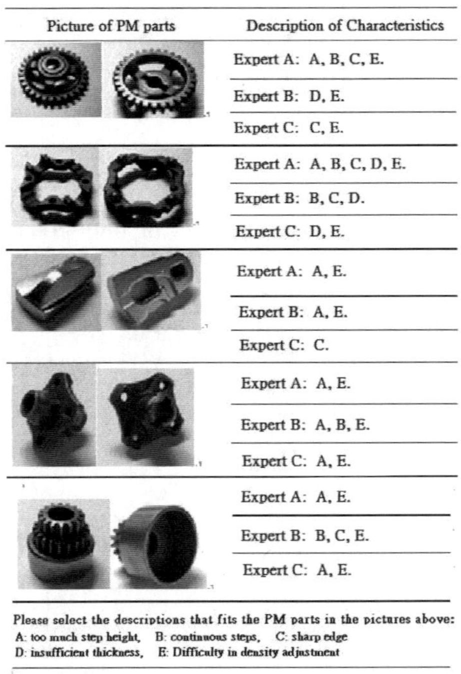

Figure 2. Categorization of Characteristics of OEM PM Parts

IOP Conf. Series: Materials Science and Engineering **542** (2019) 012074 doi:10.1088/1757-899X/542/1/012074

The result of this questionnaire survey had concluded into six design guidelines of OEM PM mechanical parts as follows: (a). Step height on the parts cannot be higher than one-third of the thickness of the next layer, (b). The factor of easy adjustment on density must be taken into account on the form design, (c). Avoid the design of long and thin shapes, (d). It is essential to keep the same thickness when there are continuous steps or curvature surfaces, (e). It is preferable to make all edges in round shape, (f). All draft angles must be 3 to 5 degrees if there are any raised ribs, columns, or surfaces.

(2) Design characteristics of cultural and creative artifacts: There were only a tiny number of previous literature concerning the application of PM technology in the field of cultural and creative artifacts in Taiwan. It was only Cheng [13] who had pioneered the application of technologies including image processing and CAD/CAM in the cultural and creative artifact. It was an elephant keychain as shown in Figure 3. However, it was a straightforward design with slight variations on the surface only. Therefore, we took reference from cultural and creative artifacts manufactured in other ways and materials and concluded that these products are prone to have more complicated surface designs such as variations curvature, steps, and thickness, which may be challenging in PM manufacturing and thus require appropriate design modifications.

Figure 3. Cheng's [6] PM Product – An Elephant Keychain

(3) Interviews of PM experts**:** As described above, we invited three PM experts from the PM manufacturer that was participated in our research to develop a questionnaire to investigate the characteristics of PM parts collaboratively. These three PM experts had also played the role of PM technical consultant for our research in the following stages.

(4) Manufacturing guidelines of PM: Additionally, the result of the questionnaire survey described earlier had also concluded into three manufacturing guidelines of OEM PM products as follows: (a). The process of forming had better take more gradual steps, (b).The adjustment of upward and downward forces of the press must be very careful, (c). Avoid having too much discrepancy on density in the product.

(5) 3D digitization design expertise: The expertise of 3D digitization design is the major strength of our research team, through which we intended to help the PM industry by combining it with the PM knowledge that we acquired in this research. Now that we had learned the experiences regarding traditional PM design and manufacturing from our experts, it should be our turn to combine our expertise with their contribution at this stage.

(6) Summarization of new design process with checklist of design quality for cultural and creative PM products: In summarization of the former stages, our research had come to the point where the resulted new design process with checklist of design quality for cultural and creative PM products can be coming to the fore. At this stage, we integrated the design process and the checklist of design quality together to become a comprehensive flowchart as shown in Figure 4 on the next page:

IOP Conf. Series: Materials Science and Engineering **542** (2019) 012074 doi:10.1088/1757-899X/542/1/012074

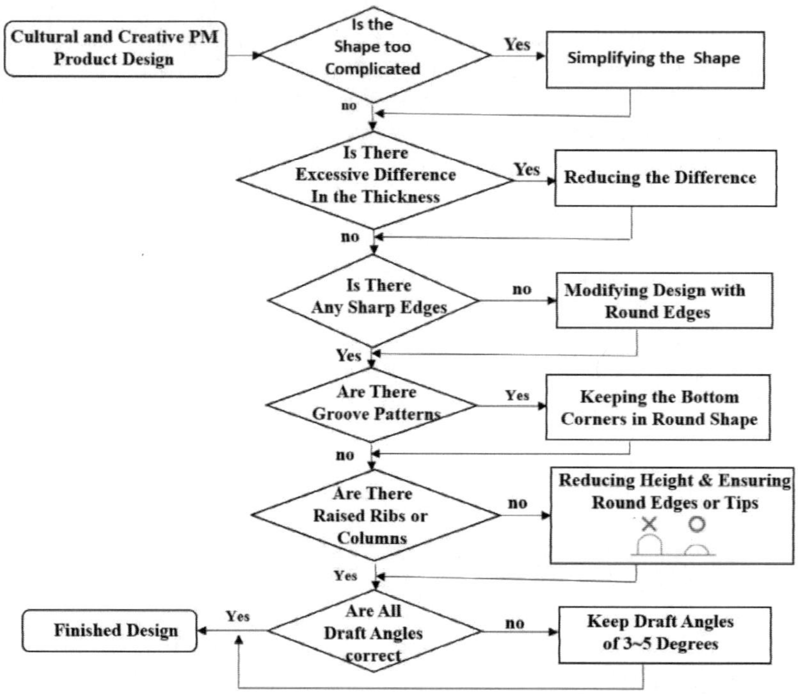

Figure 4. Design Process with Checklist of Design Quality for Cultural and Creative PM Products

(7) Real case verification:Based on the new design process with the checklist of design quality for cultural and creative PM products that we proposed above, we took Cheng's elephant keychain as a real case to verify its viability. We showed the pictures of Cheng's elephant keychain to our PM experts and consulted their opinions of required improvements, then an improved 3D design was finished, and our three PM experts had all took this design as having good design quality. Therefore, this real case has verified that the new design process with the checklist of design quality for cultural and creative PM products that we proposed is viable.

Figure 5 3D Model of the Redesigned Result of Cheng's Elephant Keychain

(8) Applications in cultural and creative PM products: After the above initial verification of our new design process with the checklist of design quality for cultural and creative PM products, we have to further verify its applicability by the design and production of a real cultural and creative PM product. As the majority of the population in Miao-li County is Hakka, thus we aimed to choose icons of Hakka culture to develop our cultural and creative product. For this purpose, we chose three icons of Hakka culture for our initial design, i.e., tong tree flower, fire dragon, and Hakka

IOP Conf. Series: Materials Science and Engineering **542** (2019) 012074 doi:10.1088/1757-899X/542/1/012074

lion. Among these three initial designs, we chose the Hakka lion to produce a test sample shows the 3D digitized model in Figure 6 below.

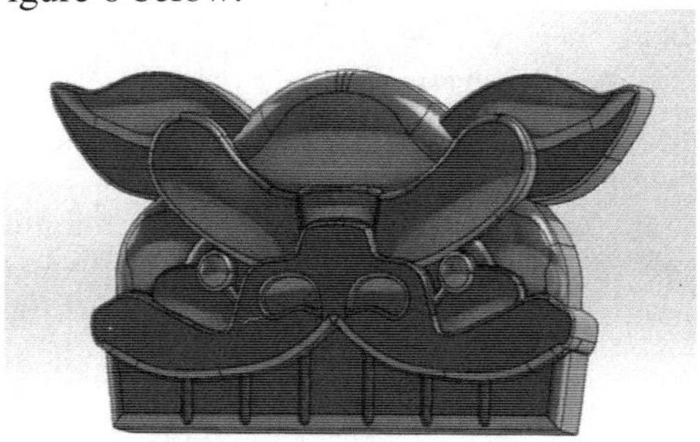

Figure 6. 3D Digitized Design of the Hakka Lion

Since cultural and creative products are usually small in size, we developed our design to fit with the powder compression model machine of 3 tons and the maximum product range under 30mm x 30mm. We pioneered the introduction of 3D printing to the PM industry by producing a plastic model using PLA, which is a convenient vehicle for enhancing the efficiency in communications in this stage. Following the plastic model, we established a CAD model of 2.5D of the embossed copper electrode, then use this CAD model to produce the CNC program with CAM software and produce it by five-axis engraving machine. The finished upper, lower and middle molds are shown in Figure 7 below.

Figure 7. Upper, Lower and Middle Molds of the Hakka Lion

The toolset finished above was then used to produce green compacts as shown in Figure 8 below.

Figure 8. Cooper and Iron Green Compacts of the Hakka Lion

4. Results and Discussions:

Finally, the results of this research had hit the aims that we set initially. They are as follows:

IOP Conf. Series: Materials Science and Engineering **542** (2019) 012074 doi:10.1088/1757-899X/542/1/012074

(1). Established a design process with the checklist for design quality of PM products, for which the viability had been verified in this research also.
(2). Introduced the technologies of 3D CAD/CAM and 3D printing in the design process of PM products and enhanced the communication efficiency significantly.
(3). Introduced the cultural and creative artifacts to the PM industry to help them secure better business opportunities and thus improved their economic value.
These research results had been proven to be effective in the PM manufacturer that collaborated with this research. It is hopeful that other PM manufacturers will be influenced to adopt these changes proposed by this research gradually.

5. **Conclusions**

This research adopted research methods of in-depth interviews and literature review, combined theories and practical experiences of PM and introduced new technologies including digitized design, CAD/CAM and 3D printing and new product of cultural and creative artifacts to the PM industry successfully. With the contributions made by this research, the PM manufacturer that collaborated with the researchers had found great prospect to prosper again in the future. We are sure that other PM manufacturers will found the hope in these changes proposed by this research and follow up to change together gradually.

References

[1] Chu C L 2008 Developments and Applications in the Powder Metallurgy Industry, *Journal of the Powder Metallurgy Association* 33(4) pp. 201-214.

[2] Chang C B 2015 Don't Look Down to the OEM Business in Taiwan *Journal of Observations* 28 pp. 34 .

[3] Executive Yuan 2018 Taiwan, *Challenging 2008—A Six-Year Plan of National Key Development,* Retrieved August 11, 2018. https://www.teg.org.tw/files/events/2002.05.31.pdf

[4] Executive Yuan 2018 Taiwan, *Creative Taiwan—An Action Plan for the Development of Cultural and Creative Industry* Retrieved August 11, 2018. http://grb-topics.stpi.narl.org.tw/app/download/4b1141c24dd3d37e014dd765ece000f6

[5] D. Newman 2017 Top 5 Digital Transformation Trends In Manufacturing *Forbes,* August 8 (2017) Retrieved August 11, 2018. https://www.forbes.com/sites/danielnewman/2017/08/08/top-5-digital-transformation-trends-in-manufacturing/#646522a249f0

[6] Wang J T 2006 A Study of the Hakka Fire Dragon Festival in Miao-li *Master Thesis* Department of Physical Education, National Ping Tung University.

[7] Chen J S 2003 A Study on the Influence of Green Compact and Sintering Temperature on the Degree of Precision of Powder Metallurgy Spur Gears *Master Thesis*, Mechanical Engineering Department, National Chiao Tung University.

[8] Gummeson P U 1972 High-Pressure Water Atomization in: *Powder Metallurgy for High-Performance Applications,* J. J. Burke and V. Weiss (Eds.), Syracuse, NY: Syracuse University Press, 1972, pp.27-55.

[9] Cheng C S, Chen Y H, Du Z F, & Huang L J 2012 The Fabrication of Powder Metallurgy Stainless Charm and the Education of Mechanical Talents *Journal of Applied Engineering Technologies* 1(2) pp. 87-96.

IOP Conf. Series: Materials Science and Engineering **542** (2019) 012075 doi:10.1088/1757-899X/542/1/012075

Performance Exploration of A Radially Layered Cymbal Piezoelectric Energy Harvester under Road Traffic Induced Low Frequency Vibration

Xinnan Liu[1] and Jianjun Wang[2*]

1 School of Civil Engineering, North China University of Technology, Beijing, China
2 Department of Applied Mechanics, University of Science and Technology Beijing, Beijing, China
E-mail: xinnanliu@139.com; jianjunwang168@ustb.edu.cn

Abstract. A type of radially layered cymbal piezoelectric energy harvester is designed and fabricated. It is made of two cymbal metal caps, two axially polarized piezoceramic rings, one solid metal disk, one common metal ring and one drilled metal ring. The mechanical couplings in radial direction are finished by using extra strong epoxy. The mechanical couplings in axial direction are finished by using screw bolts. Subsequently, the force transferring coefficient of each cymbal metal cap is approximated through both theoretical calculation and experimental calibration. Aiming at matching the road traffic induced vibration frequencies that are predominantly ranging from 5Hz to 25Hz, the approximated analytical model of the harvester is established at frequencies below 30Hz. The model' s predictions agree reasonably with the measured data. Furthermore, the experimental results indicated that the harvester was able to generate a power of about 0.92mW across a 0.8MΩ resistor under the force of 500N at a frequency of 20Hz. The present work provides a new design concept for developing the novel cymbal harvesters used in roadway systems to harvest the traffic induced vibration energy.

1. Introduction

Energy harvesting from mechanical vibration using piezoelectric devices has attracted numerous research interests over the last few decades [1, 2]. Previous literatures have reported a large volume of structure forms of piezoelectric devices in various engineering applications, such as piezoelectric stack [3-7], piezoelectric patch [8-10], piezoelectric beam [11-13], piezoelectric drum [14, 15] and piezoelectric cymbal [16-18]. Comparing to other types of piezoelectric devices, piezoelectric cymbal transducer has a unique advantage in terms of mechanical force amplification mechanism of the cymbal metal endcaps, which can increase the radial stress applied to the piezoelectric element [19, 20]. Additionally, it also has reasonable energy conversion efficiency, medium stiffness and high durability under high loading level [18, 21]. These properties of cymbal transducer match well with the characteristics of pavement structure, such as medium stiffness, high stress, low frequency response and huge repetitions [22]. Owing to its excellent properties, piezoelectric cymbal transducer is a good candidate for harvesting energy from asphalt pavement.

A recent study has proved the feasibility of energy harvesting from asphalt pavement using traditional piezoceramic cymbal [17], which was formed by sandwiching a piezoceramic disk between two cymbal metal endcaps [23] using epoxy. The results indicated that a single piezoceramic cymbal can recover up to a power of 16 μW for the pass of one heavy vehicle wheel. This harvested power is far smaller than those reported in [24-26], whose power levels can reach up to 52mW or even 100mW.

IOP Conf. Series: Materials Science and Engineering **542** (2019) 012075 doi:10.1088/1757-899X/542/1/012075

Such an obvious difference is mainly caused by the mismatch of force frequency ranging from 4Hz [17] to 100Hz or 200Hz [24-26]. In addition, the embedded location will also induce a large deviation of the produced power. It is well known that the vertical stress component in the layered pavement structure decrease with the increase of depth [27]. To improve the harvesting efficiency, it is required that the harvester must be located as close as possible to the pavement surface while ensure that the harvester will not be damaged by the passage of vehicles. However, an earlier work has reported that the maximum bearing capacity of a 0.4mm thick steel cap is 70N [26], which limit the embedded location of the energy harvester.

Meanwhile, structural design of piezoelectric cymbal is a very important internal factor for the performance of the harvester. The past decades have witnessed great improvement in piezoelectric cymbal transducer. On the one hand, the cymbal metal endcaps have evolved from the traditional ones [19, 24-26, 28] to the slotted ones [29, 30] or the curved convex ones [31] for improving the force transferring efficiency. On the other hand, the piezoelectric element has been developed from the traditional one to unimorph (a piezoelectric/steel composite disk) [32] and bimorph (a piezoelectric/steel/piezoelectric composite disk) [20] for being used in the high load application, two-layer-stacked piezoelectric disks [28] for increasing the power output. However, most of these designs have their own disadvantages, such as lower force level, smaller power output, which limit the scope of applications.

This article attempts to develop a type of new piezoelectric cymbal harvester through combining the cymbal metal endcaps and the radially layered piezoelectric structure for energy harvesting in roadway systems. The rest of the paper is organized as follows. Section 2 describes the design details of the harvester, and establishes its approximate analytical model. Section 3 introduces the experimental setup. Section 4 analyzes and discusses the theoretical and experimental results, and compares the differences of energy harvesting properties between the present and previous work. Conclusions are drawn in Section 5.

2. Design and Modeling

As shown in Figure 1, a radially layered cymbal piezoelectric energy harvester is developed. The cymbal structure consists of two cymbal metal caps, two axially polarized piezoceramic rings, two metal rings and one solid metal disk. Two piezoceramic rings are connected in parallel electrically. A drilled ring casing is designed for the coupling with the cymbal metal caps. A hole was drilled in every metal cap to thread wires. The piezoelectric material used is PZT-5H and the outer casing is made of stainless steel. Their related properties are shown in Table 1 and Table 2. All rings are coupled radially with each other by using high strength epoxy. The upper and bottom metal caps were coupled with the outmost metal ring by bolting. The geometric parameters in Figure 2 are taken as $R_1 = 10\text{mm}$, $R_2 = 15\text{mm}$, $R_3 = 20\text{mm}$, $R_4 = 25\text{mm}$, $R_5 = 35\text{mm}$, $h = 5\text{mm}$, $h_m = 3\text{mm}$, $h_C = 8\text{mm}$, $R_{m1} = 8\text{mm}$, $R_{m2} = 25\text{mm}$, $\phi_{m4} = 3\text{mm}$, $\phi_{m5} = 4\text{mm}$.

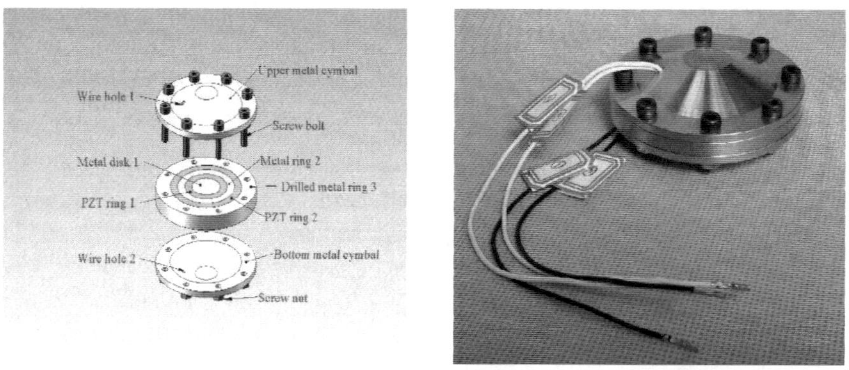

(a) (b)

Figure 1. The radially layered cymbal piezoelectric energy harvester: (a) schematic and (b) prototype.

Table 1. Material parameters of PZT-5H

Elastic constant ($\times 10^{-12}\,\mathrm{m^2N^{-1}}$)		Piezoelectric constant ($\times 10^{-12}\,\mathrm{CN^{-1}}$)	Dielectric constant	Density ($\mathrm{kgm^{-3}}$)
s_{11}^E	s_{12}^E	d_{31}	$\kappa_{33}^{\sigma} / \varepsilon_0$	ρ_P
13	-4.29	-186	4500	7450

$\varepsilon_0 = 8.85 \times 10^{-12}\,\mathrm{Fm^{-1}}$, Permittivity of free space

Table 2. Material parameters of Epoxy

Young's modulus ($\times 10^{-12}\,\mathrm{m^2N^{-1}}$)	Poisson's ratio	Density ($\mathrm{kgm^{-3}}$)
$\bar{E}$	μ	ρ_E
21	0.25	7930

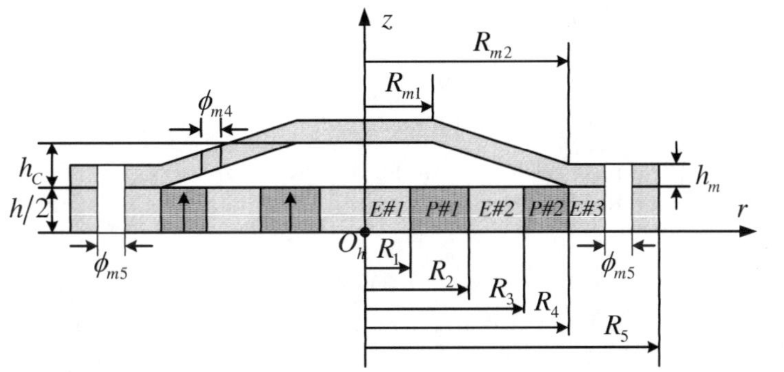

Figure 2. Geometric dimensions of the symmetric upper parts.
(E: elastic material; P: piezoelectric material; ↑: polarized direction)

Similar to the earlier cymbal transducers [23, 25, 26], the proposed devices also utilized the advantage of the cymbal structure that an axial load is transferred to the radial stress via the cymbal metal caps onto the piezoelectric element in the key region ($R_1 \le r \le R_4$), as shown in Figure 3. Under a pair of applied loads F, the radial stress is defined as

$$\bar{\sigma} = 2\frac{\chi F}{2\pi R_4 h} \tag{1}$$

where χ is the force transferring coefficient, which represents the force transferring capability of the cymbal metal cap. For the current cases, it is difficult to solve for this coefficient theoretically due to the complexity of the structure. Instead, the coefficient will be approximately obtained through experimental calibration.

Once the load transferring coefficient is determined, the next problem is to solve a layered piezoelectric/elastic composite structure subjected to a uniform radial stress, as shown in Figure 3(b). In the present work, an external mechanical force is considered as

$$F(t) = F_0 \mathrm{e}^{\mathrm{i}\omega t} \tag{2}$$

where F_0 is the force amplitude, $\mathrm{i} = \sqrt{-1}$, $\omega = 2\pi f$ is the circular frequency, and t is the time.

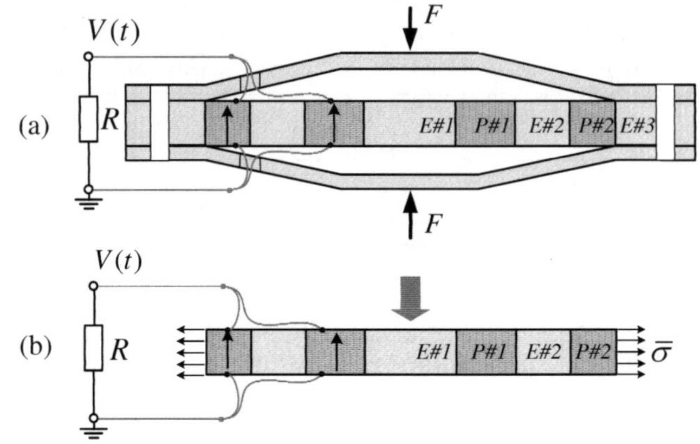

Figure 3. Force transferring mechanism.

The transferring radial stress can be expressed as

$$\bar{\sigma}(t)=\bar{\sigma}_0 \mathrm{e}^{\mathrm{i}\omega t} \tag{3}$$

where $\bar{\sigma}_0 = 2\dfrac{\chi F_0}{2\pi R_4 h}$ is the amplitude of radial stress. Furthermore, the generated voltage across the resistive load R is also written as

$$V(t)=V_0 \mathrm{e}^{\mathrm{i}\omega t} \tag{4}$$

where V_0 is the complex voltage amplitude across the resistor load R.

For harmonic and steady vibration, the electrical and mechanical components for the plain stress problem can be written as [33]

$$(u_{rPi},\sigma_{rPi},\sigma_{\theta Pi},\phi_i,E_{zi},D_{zi})=[u_{rPi}(r),\sigma_{rPi}(r),\sigma_{\theta Pi}(r),\phi_i(z),E_{zi}(z),D_{zi}(z)]\mathrm{e}^{\mathrm{i}\omega t} \quad \text{for piezoelectric layer} \tag{5}$$

$$(u_{rEi},\sigma_{rEi},\sigma_{\theta Ei})=[u_{rEi}(r),\sigma_{rEi}(r),\sigma_{\theta Ei}(r)]\mathrm{e}^{\mathrm{i}\omega t} \quad \text{for elastic layer} \tag{6}$$

where u_{rPi}, σ_{rPi}, $\sigma_{\theta Pi}$, E_{zi} and D_{zi} are the radial displacement, the radial and tangential stress components, the axial electric field and electric displacement of the *i-th* piezoelectric layer, respectively. u_{rEi}, σ_{rEi} and $\sigma_{\theta Ei}$ are the radial displacement, the radial and tangential stress components of the *i-th* elastic layer, respectively. The expressions of $u_{rPi}(r)$, $\sigma_{rPi}(r)$, $\phi_i(z)$, $D_{zi}(z)$, $u_{rEi}(r)$ and $\sigma_{rEi}(r)$ can be written as

$$u_{rPi}(r)=A_{Pi}f_1(r)+B_{Pi}f_2(r) \tag{7}$$

$$\sigma_{rPi}(r)=A_{Pi}f_3(r)+B_{Pi}f_4(r)+e_{31}\frac{V_0}{h} \tag{8}$$

$$\phi_i(z)=z\frac{V_0}{h} \tag{9}$$

$$D_{zi}(z)=A_{Pi}e_{31}k_P J_0(k_P r)+B_{Pi}e_{31}k_P Y_0(k_P r)-\kappa_{33}^{\varepsilon}\frac{V_0}{h} \tag{10}$$

$$u_{rEi}(r)=A_{Ei}f_5(r)+B_{Ei}f_6(r) \tag{11}$$

$$\sigma_{rEi}(r)=A_{Ei}f_7(r)+B_{Ei}f_8(r) \tag{12}$$

where the functions $f_1(r) \sim f_8(r)$ are the known functions defined in the previous work [33]. $J_0(k_P r)$ and $Y_0(k_P r)$ are the first and second kind Bessel functions, respectively.

As shown in Figure 3(b), this layered structure includes two mechanical boundary conditions and six continuous conditions, which can be expressed as follows

$$\begin{cases} u_{rE1}\big|_{r=0}=0 \\ \sigma_{rP2}\big|_{r=R_4}=\bar{\sigma}(t) \end{cases} \tag{13}$$

$$\begin{cases} u_{rPi}\big|_{r=R_{2i-1}} = u_{rEi}\big|_{r=R_{2i-1}} \\ \sigma_{rPi}\big|_{r=R_{2i-1}} = \sigma_{rEi}\big|_{r=R_{2i-1}} \end{cases} \quad (i=1,\ 2) \tag{14}$$

$$\begin{cases} u_{rPi}\big|_{r=R_{2i}} = u_{rE(i+1)}\big|_{r=R_{2i}} \\ \sigma_{rPi}\big|_{r=R_{2i}} = \sigma_{rE(i+1)}\big|_{r=R_{2i}} \end{cases} \quad (i=1) \tag{15}$$

Substituting equation (11) into the first condition in equation (13) yields

$$B_{E1} = 0 \tag{16}$$

Substituting equations (7), (8), (11) and (12) into equations (14) and (15), and combining equation (16), the unknown constants A_{P1}, B_{P1}, A_{E2}, B_{E2}, A_{P2} and B_{P2} can be derived as the expressions of A_{E1} and V_0. Furthermore, combining equations (8) and (13), we can obtain

$$a_1^9 A_{E1} + c_1^9 V_0 = \bar{\sigma}_0 \tag{17}$$

where

$$a_1^9 = a_1^7 f_3(R_4) + a_1^8 f_4(R_4) \tag{18}$$

$$c_1^9 = c_1^7 f_3(R_4) + c_1^8 f_4(R_4) + \frac{e_{31}}{h} \tag{19}$$

Additionally, the electrical condition can be expressed as

$$V(t) = RI(t) \tag{20}$$

where, the electrical charge $Q(t)$ and current $I(t)$ can be expressed as

$$Q(t) = \sum_{i=1}^{2} \int_0^{2\pi} \int_{R_{2i-1}}^{R_{2i}} D_{zi} r d\theta dr = (a_1^{10} A_{E1} + c_1^{10} V_0) \mathrm{e}^{\mathrm{i}\omega t} \tag{21}$$

$$I(t) = dQ(t)/dt = \mathrm{i}\omega(a_1^{10} A_{E1} + c_1^{10} V_0) \mathrm{e}^{\mathrm{i}\omega t} \tag{22}$$

in which

$$a_1^{10} = 2\pi \begin{Bmatrix} a_1^1[f_9(R_2) - f_9(R_1)] + a_1^2[f_{10}(R_2) - f_{10}(R_1)] \\ +a_1^7[f_9(R_4) - f_9(R_3)] + a_1^8[f_{10}(R_4) - f_{10}(R_3)] \end{Bmatrix} \tag{23}$$

$$c_1^{10} = 2\pi \begin{Bmatrix} c_1^1[f_9(R_2) - f_9(R_1)] + c_1^2[f_{10}(R_2) - f_{10}(R_1)] \\ -\frac{\kappa_{33}^{\varepsilon}}{2h}(R_2^2 - R_1^2) + c_1^7[f_9(R_4) - f_9(R_3)] \\ +c_1^8[f_{10}(R_4) - f_{10}(R_3)] - \frac{\kappa_{33}^{\varepsilon}}{2h}(R_4^2 - R_3^2) \end{Bmatrix} \tag{24}$$

Then, equation (20) can be simplified as

$$\mathrm{i}\omega R a_1^{10} A_{E1} + (\mathrm{i}\omega R c_1^{10} - 1) V_0 = 0 \tag{25}$$

Combining equations (17) and (25) yields

$$A_{E1} = d_{01} \bar{\sigma}_0 \tag{26}$$

where

$$d_{01} = -\frac{\mathrm{i}\omega R c_1^{10} - 1}{\mathrm{i}\omega R a_1^{10} c_1^9 - a_1^9(\mathrm{i}\omega R c_1^{10} - 1)} \tag{27}$$

When $R \to \infty$, the open circuit voltage V_{OC} can be obtained. an average power can be obtained as

$$\bar{P} = \frac{|V_0|_{,rms}^2}{R} \tag{28}$$

where the subscript *rms* denotes the root mean square.

3. Experimental Setup

IOP Conf. Series: Materials Science and Engineering **542** (2019) 012075 doi:10.1088/1757-899X/542/1/012075

As shown in Figure 4, an experimental platform is set up to measure the energy harvesting performance of the fabricated cymbal harvester. This testing system consists of a low-frequency fatigue testing machine with a computer controller, a resistor control box and a digital oscilloscope. Prior to performing the experimental calibration, a static pressure of 1100N was first preloaded to clamp the specimen. Afterward, a sinusoidal mechanical load is applied to activate the harmonic vibration, and the resistor is adjusted through the resistor control box. When the generated AC signal observed in the oscilloscope is stable, the experimental data is saved into a flash drive.

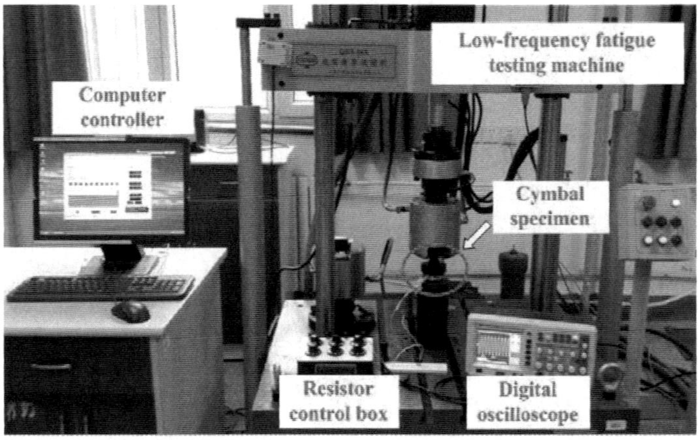

Figure 4. Photograph of experimental setup.

4. Results and Discussion

Firstly, the force transferring coefficient χ was solved for several special examples. Taking $F_0 = 500\text{N}$, f = 20, 25, 30Hz, $R = 100\text{M}\Omega$, the measured output voltage amplitudes are 52.8V, 52.4V, 53.6V, respectively.

For a typical semi-rigid asphalt pavement, the vibration frequency caused by a moving vehicle is between 10Hz and 20Hz [34]. However, road traffic tends to produce vibrations with frequencies predominantly in the range from 5Hz to 25Hz [35]. In order to match this frequency range, the frequencies below 30Hz are focused in the following analysis and discussion.

Figure 5 presents the relationship between amplitude of output voltage and the value of the load resistor. It indicates that the output voltage amplitude increases with the increase in the load resistor. Under a force of 500N at 20Hz, the experimental open circuit voltage amplitude of 52.8V is generated. Further, Figure 6 plots the curve of output average power versus the load resistor. With the increase of the value of load resistor, the output average power reached an optimal value before slowly decaying. Under a force of 500N at 20Hz in the experimental setup, the optimal average power of about 0.92mW is generated across a resistor of 0.8MΩ. Subsequently, Figure 7 shows the output voltage as a function of the applied force. It can be seen that the output voltage increases with the increase of the applied force. To investigate the frequency effects, Figure 8 gives open circuit output voltage as a function of the frequency of the applied force. It can be seen that the frequency of the applied force has slight effect on the open circuit output voltage.

In Figures 5~8, though qualitative agreement is found, there exists the difference among the predicted and experimental results, which may be caused by the following reasons: 1) a preloading was applied to clamp the specimens in the experiment, which was not considered in the approximate theoretical model and the simulation model; 2) the plain stress problem is assumed in the theoretical calculation, which cannot be satisfied in the actual situation.

IOP Conf. Series: Materials Science and Engineering **542** (2019) 012075 doi:10.1088/1757-899X/542/1/012075

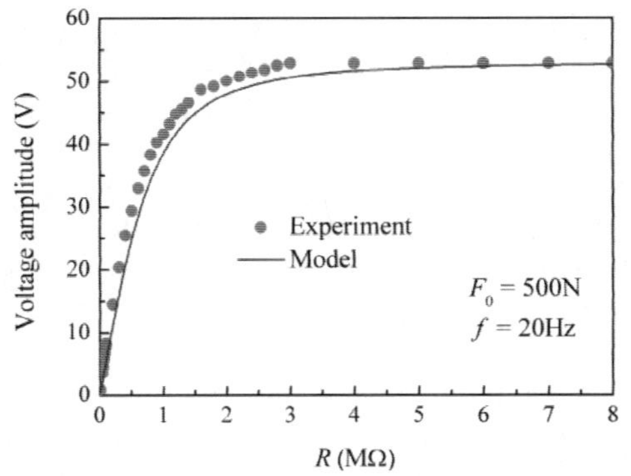

Figure 5. Output voltage amplitude versus the load resistor.

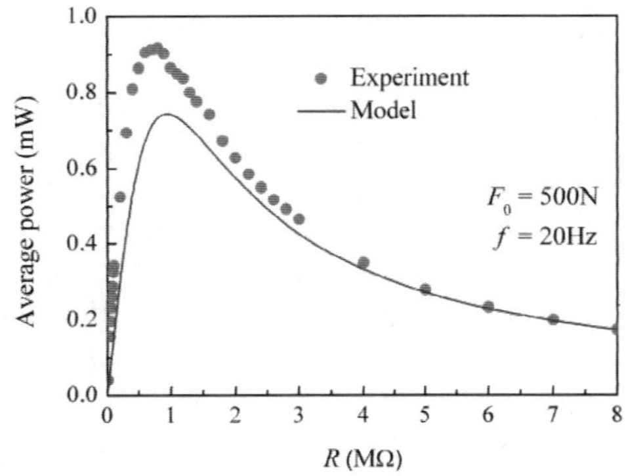

Figure 6. Average power versus the load resistor.

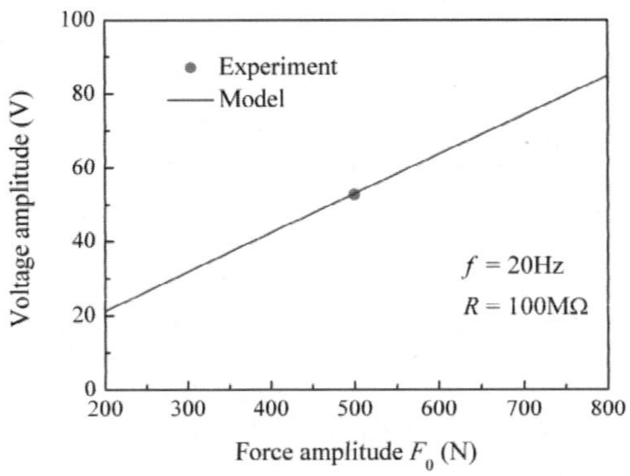

Figure 7. Output voltage versus the force amplitude.

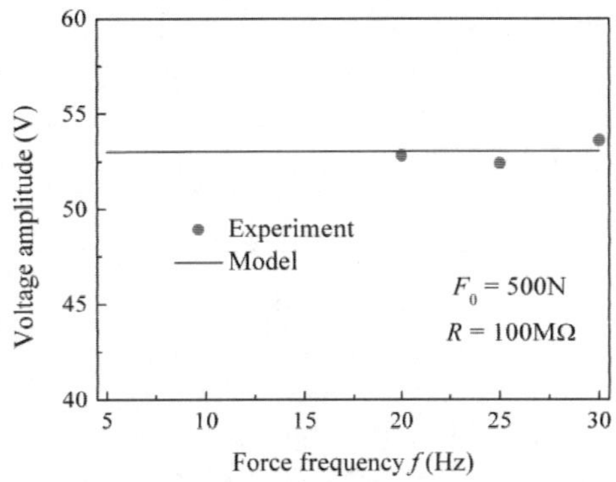

Figure 8. Open circuit output voltage versus the frequency of the applied force.

Further, Table 3 compares the differences of energy harvesting properties between the present and previous work. It can be seen clearly that most of the traditional cymbal transducers [24-26] and the alternative designs [29, 30] can harvest more power (up to 100mW) under lighter forces (less than 70N) and higher frequencies (up to 200Hz). However, the harvested power is very small at the low frequencies [19, 31, 32]. A main reason is that the power decreases dramatically with the decrease of the force frequency [25]. In addition, most of them have the lower load bearing capacity [26]. The above factors will greatly reduce their energy harvesting efficiency in the pavement systems that have the characteristics of high stress and low frequency. To match the characteristics of the pavement systems, the present work attempts to develop a type of new piezoelectric cymbal harvesters through combining the cymbal metal endcaps with the radially layered piezoelectric structure. The results showed that the harvested power is in the level of 1mW at the low frequencies. However, it proved that this type of design is feasible, which provides a new design concept. The authors envisioned that the output power can be greatly increased by adopting the multilayered piezoelectric structure, which similar to the piezoelectric stack.

5. Conclusion

This work explores a type of new radially layered cymbal piezoelectric energy harvester through combining the cymbal metal endcaps and the radially layered piezoelectric structure for energy harvesting from road traffic induced low frequency vibration. The experiment research combined with the theoretical analysis are used to investigate the effects of the load resistor, the amplitude and frequency of force on the energy harvesting performance. The differences on energy harvesting properties between the present and previous work are also highlighted. Different from the previous cymbal designs, our designed cymbal structure includes two piezoceramic layers and three metal layers in radial direction. This design attempts to adopt the layered structural style similar to piezoelectric stack that is expected to improve the power output. Although the harvested power is in the level of 1mW at the low frequencies, this study proves that this kind of design is feasible, which provides a new concept to develop the novel cymbal piezoelectric energy harvesters for energy harvesting in roadway systems. In the future, a multilayered piezoelectric structure that similar to piezoelectric stack will be examined with the hope of enhancing the output power.

Table 3. Comparisons of energy harvesting properties between the present and previous work

Type	Configuration	Literature	Force	Force frequency	Resistor	Harvested power
Traditional cymbal transducer	A piezoelectric disk is sandwiched between two concave shaped metal endcaps	[24]	7.8 N	100Hz	400 kΩ	39mW
		[26]	55 N	100Hz	400 kΩ	33mW
			70 N	100Hz	400 kΩ	52mW
		[25]	70 N	100Hz	480 kΩ	48mW
			70 N	200Hz	200 kΩ	100mW
		[19]	50 N	2Hz	10 MΩ	1.2mW
		[28]	2g acceleration	166Hz	80 kΩ	104.04μW
Alternative designs	Two curved convex steel discs	[31]	24.8N	1.19Hz	2.6 MΩ	0.66mW
	A circumferential slot between plane and conical surface of the cymbal metal endcaps	[30]	8.15N	120Hz	410 kΩ	1.4mW
	18-fringe radial slots in the cymbal endcaps edge	[29]	30N	120Hz	520 kΩ	14.5mW
	18-cone radial slots in the cymbal endcaps	[29]	30N	120Hz	500 kΩ	16mW
	Two-layer-stacked piezoelectric disks	[28]	2g acceleration	153Hz	40 kΩ	141.61μW
	Unimorph (A piezoelectric/steel composite disk)	[32]	1940N	1Hz	3.3 MΩ	121.2μW
	A radially layered piezoelectric/metal disk with the drilled ring	Present	500N	20Hz	0.8 MΩ	0.92mW

Acknowledgments
This work is supported by the National Natural Science Foundation of China (51708025 and 51808007), the Fundamental Research Funds for the Central Universities (FRF-TP-16-069A1) and the China Postdoctoral Science Foundation (2016M600038).

References

[1] Anton SR and Sodano HA 2007 *Smart Mater. Struct.* **146** R1.
[2] Erturk A and Inman DJ 2011 *Piezoelectric energy harvesting* (Chichester: Wiley).
[3] Feenstra J, Granstrom HA and Sodano HA 2008 *Mech. Syst. Signal Pr.* **22** 721-34.
[4] Jiang X, Li Y, Li J, Wang J and Yao J 2014 *J. Renew. Sustain. Ener.* **6** 043110.
[5] Wang JJ, Shi ZF and Han ZJ 2013 *J. Intell. Mater. Syst. Struct.* **24** 1626-36.
[6] Wang JJ, Shi ZF, Xiang HJ and Song G 2015 *Smart Mater. Struct.* **24** 105017.
[7] Zhao S and Erturk A 2014 *Sensor Actuat. A-Phys.***214** 58-65.
[8] Erturk A 2011 *J. Intell. Mater. Syst. Struct.* **22** 1959-73.
[9] Xiang HJ, Wang JJ, Shi ZF and Zhang ZW 2013 *Smart Mater. Struct.* **22** 095024.
[10] Zhang ZW, Xiang HJ and Shi ZF 2016 *J. Intell. Mater. Syst. Struct.* **27** 567-78.
[11] Tang L, Zhao L, Yang Y and Lefeuvre E 2014 *IEEE/ASME T. Mech.* **20** 834-44.
[12] Yang Y, Zhao L and Tang L 2013 *Appl. Phys. Lett.* **102** 064105.
[13] Zhao S and Erturk A 2013 *Smart Mater. Struct.* **22** 015002.
[14] Wang S, Lam KH, Sun KWK, Chan HLW, Guo MS and Zhao XZ 2017 *Appl. Phys. Lett.* **90** 113506.
[15] Yuan T, Yang J, Song R and Liu X 2014 *Smart Mater. Struct.* **23** 125046.
[16] Leinonen M , Palosaari J, Juuti J and Jantunen H 2014 *J. Intell. Mater. Syst. Struct.* **25** 391-400.
[17] Moure A, Rodríguez MI, Rueda SH, Gonzalo A, Rubio-Marcos F, Cuadros DU, Pérez-Lepe A and Fernández J 2016 *Convers. Manage.* **112** 246-53.
[18] Zhao HD, Yu J and Ling JM 2010 *J. Ceram. Soc. Jpn.* **118** 909-15.
[19] Daniels A, Giuliano A, Zhu M and Tiwari A 2013 *Modeling, validation and design analyses of a piezoelectric cymbal transducer for non-resonant energy harvesting* (IEEE International Conference on Green Computing and Communications and IEEE Internet of Things and IEEE Cyber, Physical and Social Computing).
[20] Mo C, Jordan S and Clark WW 2012 *Bimorph piezoelectric cymbal design in energy harvesting* (ASME 2012 Conference on Smart Materials, Adaptive Structures and Intelligent Systems).
[21] Zhao HD, Ling JM and Yu J 2012 *J. Ceram. Soc. Jpn.* **120** 317-23.
[22] Zhao HD, Liang YH and Ling JM 2011 *J. Shanghai Jiaotong Univ.* **45** 62-66.
[23] Dogan A, Uchino K and Newnham RE 1997 *IEEE Trans. Ultrason. Ferroelectr. Freq. Control* **44** 597-605.
[24] Kim H, Batra A, Priya S, Uchino K, Markley D, Newnham RE and Hofmann HF 2004 *Jpn. J. Appl. Phys.* **43** 6178-83.
[25] Kim H, Priya S and Uchino K 2006 *Jpn. J. Appl. Phys.* **45** 5836-40.
[26] Kim H, Priya S, Uchino K and Newnham RE 2005 *J. Electroceram.* **15** 27-34.
[27] Huang RB, Niu YL, Zhao HD and Chang HB 2012 *China J. Highway Transp.* **25** 1-8.
[28] Tufekcioglu E and Dogan A 2014 *Sensor Actuat. A-Phys.* **216** 355-63.
[29] Yuan J, Shan X, Xie T and Chen W 2009 *J Zhejiang Univ. Sci. A* **10** 1187-90.
[30] Yuan J, Shan X, Xie T and Chen W 2010 *J. Intell. Mater. Syst. Struct.* **21** 765-71.
[31] Palosaari J, Leinonen M, Hannu J, Juuti J and Jantunen H 2012 *J. Electroceram.* **28** 214-19.
[32] Mo C, Arnold D, Kinsel WC, Clark WW 2013 *J. Intell. Mater. Syst. Struct.* **24** 828-36.
[33] Wang JJ and Shi ZF 2013 *IEEE Trans. Ultrason. Ferroelectr. Freq. Control* **60** 2196-2203.
[34] Zhao HD, Qin L and Ling JM 2015 *Int. J. Transport. Sci. Technol.* **4** 17-28.
[35] Hunaidi O 2000 *Traffic vibrations in buildings* (Construction Technology Update: www.nrc-cnrc.gc.ca/ctu-sc/ctu_sc_n39).

IOP Conf. Series: Materials Science and Engineering **542** (2019) 012076 doi:10.1088/1757-899X/542/1/012076

Analysis of Fatigue Characteristics of Gears Based on nCode DesignLife

Suxin Chen[1,a], Yingxue Yang[2,b,*], Tao Wang[3,c], and Jie Zhang[4,d]

[1,2,3,4] School of Mechanical Engineering, Hefei University of Technology, Hefei, 230009, China

E-mail: [a]chensuxin2003@163.com, [b]1508046943@qq.com, [c]1171835691@qq.com, [d]295737015@qq.com

Abstract. For the phenomenon that the straight bevel gear transmission system is prone to fatigue failure under cyclic loading, the fatigue characteristics of gears are studied, and a fatigue damage simulation method is proposed. Taking a pair of straight bevel gear in driving line body as the research object, the dynamics analysis of the gear contact model was carried out through the ADAMS software, and the virtual load spectrum of the gear contact meshing was obtained. With the help of Workbench finite element analysis software, the gear static finite element analysis was performed to obtain the stress distribution and the weak points of fatigue of the gears. In nCode DesignLife software, the fatigue damage of the gear was estimated by loading the virtual load spectrum in the fatigue hazard area according to the nominal stress method. The stress concentration at the tooth roots of the gears obtained through simulation results in the risk of breakage of gear teeth, and the gears need to be optimized. The simulation provides reference for the fatigue reliability of the design, shortens the design cycle, and provides guarantee for the reliability of the line body.

1. Introduction

Gear transmission has advantages of stable transmission ratio, high transmission efficiency and reliable operation, which makes it widely used. Due to the structural characteristics of the gear and the transmission characteristics, the gear is prone to fatigue failure such as broken teeth and pitting on the tooth surface under repeated load for a long time. Therefore, the reliability of the gear transmission system has received much attention.

Most engineering structures or mechanical failures are caused by the accumulation of fatigue damage caused by a series of variable amplitude cyclic loads. Huang obtained the load spectrum based on the rain flow counting method and then used the Brown-Miller algorithm for fatigue damage analysis [1]. Ding based on the load spectrum simulated by the finite element method for fatigue life estimation [2]. Wang estimated the fatigue damage of herringbone gear based on rainflow counting method combined with Miner linear cumulative damage theory [3]. Li proposed a method for predicting structural fatigue life based on the multibody dynamics simulation and Finite Element Method [4]. M. Kamal [5] summarized a variety of fatigue damage estimation methods. According to the nominal stress method, based on the S-N curve of the material, the stress concentration factor and the nominal stress of the structural fatigue hazard area can be combined with the fatigue damage accumulation theory to calculate the fatigue damage [6].

2. 3-Dimensional Model

In gear transmission systems for assembly lines, metal gears and plastic gears are often paired to increase heat dissipation and reduce wear. The driving gear is made of SMF5040 powder metallurgy material

IOP Conf. Series: Materials Science and Engineering **542** (2019) 012076 doi:10.1088/1757-899X/542/1/012076

grade, and the driven gear is POM. Modeled in SolidWorks, the resulting 3D assembly model is shown in Figure 1.

Figure 1. Three-Dimensional Model of the Bevel Gears (Driven gear is a view of the left, and the right is the driving gear).

3. Dynamic Model

Gear meshing transmission mainly relies on the impact force generated by contact collision [7, 8], including normal contact impact force and tangential friction force. The normal contact impact force can be calculated by the Impact function in ADAMS:

$$F_n = \begin{cases} K(x_1 - x)^e - step(x, x_1 - d, c, x_1, 0)\dot{x} & x \leq x_1 \\ 0 & x > x_1 \end{cases} \tag{1}$$

where K is the contact stiffness coefficient, x_1 is the displacement switch to determine whether the one-side collision is active, x is displacement variables, e is the nonlinear spring force exponent, c is the damping coefficient, and d is the penetration depth of the two contact objects when the damping reaches the maximum.

Friction can be calculated by Coulomb function:

$$F_f = -F_n \times step(v, -v_s, -1, v_s, 1) \times step\big(ABS(v), v_s, c_{st}, v_d, c_{dy}\big) \tag{2}$$

where F_n is the normal force, v is the relative sliding velocity, v_s is the stiction transition vel, v_d is the friction transition vel, c_{st} is the static coefficient, and , c_{dy} is the dynamic coefficient.

According to the actual working conditions, the angular velocity of the driving gear is 6.8 $rad \cdot s^{-1}$, the load of the driven gear is 347 $N \cdot mm$, and both of them are increased gently within 0.2 seconds. The main parameter settings are shown in Table 1. The speed-time curve of the driven driven gear is shown in Figure 2. According to the rotational speed diagram of Figure. 2, it can be seen that the rotational speed of the gear meshing basically satisfies the 1:1 relationship, indicating that the established dynamic model is feasible. Gear meshing force - time curve is shown in Figure 3.

Table 1. Set of Contact Force.

K	c	v_s	v_d	d	e	c_{st}	c_{dy}
$3.0 \times 10^5\, N \cdot mm^{-1}$	$100\, N \cdot s \cdot mm^{-1}$	$10\, mm \cdot s^{-1}$	$100\, mm \cdot s^{-1}$	$0.1\, mm$	2.2	0.1	0.05

From Figure. 2 and Figure. 3, the gear rotation speed is abrupt at the beginning, and the gear meshing force is also abruptly changed, and then the rotation speed and the contact force fluctuate around a level. The fluctuation of the driven gear speed and the contact force between the two gears is due to vibration and shock during the contact engagement.

IOP Conf. Series: Materials Science and Engineering **542** (2019) 012076 doi:10.1088/1757-899X/542/1/012076

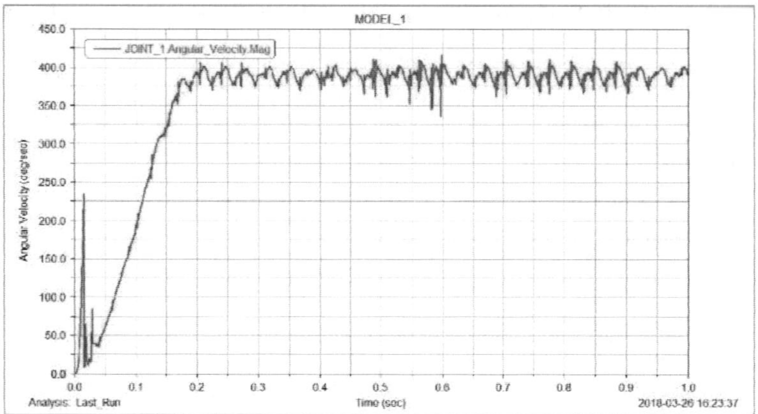

Figure 2. Rotation speed-time course curve of the driven geer is presented.

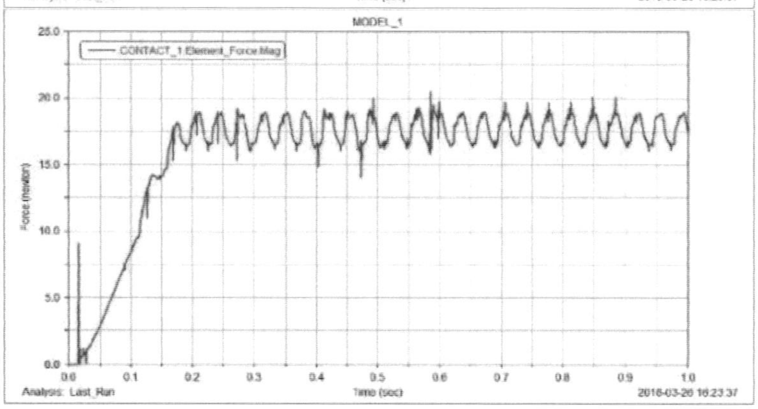

Figure 3. Meshing force between the two gears is presented.

4. Finite element model

Static contact simulation analysis can be calculated in Workbench. Considering the iterative calculations, it is necessary to speed up convergence [9]. Unreasonable solution of parameter settings and initial equilibrium condition settings will affect the convergence of the solution [10]. An angular displacement can be applied to the drive wheel and a load torque can be applied to the driven wheel. This setting facilitates solution convergence. The simulation results are shown in Figure 4.

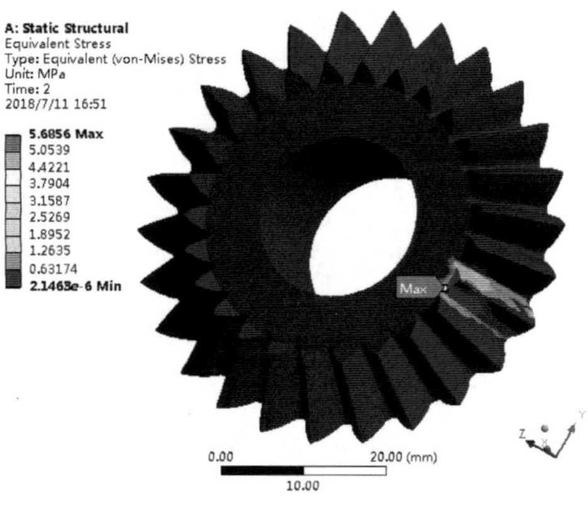

Figure 4. Stress Distributing of the gear teeth.

It can be seen from Figure. 4 that the maximum stress of the gear is located at the root of the tooth, the value is 5.6856 MPa, which is smaller than the yield strength of the POM of 60.6 MPa, indicating that the static strength of the gear meets the requirements.

5. Finite fatigue life

The fatigue damage of the gears was analyzed by using the CAE analysis software DesignLife from NCode, UK. It uses the nominal stress method as the theoretical basis to analyze the fatigue of the

IOP Conf. Series: Materials Science and Engineering **542** (2019) 012076 doi:10.1088/1757-899X/542/1/012076

component. The basic principle of the analysis is: the stress-strain results obtained by static analysis of the components using finite element analysis software are combined with the S-N curves of the component materials, and then the fatigue estimation theory is applied to calculate the distribution of fatigue damage of the components throughout the components.

Establish a fatigue analysis modal in nCode DesignLife software and calculate gear fatigue damage values. The finite element static analysis file of the powder metallurgy gear obtained in the workbench was imported into the FEInput module, and the gear meshing load spectrum file obtained by the dynamic simulation in the ADAMS was input into the SNAnalysis module through the Time Series, and the S-N curve of the material was input. Finally, the fatigue damage of the gear was obtained. The flow diagram of analysis is shown in Figure 5. And the simulation results are shown in Figure 6.

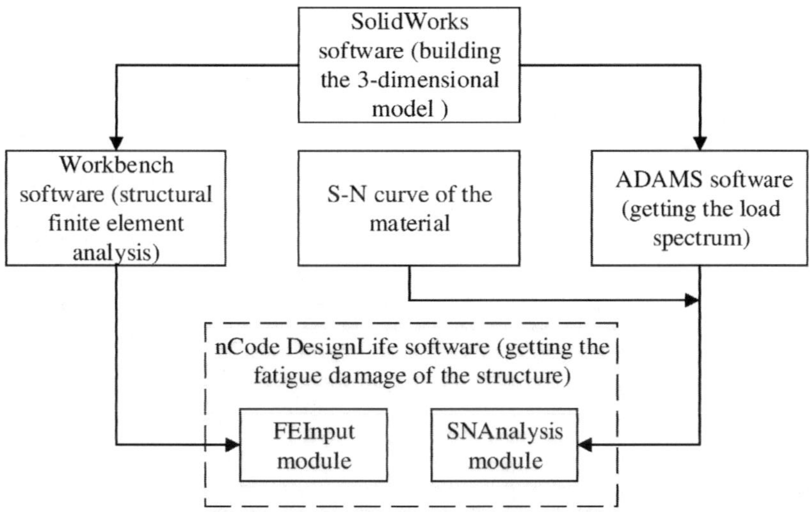

Figure 5. The flow diagram of fatigue damage analysis.

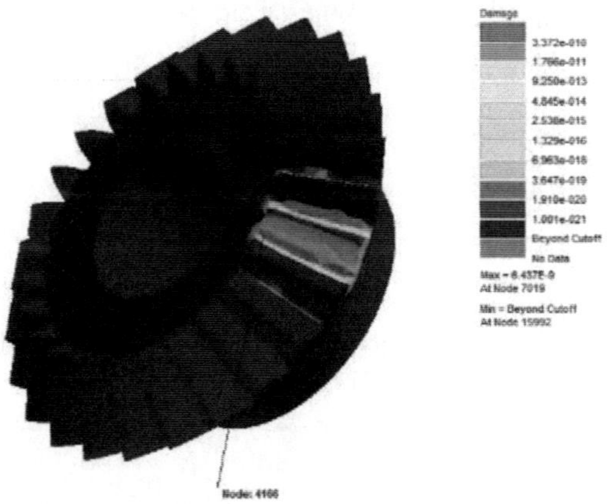

Figure 6. Fatigue damage distribution nephogram of original gear.

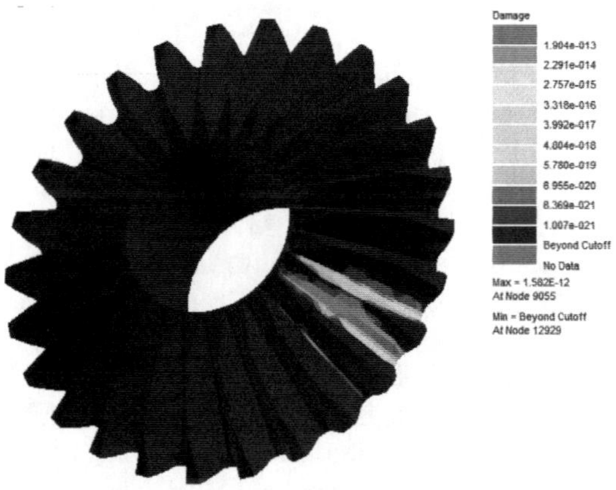

Figure 7. Fatigue damage distribution nephogram of optimized gear.

It can be seen from Figure. 6 that the root of the tooth is the weak point of the gear, and the stress concentration is easy to occur here. The position with the node number 7019 is the most dangerous, and its damage value is $6.347e^{-9}$. Therefore, measures need to be taken to improve the gear structure and reduce the stress concentration. For the root stress concentration problem, the gear is redesigned and the fatigue analysis is carried out. The fatigue damage diagram of the optimized gear is shown in Figure. 7.

It can be seen from the figure that the maximum damage value is reduced from the original $6.347e^{-9}$ to $1.582e^{-12}$, indicating that the structural optimization direction is reasonable.

6. Conclusion

In this paper, the powertrain assembly line is taken as the research object. The main research content is the fatigue damage analysis of the bevel gear transmission system in the conveyor line. The fatigue stress analysis of the gear of the transmission system is carried out by the nominal stress method. A series of simulation analysis of the gears was carried out, and the fatigue damage distribution cloud map of the parts and the specific values were obtained. The main conclusions are as follows:

(1) Based on the Adams dynamics simulation software, the simplified dynamic model of the bevel gear transmission system is established. The simulation of the transmission system under normal working conditions is carried out. The rotation speed-time course curve of the driven gear and the meshing force curve between the two gears are obtained. The correctness of the collision coefficient setting when the dynamic model is established is verified by the rotational speed diagram of the gear.

(2) The finite element analysis model of the bevel gear is established, and the stress distribution cloud diagram of the bevel gear is obtained. The dangerous part of the bevel gear is determined, which provides the directionality for the subsequent fatigue damage analysis.

(3) The process of fatigue damage analysis of bevel gears was built in nCode DesignLife. The finite element analysis results and load spectrum were imported into the input module, and the material properties and simulation parameters were set. After the analysis, the fatigue damage distribution cloud diagram of the bevel gear is obtained, which is consistent with the stress distribution cloud map. It provides an important basis for the structural design of the bevel gear. After optimizing the three-dimensional model of the gear, the gear fatigue damage value is reduced and the gear life is increased. The simulation verifies that the method can effectively verify the rationality of the gear design, shorten the design period of the gear, and guarantee the reliability of the line body.

7. References

[1] Huang Y R. Exploration of Fatigue Property of Orthogonal Face Gear based on FEM [J].Journal of Mechanical Transmission, 2014(8):51-55.

[2] Ding S, Liu S J, Hu Y. Evaluation method of fatigue life for split torque transmission spur gears in helicopter [J].Journal of Central South University (Science and Technology), 2015(1):135-140.

[3] Wang F, Z.D. Fang, S.J. Li, J.K. Jiang. Fatigue Analysis and Optimization of Marine Herringbone Gear Based on Tooth Root Dynamic Stress [J]. Engineering Mechanics, 2015(7):184-189.

[4] Li K, Tang Y, Zhao J. Fatigue simulation of transmission system of hose reel irrigators [J]. Journal of Drainage and Irrigation Machinery Engineering2017, 35(8):725-730.

[5] Kamal M, Rahman M M. Advances in fatigue life modeling: A review [J]. Renewable & Sustainable Energy Reviews, 2018, 82:940-949.

[6] Yao W X. Fatigue Life Prediction of Structures [M]. National Defense Industry Press, 2003.

[7] TANG J Y, ZHOU W, Chen S Y. Contact-impact Analysis of Gear Transmission System [J].JOURNAL OF MECHANICAL ENGINEERING, 2011, 47(7):22-30.

[8] Wei J, Zhang A Q, Qin D T, Lim T C, Shu R Z, Lin X Y and Meng F M. A coupling dynamics analysis method for a multistage planetary gear system [J]. Mechanism & Machine Theory, 2017, 110:27-49.

[9] Aköz Ahmet, Kadıoğlu Fethi, Tekin Gülçin. Quasi-static and dynamic analysis of viscoelastic plates[J]. Mechanics of Time-Dependent Materials, 2015, 19(4):483-503.

[10] Zhou C, Li Z, Hu B, Zhan H, and Han X. Analytical Solution to Bending and Contact Strength of Spiral Bevel Gears in Consideration of Friction [J]. International Journal of Mechanical Sciences, 2017, 128.

Design of Automatic Billiard Sports Service Machine

Luo Sichun, Wen Tanggen, Lan Xiang, Gong Jin and Tian Chunlai*

School of Mechanical and Electronic Engineering, Pingxiang University, Pingxiang, Jiangxi province, 337000, China
E-mail:tianchunlai@bit.edu.cn

Abstract. The billiard sports service machine which adapts to most of the billiards matches now is designed in this paper. It has strong environmental adaptability with the purposes of saving manpower, high efficiency and smooth transport of billiards. It has five main parts, such as billiards collection part, billiards sorting part, billiards push up part and billiards placement part. It is operated according with the working principles of billiard service machine. It uses the color sensor to identify the color of billiard balls. It applies the single chip microcomputer to control the mechanical actors. It makes use of the rubber sucker and the vacuum pump to grab the billiard balls. The waiting time for the new balls arrangement is shorten significantly to help players enjoy the game. The design with the feasibility plan is provided for the realization of the service for the automated billiard sports.

1. Introduction

The form of billiards is very diverse, such as snooker, pall billiard and more. Because billiards can be both competition and entertainment. In recent years, billiards sports have become more and more popular, and billiards competitions are also increasing [1]. Although scholars have a lot of research on billiard sports. However, scholars mainly focus on billiard robots, such as: visual recognition, intelligent algorithms, etc. [2, 3]. There is not much research on the billiard sports service machine. Researchers have studied the billiard tables [4, 5]. Their research has played a positive role in the development of billiards. The service of billiards sports is just in the stage of manual service. In this paper, a billiard service machine is designed to reduce the waiting time of players in entertainment or in competition.

2. Working process and overall structure

2.1. Working process

After the game, billiard balls dropped into the holes and rolled to the bottom left of the pool table. The billiard division car part is driven to the specified position according to the specified path. The baffle plate is opened under the billiard table. The billiard balls that gathered together could roll into the ball box of the car part. After all the billiard balls enter the ball box, the pneumatic push rod under the ball box will lift the wider end of the ball box. The billiard ball will roll. The billiard ball is hindered by the ball blocking device.

A color sensor is set at the narrow mouth of the ball box. The billiard ball will be identified to prepare for the next arrangement one by one. After the color is identified, the billiard ball will enter the first division plate. It divides the billiard balls into two channels according to its color. One is to handle billiard balls that needs to be rhombus or triangular, and the other one is to handle a single billiard ball.

A billiard ball is needed to be rhombus or triangular enter the second division plate. The cam mechanism divides the billiard ball in the second division plate. It is pushed onto the model board by the push plate to complete the billiard shape. However, the single billiard ball can only be raised by the pneumatic putter.

After the collection finished, the billiard division platform will be raised under the driving of the three jaw lifting platform. The three direction manipulator grabs the billiard ball and puts the billiard ball on the billiard table. Finally, the machine is restored as the initial status.

2.2. Overall structure design

Automatic billiards machine mainly consists of three parts, such as power system, transmission system and function module system. The functional modules include four parts. They are the billiard collection system, the billiard sorting system, the billiard lifting system and the billiard placement system.

The system diagram is shown in Fig.1. Firstly, the drive system sends the car to the designated position. Dropped billiard ball is collected by the collection system and will be send to the billiard sorting system. The sorting system separates the billiard ball by color. Then the billiard sorting system on the billiard division car part will put it into the desired shape box. After that, the lifting system on the billiard division car will lift the billiard division platform to achieve the position by the billiard placement system. Finally, the three direction manipulator in the billiard placement system will grab the ball and deliver it to the designated position of the table. The power system provides power support for the automatic billiards machine.

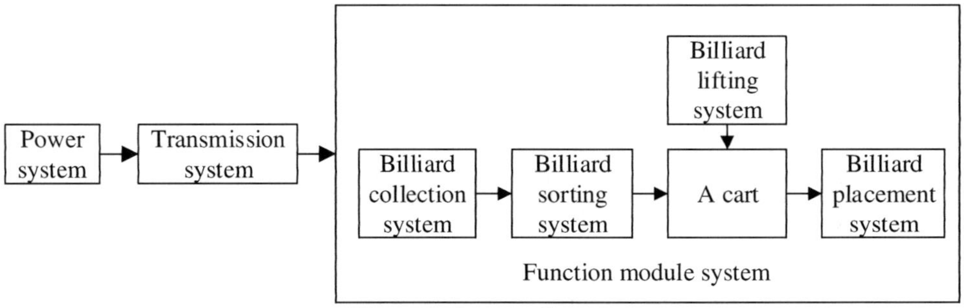

Figure 1. System diagram.

The design of the automatic billiard machine is shown in Fig.2. As shown in Fig.2, the automatic billiard machine axonometric is made up of a billiard table, a three direction manipulator and a billiard division car. It can be reformed on the basis of a traditional billiard table. The billiard collection box is set at the bottom. The billiard division car is placed on the other side of the billiard table. The three direction manipulator can be attached to the top of the billiard table.

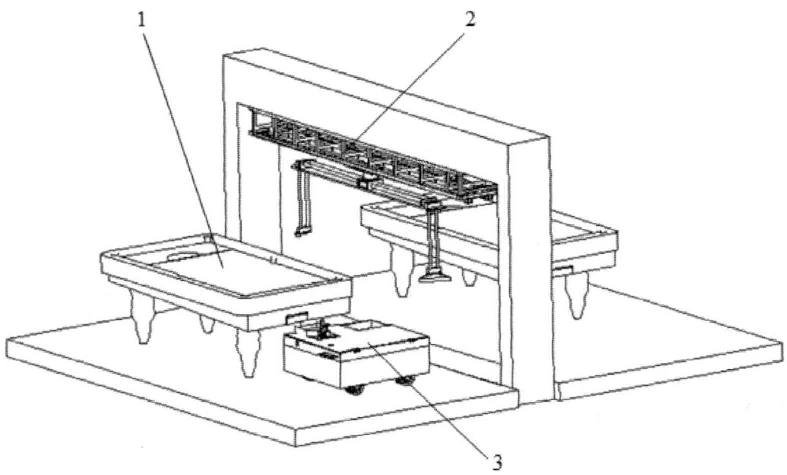

1. Billiard table; 2. Three direction manipulator 3. Billiard division car

Figure.2. Automatic billiard machine.

3. The design of the billiard division car

3.1. Design of automatic ball blocking device

The automatic ball blocking device is the key part of the billiard division car. It mainly plays the role of billiard buffer. The model is shown in Fig. 3. The power gear drives the rack moving up and down. The power shell protects the power gear. The “L” plate 1 fixed block and the “L” plate 2 fixed block are fixed on the rack. The “L” plate 1 and the “L” plate 2 are fixed on the “L” plate 1 fixed block and the “L” plate 2 fixed block, which can be moved up and down to follow the rack. The degree of engagement between the power gear and the rack is high to ensure the time of billiard ball color recognition. The power gear and rack are designed to be 7 to 8 stages of machining accuracy.

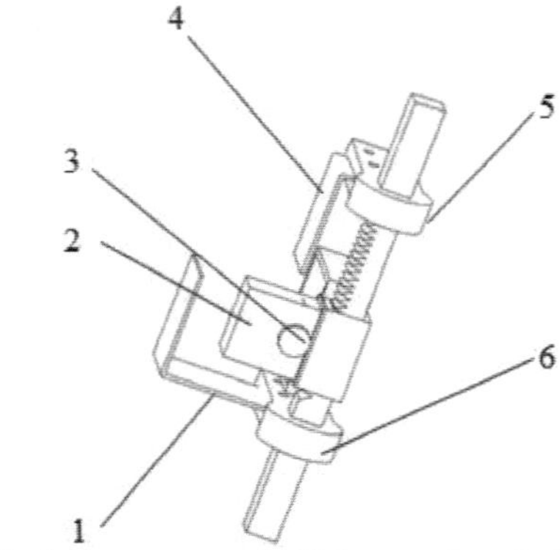

1. "L" type plate 1; 2. Power shell; 3. Power gear; 4. "L" type plate 2; 5 "L" plate 2 fixed block; 6 "L" plate 1 fixed block

Figure.3. Automatic ball blocking device.

3.2. Design of division ball box

The function of the division ball box is to store the balls and identify the color of billiard ball one by one. Its wireframe structure is shown in Fig.4. The outer rod of the pneumatic lifting rod is not fixed. The ball box is pulled by the inner lifting rod of the pneumatic hoisting rod. The color sensor, which is set at the narrow mouth of the ball box, is used to identify the color of billiard ball one by one. The color signal controls the motor rotation and rack moving up and down. In order to prevent the collision between the balls and the box, a layer of cashmere material is added on the box surface to ensure that the noise is reduced. The cylinder stroke is about 50 mm to 60 mm. There is no influence on the efficiency of the divide ball.

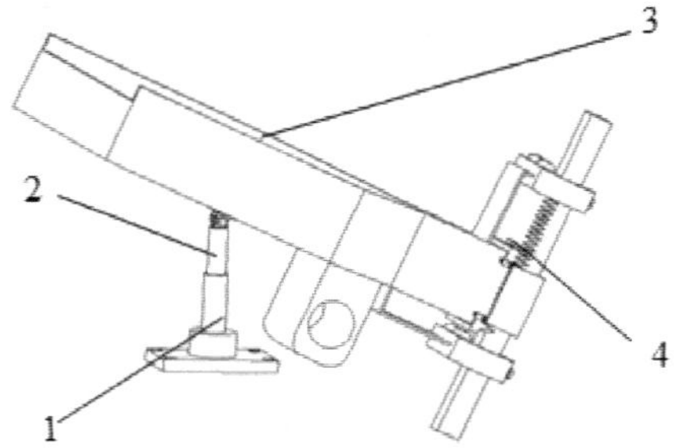

1. Pneumatic hoisting rod outer bar; 2. Pneumatic hoisting rod inner rod; 3. Ball box; 4. Identifier;

Figure.4. Division ball box.

3.3. Design of the first-level division device

As shown in Fig.5, the rocker active force drives the rocker to rotate at a certain angle. The rocker moves the dial to rotate 90 degrees to achieve the billiard lane separation function. The first channel is a passageway for balls. The core of the device is the rocker mechanism. The rocker lever accurately moves the lower part of the pick to realize the steering pick function. It realizes the lane separation of the balls.

3.4. Design of the second-level division device

As shown in Fig.6, there are four cams and baffles independent applied with each other. The baffle is raised according with the color signal. Then the billiard balls will enter different spaces, and the next step can be achieved. The mechanism uses a cam mechanism to drive the baffle so as to realize the function of the baffle blocking ball. The mechanism is characterized by the movement of each cam mechanism without interference from another one. The rotation of the cam is controlled only by the color signal. The baffle can keep their respective movements for the balls entrance through the different tracks.

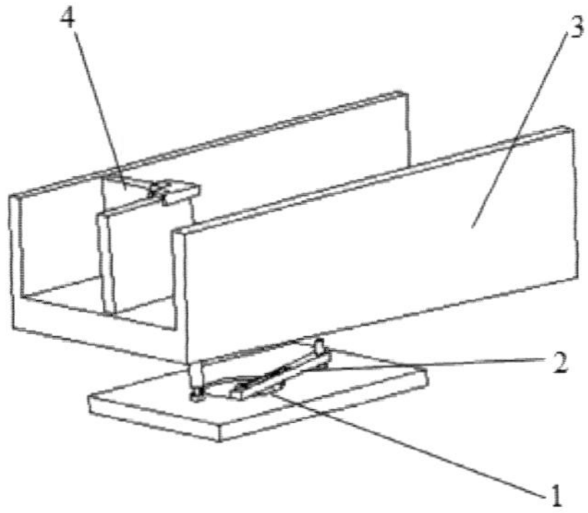

1. Rocker active force; 2. Rocker ;3. Primary lane board; 4. Pick

Figure.5. First-level division device.

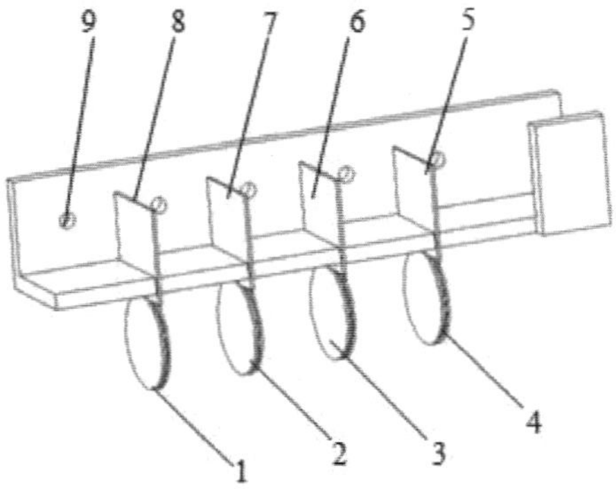

1-4. Cam; 5-8. Baffle; 9. Push rod jack

Figure.6. Second-level division device.

4. The design of the grab suction cup

The triangle suction cup is shown in Fig.7. It is fixed at the end of the manipulator. It is operated by the gas of inflating and deflating under the control of the single chip microcomputer. The ball is sucked up with the suction of the vacuum sucker, so that it is caught from the billiard division platform to the table. If it is needed by the pall billiard (Nine-Ball) arrangement, the sucker could be set as the rhombus.

Figure.7. Triangular sucker.

4.1. Design calculation of vacuum sucker

It uses a cylindrical sucker configured with an outer diameter of 60 mm, an inner diameter of 56 mm and a height of 30 mm. The material uses soft rubber, which is conducive to the adsorption of billiard balls.

Firstly, the vacuum sucker is connected with the vacuum device (vacuum pump) through the tubing when working. Then the sucker contacts with the billiard ball and starts the vacuum equipment to produce the negative pressure in the suction cup. After that the billiard ball will be sucked up and can be moved by the manipulator. When the billiard ball is moved to the destination, the vacuum pump is steadily inflated into the vacuum suction cup to make the cup from negative pressure to zero pressure or slightly positive pressure. Finally, the ball is set out of vacuum sucker and arranged well.

The advantage is that it will not damage the billiard table. The billiard material is kept good, because of the sucker is made from more resilient material. The absorbing or putting down billiard balls would not cause any damage to the appearance of billiard balls.

The formula for calculating the suction of a vacuum suction cup is

$$W = \left(pA / f\right) \times 1.778 \times 10^{-4} \tag{1}$$

where W is the adsorption force; p is the vacuum degree (relative pressure) in the suction cup; A is an effective adsorption area for the sucker and f is the safety factor.

Usually, the effective adsorption area of sucker is about 80% of the suction cup area, and the vacuum degree takes about 90% of the maximum value. The safety factor varies with the conditions of use. According to the data, the weight of billiard ball is between 154.5 g and 195 g. It gets 200 g as the weight of the target ball. It is assumed that the suction cup needs to absorb the ball's weight. In terms of the safety factor is 5 and the vacuum degree is up to 90%, the suction force of the rated suction plate is 2.52 N, which is larger than the billiard gravity. Therefore, it meets the requirements.

5. Summary

The automatic billiard sports service machine with mechanical design and operation principle research is presented here. It resolves problems of the billiard balls collections after the competitions. The billiard balls are arranged into a triangle or a diamond by the designed sucker. The manipulator could help to obtain the placement of the snooker billiards and pall billiards (Nine-Ball). The rubber sucker and vacuum pump are used as the starting points of the manipulator to avoid from causing damage to the billiard table while the general machine is swinging the ball. In all, the automatic billiards sports machine has great complement to the shortcomings in the traditional manual collecting sports. It provides a feasible scheme for the realization of automatic billiard competitions.

6. References

[1] J. Chen 2006 Analysis of the Development Status and Prospects of Billiards in China *J. SPORTS. ADULT EDU.* **22** 61-62

[2] B. R. Cheng, J. T. Li and J. S. Yang 2004 Design of the Neural-Fuzzy Compensator for a Billiard Robot *IEEE International Conference on Networking, Sensing and Control* (Taipei:IEEE) vol.2 pp 909-913

[3] S.C. Chua and E. K. Wong 2003 Pool Balls Identification and Calibration for a Pool Robot *International Conference on Robotics, Vision, Information and Signal Processing* (Changsha: IEEE) pp 312-315

[4] J.N.S. Blair, A.W. Callif, J.M. Maccarthy and J. Zebley 2009 U.S. Patent 597623

[5] Y Wang, L Zhong and H Wang 2014 P.R.C. Patent 203724745U

7. Acknowledgements

This work is supported by the Chinese National Training Program of Innovation and Entrepreneurship for Undergraduates (No. 201810895008) and the major construction project of Pingxiang University (No. 201701). And thanks to the help from the key laboratory of modern mining machinery design and intelligent manufacturing at Pingxiang University.

IOP Conf. Series: Materials Science and Engineering **542** (2019) 012078 doi:10.1088/1757-899X/542/1/012078

Design of Curling Sports Service Robot

Tian Chunlai*, Chen Qideng, Feng Quanquan and Luo Sichun

School of Mechanical and Electronic Engineering, Pingxiang University, Pingxiang, Jiangxi province, 337000, China
E-mail: tianchunlai@bit.edu.cn

Abstract. In order to implement handling and unmanned maintenance for curling in sports, a service robot is designed in this paper. It could save professional labor costs and improve work efficiency for the curling sports service with strong usability. Through analysis on the curling's feature, the mechanical model of the service robot is established by three-dimensional CAD software. The robot consists two functional sub robots. One is the curling handling robot and the other one is the curling maintenance robot. The design includes the robotic arm systems, maintenance systems, intelligent identification, industrial CT inspection systems and pneumatic control systems. Referred to the mechanical structure of the excavator, the mechanical arm movement is driven by the cylinder. The handling curling has strong stability. The maintenance system adopts a circular layout design. The simulated motion of the design is carried out with the rationality of mechanical structure analyzed. It can make reasonable serve sports and improve work efficiency significantly.

1. Introduction

The competition level of the curling sports in China has been progressed rapidly. It has made great achievements in the international competitions, which greatly gained wide recognition. The curling is expensive with its special material and high quality manufacturing [1]. As other sports equipment, the curling needs the regular maintenance in sports. The maintenance has several complicated steps with strictly rules. It is demanding with professional techniques [2]. It requires professional personnel to carry out the work. If users are lack of maintenance awareness, the service life of curling will be shortened. There are a few sport service machines for the curling, such as the decision robot, the innovation brush and others [3, 4, 5].

In this paper, a curling sports service robot is mainly designed for handling and maintenance of curling from professional manual operation to mechanical automation. It is a typical service robotics machine [6]. It could extend the service life of curling, reduces the cost of professional manual maintenance and could make the curling sports or games more popular [7]. The robot is suitable for the curling field with strong stability according with the wheel system design [8]. Its internal structure runs in an orderly flow. It could make rational use of space and improve work efficiency with saving a little cost. And it has positive significance in popularization of the curling sports.

2. Function of curling sports service robot

The robot makes handling and maintenance of curling from manual operation to mechanical automation. It consists two functional sub robots. One is the curling handling robot and the other one is the curling maintenance robot. The functional structure diagram with the main parts and system is shown in Fig.1. Firstly, the handling robot collects the curling from field. It picks up the curling and transfers it to the maintenance robot. The curling is pushed by the door conveyor and tested through an industrial CT inspector. Then, by cooperation of the lifting conveyor and the arms, the bottom is exposed. The maintenance process in carried out soon and the curling position is switched under the rotation of the table. The bottom of the curling is sequentially followed by the scrubbing arm, the waxing arm, the baking arm, the gas nitrogen cooling arm and the polishing arm. Finally, the professional maintenance work of the curling is finished.

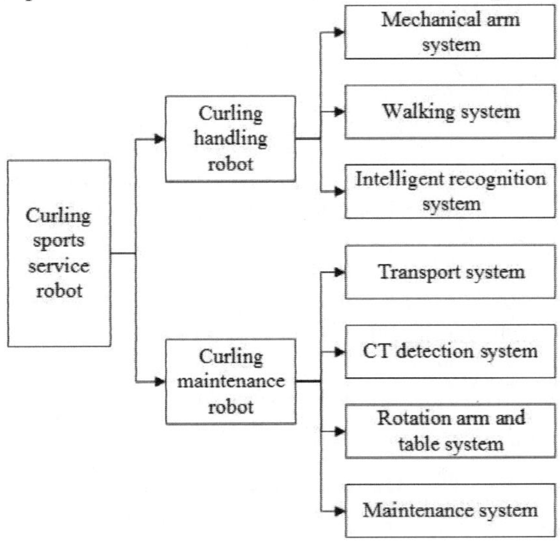

Figure.1. Functional structure diagram.

3. Design and configurations

3.1. Curling handling robot.

The curling handling robot is designed as shown in Fig.2. In the figure, Label 1 is the balance linkage which is used tomaintain horizontal levels during curling to ensure stability during handling. Label 2 is the small cylinder numbered 1. It is used to drive the balance link. Label 3 is the robotic arm. Label 4 is the manipulator drive cylinder, which powers the movement of the arm. Label 5 is the robotic arm support. It is used to fixed and supported robotic arm system. Label 6 is the Mecanum wheel. It is a kinds of universal wheel and enables the robot to move in all directions. Label 7 is the chassis. Label 8 is high definition camera with aims to position curling. Label 9 is the small cylinder numbered 2. It provides the grab power. And Label 10 is the suck for the claw. It is used to grab the curling.

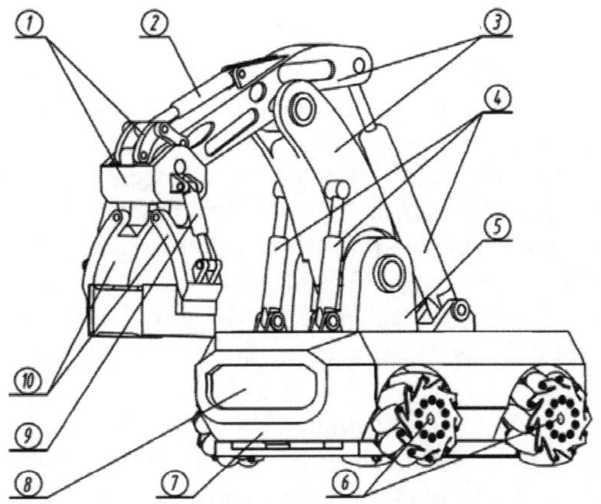

Figure.2.Curling handling robot.

The extension and contraction of the robot arm of the curling handling robot is mainly driven by the linear cylinder motion. Its mechanical structure mainly refers to the structure of the construction machinery, such as the excavator. The balance linkage mechanism ensures the stability of the transportation, which is flexible and ingenious. It adopts the cylinder as the economical power supply. The curling claws is designed according to the size of the curling. The assembly suction cup assists in the grasping, which could flexibly grab the curling without damage to the curling on the field. The machine is configurated with four Mecanum wheels, with four

IOP Conf. Series: Materials Science and Engineering **542** (2019) 012078 doi:10.1088/1757-899X/542/1/012078

secondary gearboxes and with four independent motors controlled by control and power supply system. Based on the omnidirectional motion characteristics of the Mecanum wheel, free and quickly motion can be realized with a high maneuverability of the robot.

3.2. Curling maintenance robot.

The design of the curling maintenance robot is shown in Fig.3. Its appearance resembles looks like a large curling and it is successfully integrated into the curling venue. Since the weight of curling is heavy, the curling maintenance robot is fixed on the field corner during work. The curling handling robot will carry the curling to the curling maintenance robot, and put the curling into the interior of the maintenance robot through the door conveyor. As shown in Fig.3, Label 1 is the enclosure containment of the robot. The containment is designed as similar as the curling, which makes the robot successfully integrated into the curling field. Label 2 is the industrial CT inspection system, which is used to detect the quality overall of the curling. Label 3 is the cabin landing gear. Label 4 is the door cylinder. It powers the switch of the door conveyor.

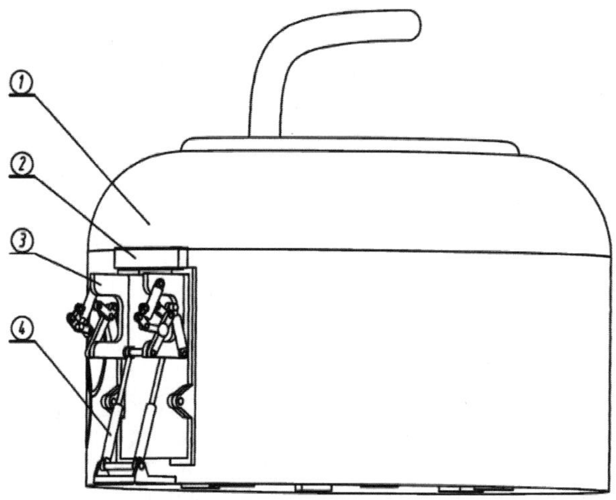

Figure.3. Curling maintenance robot.

3.2.1. Internal of the curling maintenance robot.

The internal of the curling maintenance robot is shown in Fig.4. As show in Fig.4, Label 1 is the pillar, which is used to support the suction cup rotating arm. Label 2 is the scrub arm. It wipes small portion of the ice powder remaining on the bottom. Label 3 is the polishing arm. It polishes the bottom of the curling pot. The bottom is exposed to the luster of the stone. Label 4 is the gas nitrogen cooling manipulator. It performs the cooling process in the maintenance. Label 5 is the baking arm and performs the baking wax process. Label 6 is the rotary table. The gear rotates to switch the position of the curling. Label 7 is the suction cup swivel arm. It could hold the curling, lifted by the screw drive, and can be rotated by 90 degrees. Label 8 is the waxing arm and performs the waxing process. Label 9 is the transfer lift to be used for transporting curling. Label 10 is the door conveyor. It is a part of the shell and could protect the internal structure.

The curling maintenance robot has many internal components and complicated structure. Its working process is ordered by the control system. Firstly, the hatch door is open and the cabin door conveyer receives the curling and transports it into the internal. The top of the hatch is equipped with an industrial CT detector to check the overall of the curling. Then the lifting conveyor is pneumatically close to the door conveyor and receives the curling. The curling is transferred to a suitable position with proper maintenance process. If the lifting conveyor received the curling, the rotary table will be controlled by the stepper motor to rotate 72 degrees. After that the conveyor lift will receive the next curling. If the internal has received five as full filled the positions, the lifting conveyor and the rotating arm would work corporately with each other for the maintenance. And the two parts are operated by the screw drive to lift and transfer the curling.

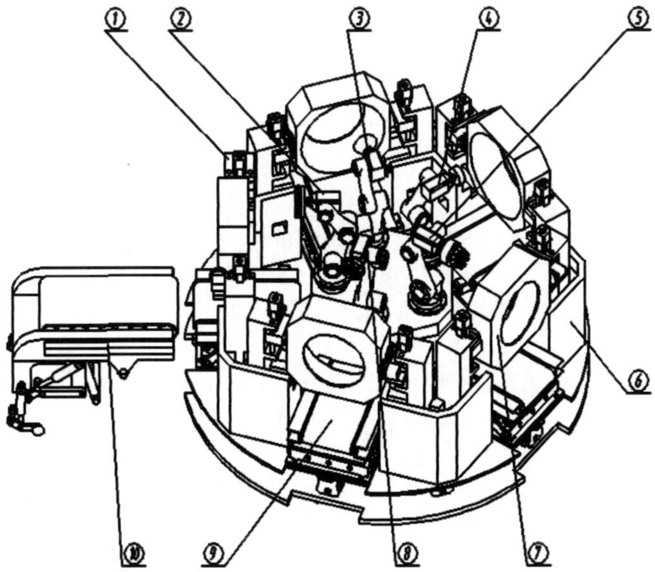

Figure.4. Curling maintenance robot internal structure.

In the later the suction disc revolving arm sucks the pot and moves upward, while the lifting machine drops two to the appropriate positions. The suction disc arm controlled by the stepper motor rotates 90 degrees and exposes the bottom of the curling to the manipulator. At the same time the lifting platform rises and holds the sucker rotating arm. Then the maintenance process on the surface of the curling is started here. The five manipulators will carry out the maintenance according to the accurate procedure. It strictly follows five steps in proper sequence, such as the scrubbing, wax, baking, cooling and polishing. It is shown in Fig.5. It is controlled by the rotary table, which is stepped rotated with 72 degrees to change the curling position.

3.2.2. Maintenance manipulators.

The maintenance manipulators system is one of the most important parts. It involves the maintenance process of the curling and conducts the professional work. As shown in Fig.5, the system consists of five five-axis manipulators. As shown from left to right, they are the scrubbing manipulator, the waxing manipulator, the baking manipulator, the gas nitrogen cooling manipulator and the polishing manipulator. The five manipulators control systems are independent and flexible scrubbed separately. The control strategy of the required professional maintenance processes is fixed and programed with operation command code written into the control program.

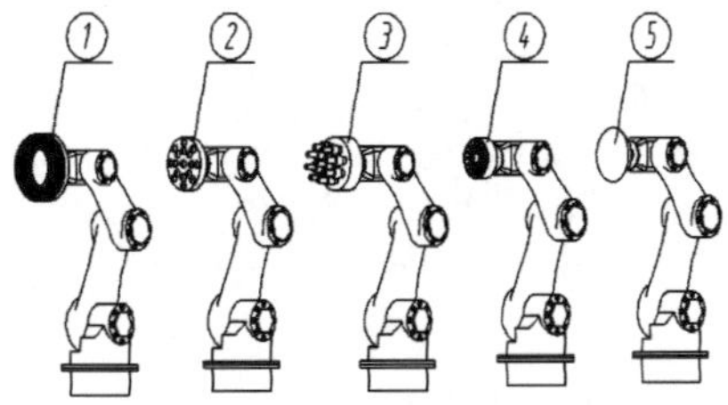

Figure. 5. Maintenance manipulators.

As shown in Fig.5, Label 1 is the scrubbing. It is a soft brush made of special soft cortical material. It is used to wipe the rubbing ring on the bottom of the curling. The aim is to wipe off the snow or impurities stuck on the ring. Label 2 is the wax manipulator, which is used to heat the wax to the proper thick shape with the internal heating tube. Through the spiral transmission, the wax is evenly spread at the bottom, so that it can be attached to the surface of the stone body. It has to be waited for its natural cooling and solidification. Label 3 is used to carry out the baking with alcohol as fuel. The torch emits uniform flames and bakes the bottom. Label 4 is used with gas-nitrogen cooling to cool the curling heated in the previous step. The curling temperature should be kept at a specify temperature that is similar to the ice surface temperature. Label 5 is the polishing. It uses a professional polishing wheel to polish the curling on a smooth sliding surface for revealing the smoothness and luster of the original.

3.3. Control system.

According to the configuration and operation management, the robot machine is applying the PLC control system to realize system control and automatic operation. The control unit of pneumatic actuator, stepper motor driver is controlled with the special operation part.

3.4. Power supply system.

The maintenance robot is static and powerred by the power supply from the field main output. The handling robot is mobile and powered by the rechargeable battery supply, which is set on the robot itself. The power charge interface is set on the maintenance robot and the handling robot with the battery could be charged by the maintenance robot during the off working time.

4. Summary

The curling sports service robot uses mechanical automation instead of professional manual operation to realize the handling and maintenance of curling. The machine replaces the humnan labor, which could reduce the cost and improve the efficiency. It aims to serve the curling sports automatic with automation robot movement. The robot has the simple mechanical structure, comprehensive function, comprehensive operation and PLC intelligent control system by the reliable operation control of various mechanisms. Each mechanism of the robot is closely coordinated, which is safe and stabe with running orderly. At present, there has been few public machine products that serves the curling sports. The curling robot in this paper can effectively serve the curling sports and so on. It could significantly improve the popularity of the curling sports.

5. References

[1] J.L. Bradley 2009 The Sports Science of Curling: A Practical Review *J. SPORT. SCI. MED.* **8** 495-500
[2] T. Kawamura, Y. Takegawa and M. Yamamoto 2017 Scientific Approach and Challenges of Curling Study *IEEJ TRANS. ELETRIC. INF. SYS.* **137** 1137-1140
[3] T. Kawamura, R. Kamimura, S. Suzuki and K. Iizuka 2015 A Study on the Curling Robot will match with Human Result of One End Game with One Human *IEEE CIG.* **1** 489-495
[4] B.A. Marmo and J.R. Blackford 2004 Friction in the Sport of Curling *ENG. SPORTS.* **1** 379-385
[5] M.P. Buckingham, B.A. Marmo and J.R. Blackford 2006 Design and Use of an Instrumented Curling Brush *PROC. INST. MECH. ENG. L-J MATER.* **220** 199-205
[6] S. Verlag 2006 *Field and Service Robotics* (Berlin: Springer) p 126
[7] Y.P. Kondratenko 2015 *Robotics Automation and Information Systems: Future Perspectives and Correlation with Culture, Sport and Life Science* (Berlin: Springer) pp 136-138
[8] C. Tian, G. Min, Q. Wen, J. Zhang and Y. Zhang 2018 P.R.C. Patent 207640084U

6. Acknowledgements

This work is supported by the Chinese National Training Program of Innovation and Entrepreneurship for Undergraduates (No. 201810895009) and the Major Construction Project of Pingxiang University (No. 201701). And thanks to the help from the key laboratory of modern mining machinery design and intelligent manufacturing at Pingxiang University.

Mechanical Design of Service Robot for Shuttlecock Sports

Tian Chunlai*, Shen Musheng, Wang Yixian and Gao Wei

School of Mechanical and Electronic Engineering, Pingxiang University, Pingxiang, Jiangxi province, 337000, China
E-mail: tianchunlai@bit.edu.cn

Abstract. Shuttlecock ball is a popular sport in modern society which including entertainment, skills, mass, fitness, popularity and education. The current status artificial ball picking during the training and sports of the shuttlecock sports was reviewed in this paper. Based on the main structure and working principle of the picking robot, it was designed and presented here. It is designed by the three-dimensional CAD software. Through motion simulation analysis, the actual working process of the shuttlecock collection robot was shown. The structure design and motion simulation analysis of the whole machine were carried out and it verified the feasibility of the design. The shuttlecock collection robot is used of retrieving, sorting and placing the shuttlecock. It could effectively solve the inconvenience cause by the manual collecting in shuttlecock ball sports.

Keywords: shuttlecock sports; sport service robot; shuttlecock collection; collection robot; mechanical design.

1. Introduction

Shuttlecock ball sport is becoming more and more popular because of its strong entertainment and not critical restricted by weather and field conditions. The shuttlecock begins to move towards the world in the exchange activities. It has developed in many countries of Europe, Asia and the United States, where the players have established the international sports organizations with the world championship system. At present, player has to pick up balls by hands in games or training. It not only costs a lot of manpower, but also delays sports time. Sports service robot is the typical service automation [1, 2]. The one designed for the sports could play an important role in promoting the scientific training and level improvement of sports players [3]. Some researchers have focused on the theoretical research on the application of the shuttlecock dynamics [4, 5]. There are some prototypes for the badminton sports, which could be referred for the shuttlecock [6, 7]. It is necessary to design a robot with the function of collecting shuttlecock automatically. This paper studies the working principle of the shuttlecock collector. The design of the shuttlecock collector and service robot is carried out by using CAD software and presented here. It has the advantages of fast collection speed and high working efficiency.

2. Working plan and principles

The shuttlecock collector robot is designed to collect the shuttlecock balls during the sports. It is a typical sports service auto machine. Before the mechanical overall design, the work plan with principles should be worked out. The work plan with operation flow is shown in Figure 1. As shown in Figure 1, it firstly collects the shuttlecock balls and transfers them into the machine inner through the collection

IOP Conf. Series: Materials Science and Engineering **542** (2019) 012079 doi:10.1088/1757-899X/542/1/012079

part. Then it adjusts the shuttlecock balls' positions through the transfer part. After that the carding and inspections part checks and combs the balls. Finally, it pushes the good shuttlecock balls into the storage device by the push rod part.

3. Mechanical overall design

The shuttlecock collection robot is design with composition of the collection part, the transfer part, the carding and inspection part, the push rod part and the storage part. The overall structure is wireframe shown in Figure 2. In the Figure 2, Label 1 is the storage part, which is used to storage the shuttlecock balls. Label 2 is designed as the push rod part. It can push the shuttlecock to the box. Label 3 is the collection part, which could collect the shuttlecock to the inside.The transfer part labelled 4 is used to move the shuttlecock ball and adjust the positions. The carding and inspection part is labelled 5 and used to check and comb the shuttlecock ball.

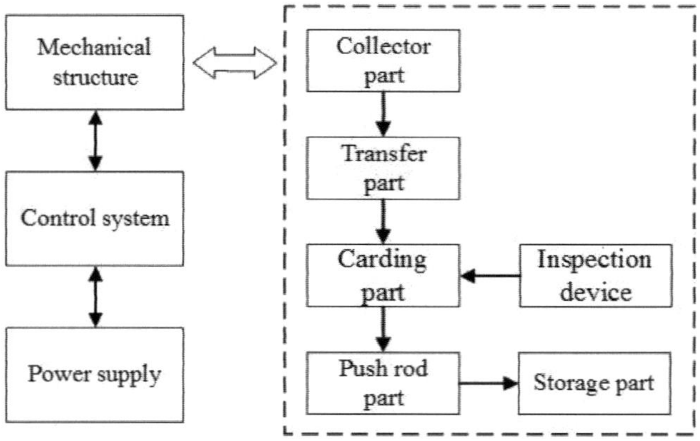

Figure 1. Work plan with operation flow.

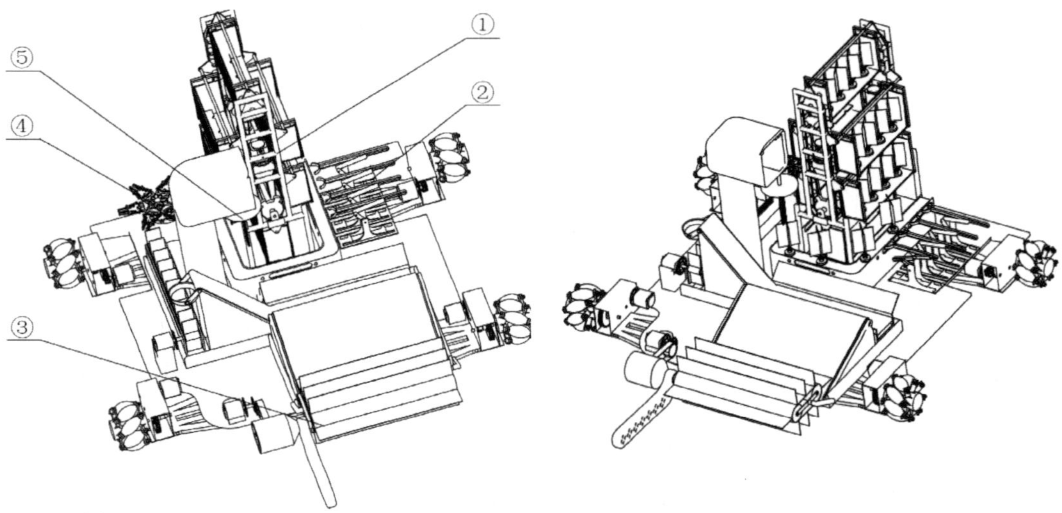

Figure 2. Overall structure of robot (the right is the symmetric view with the left).

4. Mechanical parts design

4.1.Collection part.

The collection part is mainly composed of fans and rotating collection blades. The rotating blades are rotated by the connected rollers. It can rotate the shuttlecock balls and push them into the internal of the robot. The collection part model is designed as shown in Figure 3.

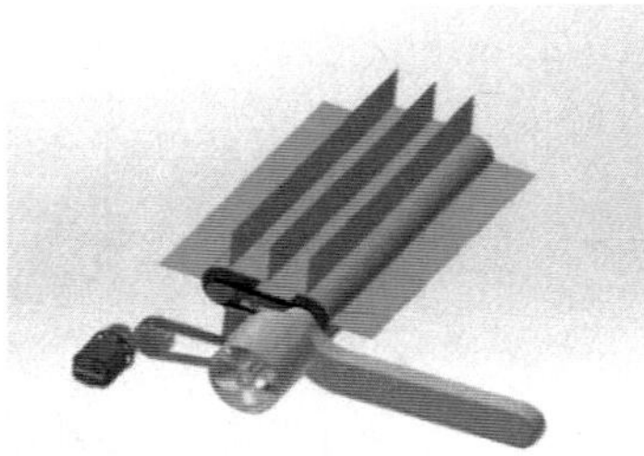

Figure 3. Collection part.

4.2.Transfer part.

The transfer part with one manipulator model is shown in Figure 4. It is composed of the groove wheel mechanism and the connecting rod mechanism. As shown in Figure 4, the hydraulic pole is labelled 1 and used to provide power for telescopic movement of manipulator. Label 2 is the screw thread and it makes the nut move in a straight line. Label 3 is the nut, which is designed to drive the connecting rod movement. The connecting rod is labelled 4 and could move the clamping arm. Label 5 is the clamp arm and used to clip and hold the shuttlecock ball.

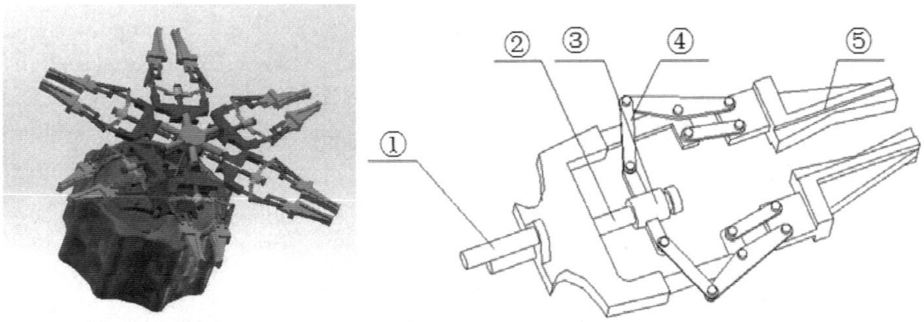

Figure 4. Transfer part with manipulator (wireframe view).

During the operation, the motor drives the slot wheel for the rotation. It makes the manipulator rotate intermittently to change the shuttlecock position. The manipulator is composed of hydraulic rod, screw thread rod, nut, connecting rod and clamping arm. The hydraulic rod and screw thread rod are connected in proper positions of the manipulator. The hydraulic cylinder drives the manipulator for the linear reciprocating motion. The screw thread rod, nut, the connecting rod and the clip arm are connected in turn. Motor drives the thread rod to arrange shuttlecock balls in a straight line. With helps of the loosening and clamping arm, the accurate positioning of shuttlecock balls is obtained.

4.3.Carding part.

The carding part is mainly composed of worm, gear and brush. The turbine worm assembly is connected with the motor and disc. The lower end of the disc is equipped with an infrared induction device to determine the position. The motor drives the worm gear component movement with the brush. It combs shuttlecock balls' feathers. The carding device is shown in Figure 5.

4.4.Push rod part.

The push rod device is mainly composed of crank rocker mechanism and push rod. The motor drives the original rod rotation. The original rod drives the connecting rod to make the push rod linear reciprocate move. It could push the shuttlecock into the collection box. The push rod part is shown in Figure 6.

Figure 5. Carding part.

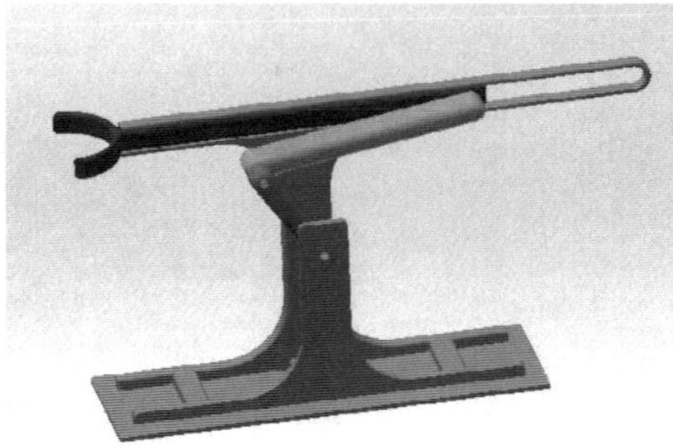

Figure 6. Push rod part.

4.5.Storage part.

The storage part is equipped with six collection components modules. Each module includes pallets and collection boxes. Each one can store four shuttlecock balls. The collection component is connected with the chain. The chain movement is driven by a motor driving gear to realize the cyclic movement of the collection component. It realizes the collection with twenty-four shuttlecock balls. The wireframe structure of the storage part is shown in Figure 7.

As shown in Figure 7, the collection box is labelled 1 and used to store the shuttlecock balls. Label 2 is the support framework, which could structural support the storage part. The gear is labelled 3 and used to drive and support the chain movement. Label 4 is the main motor. It provides the rotation with the power supply. The centre chain labelled 5 is design to drive the shuttlecock collection box circulation movement.

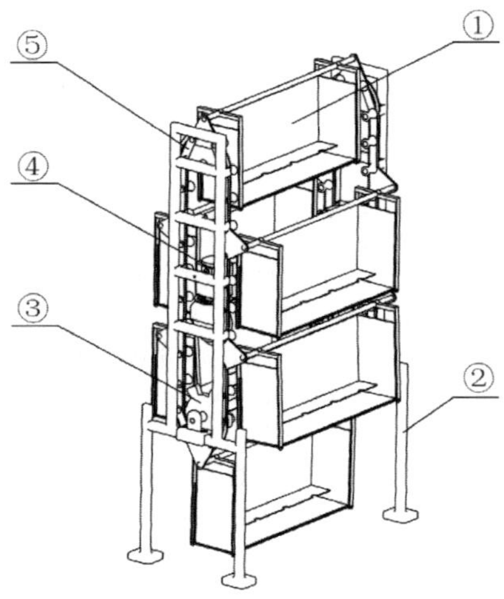

Figure 7. Storage device.

5. Walking system.

The Mecanum wheels is chosen for the robot walking system with the support plate as shown in Figure 8. As shown in Figure 8, Label 1 is the chassis, which carries the main parts of the machine. Label 2 is the space for placing the spring of the shock absorber. Label 3 is the Mecanum wheels. It could meet the design requirements, such as the volume, weight and size of the robot. Each wheel applies a brushless motor and a two-stage reducer. The rated torque of the motor is expected to be selected as 0.20 Nm with rated speed as 3000 r/min.

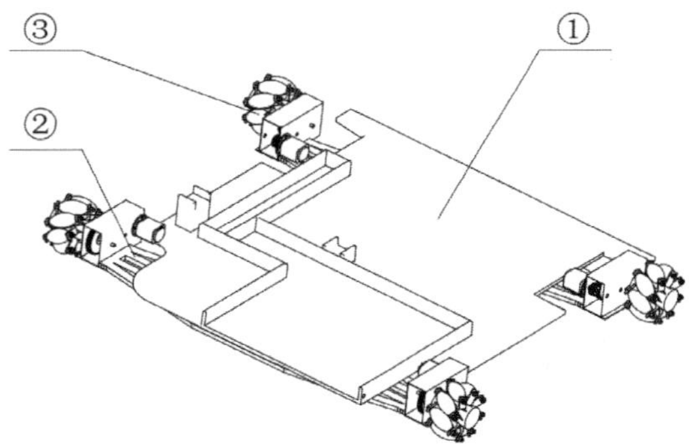

Figure 8. Walking system with support plate.

6. Control system.

It uses a microcomputer to control the system operation. The shuttlecock ball on the sports field is tested and found by a high-definition camera. After the test is completed, the images will be transmitted to the central processor of the control system. Then the robot walk route is planned to drive the robot walking along the planning path quickly. During the collection process, the function with the mechanical movement control is realized through the infrared ranging sensor, the motors and the actuators.

7. Design verification and simulation

7.1.Running speed.

The standard size of shuttlecock sports field is 11.88 m long and 6.1 m wide. The experimental analysis shows that the actual speed of the shuttlecock collection robot needs as 2.0 m/s. According to the design requirements. Calculation by formula the gear ratio as 3.06. Total transmission ratio as 9.34. The speed of the large gear after deceleration as 321.2 r/min, angular velocity as 33.6 rad/s. The maximum moving speed of the robot is calculated as 2.6 m/s. The maximum speed is larger than that of the robot, which meets the design requirements.

7.2.Torque of driving motor.

Output torque T of driving motor could be calculated by

$$T = (9.55FV) / (\eta n) \qquad (1)$$

where F is the friction resistance, reference data is 10 N, η is the transmission efficiency as 70%, V is the maximum speed of the design robot as 2.6 m/s, n is the speed of the motor as 3000 r/min. We could get T resulted as 0.22 Nm. It shows that the output torque of the driving motor is less than the rated torque, which could meet the design requirements.

7.3.Model simulation.

According to the actual size parameters of the shuttlecock collecting robot, the 3D model is built and assembled in CAD software. To ensure that the design has enough collection efficiency, the motion simulation of the robot working process is carried out through the software. The 20 shuttlecock balls were dropped randomly in the sports field before the simulation starts. Based on the actual working parameters of the robot, the drive was added as the motion simulator. The analysis of the 100 collection process was carried out. The results showed that 20 shuttlecock balls were collected and arranged well in planned operation. The time used by a robot was less than 120 s. Compared with that the time required for the manual collections under the same condition is about 5 minutes, it is far below the efficiency of the robot.

8. Summary

The robot is designed to service shuttlecock sports for the rapid collection of shuttlecock balls. It is composed of the collection part, the transfer part, the carding part, the push rod part and the storage part with the control system. After a detailed introduction to the overall structure and working principle of the robot, the three dimensional model with the motion simulation are carried out by using the three-dimensional CAD software. The working performance in the real environment of shuttlecock training field is obtained with the collection efficiency. The efficiency is much higher than the manual collection. It could provide an example for research and development of similar products in balls like sports service robots.

9. References

[1] Y.P. Kondratenko 2015 *Robotics Automation and Information Systems: Future Perspectives and Correlation with Culture, Sport and Life Science* (Berlin: Springer) pp136-138
[2] S. Verlag 2006 *Field and Service Robotics* (Berlin: Springer) p 128
[3] J.A. Ronkainen, R.F. El-Kati and P.R. Fleming 2010 Application of an Industrial Robot in the Sports Domain: Simulating the Ground Contact Phase of Running *PROC. INST. MECH. ENG. P-J SPORTS. ENG. TECH.* **224** pp 259-269
[4] Y Man, L Zhao and J Hu 2009 Application of Kalman Filter in Track Prediction of Shuttlecock *Proceeding of 2009 IEEE International Conference on Robotics and Biomimetic* (Piscataway:IEEE) pp 2205-2210
[5] A. Cooke 2002 Computer Simulation of Shuttlecock Trajectories *SPORT ENG.* **5** pp 93–105
[6] C.S. Kim, X.Y. Yap and E.S. Low 2016 A Badminton Robot - Serving Operation Design, *ARPN J ENG APP SCI* **11** pp 3968-3974.
[7] B Qiu 2008 P.R.C. Patent 101134138A

10. Acknowledgements

This work is supported by the Chinese National Training Program of Innovation and Entrepreneurship for Undergraduates (No. 201810895005) and the Major Construction Project of Pingxiang University (No. 201701). And thanks to the help from the key laboratory of modern mining machinery design and intelligent manufacturing at Pingxiang University.

IOP Conf. Series: Materials Science and Engineering **542** (2019) 012080 doi:10.1088/1757-899X/542/1/012080

Innovative accuracy design of in-site precise automatic inspecting machine of valve sleeve bore of aeronautical precision coupling

Duxi Liu* **and Mingquan Zhu**

School of Mechanical Engineering, Northwestern Polytechnical University, Xi'an 710072, China
*E-mail: liuduxi@nwpu.edu.cn

Abstract. The diameter accuracy and geometric accuracy of the valve sleeve bore of aeronautical precision coupling are both high. It is necessary to inspect the bore diameter and cylindricity error frequently in machining, but there is a lack of in-site high accuracy and low cost automatic inspection methods currently. The scheme and accuracy design of a precise automatic inspecting machine (PAIM) of valve sleeve bore is carried out. Accuracy design of PAIM by traditional error distribution method fails to meet the requirement of measurement accuracy. Based on the idea of error avoidance and TRIZ technology system evolution theory, an innovative accuracy design method by using automatic dynamic disconnected measuring chain is proposed and realized. Accuracy analysis and experiment prove the correctness and validity of the proposed method.

1.Introduction

Valve sleeve is the component of Aeronautical precision coupling, whose bore consists of multiple sections of discontinuous internal cylindrical surface, and axial width of each section is less than 3mm. it's narrow, small and deep. The tolerance grade of the valve sleeve bore diameter is higher than IT7, and its cylindricity tolerance is less than 1μm. Precision automatic measurement of valve sleeve bore is an important subject in bore machining and inspection. The measurement of small and micro internal scale is a new project in precision measurement field, which requires solving the problem of measuring accessibility and measurable depth in a certain level of measurement accuracy and range [1]. A tendency of precision machining is to develop in-site high accuracy and low cost inspecting device [2]. For the size and form error at any section of small bore in a certain depth, there is no effective measurement method presently. The goal of this paper is to explore a low cost solution of in-site precision automatic inspecting machine (PAIM) of valve sleeve bore.

IOP Conf. Series: Materials Science and Engineering **542** (2019) 012080 doi:10.1088/1757-899X/542/1/012080

Firstly scheme of PAIM based on pneumatic measurement is briefly introduced, and then accuracy design is intensively discussed. Accuracy design of PAIM by traditional error analysis and error distribution method fails to meet the requirement of measurement accuracy. Based on the idea of error avoidance and TRIZ technology system evolution theory, an innovative accuracy design by using automatic dynamic disconnected measuring chain of PAIM is proposed and realized.

2.Measurement method and scheme of PAIM

2.1.Measurement method

The measuring accuracy of on-line inspecting system mainly depends on the accuracy of the machine tool itself, but the accuracy of the grinder is difficult to meet the requirement of the measuring accuracy of valve sleeve bore. Because of the limit of geometric structure space of valve sleeve bore, it is difficult to implement on-line inspection in horning and grinding. Thus in-site inspection strategy is correct and reasonable choice.

After the comparison with various methods from the aspects of aperture measuring range, accessibility, accuracy, and environmental suitability, it is found that pneumatic measurement is most suitable for in-site high precision automatic measurement of small bore [3, 4]. Pneumatic measurement can meet the requirement of measuring accessibility and accuracy of the valve sleeve bore. Thus pneumatic measurement is most adoptable for the measuring principle of PAIM. CAG2000 type electronic column gas-electric micrometer is used as the pneumatic instrument of PAIM, its measuring range is 0.05mm, resolution is 0.5μm, and measuring accuracy is 0.2μm.

Besides the pneumatic instrument body, pneumatic measuring probe is also a key part of pneumatic measurement system. It usually has a big length-diameter ratio for small diameter deep bore, and is also called pneumatic measuring rod. Cylindrical and rectangular are common shapes of measuring nozzle of pneumatic measuring rod. The measured surface of valve sleeve bore is too narrow to be measured by cylindrical nozzles, but only by rectangular nozzles. Since the pneumatic measuring probe with opposite nozzle structure is often used for bore diameter, pneumatic measuring rod with opposite rectangular nozzles with 0.4×1.5mm is adopted in PAIM design [4]. The opposite nozzle on the pneumatic measuring rod is used as both a probe and an aiming device in pneumatic measurement. Axial aiming is used to determine the axial positions of measured cross-sections of workpiece, and radial aiming is used to determine the positions of measuring points in circumferential direction of the measured bore in the same cross-section.

2.2.Mechanism scheme of PAIM

PAIM can be used to measure the diameters and cylindricity error of the valve sleeve bore. Bore diameters can be measured directly and the cylindricity error can be approximately replaced by measuring roundness and taper. According to the requirement of automatic measuring function and referring to the motion scheme of deep bore machining machine tool, the mechanical mechanism of PAIM is designed [4-6] and shown in figure 1.

Pneumatic measuring rod 5 is mounted on supporting bracket 6, and supporting bracket 6 is mounted on slider 7 of precision linear module used for straight motion of the pneumatic measuring rod along Z coordinate axis.

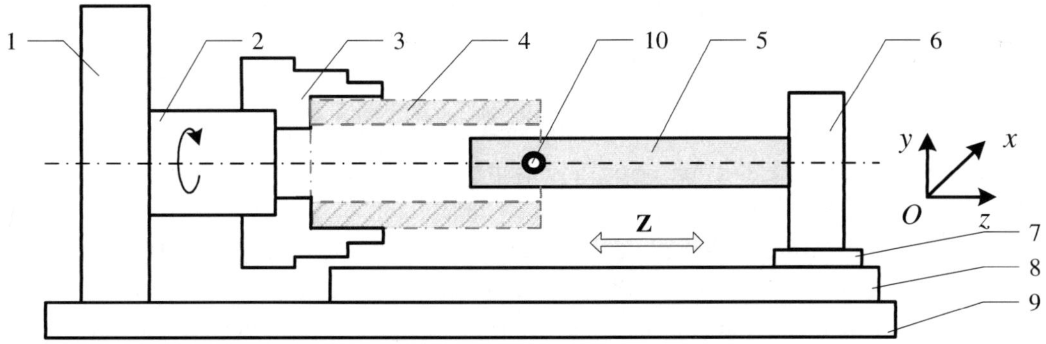

1-Column; 2-Spindle; 3-Clamp; 4-Workpiece; 5-Pneumatic measuring rod;
6-Supporting bracket; 7-Slider; 8-Guide; 9-Bed;10-Nozzle

Figure 1. Overall mechanism scheme of PAIM.

3.Error analysis and distribution of PAIM

3.1.Error analysis

Ideal position relation between the pneumatic measuring rod and the workpiece is shown in figure 2.

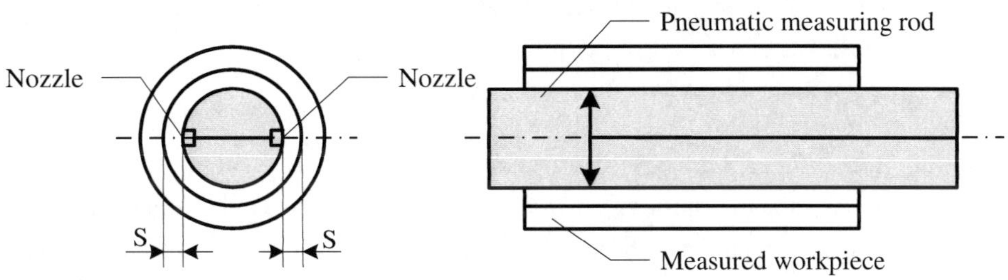

Figure 2. Ideal position relation between pneumatic measuring rod and workpiece.

Because of the influence by machining error, installation error and adjustment error of the mechanical mechanism and the pneumatic measuring rod of PAIM, the position relation between the pneumatic measuring rod and the workpiece is not ideal practically, but rather the position error of the pneumatic measuring rod exists. The position error of pneumatic measuring rod is a main error in the measuring chain of PAIM [4]. Although PAIM has many error sources, such as pneumatic instrument body, pneumatic measuring rod, mechanical mechanism, measuring standard part and environment condition, only the position error of the pneumatic measuring rod and the geometric error of the mechanical mechanism are closely related to the function scheme of PAIM. A feature of pneumatic measurement is that, the geometric error of the mechanical mechanism leads to the position error of the pneumatic measuring rod, which will affect the gap between the workpiece and the pneumatic measuring rod and result in pneumatic measurement error.

It should be noted especially that, when the workpiece hangs on the pneumatic measuring rod by self-weight, pneumatic measurement error caused by the position error of the pneumatic measuring rod can be ignored [7-9]. It provides a new idea and basis for innovative accuracy design of PAIM.

3.2.Error distribution

In accuracy design of PAIM, traditional measurement error distribution method is used firstly. It's assumed that only the pneumatic instrument, error the straightness error of guide-slider pair, the radial run-out component of spindle rotation error and the centering error of the clamp are considered. The combined error after error distribution should satisfy

$$\delta_0+\delta_1+\delta_2+\delta_3\leq\delta_\Sigma \quad (1)$$

Where δ_0 is the pneumatic instrument error, δ_1 the straightness error of guide-slider pair, δ_2 the radial run-out component of spindle rotation error, and δ_3 the centering error of the clamp. δ_Σ is the total measuring error of PAIM.

As noted before, the diameter tolerance of a certain type of valve sleeve bore is 0.015mm, and the cylindricity tolerance 1μm. Since PAIM will be used to measure the diameter and the cylindricity error, the instrument measurement accuracy must be determined by the cylindricity tolerance. The Total measuring error of PAIM should not exceed 1/3 of the tolerance of the measured parameter, and is 0.3μm, i.e. δ_Σ=0.3μm. The measuring accuracy of CAG2000 type pneumatic instrument is 0.2μm, i.e. δ_0=0.2μm, then the sum of the other three errors is only 0.1μm. Following the equal action principle, the error distributed to each item is only about 0.033μm. Even if the spindle rotation accuracy, guide accuracy and clamp centering accuracy are raised to present top technical level, it is difficult to reach the total measurement accuracy of PAIM. Obviously accuracy design of PAIM cannot perform by traditional error distribution method. Based on TRIZ theory, a serious technical conflict exists in accuracy design of PAIM. The shackles of traditional accuracy design method must be taken off, and innovative accuracy design method must be explored.

4.Innovative accuracy design of PAIM

4.1.Composition of automatic dynamic disconnected measuring chain

The principle of error avoidance employs certain technical ways to reduce, eliminate and avoid the influence of error sources on measurement system, and to make error sources ineffective under certain conditions [10]. Different from traditional accuracy design, it emphasizes active design, manipulation and avoidance of errors. To solve the technical conflict in accuracy design of PAIM, an innovative accuracy design method based on the idea of error avoidance is proposed. Its basic idea is to avoid the influence of error of PAIM on the measurement by using automatic dynamic disconnected measuring chain. The mechanical mechanism transmission chain i.e. measuring chain corresponding to the scheme of PAIM is shown in figure 3(a).

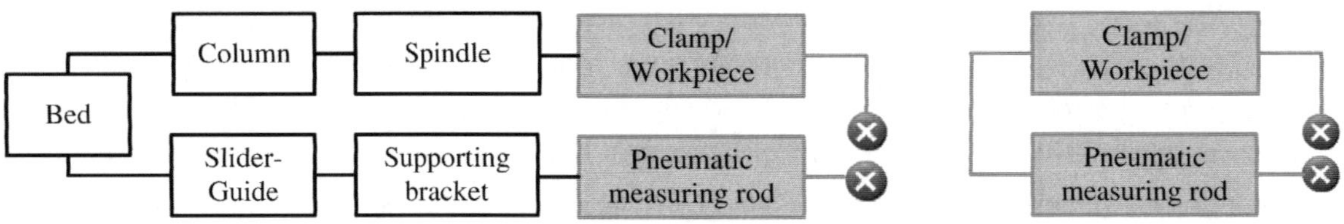

(a) Measuring chain in non-measuring state (b) Measuring chain in measuring state

Figure 3. Automatic dynamic disconnected measuring chain of PAIM.

As seen from figure 3(a), the measuring chain of PAIM corresponding to the function scheme only has seven composition rings, which includes pneumatic measuring rod, supporting bracket, slider-guide, bed, column, spindle and clamp/workpiece, which conforms to the principle of the shortest measuring chain and has Minimum number of error sources.

After the mechanism and function scheme of PAIM is determined, the measuring chain and its composition rings are usually all fixed, and they will not change even in working process. Based on the idea of error avoidance, an automatic dynamic disconnected measuring chain is proposed and established, which means that, partial composition rings of measuring chain of PAIM can be disconnected under certain conditions by automatic control in measuring process to avoid the errors of these composition rings. The number of composition rings of the measuring chain of PAIM will change dynamically with different working state and measuring action. The measuring chain in non-measuring state is shown in figure 3(a), and new measuring chain in automatic measuring state after partial composition rings are disconnected is shown in figure 3(b).

4.2.Realization of automatic dynamic disconnected measuring chain

The realization of automatic dynamic disconnected measuring chain is essentially to increase its dynamics and controllability. Based on one of the technology evolution patterns of TRIZ technology system evolution theory, by improving the flexibility, mobility, controllability and stability of the system structure, the system will evolve towards increasing dynamics and controllability [11, 12]. The evolution of the measuring chain of PAIM should be from fixed to flexible.

Composition rings of measuring chain of PAIM can be divided into static joint of mechanism and kinematic pair. Composition rings of static joint type include pneumatic measuring rod, supporting bracket, bed and column. To increase the dynamics and controllability of the measuring chain of PAIM, it is impossible to change the working state of the static joint of mechanism automatically. The breakthrough can only be found from the composition rings of kinematic pair type. Composition rings of kinematic pair type include spindle, guide-slider pair and clamp. Although the rotation motion of the spindle and the linear motion of the guide-slider pair are dynamic and controllable, they cannot make the measuring chain automatically disconnect. To realize automatic disconnection of the measuring chain and to increase the dynamics and controllability of the mechanism, the clamp can be only chose. Based on aforementioned TRIZ technology evolution pattern, the technology evolution route of the clamp is obtained as shown in figure 4.

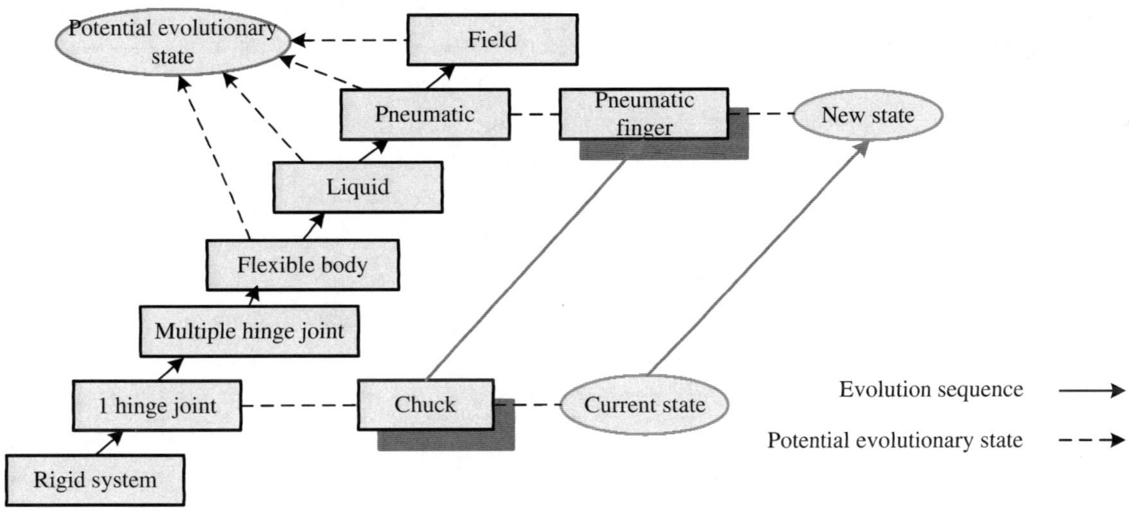

Figure 4. Technological evolution route of the clamp.

The highest evolution state is field, but there is no some kind of field with advanced principle and mature technology to be used to design clamp currently. Thus a relatively advanced evolution state, that is, pneumatic form is adopted to realize the dynamics and controllability of the clamp. Since it's easy to realize automatic control of the working state, pneumatic mechanism is used to replace mechanical clamping mechanism. Because of its simple structure and high repeatability, pneumatic finger is adopted as the clamp to enable the measuring chain of PAIM to be disconnected automatically in measuring process. Thus fixed measuring chain can evolve into flexible automatic dynamic disconnected measuring chain. Only when the pneumatic finger is loosened and the measuring chain is disconnected, can PAIM be in automatic measuring state.

4.3.Accuracy analysis

When pneumatic measurement is taken after the measuring chain is disconnected, composition rings of the measuring chain of PAIM, such as the rotary spindle, the pneumatic finger, and the guide-slider pair do not affect the position relation between the pneumatic measuring rod and the workpiece, only the pneumatic measuring rod can influence the pneumatic measurement. As noted before, when the measuring chain is disconnected, the workpiece hangs on the pneumatic measuring rod by its self-weight and remains a relative static state, and then the pneumatic measurement error caused by the position error of the pneumatic measuring rod can be ignored [7-9]. Thus the influence of the position error of the pneumatic measuring rod and the errors of all other composition rings such as the spindle rotation, the guide-slider and the clamp on the pneumatic measurement can be both avoided completely. Thus the influence of all these error sources on pneumatic measurement accuracy is equal to zero except for pneumatic measuring rod and the pneumatic instrument body. The total measurement accuracy of PAIM basically depends on the pneumatic measurement. When high accuracy pneumatic instrument is selected and the pneumatic measuring rod is correctly designed and fabricated, the measurement accuracy of PAIM can be guaranteed. Besides, after the measuring chain is disconnected, the measured workpiece and the pneumatic measuring rod are always in an isoline contact, which can ensure the consistency and repeatability of their position relationship, and then the repeatability of PAIM can be guaranteed.

In this way the dynamic automatic measurement by PAIM is essentially converted into a quasi-static pneumatic measurement. It can not only avoids the error caused by composition rings of measuring chain, but also reduce the dynamic measurement error of pneumatic measurement, thus the total measurement accuracy is ensured. The experimental results also show that proposed and designed PAIM can meet the requirement of high measurement accuracy of the valve sleeve bore.

5.Conclusions

(1) An innovative accuracy design method of PAIM of valve sleeve bore of aeronautical precision coupling based on the idea of error avoidance is proposed.

(2) Based on TRIZ technology system evolution theory, the automatic dynamic disconnected measuring chain is designed, and the pneumatic finger is used to realize automatic switching of the measuring chain. Accuracy analysis shows that, by this method the influence of the position error of the pneumatic measuring rod and the errors of all other composition rings to the pneumatic measurement accuracy can be completely avoided. The total measurement accuracy of PAIM only depends on the pneumatic measurement accuracy.

(3) The proposed and designed PAIM has been used in grinding production field. The experimental result shows that it can realize in-site precise automatic inspection of the valve sleeve bore of aeronautical precision coupling, and can reach high measurement accuracy.

6.References

IOP Conf. Series: Materials Science and Engineering **542** (2019) 012080 doi:10.1088/1757-899X/542/1/012080

[1] Tan Jiubin 2000 Some major concern questions in the research of super precision measuring technique and instrument engineering *China Mechanical Engineering* **11** pp257–61

[2] China Society of Mechanical Engineering 2009 *2008-2009 Report on advances in mechanical engineering (Mechanical manufacturing)* (Beijing: China Science and Technology Press) pp324–35

[3] Rucki M Barisic B and Varga G 2010 Air gauges as a part of the dimensional inspection systems *Measurement* **43** pp83–91

[4] Liu Duxi 2010 Research on precision-driven innovative design method for automatic inspecting machine of valve sleeve of aeronautical precision coupling (Xi'an: Northwestern Polytechnical University)

[5] Stepien K 2014 In situ measurement of cylindricity–problems and solutions *Precision Engineering* **38** pp697–701

[6] Liu J Pan X Wang G et al 2012 Design and accuracy analysis of pneumatic gauging for form error of spool valve inner hole *Flow measurement & instrumentation* **23** pp26-32

[7] Feng Defu 2007 *Practical On-line inspection technology for factories* (Bejing: National Defense Industry Press)

[8] Wang Renfu 1989 *Precision pneumatic measuring instrument* (Beijing: China Machine Press)

[9] Li Xuehai 1987 Dynamic error analysis of honing measuring head *Machine Tool* **9** pp32–34

[10] Shi Huli and Zhao Yan 2007 New idea and methods of error design 2007 *Proc. of the Academic Conference on Mechatronics* (Beijing: Publishing House of Electronics Industry) pp 355–62

[11] Ko H K and Kim K H 2012 Design evolution of an actuation system for manipulator upper arm based on TRIZ *J. of Advanced Mechanical Design Systems & Manufacturing* **6** pp131–9

[12] Roderburg A Klocke F and Koshy P 2011 Principles of technology evolutions for manufacturing process design *Procedia Engineering* 9 pp294–310

IOP Conf. Series: Materials Science and Engineering **542** (2019) 012081 doi:10.1088/1757-899X/542/1/012081

Research on dynamic characteristic of connecting rod with joint clearance

Shihua Li [1], Jilong Xu [2] and Jie Zhao [2]

1 Hebei Provincial Key Laboratory of Parallel Robot and Mechatronic System, Yanshan University, Qinhuangdao, China
2 Hebei Provincial Key Laboratory of Parallel Robot and Mechatronic System, Yanshan University, Qinhuangdao, China
E-mail: shli@ysu.edu.cn,

Abstract. In order to study the dynamic characteristics of revolving joint with clearance, the theoretical model of the revolving joint with clearance is established by using Newton Euler method and "contact-separation" state model. The kinetic equations are solved by using the Gear algorithm in MATLAB software. The change curves of the contact force and the center trajectory of the revolving joint shaft with clearance are obtained considering different clearance, contact stiffness, and rotational speed together. The research results have laid a theoretical foundation for improving the joint forces and the accuracy by using controller.

1. Introduction

With the development of modern machinery in the direction of high speed and precision, the influence of clearance on mechanism motion has attracted more and more attention from scholars both at home and abroad. Due to the existence of clearance, the working performance of the mechanism has changed, resulting in a deviation between the actual movement of the end effector and the predetermined motion. In addition, the clearance makes a violent impact and collision between the moving pairs, which speeding up the wear between the auxiliary elements of the movement, causing the mechanical vibration and noise, and seriously deteriorating the dynamic performance of the mechanical products. Thus the operation precision, stability, reliability and service life of the machine have been reduced. The adverse effects of the kinematic pair clearance are particularly serious in high speed, heavy load and micro-mechanical systems. For example, in the field of Aeronautics and Astronautics, the unstability of the extensional mechanism, the low positioning accuracy and the failure of the antenna opening often happen, due to the nonlinear effect of clearance on the mechanism, which introduced the failure of the aerospace vehicle [1-2].

Since the 70s of last century, scholars at home and abroad have carried out related research on clearance mechanism and achieved many research results[3-5]. For dealing with the clearance problem, scholars have proposed three research methods: Newton Euler method[6], Lagrange method[7], Kaine method[8]. In addition, scholars have also proposed three types of clearance models: continuous contact model, two state model and three state model. The above research methods and models have advantages and disadvantages in dealing with dynamic problems of different clearance mechanisms.

In this paper, Newton Euler method and "contact-separation" state model are used to establish the dynamic equations of revolving joint with clearances. The contact force of the shaft and the change law of the movement track of the axis center relative to the bearing bush are analyzed, with different clearance, different contact stiffness and different rotational speed. The research on dynamic modeling of clearance collision and dynamic simulation of clearance mechanism is developed, which provides an important reference for the design of the mechanism and is helpful to guide the practical application of the engineering.

2. Theoretical dynamic modeling of revolving joint with clearance

2.1 Brief description of the revolving joint with clearance

A three-dimensional model of a revolving joint with clearance is shown in Figure. 1, and a two-dimensional simplified model with clearance joints is shown in Figure. 2, where the clearance is enlarged for clarity. The length and mass of the connecting rod are L、m, S is the centroid of the connecting rod, and O_1 and O_2 are the centroid of the bearing bush and the shaft respectively. The distance between the center of mass of the connecting rod S and the center of mass of the shaft O_2 is L_S. The radius of the bearing bush and the shaft are R_1 and R_2 respectively. The distance between the two is e, and the driving torque acting on the shaft is T. θ is the angle between the central axis of the connecting rod and the counterclockwise rotation of the X-axis of the fixed coordinate system. The centroid of the bearing bush is taken as the origin of the fixed coordinate system, and the Cartesian coordinate system shown in Figure.2 is established. A clearance revolving joint is formed between the shaft and the bearing bush with clearance. A position enlarged view of clearance revolving joint is shown in Figure.3, where the parameters e_x and e_y are the projections of the axis of mass in the coordinate system OXY in the X and Y directions, and ϕ is the angle between O_1O_2 and counterclockwise rotation of the X-axis.

Figure 1. 3D model of the revolving joint with clearance

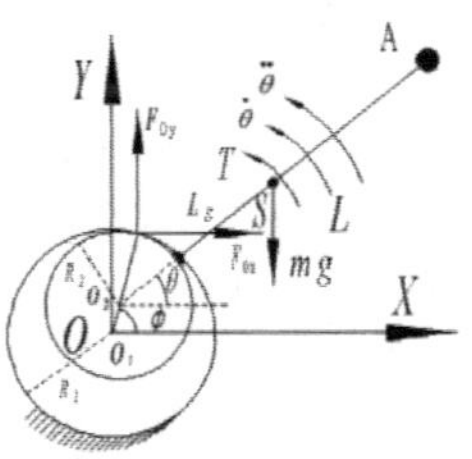

Figure 2. Two dimensional simplified model of the revolving joint with clearance

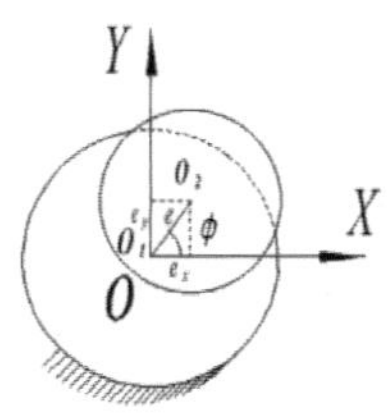

Figure 3. Magnified view of the revolving joint with clearance

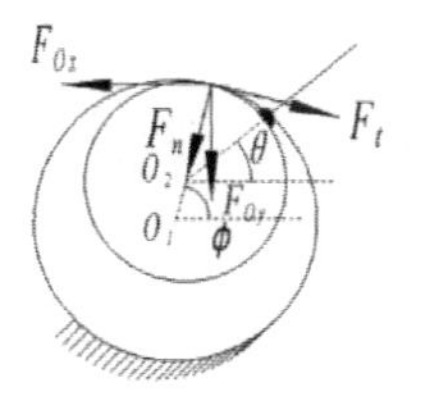

Figure 4. Force diagram of the revolving joint with clearance.

2.2 Establishment of kinematic equation

As shown in Figure. 2, the coordinate equation of the centroid of the connecting rod is:

$$\begin{bmatrix} x_s \\ y_s \end{bmatrix} = e\begin{bmatrix} \cos\phi \\ \sin\phi \end{bmatrix} + L_s\begin{bmatrix} \cos\theta \\ \sin\theta \end{bmatrix} \tag{1}$$

The acceleration formula of the link centroid can be obtained by calculating the two derivative of time on both sides of the formula (1).

$$\begin{bmatrix} \ddot{x}_S \\ \ddot{y}_S \end{bmatrix} = \ddot{e}\begin{bmatrix} \cos\phi \\ \sin\phi \end{bmatrix} + 2\dot{e}\dot{\phi}\begin{bmatrix} -\sin\phi \\ \cos\phi \end{bmatrix} + e\ddot{\phi}\begin{bmatrix} -\sin\phi \\ \cos\phi \end{bmatrix} - e\dot{\phi}^2\begin{bmatrix} \cos\phi \\ \sin\phi \end{bmatrix} - L_S\dot{\theta}^2\begin{bmatrix} \cos\theta \\ \sin\theta \end{bmatrix} + L_S\ddot{\theta}\begin{bmatrix} -\sin\theta \\ \cos\theta \end{bmatrix} \tag{2}$$

2.3 Establishment of dynamic equation

The model is established using Newton Euler equation [9] on the assumption that the shaft of the revolving joint with clearance is regarded as rigid body and the bearing bush is regarded as a flexible body. Assuming its centroid acceleration is $\boldsymbol{a}_S$, and according to Newton's second law $F_{合} = m\boldsymbol{a}_S$, the following equation are given:

$$\begin{cases} -F_{Ox} = m\ddot{x}_s \\ -F_{Oy} - mg = m\ddot{y}_s \end{cases} \tag{3}$$

According to the Euler equation $\boldsymbol{\tau} = {}^S\boldsymbol{J}\boldsymbol{\alpha} + \boldsymbol{\omega} \times ({}^S\boldsymbol{J}\boldsymbol{\omega})$, the following two equations are obtained:

$$F_{Ox}R_2 \sin\phi - F_{Oy}R_2 \cos\phi \tag{4}$$

$$-mgL_s \cos\theta + T = J_s\ddot{\theta} \tag{5}$$

The shaft collided with the bearing bush during the movements, and assume the normal deformation produced by the contact collision is δ, then there is the following relationship between δ and e and the ideal clearance c:

$$\delta = e - c \tag{6}$$

In the above formula: $c = R_1 - R_2$; $\begin{cases} \delta < 0 & \textit{Free movement} \\ \delta = 0 & \textit{Start to contact or start to separate} \cdot \\ \delta > 0 & \textit{Contact deformation} \end{cases}$

In order to solve the contact force problem in the collision hinge model, the nonlinear spring damping model proposed by Lankari-Nikravesh [10] is used to describe the contact-separation process between the motion pairs. The normal contact force F_n is expressed as:

$$F_n = u(\delta)\left(K\delta^m + C_n\dot{\delta}\right) \tag{7}$$

Where K represents the Stiffness coefficient, C_n represents the Normal damping coefficient, and m represents the Nonlinear spring contact force coefficient.

The expression of the tangential impact force is:

$$F_t = u(\delta)(-fF_n sign(v_t) - C_t v_t) \tag{8}$$

Where C_t represents the Tangential damping coefficient, f represents the Sliding friction coefficient, and v_t represents the Motion tangential relative speed.

The $u(\delta)$ in equations (7) and (8) is the step function.

$$u(\delta) = \begin{cases} 0 & \delta < 0 \\ 1 & \delta \geq 0 \end{cases} \tag{9}$$

As shown in figure 4, project F_n and F_t toward the X and Y axes to get the expressions of F_{Ox} and F_{Oy}.

$$\begin{cases} F_{Ox} = F_n \cos\phi - F_t \sin\phi \\ F_{Oy} = F_n \sin\phi + F_t \cos\phi \end{cases} \tag{10}$$

At the point of contact between the shaft and the bearing bush, the normal and tangential relative velocity expressions of the shaft relative to the bearing bush are represented as:

$$\begin{cases} v_t = -\dot{e}_x \sin\phi + \dot{e}_y \cos\phi + \dot{\theta} R_2 \\ v_n = \dot{e}_x \cos\phi + \dot{e}_y \sin\phi \end{cases} \tag{11}$$

According to equations (1) to (11), the dynamic expressions of the clearance-containing hinges are obtained.

$$\ddot{x} = M^{-1} \bullet Q \tag{12}$$

Where $\ddot{x}$ represents the Generalized acceleration array, M represents the Generalized mass matrix, and Q represents the Generalized force array.

$$\text{and } \ddot{x} = \begin{bmatrix} \ddot{e} \\ \ddot{\phi} \\ \ddot{\theta} \end{bmatrix}; \quad M = \begin{bmatrix} 0 & 0 & J_{O2} \\ \cos\phi & -e\sin\phi & -L_S \sin\theta \\ \sin\phi & e\cos\phi & L_S \cos\theta \end{bmatrix}; Q = \begin{bmatrix} F_{Ox} R_2 \sin\phi - F_{Oy} R_2 \cos\phi - mgL_S \cos\theta + T \\ \dfrac{-F_{Ox}}{m} + 2\dot{e}\dot{\phi}\sin\phi + e\dot{\phi}^2 \cos\phi + L_S \dot{\theta}^2 \cos\theta \\ \dfrac{-F_{Oy}}{m} - g - 2\dot{e}\dot{\phi}\cos\phi + e\dot{\phi}^2 \sin\phi + L_S \dot{\theta}^2 \sin\theta \end{bmatrix}. \tag{13}$$

According to the formula (12), the function expression of the time of variable parameters e, ϕ and θ is calculated by numerical calculation, in which the coordinate (e,ϕ) is the trajectory of the center of the pin axis. The expression of the contact force F_n of the axis can be obtained by taking the function expression of e, ϕ, θ about time into equation (7).

3 Analysis of dynamic characteristics of revolving joint with clearance

The Gear algorithm of MATLAB is used to solve the equation (12), and the variation of the contact force F_n of the shaft with time and the change curve of the center track $\phi - e$ are obtained.The specific parameters of the revolving joint with clearance are shown in Table 1.

Table 1. Specific parameters table with revolving joint with clearance

Connecting rod L	0.2 [m]
Connecting rod centroid L_S	0.15 [m]
Connecting rod quality m	5 [kg]
Connecting rod inertia J_{O2}	0.24 [kg•m^2]
Axial radius R_2	0.005 [m]
Gravitational acceleration g	9.8 [N/kg]

3.1 Contact force and central motion locus of shaft at different clearances

Under the condition of constant contact stiffness and speed, the contact force and central motion trajectory of the shaft under different clearance conditions are analyzed. The given simulation parameters are as follows: K=1×10^7 N/m; $C_n = 175\,N\bullet s/m$; f =0.05; $C_t = 0$; $\dot{\theta}$=2π rad/s; $t = 4s$ In the initial state, the connecting rod is stationary, and the change diagram of the contact force of the shaft with different clearance is obtained by MATLAB, as shown in figure. 5

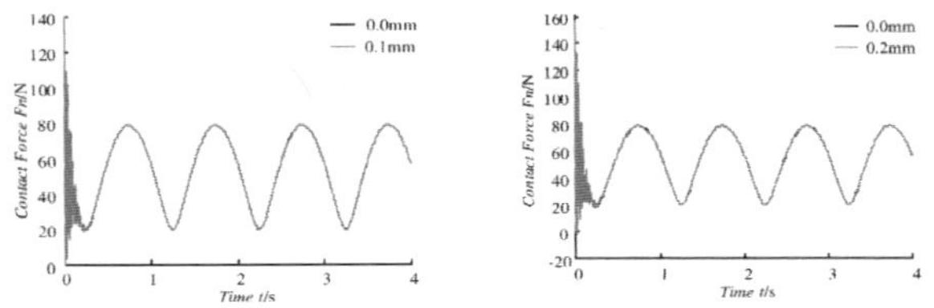

a) A change map of F_n when the clearance is $0.1mm$

b) A change map of F_n when the clearance is $0.2mm$

Figure 5. F_n change diagram at different clearance

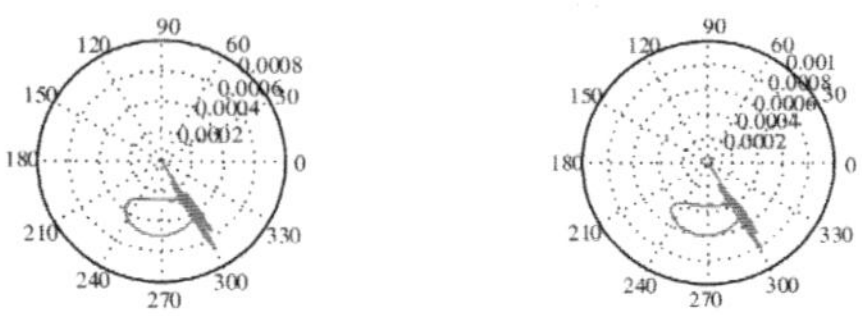

a) The $\phi - e$ coordinate diagram under $0.1mm$ clearance condition

b) The $\phi - e$ coordinate diagram under $0.2mm$ clearance condition

Figure 6. Trajectory change map with different clearance conditions

Figure 5 shows that with the increase of clearance between axle and bearing bush, the magnitude of contact force increases. The above analysis shows that the clearance is very destructive to the mechanism at the beginning stage of the movement. The greater the clearance, the greater the damage. When the connecting rod is stable, the clearance has little effect on the force of the shaft.

As shown in figure 6, with the increase of the range of the center of the axis, the larger the clearance is, the deeper the shaft is in the axle bush, the greater the contact force is. When the connecting rod moves steadily, the center of the shaft moves periodically on both sides of the gravitational direction under different clearance conditions.

3.2 Contact force and center motion trajectory of pin shaft with different contact stiffness

Under the condition of constant clearance and speed, the contact force and central motion trajectory of the shaft with different contact stiffness are analyzed. The given simulation parameters are as follows: c=0.2mm； $C_n = 175 N{\bullet}s/m$； $f = 0.05$； $C_t = 0$； $\dot{\theta} = 2\pi\, rad/s$； $t = 4s$.

Figure 7 shows that the contact force of the shaft changes abrupt at the beginning of the movement. It can be seen from the above that the instantaneous contact force increases with the increase of contact stiffness at the beginning of the movement stage. With the movement of the moving arm, the contact force of the contact shaft has no obvious difference when the connecting rod is running smoothly.

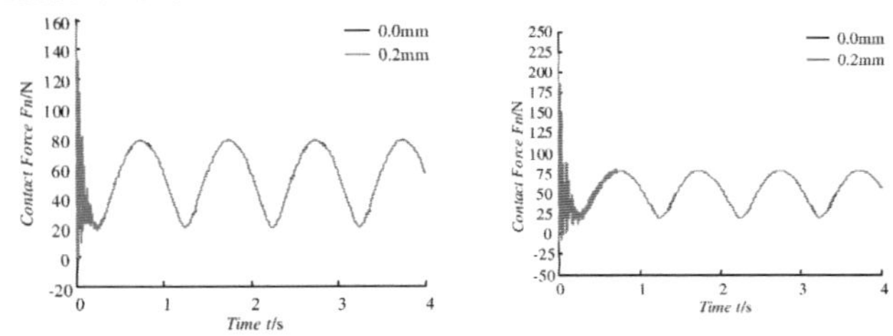

a) The change diagram of F_n under the condition of K=$1\times10^7\ N/m$

b) The change diagram of F_n under the condition of K=$5\times10^7\ N/m$

Figure 7. Change diagram of F_n fat different contact stiffness

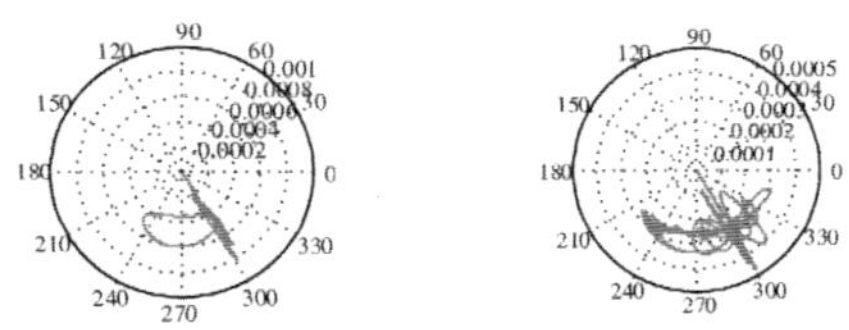

a) The $\phi - e$ coordinate diagram under K=$1\times10^7\ N/m$ condition

b) The $\phi - e$ coordinate diagram under K=$5\times10^7\ N/m$ condition

Figure 8. The $\phi - e$ coordinate diagrams under different contact stiffness

Figure 8 shows that with the increase of the contact stiffness between the shaft and the bearing bush, the range of motion of the shaft center in the horizontal direction increases, but the range of motion in the direction of gravity decreases. The reason is that the greater the contact stiffness when the shaft is subjected to the same force, the smaller the deformation of the bearing bush. When the bearing bush is deformed, the shaft will have a large slip along the inner wall of the bearing bush.

IOP Conf. Series: Materials Science and Engineering **542** (2019) 012081 doi:10.1088/1757-899X/542/1/012081

3.3 Contact force and center motion trajectory of pin at different speeds

Under the condition of constant clearance and contact stiffness, the contact force and central motion trajectory of the shaft at different speeds are analyzed. The given simulation parameters are as follows:

$K=1\times10^7\ N/m$; $C_n=175N\bullet s/m$; $f=0.05$; $C_t=0$; $c=0.2mm$; $t=4s$

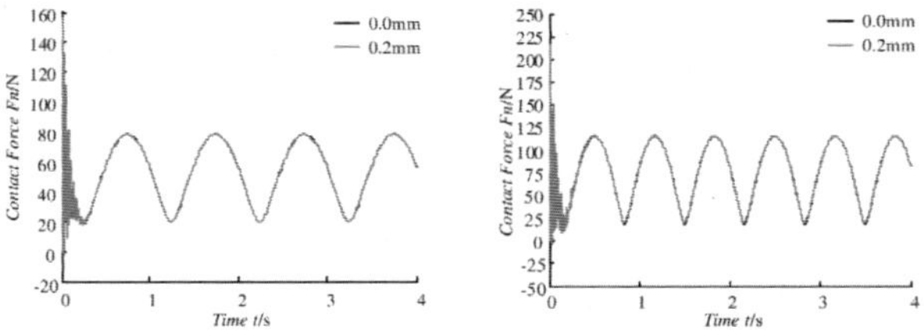

a)The change diagram of F_n under the condition of $\dot{\theta}=2\pi\ rad/s$

b)The change diagram of F_n under the condition of $\dot{\theta}=3\pi\ rad/s$

Figure 9. F_n change diagram at different rotational speeds

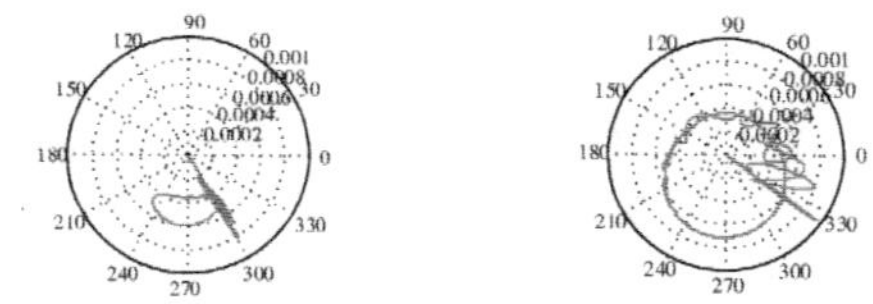

a)The $\phi-e$ coordinate diagram under $\dot{\theta}=2\pi\ rad/s$ condition

b)The $\phi-e$ coordinate diagram under $\dot{\theta}=3\pi\ rad/s$ condition

Figure 10. The $\phi-e$ coordinate diagrams under different rotational speeds

Figure 9 shows that with the increase of the rotational speed of the the connecting rod, the force applied to the shaft also increases, and the larger the rotational speed, the larger the fluctuation of F_n. In addition, the greater the rotational speed, the slower the attenuation of the contact force of the shaft, that is, the greater the rotational speed, the longer the contact time of the shaft reaches a stable value. As shown in Figure. 10, It can be seen that the maximum penetration depth of the shaft with the revolving joint with clearance relative to the bearing bush increases with the increase of the rotational speed. After the shaft is stable, its motion trajectory exhibits a periodic variation. In the gravitational field, the central trajectory of the shaft moves downward as a whole due to the effect of gravity.

4. Summary

In this paper, the dynamic model of the revolving joint with clearance with "contact-separation" state is established based on Newton Euler method. The dynamic equation is solved, and the general change law is obtained, which lays a foundation for the future analysis of dynamics and control of hinges with clearances. The specific laws are as follows:

(1) The clearance has a strong destructive effect on the mechanism in the early stage , and the greater the clearance, the greater the damage. When the connecting rod moves steadily, the clearance has little influence on the stress of the shaft.

(2) The greater the contact stiffness between the shaft and the bearing bush, the greater the force of the shaft, but the smaller the deformation of the bearing bush. When the connecting rod moves steadily, the contact stiffness between the shaft and the bearing bush has little effect on the stress of the shaft.

(3) The greater the rotational speed, the greater the contact force of the shaft, the longer the time it experienced when the contact force reaches the stable value, the higher the connecting rod' speed , the deeper the corresponding puncture' depth.

5. References

[1] YU Lixiao,LIU Caishan. The dynamic analysis of the solar array deployment with spatial clearance joint[J]. *Journal of Dynamics and Control*,2004,2(2):19-22.

[2] XU Jilong. Analysis with mechanical properties of manipulator with clearance joint in different gravity environment[D]. *Yanshan University*,2015

[3] FLORES P. Modeling and simulation of wear in revolute clearance joints in multibody systems[J]. *Mechanism and Machine Theory*,2009,44:1211-1222.

[4] OLYAEI A A,GHAZAVI M R. Stabilizing slider-crank mechanism with clearance joints[J]. *Mechanism and Machine Theory*,2012,53:17-29.

IOP Conf. Series: Materials Science and Engineering **542** (2019) 012081 doi:10.1088/1757-899X/542/1/012081

[5] LU Xuxiang,LIU Zhengqiang,HUANG Shuhong,et al. Nonlinear vibration characteristics of a vibro-impact system with clearance[J]. *Journal of Vibration Engineering*,2012,35(5):388-393.
[6] ZHANG Jing, Guo Hongwei, Liu Rongqiang, et al. Nonlinear dynamic characteristic analysis of jointed beam with clearance[J]. *Acta Astronautica*, 2016, 129:135-146.
[7] WITTENBURG J. Dynamics of systems of rigid bodies [M]. Publisher:*Stuttgart B G Teubner*,2012.
[8] SHABANA A A. Dynamics of multibody systems (3ed)[M]. *New York:Cambridge University Press*,2005.
[9] KANE T R,LEVINSON D A. Formulation of equation of motion for complex spacecraft[J]. *Journal of Guidance,Control and Dynamics*,1980,3(2):99-112.
[10] LANKARANI H M,NIKRAVESH P E. A contact force model with hysteresis damping for impact analysis of multibody systems[J]. *Journal of Mechanical Design*,1990,112:369-376.

Supported by NSFC(51775475);
Supported by Natural Science Foundation of Hebei Province of China (E2016203463).

IOP Conf. Series: Materials Science and Engineering **542** (2019) 012082 doi:10.1088/1757-899X/542/1/012082

The Control Method of Pigging Slug Flow for Multiphase Subsea Pipeline

Li Meng [1] and Xu Hao [2]

1 CNOOC Research Institute Co., Ltd., China
2 China National Oil and Gas Exploration and Development Co., Ltd., China
E-mail: peterxu2009@163.com

Abstract. This paper analyses the causes of the large number of slug flows by OLGA software and analyses its control methods. It can be seen from the simulation that closing the high-production gas well to control the pigging flow rate and venting time can effectively control the volume of the pigging slug flow, thereby reducing the volume of the slug flow trap and reducing the platform area, thereby reducing the overall engineering investment.

1. Introduction

As the main way to develop and transport production fluids in offshore oil and gas fields, submarine pipelines connect offshore oil and gas fields, oil storage facilities or land terminals into an organic whole, so that all aspects of offshore production facilities form interconnected and coordinated production systems through submarine pipelines. Submarine pipelines can be regarded as the lifeline of offshore oil and gas field development. Therefore, the flow safety of submarine pipelines is a necessary condition for maintaining oil and gas field production, and the occurrence of slug flow in submarine pipelines is one of the serious problems that threaten the safety of submarine pipelines [1-2].

The slug flow refers to the gas-liquid two-phase flow state in which a gas column and a liquid column alternately appear in the pipeline. Liquid slug flow often occurs in oil-gas two-phase flow mixing pipelines, gas long-distance transport pipelines, and riser outlets, which appear as two-phase slug flow. Due to the high flow rate and high kinetic energy, liquid slug flow can cause mechanical problems. At the same time, due to liquid level fluctuations, impacts and blockages, slug flow can also cause process problems, posing a serious threat to the safe transportation of submarine pipelines [3].

The damage caused by the slug flow to the normal operation of the pipeline system and the downstream production equipment of the pipeline is mainly manifested in: causing the pipeline pressure to fluctuate greatly; intensifying the corrosion of the pipe wall, especially the corrosion of the riser pipe wall; causing the overflow of the pipe outlet separator or interrupted flow, making the slug flow trap unstable and difficult to operate; causing mechanical damage to the pipe joints and struts, causing cavitation of the pressurized equipment (such as multiphase pumps and compressors) during pumping and reducing the efficiency of pumping and work reliability; in severe cases, it can cause the platform or even the entire oil field to be shut down [4-6].

The slug flow is mainly divided into normal working condition slug flow and pigging condition slug flow. The pigging condition pipe slug flow refers to the front end of the pigging ball due to the difference of gas-liquid flow rate in the pipe when the gas-liquid mixed pipeline is in pigging condition. The accumulation of liquids forms a pigging slug flow, and the control of pigging slug flow is particularly important when producing offshore platforms. Because the slug flow trap is generally provided at the end of the pipeline, the size of the slug flow trap cannot be too large due to the

IOP Conf. Series: Materials Science and Engineering **542** (2019) 012082 doi:10.1088/1757-899X/542/1/012082

influence of the platform area and economy. Therefore, the maximum slug flow volume in pigging condition should be controlled in production.

The method for controlling the pigging slug flow is: reducing the amount of liquid in the pipe before pig the pipe (ie, closing the source water well or the production well); controlling the pigging speed (ie, increasing the back pressure of the pipe or shutting in the high-production gas well); increasing the liquid phase outlet flow rate of slug flow trap; due to the difference in height and the difference in oil-water density, in addition to the attention to the liquid phase slug, some pipelines also need to pay attention to the change of the flow rate of the oil phase and the water phase of the pipeline outlet.

2. The cases simulation of slug flow

2.1 Oilfield overview

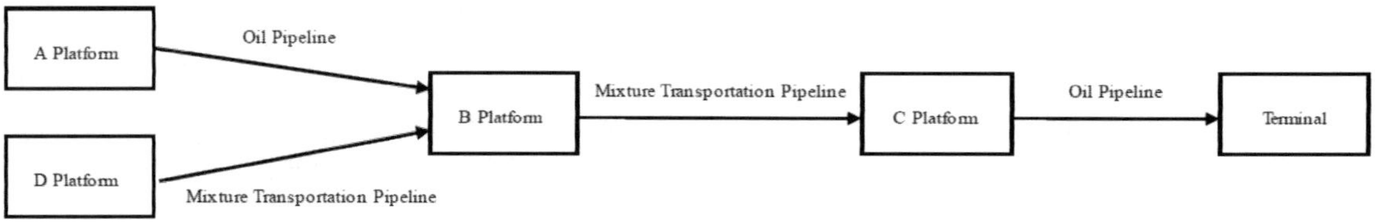

Figure 1. Schematic diagram of an offshore oilfield development

An oilfield area consists of 5 platforms and 4 pipelines. The development diagram is shown in Figure 1. This paper simulates the pigging conditions of the oil and gas-water mixed pipelines from the B platform to the C platform through software, and then analyses the pigging slug flows.

2.2 Basic data

The B to C mixed submarine pipeline is about 15.9 km long. The water depth in this area is about 25 m, the pipe diameter is 10 in, and the outlet pressure is 1500 kpaA. It is a double-layer insulation pipe with an average mud temperature of 4.8 °C. The gas-liquid-liquid ratio of the pipeline is 38.9. It belongs to low-viscosity high-condensation crude oil. In this paper, the typical working condition parameters of three years are selected for simulation calculation. The specific transmission parameters are shown in Table 1.

Table 1. Typical working condition parameters

Case	Oil m^3/d	Water m^3/d	Liquid m^3/d	Gas $10^4Sm^3/d$	Water Cut %
1	2925.1	731.3	3656.4	15.9	20.0
2	4266.1	1066.5	5332.7	19.7	20.0
3	4083.2	1020.8	5104.0	19.8	20.0

2.3 The stimulation of slug flow in pigging condition

This paper uses OLGA software to build a dynamic model of the pipeline for simulation. After simulation, the calculation results of the pipe slug flow in the typical years of B-C mixed submarine pipelines are shown in Figure 2~Figure 4. As can be seen from the figure, a large number of slug flows are generated when the pipe is in pigging condition.

IOP Conf. Series: Materials Science and Engineering **542** (2019) 012082 doi:10.1088/1757-899X/542/1/012082

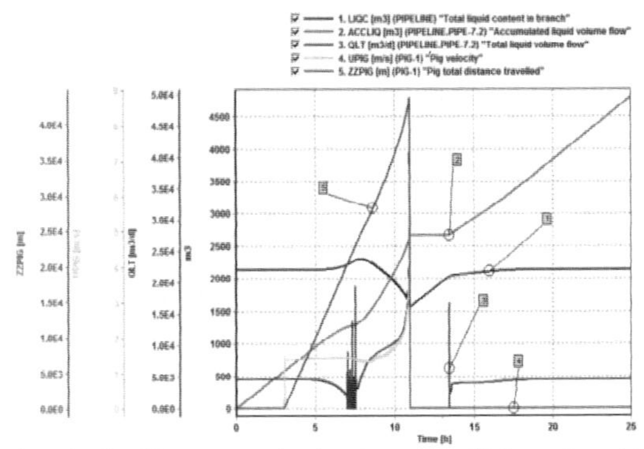

Figure 2. B to C mixed pipeline case 1 pigging condition slug flow calculation result

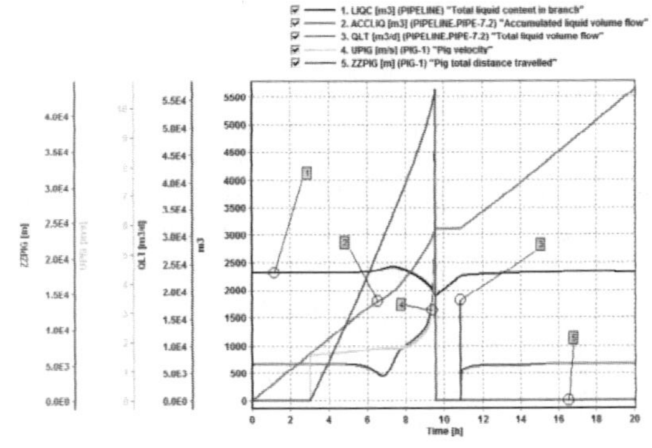

Figure 3. B to C mixed pipeline case 2 pigging condition slug flow calculation result

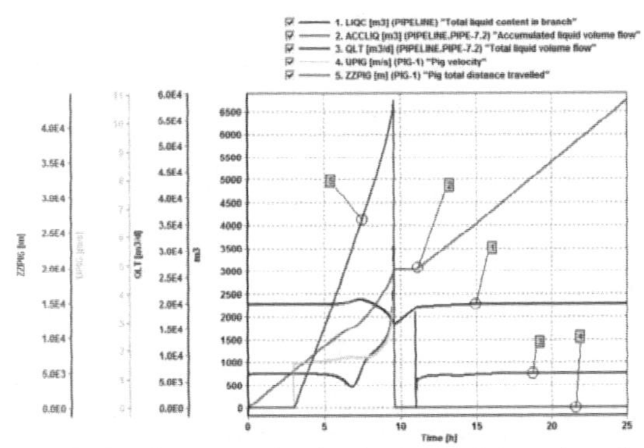

Figure 4. B to C mixed pipeline case 3 pigging condition slug flow calculation result

The results of the slug flow analysis are shown in Table 2. It can be seen from the analysis results of Table 2 that the volume of slug flow in the pigging period of the case 1, the case 2, and the case 3 is large. Therefore, it considers taking full advantage of the equipment of the downstream C platform. The plug flow trap on C platform has an effective capacity of 60 m^3 and the closed vessel capacity is 24 m^3. Therefore, the total volume of the slug flow can be stored on the C platform is 84 m^3. After deducting this volume, there is still 391 m^3 pigging slug flow that cannot be stored. If the volume of the slug flow trap is increased, the platform deck area is enlarged too much, and the overall project investment is correspondingly increased.

IOP Conf. Series: Materials Science and Engineering **542** (2019) 012082 doi:10.1088/1757-899X/542/1/012082

Table 2. B to C mixed pipeline pigging condition slug flow calculation result

Case	Year	Slug Flow Time	Slug Flow Volume	Downstream Process Equipment Processing Capacity	Downstream Process Equipment Processing Volume Capacity	The Volume Of Slug Flow That Needs To Be Stored	Slug Flow Trap Volume + Closed Vessel Capacity	Can the Slug Flow To Be Stored
		min	m^3	m^3/min	m^3	m^3	m^3	
1	2019	163	1220	5	815	405	84	No
2	2021	135	1138	5	675	463	84	No
3	2022	133	1138	5	663	475	84	No

2.4 Simulation and analysis of control method for pigging slug flow

According to the method of controlling the pigging slug flow, combined with the actual situation of no source water well in the oilfield, it is considered that the effect of controlling the outlet back pressure to control the pigging slug flow is not obvious. Therefore, it is decided to control the slug flow volume by means of closing the high-production gas well so that when the pigging is carried out, the slug venting time is prolonged, and the speed of the pigging ball is reduced. Therefore, according to the production schedule, it is considered to close three wells with a large gas production during the pigging. Figure 5 to Figure 7 show the simulation results of the plugging conditions of the B-C mixed pipeline after shut in the gas wells, and Table 3 shows the analysis results of the simulated slug flow. It can be seen from the analysis results that the slug flow can be stored after the high-production gas wells are shut in.

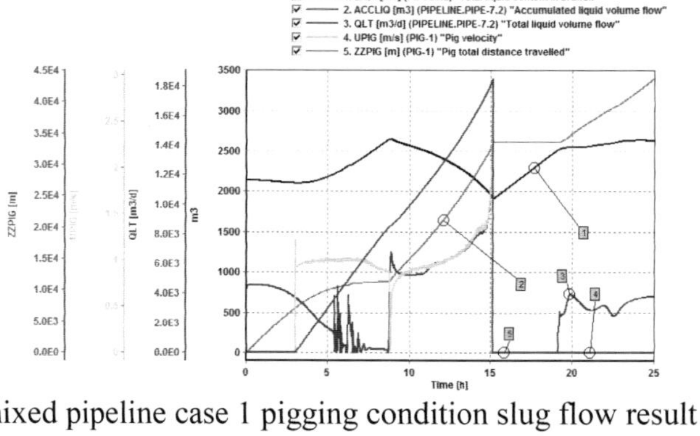

Figure 5. B to C mixed pipeline case 1 pigging condition slug flow result after shut in wells

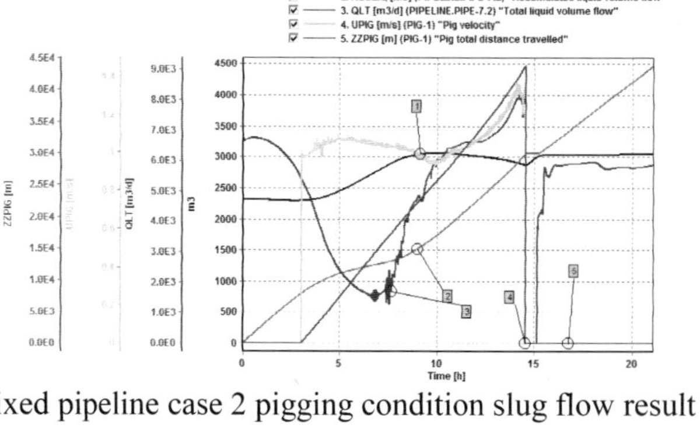

Figure 6. B to C mixed pipeline case 2 pigging condition slug flow result after shut in wells

IOP Conf. Series: Materials Science and Engineering **542** (2019) 012082 doi:10.1088/1757-899X/542/1/012082

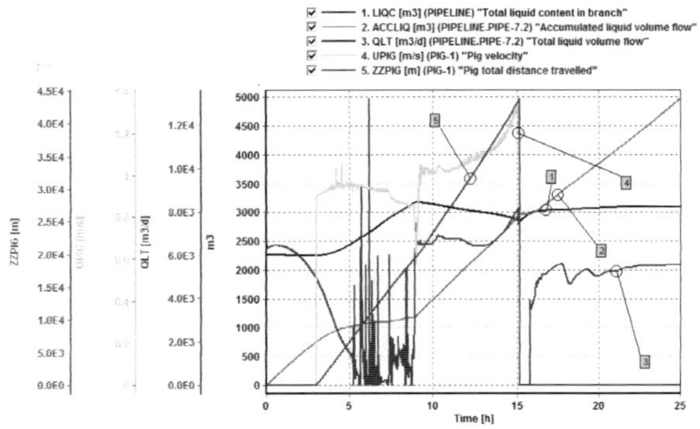

Figure 7. B to C mixed pipeline case 3 pigging condition slug flow result after shut in wells

Table 3. B to C mixed pipeline stimulation results after reducing the volume of gas production

Case	Year	Pre-pigging Flow Rate			Flow Rate During Pigging			Slug Flow		Downstream Process Equipment Processing Capacity	Downstream Process Equipment Processing Volume Capacity	The Volume of Slug Flow That Needs To Be Stored	Slug Flow Trap Volume	Can the Slug Flow To Be Stored
		Oil	Gas	Water	Oil	Gas	Water	Time	Volume					
		m^3/d	$10^4Sm^3/d$	m^3/d	m^3/d	$10^4Sm^3/d$	m^3/d	min	m^3	m^3/min	m^3	m^3	m^3	
1	2019	2925	15.9	731	2365	6.5	591	96	553	5	480	73	84	Yes
2	2021	4266	19.7	1066	3706	4.6	926	66	390	5	330	60	84	Yes
3	2022	4083	19.8	1020	3523	1.8	880	58	315	5	288	27	84	Yes

3. Conclusion

The pipe slug flow will have a great impact on the safety of the submarine pipeline. This paper analyses the causes of the large number of slug flows by OLGA software and analyses its control methods.

It can be seen from the simulation that closing the high-production gas well to control the pigging flow rate and venting time can effectively control the volume of the pigging slug flow, thereby reducing the volume of the slug flow trap and reducing the platform area, thereby reducing the overall engineering investment.

In actual production, the control method for a pipe slug flow needs to be considered in combination with actual operation conditions, operator requirements and economic investment.

References

[1] Lu Hong, Chen Jinghua, Zhu Haishan, et al. Control Method for the Pigging Slug in the Subsea Multiphase Pipeline[J]. Petroleum Planning & Engineering, 2013, 24(4):25-27

[2] Wang Bing. Structural Design and Analysis of Slug-flow Catcher[J]. Petroleum Engineering Construction, 2009, 35(S1):165-166

[3] Wang Wenguang, Wan Yufei, Qu Zhaoguang, et al. Influence factors and control method of pigging surge of subsea multiphase pipeline[J] Petroleum Engineering Construction, 2017,04(S1):44-46

[4] Ju Pengpeng, Qin Jiajun, Liu Peilin. Analysis of slug flow in multiphase subsea pipeline[J]. Ship Building of China,2012,53(S1): 79-84

[5] Chen Hongju, Zhou Xiaohong. Subsea pipeline transportation process in regional development[J]. Petroleum Planning & Engineering, 2011, 22(4):30-33

[6] Lu Hong, Chen Jinghua, Zhu Haishan, et al. Control of Pigging Slug in Multiphase Subsea Pipeline[J]. Ship Building of China, 2014, 55(S2):402-407

IOP Conf. Series: Materials Science and Engineering **542** (2019) 012083 doi:10.1088/1757-899X/542/1/012083

Effect of Local Corrosion Caused by Inclusions and Surface Scratches of Steel in High Temperature and Pressure Flue Gas Environment

Dong Wang[1], Shuzhong Wang[1,*], Jianqiao Yang[1], Zhuohang Jiang[1], Baoquan Zhang[1], Chengchao Cui[1] and Jianna Li[1].

1 Key Laboratory of Thermo-Fluid Science and Engineering of MOE, School of Energy and Power Engineering, Xi'an Jiaotong University, Xi'an, 710049, China
E-mail:szwang@aliyun.com

Abstract. By simulating the high temperature and high pressure flue gas environment in heavy oil thermal recovery, the influence of two kinds of surface defects, steel inclusions and surface scratches, on local corrosion of steel was studied. And the mechanism of local corrosion caused by inclusions and surface scratches was analysed steel #20 and 26CrMo4 were used as experimental materials, and reacted in a 9 MPa flue gas environment for 24h. The reaction temperatures were 120 °C and 250 °C. The results show that the non-metallic inclusions in the steel are weak parts of the passivation film on the surface of the sample and are easily corroded by corrosive media. The scratched area is more likely to cause aggregation of corrosive media than other parts of the steel surface, and the iron elements in the metal in this area are more likely to reach the surface and react with the corrosive medium.

1. Introduction Formatting the text

These Steam assisted gravity drainage technology is the most commonly used enhanced recovery technology in thermal recovery of heavy oil. Steam is injected through horizontal injection wells. The heated heavy oil flows along the gas cavity wall to the production well and is produced under the action of gravity. In the practical application, it is found that the steam assisted gravity drainage technology may lead to the problems of large steam consumption and low heat energy utilization rate[1-2]. Therefore, the technology of flue-steam assisted gravity drainage is developed. The flue-steam assisted gravity drainage technology can effectively enhance the oil recovery and improve the economy of oil recovery compared with the single steam injection[3].

In the process of injecting steam into a heavy oil thermal recovery well, the casing is subjected to a biggish compressive stress due to thermal expansion. Because of the relaxation and creep of metal materials, the casing is subjected to large tensile stress, and the casing bears such high tension-compression stress periodically in the process of stopping injection oil production. This causes fracture (or threaded joint slippage), deformation and casing damage in thermal recovery wells[4]. The thermal stability of ordinary American Petroleum Institute(API) casing such as N80 steel casing is poor under the condition of heavy oil thermal recovery. The Cr-Mo series low alloy steel is used as tubing material in 26CrMo4 casing, which makes the casing have good thermal stability, so it is widely used in heavy oil thermal recovery. At the same time, Steel #20 is widely used in surface pipeline.

In the oil and gas field industry, the fracture of carbon steel and alloy steel under stress and corrosion environment is one of the most dangerous forms of failure. The localized corrosion of passive metals almost always begins with the local presence of impurities, especially inclusions, as well as grain boundaries, dislocations, crack defects or mechanical damage. At present, the corrosion of steel caused by inclusions has been widely discussed[5-8]. However, there are few studies on the corrosion of casing caused by inclusions and scratches in high-temperature and high-pressure flue gas environment. In this paper, 26CrMo4 and steel #20, which are widely used in heavy oil thermal recovery, are used as experimental materials. It is of practical significance to study the corrosion law of inclusions in high temperature and high pressure flue gas environment.

2. Experimental

2.1. Material and Pretreatment

The experimental materials were steel #20 and 26CrMo4, and the chemical composition thereof is shown in Table 1. The corrosive medium is flue gas, and its composition is shown in Table 2.

Table 1. Chemical compositions of the tested steels.

Steel	C	Si	Mn	Cr	Mo	P	S	Ni	Fe
#20 (wt.%)	0.18	0.21	0.42	0.22	-	0.02	0.033	0.22	Bal.
26CrMo4(wt.%)	0.24	0.23	0.81	0.92	0.21	0.02	0.03	-	Bal.

Table 2. Chemical compositions of the corrosion fumes.

Gas	O_2	CO_2	NO_2	H_2	H_2S	N_2
Vol.%	7%	11%	0.005%	0.001%	0.001%	Bal.

The steel #20 and 26CrMo4 were respectively processed into 8mm×8mm×5mm and 6mm×6mm×6mm hexahedral samples by wire cutting. Before tests, the six surfaces of the sample were polished to the mirror surface with 200 mesh, 800 mesh,1500 mesh sandpaper to remove the outermost dirt and oxide layer. The oil on the surface of the sample is removed with acetone, and then the surface of the sample is rinsed with distilled water and placed in absolute ethanol to prevent oxidation of the material. The samples were taken out and dried with cold air and then wrapped with filter paper after all the samples were polished and cleaned. The accurately measured weight was taken after placing it in the drying dish for one hour.

2.2. Experimental method

The reactor is a tube furnace with a furnace tube size of Ø12×600mm provided by Kejing Company. Before the experiment, the tube furnace was first filled with high-purity nitrogen for 2 hours to remove oxygen. The sample is placed in the middle heating section of the tube furnace and the mixed gas fed into the furnace tube to make the pressure inside the tube reach 9MPa. The reactor was set the program to a temperature of 120°C and 250°C. The sample was taken out of the reactor after 24h and washed 3 times with distilled water and absolute ethanol. The experimental conditions are shown in Table 3.

Table 3. Chemical compositions of the tested steels.

Run No.	Sample	Temperature (°C)	Pressure (MPa)	Reaction time (h)	Corrosion medium
1	Steel #20	120	9	24	Flue gas
2	26CrMo4	120	9	24	Flue gas
3	Steel #20	250	9	24	Flue gas
4	26CrMo4	250	9	24	Flue gas

2.3. Characterization of the Corrosion Scale

The surface of the corrosion scales were observed using scanning electron microscope (SEM). The elemental compositions of the corrosion scales were analyzed using energy dispersive spectroscopy (EDS).

IOP Conf. Series: Materials Science and Engineering **542** (2019) 012083 doi:10.1088/1757-899X/542/1/012083

3. Results and discussion

3.1. Effect of inclusions on corrosion

The presence of inclusions can lead to chemical homogeneity on the surface of the sample, resulting in weak areas in the passivation film formed on the metal surface, including the boundary between the metal matrix and the non-metallic inclusions and the chemically active inclusion itself.

Sulfide inclusions are the most sensitive locations for initiation pitting of carbon steel and low alloy steel. In carbon steel, there are usually mixed Fe-Mn sulfides. The manganese-rich sulfide is easily corroded by corrosive medium， and a manganese-rich sulfide is easily attacked by a corrosive medium. Cracks are often formed around manganese-rich sulfides, and inclusions and matrix are eroded. The steel #20 and 26CrMo4 used in this experiment all contain a certain proportion of impurity elements such as Mn, S and P. This is unavoidable in the iron and steel smelting process. These materials are trapped inside the material during the formation of the metal, forming defects in the crystal. If some of these impurities are exposed during the polishing process, local defects will occur. These defects are active sites where corrosion preferentially occurs, resulting in corrosion pits. The observation of the morphology after the corrosion test did show some local corrosion aggravation behavior caused by inclusion defects. However, through the observation of the morphology after the corrosion test, it is true that some localized corrosion behaviors caused by inclusion defects exist on the surface of the sample. Figure 1 and figure 2 shows the morphology of the inclusions due to localized corrosion of inclusions in the experimental samples.

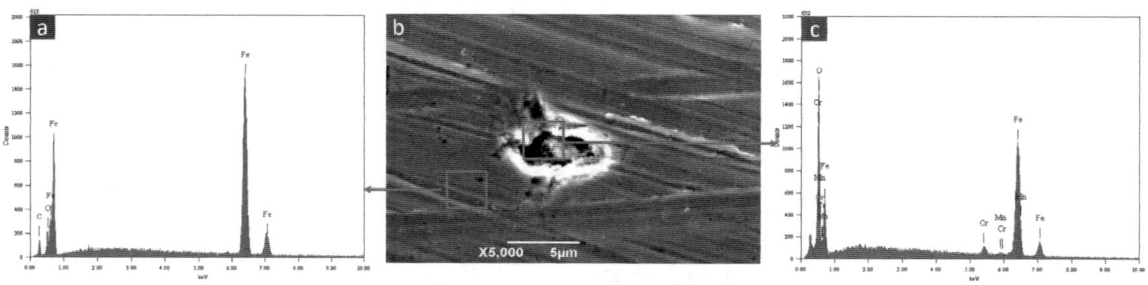

Figure 1. SEM and EDS images of steel #20 for 24h at 9MPa and 120°C:(a) EDS images of uncorroded area, (b)SEM images of sample surface,(c) EDS images of corroded area.

Figure2. SEM and EDS images of 26CrMo4 for 24h at 9MPa and 120°C:(a) EDS images of uncorroded area, (b)SEM images of sample surface,(c) EDS images of corroded area.

There are two explanations for the mechanism of steel corrosion caused by inclusions. One is that pitting occurs and developed in the micro-voids on the boundary between sulfide and metal, and the other is pitting caused by the dissolution of sulfide itself. In particular, certain composite sulfide-oxide inclusions establish a disordered state in the region immediately adjacent to them, therefore affecting the protective properties of the passivation film. At the same time, the locally disordered film is thick but often defective. The adhesion of the film is generally poor, and the physical and chemical resistance is low.

It is worth noting that the occurrence of localized corrosion caused by inclusions has little to do with the temperature, pressure and corrosion time of the experiment. The position of inclusions in the sample block cannot be predicted. The morphology of the inclusions is exactly in a certain area of the observed surface, and the surface is not covered by a thicker corrosion layer.

3.2. The effect of scratch on corrosion

The scratch is a groove left on the surface during the sanding of the sample, which determines the roughness of the surface. Methods for the rough treatment of surfaces include electrolytic polishing, grinding, pickling, and mechanical polishing. In this experiment, the surface of the sample was treated by sandpaper on the mechanical polishing of the material according to the corrosion test standard, so that scratches will be left on the surface of the sample.

In this experiment, the sample was carried out under high temperature environment of corrosion of 9MPa and 250 ° C. As shown in figure 3, such corrosion product particles grown along the scratch were observed. After 24h reaction time, there were three different morphologies on the surface of steel #20 and 26CrMo4 samples, which were flat surface, scratches and corrosion product particles growing along the scratches. At the same time, the energy analysis shows that the corrosion product is iron oxide.

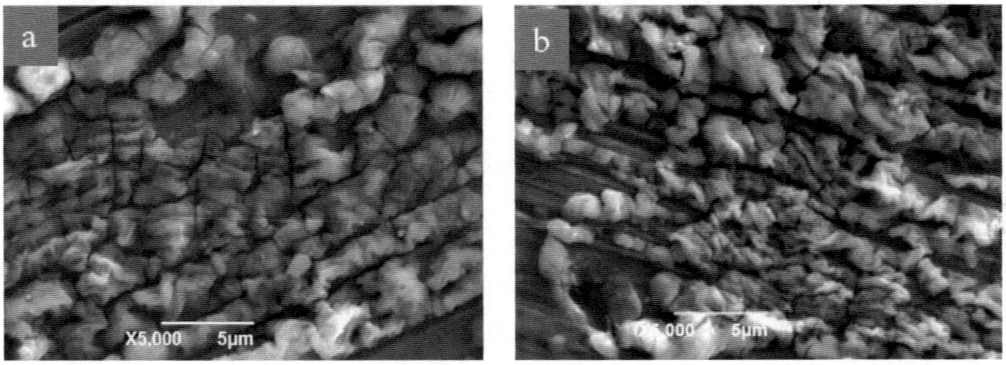

Figure 3. SEM images of corrosion for 24h at 9MPa and 250°C:(a) steel #20, (b)26CrMo4.

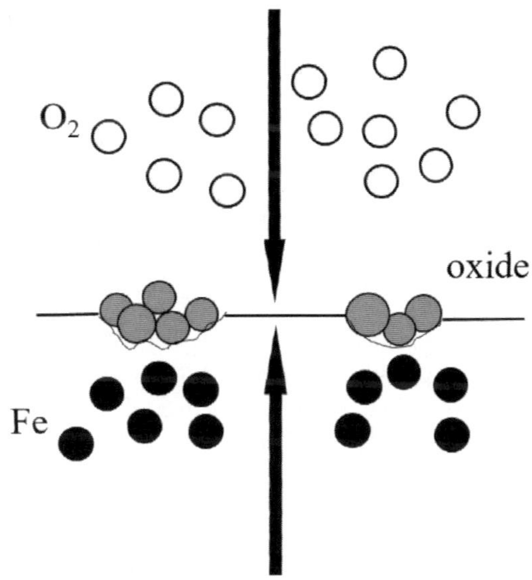

Figure 4. Schematic diagram of corrosion mechanism at scratch.

The local corrosion mechanism caused by scratches on the surface of steel is shown in figure 4. Scratched depressions provide a natural enrichment site for the enrichment of corrosive media. Since the sunken region is closer to the substrate, the iron ions diffused outward from the substrate are first brought into contact with oxygen ions adsorbed on the metal surface in these regions and undergo oxidation reaction to form a corroded particle product. The diffusion path of iron element in the matrix includes two channels, the grain boundary and the inner part of the crystal, wherein the diffusion speed at the grain boundary is higher than the diffusion speed inside the crystal. And a large number of scratches appearing on the surface during polishing can become a diffused short-circuit path. The high temperature environment helps to increase the diffusion rate and the size of product particles. This process should be more obvious in high temperature environment and occurs in the early stage of corrosion. As the corrosion progresses, the scratches will gradually be covered and the oxide growth advantage in the scratch direction will no longer exist.

4. Conclusion

In this paper, the corrosion of two common steel surface defects such as inclusions and scratches on steel surface in high temperature and high pressure corrosive media environment is studied. Carbon steel and alloy steel usually have mixed Fe-Mn sulfide inclusions. Sulfide is susceptible to corrosion by corrosive media, and a gap is formed around it, causing both inclusions and matrix to be corroded. Scratches and inclusions of materials are relatively close to the metal matrix and also become corrosive medium enrichment areas. Therefore, in order to improve the service life of the gas injection pipeline, it is necessary to avoid defects such as inclusions, scratches and mechanical damage on the steel surface in practical applications.

Acknowledgments

This research is supported by the Fundamental Research Funds for the Central Universities [xjj201820 1] and [xjj2018006], the Projects from National Natural Science Foundation of China [51871179].

References

[1] Liu S Q, Bao L C and Ma D S 1999 New Technology study for the development of a supper heavy oil reservoir in Liaohe oil field[J]. *Petroleum Exploration & Development*, **26**(4):80-81

[2] Lin R Y, Li W, Li Z M, Yang L Q and Yang J P 2012 Numerical simulation technology of flue gas-steam assisted gravity drainage[J]. *Journal of China University of Petroleum (Edition of Natural Science)*, **36**(5):136-140

[3] Tian Z Q, Li S L, Wang Z M, Xu H M and Luo R Z 2002 Gas/steam injection for heavy oil recovery: Laboratory study and field trial uses[J]. *Oilfield Chemistry*, **19**(1):47-50

[4] Lu X Q, Li Q, Li C X 2007 Development of TP110H Hi-strength Special Casing for Service in Thickened Oil Hot Well[J]. *Steel Pipe*, **36**(5):14-17

[5] Mu R S, Zhang Z Y, Cheng L and Zhao Y Y 2015 Influences by Inclusions on Hydrogen Sulfide Corrosion-resistances of Tubing/Casing and Linepipe[J]. *Steel Pipe*, **44**(4):16-21

[6] Zhang C Y, Zhang Q, Li J G, Xiong C G and Li Z 2014 Micro-corrosion test research on pitting initiation site of the inclusions ofcarbon steel and low alloy steel in chlorine ion solution[J]. *Metallurgical Analysis*, **34**(1):22-27

[7] Zhang P, Zhang G X, Bi Z Y and He S L 2014 Effect of Inclusion and Banded Microstructure on Corrosion Performance of Pipeline Steel[J]. *Hot Working Technology*, **43**(18):34-38

[8] Liang J M, Wu H B, Tang D, Zhang P C and Liu X T 2014 Influence of inclusion and grain boundary on corrosion behaviour of low alloy steel in cargo oil tank bottom plate environment[J]. British Corrosion Journal, **49**(7):633-642

IOP Conf. Series: Materials Science and Engineering **542** (2019) 012084 doi:10.1088/1757-899X/542/1/012084

Study of corrosion process of 38CrMoAl in flue gas

Zhuohang* Jiang**[1], ***Shuzhong* Wang**[1,*], ***Jianqiao* Yang**[1], ***Dong* Wang**[1],Chengchao* Cui**[1], ***Baoquan* Zhang**[1] **and *Jianna* Li**[1]

[1]Key Laboratory of Thermo-Fluid Science and Engineering of MOE, School of Energy and Power Engineering, Xi'an Jiaotong University, Xi'an 710049, PR China
E-mail:szwang@aliyun.com

Abstract. The experiments were carried out on 38CrMoAl at 3MPa and 150°C, which the temperature and pressure conditions were set to simulate the actual production conditions. The measurement methods included scanning electron microscopy analysis, X-ray energy diffraction spectrum analysis. The experimental results show that the corrosion of 38CrMoAl in the flue gas have a certain periodicity, and the corrosion process can be summarized as a cyclic process: oxidation-film formation-rupture-continuing oxidation. It can be inferred that as the exposure time increases, the corrosion process continues to circulate, and the cracking of the oxide film under the action of various stresses may even cause it to fall off, which would result in serious damage to the material.

1. Introduction

Secondary oil recovery is a kind of intensive development measure which is adopted to develop the nearly exhausted oil field by relying on the natural energy. Recently, flue gas assisted oil recovery has been widely applied in oil field as a secondary oil recovery technology, which aims to increase production and oil recovery. Meanwhile, as the specific application of CCUS (Carbon Capture, Utilization and Storage) technology, flue gas assisted oil recovery can fully capture, utilize and store a large amount of CO_2 gas generated by boilers [1], which is conducive to environmental protection and generates good economic benefits. However, the corrosive gas (CO_2, O_2) in the flue gas is more corrosive to steel, which results in corrosion of oil well pipelines and even pitting.

For the corrosion of flue gas to steel, the current focus is mainly on the corrosion of the boiler system by the flue gas generated by boiler combustion. Some scholars have studied the high temperature flue gas corrosion under normal pressure [2,3]. However, in the flue gas assisted oil recovery process, the corrosion of the pipeline is different from the flue gas corrosion under normal pressure due to a higher pressure. At present, the corrosion of oil well pipe is mainly concentrated on CO_2 corrosion, and there are few studies on mixed flue gas [4].

Hausler et al. [5] suggested that the corrosion reaction mechanism of bare metal and metal with surface film is different, and the mechanism of CO_2 corrosion is summarized as interface mechanism and phase mechanism. At present, most scholars have proposed a number of CO_2 corrosion reaction mechanisms based on the interphase mechanism, and no unified consensus has been formed. However, for the uniform corrosion in CO_2 corrosive media, the following unified conclusions are drawn: the corrosion rate and corrosion morphology are mainly determined by the flow rate and the corrosion product film [6]. Therefore, studying the corrosion process of carbon steel in flue gas is important.

The present paper aims to explain the corrosion process of pipeline material in flue gas assisted oil recovery through the simulation of the flue gas corrosion condition at 3MPa, 150°C for up to 96 h.

IOP Conf. Series: Materials Science and Engineering **542** (2019) 012084 doi:10.1088/1757-899X/542/1/012084

2. Apparatuses and procedures

2.1. Materials and apparatus

The material in this study was 38CrMoAl steel and the element contents are showed in table 1. The corrosion atmosphere (flue gas) was prepared according to the flue gas used in oilfield and the gas composition is shown in table 2.

The measurement methods included scanning electron microscopy analysis, X-ray energy diffraction spectrum analysis. The morphology of the metal surface was examined by a scanning electron microscope (SEM) instrument (JEOL 7800F). The element composition of the materials was identified by Oxford X-max Extreme energy diffraction spectrum (EDS).

Table 1. Element content of 38CrMoAl steel.

Element	C	Si	Mn	Cr	Mo	Al	Fe
Content	0.33%	0.30%	0.38%	1.45%	0.19%	0.97%	Bal

Table 2. Gas composition of corrosion atmosphere.

Composition	O_2	N_2	CO_2	NO_2	H_2	H_2S
Content	7%	Bal	11%	0.005	0.001	0.001

2.2. Experimental procedures

All experiments were carried out in a high pressure tube reactor, which integrated flue gas inlet, PID control, solenoid valve automatic control pressure, temperature control. The tube reactor was made of superalloy GH1000 to withstand a maximum temperature and pressure of 1000°C and 30MPa, respectively. All materials were first buffed to 30mm by P-1 sample polishing machine.

The inner wall of the reactor was washed with deionized water to remove the deposits, and the tube was heated to 60°C for 10 min under unsealed conditions to ensure that the surface was completely dry. Then the materials were placed in the tube reactor through sample holder. The tube reactor was pressurized to 3MPa after purging for 10 minutes. When the pressure reached to the set point, all valves were closed and the heating program was started to 150°C. Furthermore, the temperature was dropped when reach to the setting time and the materials were taken after reaching their respective exposure time.

In order to explore the effects of exposure time, all materials were placed in the tubular reactor separately, and the corresponding exposure time was 24 h, 48 h, 72 h, 96 h, respectively. The specific experimental scheme is shown in table 3.

Table 3. Specific conditions of the experiments.

Material	Pressure	Temperatur	Exposure time
38CrMoAl	3MPa	150°C	24 h
38CrMoAl	3MPa	150°C	48 h
38CrMoAl	3MPa	150°C	72 h
38CrMoAl	3MPa	150°C	96 h

3. Results and discussion

3.1. Analysis of corrosion surface morphology

The 38CrMoAl steel were conducted at 150°C, 3MPa within different exposure time. At different exposure time, the surface morphology of metal changed greatly, showing certain regularity.

3.1.1. 24h exposure time. Figure 1 shows the surface morphology on 38CrMoAl in 150°C, 3MPa after exposure in flue gas for 24 h. It is apparent from figure1 that after 24 h exposure, the material formed spherical corrosion products in the smooth and clean layer, which still have clear scratches generated when polished. That is to say, the new corrosion products on the surface of the matrix was not formed a complete layer. At the same time, the energy spectrum analysis was performed on the surface particles and the region without covered particles. As shown in figure 2, the significantly different between the two areas indicated that the peak intensity of oxygen in surface particles was relatively stronger than the no-particle region, which proves that the spherical particle is the result of oxygen element enrichment.

Figure 1. SEM images of the surface morphology on 38CrMoAl in 150°C, 3MPa for 24 h.

Figure 2. energy spectrum analysis of 38CrMoAl in 150°C, 3MPa for 24 h:(a) no-particle region; (b) particles.

3.1.2. 48h exposure time. Figure 3 shows the surface morphology on 38CrMoAl in 150°C, 3MPa for 48 hours. As shown in figure 3, obvious oxide particle stratification occurred on the surface of the 38CrMoAl steel after 48 h exposure. As the corrosion process progresses, a large amount of iron diffused outward while oxygen simultaneously enriched on the surface. Gradually, individual iron oxide balls were joined together by solid solution reaction to form a morphology as shown. The ratio of iron to oxygen was closer to 2:3, i.e., ferric oxide.

Figure 3. SEM images of the surface morphology on 38CrMoAl in 150°C, 3MPa for 48 h.

3.1.3. 72h exposure time. Figure 4 depicts the surface morphology on 38CrMoAl in 150°C, 3MPa for 24 h and 72 h. It is apparent that the ever-increasing spherical oxides on the alloy surface were mostly connected by solid solution reaction. However, the difference of iron diffusion velocity in different regions of the matrix entailed the variance of the oxide particle. At the same time, the oxide connected into a piece, and formed cracks under the action of various stresses caused by the extrusion of corrosion products and the fluctuation of temperature and pressure, which was very unfavourable to the corrosion resistance of the material.

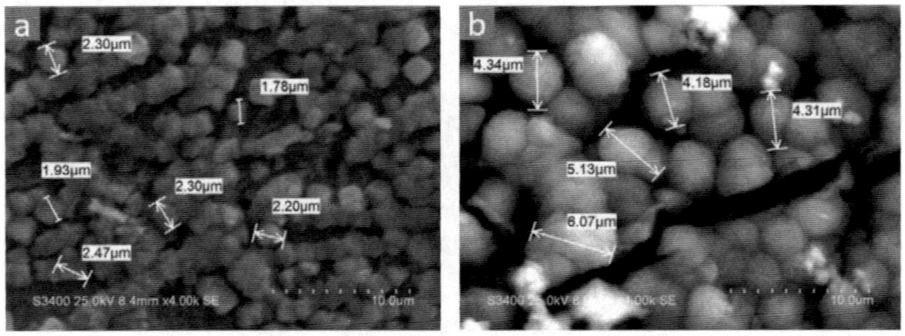

Figure 4. Surface particle size comparison of 38CrMoAl in 150°C, 3MPa: (a)for 24h (b)for 72 h.

3.1.4. 96h exposure time. Figure 5 shows the surface morphology on 38CrMoAl in 150°C, 3MPa for 48 hours. As the corrosion process continues, the iron diffusing continuously from the cracks in figure4(b) to the surface react with the oxygen to form new corrosion products. As shown in figure5, it could be seen that with the continuous corrosion and dissolution, the particle corrosion products on the surface gradually disappeared and became more flat. The outward diffusion of iron through the crevices and the reaction with oxygen resulted in the continuous thickening of the corrosion layer and the large fractured oxide layer.

As the high-pressure gas environment promoted the gas diffusion rate to the corrosion layer, the corrosion layer appeared obvious fracture to maintain the internal and external pressure difference, which indicated that the material had poor corrosion resistance in this corrosion atmosphere. The occurrence of oxidation, filming, fracture, continuous oxidation cyclic oxidation process is dangerous for the materials.

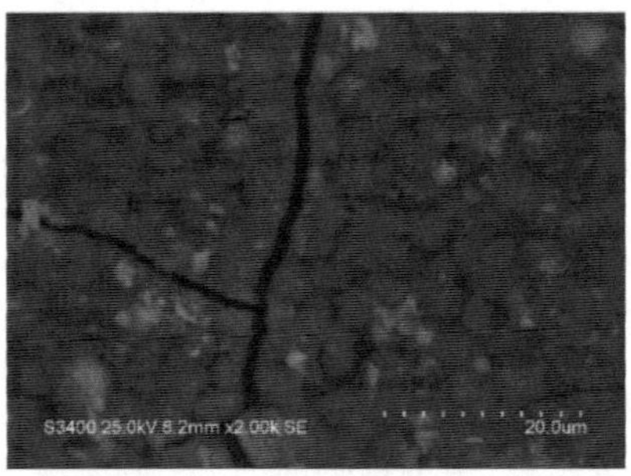

Figure 5. SEM images of the surface morphology on 38CrMoAl in 150°C, 3MPa for 96 h.

3.2. Corrosion mechanism in high pressure and temperature flue gas

3.2.1. Crystal structure of corrosion products. The main component of 38CrMoAl is iron, and EDS analysis under four exposure time showed that most of the corrosion products contained iron and oxygen. Therefore, the high temperature oxidation corrosion mechanism of 38CrMoAl can be mainly regarded as the high temperature oxidation process of iron.

When iron is oxidized in the flue gas at the experimental temperature of 150°C, it will grow into a rust layer containing Fe_3O_4 and Fe_2O_3, wherein Fe_3O_4 is close to the metal. The Fe_3O_4 phase is an inverse spinel structure in which half of Fe^{3+} is in the tetrahedral interstices, while Fe^{2+} and the remaining Fe^{3+} are in the octahedral interstices. Consequently, iron ions can diffuse through the defects at the octahedral and tetrahedral positions, which exhibits a certain degree of intrinsic semiconductor characteristics and inevitably cause electrons to diffuse out into the conduction band through electron holes, just like excess electrons.

The Fe_2O_3 phase has two forms: rhombohedral structure for α-Fe_2O_3 and cubic structure for γ-Fe_2O_3. Besides, though Fe_3O_4 is oxidized to form α- Fe_2O_3 above 400 ° C, the presence of Fe_3O_4 also needs to be considered due to the relatively low experimental temperature. In the rhombohedral structure, the combination between oxygen ions and iron ions in a gap position has the closed-packed hexagonal structure, which indicated iron ions were easily migrated.

3.2.2. Physical diffusion process and chemical reaction equation. Through the observation of the morphology of this experiment, the corrosion of most samples is mainly based on the formation of spherical corrosion particles on the surface. Therefore, it should not be the internal oxidation process caused by the inward migration of oxygen ions, but the surface oxidation process by the outward of the iron ions. The diffusion of iron ions is dominant in the oxidation process of iron based on the above introduction. Thus, the concise mechanism for the oxidation process of iron is as follows.
Iron ionized at the interface between iron and Fe_3O_4:

$$Fe \longrightarrow Fe^{2+} + 2e^- \quad (1)$$

$$Fe \longrightarrow Fe^{3+} + 3e^- \quad (2)$$

Iron ions and electrons diffused and migrated through the Fe_3O_4 layer, which went by the iron vacancies and electron holes on the tetrahedron and octahedron respectively. At the interface between Fe_3O_4 and Fe_2O_3, the reactions of Fe_2O_3 with iron ions and electrons formed Fe_3O_4:

$$Fe^{2+} + 2e^- + 4Fe_2O_3 \longrightarrow 3Fe_3O_4 \quad (3)$$

$$Fe^{3+} + 3e^- + 4Fe_2O_3 \longrightarrow 3Fe_3O_4 \quad (4)$$

In the Fe_2O_3 layer, the outward migration of iron ions through vacancies together with electrons formed new Fe_2O_3 at the Fe_2O_3/gas interface.

$$2Fe^{3+} + 6e^- + \frac{3}{2}O_2 \longrightarrow Fe_2O_3 \quad (5)$$

3.3. Corrosion process analysis
The corrosion model in flue gas was based on the outward diffusion of iron ions in dominant position. Figure 6 presents the schematic of the corrosion process. In the figure6, the corrosion process can be summarized as follows.

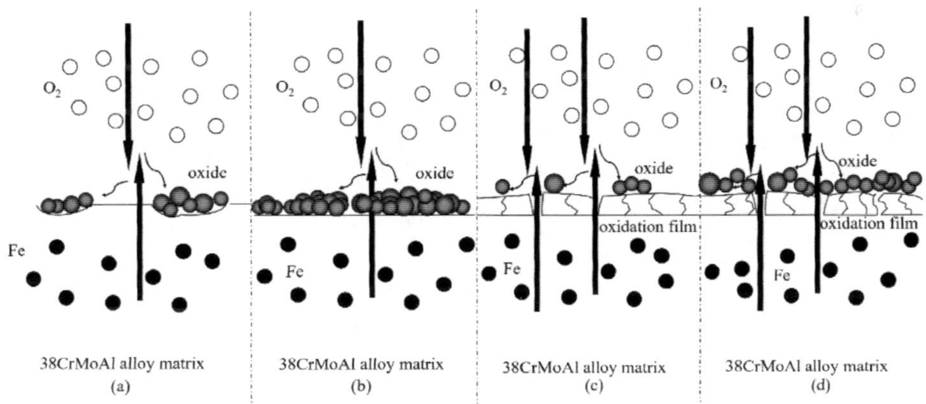

Figure 6. Schematic of the corrosion process.

First, iron ions in carbon steel matrix diffused outward while oxygen diffused inwards by ionization and adsorption on the surface of matrix, thus the contact between them on the surface of material form the spherical granular oxides. The EDS analysis of the surface particles and the uncovered substrate region showed that the oxygen content of the spherical particles was significantly higher than that of the matrix region, which proved that the spherical particles were the result of oxygen enrichment.

Second, oxygen was constantly adsorbed on the surface and reacted with iron ions which were continuously diffused from the matrix. The size of the surface oxide particles increased to form the first oxide layer by becoming a piece, in which individual iron oxide particles were connected together at this step.

Third, the surface became more flat as the continuous corrosion and dissolution reactions caused spherical particles gradually disappear. The oxide layer ruptured under the influence of various stress such as temperature stress. Due to the diffusion of iron ions from oxide layer cracks and reaction with oxygen, the size of the surface oxide particles increased further.

Last, the protectiveness of first iron oxide layer was weak due to the generation of cracks. The outward diffusion of iron through the crevices and the reaction with oxygen resulted in the continuous thickening of the corrosion layer and the forming of large fractured oxide layer.

4. Conclusions

In this work, experiment on 38CrMoAl steel was conducted to explain the corrosion process during the technology of flue gas assisted oil recovery. Experimental results indicated that the material exhibits a corrosion mechanism that iron oxide films grow periodically in high temperature and pressure flue gas. After continuous observation of 24 h, 48 h, 72 h, 96 h, the processes are concluded as a cyclic process: oxidation, film formation, rupture, continuing oxidation, which would cause relatively serious damage to the material.

Acknowledgement

This research is supported by the Fundamental Research Funds for the Central Universities [xjj2018201] and [xjj2018006], the Projects from National Natural Science Foundation of China [51871179].

References

[1] Chen W 2016 *Research on Electrochemical Corrosion Behaviour of Carbon Steel in Flue Gas Environment* (China University of Petroleum, Qingdao)

[2] Liu C 2015 *Optimization of Corrosion-Resistant Materials in Low-Temperature Boiler Flue Gas* (North China Electric Power University, Baoding)

[3] Nattawan K, Raphael I, Paitoon T and Chintana S 2011 *Energy Procedia* **4** 1761-1768

[4] Xiang Y, Wang Z, Xu MH, Li Z and Ni WD 2013 *J. Supercrit. Fluid* **82** 1-12

[5] Hausler R H and Stegmann D W 1988 *Corrosion-National Association of Corrosion Engineers Annual Conference* **1** No.363

[6] Synopsis B P A, 1984 *Corrosion-National Association of Corrosion Engineers Annual Conference* **1** 3-10

Mechanism analysis of bubbling corrosion caused by boiler flue gas on metal material

B Q Zhang[1], S Z Wang[1]2, J Q Yang[1], C C Cui[1], Z H Jiang[1], D Wang[1] and J N Li[1]

1 Key Laboratory of Thermo-Fluid Science and Engineering of MOE, School of Energy and Power Engineering, Xi'an Jiaotong University, Xi'an 710049, PR China
2 Corresponding author E-mail: szwang@aliyun.com

Abstract. Hydrogen bubbling is a typical corrosion morphology caused by hydrogen sulfide corrosion. In order to explore hydrogen sulfide corrosion of iron, the composition of flue gas under actual conditions was simulated in this research. Hydrogen sulfide corrosion in flue gas of four kinds of materials selection of ground process pipeline and underground pipe string was researched experimentally. The pressure and temperature of hydrogen bubbling corrosion morphology under actual flus gas conditions were obtained. The electrochemical corrosion mechanism between iron and hydrogen sulfide was investigated.

1. Introduction

Steam injection boiler is a special gas injection equipment for heavy oil recovery in oilfields. In the process of heavy oil recovery assisted by steam injection boiler flue gas, CO_2, SO_2, NO_2 and a small amount of H_2S in boiler flue gas are corrosive to pipeline materials, which threaten the safe operation of heavy oil recovery equipment and its auxiliary equipment. Therefore, it is necessary to study the corrosion behaviour of candidate materials in the flue gas environment at different temperatures and pressures, which is able to provide reference for materials selection of ground process pipeline and underground pipe string.

An immersion test was performed by X S Huang et al. [1] in order to analyse the influence of partial pressure of H_2S/CO_2 and sulfur deposition for the corrosion behaviour of L360QCS pipeline steel in H_2S/CO_2 environment. The results indicated that the corrosion rate of steel firstly decreased and then increased with the increasing of H_2S/CO_2 partial pressure. Jean Kittel et al. [2] studied the hydrogen induced cracking (HIC) of pipeline steel via immersion test and hydrogen permeation test. The extent of HIC depended on exposure time until it reached a stable level when the H_2S partial pressure was constant. According to Li et al. [3], the corrosion rate (CR) of 110S tube steel decreased first and then increased with increasing temperature. The minimum CR appeared at 110°C. Yin et al. [4] studied the corrosion behaviour of SM 80SS tube steel in stimulant solution containing H_2S and CO_2. Corrosion attacks increase in the initial stage and then decrease with the increase in p_{CO_2} or p_{H_2S}, and serious corrosion appears in the p_{CO_2}/p_{H_2S} range from 31 to 520.

In this study, four typical materials for ground equipment and underground pipe string are tested. According to the actual working conditions, corrosion tests were performed and the mechanism of bubbling corrosion was studied.

2. Experimental

2.1. Materials

IOP Conf. Series: Materials Science and Engineering **542** (2019) 012085 doi:10.1088/1757-899X/542/1/012085

Four different materials are steel 20#, 35CrMo steel, steel 45# and N80 steel. The standard compositions of samples are shown in table 1 [5]. Dimensions of all samples were 3*3*5mm. According to the actual flue gas detection report, the flue gas is used as corrosion medium. The compositions of the corrosion medium are shown in table 2.

Table 1. Standard compositions of materials used in the experiment

Material	C	Si	Mn	Cr	Mo	Al	P	S	Ni
Steel 20#	0.20	0.25	0.45	0.20	-	-	0.03	0.04	0.02
35CrMo steel	0.35	0.21	0.55	0.90	0.20	-	-	-	-
Steel 45#	0.46	0.27	0.65	0.22	-	-	0.04	0.04	0.21
N80 steel	0.22	1.00	1.00	13.00	-	-	0.02	0.01	0.50

Table 2. Compositions of corrosion medium

composition	CO_2	NO_2	H_2	H_2S	O_2	N_2	H_2O
Volume fraction（%）	11	0.005	0.001	0.001	7	surplus	saturation

2.2. Procedure and apparatus

The matrix was polished in the sample polishing machine, and then was cleaned in the ultrasonic material emulsifier and disperser. The matrix was cooled naturally and weighed. The experiments were performed using high pressure tubular furnace. The pressure in the adjusting furnace is 3.0MPa, 5.0MPa, 7.0MPa and 9.0MPa, respectively. The temperature is 120°C, 150°C, 200°C and 250°C, respectively, and the corrosion time is 48 hours. Samples were stored in a vacuum drying oven.

2.3. Sample analysis

The identification of the corroded sample was performed by scanning electron microscope (Model JEOL 7800F). Energy-dispersive X-ray spectroscopy (Model X-Max Extreme Oxford) was used in X-ray energy spectrum analysis of sample.

3. Results and discussion

3.1. Occurrence conditions of hydrogen bubbling corrosion

Bubbling corrosion morphology was found via scanning electron microscope (SEM) images. Figure 1 shown the typical morphology of bubbling corrosion. It can be seen that a small amount of bubbling corrosion products was produced on the surface of the matrix and most of bubbles are cracked, which can only be observed through SEM. In addition, it can be seen that the solid corrosion product layer was raised from the matrix, and there was a certain distance between the layer and matrix through the observation of the fracture of corrosion product.

It is noteworthy that this bubbling corrosion product usually occurs in the "wet hydrogen sulfide" environment that the content of $H_2S \geq 0.0075\%$ in acid gas. In this study, the content of $H_2S = 0.0075\%$ in flue gas, which didn't form the so-call "wet hydrogen sulfide" environment.

In this research, experiments have been done at different temperatures and pressures. Table 3 shown that the conditions of appearing bubbling corrosion product, it indicated that bubbling corrosion morphology mainly appeared in a certain range of experimental condition that 9MPa, 200°C and 7MPa, 120°C. It can be considered that bubbling corrosion happened mainly in high pressure region

IOP Conf. Series: Materials Science and Engineering **542** (2019) 012085 doi:10.1088/1757-899X/542/1/012085

(7MPa, 9MPa), and it happened in higher temperature region (7MPa, 120°C; 9MPa, 200°C) with the increased of pressure. In the flue gas environment containing water, the state of water in the high pressure and low temperature environment is super-cooled water, and the solubility of H_2S is high. However, the diffusion coefficient of H_2S is relatively high in the region of higher temperature, so that there is an equilibrium point between pressure and temperature. Therefore, 9MPa, 200°C and 7MPa, 120°C may be such an equilibrium points.

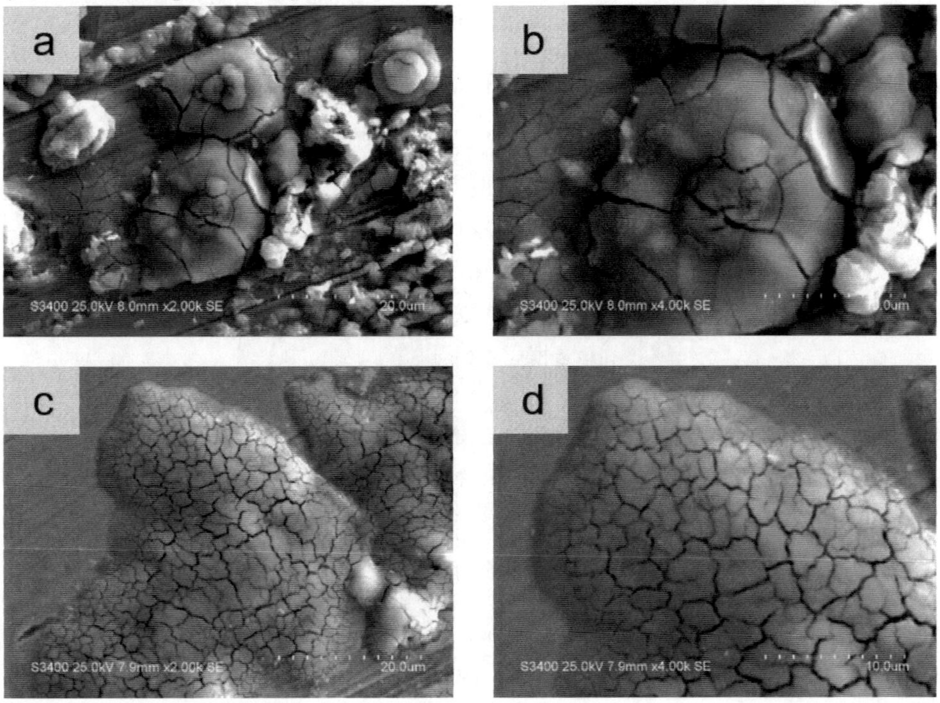

Figure 1. SEM images of two typical bubbling corrosion morphology. (a)(b). 20 steel in 9MPa and 200°C. (c)(d). 45 steel in 9MPa and 200°C.

Table 3. The conditions of appearing bubbling corrosion product in experiments

Material	Pressure	Temperature
35CrMo	9	150
All Materials	9	200
35CrMo	9	250
All Materials	7	120

3.2. Formatting the title

3.2.1. Anodic reaction mechanism. On the metal surface, zero-valent Fe lost electrons to generated divalent Fe, and then combined with HS^- (from H_2S) to formed $FeHS^+$. On the other hand, H^+ (from H_2S) combined with H_2O to formed hydronium ion (H_3O^+). $FeHS^+$ couldn't exist stably, and it will eventually be transformed into a more stable FeS [6]. The chemical reaction equations are as follows:

$$Fe \rightarrow Fe^{2+} + 2e^- \quad (1)$$

$$Fe^{2+} + HS^- \rightarrow FeHS^+ \quad (2)$$

$$H^+ + H_2O \rightarrow H_3O^+ \quad (3)$$

$$FeHS^+ + H_2O \rightarrow FeS + H_3O^+ \quad (4)$$

In the case of the concentration of H_2S is higher than 0.02% and the temperature is higher than 40°C, iron- rich sulfide $FeS_{(1-x)}$ will be formed; On the contrary, sulfur-rich sulfide $FeS_{(1+x)}$ will be formed when both the concentration and temperature are low [7]. In this study, the concentration of H_2S was 0.001%, and the temperature of environment was 120-250°C. It was considered that the corrosion product film should contained FeS, it can protected metal materials from further corrosion in a way due to their compactness [6]. The energy spectrum analysis of corrosion product films and cracks is shown in figure 2. It can be seen that the content of sulfur in corrosion products was very low, but oxygen was relatively high. It indicated that the main component of corrosion product film was still iron oxide. The content of H_2S was relatively low, therefore, corrosion product film was not as compact as FeS, so it was prone to rupture.

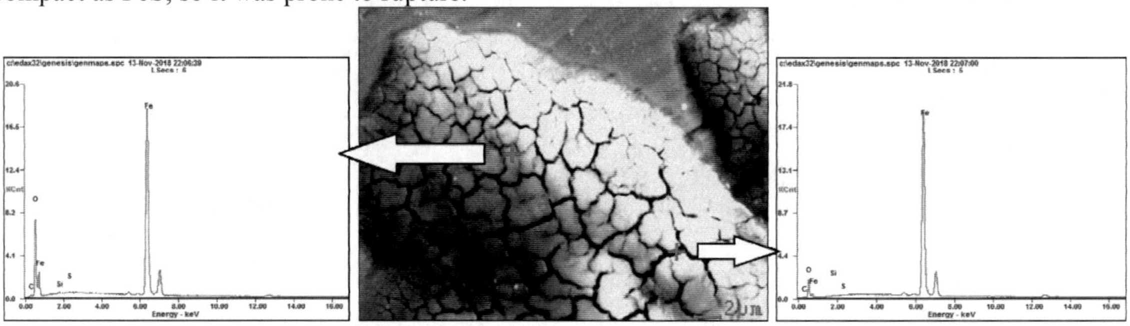

Figure 2. Spectral analysis of bubbling region. (a)(b). Position of analysis. (c)(d). Spectrum analysis spectrogram at specified locations.

3.2.2. Anodic reaction mechanism. The hydrolysis process of H_2S was carried out in two steps, in water, with the production of H_3O^+, HS^- and S^{2-}. The chemical reaction equations are as follows:

$$H_2S + H_2O \rightarrow H_3O^+ + HS^- \tag{5}$$

$$HS^- + H_2O \rightarrow H_3O^+ + S^{2-} \tag{6}$$

Hydronium ions that produced on metal surfaces were gave to electrons and generated hydrogen atoms adsorbed on metal surfaces. The chemical reaction equations are as follows:

$$H_3O^+ + e^- \rightarrow H + H_2O \tag{7}$$

$$Fe + H \rightarrow Fe \cdot H_{(adsorption)} \tag{8}$$

There were two trends for hydrogen atoms adsorbed on metal surfaces. Some of the hydrogen atoms were further transformed into dissolved hydrogen atoms and diffused into metal materials, it were captured by steel defects such as plane defects, crystal lattice dislocation and stress concentration in the process of diffusion, and then became interstitial atoms in metal. In this case, the pressure in the position of defects increasing, which led to embrittlement of steel, flaws are initiated and propagated to form cracks. The chemical reaction equations are as follows:

$$Fe \cdot H_{(adsorption)} \rightarrow Fe \cdot H_{(dissolution)} \tag{9}$$

The other part of hydrogen atoms compounded formed gaseous hydrogen molecules and escaped from the metal surface. The chemical reaction equations are as follows:

$$2Fe \cdot H_{(adsorption)} \rightarrow Fe \cdot H_2 + Fe \tag{10}$$

$$Fe \cdot H_2 \rightarrow Fe + H_2 \uparrow \tag{11}$$

Hydrogen gas accumulated continuously between the matrix and the compact solid corrosion products, so that the surface of the material gradually bubbled up with a growing pressure difference between two sides. When the pressure difference reached a critical value, the corrosion product film fractured and the surface of matrix contacted newly with corrosion medium. In such a case, matrix began to corrode again.

The general chemical reaction equation could be expressed as follows:

$$Fe + H_2S \rightarrow FeS + H_2 \uparrow \quad (12)$$

In this study, the concentration of H_2S was lower than the standard of "wet hydrogen sulfide" environment, but there was still bubbling corrosion morphology. The possible reason is that there were a lot of CO_2 and NO_2 in flue gas, which lead to the increase of pH value of corrosion medium. To some extent, hydrogen bubbling corrosion was promoted.

4. Conclusions

In this research, the hydrogen bubbling corrosion of four metal materials caused by hydrogen sulfide was studied experimentally in the flue gas of boilers under the conditions of 3MPa - 9MPa and 120°C - 250°C. The temperature and pressure conditions of hydrogen bubbling corrosion morphology were analyzed. The results indicated that hydrogen bubbling analysis mainly occurs under corrosion conditions of high pressure region (7MPa, 120°C; 9MPa, 200°C), but the higher pressure has not been studied. According to the corrosion medium of metal, the mechanism of hydrogen bubbling corrosion caused by hydrogen sulfide was proposed. In the process of corrosion, corrosion cell were formed with steel as anode and hydrogen sulfide as cathode. The results of this research provided a basis for the further study of corrosion mechanism of iron in boiler flue gas.

5. Acknowledgement

This research is supported by the Fundamental Research Funds for the Central Universities [xjj2018201] and [xjj2018006], the Projects from National Natural Science Foundation of China [51871179].

6. References

[1] Huang X S, Qi Y M, Chen C F, Yu H B and Lu G W, *Corros. Eng. Sci. Techn.*, **50**, 169-177
[2] Kittel J, Smanio V, Fregonese M, Garnier L and Lefebvre X, *Corros. Sci.*, **52**, 1386-1392
[3] Li W F, Zhou Y J, Xue Y, *J. Iron Steel Res.*, 19, 59–65(2012)
[4] Yin Z F, Zhao W Z, Bai Z Q, Feng Y R, Zhou W J, *Electrochim. Acta.*, 53, 3690-3700(2008)
[5] Shi S Z, Jun D B, Zeng D Z, Yu H Y, Chen Y X, *J. Sou. Pet.*, 40,162-168
[6] Ma H Y, Cheng X L, Li G Q, Chen S H, Quan Z L, Zhao S Y and Niu L, *Corros. Sci.*, **42**, 1669-1683
[7] Chen X Y, Zheng Y P, Li L, Ge T Z, Yu Q X, *Pipe. Techn. Equ.* , no.2, 49-51

IOP Conf. Series: Materials Science and Engineering **542** (2019) 012086 doi:10.1088/1757-899X/542/1/012086

The Inhibitory effect of Natural materials Black Tea Theaflavins on MMPs-2 and MMPs-9 activities in Dentine

Hong Li, Yina Wang and Jiongyuan Xu

Hangzhou Dental Hospital, School of Stomatology, Zhejiang Chinese Medical University, Hangzhou, China
E-mail:lihong8689@163.com

Abstract. The aim of the present study is to investigate the inhibitory effect of black tea theaflavins(TF) on gelatinases (MMP-2, MMP-9) in human dentine. Protein extracts from demineralized human dentine powder were treated with TFs concentration range from 0.1-1.6 mg/mL; Samples treated with EDTA was used as a positive control while untreated sample as a negative control; Zymography was utilized to assess MMP-2 and MMP-9 activities in dentin. The results demonstrated that untreated demineralized dentine showed the strongest band of MMPs activity on zymography, while EDTA treatment completely inhibited MMPs activity; MMPs activity was inhibited by TFs in a concentration dependent manner. We conclude that theaflavins inhibited MMP-2 and MMP-9 activities, suggesting that natural product theaflavins could be a potential supplementary therapy both for prevention and treatment of dentine caries.

1. Introduction

Caries is defined as a chronic and infectious disease caused by certain cariogenic bacteria. The bacteria aids dissolving the hard tissues of teeth like enamel, dentine and cementum. Being different from enamel, dentine contains less mineral and more organic material. Dentine carious process also differs from enamel caries for its construction. Dentine mineral dissolved and organic dentine extracellular matrix (ECM) exposed, followed by ECM degradation by proteases, cariogenic bacteria aggress toward pulp tissues. In the past, people thought proteases were secreted by bacteria. however, host-derived protease especially matrix metalloproteinases (MMPs) are recently found to play an important role in caries as well [1].

MMPs participate in normal cell functions and regulate many physiological processes. However, they also implicate in pathological conditions such as inflammation, tumor development, and arthritis. In the oral region, MMPs have implications in tooth development, pulp and periradicular pathoses, periodontal disease and dental caries. The dentine extracellular matrix contains type I collagen mainly,

IOP Conf. Series: Materials Science and Engineering **542** (2019) 012086 doi:10.1088/1757-899X/542/1/012086

which can be degraded by gelatinases (MMP-2 and MMP-9). Several studies have suggested that MMP-2 and MMP-9 can be found in caries lesions and have contribution to dentine matrix degradation [2]. So people try to find out different ways to inhibit MMPs for decreasing caries. Recently nature MMPs inhibitors becomes popular for their nontoxicity and less side effects.

The black tea is one of the most popular drink in the world. It's fermented and contains four different theaflavins which are the main and active components in black tea. Some researches indicated that the extracts of black tea could inhibit P. gingivalis and inflammatory facts secreted by epithelial cells. Mughal et.al found that the MIC of methanolic extracts of black tea diluted in methanol was 0.6 mg/ml[3]. Our previous study showed the MIC and MBC of theaflavins were 0.5 mg/ml and 1.0 mg/ml respectively.

The black tea exhibits its antibacterial character. However, to our knowledge there are few studies focus on the influence of theaflavins on gelatinases in dentine. Therefore, the aim of this study is to determine the capacity of theaflavins to inhibit MMP-2 and MMP-9 in dentine caries. [1-3]

2. Methods and materials

2.1. Analysis of theaflavins

Commercial TFs (DeHe Bio-technology Company, Jiangsu, China) were analyzed on a high performance liquid chromatography (HPLC) system (Waters 1525 pump, 2487 dual γ absorbance detector, 717-plus injector). A Phenyl-Hexyl C18 reversed-phase column (250 mm × 4.6 mm inner diameter) was used for applications. The flow rate was 1mL/min.

2.2. Specimens preparation and demineralization

25 sound human molars were selected after informed content had been obtained under a protocol approved by Zhejiang Chinese Medical University. Teeth were cleaned in distilled water and attached soft tissue was removed by a scalpel carefully. Enamel, roots and pulp tissue were removed with diamond burs under flowing water. Dentine powder was prepared by pulverizing liquid nitrogen-frozen coronal dentine with a hammer and pestle. The commercial 80% theaflavins (DeHe Bio-technology Company, Jiangsu, China) were diluted to concentrations of 1.6, 0.8, 0.4, 0.2 and 0.1 mg/mL using two-fold dilution method. Seven 500mg of aliquots of dentine powder were treated as follow. Group 1: dentine powder demineralized in 37% aqueous PA for 1 min and 10mM EDTA for 5 min; Group 2: dentine powder demineralized in 37% aqueous PA for 1 min; Group 3-7: dentine powder demineralized in 37% aqueous PA for 1 min and different concentrations of TFs solution for 5 min.

2.3. Enzyme extraction of dentine specimens

After treatment, all specimens were rinsed with distilled water for 2 times and centrifuged for 10 min (13000rpm) at 4℃. Supernatants were discarded and specimens were resuspended in 1mL extraction buffer (50mM Tris-HCl, pH 6.4, 50mM $CaCl_2$, 100mM NaCl, 0.1% Triton X-100, 0.1% NP-40, 0.1mM $ZnCl_2$, 0.02%NaN_3) and EDTA-free protease inhibitor cocktail (P8340, Sigma company) for 24h at 4℃. Supernatants were collected after centrifuging for 10 min at 4℃ and protein was precipitated with cold acetone at -20℃ for 2 h. The protein content was then centrifuged for 10 min and dissolved in 1X loading buffer. Total protein concentration of the extracts of treated dentine powder was then quantified (Enhanced BCA Protein Assay Kit, Beyotime, Jiangsu, China).

2.4. Gelatin zymography

Gel preparation was performed according to the protocol described by Marta et al [4]. Proteins were electrophorized at 4℃ under non-reducing conditions on 10% SDS-polyacrylamide gels copolymerized with 1g/L gelatin (porcine skin, Sigma company) at constant voltage (120V, 40mA, 120 min). Carefully removed the gel and placed it in renaturing buffer. Incubating the gel at room temperature for 30 min with gentle agitation and rinsed the gel with distilled water for 30 min. The gel was incubated in developing buffer at room temperature for 30 min and then replaced in another fresh developing buffer at 37℃ for approximately 18h. Staining the gel with 0.5% Goomassie Brilliant Blue for 1h and destained in destaning buffer (50% methanol, 10% acetic acid, 40% water).

3. Results

The HPLC analysis demonstrated that TFs were mainly composed of TF, TF-3-G, TF-3'-G and TF-D-G (Fig.1). The concentration of the commercial theaflavins was 71.4% (Tab.1).

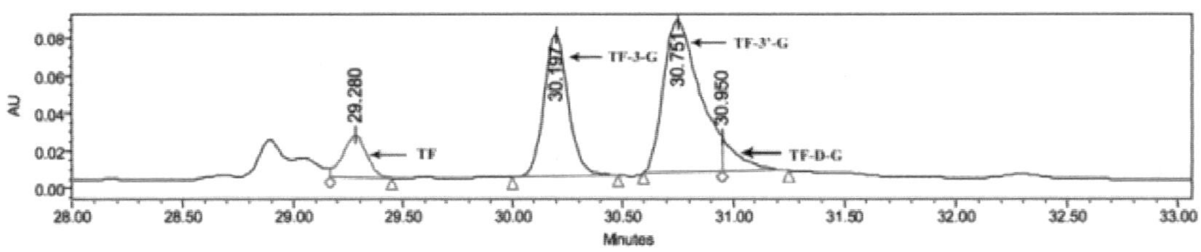

Fig.1-The retention times of theaflavins (TF), TF-3-G, TF-3'-G and TF-D-G were 29.3, 30.1, 30.7 and 30.9 min, respectively.

Name	Retention Time	Area	Area Percent	Concentration
TF	29.269	170075	9.31	12.925
TF-3-G	30.184	590574	32.34	44.174
TF-3'-G	30.736	949802	52.02	76.744
TF-D-G	30.950	115672	6.33	8.976
		1826123	100.00	71.410

Tab.1-The main components of theaflavins (TF) and their concentrations respectively. The total concentration of the commercial TFs was 71.4%.

Gelatin zymography showed four kinds of zymographic lytic bands ranged from 62 kDa to 130 kDa (Fig.2). The latent proforms (92 kDa) and activated enzyme forms (82 kDa) of MMP-9 as well as activated forms of MMP-2 (62 kDa) were detected. 130 kDa form was probably complexes of MMPs. Gelatinolytic activities varied amongst specimens, ranging from a strong band to trace amounts. With the increasing concentration of TFs, MMPs decreased activity. Demineralized dentin powder presented the strongest band while the positive control showed nothing since they were inhibited by EDTA completely.

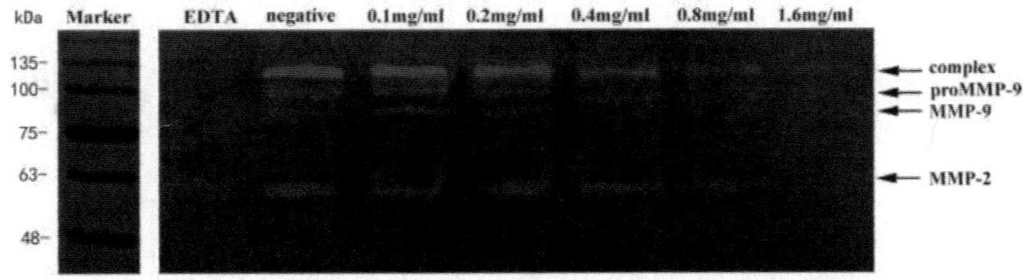

Fig.2- The zymography of MMP-2 and -9. Lane 1: protein extract of EDTA-treated demineralized dentine powder; Lane 2: protein extract of demineralized dentine powder without treatment; Lane 3-7: protein extracts of TFs-treated demineralized dentine powder.

4. Discussion

As we all know, the cariogenic biofilm adheres to teeth surface resulting in caries. The bacteria especially streptococcus mutans lead to localized demineralization and destruction of dentine hard tissue by producing acids. Except for the influence of bacteria, MMPs are found to play a role in caries as well. This proteinase family contribute to homeostasis of tissues and participate in physiological processes like angiogenesis and bone remodeling. However, they also lead to a serials of pathological processes(2). MMP-2 and MMP-9, known as gelatinases, can degrade type Ⅳ collagen which is the main component of ECM. Therefore, they can always be detected in inflammatory periodontal tissues. Type I collagen constitutes nearly 90% of the dentine organic matrix. In general, collagens are degraded by collagenases and then lose the triple-helical conformation. After that, they can be further degraded by MMP-2 and MMP-9. So gelatinases activities can be observed in caries lesions. Different with MMP-9, MMP-2 also has the ability to degrade collagen [4].

Early research has found that 0.03% chlorhexidine can completely inhibit MMP-2 and MMP-9. The inhibition of gelatinases by sodium fluoride is reversible at low concentrations (250-1500 ppm), but irreversible at 5000 ppm. However, CHX and fluoride have limitations for their side effects. For instance, desquamations in the oral mucosa and discolored tongues and tooth surfaces are observed when using CHX mouth washes. Excess fluoride ingestion during tooth formation can result in dental fluorosis which leading to a loss of enamel surface [5]. Because of these side effects, many researchers begin to focus on natural plants. Tea is a popular drink in the world. Lots of studies indicate that green tea has a significant effect on cariogenic bacteria and MMPs[6]. EGCG, the main component of green tea, can inhibit MMP-2 and MMP-9 directly. The mechanism may be that EGCG has metal- chelating property or it can directly bind to gelatinases. Black tea accounts for 78% of the tea production worldwide. It has more active components than green tea for its oxidation during fermentation procedure. Theaflavins are the main polyphenols which are the standard to evaluate black tea quality and the important bioactive substance in black tea. Studies of TFs range from antioxidant to anticancer[6], while small amounts of them focus on the potential effects in oral diseases. Among these studies, antibacteria effect is observed the most. Black tea extract can inhibit S. mutans and the inhibitory effect of black tea is even in lower concentration than green tea[7]. However, few studies observe the inhibitory effect of TFs on MMPs, which are also the important facts in dentine caries. Therefore, it's necessary to evaluate if TFs could suppress MMP-2 and MMP-9 efficiently.

This study revealed four different kinds of zymographic lytic bands. MMP-9 was detected at two molecular weight. 92 kDa indicated the latent proforms and 82 kDa showed the activated enzyme forms. The activated forms of MMP-2 were detected as well while the latent proforms of it were not observed. There were 130 kDa forms which might probably be complexes of MMPs. But in some studies, they were thought to be homodimers of MMP-9[8]. The relation between gelatinolytic activities and concentrations of TFs was obvious. Demineralized dentin powder without TFs treatment presented the strongest band. With the increasing of concentrations, the activities of MMPs decreased. The band could nearly not be seen when the dentine powder was treated with 1.6mg/mL of TFs. The positive control showed nothing since they were inhibited by EDTA completely. So we could conclude that TFs could inhibit the activities of MMP-2 and MMP-9 especially in a high concentration. The reason why TFs can inhibit gelatinases may because they are flavonoids. Flavonoid is proved to chelate metal efficiently. So the inhibition of MMP-2 and MMP-9 activities by TFs may be related to the ability to chelate zinc metal which is essential for enzymatic activity. Otherwise, the 130 kDa forms and MMP-2 were easy to observe indicating their high concentrations in enzyme extract. Compared with MMP-9, MMP-2 is suggested to be the major gelatinase in human dentin activing in the organization of predentine, initiation of dentine formation and lateral extension of caries along DEJ[9]. Charadram also found the activity of MMP-2 in dentine caries was higher than those in healthy dentine[10]. This suggest MMP-2 plays a very active role in the dentine caries procedure.

In summary, the present study shows that TFs can efficiently inhibit the activity of gelatinases. This inhibition suggests that TFs may be a promising method both in the prevention and treatment in dentine caries.

Acknowledgement

The project was supported by the National Natural Science Foundation of China (81571003 and 51290291), the project of Hangzhou Scientific Research Funding (20140633B46)

Corresponding author: (1) Prof. Hong Li. DMD. Email: lihong8689@163.com .(2) Yina Wang Master students of Prof. Hong Li. Made the same contribution. Email: beccaria@msn.cn.

Reference

[1] Sambandam V, Neelakantan P. *Matrix Metalloproteinases (Mmp) in Restorative Dentistry and Endodontics.* J. Clin. Pediatr. Dent. 2014;39(1):57-59

[2] Vidal C, Tjäderhane L, Scaffa P, Tersariol I, Pashley D, Nader H, et al. *Abundance of MMPs and cysteine cathepsins in caries-affected dentin.* J. Dent. Res. 2014;93(3):269-274

[3] Kong L, Qi X, Huang S, Chen S, Wu Y, Zhao L. *Theaflavins inhibit pathogenic properties of P. gingivalis and MMPs production in P. gingivalis-stimulated human gingival fibroblasts.* Arch. Oral

Biol. 2015;60(1):12-22

[4] Toth M, Sohail A, Fridman R. *Assessment of gelatinases (MMP-2 and MMP-9) by gelatin zymography.* Metastasis Research Protocols 2012:121-135

[5] DenBesten P, Li W. *Chronic fluoride toxicity: dental fluorosis. In: Fluoride and the Oral Environment.* Karger Publishers; 2011. p. 81-96

[6] Jazaeri M, Pakdek F, Rezaei-Soufi L, Abdolsamadi H, Rafieian N. *Cariostatic Effect of Green Tea in Comparison with Common Anticariogenic Agents: An in Vitro Study.* J Dent Res Dent Clin Dent Prospects 2015;9(1):44-48

[7] Allah AA, Ibrahium M, Allah SA, Amin M. *Antimicrobial effect of tea and tea with milk beverages on oral streptococcus mutans and lactobacilli.* World Applied Sciences Journal 2012;19(9):1327-1334

[8] Katchman BA, Antwi K, Hostetter G, Demeure MJ, Watanabe A, Decker GA, et al. *Quiescin sulfhydryl oxidase 1 promotes invasion of pancreatic tumor cells mediated by matrix metalloproteinases.* Molecular Cancer Research 2011;9(12):1621-1631 .

[9] Niu L, Zhang L, Jiao K, Li F, Ding Y, Wang D, et al. *Localization of MMP-2, MMP-9, TIMP-1, and TIMP-2 in human coronal dentine.* J. Dent. 2011;39(8):536-542.

[10] Charadram N, Farahani RM, Harty D, Rathsam C, Swain MV, Hunter N. *Regulation of reactionary dentin formation by odontoblasts in response to polymicrobial invasion of dentin matrix.* Bone 2012;50(1):265-275

Institute of Physics
Dirac House, Temple Back
Bristol BS1 6BE UK

ISSN: 1757-8981
ISBN 978-1-5108-9087-9